THE PHYSICAL

SCIENCES

THE PHYSICAL SCIENCES

AN INTEGRATED

APPROACH

Robert M. Hazen

James Trefil

GEORGE MASON UNIVERSITY

JOHN WILEY & SONS, INC.

New York   Chichester   Brisbane   Toronto   Singapore

Acquisitions Editors *Cliff Mills, Stuart Johnson*
Development Editor *Barbara Heaney*
Marketing Manager *Catherine Faduska*
Assistant Marketing Manager *Ethan Goodman*
Production Manager *Lucille Buonocore*
Sr. Production Editors *Nancy Prinz, Tracey Kuehn*
Designer *Madelyn Lesure*
Cover Photo *John Roberts/The Stock Market*
Manufacturing Manager *Mark Cirillo*
Photo Editor *Lisa Passmore*
Photo Researcher *Ramón Rivera Moret*
Illustration Coordinator *Edward Starr*

This book was set in 10/12 Stone Serif by Bi-Comp, Inc., and printed and bound by Von Hoffmann Press. The cover was printed by Phoenix Color.

Recognizing the importance of preserving what has been written, it is a policy of John Wiley & Sons, Inc., to have books of enduring value published in the United States printed on acid-free paper, and we exert our best efforts to that end.

Library of Congress Cataloging-in-Publication Data:

Hazen, Robert M., 1948–
    The physical sciences : an integrated approach / Robert M. Hazen
and James Trefil.
        p.   cm.
    Includes index.
    ISBN 0-471-15440-7 (cloth : alk. paper). — ISBN 0-471-00249-6 (pbk. : alk. paper)
    1. Physical sciences. I. Trefil, James S., 1938– . II. Title.
Q158.5.H39 1995
500.2--dc20   95-40786
                                                                                    CIP

Printed in the United States of America

10 9 8 7 6 5 4 3 2 1

PREFACE

WHY STUDY SCIENCE?

Why should every college student study science? Most students are not going to be professional scientists—only about one in every hundred Americans will. Most people won't ever have to predict the weather or cure a disease, much less calculate the orbit of a planet or the acidity of a solution. So why not just leave science to the experts? The answer is simple: learning about the central ideas and methods of science empowers everyone to make informed decisions about issues that affect work, family, and other aspects of daily life.

Most existing jobs, and the vast majority of new jobs that will be created in the coming decades, depend on advances in science and technology. New technologies are a driving force in economics, business, medicine, and even many aspects of law. Semiconductor advances, agricultural methods, and information processing have altered our world. New materials and manufacturing techniques drive our industries and present constant challenges to workers in sales and advertising. Plastics, superconductors, and composite materials have changed the way we shape our environment. Even professional athletes must constantly evaluate and use new and improved gear and rely on improved medical treatments and therapies. And, of course, future teachers will be called upon more and more to provide their students with confident command of scientific methods and content. By studying science, students will not only be able to incorporate these advances into their professional lives—they will also understand some of the processes by which the advances were made.

Science is no less central to daily life. As consumers, students are besieged by new products and processes, not to mention a bewildering variety of warnings about health and safety. As taxpayers, they must vote on issues that directly affect our communities—energy taxes, recycling proposals, government spending on research, to name a few. And as parents, they will have to nurture and guide children through an ever more complex world. A firm grasp of the principles and methods of science will help students make life's important decisions in a more informed way.

By studying science, students become prepared to understand and deal with the changing world, both on the job and at home. But there's more to science than this. The scientifically literate student will be poised to share in the excitement of the scientific discoveries that, week by week, transform our understanding of the universe and our place in it. Science opens up astonishing, unimagined worlds—bizarre life forms in the deep oceans, exploding stars in deep space, and aspects of the history of life and the universe more wondrous than any fiction. Dinosaurs, black holes, superconductors, mass extinctions, buckeyballs, space travel, and much more await the informed science reader.

With this book as an introduction, every student can share in the greatest ongoing adventure of the human race—the adventure of science.

REASSESSING TRADITIONAL SCIENCE EDUCATION

Pick up a newspaper any day of the week and you will find a dozen articles that relate to science and technology. Every issue has stories on the weather, energy, the environment, medical advances—the list goes on and on. Is the average U.S. college graduate prepared to understand the scientific component of these issues? In most cases the answer is no.

A week prior to graduation at a major U.S. university twenty-five seniors, selected at random, were asked a simple question: "What is the difference between an atom and a molecule?" Only one-third of the students could answer the question correctly. This result does not instill confidence in our ability to produce students who are in command of rudimentary facts about the physical universe. There is little doubt that we are faced with a generation of Americans who complete their education without learning even the most basic concepts about science. These citizens lack the critical knowledge to make informed decisions regarding environmental issues, resource management, medical advances, and the myriad other complex issues that face our society.

We have written this book to address two problems that pervade the organization and presentation of science at U.S. colleges and universities. First, many introductory science courses are geared toward science majors. Specialization is vital for these students—typically a few percent of all college students—who must learn an appropriate vocabulary and develop skills in experimental method and mathematical manipulations to solve problems. Unfortunately, few of these science majors ever gain a broad overview of the sciences, nor do they understand how their chosen specialty fits into the larger scheme of science and society. A physics major may never learn about the modern revolution in plate tectonics, while the geology major is unlikely to appreciate the nature and importance of semiconductors in our technological society.

If such specialization is unfortunate for science majors, it is totally inappropriate for the majority of students—nonscience majors, for whom experimental technique and mathematical rigor often divorce science from its familiar day-to-day context. Introductory science courses designed for science majors fail to foster scientific understanding among these other students. Without a broad context, many students neither understand the distinctive process of science, nor do they retain the abstract content of what they have been taught. Ultimately, this needlessly narrow approach to science education alienates most nonmajors, who graduate with the perception that science is difficult, boring, and irrelevant to their everyday interests.

The second problem with most introductory science courses at the college level, even among those science courses specifically designed for nonscientists, is that they rarely integrate physics, astronomy, chemistry, and Earth science. Such departmentally-based courses cannot produce graduates who are broadly literate in science. The students who take introductory geology learn nothing about lasers or nuclear reactions, while those who take "Physics for Poets" remain uninformed about the underlying causes of earthquakes and volcanoes. And neither physics nor geology classes touch on such vital modern fields as chemical reactions, environmental chemistry, space exploration, or materials science. Students, therefore, must take courses in at least four departments to gain a basic overview of the physical sciences.

Perhaps most disturbing, few students—science majors or nonmajors—ever learn how the often arbitrary divisions of specialized knowledge fit into the overall sweep of the sciences. In short, traditional science curricula of most colleges and universities fails to provide that basic science education that is necessary to understand the many scientific and technological issues facing our society.

ACHIEVING SCIENTIFIC LITERACY FOR ALL AMERICANS

This first edition of *The Physical Sciences: An Integrated Approach* is the culmination of a collective effort of dozens of faculty and thousands of students. The text is an expansion of the first two-thirds of our textbook, *The Sciences: An Integrated Approach* (Wiley, 1995), which was introduced in a preliminary edition during the 1993–1994 academic year. The content and scope of this book were shaped by the constructive comments and suggestions of scholars from across the United States and Canada—an integrated effort that is reflected throughout the text.

The central goal of science education must be to give every student the ability to place important public issues such as the environment, energy, and medical advances in a scientific context. Students should understand the scientific process, be familiar with the role of experiments in probing nature, and recognize the importance of mathematics in describing its behavior. They should be able to read and appreciate popular accounts of major discoveries in physics, astronomy, chemistry, and geology, as well as advances in medicine, information technology, and new materials. And, most importantly, they should understand that a few universal laws describe the behavior of our physical surroundings—laws that operate every day, in every action of our lives.

Most societal issues concerning science and technology draw on a broad range of knowledge. To understand the debate over nuclear waste disposal, for example, one needs to know how nuclei decay to produce radiation (physics), how radioactive atoms interact with their environment (chemistry), and how radioactive elements from waste can enter the biosphere (Earth science). These scientific aspects must be weighed with other societal issues—economics, energy demand, perceptions of risk, and demographics, for example. Other public issues, such as global warming, space research, alternative energy sources, and AIDS prevention, likewise depend on a spectrum of scientific and technological concepts, as well as other social concerns.

Many recent studies recognize the urgent need for reform in science education and advocate an interdisciplinary approach—see, for example, reports by the American Association for the Advancement of Science (*Science for All Americans—Project 2061*, Washington, 1990; *Benchmarks for Scientific Literacy*, Washington, 1993; *The Liberal Art of Science*, Washington, 1990), by the National Research Council (*National Science Education Standards,* Washington, 1994), and by the White House (*Science in the National Interest*, Washington, 1994). Our approach recognizes that science forms a seamless web of knowledge about the universe. Our integrated physical science course encompasses physics, chemistry, Earth sciences, and astronomy, and emphasizes general principles and their application to real-world situations rather than esoteric detail.

There are many ways to achieve this synthesis, but any general treatment should take advantage of the fact that virtually everything in science is based on a few simple overarching principles. Newton's laws, the atomic model, chemical bonding, and plate tectonics would make every scientist's list of these "great ideas." By returning to general science courses for all students, we have the means to achieve our goal—to produce college graduates who appreciate that scientific understanding is one of the crowning achievements of the human mind, that the universe is a place of magnificent order, and that science provides the most powerful means for discovering knowledge that can help us understand and shape our world.

THE ORGANIZATION OF THIS BOOK

This text adopts a distinctive and widely acclaimed approach to science education, based on the principle that general science courses are a key to a balanced and effective college-level science education for nonmajors and a broadening experience for science majors. We organize the text around a series of 22 scientific concepts. The most basic principle, the starting point of all science, is the idea that the universe can be studied by observation and experiment (Chapter 1). A surprising number of students—even science majors—have no clear idea how this central concept sets science apart from religion, philosophy, and the arts as a way of understanding our place in the cosmos. This distinctive way of knowing requires a specialized language (Chapter 2), including specialized vocabulary and mathematical models.

Once students understand the nature of science and its practice, they can appreciate some of the basic principles shared by all sciences: Newton's laws governing force and motion (Chapters 3 and 4), the laws of thermodynamics that govern energy and entropy (Chapters 5 and 6), the eqivalence of electricity and magnetism (Chapters 7 and 8), and the atomic structure of all matter (Chapters 9 through 13). These concepts apply to everyday life and explain, for example, the compelling reasons for wearing seat belts, the circulation of the blood, the dynamics of a pot of soup, the regulation of public airwaves, and the rationale for dieting. In one form or another, all of these ideas appear in virtually every elementary science textbook, but often in abstract form. As educators we must strive to make them part of every student's day-to-day experience.

Once these general principles have been laid down, we can examine specific natural systems such as atoms, the Earth, or stars. The realm of the nucleus (Chapter 14) and subatomic particles (Chapter 15), for example, must follow the basic rules governing all matter and energy. An optional chapter on the theory of relativity (Chapter 16) examines the consequences of a universe in which all observers discern the same laws of nature.

Plate tectonics and the cycles of rocks, water, and the atmosphere unify the Earth sciences (Chapters 17 through 20). The laws of thermodynamics, which decree that no feature on the Earth's surface is permanent, can be used to explain geologic time, gradualism, and the causes of earthquakes and volcanoes. The fact that matter is composed of atoms tells us that individual atoms in the Earth system—for example, in a grain of sand, a gold ring, or a student's most recent breath—have been recycling for billions of years.

Finally, in sections on astronomy and cosmology (Chapters 21 and 22), students learn that stars and planets form and move as predicted by Newton's laws, that stars eventually burn up according to the laws of thermodynamics, that nuclear reactions fuel stars by the conversion of mass into energy, and that stars produce light as a consequence of electromagnetism.

The text has been designed so that four of these chapters—"Quantum Mechanics" (10), "The Ultimate Structure of Matter" (15), "Albert Einstein and the Theory of Relativity" (16), and "Cosmology" (22)—may be skipped without loss of continuity.

SPECIAL FEATURES

In an effort to aid student learning and underscore the interconnections among the sciences, we have attempted to relate scientific principles to each student's everyday life. To this end, we have incorporated several distinctive features throughout the book.

Great Ideas.   Each chapter begins with a statement of a great unifying idea or theme in science, so that students immediately grasp the chief concept of that chapter. These statements are not intended to be recited or memorized, but to provide a framework for placing everyday experience in a broad context.

Random Walks.   Each chapter begins with "A Random Walk," in which we tie the chapter's main theme to common experiences, such as eating, driving a car, or suntanning. These "Random Walks" grew out of our idea of the perfect class: during every class period, we would meet outdoors and walk until we saw something that would illustrate that day's topic.

Technology.   The application of scientific ideas to commerce, industry, and other modern technological concerns is perhaps the most immediate way students encounter science. In almost every chapter, we include examples of these technologies, such as petroleum refining, microwave ovens, and nuclear medicine.

Environment.   Many environmental concerns face our society. Most chapters include sections that examine ways to model and modify our environment through our understanding of scientific principles. Features on the environment, furthermore, underscore the interdisciplinary nature of science and its application to issues that affect all people.

Developing Your Intuition.   Everyone has practical experience with the physical world and how it works. In the "Developing Your Intuition" features, students are asked to think about how events in their daily lives relate to key scientific principles. Sections on falling objects, the behavior of ice cubes, and the information content of books, for example, amplify key ideas in the physical sciences.

Science in the Making.   These historical episodes trace the process of scientific discovery as well as portray the lives of central figures in science. In them we have tried to illustrate the scientific method, examine the interplay of science and society, and reveal the role of serendipity in scientific discovery.

Thinking More About.   Each chapter ends with a section that addresses a social or philosophical issue tied to science, such as federal funding, nuclear waste disposal, the teaching of evolution, and dietary fads.

Mathematical Equations and Worked Examples.   Unlike the content of many science texts, formulae and mathematical derivations play a subsidiary role in our treatment. We rely much more on real-world experience and everyday vocabulary. We believe, however, that every student should understand the role of mathematics in science. We have, therefore, included a few key equations and appropriate worked examples in many chapters. Whenever an equation is introduced, it is presented in three steps: first as an English sentence, then as a word equation, and finally in its traditional symbolic form. Many equations are also treated graphically in order to demonstrate more clearly the process of scientific investigation (for example, the scatter of real measurements) and the power of being able to visualize data. In this way, students can focus on the meaning, rather than the abstraction, of the mathematics. We also include an appendix on English and SI units.

Science by the Numbers.   We also think that students should understand the importance of simple mathematical calculations in making estimates and determining orders of magnitude. We have, therefore, incorporated many nontraditional calculations of this kind. For example, readers are asked to determine how much solid waste is generated in the United States, how long it would take to erode a mountain, and how many people were required to build Stonehenge.

Key Words.   We believe that most science texts suffer from too much complex vocabulary, and we have avoided any unnecessary jargon. Nevertheless, the scientifically literate student must be familiar with many words and concepts that appear regularly in newspaper articles or other material for general readers. In each chapter a number of these words appear in **boldface,** and they are included in a fill-in-the-blank exercise at the end of the chapter. In Chapter 14 on nuclear physics, for example, key words include proton, neutron, isotope, radioactivity, half-life, radiometric dating, fission, fusion, and nuclear reactor—all terms likely to appear in the newspaper.

Many other scientific terms are important, though more specialized; we have highlighted these terms in *italics*. We strongly recommend that students be expected to know the meaning and context of key words, but not be expected to memorize these italicized words. We encourage all adopters of this text to provide their own lists of key words and other terms—both ones we have omitted and ones you would eliminate from our list.

Questions.   We feature five levels of end-of-chapter questions. *Review Questions* test the most important factual information covered in the text, while *Fill-in-the-Blanks* emphasize important words and phrases that every student should know. *Discussion Questions* are also based on material in the text, but they examine student comprehension, and they explore applications and analysis of the scientific concepts. *Problems* are quantitative questions that require students to use mathematical operations. Answers to odd-numbered problems are included in an appendix. Finally, *Investigations* require additional research outside the classroom. Each instructor should decide which levels of questions are most appropriate for his or her students. We welcome suggestions for additional questions, which will be added to the next edition of this text.

Additional Reading.   Each chapter concludes with a short reading list of popular books and articles that expand on themes introduced in that chapter.

Illustrations.   Students come to any science class with years of experience in dealing with the physical universe. Everyday life provides an invaluable science laboratory—the physics of sports, the chemistry of cooking, and the biology of being alive. This book has thus been extensively illustrated with familiar color images in an effort to amplify the key ideas and principles. All diagrams and graphs have been designed for maximum clarity and impact.

SUPPLEMENTS

The supplements that accompany *The Physical Sciences: An Integrated Approach* assist both the instructor and the student. The **Instructor's Manual** was prepared by Gail Steehler of Roanoke College. The manual contains teaching suggestions, lecture notes, answers to problems in the text, practice questions and problems, a list of supplemental readings and films, as well as ideas for beginning and ending lectures. In addition, there are sections called "Connecting Back" and "Connecting Ahead" that offer suggestions on how to connect the material from different chapters together.

The **Active Learning Guide** was prepared by Ron Canterna of the University of Wyoming. The Guide contains contextual examples, applications of conceptual experiments, and quantitative problem solving related to conceptual examples. In addition, there is a brief math review, and answers to 25% of the discussion questions in the text, as well as solutions to 25% of the problems in the text.

The **Test Bank** contains approximately 1250 multiple-choice, true-false, and short answer conceptual questions. There are also essay questions for each chapter.

The **Laboratory Manual,** prepared by Ron Canterna of the University of Wyoming and Kelly V. Beck, contains approximately 25 labs using the predictive learning cycle.

ACKNOWLEDGMENTS

The development of this text has benefitted immensely from the help and advice of numerous people.

Student Involvement

Students in our "Great Ideas in Science" course at George Mason University have played a central role in designing this text. Approximately 1,500 students—the majority of whom were nonscience majors—have enrolled in the course over the past seven years. They represent a diverse cross-section of U.S. students: more than half were women, while many minority, foreign-born, and adult learners were included. Their candid assessments of course content and objectives and their constructive suggestions for improvements have shaped our text.

Faculty Input

We are also grateful to members of the Core Science Course Committee at George Mason University, including Richard Diecchio (Earth Systems Science), Don Kelso and Harold Morowitz (Biology), Minos Kafatos and Jean Toth-Allen (Physics), and Suzanne Slaydon (Chemistry), who helped to design many aspects of the original treatment.

We thank the many teachers across the country who are developing integrated science courses. Their letters to us and responses to our publisher's survey inspired us to develop new curricula for nonscience majors. In particular, we thank those who used the preliminary edition of *The Sciences* and provided comments; these individuals include:

Lauretta Bushar, *Beaver College*
Tim Champion, *Johnson C. Smith University*
Ben Chastain, *Samford University*
Marvin Goldberg, *Syracuse University*
James Grant and Michael Held, *St. Peter's College*
Jim Holler, *University of Kentucky*
Patricia Hughey, *Lansing Community College*
Joseph Ledbetter and Rick Saparano, *Contra Costa College*
Leigh Mazany, *Dalhousie University*
Donald Miller, *University of Michigan at Dearborn*
Lynn Narasimhan, *DePaul University*
Ervin Poduska, *Kirkwood Community College*
Gail Steehler, *Roanoke College*
Jim Yoder, *Hesston College*

We especially thank those professors who have provided detailed reviews of this physical sciences text. Their expertise in science and pedagogy have shaped our treatment and content.

Robert Backes, *Pittsburgh State University*
Edward Borchardt, *Mankato State University*
Robert Boram, *Morehead State University*
John Breshears, *Southeastern Community College*
Ron Canterna, *University of Wyoming*
Paul Deaton, *Asbury College*
Paul Fishbane, *University of Virginia*
Allan Gelfand, *New York City Technical College*
Simon George, *California State University–Long Beach*
John Graham, *Carnegie Institution of Washington*
Joseph Heafner, *Catawba Valley Community College*
Paul Jasien, *California State University–San Marcos*
Stan Jones, *University of Alabama*
Glenn Julian, *Miami University*
Rudi Kiefer, *University of North Carolina–Wilmington*
Mary Lu Larsen, *Towson State University*
Kenneth Mendelson, *Marquette University*
Harold Morowitz, *George Mason University*
Victor Motz, *Northern Essex Community College (MS)*
J. Ronald Mowery, *Harrisburg Area Community College*
Bjorn Mysen, *Geophysical Laboratory*
J. William Nelson, *Florida State University*
Charles O'Neill, *Edison Community College (FL)*
Charles Ostrander, *Hudson Valley Community College*
Donald Rickard, *Arkansas Technical University*
Stephen Roeder, *San Diego State University*
Selwyn Sacks, *Carnegie Institution of Washington*
Stanley Shepherd, *Pennsylvania State University*
Gregory Smith, *Edison Community College (FL)*
Harry Smith, *Tallahassee Community College*
Fred Thomas, *Sinclair Community College*
John Thompson, *Texas A&M University–Kingsville*
Don Wieber, *Contra Costa College*

Publisher Support

Finally, we applaud the many people at John Wiley who proposed *The Sciences: An Integrated Approach,* and this book, *The Physical Sciences.* In our combined 50 years of book writing, we have never experienced a more dedicated, knowledgeable, enthusiastic, or capable editorial and production team.

Acquisitions editor Clifford Mills and publisher Kaye Pace formulated the idea for this text, and provided persuasive encouragement for us to pursue the project. Carl Kulo designed and analyzed all the marketing surveys.

We are indebted to developmental editor Barbara Heaney, who has worked closely with us on *The Physical Sciences,* as she did with our previous text. She has played a key role in helping us to craft a unique approach to presenting science to nonscientists. Her many substantive contributions, along with those of freelance editor Maxine-Effenson Chuck, as well as her cheerful attention to the innumerable details associated with such an effort, are reflected on every page.

Nancy Prinz and Tracey Kuehn supervised the accelerated production schedule, handling the countless technical details with efficiency and skill. We thank

Lisa Passmore and Ramón Rivera-Moret, who displayed great resourcefulness in locating the numerous photos for this edition. Madelyn Lesure designed the text with great skill. Edward Starr coordinated the development of the illustrations. Catherine Faduska and Ethan Goodman supervised the marketing of *The Physical Sciences* with creativity and good humor. Jennifer Bruer coordinated the supplements. Brent Peich, Matt Van Hattem, and Monica Stipanov provided administrative support throughout the development of the text.

To all the staff at John Wiley we owe a great debt for their enthusiastic support, constant encouragement, and sincere dedication to science education reform.

The readers of this book can help us to achieve additional objectives. Please take some time to let us know, by writing to Robert Hazen and James Trefil, Robinson Professors, George Mason University, Fairfax, VA 22030. We are eager to incorporate your comments and suggestions in subsequent editions, and thank you in advance for sharing in this development.

Robert M. Hazen
James Trefil

George Mason University

BRIEF CONTENTS

CONTENTS

8 WAVES AND ELECTROMAGNETIC RADIATION

9 THE ATOM

17 THE EARTH AND OTHER PLANETS

Great Idea   *The Earth, one of nine planets that orbit the Sun, formed 4.5 billion years ago from a great cloud of dust.*

18 THE DYNAMIC EARTH

Great Idea   *The entire Earth is still changing, due to the slow convection of soft, hot rocks deep within the planet.*

THE PHYSICAL SCIENCES

1

SCIENCE: A WAY OF KNOWING

SCIENCE IS A WAY OF ANSWERING QUESTIONS ABOUT
THE PHYSICAL UNIVERSE.

Making Choices

Our lives are filled with choices. What should I wear? What should I eat? Should I bother to recycle an aluminum can, or just throw it in the trash? Which road should I take? Every day we have to make dozens of decisions; each choice is based, in part, on the knowledge that actions in a physical world have predictable consequences.

When you pull into a gas station, for example, you have to decide what sort of gasoline to put into your car. Over a period of time, you may try several different types, observing how your car responds to each. In the end, you may conclude that a particular brand and grade suits your car best, and decide to buy that one in the future. You engage in a similar process when you buy shampoo, pain relievers, athletic shoes, and scores of other products.

This simple example illustrates one way we learn about the universe. First, we look at the world to see what is there and learn how it works. Then we generalize, making rules that seem to fit what we see. Finally, we apply those rules to situations we've never encountered, fully expecting them to work.

There doesn't seem to be anything earth-shattering about choosing a brand of gasoline or shampoo. But the same basic procedure can be applied in a more formal and quantitative way when we want to understand the workings of a distant star or a living cell. In these cases, the enterprise is called **science,** and the people who study these questions for a living are called **scientists.**

science The discipline that uses the scientific method to ask and answer questions about the physical world.

scientists Women and men who work in one or more areas of science.

THE SCIENTIFIC METHOD

scientific method   A continuous process used to collect observations, form and test hypotheses, make predictions, and identify patterns in the physical world.

Science is a discipline that asks and answers questions about the physical world. It is not primarily a set of facts or a catalog of answers, but rather a way of conducting an ongoing dialogue with our physical surroundings. Like any human activity, science is enormously varied and rich in subtleties. Nevertheless, there are a few basic steps that, taken together, can be said to comprise the **scientific method.**

Observation

If our goal is to learn about the world, the first thing we have to do is look around us and see what's there. This statement may seem obvious to us; yet, throughout much of history, learned men and women rejected the idea that you can understand the world simply by observing it.

The Greek philosopher Plato, living during the Golden Age of Athens, would have argued that one cannot deduce the true nature of the universe by trusting to the senses. The senses lie, he would have said. Only the use of reason and the insights of the human mind can lead us to true understandings. In his famous book *The Republic*, Plato compared human beings to people living in a cave, watching shadows on a wall, but unable to see the objects causing the shadows. In just the same way, he argued, observing the physical world will never put us in contact with reality, but will doom us to a lifetime of wrestling with the shadows. Only with the "eye of the mind" can we break free from illusion and arrive at the truth.

During the Middle Ages in Europe, a similar frame of mind existed, but with a devout trust in wisdom passed down from classical scholars and theologians replacing human reason as the ultimate tool in the search for truth. There is a story (probably apocryphal) about a debate in an Oxford college on the question of how many teeth a horse has. One learned scholar quoted the Greek scientist Aristotle on the subject, and another quoted the theologian St. Augustine to put forward a different answer. Finally, a young monk at the back of the hall got up and noted that since there was a horse outside, they could settle the question by looking in its mouth. At this point, the manuscript states, the assembled scholars "fell upon him, smote him hip and thigh, and cast him from the company of educated men."

As both of these examples illustrate, we can develop strategies to learn about the physical world by using our reasoning powers alone, without actually making observations. Such approaches are not, however, what we call the scientific method, nor did they produce the kinds of advanced technologies and knowledge we associate with modern societies. These other attempts to understand our place in the cosmos were, however, perfectly serious, and were pursued by people every bit as intelligent as we are. In Chapter 3, we will see how human beings gradually came to understand that observation complements pure reasoning, and thus has an important role to play in learning about the universe.

observation   The act of observing nature without manipulating it.

experiment   The manipulation of some aspect of nature to observe an outcome.

In the remainder of this book, we will differentiate between **observations,** in which we observe nature without manipulating it, and **experiments,** where we manipulate some aspect of nature and observe the

Table 1–1 • Measurements of Falling Objects

Time of Fall (seconds)	Distance of Fall (meters)
0	0
1	16
2	64
3	144

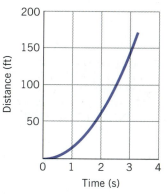

Figure 1–1
Measurements of a falling object can be presented visually in the form of a graph. Distance of fall (on the vertical axis) is plotted versus time of fall (on the horizontal axis).

outcome. An astronomer, for example, observes distant stars without changing them, while a chemist experiments by mixing materials together and seeing what happens.

Identifying Patterns and Regularities

When we observe a particular phenomenon over and over again, we begin to get a sense of how nature behaves. We start to recognize patterns in nature. Eventually, we will generalize our experience into a synthesis that summarizes what we have learned about the way the world works. We may, for example, notice that whenever we drop something, it falls. This statement would represent a summary of the results of many observations.

Scientists often summarize the results of their observations in mathematical form, particularly if they have been making quantitative measurements. In the case of a falling object, for example, they might be measuring the time it takes an object to fall a certain distance, rather than just noticing that the object falls. The next step would be to collect data in the form of a table (see Table 1–1).

These data could also be presented in the form of a graph, in which distance is plotted against time (see Figure 1–1). After preparing tables and graphs of their data, scientists would notice that the longer the time something falls, the greater the distance it travels. Furthermore, the distance isn't simply proportional to the time of fall. If the object falls for twice as long, it does not travel twice as far. Rather, if one object falls for twice as long as another, it will travel four times as far; if it falls three time longer, it will travel nine times as far; and so on. This statement can be summarized in three ways (a format that we'll use throughout this book):

▶ **In words:**

The distance traveled is proportional to the square of the time of travel (i.e., time × time).

▶ **In equation form:**

distance = constant × time × time
= constant × (time)^2

▶ **In symbols:**

$D = k \times T^2$

Time lapse image of a falling light bulb.

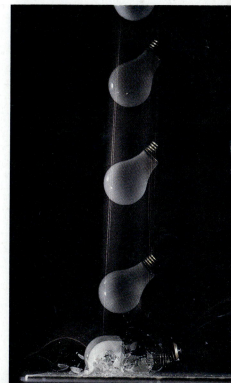

The symbol, *k*, is a *constant* that defines the quantitative mathematical relationship between distance and time squared. This constant has to be determined from the measurements. (We'll return to the subject of constants in Chapter 2.)

Mathematics is a concise language that allows scientists to communicate their results and to make very precise predictions, but anything that can be said in an equation can also be said (though in a less concise way) in a plain English sentence. When you encounter equations in your science courses, you should always ask, "What English sentence does this equation represent?" This routine will keep the mathematics from obscuring the simple ideas that lie behind most equations.

Not every scientific idea can be or has to be stated this precisely, though. A scientist studying the gradual encroachment of a forest into an abandoned field, for example, might notice that certain plants seem to follow each other—weeds, followed by scrub trees, followed by pines, followed by hardwoods, for example. We can conclude that a succession of plant types will be observed everywhere in a particular climate zone. This conclusion can be tested, and so it is a part of scientific inquiry.

Hypothesis and Theory

hypothesis   A tentative guess about how the world works, based on a summary of experimental or observational results and phrased so that it can be tested by experimentation.

Once we have summarized experimental and observational results, we can form a **hypothesis**—a tentative, educated guess—about how the world works. In the case of our everyday experience with falling objects, a hypothesis could be very easily formulated. We could say, "When I drop something, it falls." In other cases, the formation of the hypothesis may be more complicated, and the hypothesis may be stated in the form of mathematical equations. When confronted with a new phenomenon, scientists often weigh several different hypotheses at once, much as a detective in a murder mystery may consider several different suspects.

theory   A description of the world that covers a relatively large number of phenomena and has met many observational and experimental tests. A conclusion based upon observations of nature.

The word **theory** refers to a description of the world that covers a relatively large number of phenomena and has met many observational and experimental tests. After observing hundreds of dropped objects, for example, we could state a theory such as, "In the absence of wind resistance, all objects fall a distance proportional to the square of the time of the fall." Just as a detective announces a theory at the solution of a murder mystery, so, too, scientists reach a conclusion based on their observations of nature.

Prediction and Testing

prediction   A guess about how a particular system will behave, followed by observations to see if the system did behave as expected within a specified range of situations.

In science, every hypothesis must be tested. We test hypotheses by using them to make **predictions** about how a particular system will behave, then by observing nature to see if the system behaves as predicted. For example, if we hypothesize that all objects fall when they are dropped, then that idea can be tested by dropping all sorts of objects. Each drop constitutes a test of our prediction, and the more successful tests we conduct, the more confidence we have that the hypothesis is correct.

As long as we restrict our tests to solids or liquids on the Earth's surface, the hypothesis is consistently confirmed. Test a helium-filled balloon, however, and we discover a clear exception to the rule. The balloon "falls" up. The original hypothesis, which worked so well for

In orbit, an object will not fall if you drop it but continue floating. This is an example in which the simple hypothesis that "things fall to the ground when released" doesn't work.

most objects, fails for certain gases. And more tests would show that that's not the only limitation. If you were an astronaut in the space shuttle, every time you held something out and let it go it would not fall or rise. It would float in space. Evidently, our hypothesis is invalid in orbit, as well.

This example illustrates an important aspect about testing hypotheses. Tests do not necessarily prove or disprove a hypothesis; instead, they often serve to define the range of situations under which the hypothesis is valid. We may, for example, observe that nature behaves in a certain way only at high temperatures or only at low ones; only at low velocities or only at high ones. Such limitations indicate that the original hypothesis doesn't cover enough ground and has to be replaced by something more general. In the example of falling objects, we will see that the hypothesis that "objects fall when dropped" has to be replaced by a more sophisticated and general set of hypotheses called Newton's laws of motion and the law of universal gravitation. These laws describe and predict the motion of dropped objects both on the Earth and in space and are, therefore, a more successful set of statements than the original hypothesis. We will discuss them in more detail in Chapters 3 and 4.

When a hypothesis has been tested extensively and seems to apply everywhere in the universe—when we have had enough experience with it to have a lot of confidence that it is true—we generally elevate the hypothesis to a new status. We call it a **law of nature.** We will encounter a number of such laws in this book, all backed by countless observations and measurements. It is important, however, to remember where these laws come from. They are not written on tablets of stone, nor are they simply good ideas that someone once had. They arise from repeated and rigorous observation and testing. They represent our best understanding of how nature works.

law of nature An overarching statement of how the universe works, following repeated and rigorous observation and testing of a theory or group of related theories to show that the theory seems to apply everywhere in the universe.

Remember—we never stop questioning the validity of our hypotheses, even after we call them laws. Scientists constantly think up new, more rigorous experiments to test the limits of our theories. In fact, one of the central tenets of science is that

> **Every law of nature is subject to change, based on new observations.**

The Scientific Method in Operation

The elements of observation, hypothesis formation, prediction, and testing together comprise the scientific method. In practice, you can think of the method as working as shown in Figure 1–2. In this never-ending cycle, observations lead to hypotheses, which lead to more observations.

Figure 1–2
The scientific method can be represented as an endless cycle of collecting observations (data), identifying patterns and regularities in the data (synthesis), forming hypotheses, and making predictions, which lead to more observations.

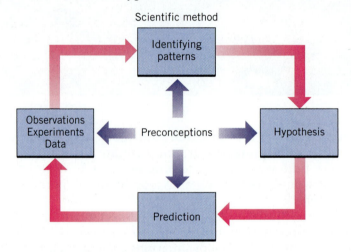

If observations confirm a hypothesis, then more tests may be devised. If the hypothesis fails, then the new observations are used to revise it, after which the revised hypothesis is tested again. Scientists continue this process until the limits of existing equipment are reached, in which case researchers often try to develop better instruments to do even more tests. If it appears that there's just no point in going further—when numerous experiments confirm a given hypothesis—that hypothesis may eventually be elevated to a law of nature.

Several important points should be made about the scientific method.

1. While scientists attempt to be objective, they often observe nature with preconceptions about what they are going to find. Most experiments and observations are designed and undertaken with a specific hypothesis in mind, and most researchers have a strong hunch about whether that hypothesis is right or wrong. Perhaps the most important point about the scientific method is that scientists have to believe the results of their experiments and observations, whether they fit preconceived notions or not. Science does not demand that we have no ideas when we enter the cycle (Figure 1–2)—only that we be ready to change those ideas if the evidence forces us to.

2. There is no "right" place to enter the cycle. Scientists often start their work by making extensive observations, but they can also start with a theory and test it. Wherever they enter the cycle, the scientific process will take them all the way around.

3. Observations and experiments must be reported in such a way that anyone with the proper equipment can verify the results. Scientific results, in other words, must be **reproducible.**

4. There is no end to the cycle, but each cycle lifts us to a new level of understanding. Science does not always provide final answers, nor is it always a search for ultimate truth. Science is a way of producing successively more detailed and exact descriptions of the physical world—descriptions that allow us to predict the behavior of that world with higher and higher levels of confidence.

5. Finally, while the orderly cycle shown in Figure 1–2 provides a useful framework to help us think about science, science is not a rigid, cookbook-style set of steps to follow. Because science is undertaken by human beings, it involves occasional bursts of intuition, sudden leaps, a joyful breaking of the rules, and all the other sorts of things we associate with human activities.

reproducible A criterion for the results of an experiment. In the scientific method, observations and experiments must be reported so that anyone with the proper equipment can verify the results.

Science in the Making

Dimitri Mendeleev and the Periodic Table

The discoveries of previously unrecognized patterns in nature provide scientists with some of their most exhilarating moments. Dimitri Mendeleev (1834–1907), a popular chemistry professor at the Technological Institute of St. Petersburg in Russia, experienced such a breakthrough in 1869 as he was tabulating data for a new chemistry textbook.

The mid-nineteenth century was a time of great excitement in chemistry. Almost every year saw the discovery of one or two new chemical elements, and new apparatus and processes were greatly expanding the repertoire of laboratory and industrial chemists. In such a stimulating field, it was no easy job to keep up to date with all the developments and summarize them in a textbook. In an effort to consolidate the current state of knowledge about the most basic chemical building blocks, Mendeleev listed various properties of the 63 known chemical elements (substances that could not be divided by chemical means). He arranged his list in order of increasing atomic mass, and then noted the distinctive chemical behavior of each element.

Examining his list, Mendeleev realized an extraordinary thing: elements with similar chemical properties appeared at regular, or *periodic*, intervals. One group of elements, including lithium, sodium, potassium, and rubidium (he called them group-one elements), formed compounds with chlorine in a one-to-one ratio. Immediately following the group-one elements in the list were beryllium, magnesium, calcium, and barium—group-two elements that form compounds with chlorine in a one-to-two ratio, and so on.

Dimitri Mendeleev recognized regular patterns in the properties of known chemical elements, and thereby devised the first periodic table of elements.

Figure 1–3
Early published versions of Dimitri Mendeleev's periodic table of the elements revealed regular patterns in the chemical behavior of known elements, as well as obvious gaps where as yet undiscovered elements must lie. Elements in the same row (which Mendeleev called Groups I, II, and so on) have similar chemical properties.

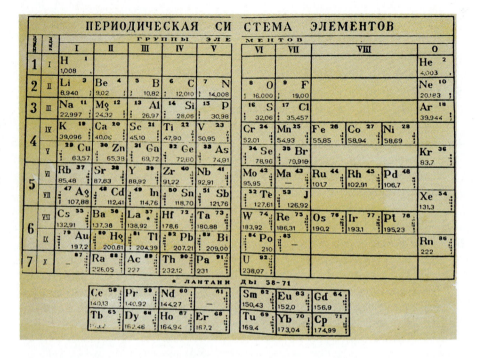

TABLE I Distribution of the Elements in Groups and Series										
Group	**I.**	**II.**	**III.**	**IV.**	**V.**	**VI.**	**VII.**	**VIII.**		
Series 1	H	—	—	—	—	—	—			
" 2	Li	Be	B	C	N	O	F			
" 3	Na	Mg	Al	Si	P	S	Cl			
" 4	K	Ca	Sc	Ti	V	Cr	Mn	Fe . Co . Ni .		Cu
" 5	(Cu)	Zn	Ga	Ge	As	Se	Br			
" 6	Rb	Sr	Y	Zr	Nb	Mo	—	Ru . Rh . Pd .		Ag
" 7	(Ag)	Cd	In	Sn	Sb	Te	I			
" 8	Cs	Ba	La	Ce	Di?	—	—	— . — . — .		—
" 9										
" 10	—	—	Yb	—	Ta	W	—	Os . Ir . Pt .		Au
" 11	(Au)	Hg	Tl	Pb	Bi	—	—			
" 12	—	—	—	Th	—	U	—			
	R_2O	R_2O_2	R_2O_3	R_2O_4	R_2O_5	R_2O_6	R_2O_7	Higher oxides		
	—	RO	—	RO_2	—	RO_3	—	RO_4		
	—	—	—	RH_4	RH_3	RH_2	RH	Hydrogen compounds		

As other similar patterns emerged from his list, Mendeleev realized that the elements could be arranged in the form of a table (Figure 1–3). Not only did this so-called periodic table highlight previously unrecognized relationships among the elements, it also revealed obvious gaps—places where as yet undiscovered elements must lie.

The power of Mendeleev's periodic table of the elements was underscored when several new elements, with atomic masses and chemical properties just as he had predicted, were discovered in the following years.

The discovery of the periodic table ranks as one of the great achievements of science. It was so important, in fact, that Mendeleev's students carried a copy of it behind his coffin in his funeral procession. ●

OTHER WAYS OF KNOWING

The central idea of science revolves around the notion that we can discover laws that describe how nature works by observing and measuring. *Every* idea in science is subject to testing. If an idea cannot be tested, it may not be wrong, but it isn't a part of science.

For example, a scientist can hypothesize whether a particular painting was executed in the seventeenth century. He or she could use various chemical tests to find the age of the paint, study the canvas, X-ray the painting, and so on. It may turn out that the statement about the age of the painting is wrong (it may, for example, be a modern forgery). The key here is that the statement can be tested. It is, therefore, an acceptable scientific hypothesis.

But some questions cannot be answered by the methods of science. No physical or chemical test will tell us whether or not the painting is beautiful, nor can any test be devised that will tell us how we are to respond to it. These questions are simply outside the realm of science. In fact, the methods of science are not the only way to answer many questions that matter in our lives. While science provides us with a way of tackling questions about the physical world, such as how it works and how we can shape it to our needs, many questions—some would say the most important questions—lie beyond the scope of science and the scientific method. Some of these questions are deeply philosophical: What is the meaning of life? Why does the world hold so much suffering? Is there a God? Other important questions are more personal: What career should I choose? Whom should I marry? Should I have children? These are not questions that can be answered by the cycle of observaton, hypothesis, and testing. For answers, we turn instead to religion, philosophy, and the arts.

A symphony, a poem, and a painting are not, in the end, objects to be studied scientifically. These art forms address different human needs and they use different methods than science. The same can be said about religious faith. Strictly speaking, there should be no conflict between science and religion, because they deal with different aspects of life. Conflicts arise only when zealots on either side try to push their methods into areas where they aren't applicable.

Pseudoscience

Many kinds of inquiry—extrasensory perception (ESP), unidentified flying objects (UFOs), astrology, crystal power, reincarnation, or any of the myriad notions you see in magazines at supermarket checkout counters—fail the elementary test that defines science. None of these subjects, collectively labeled **pseudoscience,** is subject to testing in the sense we are using that term. There is no test you can imagine that will convince those who believe in these notions that their ideas are wrong. Yet, as we have

pseudoscience A kind of inquiry, falling in the realm of belief or dogma, that includes subjects that cannot be proved or disproved with a reproducible test. The subjects include creationism, extrasensory perception (ESP), unidentified flying objects (UFOs), astrology, crystal power, and reincarnation.

seen, the central property of scientific ideas is that they can be tested and that they may, at least in principle, be wrong. Pseudoscience lies outside the domain of science, within the realm of belief or dogma.

Science by the Numbers

Astrology

Astrology is a very old system of beliefs that most modern scientists would call a pseudoscience. The central belief of astrology is that the positions of objects in the sky at a given time (at the moment of a person's birth, for example) determine a person's future. Astrology was part of a complex set of omen systems developed by the Babylonians thousands of years ago, and was practiced by many famous astronomers well into modern times.

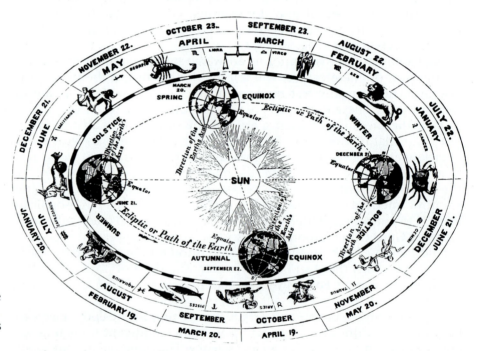

A horoscope such as that shown at right describes the positions of the Sun, Moon, and planets among the stars at the moment of birth.

If you were in a spaceship above the Earth's atmosphere, you could see the Sun and the stars at the same time. As the Earth traveled around the Sun, you would see the backdrop of stars change. The band of background stars through which the Sun appears to move is called the zodiac. The stars of the zodiac are customarily divided into twelve constellations, called "signs" or "houses." At any time, the Sun, the Moon, and the planets all appear in one of these constellations, and a diagram showing these positions is called a horoscope. The constellation in which the Sun appeared at the time of your birth is your "Sun sign," or, simply, your "sign."

Astrologers have a complex (and far from unified) system in which each combination of heavenly bodies and signs is believed to signify particular things. The Sun, for example, is thought to indicate the out-going, expressive aspects of one's character, the Moon represents the inner-directed ones, and so on.

Scientists reject astrology for two reasons. First, there is no known way that planets and stars could exert a significant influence on a child at birth. It is true, as we shall learn in Chapter 4, that stars and planets exert a tiny gravitational force on the infant, but the gravitational force exerted by the delivering physician (who is smaller but much closer) is much greater than that exerted by any celestial object.

But, more importantly, scientists reject astrology because it just doesn't work. Over the millennia, there has been no evidence at all that the stars can predict the future.

You can test the ideas of astrology for yourself, if you like. Try this: Have a member of your class take the horoscopes from yesterday's newspaper and type them on a sheet of paper without indicating which horoscope goes with which sign. Then ask members of your class to indicate the horoscope that best matches the day they actually had. Have them write their birthday (or sign) on the paper as well.

If people just picked horoscopes at random, you would expect about 1 person in 12 to pick the horoscope corresponding to his or her sign. Are the results of your survey any better than that? What does this tell you about the predictive power of astrology? ●

THE ORGANIZATION OF SCIENCE

Scientists investigate all sorts of natural objects and phenomena, from the tiniest elementary particles to living cells to the human body to forests to the Earth to stars to the entire cosmos. In all of this vast sweep, the same scientific method can be applied. Men and women have been carrying out this task for hundreds of years, and by now we have a pretty good idea about how many parts of our universe work. And, in the process, we have found a kind of overall organization to scientific knowledge.

The Web of Knowledge

The organization of science is analogous to a large spiderweb. Around the periphery of the web are all the phenomena examined by scientists, from atoms to fish to comets. Moving toward the web's center, we find the cross-linking hypotheses that scientists have developed to explain how these phenomena work. The farther in we move, the more general these hypotheses become and the more they explain. Radiating out from the center of the web, connecting all the parts and holding the entire structure together, we find a small number of very general principles that have attained the rank of laws of nature.

No matter where you start on the web, no matter what part of nature you are investigating, you will eventually come to one of the ideas that intersect at the central core. Everything that happens in the universe happens because one or more of these laws is operating (Figure 1–4).

The hierarchical organization of scientific knowledge provides an ideal way to approach the study of nature. At the center of any scientific question are a few laws of nature—we prefer to call them "great ideas of science.'' We begin by looking at the laws that describe everyday forces and motions in the universe. These overarching principles of science are accepted and shared by all scientists, no matter what their field of research. These ideas recur over and over again as we study different parts of the world. You will find that many of these ideas and their consequences seem quite simple, perhaps even obvious, because you are intimately familiar with the physical world in which these laws of nature are continuously operating.

After an introduction to these general principles, we look at how the scientific method is applied in specific areas of nature. We examine the nature of materials and the atoms that make them, for example, and look at the chemical reactions that form them. We explore the planet on which we live and discover how mountains and oceans, rivers and plains are formed and evolve over time. And we move beyond our Earth to consider the stars and galaxies and ask how they came to be.

By the time you have finished this journey, you will have touched on many of the great truths about the physical universe that scientists have deduced over the centuries. You will discover how the different parts of our universe operate and how all the parts fit together, and you will know that there are still great unanswered questions that drive scientists today. You will understand some of the great scientific and technological challenges that face our society, and, more importantly,

Figure 1–4
The interconnected web of scientific knowledge.

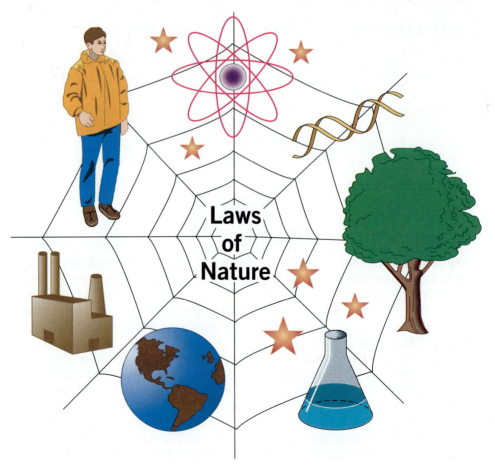

Laws
of
Nature

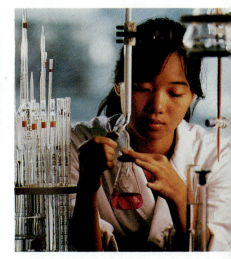

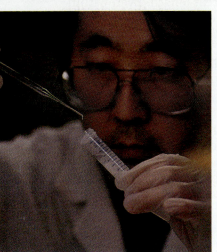
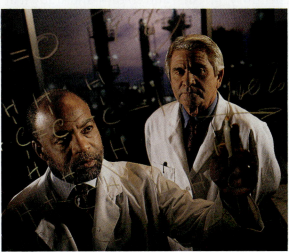

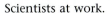

Scientists at work.

you will know enough about how the world works to deal with many of the new problems that will arise in the future.

The Organization of Scientists

Scientists do science. Hollywood films have created many stereotypes of scientists—the absent-minded professor with frizzy white hair; the mad scientist intent on destroying the world; the bespectacled nerd in a white lab coat. Popular fiction seems to portray the scientist, alternately, as an eccentric genius with solutions to every human crisis, or an arrogant, self-centered villain who pursues forbidden knowledge. Some people still mistakenly think of all scientists as white and male.

Our own experience with our colleagues does not support these stereotypes. Scientists include women and men of every race, religion, and national heritage. Some are scholarly types and some are party animals. Some are honest men and women, while others you wouldn't trust as far as you could throw them. Scientists, in short, come in pretty much the same spectrum of personality types as accountants, butchers, and automobile mechanics. And as far as appearances go, all we can say is that neither of the authors, despite a combined total of 50 years in research careers, has *ever* worn a white lab coat.

Scientific Specialties

When modern science first started in the seventeenth century, it was possible for one person to know almost all there was to know about the physical world and the "three kingdoms" of animals, vegetables, and minerals. In the seventeenth century, Isaac Newton could do forefront research in astronomy, in the physics of moving objects, in the behavior of light, and in mathematics. Today, our knowledge and understanding of the world is so much more sophisticated and complex that no one person could possibly be at the frontier in such a wide variety of fields. Scientists today must specialize. They must choose a field—biology, chemistry, physics, and so on—and study it at great length. Even within these disciplines, there are different subspecialties.

In physics, for example, a student may elect to study the properties of materials, semiconductors or superconductors, the nucleus of the atom, elementary particles, or the origin of the universe. The amount of information and expertise required to get to the frontier in any of these fields is so large that most students have to ignore almost everything else to learn their specialty.

Scientists within each subspecialty approach problems in different ways. Some scientists, called *theorists*, spend their time imagining universes that might exist. Other scientists, called *experimentalists*, observe and experiment in order to determine in which of the possible universes we actually live. Both kinds of scientists need to work together to make progress.

Basic Research, Applied Research, and Technology

basic research The type of research performed by scientists who are interested in simply finding out how the world works, in knowledge for its own sake.

technology The application of the results of science to specific commercial or industrial goals.

applied research The type of research performed by scientists with specific and practical goals in mind. This research is often translated into practical systems by large-scale research and development projects.

research and development A kind of research, aimed at specific problems, usually performed in government and industry laboratories.

There are many ways to study the physical universe, and many reasons for doing so. Many scientists are simply interested in finding out how the world works—in knowledge for its own sake. They are engaged in **basic research,** and may be found studying the behavior of distant stars or subatomic particles. Although discoveries made by basic researchers may have profound effects on society (see the discussion of the discovery of the electric generator in Chapter 7, for example), that is not the primary goal of these scientists.

Many other scientists approach their work with specific goals in mind. They wish to develop **technology,** in which they apply the results of science to specific commercial or industrial goals. These scientists are said to be doing **applied research,** and their ideas are often translated into practical systems by large-scale **research and development (R & D)** projects.

Government laboratories, colleges and universities, and private industries all support both basic and applied research; however, most large scale R & D (as well as most applied research) is done in government laboratories and by private industry (Table 1–2).

Table 1–2 • Some Important Research Laboratories in the United States

Facility	Type	Location
Argonne National Laboratory	Govt/Univ	Near Chicago, IL
AT&T Bell Laboratories	Industrial	Murray Hill, NJ
Brookhaven National Laboratory	Government	Long Island, NY
Dupont R & D Center	Industrial	Wilmington, DL
Fermi National Accelerator Lab	Govt/Univ	Near Chicago, IL
IBM Watson Research Laboratory	Industrial	Yorktown Hts, NY
Keck Telescope	University	Mauna Kea, HI
Los Alamos National Laboratory	Government	Los Alamos, NM
National Institutes of Health	Government	Bethesda, MD
Oak Ridge National Laboratory	Government	Oak Ridge, TN
Stanford Linear Accelerator	Govt/Univ	Stanford, CA
Texas Center for Superconductivity	University	Houston, TX
United States Geological Survey	Government	Reston, VA
Upjohn Pharmaceuticals Laboratory	Industrial	Kalamazoo, MI
Woods Hole Oceanographic Institution	University	Woods Hole, MA

Communication Among Scientists

Sometimes it's easier to do your homework with other students than by yourself; scientists also find this to be the case. It is very hard to work in isolation, and scientists often seek out other people with whom to converse and collaborate. The popular stereotype of the lonely genius changing the course of history just doesn't describe the world of the working scientist. The next time you walk down the hall of a science department at your university, you will probably see faculty and students deep in conversation, talking and scribbling on blackboards. This is the simplest type of scientific communication—direct contact between colleagues.

Scientific meetings provide a more formal and structured forum for communication. Every week of the year, at conference retreats and convention centers across the country, groups of scientists gather to trade

Sometimes scientist trade information at meetings, like this conference on AIDS in Berlin, Germany.

An extraordinary range of scientific periodicals keeps professionals, as well as non-scientists, informed about the latest important research.

ideas. You may notice that science stories in the media often originate in the largest of these meetings, where thousands of scientists converge at one time, and where a cadre of science reporters, with their own special briefing room, is poised to publicize exciting results. Scientists often hold off announcing important discoveries until they can make a splash at such a well-attended meeting and press conference.

At almost any scientific gathering, you will find people from many different countries. Electrons in Japan, Brazil, or the United States are exactly the same, and this universality is reflected in the composition of the scientific community.

Finally, scientists communicate with each other in writing. In addition, to informal communications by letter, fax, and electronic mail, all scientific fields have specialized journals to publish the results of research. The system works like this: When a group of scientists has finished a piece of research and wants to communicate the results, they write a concise paper describing exactly what they've done, giving the technical details of their method so that others can reproduce the data, and stating their results and conclusions. There are many thousands of specialty journals in science and technology to choose from; however, if the results are especially newsworthy, scientists often submit their work to the editors of one of a few high-profile periodicals such as *Science* or *Nature*, which are read weekly by hundreds of thousands of scientists. More specialized research will find its way to journals with more limited readership. Physicists, for example, might submit papers to a journal called *Physical Review*, chemists to *Journal of the American Chemical Society,* earth scientists to *American Mineralogist*, and medical researchers to *The New England Journal of Medicine* or *Cell*.

The journal editor sends the submitted manuscript to one or more knowledgeable scientists who act as referees. These reviewers, whose identity is not revealed to the authors, read the paper carefully, checking for mistakes, misstatements, or shoddy procedures. If they tell the editor that the work passes muster, it will probably be published. Often, such a recommendation is accompanied by a list of necessary modifications and corrections. This system, called **peer review,** is one of the cornerstones of modern science.

The presence of peer review and a clear protocol for entering new results into the scientific literature explain why scientists get so upset when one of their colleagues tries to bypass the system and announce results at a press conference. Such work has not been subject to the thorough review process, and no one can be sure that it meets established standards. When the results turn out to be unreproducible, overstated, or just plain wrong, as they were in the case of cold fusion (see Chapter 15), it damages the credibility of the entire scientific community.

Scientific Societies

Professional organizations are an integral part of the life of a scientist, and they play a central role in communicating scientific ideas and discoveries to the general public. Since 1660, the year the Royal Society of London for the Promotion of Natural Knowledge was founded, scientists have recognized the need for organizations devoted to the promotion of their craft. Scientists in America established many scientific societies of their own, beginning in 1769 with the American Philosophical Society, "held at Philadelphia, for promoting useful knowledge." Benjamin Franklin and Thomas Jefferson were both active members.

Many American scientists now join the **American Association for the Advancement of Science,** universally referred to as the AAAS ("triple-A S"). This Washington-based organization represents all branches of the physical, biological, and social sciences. With approximately 133,000 members, the AAAS is among the largest scientific societies in the world. It holds a national meeting and several regional meetings every year, publishes the weekly periodical *Science*, and is a strong force in establishing science policy and promoting science education.

The **National Academy of Sciences,** based in Washington, DC, serves a very different function. Founded by President Abraham Lincoln and Congress during the Civil War to help the government with scientific and technical problems, the National Academy provides professional advice for the government on policy issues ranging from environmental risks and natural resource management to education and funding for science research. Membership in the Academy is limited to just 60 new members per year—approximately 2,000 scientists in all—who are elected by existing members in semiannual ballots. Election to fellowship in the National Academy is a mark of high distinction among American scientists, as are major scientific prizes and awards (see Table 1–3).

peer review A system by which the editor of a scientific journal submits manuscripts considered for publication to a panel of knowledgeable scientists who, in confidence, evaluate the manuscripts for mistakes, misstatements, or shoddy procedures. Following the review, if a manuscript is to be published, it is returned to the author with a list of modifications and corrections to be completed.

American Association for the Advancement of Science One of the largest scientific societies, representing all branches of the physical, biological, and social sciences. AAAS is a strong force in establishing science policy and promoting science education.

National Academy of Sciences A nationally recognized association of scientists, elected to membership by their peers to provide professional advice for the government on policy issues ranging from environmental risks and natural resource management to education and funding for science research.

Table 1–3 • A Few Major Scientific Awards

Name of Prize	Subject Area
Arthur Day Medal	Earth Science
Field Prize	Mathematics
Japan Prize	A subject not covered by Nobel Prizes
Nobel Prize	Chemistry, Medicine, Physics
Priestly Award	Chemistry

Table 1–4 • **Major Scientific Societies in the United States**

Name of Society	Membership
American Chemical Society	144,000
American Geophysical Union	30,000
American Institute of Biological Sciences	90,000
American Institute of Physics	90,000
American Physical Society	40,000
Geological Society of America	17,000

In addition to these general science organizations, hundreds of societies are devoted to specific fields of study. A few of the largest are summarized in Table 1–4.

Figure 1–5
A graph showing the spending on scientific research (in constant dollars) from 1980–1994.

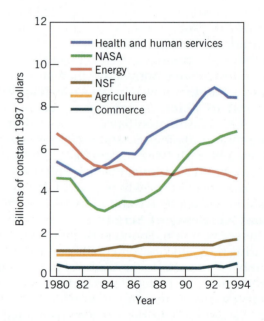

FUNDING FOR SCIENCE

An overwhelming proportion of funding for American scientific research comes from various agencies of the federal government—your tax dollars at work (see Table 1–5 and Figure 1–5). In 1995, the United States government's total research and development budget was almost 80 billion dollars. The **National Science Foundation,** with an annual budget of about three billion dollars, supports research and education in all areas of science. Other agencies, including the Department of Energy, the Department of Defense, the Environmental Protection Agency, and the National Aeronautics and Space Administration, fund research and science education in their own particular areas of interest, while Congress may appropriate additional money for special projects. Funding for basic

National Science Foundation   A federal agency that funds American scientific research and education in all areas of science.

Table 1–5 • Your Tax Dollars: 1995 Federal Science Funding

Agency or Department	Funding (in millions of $)
Department of Agriculture	1,436
Department of Defense	35,870
Department of Energy	
General science and research	5,208
Environmental restoration; nuclear waste management	5,837
Environmental Protection Agency	350
National Aeronautics and Space Administration	
Research and development	5,891
Human space flight and mission support	8,129
National Institutes of Health	11,334
National Institutes of Standards and Technology	790
National Oceanographic and Atmospheric Administration	259
National Science Foundation	3,360

and applied research in medicine and biology comes from the **National Institutes of Health,** headquartered in Bethesda, Maryland.

An individual scientist seeking funding for research will usually submit a grant proposal to the appropriate federal agency. Such a proposal will include an outline of the planned research together with a statement about why the work is important. The agency evaluating the proposals asks panels of independent scientists to rank them in order of importance, and funds as many as it can. Depending on the field, a proposal has anywhere from a 10% to 40% chance of being successful. This money from federal grants buys experimental equipment and computer time, pays the salaries of researchers, and supports advanced graduate students. Without this support, science in the United States would all but come to a halt. The funding of science by the federal government is one place where the opinions and ideas of citizens, through their elected representatives, have a direct effect on the development of science.

As you might expect, scientists and politicians engage in many debates about how this research money should be spent. One constant point of contention, for example, concerns the question of basic versus applied research. How much money should we put into the latter, which can be expected to show a quick payoff, as opposed to the former, which may not have a payoff for years (if at all)?

Another continuing debate concerns the division of money between "big" and "little" science. Most scientists work in small groups, concentrating on specialized research. This method of operation is called little science, and many of the important discoveries we'll be talking about in later chapters resulted from this tradition.

On the other hand, sometimes science is done at huge, centralized (and very expensive) installations. Today, there are proposals to build a $35 billion dollar Space Station and conduct a $5 billion dollar Human Genome Project, while plans to construct a $12 billion machine called

National Institutes of Health
A federal agency that provides funding for basic and applied research in medicine and biology.

the Superconducting Supercollider were canceled by Congress after it was partially built. These projects epitomize big science. Many of the important discoveries we'll discuss later were made at these sorts of establishments.

The conflict arises when we ask how many small science projects are to be canceled to pay for the expensive instruments needed to conduct big science. This sort of conflict is not easy to resolve—in fact, the two authors of this book often find themselves in disagreement on specific policy issues related to the question of big versus little science.

Technology

Buckeyballs: A Technology of the Future?

An extraordinary discovery, announced in 1990, reveals the close relationships among pure scientific research, applied research, and technology. For centuries, the element carbon was known in only two basic forms—a soft black mineral, graphite, and a hard transparent gemstone, diamond. But in 1985, chemists at Sussex University in England and Rice University in Texas found evidence for a totally new form of carbon, in which 60 carbon atoms bond together in a ball-shaped molecule (Figure 1–6). The distinctive linkage of the atoms, much like the geodesic domes of architect-inventor Buckminster Fuller, led scientists to dub the new material buckminsterfuller—or "buckeyballs" for short. Buckeyballs, though completely unexpected, at first excited little attention outside a small research community, for the material had no known uses and could be produced only in minute quantities. Nevertheless, the lure of the new form of an important chemical element kept several research groups busy studying the stuff. With no obvious practical applications, these early buckeyball studies were examples of basic research.

A major advance came in May of 1990, when a small team of German chemists discovered a way to produce and isolate large quantities of buckeyball crystals in a simple and inexpensive device. With the possibility of commercial-scale production, an explosion of applied

Figure 1–6
Buckeyballs are soccer-ball-like molecules of 60 carbon atoms (red spheres) that form crystals in which these round molecules stack together like oranges at the grocery store.

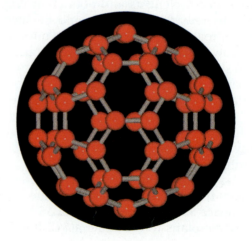

buckeyball research followed. Thousands of scientists, including teams at most major industrial and government laboratories, jumped on the buckeyball bandwagon, and hundreds of scientific articles documented an astonishing range of chemical and physical properties for the new carbon.

Among the preliminary findings: buckeyballs and closely related materials may contribute to a new generation of versatile electronic materials, powerful lightweight magnets, atom-sized ball bearings, and super-strong building materials. With such extraordinary prospects on the horizon, buckeyball investigations may soon become the domain of engineers developing new technologies—new kinds of batteries for automobiles, carbon-based girders for skyscrapers, unparalleled lubricants, and other products as yet undreamed.

Within the next few years, the first buckeyball products may appear at your hardware store, as engineers take the results of applied scientific research and use them to design large-scale production facilities. When the discovery is big enough, the transition from small, basic research to new technologies may be rapid indeed! ●

THINKING MORE ABOUT SCIENCE

Research Priorities

Sometimes questions of research funding get caught up in questions of public policy. For example, there have been 800,000 cases of AIDS diagnosed in the United States since 1980, primarily among male homosexuals and intravenous drug users. Over that same period, more than 5 million Americans have died of cancer. Yet by 1990, the budget for AIDS research at the National Institutes of Health exceeded that for cancer.

Critics of this policy argue that research money should be spent on those diseases that affect the greatest number of people, and that the federal policy has been distorted by a vocal minority. Supporters argue that AIDS, an incurable and invariably fatal disease, is a *potential* threat to many more people than cancer, and point to the tens of millions of heterosexual men and women who have died of the disease in Africa as a portent of what could happen if a cure or vaccine for the disease is not found soon.

As so often happens, there is no "scientific" solution to this problem. What do you think the proper course for the government ought to be? Should we spend more to combat a disease that is already killing many people, or one that is relatively minor now but could be even more deadly? What nonscientific arguments have to be brought to bear in making decisions like these?

▶ Summary

Science, a way of learning about our physical universe, is undertaken by women and men called *scientists*. The *scientific method* relies on making *reproducible observations* and *experiments*, which may suggest general trends and *hypotheses*, or *theories*. Hypotheses, in turn, lead to *predictions* that can be treated with more observations and experiments. Successful hypotheses may, after extensive testing, be elevated to the status of *laws of nature*, but are always subject to further testing. Science and the scientific method differ from other ways of knowing, including religion, philosophy, and the arts, and differ from *pseudosciences*.

Science is organized around a hierarchy of fundamental principles. Overarching concepts about forces, motion, matter, and energy apply to all scientific disciplines. Additional great ideas relate to specific systems—molecules, cells, planets, or stars. This body of scientific knowledge forms a seamless web, in which every detail fits into a larger, integrated picture of our universe.

Scientists engage in *basic research*, whose goal is solely the acquisition of knowledge; in *applied research*; and in *research and development (R & D)* aimed at specific problems. *Technology* is developed by this process.

Scientific societies like the *American Association for the Advancement of Science* and the *National Academy of Science* facilitate the scientific process by organizing professional meetings, publishing research results, and fostering science education. Scientific results are communicated in *peer reviewed publications*. The federal government plays the important role of funding most scientific research and advanced science education in the United States. Much basic research is funded by the *National Science Foundation* and the *National Institutes of Health*.

▶ Review Questions

1. Describe the steps in the scientific method.

2. What distinguishes a theory from a natural law?

3. What is the difference between an experiment and an observation?

4. What kind of experiment might a chemist perform?

5. What kind of observation might an astronomer make?

6. What are ways of knowing that are not science?

7. Why do scientists use equations?

8. How does a hypothesis differ from a guess?

9. How does a scientist choose between competing hypotheses?

10. Must scientists always conduct their research without preconceptions?

11. What does it mean that scientific experiments must be reproducible?

12. What is the central idea of science?

13. What is pseudoscience? Give an example.

14. What is meant by the interconnected web of scientific knowledge?

15. What is the difference between basic and applied research?

16. What is peer review and how does it work?

17. Name three important scientific organizations.

18. What is the National Science Foundation?

19. Who pays for most scientific research in the United States?

20. Give examples of big and little science.

▶ Fill in the Blanks

Complete the following paragraph with words and phrases from the list of key terms.

applied research
basic research
experiment
hypothesis
law of nature
National Academy of Science
National Institutes of Health
National Science Foundation

observation
peer review
research and development (R & D)
science
scientist
scientific method
technology
theory

_____ is a process of asking and answering questions about the physical universe. The _____ usually begins by observing the physical world. An _____ does not change things in that world, while an _____ involves manipulation. From data, a _____ can be formed and tested by the _____. A hypothesis that has been widely tested is called a _____, or may even be elevated to a _____.

Research aimed at acquiring knowledge for its own sake is called _____, in contrast to _____, which is usually performed with a definite goal in mind, or _____, which is meant to lead to a particular product. When science is applied to achieve practical goals, the result is often referred to as _____.

Scientific advances are reported in specialized journals, whose entries are subject to _____ to ensure correctness. Science in the United States is largely funded by the federal government, and some agencies that fund science are _____, _____, and _____.

▶ Discussion Questions

1. Which of the following statements could be tested scientifically? Explain your reasoning.

    a. Most of the energy coming from the Sun is in the form of visible light.

    b. Unicorns exist.

    c. Shelley wrote beautiful poetry.

    d. The Earth was created over four billion years ago.

    e. Diamond is harder than steel.

    f. Diamond is more beautiful than ruby.

    g. A virus causes the flu.

    h. Chocolate ice cream tastes better than strawberry ice cream.

2. The claim is sometimes made that the cycle of the scientific method produces closer and closer approximations to "reality." Is this a scientific statement? Why or why not?

3. Scientists are currently investigating whether certain microscopic organisms can clean up toxic wastes. How might you set up an experiment to determine that you had found such an organism?

4. Categorize the following examples as basic research or applied research.

    a. The discovery of a new galaxy

    b. The development of a better method to fabricate rubber tires

    c. The breeding of a new variety of disease-resistant chicken

    d. A study of the diet of parrots in a tropical rain forest

    e. The identification of a new chemical element

    f. The improvement of a method to extract the element gold from stream gravels

5. Following World War II, in the 1950s many movies portrayed evil physicists and rocket scientists, who seemed ready to destroy the world. In more recent movies biologists are more often the villains. What discoveries and events may have influenced this change?

6. A recent television commercial claimed that an antacid consumed "47 times its own weight in excess stomach acid." How might you test this statement in the laboratory? As a consumer, what additional questions would you ask before deciding to buy this product? Are all of these questions subject to the scientific method?

7. State if the following activities are an example of an observation or an experiment and explain the reason(s) for your choice.

    a. Recording the position of the setting sun on the horizon during the year

    b. Recording the changes in atmospheric pressure while a cold front moves through your hometown

    c. Measuring the amount of snowfall or rain on the roof of your physical science building

    d. Recording the boiling point for a beaker of water while you add different amounts of salt to the water

    e. Recording the times of free-fall for different objects that were thrown off the roof of your science building

    f. Measuring the difference between the temperature inside and outside your dormitory room window during the winter and relating this difference to the presence and absence of fog on the window

8. Why is reproducibility significant to the process of valid science?

9. What are the parts or major features of the scientific method? Outline this process using these major features.

10. Outline, diagram, or sketch the interconnected web of scientific knowledge. How does this compare with your personal web of scientific knowledge?

▶ Problems

1. Susan has kept careful records of driving speed versus fuel efficiency. She has noted that traveling 10 miles per hour (mph) she averages 22 miles per gallon (mpg) of gasoline. Similarly, she gets 26 mpg at 20 mph, 29 mpg at 30 mph, 31 mpg at 40 mph, 32 mpg at 50 mph, 28 mpg at 60 mph, and 24 mgp at 65 mph. Describe and illustrate some of the ways you might present these data. What additional data would you like to obtain to improve your description?

2. Measure the height and weight of ten friends and present these data both in a table and graphically. What trends do you observe? Why might physicians find such a table useful?

3. Follow these instructions for the next six experiments or observations.

    a. Describe and identify in words the pattern that you observe.

    b. Using a graph, plot the data that will best illustrate this pattern.

    A. Your mother is at a stop sign in your hometown before she continues your trip to grandma's house. You record the speedometer readings every 2 seconds as she travels to the next stop sign. The following table gives the times and speedometer readings for your mother's car.

Time (sec)	Speed (miles per hour)
0	0
2	10
4	20
6	30
8	40
10	50
12	50
14	50
16	50
18	35
20	20

    B. The following table gives the measured times for a simple pendulum (a washer at the end of a string) to complete one complete oscillation (period), and the length of the string. (*Hint:* you may want to consider the square of the period for your pattern.)

String Length	Period
5 cm	0.45 sec
10 cm	0.63 sec
15 cm	0.78 sec
20 cm	0.90 sec
25 cm	1.00 sec
50 cm	1.42 sec

C. An exercise therapist has recorded the maximum heart rate for a sample of typical human beings. The maximum heart rate and age of this sample of human beings is given in the following table.

Heart Rate (beats per minute)	Age (years)
200	20
195	25
190	30
180	40
170	50
155	65
140	80

D. While waiting for the gas station attendant to fill up your car's 10-gallon tank, you record the time it takes for the pump to reach every 2 gallons. A table of your findings is given below.

Volume (gallons)	Time (sec)
0	0.0
2	2.5
4	5.0
6	7.5
8	10.0
10	12.5

E. Everyday Cowboy Joe used to set his clock to noon by noting the time when the shadow of the corral gate was the shortest during the day. Cowboy Joe, taking a keen interest in geometry, used this information to calculate the altitude of the sun at noon (i.e., the angle above the horizon the sun is at noon). Joe also noticed that the length of this noontime shadow and the altitude of the sun changed throughout the year. The table below gives the noontime altitude of the sun and the length of the corral gate's shadow during the year.

Date	Altitude of Sun	Shadow Length
Jan 22	19 degrees	29 ft
Feb 20	28 "	19 "
Mar 21	39 "	12 "
April 21	51 "	8 "
May 20	58 "	6 "
June 21	62 "	5 "
July 21	59 "	6 "
August 21	50 "	8 "
September 21	39 "	12 "
October 24	27 "	19 "
November 22	19 "	29 "
December 21	16 "	35 "

F. The estimated world population is given in the following table.

Decade	Population
A.D. 1650	0.5 billion
A.D. 1850	1.0 "
A.D. 1920	2.0 "
A.D. 1990	5.5 "

4. Make the following predictions based on the data presented in Problem 3 and the trends that you observed. Use the six tables in Problem 3.

a. What would be the speed of your mother's car at the time 22 sec? 24 sec? (See Table 3A)

b. What would be the period of the simple pendulum for a string of length 75 cm? (See Table 3B)

What would be the string length if the period of the pendulum was 2 sec? (See Table 3B)

c. What would be the maximum heart rate for a 90-year-old women? A 15-year-old boy? (See Table 3C)

d. How long would it take you to pump 15 gallons of gas using the pump in example 3D?

e. How long would the shadow of the corral gate be on May 1? (See Table 3E)

What would the altitude of the sun be on August 1? (See Table 3E)

f. What will the world population be in the year A.D. 2010? (See Table 3F)

▶ Investigations

1. What is the closest major government research laboratory to your school? The closest industrial laboratory?

2. How did your representatives in Congress vote on the Superconducting Supercollider? How did they vote on funding for the Space Station? Why did they vote that way?

3. How are animals used in scientific experimentation? What limits should scientists accept in research using animals?

4. Malaria, the deadliest infectious disease in the world, kills more than 2 million people (mostly children in poor countries) every year. The annual malaria research budget in the United States is less than a million dollars—a minuscule fraction of the spending on cancer, heart disease, and AIDS. Should the United States devote more research funds to this disease, which does not occur in North America? Why?

5. Describe a program of scientific research carried out by a member of your school's faculty. How is the scientific method employed in this research?

6. Identify a current piece of legislation relating to science or technology (perhaps an environmental or energy bill). How did your representatives in Congress vote on this issue?

7. Look at a recent newspaper article about science funding. Are the projects described big or little science?

8. Find a science story in a newspaper or popular magazine. Did it originate at a scientific meeting? Which one?

9. How were scientists depicted in the novel and film of *Jurassic Park* by Michael Crichton? Were you convinced by these portrayals? Why?

10. Design an experiment to test the relative ability of three different brands of paper towel to absorb water. What data would you need to collect?

11. Take the published height and weights of 10 members of your school's women's volleyball team and 10 members of the men's basketball team and present these data in a table and graphically. Do find the same trends or different ones compared to those for your friends? What explanations can you give for your comparisons?

▶ Additional Reading

Alvarez, Luis. *Alvarez: Adventures of a Physicist.* New York: Basic Books, 1987.

Gross, Paul, and Norman Leavitt. *Higher Superstition.* Baltimore: Johns Hopkins University Press, 1994.

Harre, Rom. *Great Scientific Experiments.* New York: Oxford University Press, 1981.

Holton, Gerald. *Thematic Origins of Scientific Thought: Kepler to Einstein.* Cambridge, Mass.: Harvard University Press, 1988.

Kevles, Daniel. *The Physicists.* Cambridge, Mass.: Harvard University Press, 1971.

Kuhn, Thomas. *The Structure of Scientific Revolutions.* University of Chicago Press, 1962.

2

THE LANGUAGE

OF SCIENCE

MATHEMATICS IS THE UNIVERSAL LANGUAGE
OF SCIENCE.

The Undecipherable Blackboard

To many people, science means obscure equations written in strange, undecipherable symbols. Next time you're in the science area of your college or university, look into an advanced classroom. Chances are you will see a confusing jumble of unintelligible formulas on the chalkboard. Alongside the equations there might be lists of chemicals, tables crammed with numbers, and assorted graphs and charts.

Have you ever wondered why scientists need all those complex mathematical equations? Science is supposed to help us understand the physical world around us, so why can't scientists just use plain English?

"I THINK YOU SHOULD BE MORE EXPLICIT HERE IN STEP TWO."

© Sidney Harris

QUANTIFYING NATURE

Take a stroll outside and look carefully at a favorite tree. Think about how you might describe the tree in as much detail as possible so that a distant friend could envison exactly what you see, and distinguish that tree from all others.

A cursory description would note the rough brown bark, branching limbs, and canopy of green leaves, but that description would do little to distinguish your tree from most others. You might use adjectives like lofty, graceful, or stately to convey an overall impression of the tree. Better yet, you could identify the exact kind of tree and specify its stage of growth—a sugar maple at the peak of autumn color, for example—but even then your friend has relatively little to go on.

There are many ways to describe a tree, from focusing on its shape to a discussion of the complex chemistry that goes on in its leaves.

Your description would be enhanced by giving exact dimensions of the tree—how tall it is, the distance spanned by its branches, or the diameter of the trunk. You could document the shape and size of leaves, the texture of the bark, the angles and spacing of the branching limbs, and the tree's approximate age. You could be even more exact and calculate the number of board feet of lumber the tree could produce.

Furthermore, you might examine the tree for moss and insects living on the trunk and for evidence of disease on the leaves. The more detailed your description, the more varied the vocabulary you would need to command, and the more precise your measurements of the tree's many parts would have to be. Photographs and other illustrations might be included to supplement your written report. Ultimately, you might even probe the tree at the microscopic level, examining the cells and molecules that give the tree its unique characteristics.

For each different kind of description of the tree, there is an appropriate language. For some uses, words might be sufficient. If, for example, you were doing a census of a particular group of trees, a simple "oak tree" might be enough to get the job done. For other uses—for instance, including an image of the tree in a decorating scheme, you might want a picture or a geometrical shape. For still others, such as a quantitative description of the tree's energy balance or its economic value as lumber, you would probably need to use numbers. All of these are useful descriptions of the tree, but each is appropriate for answering a different question about the tree.

Scientists are constantly faced with the challenge of describing our world. Their solution to the problem invariably involves developing a complex vocabulary, coupled with appropriate mathematical expressions—the language of science.

One way to look at a tree is in terms of how much lumber it will produce.

Language and Science

What makes a language useful? First and foremost, a language must be able to communicate a wide range of expression without ambiguity or confusion. In most day-to-day activities, two or three thousand words suffice for basic communication, but as soon as you deal with a complex system, such as an automobile, the vocabulary may increase dramatically. Think about the last time you had to have your car repaired, for example. The mechanic first had to know which of the hundreds of makes, models, years, and engine types you own. The mechanic then had to identify which of the thousand or so automobile parts was defective. Just to describe the problem, the mechanic has to master thousands of words—the specialized vocabulary of automobiles.

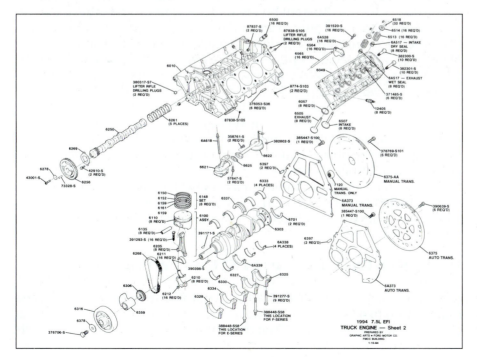

To be a mechanic, you have to learn a specialized vocabulary that you don't need just to operate a car.

A catalog of parts, alone, however, is insufficient to describe your automobile and what it does. Other statements are needed to describe the car's operation. An engine must idle at a prescribed speed, for example, and the tires must be inflated to a safe pressure. All of these conditions and hundreds more are measured by various gauges and sensors, which are critical to the operation of your car. These quantities are best described by numbers, not words. Indeed, almost everything to do with the mechanics of driving—speed, acceleration, distance, and time—is expressed by numbers.

The same situation applies to many other things we do in everyday life. To prepare your meals you must contend with the complex vocabulary of food, including numerous varieties of fruits and vegetables, dozens of cuts of meat and types of seafood, shelves of herbs and spices, and so on. But any cook needs numbers—quantifiable information—as well, to communicate the details of a recipe: how much, how hot, how long? Similarly, virtually all sports have evolved specialized vocabularies, and they often employ sophisticated mathematical scales to measure performance: earned run average, third down efficiency, serving percentage, and a host of other parameters, which enliven sports reporting.

Communication in science poses special challenges because, like your automobile, natural systems are complex in design and operate according to strict quantitative guidelines. And, like cooking and sports, science involves complex procedures that must be documented with precision so that others can try the activity for themselves.

Learning the Language of Science

Memorizing complex vocabulary is an integral part of learning *to do* science. Doctors and medical researchers, for example, must be able to refer to thousands of different bones, muscles, nerves, and other anatomical features. Chemists must have command of the names of more than a hundred elements and countless chemical compounds. The vocabulary of every geologist includes minerals, rocks, fossils, and ancient time periods. Without a specialized vocabulary, communication between specialists would be all but impossible.

The average person doesn't need to know all of the specialized vocabulary to deal with the science that typically comes his or her way. You don't have to learn all the mechanic's jargon to know if your car is running properly; however, if you decide to become a mechanic yourself, you'll need a lot of specialized training, which includes the vocabulary. Similarly, you don't have to be a master chef to enjoy good food, or be a star athlete to appreciate sports. The same is true of science—you can appreciate science without having to become a scientist.

DESCRIBING THE PHYSICAL WORLD

The challenge of describing the vast and complex universe may be divided roughly into two tasks. First, scientists must describe all kinds of physical objects, from atoms to stars. Then they must document how these objects interact and change over time. Both of these jobs rely, in large measure, on mathematics.

Describing Objects: What Is It?

We can't understand how the universe works unless we know its components. For hundreds of years, astronomers plotted the position of every visible star, while geographers mapped the features of our globe. Naturalists travelled to the ends of the Earth collecting every possible rock, shell, flower, and other curio for their museum collections. In our own century, the discovery of vast numbers of galaxies, disease-causing viruses, and a complex zoo of subatomic particles have transformed our understanding of the universe.

Describing new objects requires the ability to identify the features that distinguish one object from all others. To a certain extent, these descriptions rely on words, which is why the vocabulary of science has become so complex. A rock, for example, might be described as rose pink, fine-grained, silica-rich, and intrusive. But, eventually, such a description has to incorporate numbers for added precision. What is the average size of the grains? How much silica is contained in the rock? What are the light-absorbing properties that give the rock a pink color?

Scalars and Vectors

All the descriptions and units we've discussed so far can be expressed as a single number—you buy one gallon of paint, or you drive 10 miles to work. Any quantity that can be expressed as a single number is called a **scalar.** Scalars are crucial to the description of the physical world. As a consumer you are besieged by scalars: the wattage of a light bulb, the octane rating of gasoline, the efficiency of appliances, and the voltage of your car's battery. You pay for coffee by the pound, fabric by the yard, milk by the gallon, and electricity by the kilowatt-hour. In science we measure the size of microbes, the mass of stars, the density of crystals, and the temperature of our bodies. We will even find that the colors of light may be represented as a scalar quantity (see Chapter 8).

Sometimes you can't give a description in terms of a single number. If you were giving a friend instructions to your favorite restaurant, for example, you might say something like "Go north three miles on Main Street." Here you have to give a scalar ("three miles") and some additional information (in this case, the direction "north"). When a definition requires two numbers, such as a magnitude and a direction, it is called a **vector.** We will discuss vectors in more detail in Chapter 3.

The Dynamics of the Physical World: How Does It Change?

If scientists just described objects in the universe, science would seem pretty boring. What makes science fascinating and useful is that systems change. Science is a search to understand and predict these changes—the dynamics of our physical world.

Change can be described in words, in tables of numbers, or visually through the use of graphs. Of special importance are *equations*, which define a precise mathematical relationship among two or more measurements. Let's look at an example to see how the same physical phenomenon can be described in three different ways.

A biologist examining the forest canopy in Panama.

scalar Any number which measures a physical quantity such as weight, length, or time.

vector A quantity measured with two scalar numbers that also has a direction, such as velocity, acceleration, or momentum.

As you will see in Chapter 11, a bar of iron will expand when heated. A researcher might carefully measure the length of a 1-meter iron bar at a series of temperatures and prepare a table as follows:

Table 2–1 • Thermal Expansion of an Iron Bar

Temperature (°C)	Length (meters)
0	1.0000
100	1.0056
200	1.0115
300	1.0179
400	1.0244

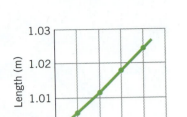

Figure 2–1
A graph of temperature versus the length of an iron bar illustrates how two scalar properties are related.

These data can be described in several different ways.

1. We can describe what happens to the iron bar in words:
    When we heat a 1-meter iron bar, it gets longer.

2. We can express this idea as an equation with words:
    The length of the bar equals the original 1-meter length plus a constant times temperature.

3. We can use an equation with symbols and numbers (approximately):
    $L = 1 + (0.00006 \times T)$

4. Finally, the data might be displayed in graphical form, as a plot of temperature versus length (see Figure 2–1).

You may have noticed that the data points in Figure 2–1 don't all fall exactly on the line given by the equation; they display some *scatter*. All experimental measurements, such as the length or temperature of an iron bar, contain such errors. In the case of thermal expansion, furthermore, the simple equation represented by the line in Figure 2–1 only approximates iron's more complex behavior. At higher temperatures, for example, iron expands at a faster and faster rate, causing the line in Figure 2–1 to curve upward.

Similar relationships are found in all scientific literature. Researchers graphically document changes in the volume of a gas with pressure, changes in the distance objects fall with time, changes in the growth rate of bacteria with concentration of nutrients, and countless other trends.

The example of the expanding iron bar illustrates why scientists use the language of mathematics. Table 2–1 certainly represents the data, and in a modern experiment that list of numbers could easily run to thousands or even millions of entries. Yet all of that information may be packaged into a one-line equation. Thus the use of mathematics allows us to express the results of experiments in a highly compressed and convenient form and to predict or extrapolate beyond what has been measured.

As we shall see later, equations have the added advantage of providing us with the best way to make predictions about the behavior of our surroundings. And, finally, they transcend national barriers in that they have exactly the same meaning all over the world.

Blacksmith with a hot iron bar.

Equations allow us to describe with precision the behavior of objects in our physical world. One such equation predicts the behavior of falling objects.

Developing Your Intuition

Fuel Efficiency

How would you describe the gas efficiency of your automobile? A colloquial answer might be, "I get pretty good mileage, especially on Interstates."

Most people would accept that answer, but it wouldn't be very useful in trying to compare two different cars. To give a more accurate answer, you could keep exact records of your car's mileage and the amount of gas purchased each time you fill up the tank. By dividing the total miles driven by the number of gallons purchased, you could calculate:

$$\text{miles per gallon} = \frac{\text{total miles driven}}{\text{gallons of gas purchased}}$$

Now you could reply with a scalar quantity, "I get about 30 miles per gallon."

Your answer could be even more precise if you record additional notes. What was the brand and grade of fuel? Was the driving between each fillup in the city or on high-speed roads? Was the air conditioner used? Did you recently have an oil change? Were the tires properly inflated? What were the weather and road conditions? With sufficiently detailed records you might be able to say, "My car, when properly serviced and fueled with regular unleaded gas, averages 33.7 miles per gallon when traveling 55 miles per hour on level, dry interstate highways, and approximately 25.5 miles per gallon in city traffic. Use of the air conditioner reduces these values by about 2.5 miles per gallon."

Automobile manufacturers, who must document the fuel efficiency of their vehicles, carry this process a step further by running carefully controlled mileage experiments on dozens of cars in special labora-

Even buying gasoline involves using numbers and units.

tories. There, engineers develop graphs and equations that relate fuel consumption to numerous other variables. Many of these tests are now mandated by the Environmental Protection Agency to provide consumers with an accurate measure of each brand's fuel efficiency. ●

Equations: Modeling the World

Scientists have devised many ways to describe the natural world. As shown in the example of the expanding iron bar, the behavior of a physical system may be documented in words, in tables of numbers, or in graphs. But no description is more compact and efficient than an equation. A brief survey will help you visualize the everyday reality underlying four common kinds of equations that are used in this book: direct, inverse, power law, and inverse square.

1. Direct Relationships.   The simplest equations consist of a *direct relationship* between two variables, *A* and *B*, in the form

$$A = k \times B,$$

where *k* is called a *constant of proportionality*. You use a direct relationship every time you buy gas by the gallon, or food by the pound:

$$(\text{cost}) = (\text{price per pound}) \times (\text{weight}).$$

In this case, two variables, the weight and the cost, are related by a constant of proportionality called the price per pound.

In a direct relationship, the two variables change together: if weight doubles, so does the cost; if weight triples, so does the cost. We say that the cost is *proportional* to the weight (Figure 2–2). In subsequent chapters, we will find many direct relationships between pairs of variables, including:

- Acceleration is proportional to force (Chapter 4).
- Electric power is proportional to electrical current (Chapter 7).
- Wave frequency is proportional to wave velocity (Chapter 8).

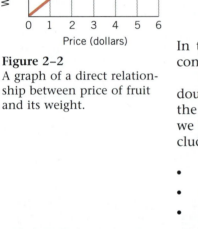

Price = $1.25 per pound × weight

Figure 2–2
A graph of a direct relationship between price of fruit and its weight.

The price per pound of produce is a constant of proportionality between weight and cost.

2. Inverse Relationships.

In many everyday situations, one variable increases as another decreases—a situation called an *inverse relationship*.

$$A = \frac{k}{B}$$

Think, for example, about assembly lines at which workers who work at different speeds produce toaster ovens. The shorter the time, T, it takes for a worker to produce one oven, the greater the number, N, of ovens produced per hour (Figure 2–3). We say that the number of ovens produced is *inversely proportional* to the production time:

$$(\text{output of ovens}) = \frac{(\text{constant})}{(\text{production time per oven})},$$

or,

$$N = \frac{k}{T}.$$

In this case, the constant k is a time, perhaps eight hours or one week.

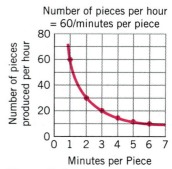

Number of pieces per hour
= 60/minutes per piece

Figure 2–3
A graph of an inverse relationship between the time it takes a worker to assemble an item, and the worker's hourly output. The shorter the assembly time per item, the greater the hourly output.

The output of automobiles is inversely proportional to the production time per automobile.

Think about the behavior of the two variables, production time and output in this case. If you make an oven in half the time of a fellow worker, you will produce twice as many ovens in any given time period. If you produce an oven in a third of the time, you'll produce three times as many, and so forth. Inverse relationships thus lead to the distinctive kind of curving graph illustrated in Figure 2–3. We will encounter many examples of such inverse relationships, such as:

• For a given force, acceleration is inversely proportional to the mass being accelerated (Chapter 4).

• The wavelength of light is inversely proportional to the frequency of light (Chapter 8).

This is a microphotograph of rod-shaped bacteria on the head of a household pin seen at progressively greater magnifications: ×9, ×36, ×875, and ×4,375.

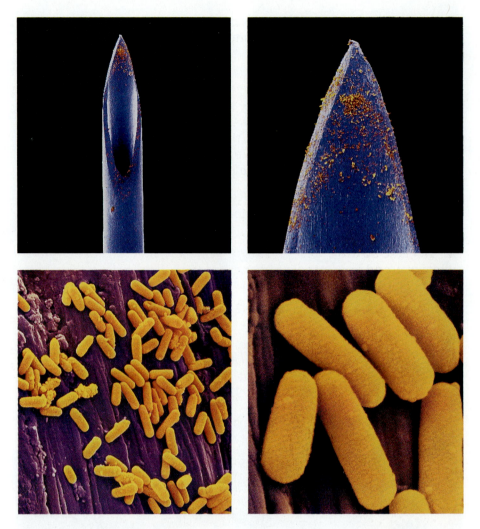

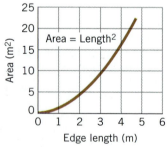

Figure 2–4
A graph of a squared relationship between the area and edge length of a square.

3. Power Law Relationships. You may recall from a geometry class the equation that defines the area of a square, A, in terms of the edge length, L:

$$A = L \times L,$$

which can be rewritten as

$$A = L^2.$$

Area is said to be equal to length *squared* (Figure 2–4). Squared relationships are common in our daily lives. For example, in Chapter 3 we will see that a dropped object falls a distance that is proportional to the time of fall squared.

A similar kind of relationship is found between the *volume* of a cube, V, and its edge length, L:

$$V = L \times L \times L,$$

or, in mathematical notation,

$$V = L^3.$$

Note that volume is measured in units of distance *cubed*, such as cubic meters. Many systems of measurement adopt special volume units, such as liters or gallons.

Squared-type and cubed-type equations are special cases of a more general class, called *power law equations*. These common equations have the form

$$A = B^n,$$

where *n* is any number. A general feature of these relationships is that one variable changes much more quickly than another; double the edge of a cube, for example, and the volume increases eightfold; triple the edge length, and the volume becomes 27 times larger. The result is a steeply rising graph, as shown in Figure 2–4.

4. Inverse Square Relationship. Imagine shining a flashlight on a white wall and measuring the brightness with a photographic light meter. If you move the flashlight twice the distance from the wall and make the same measurement, what will you observe? The flashlight will illuminate an area twice as high and twice as wide, so the light will be spread out over an area four times as large as before (see Figure 2–5*a*). Consequently, the light meter will read an intensity only one fourth as strong as before. This relationship between light intensity, *I*, and distance, *D*, is called an *inverse square relationship*:

$$I = \frac{k}{D^2}$$

where *k* is a constant.

A graph of an inverse square relationship (Figure 2–5*b*) reveals the sharp fall-off of one variable as the other changes gradually. We will find in later chapters that inverse square relationships are common in nature; for example, an inverse square relationship describes the change of magnitude of everyday electrostatic and gravitational forces as a function of the distance between two objects (see Chapters 4 and 7).

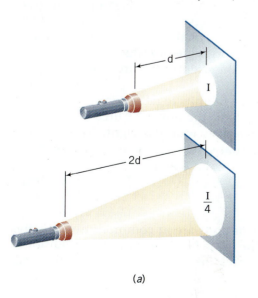

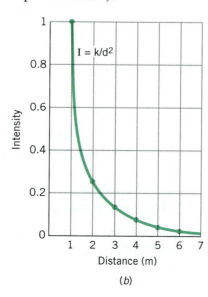

(a)   (b)

Figure 2–5
(*a*) A light shines on a white wall. As the distance from the light to the wall doubles, the light intensity decreases by $\frac{1}{4}$, in an inverse square relationship. (*b*) A graph of distance versus intensity for an inverse square relationship.

Different products are sold in different units at a hardware store.

UNITS AND MEASURES

As soon as we begin to talk about using numbers to describe any physical system, we have to deal with the question of units. Walk into any hardware store in the United States and you will notice immediately that the things for sale are measured in many different ways. You buy paint by the gallon, insulation in terms of how many BTUs will leak through it, grass seed by the pound. In some cases, the units are strange indeed—nails, for example, are ranked by "penny" (abbreviated "d"). A 16d nail is a fairly substantial thing, perfect for holding the framework of a house together, while a 6d nail might find use tacking up a wall shelf.

No matter what the material, there is a unit to measure how much is being sold. In the same way, in all areas of science, systems of units have been developed to measure how much of a given quantity there is. We will encounter many of these units in the text—the newton as a measure of force, for example, and the degree Celsius as a measure of temperature. Every quantity used in the sciences has an appropriate unit associated with it.

system of units A method of measuring objects using a common set of standard quantities.

We customarily use certain kinds of units together, in what is called a **system of units.** In a given system, there will be units assigned to fundamental quantities such as mass (as well as weight, length, time, and temperature). Someone using that system will use only those units and ignore the units associated with other systems.

The International System

In the United States, two systems of units are in common use. The one encountered most often in daily life is the *English system.* This traditional system of units has roots that go back into the Middle Ages. The basic unit of length is the foot (which was actually defined in terms of the average lengths of the shoes of men outside a certain church on a certain day), and the basic unit of weight is the pound.

Throughout this book, and throughout most of the world outside of the United States, the so-called **metric system** or, more correctly, **International System** (abbreviated **SI,** for Système Internationale), is preferred. In this system, the unit of length is the meter and the unit of mass is the kilogram. In both the SI and English system, the basic unit of time is the second.

In all probability, the unit from the metric system with which you are most familiar is the liter, a measure of volume. A liter is the volume enclosed by a cube ten centimeters on a side, or 1,000 cubes one centimeter on a side. Soft drinks and other liquids are routinely sold in one- and two-liter bottles in the United States. The cubic meter—the volume contained in a cube one meter on a side—is also often used as a volume measure in the metric system. The measure of volume in the English system is the cubic foot, but liquids are commonly measured in gallons (3.79 liters), quarts ($\frac{1}{4}$ gallon) and pints ($\frac{1}{8}$ gallon).

Within the SI system, units are based on multiples of 10. Thus the centimeter is 1 one-hundredth ($\frac{1}{100}$) the length of a meter, the millimeter 1 one-thousandth ($\frac{1}{1000}$) and so on. In the same way, a kilometer is 1000 meters, a kilogram is 1000 grams, and so on. This organization differs from that of the English system, in which a foot equals 12 inches, and 3 feet make a yard. A list of metric prefixes follows.

metric system or international system (SI) An internationally recognized system of measurement comprised of units based on multiples of ten; standard units include the meter, kilogram, liter, and second.

The most commonly seen metric unit in the United States is the liter, which is used to measure volumes of soft drinks.

Metric Prefixes

If the prefix is:	Multiply the basic unit by:
giga-	billion (thousand million)
mega-	million
kilo-	thousand
hecto-	hundred
deka-	ten

If the prefix is:	Divide the basic unit by:
deci-	ten
centi-	hundred
milli-	thousand
micro-	million
nano-	billion

Science in the Making

The History of Systems of Units

Ever since humans started engaging in commerce, there has been a need for agreements on weights and measures. Merchants needed to be assured that they were buying and selling the same thing, that buyers were getting what they paid for, and that sellers were receiving full value for their wares. This meant that someone (often a government) had to set up and maintain a system of standard weights and lengths.

An Egyptian tomb painting c. 2000 B.C. A scribe keeps tally as the balance operator weighs against stone weights.

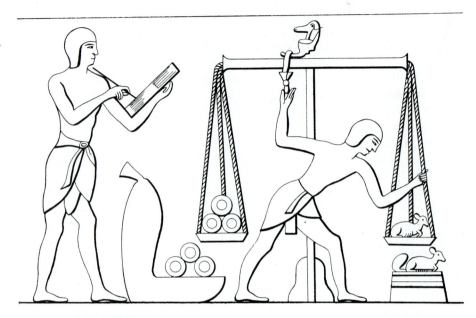

King John grants the "Magna Carta" (Great Charter) to his barons in England. The charter required the establishment of common standards for weights and lengths throughout the kingdom.

The oldest weight standard we know about is the Babylonian *mina*, which weighed between one and two pounds. Archaeologists have found standard stone weights carved in the shapes of ducks (5 mina) and swans (10 mina). In medieval Europe, almost every town maintained its own system of weights and measures, and the only institutions pushing for universal standards were the great trade fairs. The Keeper of the Fair in Champagne, France, for example, kept an iron bar against which all bolts of cloth sold at the fair had to be measured. The Magna Carta, signed by King John of England in 1215 and generally reckoned to be one of the key documents in the history of democracy, required that "There shall be standard measures of wine, ale, and corn throughout the kingdom." The English system of units eventually evolved from the welter of medieval systems.

The metric system, on the other hand, was a product of the French Revolution at the end of the eighteenth century. In 1799, the French Academy recommended that the length standard be the meter, then defined to be 1 ten-millionth of the distance between the equator and the North Pole at the longitude of Paris, and that the gram be defined to be the mass of a cubic centimeter of water at 4°C. In the Technology section in this chapter, we discuss the modern definitions of these quantities. ●

Conversion Factors

All systems of units facilitate the description of physical objects and events. Confusion may arise, however, when switching back and forth between two different systems. *Conversion factors*, which are used to shift from one system of units to another, are thus vital in both science and commerce. If you have ever visited a foreign country you have had direct experience with this process; you had to use conversion factors all the time, when converting dollars to some other currency.

Hundreds of conversion factors apply to the physical world. One person may give temperature in degrees Fahrenheit, another in degrees Celsius. Distance may be recorded in centimeters or in inches. A number of important conversion factors are tabulated in Appendix A.

Example 2–1:   Driving in North America

Suppose you are driving in Canada. The odometer on your car reads 20,580 miles. You see a sign that says "Toronto 87 kilometers." What will your odometer read when you get to that city?

▶ **Reasoning:**   Since the odometer reads in miles, the first thing to do is convert 87 kilometers to miles by using the conversion tables in Appendix A. We will then add that mileage to the current reading to get our answer.

▶ **Solution:**   From Appendix A, the conversion factor from kilometers to miles is 0.6214. When you see the sign, then, the distance to Toronto is

$$87 \text{ kilometers} \times 0.6214 \text{ kilometers/mile} = 54 \text{ miles}.$$

When you have traveled this far the odometer will read

$$20,580 + 54 = 20,634 \text{ miles}. \quad ▲$$

What are the advantages and disadvantages of conversion to metric units in the United States?

Example 2–2:   Running the Dash

American athletes used to run an event called the 100-yard dash. If an athlete could run the 100-yard dash in 10 seconds, what time would you expect her to have in the 100-meter dash?

▶ **Reasoning and Solution:** The first step is to use the conversion factor in Appendix A to convert 100 meters to a distance in feet. From Appendix A, the conversion factor for meters to feet is 3.281. Consequently, 100 meter is

$$100 \text{ meters} \times 3.281 \text{ feet/meters} = 328.1 \text{ feet}.$$

We then have to divide the number of feet by 3 to convert to yards.

$$\frac{328.1 \text{ feet}}{3 \text{ feet/yard}} = 109.4 \text{ yards}$$

If we assume that the runner travels at the same speed in the two races, the 100-meter race (equal to a 109.4-yard race) will take longer than the 100-yard race. The time of the 100-meter race is proportional to the distance of the two races multiplied by 10 seconds.

$$\text{time} = 10 \text{ seconds} \times (109.4 \text{ yards}/100 \text{ yards})$$
$$= 10.94 \text{ second} \ \blacktriangle$$

Science by the Numbers

Powers of 10

Very large or very small numbers may be written conveniently in a compact way that doesn't involve writing down a lot of zeroes. The so-called "powers of 10" notation accomplishes this goal. The basic rules for the notation are:

1. Every number is written as a number between 1 and 10 followed by 10 raised to a power, or an exponent.
2. If the power of 10 is positive, it means "move the decimal point this many places to the right."
3. If the power of 10 is negative, it means "move the decimal point this many places to the left."

Thus, for example, $3.56 \times 10^3 = 3560$, and $7.87 \times 10^{-4} = 0.000787$.

Multiplying or dividing numbers with powers of 10 requires special care. If you are multiplying two numbers, such 2.5×10^3 and 4.3×10^5, you multiply 2.5 and 4.3, but you add the two exponents.

$$(2.5 \times 10^3) \times (4.3 \times 10^5) = (2.5 \times 4.3) \times 10^{3+5}$$
$$= 10.75 \times 10^8$$
$$= 1.075 \times 10^9$$

When dividing two numbers, such as 4.3×10^5 divided by 2.5×10^3, you divide 4.3 by 2.5, but you subtract the denominator exponent from the numerator exponent.

$$\frac{(4.3 \times 10^5)}{(2.5 \times 10^3)} = \frac{4.3}{2.5} \times 10^{5-3}$$
$$= 1.72 \times 10^2$$
$$= 172 \ \bullet$$

Technology

Maintaining Standards

Systems of units are one place where governments become intimately involved with science, since the maintenance of standards has traditionally been the task of governments. When you buy a pound of meat in a supermarket, for example, you know that you are getting full weight for your money because the scale is certified by a state agency, which relies, ultimately, on international standards of weight maintained by a treaty among all nations.

Originally, the standards were kept in sealed vaults at the International Bureau of Weights and Measures near Paris, with secondary copies at places such as the National Institutes of Standards and Technology (formerly National Bureau of Standards) in the United States. The meter, for example, was defined as the distance between two marks on a particular bar of metal, and the kilogram was defined as the mass of a particular block of iridium-platinum alloy. The second was defined as a certain fraction of the length of the year.

Today, however, only the kilogram is still defined in this way. Since 1967, the second has been defined as the time it takes for 9,192,631,770 wave crests of a certain type of the light emitted by a cesium atom to pass by a given point. In 1960, the meter was defined as the length of 1,650,763.73 wavelengths of the radiation from a krypton atom, and in 1983 redefined to be the distance light travels in [1/299,792,485] second. In both these cases, the old standards have been replaced by numbers relating to atoms—standards that any reasonably equipped laboratory can maintain for itself. Atomic standards have the additional advantage of being truly universal—every cesium atom in the universe is equivalent to any other. Only mass is still defined in the old way, in relation to a specific block of material kept in a vault, and scientists are working hard to replace that standard by one based on the mass of individual atoms. ●

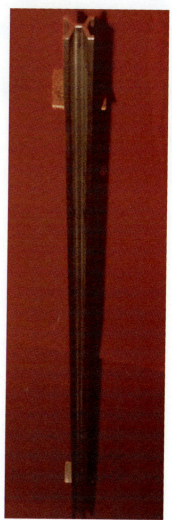

The meter, which used to be defined in terms of the distance between marks on this bar, is now defined in terms of measurements on atoms.

National Institutes of Standards and Technologies headquarters in Gaithersburg, MD.

In the United States, land area is commonly measured in a unit called the acre. Gemstones are usually ranked in terms of a weight unit called the carat, equal to a fifth of a gram. Firewood is usually measured in terms of a volume unit called the cord.

Units You Use in Your Life

The metric and English systems each give a comprehensive set of units that could, in principle, be used to measure everything we encounter in our lives. In point of fact, for various historical and technical reasons, we use units that don't fit easily into either system all the time. How many of the following do you recognize?

Acre. Used to measure land area in United States (43,560 square feet, equal to the area of a square 640 feet on a side).

Acre-foot. The amount of water that produces a depth of one foot on an acre of land (hence 640 cubic feet). This unit is widely employed in the western United States to measure water used in irrigation systems.

Barrel. International unit for oil production (42 gallons, though many different specialized definitions of barrel exist for other commodities, including wine, spirits, and cranberries).

Bushel. Used to measure production of grains in the United States (1.24 cubic feet).

Caliber. Used to measure diameter of bullets and gun barrels (0.01 inches).

Carat. Used to measure size of gemstones (0.2 grams).

Cord. Used to measure firewood (128 cubic feet).

Fathom. Used to measure depth of navigable water (6 feet).

Furlong. Used to mark distances for certain horse races (660 feet).

Grain. Used to measure weight of bullets or arrowheads ($\frac{1}{7000}$ pound).

Hand. Used to measure height of horses (4 inches).

Knot. Used to measure speed of ships (1.85 kilometers per hour).

Ounce. Used to measure the weight of produce ($\frac{1}{16}$ pound).

Troy ounce. Used to measure precious metals ($\frac{1}{12}$ pound).

THINKING MORE ABOUT SCIENTIFIC LANGUAGE

Conversion to Metric Units

Why does the United States still use English units long after most of the rest of the world has converted to SI units? It may have to do wih nonscientific factors such as the geographical isolation of the country, the size of our economy (the world's largest), and, perhaps most importantly, the expense of making the conversion. (Think, for example, of what it would cost to change all of the road signs on the Interstate Highway System so that the distances read in kilometers instead of miles.)

To understand the debate over conversion, you have to realize one important point about units. There is no such thing as a "right" or "scientific" system of units. Units can only be convenient or inconvenient. Thus U.S. manufacturers who sell significant quantities of goods in foreign markets long ago converted to metric standards to make those sales easier. Builders, on the other hand, whose market is largely restricted to the United States, have not.

By the same token, very few scientists actually use the SI exclusively in their work. Almost every discipline, including physics, chemistry, geology, biology, and astronomy, has its own preferred non-SI units for some measurement. Astronomers, for example, often measure distance in light years or in parsecs, geologists usually measure pressure in kilobars, and many physicists prefer to record energy in electron volts. In the United States, many engineers use English units almost exclusively—indeed, when the federal government was considering a tax on energy use in 1993, it was referred to as a "BTU tax" (the BTU, or British thermal unit, is the unit for energy in the English system). Hospital and medical professionals routinely use the so-called cgs system, in which the unit of length is the centimeter and the unit of mass is the gram. Next time you have blood drawn, take a look at the needle. It will be calibrated in cc—cubic centimeters.

Given this wide range of units actually in use, how much emphasis should the U.S. government give to metric conversion? How much should the government be willing to spend on the conversion process: how many new signs as opposed to how many repaired potholes on the roads?

▶ Summary

Language allows people to communicate information and ideas. In their efforts to describe the physical world with accuracy and efficiency, scientists have created many new words to distinguish the many different kinds of objects in the universe.

These descriptive terms are amplified by mathematics, which allows scientists to quantify their observations. *Scalars* are numbers that indicate a quantity—mass, length, temperature, and time are familiar scalar quantities. Objects in motion must be described with a *vector*, which combines information on both speed (a scalar) and direction.

Many scientists work to find mathematical relationships between two or more properties—temperature and the length of a metal bar, for example. Such relationships may be presented in the form of an equation or graph.

Scientific measurements rely on a *system of units*, particularly the *metric system* or *international system (SI)*.

▶ Review Questions

1. Give three examples of scientific fields that require specialized vocabulary.

2. What is a scalar quantity? Give an everyday example.

3. What is a vector quantity? Give an everyday example.

4. What is a system of units? What system do scientists use?

5. What are units of length, mass, and volume in the English system of units?

6. What are units of length, mass, and volume in the metric system of units?

7. What is a conversion factor?.

8. Identify three units of measurement that you might encounter at a grocery store.

9. Which branch of the United States government regulates standard weights and measures?

▶ Fill in the Blanks

English system	pound
foot	scalar
kilogram	second
meter	vector
metric system, or international system (SI)	

Mathematics and numbers are the language of science. A quantity described by a single number is called a _____, while one that requires two numbers is called a _____. In the United States, the most commonly used system of units is called the _____, while most of the rest of the world uses the _____. The units of length in these two systems are the _____ and the _____, respectively, but both use the _____ to measure time. The unit of mass in the SI system is the _____, while the unit of weight in the English system is called the _____.

▶ Discussion Questions

1. Why is scientific vocabulary still growing?

2. What is the role of equations in science?

3. Discuss the relative advantages of the English and SI systems of units and measurements.

4. Why do we need standards of weights and measures?

5. Describe the ways a scientist might present quantitative data.

6. Categorize the following as either a scalar or vector quantity. Explain your reasoning for each choice.

    a. a 100-yard dash
    b. 20,000 leagues under the sea
    c. a blue dress
    d. 4 days
    e. 100 degrees Celsius
    f. A.D. 1996
    g. one mile northeast
    h. the lower 40 acres
    i. 12:45 P.M.
    j. a cubic yard of cement
    k. a force of 20 newtons
    l. 40 m/sec west
    m. 20 kg
    n. 30 revolutions per minute

7. How are conversion factors related to units?

8. What is the principal difference between a vector and scalar quantity, and why is that important in science?

9. Why is powers of 10 notation important in modern science?

▶ Problems

1. In Canada a speed limit sign says 70 kilometers per hour. What is the legal speed in miles per hour?

2. How many liters are in a half-gallon container of milk?

3. A runner consistently completes a 1-mile race in

4 minutes. What is his expected time in a 1500-meter race?

4. Write the following numbers in powers of 10 notation.

    a. 1,000,000

    b. $\frac{1}{1,000,000}$

    c. 2.5

    d. $\frac{1}{2.5}$

5. Convert the following numbers into decimal notation.

    a. 7×10^4

    b. 7×10^{-4}

    c. 6.41×10^6

    d. 6.41×10^{-6}

6. The following table gives the volume of a gallon (4 quarts) of antifreeze for various temperatures initially at a temperature of 6 degrees Celsius. This antifreeze is typically used in American cars.

Volume (qts)	Temperature (°C)
4.0000	6
4.0096	12
4.0288	24
4.0672	48
4.1440	96

    a. Express any trends or patterns in words.

    b. Display the data in graphic form.

    c. Express any trends or patterns in an equation with words.

    d. Express any trends or patterns in an equation with symbols.

7. Veronica performed a laboratory assignment in which she investigated the relationship between the time of descent, final speed, acceleration, and incline angle of a marble rolling down an inclined track. Her data are given below.

Incline Angle (degrees)	Time of Descent (s)	Final Speed (m/s)	Acceleration [(m/s)/s]
5	3.75	3.2	0.85
10	2.65	4.5	1.70
15	2.17	5.5	2.54
20	1.89	6.3	3.40
30	1.56	7.6	5.08

    a. Consider only the incline angle and the acceleration.

      1. Express any trends or patterns in words.

      2. Display the data in graphic form.

      3. Express any trends or patterns in an equation with words.

      4. Express any trends or patterns in an equation with symbols.

    b. Consider only the incline angle and the time of descent.

      1. Express any trends or patterns in words.

      2. Display the data in graphic form.

      3. Express any trends or patterns in an equation with words.

      4. Express any trends or patterns in an equation with symbols.

    c. Consider only the incline angle and the final speed.

      1. Express any trends or patterns in words.

      2. Display the data in graphic form.

      3. Express any trends or patterns in an equation with words.

      4. Express any trends or patterns in an equation with symbols.

8. An industrious student named Sarah decided that she wanted to prove certain laws about gases and the relationships between pressure, volume, and temperature. In her science laboratory, she collected the following data.

Temperature (kelvins)	Volume (liters)	Pressure (atmospheres)
100	1000	1.0
100	500	2.0
100	250	4.0
100	125	8.0
200	2000	1.0
200	1000	2.0
200	500	4.0
300	750	4.0
600	1500	4.0

    a. Show by using a graph, an equation, or a written statement that the volume is directly proportional to the temperature if the pressure is held constant.

    b. Use this data to show Boyle's law, which states that at a constant temperature the pressure and volume vary inversely.

9. The brightness of a light bulb can be measured by a light meter in a unit named lumens. Jeremy decided to investigate how the brightness of a certain light bulb will change with the distance from the light bulb. He recorded the following data.

Distance from the Bulb (ft)	Brightness (lumens)
1	1600
2	400
3	178
4	100
5	64
10	16
20	4

a. Express any trends or patterns in words.

b. Display the data in graphic form.

c. Express any trends or patterns in an equation with words.

d. Express any trends or patterns in an equation with symbols.

10. Ron used 6980 kilowatt-hours (kwh) of electricity last year. Jennifer used 5235 kwh in only 10 months; the other two months she lived with her parents and did not have to pay for electricity.

a. Who used the most electricity per month?

b. What were Ron and Jennifer's daily use of electricity? (Assume a 30-day month.)

c. If Ron paid 8 cents per kwh and Jennifer paid 10 cents per kwh, who paid more for their electricity last year?

11. P.J. would follow the following exercise schedule. Every Sunday he would run for 30 minutes using 12 calories per minute. Monday, Wednesday, and Friday, P.J. would take brisk walks for one hour each day (4 calories per minute), and then play volleyball every Tuesday and Thursday for 1.5 hours (5.5 calories per minute). On Saturday she would not exercise.

a. Calculate the total amount of calories P.J. expends during the week.

b. How many calories per day (include Saturday) does P.J. expend?

12. The G-7 countries (U.S.A., Canada, Japan, Germany, the United Kingdom, France, and Italy) are the leading industrialized countries in the world. Their populations, total energy use, total wilderness areas, and total solid waste disposed for 1991 are given in the following table.

Name	Population (millions)	Energy Use (10^9 BTU)	Wilderness Area (sq miles)	Solid Waste Disposed (10^6 tons)
U.S.A.	249.2	76,355	379,698	230.1
Canada	26.5	10,309	2,473,308	18.1
Japan	123.5	15,707	9,276	53.2
Germany	77.5	13,881	19,128	21.0
U.K.	57.2	8,575	17,912	22.0
France	56.1	8,355	18,454	30.2
Italy	57.1	6,579	5,021	19.1

a. Which country uses the most energy per person? The least energy per person?

b. Which country has the most wilderness area per person? The least per person?

c. Which country disposes the most solid waste per person? The least per person?

d. If you combined the four European countries into one individual country, would your answers to a, b, and c change?

13. Express the following in powers of 10 notation.

a. 150 giga dollars

b. 43 hecto feet

c. 23 micrometers

d. 92 nanoseconds

e. 74 milligrams

f. 617 kilo bucks!

g. 43 micro breweries

14. Complete the following multiplications.

a. $(4.3 \times 10^6) \times (7.4 \times 10^{-7})$

b. $(1.2 \times 10^{-8}) \times (3.4 \times 10^{-5})$

c. $(5.5 \times 10^3) \times (6.7 \times 10^7)$

d. $(6.6 \times 10^2) \times 120$

e. $(2.3 \times 10^{12}) \times (4.9 \times 10^8)$

15. Complete the following divisions.

a. $(3.3 \times 10^{12}) \div (3.0 \times 10^{-4})$

b. $(7.6 \times 10^{-6}) \div (8.2 \times 10^8)$

c. $(1.5 \times 10^2) \div (5.0 \times 10^7)$

d. $(2.2 \times 10^{11}) \div (4.5 \times 10^8)$

16. Convert the quantity in the first column to the units in the second column.

a. 40 acres   (square miles)

b. 23,000 acre-feet   (cubic yards)

c. 50 barrels   (liters)

d. 125 bushels   (cubic meters)

e. 50 caliber   (millimeters)

f. 50,000 carats   (grams)

g. 20 fathoms   (meters)

h. 10 furlongs   (kilometers)

i. 30 hands   (centimeters)

j. 600 knots   (kilometers per second)

k. 540 knots   (meters per second)

► Investigations

1. Identify 20 specialized terms that relate to your favorite sport. What statistics are commonly recorded, and how are they calculated?

2. Investigate the history of temperature scales. Why are two different scales, Celsius and Fahrenheit, still in use?

3. Are scientists the only people who have devised a specialized vocabulary? What other fields have their own jargon?

4. Describe your favorite tree so that another student could identify it.

5. Read a history of the French Revolution. Why should this political movement have led to a new system of weights and measures?

▶ Additional Reading

Hardy, G. H. *A Mathematician's Apology*. Cambridge, England: Cambridge University Press, 1940 (reprinted 1993).

Klein, Herbert Arthur. *The Science of Measurement: A Historical Survey*. New York: Dover, 1974 (reprinted 1988).

Paulos, John Allen. *Innumeracy: Mathematical Illiteracy and its Consequences*. New York: Hill and Wang, 1988.

3 THE ORDERED UNIVERSE

THE REGULARITIES OF THE UNIVERSE CAN BE
DISCOVERED BY OBSERVATION AND OFTEN CAN BE
SUMMARIZED IN MATHEMATICAL FORM.

Calculated Moves

How many times have you crossed a street today? Once? A dozen times? More? It's such an ordinary thing to do that you probably don't even keep track. Yet a simple act like crossing a street contains, within itself, a very important lesson about the way the universe works.

Think about what happens when you see a car approaching. You watch it for a while, estimate its speed, make an unconscious calculation about how long it will take before the car gets to your corner, and only then do you make a decision about whether to start across the street. You couldn't carry out this ordinary process if you didn't have an understanding of how objects move—an understanding born of long experience with cars and their behavior. Early in your life you observed cars in motion, came to some conclusions about their properties, and have used (and tested) that knowledge ever since.

In the same way, scientists observe the world and summarize their conclusions, often using a series of mathematical laws. This process forms the core of what we call science, a method first applied in a systematic way to a study of the sky—the science we now call astronomy.

THE NIGHT SKY

Among the most predictable things in the universe are the lights we see in the sky at night; these are the stars and planets. People who live in large metropolitan areas no longer pay much attention to the richness of the night sky's shifting patterns. But think about the last time you were out in the country on a clear, moonless night, far from the lights of town. There, the stars seem very close, very real. Now try to imagine what it was like in the nineteenth century, before the development of artificial lighting. Human beings often experienced jet black skies that were filled with brilliant pinpoint stars.

Predictability

The sky changes; it's never quite the same from one night to the next. Our ancestors noticed regularities in the arrangement and movements of stars and planets, and they wove these patterns into their religion and mythology. They knew, based on their observations, that when the Sun rose in a certain place, it was time to plant crops because spring was on its way. They came to know that there were certain times of the month when a full Moon would illuminate the ground, allowing them to continue harvesting and hunting after sunset. To these people, knowing the behavior of the sky was not an intellectual game or an educational frill; it was an essential part of their lives. It is no wonder, then, that astronomy, the study of the heavens, was one of the first sciences to develop.

By relying on their observations and records of the regular motions of the stars and planets, ancient observers of the sky were perhaps the first humans to accept the most basic tenet of science:

> **The universe is predictable and quantifiable.**

Without the predictability of physical events as we saw in Chapter 1, and our ability to quantify what we observe as discussed in Chapter 2, the scientific method could not proceed.

The stars in the night sky, showing constellations. Most of these stars are not visible in the glare of city lights.

Stonehenge

There exists no better symbol of humankind's early preoccupation with astronomy than Stonehenge, the great prehistoric stone monument on Salisbury Plain in southern England. The structure consists of a large circular bank of earth, surrounding a ring of single upright stones, which, in turn, encircle a horseshoe-shaped structure of five giant stone archways. Each arch is constructed from three massive blocks—two vertical supports several meters tall capped by a great stone lintel. The open end of the horseshoe aligns with an avenue that leads northeast to another large stone, called the "heel stone" (see Figure 3–1).

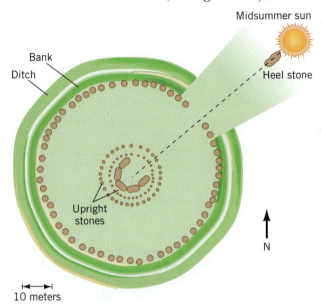

Figure 3–1
Stonehenge is built so that someone standing at the center will see the Sun rise over the heel stone on midsummer's morning.

Stonehenge was built in spurts over a long period of time, starting in about 2800 B.C. Despite various legends assigning it to the Druids, Julius Caesar, the magician Merlin (who was supposed to have levitated the stones from Ireland), or mysterious unknown races, archaeologists have shown that it was built by several groups of people, none of whom had a written language, and some of whom even lacked metal tools. Why would these people expend such a tremendous effort to erect one of the world's great monuments?

Stonehenge, like many similar structures scattered around the world, was built to mark the passing of time. It was a calendar based on the movement of objects in the sky. The most famous astronomical function of Stonehenge was to measure the seasons, because in an agricultural society people needed to know when it was time to plant the crops. At Stonehenge, this job was done by using the stones as markers. On midsummer's morning, for example, someone standing in the center of the monument would see the Sun rising directly over the heel stone, and this event would mark the beginning of a new year.

Building a structure like Stonehenge required the accumulation of a great deal of knowledge about the sky. This knowledge could have been gained only through many years of observation. Without a written language, people needed to pass complex information about the movements of the Sun, the Moon, and the planets from one generation to the next.

How else could they have aligned their stones so perfectly that modern day Druids in England could still greet the midsummer sunrise over the heel stone?

But as impressive as Stonehenge the monument, might be, Stonehenge the symbol of universal regularity and predictability is even more impressive. If the universe were not regular and predictable—if repeated observation could not show us patterns that occur over and over again—the very concept of a monument like Stonehenge would be impossible. And yet, it continues to stand after 4000 years, a testament to human ingenuity, and to the possibility of predicting the behavior of the universe in which we live.

Developing Your Intuition

Ancient Astronauts

Who built Stonehenge? Confronted by such an awesome stone monument, with its precise orientation and epic proportions, some writers evoke outside intervention by extraterrestrial visitors. Many ancient monuments, including the pyramids of Egypt, the Mayan temples of Central America, and the giant statues of Easter Island, have been ascribed to mysterious aliens. These authors and their followers refuse to accept the notion that such structures could have been built by the ingenuity and hard work of ancient people.

Such conjecture is unconvincing unless you first show that building the monument was beyond the capabilities of the indigenous people. Suppose, for example, that Columbus had found a building like the World Trade Center when he landed in America in 1492. Both the technology to produce the materials (steel, glass, and plastic, for example) and the techniques to construct a building more than 100 stories tall were beyond the abilities of Native Americans at that time. A reasonable case could have been made for the intervention of ancient astronauts or some other advanced intelligence.

Is Stonehenge a similar case? How might you discover the answer to such a question? The materials, all local stone, were certainly available to anyone who wanted to use them. Working and shaping stone was also a skill, albeit a laborious one, that was available to early civilizations. The key question, then, is whether people without steel tools or wheeled vehicles could have moved the stones from the quarry to the construction site.

Giant statues lined up on a grassy hillside on Easter Island, Chile. There's no reason to think they were made by ancient astronauts.

The largest stone, about 10 meters (more than 30 feet) in length, weighs about 50 metric tons (50,000 kilograms or about 100,000 pounds) and had to be moved overland some 30 kilometers (20 miles) from quarries to the north. Could this massive block have been moved by primitive people, equipped only with wood and ropes?

While Stonehenge was being built, the climate was cooler than now and it snowed frequently in southern England. This means that the stones could have been hauled on sleds. A single person can easily haul 100 kilograms on a sled (think of pulling a couple of your friends). How many people would it take to haul a 50,000-kilogram stone? To find the answer, we divide the total weight of the largest stone by the weight an individual could move:

$$\frac{50{,}000 \text{ kilograms}}{100 \text{ kilograms pulled by each person}} = 500 \text{ people}$$

Organizing 500 people for the job would have been a major social achievement in ancient times, but it was certainly physically possible (see Figure 3–2).

Figure 3–2
Perhaps the most puzzling aspect of the construction of Stonehenge is the raising of the giant lintel stones. Three steps in the process were probably (a) to dig a pit for each of the upright stones; (b) to pile dirt into a long, sloping ramp up to the level of the two up-rights so that the lintel stones could be rolled into place; and (c) cart away the dirt, thus leaving the stone archway.

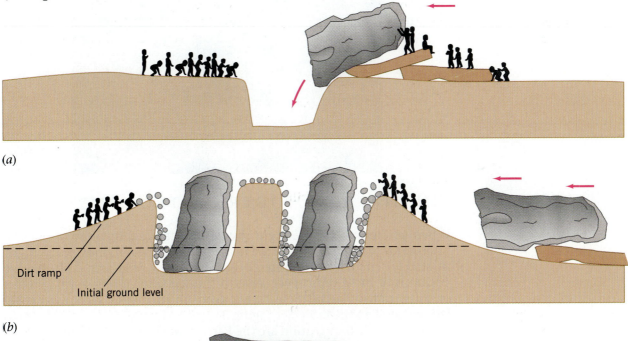

(a)

(b)

Dirt ramp

Initial ground level

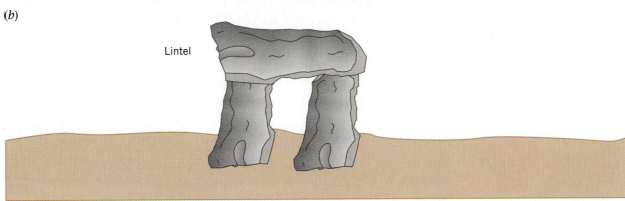

Lintel

(c)

Scientists cannot absolutely disprove the possibility that Stonehenge was constructed by some strange, forgotten technology. All of us are fascinated and awed by the mysterious and unknown, and an ancient structure like Stonehenge, standing stark and bold on the Salisbury Plain, certainly evokes these feelings. But why invoke such alien intervention when the concerted actions of a dedicated, hard-working human society would have sufficed?

When confronted with phenomena in a physical world, we should accept the simplest explanation as the most likely. This procedure is called *Occam's razor*, after William of Occam, a fourteenth-century English philosopher who argued that "postulates must not be multiplied without necessity." That is, given the choice, the simplest solution to a problem is most likely to be right. Scientists thus reject the notion of ancient astronauts building Stonehenge, and they relegate such speculation to the realm of pseudoscience. ●

Figure 3–3
This drawing of the northern sky and horizon shows the Big Dipper and Little Dipper contsellations. Polaris is also known as the North Star.

THE BIRTH OF MODERN ASTRONOMY

When you look up at the night sky, you see a dazzling array of objects. Thousands of visible stars fill the heavens and appear to move each night in stately, circular arcs centered on the north pole star. The relative positions of these stars never seem to change, and closely spaced groups of stars, called *constellations*, have been given names such as the Big Dipper and Leo the Lion (See Figure 3–3 and Appendix E.) Moving across this fixed starry background are the Earth's Moon, with its regular succession of phases, and half a dozen planets that wander through the sky. You might also see swift, streaking meteors or, more rarely, long-tailed comets, transient objects that grace the night sky from time to time.

What causes these objects to move, and what do their motions tell us about the universe in which we live?

The Historical Background: Ptolemy and Copernicus

Since before recorded history, people have observed the characteristic motions of objects in the sky and have tried to explain them. Most societies created legends and myths tied to these movements, and some (the Babylonians, for example) had long records of sophisticated astro-

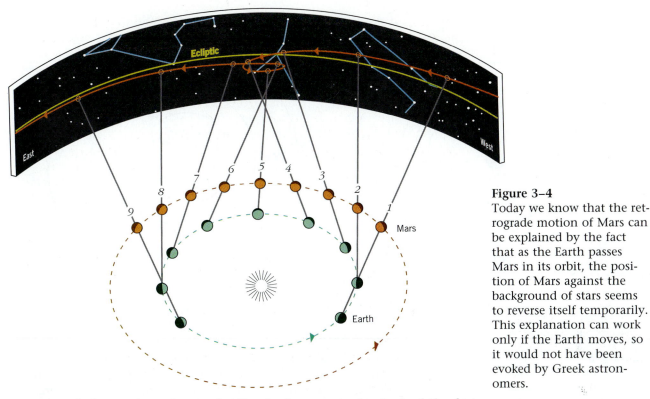

Figure 3–4
Today we know that the retrograde motion of Mars can be explained by the fact that as the Earth passes Mars in its orbit, the position of Mars against the background of stars seems to reverse itself temporarily. This explanation can work only if the Earth moves, so it would not have been evoked by Greek astronomers.

nomical observations. It was the Greeks, however, who devised the first astronomical explanations that incorporated elements of modern science.

Ancient scholars believed in the perfection of the heavens. The universe must be finite and spherical, they said, and all motions in the heavens must be represented by uniform motions of heavenly objects on spheres. Given these constraints, perhaps the greatest challenge in modeling the heavens was the occasional *retrograde* motion of planets. Most of the time a planet—Mars, for example—appears to move from east to west across the fixed background of stars. But every so often the planet's motion is retrograde: that is, for a few weeks it slows, stops, and reverses direction with respect to the stars (see Figure 3–4). What combination of circular motions could cause such behavior?

Claudius Ptolemy, an Egyptian-born Greek astronomer and geographer who lived in Alexandria in the second century A.D., proposed the first plausible explanation for complex celestial motions. Working with the accumulated observations of earlier Babylonian and Greek astronomers, he put together a singularly successful model—a theory, to use the modern term—about how the heavens had to be arranged to produce the display we see every night. In the Ptolemaic description of the universe, the Earth sat unmoved at the center. Around it, on a concentric series of rotating spheres, moved the stars and planets. The model was carefully crafted to take account of observations. The planets, for example, were attached to small spheres rolling inside of the larger spheres so that their uneven retrograde motion across the sky could be understood.

This system remained the best explanation of the universe for almost 1500 years. It successfully predicted planetary motions, eclipses, and a host of other heavenly phenomena, and was one of the longest-lived scientific theories ever devised.

A portrait of Claudius Ptolemy, painted centuries after his death. Such portraits of Greek scholars were commonly done during the Middle Ages and Renaissance.

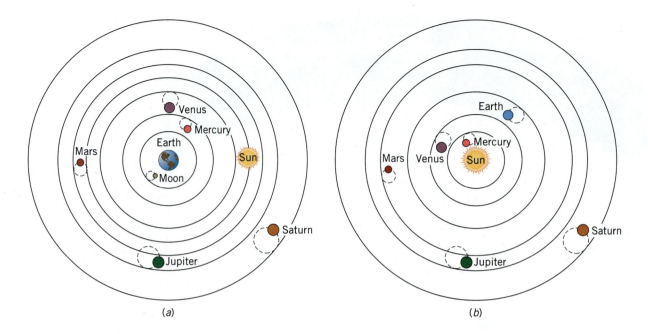

(a)   (b)

Figure 3–5
The Ptolemaic (*a*) and Copernican (*b*) systems. Both systems used circular orbits. The fundamental difference is that Copernicus placed the Sun at the center.

During the first decades of the sixteenth century, however, a Polish cleric by the name of Nicolas Copernicus (1473–1543) proposed a competing hypothesis that was to herald the end of Ptolemy's crystal spheres. His ideas were published in 1543 under the title *On the Revolutions of the Spheres*. Copernicus retained the notions of a spherical universe with circular orbits, and even kept the ideas of spheres rolling within spheres, but he asked a simple and extraordinary question: "Is it possible to construct a model of the heavens whose predictions are as accurate as Ptolemy's, but in which the Sun, rather than the Earth, is at the center?" We do not know how Copernicus, a busy many of affairs in medieval Poland, conceived this question, nor do we know why he devoted his spare time for most of his adult life to answering it. We do know, however, that in 1543, for the first time in over a millenium, a serious alternative to the Ptolemaic system was presented (see Figure 3–5).

Observations: Tycho Brahe and Johannes Kepler

With the publication of the Copernican theory, astronomers were confronted by two competing models of the universe. The Ptolemaic and Copernican systems differed in a fundamental way that had far-reaching implications about the place of humanity in the universe. They both described possible universes, but in one the Earth, and by implication humankind, was no longer at the center. The astronomers' task was to decide which model best describes our universe.

To resolve this question, astronomers had to compare the predictions of the two competing hypotheses. These predictions required keeping track of the exact time a planet or the Moon would appear in an exact location in the heavens. Many astronomers had spent their lives recording accurate observations of the times of positions of stars and planets. When they considered these observations, a fundamental problem became ap-

A portrait of Nicolas Copernicus (1473–1543).

parent. Although the models of Ptolemy and Copernicus made different predictions about various things, such as the position of a planet at midnight, or the time of moonrise, the differences were too small to be measured with equipment that was available at the time. The telescope had not yet been invented, and astronomers had to record planetary positions by depending entirely on naked-eye measurements with awkward instruments. Until the accuracy of measurement was improved, the question of whether or not the Earth was at the center of the universe couldn't be decided.

Some scientists thrive on experimental challenges, and they revel in devising new tricks for making measurements better than anyone else before. The Danish nobleman Tycho Brahe (1546–1601) was such a scientist. Though abducted in infancy by his uncle, Tycho was raised in comfort and given the best possible education. His scientific reputation was firmly established at the age of 25, when he observed and described a new star in the sky (in fact, a supernova—see Chapter 21). By the age of 30, he had been given the island of Hveen off the coast of Denmark and funds to build an observatory there by the Danish king.

An engraving of Tycho Brahe surrounded by his instruments and assistants in his castle in Uraniborg, Denmark.

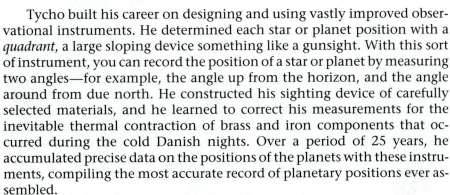

This instrument, called a quadrant, measures the angular position of stars and planets.

Tycho built his career on designing and using vastly improved observational instruments. He determined each star or planet position with a *quadrant*, a large sloping device something like a gunsight. With this sort of instrument, you can record the position of a star or planet by measuring two angles—for example, the angle up from the horizon, and the angle around from due north. He constructed his sighting device of carefully selected materials, and he learned to correct his measurements for the inevitable thermal contraction of brass and iron components that occurred during the cold Danish nights. Over a period of 25 years, he accumulated precise data on the positions of the planets with these instruments, compiling the most accurate record of planetary positions ever assembled.

Johannes Kepler (1571–1630)

Kepler's Laws

When Tycho died in 1601, his data passed into the hands of his assistant, Johannes Kepler (1571–1630), a German mathematician who had joined Tycho two years before. Kepler was skilled in mathematics, and he was able to analyze Tycho Brahe's decades of planetary data in new ways. In the end, Kepler found that the data could be summarized in three basic mathematical statements about the solar system, known as **Kepler's laws of planetary motion.** The most important of these (shown in Figure 3–6) stated that all planets, including the Earth, orbit the Sun in elliptical paths, not perfect circles as had been previously assumed. In this picture, the spheres-within-spheres are gone. Not only do Kepler's laws give a better description of what is observed in the sky, but they present a simpler picture of the solar system as well.

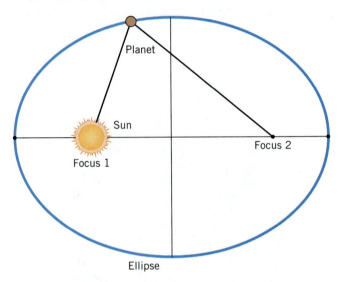

Figure 3–6
Kepler's first law shown schematically. An ellipse is a geometrical figure in which the sum of the distances to two fixed points (each of which is called a focus) is always the same. For the planets, the Sun is at one focus of the ellipse.

An *ellipse* is defined as a curve drawn so that the sum of the distances from any point on the curve to two fixed points is always the same. Imagine tracking a length of loose string down at two points, then drawing a curve by holding a pencil in the string as shown in Figure 3–7. Each fixed point is called a *focus*. What Kepler found was that the orbits of all known planets, from Mercury to Saturn, have one focus at the Sun. The statement that *the planets have elliptical orbits with one focus at the Sun* is known as Kepler's first law of planetary motion.

Kepler's second law describes the speed at which the planets move in their elliptical orbits. If you remember the playground game where you run toward a post, then grab it on the fly and swing around, you have a pretty good notion of the way the planets move. They speed up as they get closer to the Sun, then slow down in the farther parts of their arcs. Kepler's second law is usually stated in terms of "equal areas," a

Kepler's laws of planetary motion Three mathematical laws about the solar system derived by German mathematician Johannes Kepler. 1. All planets orbit the Sun in an elliptical path. 2. A planet's velocity will increase as it moves closer to the Sun. 3. The farther a planet is from the Sun, the longer its year will be.

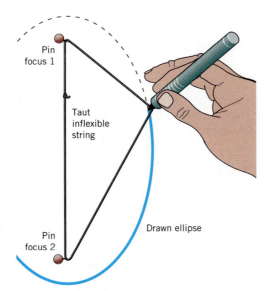

Figure 3–7
You can drawn an ellipse by tying a loop of string to a pencil, looping the string around two nails, and drawing a curve while keeping the string taut. The two nails are at the foci of the ellipse. Note that if the two nails are brought close together, the ellipse more nearly approximates a circle.

concept illustrated in Figure 3–8. Imagine that a line drawn from the Sun to an orbiting planet sweeps out an area in a fixed period of time. Kepler's second law says that *for a given time interval, this swept-out area is the same, no matter where the planet is in its orbit.* A glance at Figure 3–7 should convince you that this means that planets move faster when they are nearest the Sun, and slower when farther away.

Finally, Kepler turned his attention to the *period* of a planet's revolution: the time it takes for one complete orbit, or its "year." Planets farther from the Sun have a longer year than those closer in for two reasons: (1) they have farther to go as they make their circuit, and (2) the outer planets travel more slowly than the inner ones. The net effect is a regularity between the length of a planet's year, or its period, and the distance of its orbit from the Sun. This relationship between a planet's distance from the Sun and its period may be expressed as a simple equation that allows us to predict the behavior of orbiting objects. This expression is known as Kepler's third law. (See Figure 3–9.)

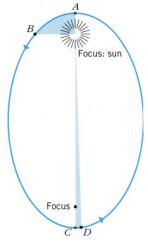

Figure 3–8
Kepler's second law of equal areas states that planets move fastest when they are closest to the Sun, and slowest when at the farthest point in their orbits. Thus a planet in an elliptical orbit sweeps out equal areas in equal times.

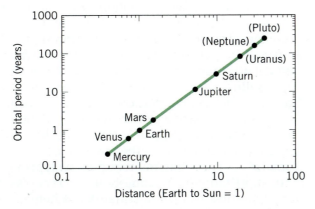

Figure 3–9
Kepler's third law states that the square of the period of a planet's orbit is proportional to the cube of its average distance from the Sun.

▶ **In words:**

The farther a planet is from the Sun, the longer is its year.

▶ **In equation form:**

The square of the period of a planet's orbit is proportional to the cube of its average distance from the Sun.

▶ **In symbols:**

$T^2 = \text{constant} \times R^3$

where T is the time it takes a planet to go around the Sun and R is the long axis of the elliptical orbit. If we approximate the orbit by a circle (which works pretty well for the planets), then R is the radius of that circle.

If T is measured in Earth years and R is measured in terms of the distance between the Sun and the Earth, then the constant in Kepler's law is equal to 1.

Example 3–1: Planetary Years

How long would a planet's years be if the planet were three times as far from the Sun as the Earth?

▶ **Reasoning and Solution:** Kepler's third law relates a planet's year, T, and its distance from the Sun, R. In this case, $R = 3$, and we want to find T by substituting in the equation.

$$\frac{T^2}{R^3} = 1$$

Thus

$$T = \sqrt{27} = 5.2 \text{ Earth years}$$

Thus this mythical planet, three times as far from the Sun as we are, would take five times as long to make one orbit. ▲

Decades of careful observations by Brahe and mathematical analysis by Kepler firmly established that the Earth is not at the center of the universe, that planetary orbits are not circular, and that neither Ptolemy nor Copernicus was correct in their models of the universe (though the Copernican model was much closer to the modern view than Ptolemy's). This research also illustrates a recurrent point about scientific progress. The ability to answer scientific questions dealing with the most fundamental aspects of human existence often depends on the kinds of instruments scientists have at their disposal. A person dealing with the grubby details of how a telescope measures the angle to a star may not appear to be doing something glamorous, but it is often such people who provide insights into the most fundamental questions. In Tycho's case, for example, meticulous attention to experimental detail helped us answer the question "What is the place of humanity in the universe?"

At the end of this historical episode, astronomers had Kepler's laws to describe *how* the planets in the solar system move; however, they had no idea *why* planets behave the way they do. Kepler's laws, as important as they are, give no insight into the basic mechanisms that make a solar system operate. The answer to that question was to come from an unexpected source.

THE BIRTH OF MECHANICS AND EXPERIMENTAL SCIENCE: GALILEO GALILEI

Mechanics is an old word for the branch of science that deals with the motions of material objects. A rock rolling down a hill, a ball thrown into the air, and a sailboat skimming over the waves are all fit subjects for this science. For sixteenth-century military leaders concerned with the behavior of cannonballs and other projectiles, mechanics was becoming a science of more than abstract interest. Since ancient times, philosophers had speculated on why things move the way they do, but it wasn't until about 1600 that our modern understanding of the subject began to emerge.

The Italian physicist and philosopher Galileo Galilei (1564–1642) was in many ways a forerunner of the twentieth-century scientist. A professor of mathematics at the University of Padua, he quickly became an advisor to the powerful court of the Medici at Florence, as well as a consultant at the Arsenal of Venice, the most advanced naval construction center in the world. He invented many practical devices, such as the first thermometer to measure temperature, the pendulum clock, and the proportional compass that draftsmen still use today. Galileo gained fame as the first to observe the heavens with a telescope, which he built after hearing of the instrument from others. He was the first to see the moons of Jupiter (now called the Galilean moons) and the rings of Saturn.

> **mechanics** The branch of science that deals with the motions of material objects and the forces that act on them; for example, a rolling rock or a thrown ball.

Science in the Making

The Heresy Trial of Galileo

Galileo is famous for the wrong reason. Despite the fact that he was a founder of modern experimental science and was the first to make a systematic survey of the sky with a telescope, he is remembered primarily because of his trial on suspicion of heresy in 1633.

Galileo published a summary of his telescopic observations in a book called *The Starry Messenger*. This book was written in Italian, the language of the people, rather than Latin, the language of scholars. Thus Copernican ideas, including the disturbing concept that the Earth is not the center of the universe, became available to the educated public. Some readers complained that these ideas violated Church doctrine and, in 1616, Galileo was called before the College of Cardinals. What happened at this meeting is not clear. The Church later claimed that Galileo had been warned not to discuss Copernican ideas unless he treated them, as Copernicus had, as a hypothesis. Galileo, on the other hand, claimed he had not been given any such warning.

Galileo Galilei (1564–1642)

A modern painting of the trial of Galileo.

In any case, the situation remained in this unsettled state until 1632, when Galileo published a book called *A Dialogue Concerning Two World Systems*, which was a long defense of the Copernican system. This publication led to the famous trial, in which Galileo purged himself of charges of heresy by denying that he held the views in his book. He was already an old man by this time, and spent his last few years under virtual house arrest in his villa near Florence.

The legend of the trial of Galileo, in which an earnest seeker after truth is crushed by a rigid hierarchy, bears little resemblance to the historical events. The Catholic Church had not banned Copernican ideas. Copernicus, after all, was a savvy Church politician who knew how to get his ideas across without ruffling feathers. Indeed, seminars on the Copernican system were given at the Vatican in the years before Galileo. Furthermore, Galileo's arguments in favor of the system were not very convincing. Much of the *Dialogues*, for example, is taken up by a completely incorrect discussion of the tides. His confrontational tactics of putting the Pope's favorite arguments into the mouth of a foolish character in the book brought a predictable reaction that earlier, more reasonable approaches had not. As often happens, under close inspection the simple myth associated with a historical event dissolves into something much more complex.

A footnote: In 1992, the Roman Catholic Church reopened the case and, in effect, issued a retroactive "not guilty" in the case of Galileo. The grounds for the reversal were that the original judges had not separated questions of faith from questions of scientific fact. ●

Telescope used by Galileo Galilei in his astronomical studies.

Speed, Velocity, and Acceleration

From a scientist's point of view, Galileo's greatest achievement was his work on experimental technique. You can see why by considering his work on the question of the behavior of objects thrown or dropped on the surface of the Earth. Greek philosophers, using pure reason, had taught the reasonable idea that heavier objects must fall faster than light ones. In a series of classic experiments, Galileo showed that this was not the case. He showed that at the surface of the Earth all objects fall at the same rate.

To understand Galileo's study of moving bodies, you must first understand the distinction between three terms: speed, velocity, and acceleration.

Speed

► **In words:**

> **Speed** is the distance an object travels divided by the time that it takes to travel that distance.

speed   The distance an object travels divided by the time that it takes to travel that distance.

► **In equation form:**

> speed = distance traveled ÷ time of travel .

► **In symbols:**

$$s = \frac{d}{t}$$

This *direct relationship* between speed and distance is illustrated in graphical form in Figure 3–10. If you know the distance traveled and the time elapsed during the travel, you can calculate the speed. From time to time we will need to use two variations of this equation:

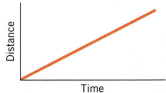

Figure 3–10
The distance traveled by an object moving at a constant velocity.

1. If you know the average speed, *s*, of an object, and the time of travel, *t*, you can calculate how far the object has traveled, *d*:

   distance traveled = average speed × time of travel

   or

   $$d = s \times t$$

2. If you know the total distance traveled and the average speed of travel, you can calculate how long the journey takes:

   $$\text{time of travel} = \frac{\text{distance traveled}}{\text{average speed}}$$

   or

   $$t = \frac{d}{s}$$

Example 3–2:   Driving

If the speedometer on your car reads 30 miles per hour, how far will you go in 15 minutes?

▶ **Reasoning and Solution:** This question involves changing units, as well as applying the equation that relates time, speed, and distance. First, we must know the travel time in hours:

$$\frac{15 \text{ min}}{60 \text{ min/h}} = \tfrac{1}{4}\text{ h}$$

Then, using the relationship between distance and time given above, we find

$$\text{distance} = 30 \text{ miles/h} \times \tfrac{1}{4}\text{ h}$$
$$= 7.5 \text{ miles}$$

It would take the average person about two hours to walk this far.

You may have noticed that we put $\tfrac{1}{4}$ hour into the equation for the time instead of 15 minutes. The reason we did this was to be consistent with the units in which an automobile speedometer measures speed. Since the automobile dial reads in miles per hour, we also put the time in hours to make the equation balance. A useful way to deal with situations like this is to imagine that the units are quantities that can be canceled in fractions, just like numbers. In this case, we would have

$$\text{distance} = (\text{miles/h}) \times \text{h}$$
$$= \text{miles}$$

If, however, we put the time in minutes, we'd have

$$\text{distance} = (\text{miles/h}) \times \text{min}$$

and there would be no cancellation.

Whenever you do a problem like this, it's a good idea to check to make sure the units come out correctly. This process is known as *dimensional analysis*. ▲

Velocity

velocity The distance an object travels divided by the time it takes to travel that distance, including the direction of travel. The velocity of a falling object is proportional to the length of time that it has been falling.

Velocity has the same numerical value as speed, but it is a vector quantity that also includes information on the direction of travel (see Chapter 2). The speed of a car might be 40 miles per hour, for example, while the velocity is 40 miles per hour due west. Velocity and speed are measured in units of distance per time, such as meters per second, feet per second, or miles per hour.

Acceleration

acceleration The amount of change in velocity divided by the time it takes the change to occur. Acceleration can involve changes of speed, or changes in direction, or both.

Acceleration measures the rate of change of velocity.

▶ **In words:**

Acceleration is the change in velocity divided by the time it takes that change to occur.

▶ **In equation form:**

Acceleration = (final velocity − initial velocity) ÷ time

▶ **In symbols:**

$$a = \frac{(v_f - v_i)}{t}$$

Like velocity, acceleration requires information about the direction, and it is therefore a vector.

When velocity changes, it may be by a certain number of feet per second or meters per second in each second. Consequently, the units of acceleration are meter/second/second (described as "meters per second per second," and usually abbreviated m/s², where the first "meters per second" refers to the velocity, and the last "per second" to the time the velocity takes to change.

To understand the difference between acceleration and velocity, think about the last time you were behind the wheel of a car, glancing at the speedometer. If the needle is unmoving (at 30 miles per hour, for example), you are moving at a constant speed. Suppose, however, that the needle isn't stationary on the speedometer scale (perhaps because you have your foot on the gas or on the brake). This means your speed is changing and, by the definition above, you are accelerating. The higher the acceleration, the faster the needle moves. If the needle doesn't move, however, *this doesn't mean you and the car aren't moving.* As we saw above, an unmoving needle simply means that you are traveling at a constant speed without acceleration. Motion at a constant speed in a single direction is called **uniform motion.**

If we want to measure a velocity, then we have to specify the distance we plan to go and see how long it takes to travel that distance. If the distance were 30 miles and it took a car an hour to go that distance, then the car would be traveling 30 miles per hour. If the distance were a meter and it took something a second to move that distance, the velocity would be one meter per second. In your car's speedometer, what is actually measured is the time it takes a particular gear in the transmission to make one revolution. When the wheels turn a known number of times and the car moves a known distance forward, this gear turns and the motion registers the number actually displayed on your speedometer.

While it is fairly straightforward to measure speeds and velocities for things that aren't accelerating, measuring becomes more complicated when acceleration occurs. If your speedometer needle climbs steadily from 30 to 40 miles per hour, the speed of your car is constantly changing. At no time during the interval is speed (or velocity) a constant. So how do we deal with objects whose speed is changing?

uniform motion   The motion of an object if it travels in a straight line at a constant speed. All other motions involve acceleration.

Technology

Measuring Time without a Watch

In our age of stopwatches, digital timers, and atomic clocks, with Olympic races routinely measured to a thousandth of a second, it's hard to imagine measuring time without an accurate device. But when Galileo set out to study accelerated motion in the 1600s, measuring time was a formidable technological challenge. Think about how you might determine small time intervals if all you had was a clock that ticked off minutes, but could record no shorter times.

Galileo wanted to document the way falling objects accelerate as accurately as possible, but these were measurements that required a knowledge of distance and time. Since objects that fall straight down moved much too fast for him to measure, he devised an experiment in which balls rolled down a gently inclined plane. However, even these balls moved too fast to time with available clocks (Figure 3–11).

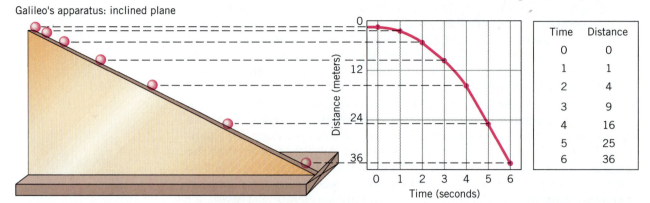

Galileo's apparatus: inclined plane

Time	Distance
0	0
1	1
2	4
3	9
4	16
5	25
6	36

Figure 3–11
Galileo's falling-ball apparatus with a table of measurements and a graph of distance versus time.

Galileo tried a variety of different methods to measure these small increments of time. He experimented with his own heartbeat, but his pulse proved too irregular. He tested rapidly swinging pendulums, but they were difficult to start and stop precisely. He had more success measuring the weight of water accumulated when a steady flow was started and stopped to coincide with the period of an object's fall, but irregularities in the flow and uncertainties in starting and stopping the water limited the accuracy of that method.

Galileo's most ingenious solution for measuring time intervals relied on his musical training. Galileo stretched lute strings across the rolling ball's path, so that there was a discernible "twang" when the ball passed the string. He then adjusted the distance between the strings until he heard the notes coming at precisely equal intervals. A musician with a good ear can tell if notes in a series are off by as little as $\frac{1}{64}$ of a second. Thus, even though Galileo did not have clocks capable of measuring time intervals better than a second or so, with this scheme he could be certain that several very short time intervals were the same.

When he got the time intervals between twangs just right, Galileo measured the distances between lute strings. He found that if a ball had traveled an inch in the first twang, then it traveled 4 inches by the end of two twangs, 9 inches by the end of three twangs, 16 inches by the end of four twangs, and so on. In other words, he found that the distance traveled by an object that is being accelerated depends on the *square* of the time, not just on the time itself.

Modern versions of the rolling-ball experiment, which rely on laser beams and electronic timers, have greatly improved the accuracy of Galileo's experiment, but they produce exactly the same result. ●

The Velocity of an Accelerating Object

Galileo's experiments revealed a simple relationship between an object's acceleration and the distance it travels.

► **In words:**

When an object is accelerated in a uniform way, the distance it covers in a given time depends on the square of the time. It goes nine times as far in three minutes as in one, for example.

► **In equation form:**

distance traveled $= \frac{1}{2} \times$ acceleration $\times$ time^2

► **In symbols:**

$d = \frac{1}{2} at^2$

This relationship between distance traveled and time can also be represented in a graph, as shown in Figure 3–12.

If you think about the moving needle on the speedometer of your car, you realize that it is possible to read off a velocity for the car at each moment. Given what we've just said about the difficulty of defining the velocity of an accelerating object, what does that speedometer reading mean?

Think of the process this way: Suppose you had marked out a short distance on the pavement; a few feet, for example, or even a fraction of an inch. As the accelerating car goes by, you can time how long it takes to cross the interval. If the interval is short enough, the velocity won't change very much as the car crosses it. The *average velocity* of the car is just the length of the small interval divided by the length of time it takes the car to cross it. The magnitude of this velocity is the number that's registering on your speedometer. The *instantaneous velocity* is the average velocity measured over a very small time interval. It is the instantaneous velocity to which we usually refer when we ask the question, "How fast is it going *now*?"

Think about the instantaneous velocity of a falling object as an example. At the instant the object is released (time equals zero), the velocity is also zero, but the object begins to accelerate at a uniform rate. Thereafter, the instantaneous velocity of the falling body is given by the following equation:

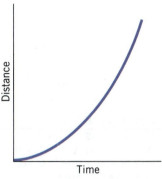

Figure 3–12
The distance traveled by an object accelerating at a constant rate.

► **In words:**

The instantaneous velocity of a uniformly accelerating object that started at rest equals the acceleration multiplied by the total time of acceleration.

► **In equation form:**

velocity $=$ acceleration $\times$ time

► **In symbols:**

$v = at$

The *direct relationship* between velocity and time of travel for a given distance is illustrated in Figure 3–13.

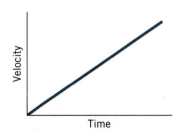

Figure 3–13
The velocity of an object accelerating at a constant rate.

Example 3–3: Out of the Blocks

A sprinter accelerates from the starting blocks to a speed of 10 meters per second in one second. Answer the following questions about the sprinter's speed, acceleration, time, and distance run.

▶ **Reasoning:**   In each case, answer the question by substituting into the appropriate motion equation.

1. What is the sprinter's acceleration?

$$\text{acceleration} = \frac{\text{final velocity} - \text{initial velocity}}{\text{time}}$$

In this case, the sprinter starts at rest at the beginning of the race, so the initial velocity is zero. Thus

$$\text{acceleration} = \frac{(10 \text{ m/s})}{1 \text{ s}}$$

$$= 10 \text{ m/s}^2$$

2. How far does the sprinter travel during this acceleration?

$$\text{distance} = \tfrac{1}{2} \text{ acceleration} \times \text{time}^2$$
$$= \tfrac{1}{2} \times (10 \text{ m/s}^2) \times (1 \text{ s})^2$$
$$= 5 \text{ m}$$

3. How fast is the sprinter going halfway through the period of acceleration?

$$\text{instantaneous velocity} = \text{acceleration} \times \text{time}$$

At $\tfrac{1}{2}$ second, the instaneous velocity will be

$$\text{velocity} = 10 \text{ m/s}^2 \times \tfrac{1}{2} \text{ s}$$
$$= \ \ 5 \text{ m/s}$$

4. How far has the sprinter traveled in the first half second?

$$\text{distance} = \tfrac{1}{2} \times \text{acceleration} \times \text{time}^2$$
$$= \tfrac{1}{2} \times 10 \text{ m/s}^2 \times \tfrac{1}{2} \text{ s}^2$$
$$= \tfrac{1}{2} \times 10 \text{ m/s}^2 \times \tfrac{1}{4} \text{ s}$$
$$= 1.25 \text{ m}$$

Notice that halfway through the period of acceleration the sprinter has not covered half of the 5 meters (see part 2 above). This feature is common to accelerated motion. The accelerating sprinter moves much faster during the second half-second, and therefore covers more ground.

5. Assuming the sprinter covers the remaining 95 meters at a speed of 10 meters per second, what will be the sprinter's time for the event? We have already calculated that the time to cover the first 5 meters is one second. The time required to cover the remaining 95 meters

at a constant velocity of 10 meters per second is

$$\text{time} = \frac{\text{distance}}{\text{velocity}}$$

$$= \frac{95 \text{ m}}{10 \text{ m/s}}$$

$$= 9.5 \text{ s}$$

Thus   total time $= 1 + 9.5 = 10.5$ s

For reference, the world's record for the 100-meter dash, set by Carl Lewis of the United States in 1991, is 9.86 seconds. ▲

Science by the Numbers

Instantaneous Velocity

When we defined instantaneous velocity, we mentioned that the acceleration wouldn't have much of an effect over a small distance. Let's see how valid this assumption is in Example 3–3 when the sprinter is halfway through the acceleration period.

Suppose we mark out a 1-centimeter (0.01-meter) space on the ground 1.25 meters from the starting blocks, which the sprinter will reach a half second into the race. The instantaneous velocity at this point is 5 meters per second, so the sprinter should cross the 1-centimeter distance in approximately

$$\text{time} = \frac{\text{distance}}{\text{velocity}}$$

$$= \frac{0.01 \text{ m}}{5 \text{ m/s}}$$

$$= 0.002 \text{ s}$$

How much does the velocity change during this time interval? The sprinter accelerates at the rate of 10 meters per second every second.

From the definition of acceleration, $$\text{Acceleration} = \frac{\text{change in velocity}}{\text{time during which change occurs}}$$

Therefore, $$10 \text{ m/s}^2 = \frac{\text{change in velocity}}{0.002 \text{ s}}$$

Thus   change in velocity $= 10 \text{ m/s}^2 \times 0.002 \text{ s}$

$$= 0.02 \text{ m/s}$$

This change is less than 0.1% of the instantaneous velocity, so our assumption that the velocity wouldn't change much over the small distance interval is valid to that level of accuracy. We leave it to the problems at the end of the chapter to show that the velocity change would be even smaller if, instead of a 1-centimeter distance, we had chosen a smaller interval. ●

Figure 3–14
Galileo did his experiment by rolling a ball down an inclined plane. The steeper the angle of the plane, the faster the fall. A plane at right angles to the ground corresponds to a freely falling object.

acceleration due to gravity
(*g*)   A constant numerical value for the specific acceleration that all objects experience at the Earth's surface; determined by measuring the actual fall rate of objects in a laboratory: 9.8 m/s² = 32 feet/s.

Acceleration Due to Gravity

In Galileo's experimental scheme, he would produce greater accelerations by increasing the angle of the incline down which his balls rolled (Figure 3–14). Eventually, at very steep angles, the motion of the balls became too fast for his ear to differentiate the time intervals. In our everyday lives, most falling objects don't roll down a tilted plane. The situation where an object falls straight down (equivalent to a plane tilted at an angle of 90 degrees) is most common.

You can verify that an acceleration is involved by dropping something and watching it fall. Notice that at the instant you release the object it barely moves. Can you see that it's moving faster at the end of its fall than at the beginning?

The acceleration of a freely falling body is so important that physicists give it a special name and assign it a special letter of the alphabet. It is called the **acceleration due to gravity** and is denoted by the letter *g*. Galileo's experiments led him to the hypothesis that any object dropped near the Earth's surface, no matter how heavy or light, falls with exactly the same constant acceleration. In the absence of complications such as wind, which may slow down or alter the direction of a falling object, it makes no difference how massive an object is. All objects experience exactly the same acceleration. (In Chapter 4 you will learn why this is so.)

The numerical value of *g* can be determined by measuring the actual motion of objects. The modern value of *g* is given by

$$g = 32 \text{ ft/s}^2 = 9.8 \text{ m/s}^2$$

This equation tells us that in the first second, a falling object accelerates from a stationary position to a velocity of 9.8 m/s, (about 22 miles per hour) straight down. After 2 seconds, the velocity doubles to 19.6 m/s; after 3 seconds, it triples to 29.4 m/s, and so on.

Ironically, Galileo probably never performed the one experiment for which he is most famous, dropping two different weights from the leaning Tower of Pisa to see which would land first. Had he done so, the effects of friction between the air and the falling bodies probably would have slowed the lighter object slightly more than the heavier one, so an observer would perceive the two objects to fall at slightly different rates.

Developing Your Intuition

Freely Falling Objects

According to Galileo's results, every object at the surface of the Earth, once released, should accelerate downward at the same rate (Figure 3–15). But we know that if a leaf and an acorn fall from a branch at the same time, the acorn will get to the ground well before the leaf, even if they have equal mass. Was Galileo wrong?

To understand this problem, think for a minute about *how* the leaf and the acorn fall. The acorn plummets straight down to the ground, accelerating as it goes. The leaf, on the other hand, flutters down slowly. It appears that the air has a much greater effect on the leaf than on the acorn. The acorn, being compact, has a much smaller force exerted on it by the air than does the leaf. In fact, if we repeated this experiment on the Moon (where there is no air) or in a tube from which air has been removed, the leaf would not flutter and the two objects would fall at exactly the same rate. ●

Example 3–4: Dropping a Penny from the Sears Tower

The tallest building in the world is the Sears Tower in Chicago, with a height of 1454 feet. Ignoring wind resistance, how fast would a penny dropped from the top be moving when it hit the ground?

▶ **Reasoning:** The penny is dropped with zero initial velocity. We first need to calculate the time it takes to fall 1454 feet. From this time we can calculate the velocity at impact.

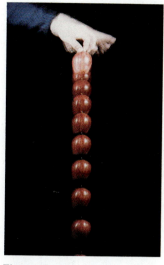

Step 1. Time of fall. The distance travelled by an accelerating object is

$$\text{distance} = \tfrac{1}{2} \times \text{acceleration} \times \text{time}^2$$
$$= \tfrac{1}{2} \times 32 \text{ ft/s}^2 \times t^2$$
$$= 16 \text{ ft/s}^2 \times t^2$$

Recall that distance equals 1454 feet, so rearranging gives

$$t^2 = \frac{1454 \text{ ft}}{16 \text{ ft/s}^2}$$
$$= 90.88 \text{ s}^2$$

Taking the square root of both sides gives time:

$$\sqrt{t^2} = \sqrt{\frac{1454}{16 \text{ ft/s}^2}}$$
$$\sqrt{t^2} = \sqrt{90.9}$$
$$t = 9.5^2 \text{ s}$$

Step 2. Velocity at impact. The velocity of an accelerating object is

$$\text{velocity} = \text{acceleration} \times \text{time}$$
$$= 32 \text{ ft/s}^2 \times 9.5 \text{ s}$$
$$= 304 \text{ ft/s}$$

Figure 3–15
The accelerated motion of a falling apple is captured by a multiple-exposure photograph. In each successive time interval, the apple falls farther.

This velocity is about 200 miles per hour, a high speed indeed. A penny traveling at such a velocity could easily kill a person on the sidewalk, so *don't* try this experiment!

In fact, most objects dropped in air will not accelerate indefinitely. Because of air resistance, an object will accelerate until it reaches its *terminal velocity*, and will continue falling at a constant velocity after that point. Terminal velocity for a penny would be somewhat less than 200 miles per hour, though still fast enough to kill. ▲

Movement in Two Dimensions: The Law of Compound Motion

Up to now, we have talked only about motion along a straight line—what we call motion in one dimension. However, if you throw a baseball or ride in a car on a highway, you experience motion in more than one dimension. The baseball, for example, follows an arching path as it travels outward. The car goes around curves and turns corners as well as moving forward. In Galileo's time, analyzing motion of projectiles like baseballs

A feather and an apple falling in a vacuum will accelerate at the same rate.

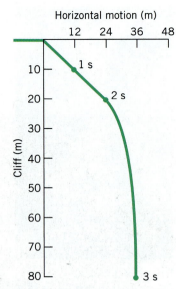

Figure 3–16
The trajectory of a rock thrown off a cliff as described. The motion in the horizontal direction is independent of the motion in the vertical direction.

was particularly important because the cannon had just been introduced into warfare, and cannonballs, like baseballs, move in two-dimensional arcs.

The central question in analyzing two-dimensional motion is this: How does the motion in one direction affect the motion in the other? When a baseball is thrown, for example, does it speed in the horizontal direction affect how high it will go? Galileo proposed what we now call *the law of compound motion*. It states that

> **Motion in one dimension has no effect on motion in another.**

Let's take a simple example to see how this law works. Suppose you are standing on a high cliff and throw a rock outward, as shown in Figure 3–16. The law of compound motion tells us that we should think of the rock's path as being made up of two separate parts or components. In the horizontal direction, the rock moves at a constant velocity imparted to it by your hand. In the downward direction, the rock accelerates like any other falling object. See Figure 3–17 for an illustration of motion in the vertical direction only.

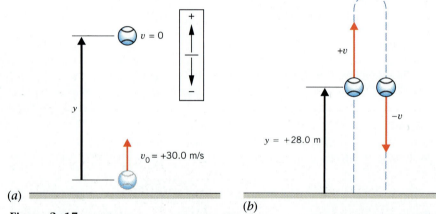

(a)

(b)

Figure 3–17
(*a*) A ball thrown upward, here with an initial velocity of 30.0 m/s, reaches a velocity of zero at its maximum height. (*b*) When it falls, the ball's downward velocity is equal to its upward velocity at any given horizontal location. Note the convention that the upward direction is positive, the downward negative.

Suppose, for example, that the rock is moving 12 meters per second in the horizontal direction. At the end of 1 second, the rock will be 12 meters from the cliff. At the same time, however, the rock is falling with an acceleration of 9.8 meters per second per second, so at the end of one second it will be 4.9 meters down (remember: distance $= \frac{1}{2}gt^2$). The rock's position, then, will be 12 meters out and 4.9 meters down, as shown. A second later, it will be 24 meters out and 19.6 meters down, and so on. The path the rock follows is called a *parabola*. A full parabola is illustrated in Figure 3–18.

The law of compound motion also applies to projectiles thrown up from the ground, such as a baseball leaving a bat or a football being thrown. In this case, the vertical motion of the ball is an ever-slowing motion upward, followed by a fall similar to what is shown in Figure

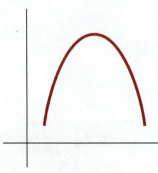

Figure 3–18
An object thrown upward at the Earth's surface follows an arching path, called a parabola.

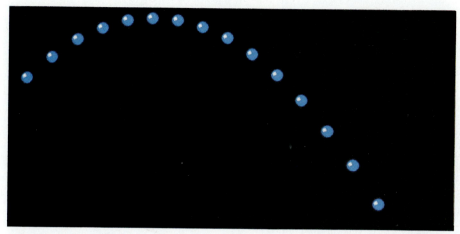

A computer-generated time-lapse image of a ball thrown upward at a 30 degree angle and initial velocity of 6 meters/second.

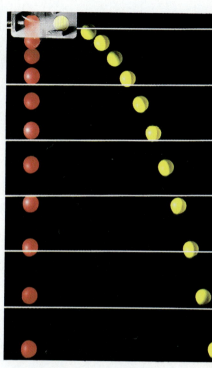

3–16 (if the effects of wind are negligible). The horizontal motion is, as in Figure 3–16, movement at a constant velocity to the right. The combination will be an arc, also in the shape of a parabola, that begins at the ground, goes up to the peak in a path that is the mirror image of that in Figure 3–16, and then goes back to the ground exactly as in Figure 3–16.

Example 3–5: The Human Cannonball

One of the attractions you may see at the circus is the "human cannonball," a person who is lowered into the barrel of a giant cannon and then shot out, traveling through the air and landing in a safety net to the sound of applause from the audience. Suppose he emerges from the cannon's, mouth with a vertical velocity of 19.6 m/s and a horizontal velocity of 8 m/s. How far away would you have to place the net to make sure he landed safely?

A ball dropped down and one thrown sideways hit the ground at the same time, in accordance with the law of compound motion.

The motion of a human cannonball illustrates the law of compound motion. The accelerated up-and-down motion in the vertical direction is completely independent of the uniform motion in the horizontal direction.

▶ **Reasoning:** The way to approach this problem is to remember that the horizontal and vertical motions are independent of each other. In the vertical direction, we have a situation in which an object is thrown upward, but in which there is a constant downward acceleration equal to *g*. In this direction, the object will slow down as it moves up, then stop and fall back down. The time for the up-and-down trip will be twice the time it takes to get from the ground to the top of the arc. The total time of up-and-down travel, multiplied by 8 m/s, will tell us the distance the object travels horizontally, and hence where to place the net. We will use the convention that acceleration in the downward direction is negative: -9.8 m/s^2.

Step 1. How long will it take to get to the top of the arc? The velocity in the vertical direction is

$$\text{final velocity} = \text{initial velocity} + \text{acceleration} \times \text{time}$$

where initial velocity is 19.6 m/s and final velocity at the top of the arc is 0. Thus

$$0 = 19.6 \text{ m/s} + (-9.8 \text{ m/s}^2) \times t$$

where all values are in SI units. Solving for time, *t*,

$$(9.8 \text{ m/s}^2) \times t = 19.6 \text{ m/s}$$

$$t = \frac{19.6 \text{ m/s}}{9.8 \text{ m/s}^2}$$

$$= 2 \text{ s}$$

In other words, the human cannonball takes 2 seconds to get to the top of the arc, and 2 more seconds to come down, for a total flight time of 4 seconds.

Step 2. Where do you place the net? In the horizontal direction, the human cannonball is travelling at a steady 8 m/s. In 4 seconds the horizontal distance traveled is

$$\text{distance} = \text{velocity} \times \text{time}$$

$$= 8 \text{ m/s} \times 4 \text{ s}$$

$$= 32 \text{ m}$$

The net should be placed 32 meters from the mouth of the cannon. ▲

Motion in a Circle

The motion of an object moving in an arc or circle presents a more difficult challenge to analyze than a baseball or a rock thrown from a cliff, but the principle is very much the same. For simplicity, let's talk about an object—perhaps a car on a track or a discus in the hands of an athlete in the process of launching it—moving around a circular path at a constant speed. In both cases, the object is being accelerated.

If you have difficulty envisioning this acceleration, you are in good company. Many famous scientists, including Galileo, tried to work out the properties of uniform circular motion and failed. But if you remember the definition of velocity, you will realize that it involves *both speed and direction*. The ordinary kind of accelerated motion we discussed in the last section, like stepping on your car's gas pedal or brake, involves a change of speed without a change of direction. We have no difficulty recognizing this motion as acceleration. Uniform circular motion, on the other hand, involves a change of direction without a change of speed.

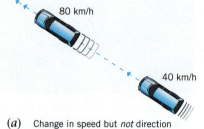

80 km/h

40 km/h

(*a*) Change in speed but *not* direction

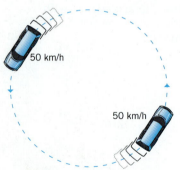

50 km/h

50 km/h

(*b*) Change in direction but *not* speed

Figure 3–19
(*a*) A car accelerates if its speed changes but not its direction. (*b*) A car also accelerates if it travels in a circle at constant speed.

It, too, involves a change in the velocity, and therefore is an acceleration (see Figure 3–19).

In fact, the acceleration of an object moving in a circle of radius r with a velocity whose magnitude (speed) is v, is just

$$\text{acceleration} = \frac{\text{velocity}^2}{\text{radius}}$$

or $a_c = \dfrac{v^2}{r}$

The acceleration of an object moving in a circle is illustrated in Figure 3–20. This acceleration is called *centripetal*, or "center seeking," acceleration, written here as a_c. We will encounter it again in the next chapter when we discuss the orbits of satellites.

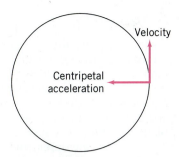

Figure 3–20
The acceleration of an object moving in a circle. The object moves at the same speed, and only the direction changes.

THINKING MORE ABOUT THE ORDERED UNIVERSE

Zeno's Paradox

Zeno of Eleas was a Greek philosopher who lived in the fifth century B.C. He is most famous for stating a paradox that, on the face of it, seems to urge that motion of any sort is logically impossible. Here's a statement of the paradox from an old manuscript.

> If anything is moving, it must be moving either in the place where it is or in the place where it is not. It cannot move in the place where it is (for that place is the same size as itself) and it cannot move in the place where it is not. Hence, movement is impossible.

Two common versions of the paradox are often cited.

1. In order to cross a room, you must first cross half, then half of what's left, then half of *that* remainder, and so on. Since there is an infinite number of divisions, you can never cross them all. Hence you cannot cross the room.
2. If Achilles were to race a tortoise and give it a short head start, he would have to get to the spot where the tortoise is now. But by then, the tortoise would have moved on. Thus Achilles could never catch the tortoise.

Both of these paradoxes are related to the notion of instantaneous velocity that we developed earlier. While it is true that you must cross shorter and shorter intervals to get across a room, it is also true that the time it takes to cross each interval gets shorter as well. For example, suppose the room is 10 feet across and we walk at a rate of 5 feet per second.

We will cross the first half in

$$\text{time} = \frac{\text{distance}}{\text{velocity}}$$
$$= \frac{5 \text{ ft}}{5 \text{ ft/s}}$$
$$= 1 \text{ s}$$

We will then cross half of the remainder in

$$\text{time} = \frac{2.5 \text{ ft}}{5 \text{ ft/s}}$$
$$= 0.5 \text{ s}$$

and half of that in

$$\text{time} = \frac{1.25 \text{ ft}}{5 \text{ ft/s}}$$
$$= 0.25 \text{ s}$$

and so on. If you add up all the increasingly shorter times, they come to 2 seconds, which is just what you'd expect from applying the equations you just learned.

This resolution of Zeno's paradox follows from concepts that are central to the branch of mathematics known as *calculus*, which was developed by Isaac Newton in the seventeenth century.

► Summary

Since before recorded history, astronomers have observed regularities in the heavens and have built monuments such as Stonehenge to help establish order in their lives. Models, such as the Earth-centered system of Ptolemy and the Sun-centered system of Copernicus, attempted to explain these regular motions of stars and planets. *Astronomers* such as Tycho Brahe made ever more precise measurements of star and planet positions. These data led mathematician Johannes Kepler to propose his *laws of planetary motion*, which, among other things, state that planets move around the Sun in elliptical orbits, not circular orbits as had been previously assumed.

Meanwhile, Galileo Galilei and other scientists investigated the science of *mechanics*, which is the study of how things move near the Earth's surface. These workers recognized two fundamental different kinds of motion: *uniform motion* at a constant *speed* in one direction (*velocity*), and *acceleration*, which entails a change in either speed or direction of travel. Galileo devised experiments to study falling objects, and he discovered that all things fall at the constant rate of acceleration of 9.8 meters per second per second, which is called the *acceleration due to gravity* or *gravitational constant (g)*. He also discovered the *law of compound motion*, which states that the motion in one dimension has no effect on motion in another dimension.

► Key Equations

$$\text{speed} = \frac{\text{distance}}{\text{time}}$$

$$\text{acceleration} = \frac{(\text{final velocity} - \text{initial velocity})}{\text{time}}$$

For an object starting at rest and experiencing constant acceleration:

$\text{distance} = \frac{1}{2} \times \text{acceleration} \times \text{time}^2$

$\text{velocity of falling object} = \text{the constant } g \times \text{time}$

$\text{acceleration due to gravity} = g = 9.8 \text{ m/s}^2$

For an object in circular motion:

$$\text{acceleration} = \frac{\text{velocity}^2}{\text{radius}}$$

► Review Questions

1. How did Stonehenge allow ancient people to make predictions?

2. Why do scientists argue that Stonehenge was not built by ancient astronauts?

3. What are the characteristic movements of some of the objects you see in the night sky?

4. Describe the main features of the Ptolemaic and Copernican systems of the universe. In what ways are they similar?

5. What did Tycho Brahe try to do to resolve the question of the structure of the universe?

6. What was Kepler's role in interpreting Tycho Brahe's data?

7. What is Kepler's first law of planetary motion? What assumption of the Copernican system did this law refute?

8. What is Kepler's second law of planetary motion? According to this law, at what point in its orbit does a planet move fastest?

9. What is Kepler's third law of planetary motion? Given this law, what are the relative lengths of the year for Earth, Venus (next closest planet to the Sun), and Mars (next farthest planet from the Sun)? (*Hint:* See the appendices for more information.)

10. What is mechanics? Provide an example of an event that might be studied by this discipline.

11. How did Galileo slow down the rate at which objects fall in the laboratory?

12. How did Galileo measure time?

13. Define speed. How can you calculate speed from a known distance traveled and time of travel?

14. What is the difference between speed and velocity?

15. Define acceleration. How can you tell if your car is accelerating by looking at the speedometer?

16. What is *g*?

17. Why does any circular motion involve an acceleration?

18. What is instantaneous velocity?

► Fill in the Blanks

Complete the following paragraph by inserting words or phrases from the following list of key terms.

acceleration	law of compound
acceleration due to grav-	motion
ity (*g*)	mechanics
astronomy	speed
Kepler's laws of plane-	velocity
tary motion	

_____ is the branch of science devoted to the study of the motion of objects in the heavens. It took thousands of years for people to understand that the planets move according to _____. At that time, there was no obvious connection between the planets and _____, the branch of science devoted to the study of motion. In this branch of science, the _____ of an object is the distance it travels divided by the time it takes to travel that distance, while the _____ is that number plus the direction in which the motion occurs. A change in either the number or direction is called an _____. Something dropped at the surface of the Earth falls with the _____, equal to 32 feet per second per second. The _____ is used to describe the motion of objects in more than one dimension.

▶ Discussion Questions

1. The asteroid belt is a large collection of rocks and boulders that lies about three times as far from the Sun as the Earth. How long is the orbital period, or "year," for one of these rocks?

2. Which of the following are in uniform motion, and which are in accelerated motion? Explain each response.

    a. a car heading north at 35 mph

    b. a car going around a curve at 50 mph

    c. a dolphin leaping out of the water

    d. an airplane cruising at 30,000 feet at 500 mph

    e. a book resting on your desk

    f. the Moon

3. People who put oversized wheels on their cars often find that their actual speed is significantly greater than the speedometer indicates. Can you explain why this is so?

4. What two quantities did Galileo have to measure in his rolling-ball experiment? How might you improve on his experiment using modern technology?

5. List at least three objects that exhibit recurring patterns that relate to the Earth's physical environment. (For example, the leaves of a tree change with the seasons.)

6. By extending the logic used to define instantaneous velocity, define instantaneous acceleration.

7. Consider a comet with a very elongated, elliptical orbit around the Sun. Using Kepler's three laws of motion, describe the speed of the comet as it orbits the Sun.

8. How are the works of Tycho Brahe and Kepler an example of the scientific method?

9. Why is the measurement of time important in mechanics?

▶ Problems

1. If a race car completes a 3-mile oval track in 58 seconds, what is its average speed? Did the car accelerate during the 58 seconds?

2. If your car goes from 0 to 60 miles per hour in 6 seconds, what is your average acceleration?

3. The hare and the tortoise are at the starting line together. When the gun goes off, the hare moves off at a constant speed of 10 meters per second. The tortoise starts more slowly, but accelerates at the rate of 1 meter per second per second. Make a table showing the positions of the two racers after 1 second, 2 seconds, 3 seconds, and so forth. How long will it be before the tortoise passes the hare?

4. Work out the change in velocity in a small interval as in the Science by the Numbers box on page 71 for an interval of 1 mm (0.001 m); for 0.01 mm (0.00001 m); and for 0.0001 mm (0.000001 m). Does making the interval shorter improve the definition of instantaneous velocity?

5. Someone in a car going past you at a speed of 20 meters per second drops a small rock from a height of 2 meters. How far from the point will the rock hit the ground? (*Hint:* Find how long it will take the rock to fall and then apply the law of compound motion.)

6. The Statue of Liberty weighs nearly 205 metric tons. If a person can pull an average of 100 kg, how many people will it take to move the Statue of Liberty? (*Hint:* 1 metric ton is equal to 1000 kg.)

7. The weight of the space shuttle is about 4.5 million pounds. How many people would it take to move it? (See problem 6.)

8. Eccentricity is a measure of how elliptical or oval an ellipse can be. It is defined (see Figure 3–21) for a planet as the ratio of the distance between the two foci (f_1 to f_2) divided by twice the average distance from the sun (d_1 to d_2). A perfect circle has an eccentricity of 0.0 because the two foci are at the same position.

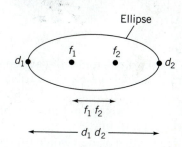

Figure 3–21

    a. Calculate the eccentricities for the following solar system objects. All data are in terms of the average distance of the Earth from the Sun, called the astronomical unit (AU).

Object	f_1 to f_2 (AU)	d_1 to d_2 (AU)
Earth	0.017	1.0
Mars	0.14	1.52
Pluto	9.8	39.5
Halley's comet	17.4	17.9

    b. Which object has the most nearly circular orbit? The most elliptical orbit?

9. For the planets and comet in the above list from problem 8, calculate the orbital periods using Kepler's third law of planetary motion.

10. Consider the orbits of two fictitious solar system objects given in Figure 3–22. Which object (*A* or *B*) has the longest orbital period? Explain your reasoning.

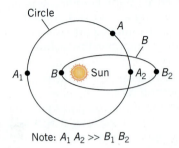

Note: $A_1 A_2 \gg B_1 B_2$

Figure 3–22

11. Consider the orbit of a typical comet around the Sun, given in Figure 3–23, which is marked at five different positions: *A, B, C, D,* and *E.* Using Kepler's second law of planetary motion, rank those positions in order of their relative speeds, the position for the fastest speed first.

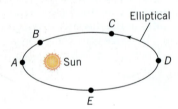

Figure 3–23

12. A new, mythical asteroid is discovered in the solar system, with a circular orbit and an orbital period of 8 years.
  a. What is the average distance of this mythical object in Earth units?
  b. Between which planets would this new asteroid be located?

13. The four Galilean satellites of Jupiter are Io, Europa, Ganymede, and Callisto. Their average distances from Jupiter and orbital periods are listed below in terms of Io's values.

Satellite	Relative Average Distance	Relative Orbital Period
Io	1.00	1.00
Europa	1.59	2.00
Ganymede	2.54	4.05
Callisto	4.46	9.42

  a. Using a graph, plot the cube of the relative average distance versus the square of the relative orbital period for each satellite. In words, state the pattern in your graph.
  b. From this information, do you agree or disagree that Kepler's third law (modified version) holds for Jupiter's four Galilean satellites? Explain.

14. A racewalker can walk 1 mile in 10 minutes.
  a. What is the speed in miles per hour? In km per hour?
  b. How long would it take a racewalker to walk 3.5 miles? 10 km?
  c. How far can a racewalker walk in 45 minutes? In 1.5 hours?

15. The North American continental plate is moving away from the European continental plate at a constant speed of 4.2 cm per year.
  a. If the average distance between the two plates is 7000 km, and the two plates have maintained their constant speeds, how long ago were the two continental plates together?
  b. In 1 million years (10^6 years), how large will the separation be between the two plates?

16. A typical motorist in the United States travels 25,000 miles in his or her car every year. If you assume that the average speed of the car while traveling is 45 miles per hour (remember that the car is not always moving), calculate the total number of hours an average motorist spends in his or her car. How many hours per day is this, averaged over one year? Do you think the average speed is a reasonable estimate? Explain.

17. The typical airborne speed of an intercontinental 747 jet is 530 miles per hour, while the airborne speed of the supersonic Concorde is 1500 miles per hour. If each airliner were to circumnavigate the Earth (25,000 miles) what would be the difference in air time spent by the two aircraft?

18. It takes light (speed = 3.0×10^8 m/s) 8.33 minutes to travel from the Sun to the Earth, and 1.3 seconds from the Moon to the Earth. What is the average distance of the Sun and Moon from the Earth?

19. While traveling out in the country at 50 miles per hour, your car's engine and brakes stop working and you coast to a stop in 25 seconds. What was your average deceleration during the time after the motor shut off?

20. Starting from rest, a train will reach a final, constant speed in 35 seconds while accelerating at a constant rate of 3 km/h/sec.
  a. What is the final speed of the train?
  b. What is the total distance traveled by the train during this period of constant acceleration? (Be careful here with your units.)

21. A rock falls to the bottom of a tall canyon, falling freely with no air resistance, for 4.5 seconds. Make a table of the distance traveled by the rock and its velocity after 1.0 s, 2.0 s, 3.0 s, 3.5 s, and 4.5 s.

22. A car traveling at 65 miles per hour crashes directly into a wall, coming to a complete stop. The time of contact for the crash was 0.25 s. What is the deceleration of the car in terms of the acceleration of gravity (i.e., the number of g's)?

23. A baseball is hit off the edge of a cliff horizontally at a speed of 30 m/s. It takes the ball 3 seconds to reach the ground, with no air resistance.

    a. How far from the cliff wall does the ball land?

    b. How high is the cliff wall?

24. Michael Jordan pops up a baseball directly over the batter's box. It takes the ball 5.0 seconds to return to the waiting glove of the catcher.

    a. What is the instantaneous speed of the ball at the top of the ball's path?

    b. What is the instantaneous speed of the ball immediately after it was in contact with the bat?

    c. How far above the ground did the ball travel? (Assume that the ball was caught at the same height it was hit.)

25. Two balls are released simultaneously from the same height, 10 meters above the ground. The first ball is released at rest and the second ball is released with a horizontal velocity of 15 m/s. Which ball reaches the ground first? Why?

26. A girl grabs a bucket of water and swings it around her in a horizontal circle, at a constant speed of 20 m/s at an arm's length of 0.7 meters. What is the centripetal acceleration of the bucket of water?

27. The space shuttle orbits 100 miles above the Earth's surface in a near-circular orbit at a constant speed. Assume that the centripetal acceleration is equal to the acceleration due to gravity at sea level (9.8 m/s²), and the orbital radius is equal to the radius of the Earth (6380 km),

    a. What is the average speed of the space shuttle?

    b. How long does the space shuttle take to make one orbit around the Earth?

28. The Moon moves in a nearly circular orbit around the Earth of radius of 3.84×10^8 m in 27.3 days. What is the centripetal acceleration of the Moon in m/s²?

▶ Investigations

1. Read the Bertold Brecht play *Galileo*, which dramatizes Galileo Galilei's heresy trial. Discuss the dilemma faced by scientists whose discoveries offend conventional ideas. What areas of scientific research does today's society find offensive or immoral? Why?

2. What other kinds of models of the universe did old civilizations develop? Look up those of the Mayans, the Chinese, and the Indians of the American Southwest, and describe some of their models. What features do these models have in common?

3. Find out how Galileo came to the idea of the pendulum clock. What did he actually observe that led him to this development?

4. Drop a wadded-up sheet of paper and a flat one side by side. Which reaches the ground first? Why? What do you think would happen if this experiment were done in a vacuum?

5. When you are in a car traveling at a constant speed, throw a ball up and describe its motion as you see it. Is there a difference when the car is being accelerated?

6. Drop a helium-filled balloon. Does it fall with acceleration g? Why? What do you think would happen if you dropped the balloon in a vacuum?

▶ Additional Reading

Drake, Stillman. *Galileo: Pioneer Scientist*. Toronto: University of Toronto Press, 1990.

Finnochairo, Maurice A. *The Galileo Affair*. Berkeley: University of California Press, 1989.

Gingerich, Owen. *The Eye of Heaven: Ptolemy, Copernicus, Kepler*. American Institute of Physics, 1993.

Hawking, Gerald S. *Stonehenge Decoded*. New York: Doubleday, 1965.

4 THE CLOCKWORK UNIVERSE

ONE SET OF LAWS—NEWTON'S LAWS OF MOTION AND
GRAVITY—PREDICT THE BEHAVIOR OF OBJECTS ON
EARTH AND IN SPACE.

Summer Vacation

The year 1665 was not a good one for Cambridge University. Bubonic plague (the "Black Death") was making one of its periodic appearances in England, and people fleeing the cities were spreading the disease everywhere. In some towns the situation got so bad that convicts who had been sentenced to be hanged were given the chance to escape their fate by burning the bodies of plague victims. Each morning they would wheel their carts through the streets, calling "Bring out your dead," then take the bodies outside the city for cremation.

In situations like these, even university administrations take action. They closed Cambridge University, sending the students back to their homes until the plague abated. For most students, it was probably just a vacation, but for one young man named Isaac Newton, the break provided a chance to think on his own, without the distractions of the university. By the time the university opened again, 18 months later, he had developed (1) a new branch of mathematics known as calculus, (2) the basis for the modern theory of color, and (3) the law of universal gravitation (which, as we'll see, connected the sciences of physics and astronomy). In his spare time, he polished off a few mathematical problems that had eluded solution up to that time. In his words, "In those days I was in the prime of my age for invention, and minded Mathematics and Philosophy more than at any time since."

What did *you* do on your last summer vacation?

Isaac Newton (1642–1727)

Newton's laws of motion
Three basic principles, expressed as laws, that govern the motion of everything in the universe, from stars and planets to cannonballs and muscles. *The first law* states that a moving object will continue moving in a straight line at a constant speed, and a stationary object will remain at rest, unless acted on by an unbalanced force. *The second law* states that the acceleration produced on a body by a force is proportional to the magnitude of the force and inversely proportional to the mass of the object. *The third law* states that for every action there is an equal and opposite reaction.

force A push or pull that, acting alone, causes a change in acceleration of the object on which it acts.

ISAAC NEWTON AND THE UNIVERSAL LAWS OF MOTION

Kepler's laws (see Chapter 3) describe the way that planets move in their orbits around the Sun. Galileo's discoveries, including the law of compound motion and the equations for accelerated motion, describe the way that objects move in a laboratory here on the surface of the Earth. Neither of these sets of laws, however, contains any suggestion that there is any way in which a planet might be similar to a cannonball, or that the laws that operate on the Earth might also operate in the heavens. In other words, after Kepler and Galileo had completed their work, physics and astronomy were two separate, unconnected areas of science. It was Isaac Newton (1642–1727) who first showed us that what happens on the Earth is in no way different from what happens anywhere else in the universe.

Newton, perhaps the most brilliant scientist who ever lived, synthesized the work of Galileo and others into a statement of the basic principles that govern the motion of everything in the universe, from stars and planets to clouds, cannonballs, and the muscles in your body. We call these results **Newton's laws of motion.** They sound so simple and obvious that it's hard to realize that they represent the results of centuries of experiment and observation, and even harder to appreciate what an extraordinary effect they had on the development of science.

Three basic laws formed the centerpiece of Newton's description of motions.

The First Law

Newton's first law of motion links the key concepts of motion and force.

> **A moving object will continue moving in a straight line at a constant speed, and a stationary object will remain at rest, unless acted on by an unbalanced force.**

Newton's first law seems so simple that it scarcely needs to be stated. It seems obvious to us that if you leave something alone, it won't change its state of motion. Yet, virtually all scientists from the Greeks to Copernicus would have argued that this statement is wrong. They believed that since the circle is the most perfect geometrical shape, objects will move in circles unless something interferes. They believed that their heavenly spheres (see Chapter 3) would keep turning without the action of any outside force.

Newton, basing his arguments on observations and the work of his predecessors, turned this notion around. A moving object left to itself will travel in a straight line, and if you want to get it to move in a circle, you have to apply a force. You know this is true. If you swing something around your head, it will move in a circle only so long as you hold on to it. Let go, and off it goes in a straight line.

Newton's first law tells us that when we see an acceleration, something must have acted to produce that change. We define a **force** as something that produces a change in the state of motion of an object. In fact, we will use the first law of motion extensively in this book to tell us how to recognize when a force, particularly a new kind of force, is acting.

We sometimes give the tendency of an object to remain in uniform motion a special name: **inertia.** This term is often used in everyday speech. We may, for example, speak of the inertia in a company or government organization when we want to get across the idea that it is resistant to change.

The Second Law

If Newton's first law of motion tells you when a force is acting, then the second law of motion tells you what the force does when it acts.

> **The acceleration produced on a body by a force is proportional to the magnitude of the force, and inversely proportional to the mass of the object.**

This law conforms to our everyday experience that it is easier to move a bicycle than a car, easier to lift a child than an adult, and easier to push a ballerina than a defensive tackle. The second law of motion can be described as an equation.

▶ **In words:**

The bigger the force, the greater the acceleration; the greater the mass, the less the acceleration.

▶ **In equation form:**

force (in newtons) = mass (in kilograms) × acceleration
(in meters per second²)

▶ **In symbols:**

$F = ma$

This equation, well known to generations of physics majors, tells us that if we know the forces acting on a system of known mass, we can predict its future motion. The equation conforms to our intuition that an object's acceleration is a balance between two factors.

A pool game illustrates Newton's laws of motion.

inertia The tendency of a body to remain in uniform motion; the resistance to change.

A small force can cause a change of motion in a large object.

On the one hand, a force causes the acceleration; therefore, the greater the force, the greater the acceleration. The harder you throw a ball, the farther it goes. Acceleration is directly proportional to forces (see Chapter 2).

On the other hand, the greater the object's mass—the more "stuff" you have to accelerate—the less effect a given force is going to have. A given force will accelerate a golf ball more than a bowling ball, for example. Newton's second law of motion thus defines the balance between force and mass in producing an acceleration. Recasting the equation, we can see that acceleration is inversely proportional to mass:

$$\text{acceleration} = \text{force} \div \text{mass}$$

or

$$a = \frac{F}{m}$$

The second law of motion does not imply that every time a force acts, motion must result. A book placed on a table remains stationary even though it feels the force of gravity, and you can push against a wall without moving it. In these situations, the atoms in the table or the wall shift around and exert their own force that balances the force that acts on them. Only unbalanced forces actually give rise to acceleration.

newton (N) The unit of force. The force required to give an acceleration of 1 m/s² to a mass of 1 kg.

In the English system, force is measured in the familiar unit of the pound. In the SI system, the unit of force is called the **newton** (abbreviated N). A newton is defined to be the force that will accelerate a 1-kilogram mass at the rate of 1 meter per second². A force of one pound is equal to about 4.45 N.

Example 4–1: From 0 to 100 in 8 Seconds

Your car, which has a mass of 750 kilograms when you are driving, accelerates steadily from 0 to 100 kilometers per hour (about 60 miles per hour) in 8 seconds. How much force must be applied during this acceleration?

▶ **Reasoning:** Newton's second law says that force equals mass (in kilograms) times acceleration (in meters per second²). We know the mass in kilograms, and the acceleration in kilometers per hour per second. Therefore, we must convert acceleration to the proper units of meters per second².

▶ **Solution:** The acceleration is 100 kilometers per hour in 8 seconds.

$$\text{acceleration} = \frac{100 \text{ km/h}}{8 \text{ s}}$$

$$= \frac{100}{8} \text{ km/h/s}$$

$$= 12.5 \text{ km/h/s}$$

Thus, each second during this acceleration the car's speed increases by 12.5 kilometers per hour.

To convert this acceleration to the proper units, remember that a kilometer is 1000 meters. So

$$\text{acceleration} = (12.5 \text{ km/h})/\text{s} \times (1000 \text{ m/km})$$
$$= (12,500 \text{ m/h})/\text{s}$$

Furthermore, a second is 1/3600th of an hour, so:

$$\text{acceleration} = (12,500 \text{ m/h})/\text{s} \times (1/3600 \text{ h})/\text{s}$$
$$= 3.47 \text{ m/s}^2$$

Now that we have converted to the proper unit, the force can be calculated:

$$\text{force} = \text{mass} \times \text{acceleration}$$
$$= 750 \text{ kg} \times 3.47 \text{ m/s}^2$$
$$= 2602.5 \text{ N} \; \blacktriangle$$

Example 4–2   Physics on the Football Field

A football player has a mass of 120 kilograms. On a particular play, he has to be moved aside 1 meter in a half second. Ignoring any strategy that the player may take, what force must be applied to accomplish this?

▶ **Reasoning and Solution:**   The first step in the solution is to figure out what the player's acceleration must be. From Chapter 3, we know that if the distance traveled by an accelerating body is d, the acceleration is a, and the time is t, then

$$d = \tfrac{1}{2} at^2$$

In this case, the distance we wish to move the player is 1 meter, the time $\tfrac{1}{2}$ second. Thus, we have:

$$1 \text{ meter} = \tfrac{1}{2}a \times (\tfrac{1}{2} \text{ s})^2$$
$$= \tfrac{1}{2}a \times \tfrac{1}{4} \text{ s}^2$$
$$= (\tfrac{1}{8}a) \text{ s}^2$$

so that:
$$a = 1 \text{ m} \times 8/\text{s}^2$$
$$= 8 \text{ m/s}^2$$

The next step in the solution is to use Newton's second law to calculate the force needed to produce this acceleration.

$$F = ma$$
$$= 120 \text{ kg} \times 8 \text{ m/s}^2$$
$$= 960 \text{ N} \; \blacktriangle$$

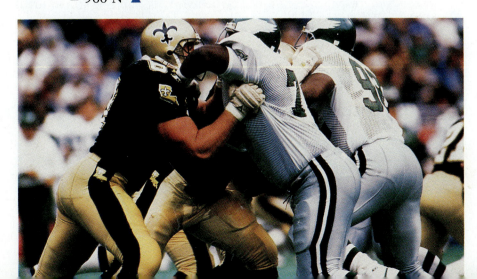

To move a football player, Newton's second law says that a force must be exerted.

The Third Law

Newton's third law of motion tells us that whenever a force is applied to an object, that object simultaneously exerts an equal and opposite force.

> **For every action (force) there is an equal and opposite reaction (force).**

When you push on a wall, for example, it instantaneously pushes back on you; you can feel the force on the palm of your hand. In fact, the force the wall exerts on you is equal in magnitude, but opposite in direction, to the force you exert on it.

The third law of motion is perhaps the least intuitive of the three. We tend to think of our world in terms of causes and effects, in which big or fast objects exert forces on smaller, slower ones: a car slams into a tree, a batter drives the ball into deep left field, a boxer punches his opponent's eye. But in terms of Newton's third law, it is equally valid to think of these events the "other way around." The tree stops the car's motion, the baseball alters the swing of the bat, and the opponent's eye blocks the thrust of the boxer's glove, thus exerting a force and changing the direction and speed of the punch. Forces *always* act simultaneously in pairs (see Figure 4–1).

The interplay of Newton's three laws of motion can be envisioned by a simple example. Imagine a child standing on roller skates holding a stack of baseballs. She throws the balls, one by one. Each time she throws a baseball, the first law tells us that she has to exert a force so that the ball accelerates. The third law then tells us that the baseball will exert an equal and opposite force on her. This force acting on the child, according to the second law, will cause her to recoil backward.

And while the example of the child and the baseballs may seem a bit contrived, it illustrates the exact principle by which rockets work. In the rocket motor, forces are exerted on hot gases, accelerating them out the tail end. Using the argument just given above, we see that an equal

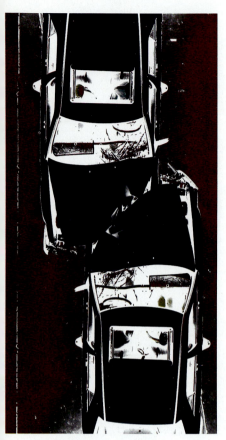

Figure 4–1
Two automobiles in a collision exert equal and opposite forces on each other.

The space shuttle *Discovery* rises from its launch pad at Cape Canaveral, Florida. As hot gases accelerate violently out the rocket's engines, the shuttle experiences an equal and opposite force that lifts it into orbit.

and opposite force must be exerted on the rocket, propelling it forward. Every rocket, from Fourth of July fireworks to the space shuttle, works this way.

Isaac Newton's three laws of motion form a comprehensive description of all possible motions, as well as the forces that cause them. In and of themselves, however, Newton's laws do not say anything about the nature of those forces. In fact, much of the progress of science during and after Newton's times has been associated with the discovery and elucidation of the forces of nature. But the laws contain even more information, as we shall now see.

Developing Your Intuition

Isaac Newton Plays Volleyball

Newton's three laws of motion apply to every action of our lives. Think about a game of doubles volleyball at the beach. It's a beautiful sunny day with a light breeze off the ocean. The opposing team serves the ball low and hard but your partner gives you a perfect pass. You set the ball high and outside, a foot off the net. She makes her approach, leaps high, and hits down the line for a side out.

During this one brief play dozens of forces and motions took place—the ball changed direction and speed, players ran and jumped, sand grains shifted, and the wind blew while waves crashed in the background. Think about how Newton's laws of motion apply to these events.

The first law says that nothing happens without an unbalanced force. When a ball flies through the air or a person jumps high off the ground, a force must be involved. One everyday force is gravity, which causes the ball (and the players) to fall back to Earth. Contact forces, such as occur when a player's hand strikes the ball or when she leaps into the air by pushing against the sand, are also all around us. The origin of other forces, such as those that cause the wind to blow or waves to crash onto the shore, are less obvious. We'll learn more about them in Chapter 7. But whatever the event, Newton's first law tells us that a force must exist.

Newton's second law allows us to calculate exactly how much force must be applied to achieve a given result (force equals mass times acceleration). A hard serve requires more force than a dink serve, just as a 36-inch vertical jump demands more force than a 12-inch jump. Volleyball players can't stop to perform complex mathematical calculations in the middle of their game. Nevertheless, our cumulative experience, living in a world of forces and motions, gives us a pretty good idea how much force to use in a given situation. And the more you practice a sport, the better your intuition about forces, motions, and Newton's second law becomes.

Newton's subtle third law comes into play throughout the volleyball game: all forces occur in pairs. As the player hits the ball, applying a force to propel it forward, the player's hand feels the force of the ball pushing back. (If you've ever received a painfully hard spike you'll know about forces acting in pairs.) When you push down to jump

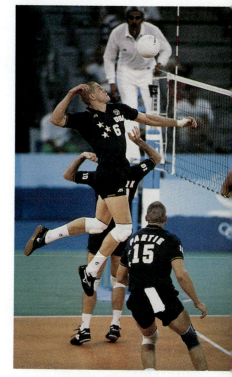

Newton's laws of motion apply everywhere, including the volleyball court.

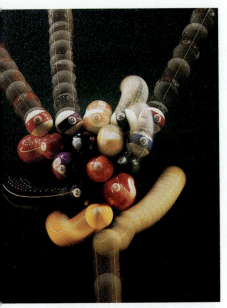

Conservation of momentum is illustrated in the collision of billiard balls.

momentum   The product of the mass of an object times its velocity.

high off the sand, the sand pushes with an equal and opposite force against you. As the wind blows against your hair, your hair simultaneously changes the path of the wind.

Next time you're at the beach, think about Newton's three laws (but don't think about them too hard). ●

MOMENTUM

Have you played a game of pool recently? If so, you remember shooting the white cue ball so that it collided with a colored ball, propelling it toward a pocket. Before the collision, the colored ball was stationary, while the cue ball was moving with a (more or less) constant velocity. After the collision, both balls were moving, though less swiftly than the cue ball had moved before. You can see the same sort of thing in many games: bowling, marbles, and croquet are examples. And, as we shall see in Chapter 5, such collisions are constantly taking place between the atoms that make up material objects. Since they involve the motion of material objects, these collisions ought to be explainable by Newton's laws.

It often happens that when we understand phenomena in terms of a comprehensive scheme such as Newton's laws, we find that these laws give us a new way of looking at familiar things. The analysis of collisions provides a good example.

Newton's laws tell us that when two billiard balls (or two atoms) collide, no external forces are at work. One billiard ball or atom may exert a force on the other, but the other billiard ball or atom must then exert an equal and opposite force on the first. These internally generated forces cause each object to change speed and direction, but when we add up all forces on the entire system, they cancel each other. Thus, although each individual object feels an unbalanced force (and therefore accelerates), the entire system feels no net force acting.

The motion of each pool ball can be described by two quantities, its mass and its velocity. **Momentum** is defined as the product of the mass of an object times its velocity. Every moving object has a momentum. The momentum can be large if the object has a small mass and a high velocity (a rifle bullet, for example), or a large mass and a low velocity (a freight train creeping to a stop). Like velocity, momentum has a direction, and is therefore a vector quantity. If, for example, a ball rolling to the right has a positive momentum, then a ball rolling to the left with the same speed will have the same momentum value, but with a negative sign in front of it, a convenient way to show that the motion is in the opposite direction.

total momentum   The sum of all the momenta in a system.

The sum of all the momenta in a system (the sum of the momenta of the cue ball and colored ball before the collision, for example) is called the **total momentum** of the system. We can use Newton's laws of motion to derive an important fact about total momentum.

▶ **In words:**

In the absence of external forces, the total momentum of a system does not change; so when two objects collide, their velocities after the collision must be arranged so that the total momentum is exactly the same after the collision as it was before.

▶ **In equation form:**

(mass × velocity)$_{Object 1}$ + (mass × velocity)$_{Object 1}$ = constant

▶ **In symbols:**

$(m_1 \times v_1) + (m_2 \times v_2) = k$

In the language of physics, something that doesn't change during an interaction is said to be *conserved*. Therefore, this rule is sometimes called the law of **conservation of momentum.**

In the case of the pool balls, for example, suppose the cue ball has a mass m_1 and travels with a velocity v_1, and the colored ball has a mass m_2 (before the collision, the velocity of the second ball is zero). If the two balls have velocities u_1 and u_2 after the collision, then the law of conservation of momentum tells us, in symbols:

$(m_1 \times v_1) + (m_2 \times 0) = (m_1 \times u_1) + (m_2 \times u_2)$

Let's go through a little exercise to see how this works.

Suppose that a speeding bullet (mass m_1, velocity $-v_1$) and a powerful locomotive (mass m_2, velocity v_2 in the opposite direction) are approaching each other. The total momentum of the system before impact is just

$(m_1 \times (-v_1)) + (m_2 \times v_2) = -(m_1 + v_1) + (m_2 \times v_2)$

The minus sign in this expression represents the facts that the bullet and the train are moving in opposite directions before the collision. Suppose, for the sake of argument, that after the collision the bullet recoils and is moving in the same direction as the train, but with velocity u_1, while the train keeps moving forward with velocity u_2 (where u_2 is not the same as u_1). The new total momentum is

$(m_1 \times u_1) + (m_2 \times u_2)$ = total momentum after collision

The conservation of momentum tells us that the total momenta before and after the collision have to be the same. Since the momentum of the bullet is reversed, the momentum of the train has to decrease, and the train has to slow down. You can, in fact, think of the collision as a transfer of momentum from the train to the bullet, with the train losing

conservation of momentum
A law which states that the total momentum of a system does not change unless an outside force is applied.

The explosion of fireworks demonstrates the conservation of momentum.

positive (rightward directed) momentum and the bullet gaining this positive momentum as it reverses direction.

It's important to keep in mind that the law of conservation of momentum doesn't say that momentum can never change. It just says that it won't change unless an outside force is applied. If a soccer ball is rolling across a field and a player kicks it, a force is applied to the ball as soon as her foot touches it. At that moment, the momentum of the ball changes, and that change is reflected in its change of direction and speed.

You may not have realized it at the time, but you were seeing the consequences of the conservation of momentum the last time you watched a fireworks display. Think about a rocket shooting straight up and, just to make things simple, imagine that the firework explodes just at the moment that the rocket is stationary at the top of its path, at the instant when its total momentum is zero. After the explosion, brightly colored burning bits of material fly out in all directions. Each of these pieces has a mass and a velocity, so each has some momentum. Conservation of momentum, however, tells us that when we add up all the momenta of the pieces, they should cancel each other out and give a total momentum of zero. Thus, for example, if there is a 1-gram piece moving to the right at 10 meters per second, there has to be the equivalent of a 1-gram piece moving to the left at the same velocity. This conservation of momentum gives fireworks their characteristic symmetric starburst pattern.

In 1880, engines like this converted stored energy in coal to kinetic energy.

Angular Momentum

angular momentum The momentum of rotating objects.

The kind of momentum we have been talking about so far is often referred to as *linear momentum*, which is characteristic of objects moving in a straight line. **Angular momentum** is a similar concept that applies to rotating objects. Just as an object moving in a straight line will keep moving unless a force acts, an object that is rotating will keep rotating until something acts to make it stop. Thus a spinning top will keep spinning until the friction between its point of contact and the floor slows it down.

Like its counterpart, linear momentum, angular momentum is conserved. It does not change unless a force acts to make it do so. The Earth, the largest spinning object with which we come into contact, now rotates once every 24 hours. Early in its history, however, the Earth rotated much faster; some scientists claim days were as short as 14 hours. External forces, especially the Moon's gravitational pull, have gradually slowed down the Earth and reduced its angular momentum.

THE UNIVERSAL FORCE OF GRAVITY

gravity A force that acts on every object in the universe.

Gravity is the most familiar (and insistent) force in our daily lives. It holds you down in your chair and keeps you from floating off into space. It guarantees that when we drop things, they fall. The effects of what we call gravity were known to the ancients, and its quantitative properties were studied by Galileo and many of his contemporaries. It was Isaac Newton, however, who revealed its universality.

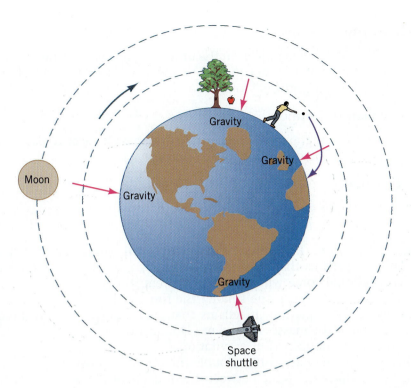

Space
shuttle

Figure 4–2
An apple falling, a ball being thrown, and the or-
biting Moon all display the
influence of the force of
gravity.

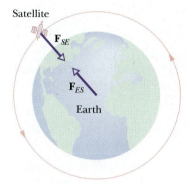

Figure 4–3
The Earth exerts an attractive gravitational force on
this satellite, F_{ES}, and the
satellite exerts an equal and
opposite gravitational force
on the Earth, F_{SE}.

Everyone knew that gravity pulled things toward the Earth, but until Newton, most people assumed that the phenomenon was local in that it appeared only near the planet's surface. People believed that farther out, in the realm of the stars and planets, different rules applied to the turning of the celestial spheres. They would say that "terrestrial gravity" operated on the Earth and "celestial gravity" operated in the heavens, but the two had little to do with each other. Isaac Newton discovered that these two seemingly different kinds of gravity were, in fact, one and the same. In modern language, in a remarkable union of seemingly disparate elements, he unified earthly and heavenly gravity.

By Newton's account, his great insight came to him in an apple orchard. He saw an apple fall and, at the same time, saw the Moon in the sky behind it. He wondered whether the gravity that caused the apple to fall could extend far outward to the Moon, supplying the force that kept it in its orbit.

If you think about the Earth–Moon system in terms of his laws of motion, you can see why this sort of thought would have occurred to Newton (see Figure 4–2). The Moon goes around the Earth in a roughly circular orbit, so it isn't moving in a straight line. From the first law of motion, it follows that some sort of force must be acting on the Moon to keep it in orbit. The question was the exact nature of that force. Newton's insight in his apple orchard was that the force holding the Moon in its orbit could be the same force that made the apple fall—the familiar force of gravity.

Eventually, he realized that the orbits of all the planets could be understood if gravity was not restricted to the surface of the Earth but was a force found throughout the universe (see Figure 4–3). He formulated this insight (an insight that has been overwhelmingly confirmed by observations) in what is called **Newton's law of universal gravitation.**

Newton's law of universal gravitation Between any two objects in the universe there is an attractive force (gravity) that is proportional to the masses of the objects and inversely proportional to the square of the distance between them. In other words, the more massive two objects are, the greater the force between them will be, and the farther apart they are, the less the force will be.

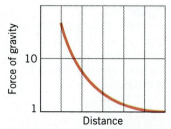

Figure 4–4
The force of gravity falls off rapidly as objects get farther apart.

gravitational constant (G)
The constant of proportionality in Newton's law of universal gravitation.

▶ **In words:**

Between any two objects in the universe there is an attractive force (gravity) that is proportional to the masses of the objects and inversely proportional to the square of the distance between them.

▶ **In equation form:**

The force of gravity = constant × (first mass × second mass ÷ the square of the distance between masses).

▶ **In symbols:**

$$F = G \times \frac{m_1 \times m_2}{d^2}$$

where G is a number known as the **gravitational constant** (see below).

This law tells us that the more massive two objects are, the greater the gravitational force will be between them; the greater the distance between them, however, the smaller the force will be. Think about the meaning of this equation. It tells us that gravitational force is directly proportional to the masses m_1 and m_2. Double one of the masses and the force of gravity doubles. But gravitational force varies with distance by an inverse square relationship: double the distance between two objects and the gravitational force between them decreases by a factor of four (see Figure 4–4).

The equation for gravitational force incorporates a number, G, that is a *constant of proportionality* (see Chapter 2). In the equation for gravity, G expresses the exact numerical relation between the masses of two objects and the distance between them, on the one hand, and the force between them on the other. Henry Cavendish, a scientist at Oxford University in England, first measured G in 1798 by using the apparatus shown in Figure 4–5. A dumbbell made of two small balls was suspended from a wire. Two large lead spheres were brought near the balls, and the resulting gravitational attraction between the small balls and the lead spheres turned the dumbbell until the twisted wire exerted a counterbalancing force strong enough to stop the rotation. By measuring the amount of twisting force (also known as a torque) on the wire, Cavendish measured

Figure 4–5
The Cavendish balance measures the gravitational constant G by balancing the gravitational force against the force exerted by a twisted wire.

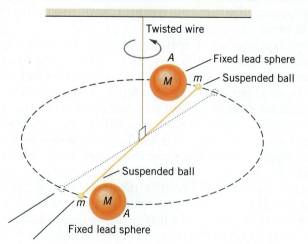

the force on the dumbbells. This measured force, together with the known masses of the dumbbells, the masses of the heavy spheres, and knowledge of the distance between them, gave him the numerical value of everything in Newton's law of universal gravitation except G, which he then calculated using simple arithmetic.

In metric units, the value of G is 6.67×10^{-11} m³/s²-kg, or 6.67×10^{-11} N-m²/kg². We expect that this constant is universal, and has the same value everywhere in our universe. Modern descendants of the Cavendish balance have succeeded in measuring G to an accuracy of 5 decimal places.

Weight and Mass

Recall that the law of universal gravitation says that there is a force between *any* two objects in the universe. Between you and the Earth, for example, there exists a force proportional to your mass and the much larger mass of the Earth. The distance between you and the center of the Earth is the radius of the Earth. This distance, which we will denote by R_E, is about 6400 km (4000 miles), and it is the number you would put in for the distance in the law if you were calculating the force with which you are attracted to the Earth.

A scale measures the force needed to cancel the downward pull of gravity, and hence tells you your weight.

The gravitational attraction between you and the Earth would accelerate you downward if you weren't standing on the ground. As it is, the ground exerts an equal and opposite force to cancel gravity, a force you can feel in the soles of your feet.

An ordinary bathroom scale makes use of this interplay of forces. Inside the scale there is a spring or some other mechanism that, when compressed, exerts an upward force. This upward force exerted by the scale keeps you from falling, and the size of this counterbalancing force registers on a display, allowing you to measure your weight.

Weight, in fact, is just the force of gravity on an object located at a particular point. Weight depends on where you are; on the surface of the Earth you weigh one thing, on the surface of the Moon another, and in the depths of interstellar space you would weigh next to nothing. You even weigh a little less on a mountaintop than you do at sea level. Weight contrasts with your **mass,** which is related to the amount of matter—a quantity that stays the same no matter where you go. If you stood on a scale in an elevator, for example, you would see that your weight changed depending on the elevator's acceleration, but your mass would remain constant.

weight The force of gravity on an object located at particular point.

mass A quantity related to the amount of matter contained in an object, independent of where that object is found.

If we call M_E and R_E the mass and radius of the Earth, and m is your own mass, then the force that the Earth is exerting on you right now is given by the equation of gravity:

$$\text{force} = G \times \frac{(\text{first mass} \times \text{second mass})}{\text{distance}^2}$$

Thus

$$\text{your weight} = G \times \frac{(\text{mass}_{\text{Earth}} \times \text{your mass})}{\text{radius}_{\text{Earth}}^2}$$

$$= G \times \frac{(M_E \times m)}{R_E^2}$$

or, rearranging the terms:

$$F = m \times \left[\frac{G \times M_E}{R_E^2} \right]$$

Note that this equation is in the form "force = mass × [something in square brackets]." If we compare this to Newton's second law, "force = mass × acceleration," we see that what's in square brackets must be the acceleration you would feel if there were no force countering gravity. This missing number is identical to the quantity we called g in Chapter 3. In other words, on the Earth's surface

$$\text{force} = \text{mass} \times g = \text{weight}$$

and

$$g = G \times \frac{M_E}{R_E^2}$$

This result is extremely important. For Galileo, g was a number to be measured, but whose value he could not predict. For Newton, on the other hand, g is a number that can be calculated purely from the size and mass of the Earth.

One way of keeping weight and mass distinct in your mind is to remember that your weight can change from one place to another. For instance, you would weigh less on the Moon than you do on Earth. Your mass, however, which measures the amount of material in your body, is the same everywhere in the universe.

Example 4–3:   The Earth's Gravity

Given that the mass of the Earth is 6×10^{24} kilograms, what is the acceleration due to gravity at the Earth's surface?

▶ **Reasoning and Solution:**   To answer this question, we put the Earth's mass and radius into the expression for g given above.

$$g = \frac{(6.67 \times 10^{-11}\,\text{m}^3/\text{s}^2\text{-kg}) \times (6 \times 10^{24}\,\text{kg})}{(6.4 \times 10^6\,\text{m})^2}$$

$$= \frac{40.02 \times 10^{13}\,\text{m}^3/\text{s}^2}{40.96 \times 10^{12}\,\text{m}^2}$$

$$= 9.8\,\text{m/s}^2$$

This number is the same constant that Galileo and others have measured. ▲

Because we now understand where g comes from, we can predict the appropriate value of g not only for the Earth, but for any body in the universe, provided we know its mass and radius.

Example 4–4:   Weighty Matters

A cantaloupe has a mass of 0.5 kilograms. What does it weigh?

▶ **Reasoning and Solution:**   To answer this question, we have to calculate the force of gravity exerted on the cantaloupe at the Earth's surface. The relation between mass and weight is

$$\text{weight} = \text{mass} \times g$$
$$= 0.5 \text{ kg} \times 9.8 \text{ m/s}^2$$
$$= 4.9 \text{ kg m/s}^2$$
$$= 4.9 \text{ N}$$

This value is the weight of the cantaloupe. Remember that kilogram is not a unit of weight, despite popular usage to the contrary. ▲

Example 4–5: Football Player Mass

Earlier in this chapter we talked about a football player with a mass of 120 kg. What is his weight in newtons? In pounds?

▶ **Reasoning and Solution:**   To find the weight in newtons, use the same equation as in the previous example.

$$\text{weight} = \text{mass} \times g$$
$$= 120 \text{ kg} \times 9.8 \text{ m/s}^2$$
$$= 1176 \text{ N}$$

From the table in Appendix A we multiply pounds by 4.45 to get newtons, so that

$$\text{weight (lb)} = \frac{\text{weight (N)}}{4.45}$$
$$= \frac{1176}{4.45}$$
$$= 264 \text{ lb}$$

This weight shows the person to be large, but not unusually so for a football player. ▲

Example 4–6: Weight on the Moon

The mass of the Moon, M_M, is 7.18×10^{22} kg, and its radius, R_M, is 1738 km. If your mass is 60 kg, what would you weigh on the Moon?

Because the Moon is less massive than the Earth, people weigh less on its surface than they do at home.

▶ **Reasoning:** Once again we have to calculate the force exerted on an object at the surface of an astronomical body. This time, though, both the mass and the radius of the body are different from that of the Earth, while G is the same.

▶ **Solution:** From the equation that defines weight, we have

weight = force due to gravity

$$= G \times \frac{M_M \times 60 \text{ kg}}{R_M{}^2}$$

$$= (6.67 \times 10^{-11} \text{ m}^3/\text{s}^2 = \text{kg}) \times \frac{(7.18 \times 10^{22} \text{ kg}) \times 60 \text{ kg}}{(1.74 \times 10^6 \text{ m})^2}$$

$$= 95 \text{ N}$$

This weight is about a sixth of the weight the same object would have on the Earth, *even though its mass is the same in both places* (see Figure 4–6). ▲

Figure 4–6
A person who weighs 570 N on the Earth would weigh only 95 N on the Moon, although the mass doesn't change. The reason is that the Moon's gravitational attraction is only one-sixth that of the Earth.

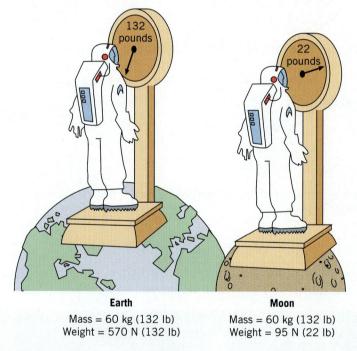

Earth	**Moon**
Mass = 60 kg (132 lb)	Mass = 60 kg (132 lb)
Weight = 570 N (132 lb)	Weight = 95 N (22 lb)

GRAVITY AND ORBITS

Perhaps the most striking result of Newton's insight about universal gravitation was the ability it gave scientists to understand the orbits of objects under the influence of gravity. In Chapter 3, we saw that the acceleration, a_c, required to keep an object like the Moon moving around in a circular path of radius r with a velocity whose magnitude is v is

$$a_c = \frac{v^2}{r}$$

Isaac Newton used this expression to calculate the behavior of orbiting objects.

Centripetal Force

Now that we have Newton's laws at our disposal, we know that to produce orbital acceleration a force must act. An object will stay in a circular path only as long as the force acts. If the force disappears, Newton's first law tells us that the object will move off in a straight line. Think of twirling a weight on a string around your head. The force that keeps the weight circling is the tension in the string; you can feel it in your fingers. If you let go of the string, however, the weight doesn't keep circling around, but flies off in a straight line in whatever direction it happened to be going at the moment of release. Discus throwers operate in exactly this way. First they spin around so that the discus, held at arm's length, is moving in a circular path. Then they release it so that it flies out over the field.

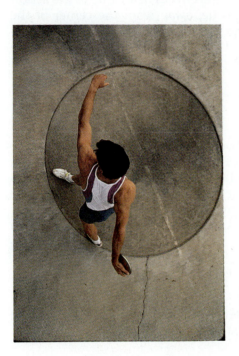

A discus moves in a circular path as long as the athlete holds on to it, but in a straight line when it is released.

From Newton's second law, we know that the force needed to produce the acceleration that keeps something moving in a circle is

$$\text{force} = \text{mass} \times \text{acceleration}$$
$$= ma_c$$

and, substituting the expression for a_c above,

$$F = ma_c$$
$$= \frac{mv^2}{r}$$

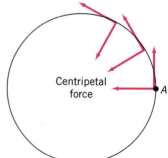

Figure 4–7
Schematic of centripetal force with weight, A, on twirling string.

When you swing a weight around your head on a string, for example, this expression defines the force the string has to exert to keep the motion going. The force is directed toward the center of the circle, and hence is often called the **centripetal force,** which means "center-seeking" force.

The action of the centripetal force is illustrated for a weight on a string in Figure 4–7. The natural tendency of the weight is to move off in a

centripetal force The force needed to keep an object moving in a circle; the force is directed inward, toward the center of the circle.

centrifugal force The force apparently produced by an object moving in a circle, the force is directed outward, away from the center of the circle.

The orbit equation applies to all satellites, whether they are moons, planets, or manufactured.

straight line, as shown at point *A*. To keep it from flying off, you have to exert a force to pull it back into the circle. It then wants to fly off again, so that you have to pull it back again, and so on. The tug you feel in your hand as the object goes around is the reaction to your constantly pulling on the string, constantly tugging the object back.

The centripetal force, which is a real force exerted on an object to keep it in a circular path, has to be distinguished from the **centrifugal force**, or "center-fleeing" force. Centrifugal force is the name we apply to the force you feel when you are inside an object that is being accelerated in circular motion, but this force is not a real force. When you go around a curve in a car, for example, your body will move in a straight line (Newton's first law) as the car turns underneath you. The car will exert a force on you (Newton's second law) and you, too, will start to move in a circle. We perceive this sequence of events as being due to the action of a force that pushes us to the outside, away from the center of the circle, although that is not what happened. We call this force the centrifugal force, but it is a force that is felt only by someone actually in the accelerating system. For this reason, physicists often refer to centrifugal force as a "fictitious force."

The Orbit Equation

In the case of the circling weight, you can feel the force pulling on the string, so identifying centripetal force is simple. In the case of the Moon, however, the origin of the centripetal force isn't so obvious. Newton's insight was so important because he realized that the force that binds the solar system together by keeping the planets and moons in their orbits is gravity. In fact, the basic equation that defines the orbit of a satellite, be it a planet, a moon, or a space shuttle, follows from this statement:

▶ **In words:**

> The centripetal force on a satellite in circular orbit is equal to the force of gravity exerted on that object.

▶ **In equation form:**

$$\frac{\text{satellite mass} \times \text{satellite velocity squared}}{\text{orbital distance}}$$
$$= \frac{\text{constant } G \times \text{satellite mass} \times \text{central mass}}{\text{distance squared}}$$

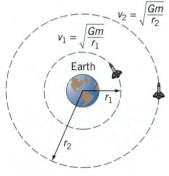

Figure 4–8
For each distance r_1 or r_2 from the Earth, there is one velocity (v_1 or v_2) that produces a stable orbit.

▶ **In symbols:**

$$\frac{mv^2}{r} = \frac{GMm}{r^2}$$

where *m* is the mass of the satellite, *M* is the mass of the central body, *r* is the distance from the satellite to the center of the central body, and *v* is the speed of the satellite in its orbit.

If we multiply both sides of this equation by *r* and divide by *m*, we find that it reduces to:

$$v^2 = \frac{GM}{r}$$

This equation is a good one to examine because it tells us that for a given distance, *r*, between a satellite and its central body there is one and only one speed, *v*, at which the satellite can move and remain in circular orbit (Figure 4–8). It also tells us that the speed at which a satellite has to move doesn't depend in any way on its mass. For example, a grapefruit orbiting the Earth at the same distance as the Moon would circle the Earth every 29 days, just as the Moon does.

Science by the Numbers

The Space Shuttle

We can get an idea of how the orbit equation works by thinking about the space shuttle. A typical shuttle orbit might be 150 kilometers (a bit less than 100 miles) above the Earth. How fast does a satellite at this distance have to move to stay in orbit?

The equation for a satellite's velocity, derived above, is

$$v^2 = \frac{GM}{r}$$

As we saw earlier, the mass of the Earth is 6×10^{24} kg, and its radius is 6.4×10^6 meters. The shuttle orbit, 150,000 (0.15×10^6) meters above the surface, must therefore correspond to a value of *r* of about 6.6×10^6 meters. If we put these values into the equation, we find that the velocity of the shuttle in its orbit is

$$v^2 = \frac{(6.67 \times 10^{-11}\,\text{N-m}^2/\text{kg}) \times (6 \times 10^{24}\,\text{kg})}{6.6 \times 10^6\,\text{m}}$$

$$= 6 \times 10^7\,\text{m}^2/\text{s}^2,$$

which means that the velocity is

$$v = 7.8 \times 10^3\,\text{m/s}$$

This velocity, equal to almost 8 km (5 miles) per second, is many times faster than commercial jet aircraft, which cruise at about 0.3 km (0.2 miles) per second. One way of getting an idea of this velocity is to ask how long it would take the shuttle to complete one orbit. The length of the orbit is just the circumference of a circle of radius *r*, or $2\pi r$. Thus the orbital period, or time for one revolution, is

$$\text{orbital period} = 2\pi \frac{r}{v}$$

$$= 2\pi \frac{6.6 \times 10^6\,\text{m}}{7.8 \times 10^3\,\text{m/s}}$$

$$= 5300\,\text{s}$$

$$= 89\,\text{min}$$

The space shuttle completes an orbit of the Earth every hour and a half. Why does the shuttle stay in orbit? The force of gravity is not much different at the shuttle orbit (r = 6400 km + 150 km) than it is where you are sitting (r = 6400 km). The shuttle stays in orbit because enormous amounts of energy have been expended to get it moving very fast, so that its tendency to fly off in a straight line is just balanced by gravity. In that sense, the only reason you aren't in orbit at this moment is that you're not moving fast enough! ●

Kepler and Newton

Newton's laws of motion, taken together with the law of universal gravitation, enable us to derive mathematically the results of Galileo's experimental investigations into mechanics. In the same way, these two great organizing principles of nature allow us to predict Kepler's laws of planetary motion, rather than just accept them.

You now know enough to understand how Newton's work can be used to derive Kepler's. Think about a satellite of mass m moving with speed v around an orbit of radius r centered on a central body of mass M.

As we saw in the Science by the Numbers exercise, the time T it takes for the satellite to make one revolution is

$$T = 2\pi \frac{r}{v}$$

so that

$$v = 2\pi \frac{r}{T}$$

If we square both sides of this equation, we find that

$$v^2 = 4\pi^2 \frac{r^2}{T^2}$$

But from the satellite equation we have

$$v^2 = \frac{GM}{r}$$

so that, equating the two expressions for v^2, we obtain

$$\frac{GM}{r} = 4\pi^2 \frac{r^2}{T^2}$$

While this expression may not seem familiar, it is actually a statement of Kepler's third law of planetary motion. To see this, multiply both sides of the equation by $r/4\pi^2$. It then reads

$$\frac{GM}{4\pi^2} = \frac{r^3}{T^2}$$

This equation says that the cube of the radius of the orbit divided by the square of the time it takes to go around the orbit (the part on the right side) equals a constant (the part on the left). This equation is an exact statement of Kepler's third law (see Chapter 3).

What Kepler didn't know, however, is that his third law is just a special case of a general set of laws that allows us to transcend the observations that first led to them. Whereas Kepler's law was a discovery about a specific set of satellites (the planets of our Sun), it also applies to planets around other suns, to the moons that circle the outer planets of our own solar system, and to the collection of artificial satellites that circle the Earth. Furthermore, where Kepler could simply observe the numerical value of the constant in his law, the above expression allows us to *predict what that constant will be.*

Science in the Making

The Recovery of Halley's Comet

Of all celestial phenomena, none seemed more portentous and magical to the ancients than comets (Figure 4–9). These glorious lights in the sky, with their luminous sweeping tails, are not like the planets. They appear sporadically and, in Newton's day, appeared unpredictably. Yet even comets are subject to Newton's laws.

Figure 4–9
Halley's comet in 1066 A.D., from a section of the Bayeux tapestry.

We associate the discovery of the orbital nature of comets with the British astronomer Edmond Halley (1656–1742). Unlike most of the scientists you'll read about in this book, Halley led an adventurous (even swashbuckling) life before he settled down as Britain's Astronomer Royal. At various points in his life, he ran a diving company, captained a Royal Navy survey ship (facing down a mutiny in the process), and, if we are to believe the legend, used his growing reputation as an astronomer to travel around European capitals as a secret agent for his country. He was the first European astronomer to produce modern maps of the skies of the Southern Hemisphere, and his navigational maps of the Earth's magnetic field were used well into the nineteenth century.

So wedded was he to the adventurous life that when he was proposed for a professorship at Oxford, one colleague wrote:

> Mr. Halley expects [the professorship], who now talks, swears, and drinks brandy like a sea captain; so much that I fear that his ill behavior will deprive him of the advantage of this vacancy.

Nevertheless, he got the post at Oxford (the house he lived in is still there).

About the time of Halley's appointment, scientists were starting to think about explaining comets. Even though they knew about universal gravitation, they didn't have the mathematical tools to solve the problem of comets' elongated orbits. The reason for this difficulty is that, except for orbits that are nearly circular like those of the planets, the distance between the central body and a satellite will vary considerably from one point to the next, and hence the gravitational force will not be the same at each point around the orbit. The problem of deducing the shape of an orbit under these circumstances is a difficult one, but it could be dealt with using the new mathematics of calculus, which was invented independently by Isaac Newton and the German mathematician Gottfried Leibniz.

In 1684, Halley visited Newton at Cambridge. Newton told him over dinner that, according to his calculations, all bodies subject to a gravitational force would move in orbits shaped like ellipses. Bolstered by this information, Halley analyzed the historical records of some 24 comets. Knowing that the orbits had to be elliptical, he was able to use the observations to determine exactly the elliptical path along which each comet moved.

He found that three recorded comets—those that had appeared in 1531, 1607, and 1682—seemed to be following the same orbit. He realized that the sightings represented not three separate comets, but one comet that was appearing over and over again.

Predicting the next appearance of the comet wasn't as simple as you might think, because the gravitational effects of Jupiter and Saturn could change the period of the comet by several years. After some work, Halley predicted that the comet would reappear in 1758. On Christmas day, 1758, an amateur astronomer in Germany sighted the comet coming back toward the Earth. This so-called "recovery" of what is now known as Halley's comet marked a great triumph for the Newtonian picture of the world.

With characteristic aplomb, Halley had the last word on his prediction of the comet's return:

> Wherefore if [the comet] should return again about the year 1758, candid posterity will not refuse to acknowledge that this was first discovered by an Englishman. ●

Comets are distinguished by their bright gaseous tails, which stream away from the Sun.

THINKING MORE ABOUT THE CLOCKWORK UNIVERSE

Predictability

Newton bequeathed a picture of the universe that is beautiful and ordered. The planets orbit the Sun in stately paths, forever trying to move off in straight lines, forever prevented from doing so by the inward tug of gravity. The same laws that operate in the cosmos operate on Earth, and these laws were discovered by applying the scientific method. To an observer with Newton's perspective, the universe was like a clock. It had been wound up by God and was ticking along according to His laws. Newton and his followers were persuaded that in carrying out their work, they were discovering what was in the mind of God when He created the universe.

The Newtonian universe seemed regular and predictable. If you knew the present state of a system and the forces acting on it, the laws of motion would allow you to predict its entire future. This notion was taken to the extreme by the French mathematician Pierre Simon Laplace (1749–1827), who proposed the notion of the *Divine Calculator*. His argument (in modern language) was this: If we knew the position and velocity of every atom in the universe and we had infinite computational power, then we could predict the position and velocity of every atom in the universe for all times. He made no distinction between an atom in a rock and an atom in your hand. According to the argument, everyone's movements are completely determined by the laws of physics to the end of time. You cannot choose your future. What is to be was determined from the very beginning.

While this argument raises many interesting questions, it has been rendered moot by two modern developments in science. One of these, the Heisenberg uncertainty principle (see Chapter 8), tells us that at the level of the atom, it is impossible to know simultaneously and exactly both the position and velocity of any particle. (Any measurement of position, Heisenberg showed, alters velocity, and vice versa.) Thus you can never get all the information the Divine Calculator needs to begin working.

More recently, scientists working with computer models have discovered that there are many systems in nature that can be described in simple Newtonian terms but whose futures are, to all intents and purposes, unpredictable. These are called "chaotic systems," and the field of study devoted to them is called *chaos*.

White water in a mountain stream provides a familiar chaotic system. Imagine putting two chips of wood in water on the upstream side of the rapids. No matter how small you make the chips, or how close together they are at the beginning, those chips (and the water on which they ride) may be widely separated by the time they

White water is an example of a chaotic system, in which a small change in the initial position can produce a large difference in the outcome.

get to the end of the rapids. If you know the exact initial position of a chip and every detail of the waterway's shape and other characteristics with complete mathematical precision, you can, in principle, predict where the chip will come out downstream. But if there is the slightest error in your initial description, no matter how small, the actual position of the chip and your prediction will differ, often wildly. Every measurement in the real world has some error associated with it, so it is never possible to determine the exact position of the chip at the start of its trip. For all practical purposes, you cannot predict where it will come out *even if you know all the forces acting on it.*

The existence of chaos tells us that the philosophical conclusions drawn from the Newtonian vision of the universe simply don't apply to some systems in nature. Of course, most situations with which we are familiar are not chaotic. The fact that we can guide a spacecraft to Jupiter or predict the orbits of comets tells us that most of our world is comfortably predictable, at least on the time scales relevant to most human activity. But systems like the turbulent flow of fluids are not.

Many scientists believe that the flow of the Earth's atmosphere and the long-term development of the weather is chaotic, and it may turn out that complex communities of living things and their surroundings, called ecosystems, behave this way as well. If this is true, governments will be faced with serious implications when they have to deal with issues such as global warming (see Chapter 19) and the preservation of endangered species, because scientists will never be able to be certain about their predictions. How confident do you have to be that something is going to happen before you start taking steps to avoid it?

▶ Summary

Isaac Newton combined the work of Kepler, Galileo, and others in his *laws of motion* and the *law of universal gravitation*. The laws of motion state that (1) nothing accelerates without a *force* acting, (2) the amount of acceleration is proportional to the force applied and inversely proportional to the *mass*, and (3) forces always act in equal and opposite pairs. *Momentum*, defined as an object's mass times its velocity, provides a measure of how difficult it is to change the motion of the object. One consequence of Newton's laws is *conservation of momentum*—that is, in any system in which no external forces are acting, momentum is conserved.

Newton's laws apply to *gravity*, the most prevalent force in our daily lives. At the Earth's surface, the gravitational force exerted on an object is called its *weight*. The same force that pulls a falling apple to Earth supplies the *centripetal force* that causes the Moon to curve around the Earth in its orbit. Indeed, the force of gravity (with the same *gravitational constant*) operates everywhere, with pairs of forces between every pair of masses in the universe. Newton's laws of motion, together with the law of universal gravitation, describe the orbits of planets, moons, comets, and satellites. They also allow us to derive Galileo's results and Kepler's laws of planetary motion, thereby unifying the sciencies of mechanics and astronomy.

▶ Key Equations

force = mass × acceleration

$$\text{force} = G \times \frac{(\text{first mass} \times \text{second mass})}{\text{distance}^2}$$

force = mass × g = weight

$$(\text{velocity of a satellite})^2 = G \times \frac{\text{mass of central body}}{\text{radius of orbit}}$$

▶ Review Questions

1. What is Isaac Newton's first law of motion? Give an everyday example of this law in action.

2. What is Isaac Newton's second law of motion? Give an everyday example of this law in action.

3. What is Isaac Newton's third law of motion? Give an everyday example of this law in action.

4. What is a force?

5. According to Newton, what are the two kinds of motion in the universe?

6. What is momentum?

7. What is the law of conservation of momentum? Give examples of this law in operation.

8. Why is gravity called a "universal" force?

9. State the law of universal gravitation.

10. What similarity did Newton see between the Moon and an apple?

11. What is the difference between weight and mass?

12. What is centripetal force? Centrifugal force?

13. What supplies the centripetal force that keeps the planets in their orbits?

14. What is the relation between the velocity of a satellite, the radius of its orbit, and the mass of the central body?

15. How did the work of Edmond Halley support Newton's theories?

16. What is a chaotic system?

▶ Fill in the Blanks

Complete the following paragraph using words or phrases from the list of key terms.

centripetal force	momentum
conservation of mo-	Newton's law of univer-
mentum	sal gravitation
force	Newton's laws of motion
gravitational constant	weight
mass	

Isaac Newton summarized the rules that govern the motion of all objects in _____. The first law defines _____, the quantity that causes acceleration. The second law relates the values of acceleration, _____, and _____. The third law says that for every action, there is an equal and opposite reaction. The quantity of mass times velocity is given a special name, _____. In the absence of an external force, the law of _____ states that this quantity does not change.

_____ tells us that the force of gravity acts everywhere in the universe. The quantity G that appears in this law is called the _____. The force of gravity that acts on an object is called its _____, while the quantity that measures the amount of material in the object is called its _____. An object moving in a circle has to be accelerated, and the force that produces this acceleration is called the _____.

▶ Discussion Questions

1. Pick an everyday activity and discuss how Newton's three laws of motion come into play.

2. Think about your favorite sport. In what circumstances is being small an advantage? In what circumstances is being large an advantage? Why?

3. Which of the following does not exert a gravitational force on you?

a. this book

b. the Sun

c. the nearest star

d. a distant galaxy

4. What pairs of forces act in the following situations?

a. A pitcher throws a fast ball.

b. An apple falls to the ground.

c. The Moon orbits the Earth.

d. A pencil rests on your desk.

5. When Galileo first observed the five largest moons orbiting the planet Jupiter, he quickly determined the time it took each to complete one orbit. Why won't this measurement allow us to determine the masses of the moons? Could such a measurement allow us to determine the mass of Jupiter?

6. Retell Newton's second law of motion in your own words.

7. Why is mass significant in Newton's second law?

8. An example of Newton's third law is a rocket. Outline the forces acting on the rocket and the propellants.

9. How is weight related to mass?

10. How is inertia related to mass?

11. What is the main idea of Newton's universal law of gravitation?

12. List some of the accomplishments of Sir Isaac Newton during the plague summer of 1665.

13. How is Newton's law of gravitation related to Kepler's third law of planetary motion?

14. What evidence can you give that relates the conservation of momentum to Newton's third law of motion?

15. Does the Moon fall toward the Earth? Explain.

▶ Problems

1. Calculate the momenta of the following:

a. a 200-gram rifle bullet traveling 300 m/s

b. a 1000-kg automobile traveling 0.1 m/s (a few miles per hour)

c. a 70-kg person running 10 m/s (a fast sprint)

d. a 10,000-kg truck traveling 0.01 m/s (a slow roll)

2. A 5-kg bowling ball is moving at 1 m/s. It hits a 1-kg pin that moves off at a speed of 2 m/s after the collision. What is the speed of the bowling ball after the collision?

3. What do you weigh in pounds? In newtons?

4. What is your mass in kilograms?

5. What would you weigh if the Earth were four times as heavy as it is and its radius were twice its present value?

6. How much force must be applied to accelerate a 20-kg satellite to counter the Earth's gravitational acceleration of 9.8 m/s²?

7. How long would the year be (in terms of the present year) if the Sun were half its present mass and the Earth's orbit were in the same place as it is now?

8. How much would you weigh if you were standing on a tower 150 km tall (i.e., if you were standing still at about the altitude of a space shuttle orbit)? How much does this differ from your weight on the surface of the Earth? Would you be able to detect this weight difference on a bathroom scale?

9. Which of these objects has the greatest inertia: a mosquito, a Volkswagen "Beetle," or an ocean liner? Justify your response by creating a mental experiment that uses Newton's second law, which relates mass to inertia.

10. You are driving a car down a straight road at 55 mph.

　a. Are there any forces acting on the car? If so, list them.

　b. Is there a net, unbalanced force acting on the car? Explain.

　c. If the car started to go around a long bend, still maintaining its constant speed of 55 mph, is there a net force acting on the car? Explain.

11. Johnnie pushes a wagon, which is initially at rest, with a constant force of 10 N. The mass of the wagon is 15 kg.

　a. What is the constant acceleration of the wagon?

　b. If Johnnie pushes the wagon for 3 s, what distance is covered by the wagon during this time?

　c. What is the speed of the wagon after 3 seconds?

12. Suzie (50 kg) is roller blading down the sidewalk going 20 miles per hour. She notices a group of workers who have unexpectedly blocked her path, and she makes a quick stop in 0.5 s.

　a. What is Suzie's average acceleration in m/s²?

　b. What force in newtons was exerted to stop Suzie?

　c. Where did this force come from?

13. Billie (35 kg) and Joannie (50 kg), both with brand new roller blades, are at rest facing each other in the parking lot. They push off each other and move in opposite directions, Joannie moving at a constant speed of 14 ft/s. What speed is Billie moving?

14. Tracy (50 kg) and Tommie (75 kg) are standing at rest in the center of the roller rink, facing each other, free to move. Tracy pushes off Tommie with her hands and remains in contact with Tommie's hands, applying

a constant force for 0.75 s. Tracy moves 0.5 m during this time. When she stops pushing off Tommie, she moves at a constant speed.

　a. What is Tracy's constant acceleration during her time of contact with Tommie?

　b. What is Tracy's final speed after this contact?

　c. What force is applied to Tracy during this time? What is its origin?

　d. What happens to Tommie? If Tommie moves, describe his motion, its force, acceleration, or Tommie's final velocity.

15. A mosquito hovering at rest above a riverside road is instantly hit by a fast-moving Volkswagen "Beetle" moving at 60 mph.

　a. Which bug (insect or VW) experiences the largest force?

　b. Which bug encountered the greatest acceleration?

　c. Which bug experiences the greatest change in momentum?

16. Allison (30 kg) is coasting in her wagon (10 kg) at a constant speed of 5 m/s. She passes her mother, who drops a bag of toys (5 kg) in the wagon.

　a. What is the initial momentum of Allison and the wagon before her mother drops the toys?

　b. What is the final momentum of Allison, the wagon, and the toys?

　c. What is the final speed of the wagon after Allison's mother drops the toys?

17. Tony (60 kg) coasts on his bicycle (10 kg) at a constant speed of 30 m/s, carrying a 5-kg pack. Tony throws his pack forward, in the direction of his motion, at 5 m/s relative to his bicycle.

　a. What is the initial momentum of the system (Tony, the bicycle, and the pack)?

　b. What is the final momentum of the system immediately after the pack leaves Tony's hand?

　c. Is there a change in the speed of Tony's bicycle? If so, what is that speed?

18. Calculate the force of gravity on a 65-kg person under the following conditions.

　a. at the surface of the Earth (R = 6400 km)

　b. at twice the Earth's radius

　c. at four times the Earth's radius

　d. Plot this gravitational force with distance. What pattern or relationship do you expect to obtain? Confirm your answer.

19. Compare the gravitational force on a 1-kg mass at the surface of the Earth with that on the surface of the Moon ($M_{Moon} = \frac{1}{81.3}$ mass of the Earth; R = 3475 km).

20. Calculate the weight in pounds and newtons of

the following objects.

    a. the Statue of Liberty (205 metric tons; 1 metric ton is equal to 1000 kg)

    b. A 40-oz softball bat

    c. a solid rocket booster of the Space Shuttle (5.9×10^5 kg)

21. Calculate the weight in pounds and newtons of the three objects in problem 13 if they were on the surface of the Moon (see problem 12), and the surface of Mars (M = 0.11 mass of the Earth; R = 0.53 Earth radius).

22. Calculate the speed and period of a ball tied to a string of length of 0.3 meters, making 2.5 revolutions per second.

23. Calculate the average speed of the Moon in km/s around the Earth. The period of revolution is 27.3 days, and the average distance is 3.84×10^8 m.

24. Calculate the speed at the edge of a compact disc (radius = 6 cm) that rotates 3.5 revolutions per second.

25. Calculate the centripetal force on the Earth by the Sun. Assume that the period of revolution for the Earth is 364.25 days, and the average distance is 1.5×10^8 km.

▶ Investigations

1. The concept of predestination plays an important role in some kinds of theology. What is it? How does it relate to Laplace? To chaos?

2. In Chapter 1, we talked about astrology and whether a planet or star can influence our lives. Calculate the gravitational force on a newborn infant exerted by a star the size of the Sun 1 light year (9.5×10^{15} m) away. Compare it to the gravitational force exerted by a 100-kg physician 0.1 m away.

3. One objection that Copernicus's contemporaries raised to his theory was that if the Earth were really turning, we would all be thrown off, the way that clay is thrown off a spinning potter's wheel. Use Newton's laws of motion and the law of universal gravitation to counter this argument.

4. In what sense is the Newtonian universe simpler than Ptolemy's? Suppose observations had shown that the two did equally well at explaining the data. Construct an argument you would make to say that Newton's universe should still be preferred.

5. If Kepler had been transported to another solar system, what would he have had to do in order to show that his laws applied there? What would Newton have had to do?

6. Read a biography of Isaac Newton. Were all of his ideas about the physical world correct? What were his views on religion and mysticism?

▶ Additional Reading

Christiansen, Gale E. *In the Presence of the Creator: Isaac Newton and His Times.* New York: Macmillan/Free Press, 1984.

Cohen, I. Bernard. *The Birth of the New Physics.* New York: W.W. Norton, 1985.

Gleick, James. *Chaos.* New York: Viking, 1987.

5 ENERGY

THE MANY DIFFERENT KINDS OF ENERGY ARE
INTERCHANGEABLE, AND THE TOTAL AMOUNT OF
ENERGY IN AN ISOLATED SYSTEM IS CONSERVED.

The Great Chain of Energy

A few days ago, you may have gone to your local gas station and filled up the tank of your car with gasoline. Have you ever wondered where that gasoline came from?

Hundreds of millions of years ago, the energy in that gasoline was generated in the flaming core of the Sun. In between the Sun and the gasoline, that energy has flowed through space and plants, eventually being buried and transformed deep within the Earth until just a few months ago, when the stored energy was brought up out of the ground in the form of petroleum.

The last time you drove, you burned that gasoline, freeing the stored energy and using it to move your car. When you parked the car, the hot engine slowly cooled, and that bit of heat energy, after having been delayed for a few hundred million years, was radiated into space to continue its voyage away from the solar system. As you read these words, the energy you freed yesterday has long since left the solar system, and is out in the depths of interstellar space.

All the energy you use every day, like the energy in our story, is in transit. But no matter how complicated the story, no matter how many different forms that energy takes, there remains one central fact: the total amount of energy doesn't change. Energy is neither created nor destroyed, but just changes form. This fact, as we shall see, is one of the great laws of nature that governs our universe.

Figure 5–1
Work is done when a force is exerted over a distance.

SCIENTIFICALLY SPEAKING

Whenever you ride in a car, climb a flight of stairs, or just take a breath, you use energy. At this moment, trillions of cells in your body are hard at work turning yesterday's food into the chemical energy that is keeping you alive today. Energy in the atmosphere generates sweeping winds and powerful storms, while the ocean's energy drives mighty currents and incessant tides. Meanwhile, deep within the Earth, energy in the form of heat is moving the continent on which you are standing.

Energy is all around you—in the ever-shifting atmosphere and restless seas, in simple bacteria and mighty redwood trees, in brilliant sunlight and shimmering moonlight. Energy affects everything in the physical world, and the laws that govern its behavior are among the most important concepts in science.

In any situation where energy is expended, you will find one thing in common. If you look at the event closely enough you will find that, in accord with Newton's laws of motion (Chapter 4), a force is being exerted on an object to make it move. When your car burns gasoline, some of the fuel's energy ultimately turns the wheels of your car, which then exert a force on the road; the road exerts an equal and opposite force on the car, pushing it forward. When you climb stairs, your muscles exert a force that lifts you upward against gravity. When you breathe, your lungs exert a force as they pull air in and push air out. Even in your body's cells, a force is exerted on molecules in chemical reactions. Energy, then, is intimately connected with the application of a force.

In everyday conversation, we may speak of the seemingly inexhaustible energy of a small child, of a song that sounds energetic, or of an athlete being energized. In science, the term energy has a precise definition that is somewhat different from the ordinary meaning. To see what scientists mean when they talk about energy, we must first introduce the concept of work.

Work

work   A force that is exerted times the distance over which it is exerted; measured in joules in the metric system, in foot-pounds in the English.

In the lexicon of physics, we say that **work** is done whenever a force is exerted over a distance. When you picked up this book, for example, your muscles applied a force equal to the weight of the book over a distance of a foot or so. You did work (Figure 5–1).

This definition of work differs considerably from everyday usage. From a physicist's point of view, if you accidently drive into a stone wall and smash your fender, work has been done by the wall, because a force deformed the car's metal a measurable distance. On the other hand, a physicist would say that you haven't done any work if you spent an hour in a futile effort to move a large boulder, no matter how tired you get. Even though you exerted a considerable force, the distance over which you exerted it is zero.

Physicists provide an exact mathematical definition to their notion of work.

▶   **In words:**

Work is equal to the force that is exerted times the distance over which it is exerted.

► **In equation form:**

work (in joules) = force (in newtons) × distance (in meters) where a joule is the unit of work, as defined below.

► **In symbols:**

$$W = F \times d$$

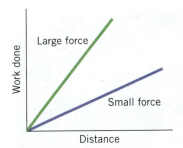

Figure 5–2
The work done increases proportionally to the distance over which the force is applied.

The *direct relationship* between work and distance is illustrated in Figure 5–2. In practical terms, even a small force can do a lot of work if it is exerted over a long distance.

Using the equation above, we can see that the units of work are equal to a force times a distance. In the metric system of units, where force is measured in newtons and distance in meters, work is measured in newton-meters. This unit is given the special name joule, after the English scientist James Prescott Joule (1818–1889), one of the first people to understand the properties of energy. One **joule** (usually abbreviated J) is defined to be the amount of work done when you exert a force of 1 newton through a distance of 1 meter.

joule The amount of work done when you exert a force of one newton through a distance of one meter.

1 joule of work = 1 newton of force × 1 meter of distance.

James Prescott Joule (1818–1889)

Example 5–1: Working Against Gravity

How much work do you do when you lift a 10-kg suitcase 1 m off the ground?

The work done by a weight lifter is equal to the magnitude of the weight multiplied by the distance it is lifted.

▶ **Reasoning and Solution:** We must first calculate the force needed to lift a 10-kg mass before we can determine work. From the previous chapter, we know that to lift a 10-kg mass against the force of gravity (whose acceleration is 9.8 m/s²) requires a force given by

$$\text{force} = \text{mass} \times g$$
$$= 10 \text{ kg} \times 9.8 \text{ m/s}^2$$
$$= 98 \text{ N}$$

Then, from the equation for work,
$$\text{work} = \text{force} \times \text{distance}$$
$$= 98 \text{ N} \times 1 \text{ m}$$
$$= 98 \text{ J} \blacktriangle$$

In North America, you may notice that work is often measured in the English system of units (see Appendix A), where force is recorded in pounds and distance in feet. Work is thus measured in a unit called the *foot-pound* (usually abbreviated ft-lb), which corresponds to the work done in lifting a weight of 1 pound 1 foot upward against the force of gravity.

Example 5–2: Lifting Weights

It's not unusual for a professional athlete to be able to lift as much as 400 lb. Suppose that such an athlete, lying on his back, pushes a 400-lb weight from his chest to a position in which his arms are fully extended upward (this is called a "bench press," and gives rise to the sporting slang in which an athlete is said to be able to "bench" a certain weight). Estimate the amount of work he does.

▶ **Reasoning:** To calculate the work done, we need to multiply force times distance. The force in this problem is clearly 400 lb, but the distance isn't given. That means we have to estimate it. The athlete starts with the weight on his chest and keeps pushing until his arms are extended. This means that the distance over which the force is applied must be about the length of his arms. The length of a large man's arms is about 3 ft (in fact, the "yard" in the English system of units was originally defined as the distance from a man's nose to the end of his hand). Let's take the distance in the problem as 3 ft.

▶ **Solution:** Work is force times distance, so

$$W = F \text{ (in pounds)} \times d \text{ (in feet)}$$
$$= 400 \text{ lb} \times 3 \text{ ft}$$
$$= 1200 \text{ ft-lb} \blacktriangle$$

Energy

energy The ability to do work; the capacity to exert a force over a distance. A system's energy can be measured in joules or foot-pounds.

Energy is defined as the ability to do work. If a system is capable of exerting a force over distance, then it possesses energy. The amount of a system's energy, which can be recorded in joules or foot-pounds (the same units used for work), is a measure of how much work the system might do. When you run out of energy, you simply can't do any work.

Power

Scientists define **power** as the rate at which work is done, or, equivalently, the rate at which energy is expended. In order to complete a physical task quickly, you must generate more power than if the same amount of work is done more slowly (Figure 5–3). If you run up a flight of stairs your muscles need to generate more power than they would if you were to walk up that same flight. A power hitter in baseball swings the bat faster, converting the energy in his muscles more quickly than most other players.

Power can be expressed in an exact form.

▶ **In words:**

> Power is the amount of work done, divided by the time it takes to do it; or, power is the energy expended, divided by the time it takes to expend it.

▶ **In equation form:**

$$\text{power (in watts)} = \frac{\text{work (in joules)}}{\text{time (in seconds)}}$$

or

$$\text{power (in watts)} = \frac{\text{energy (in joules)}}{\text{time (in seconds)}}$$

where the watt is the unit of power, as defined below.

▶ **In symbols:**

$$P = \frac{W}{t}$$

or

$$P = \frac{E}{t}$$

In the metric system, power is measured in watts, after James Watt (1736–1819), the Scottish inventor who developed the modern steam

power The rate at which work is done, or the rate at which energy is expended. The amount of work done, divided by the time it takes to do it. Power is measured in watts in the metric system, horsepower in the English.

Figure 5–3
Athletes must release their energy quickly and with great power to succeed in events such as the javelin.

James Watt (1736–1819) performing a steam experiment at home.

An old steam engine at Golden Spike National Historic Site, Promontory, Utah.

watt A unit of power that is the expenditure of 1 joule of energy in 1 second.

engine that powered the Industrial Revolution. The **watt** (abbreviated W), a unit of measurement that you probably encounter every day, is defined as the expenditure of 1 joule of energy in 1 second:

$$1 \text{ watt of power} = \frac{(1 \text{ joule of energy})}{(1 \text{ second of time})}$$

When you change a light bulb, for example, you look at the rating of the new bulb to see whether it's 60, 75, or 100 watts. This number provides a measure of the rate of energy that the light bulb consumes when it is operating. Almost any hand tool or appliance in your home will also be labeled with a power rating in watts. The unit of 1000 watts (corresponding to an expenditure of 1000 joules per second) is called a **kilowatt** (abbreviated kW), a commonly used measurement of electrical power. The English system, on the other hand, uses *horsepower* (usually abbreviated hp), which is defined as 550 foot-pounds per second (see Science in the Making, which follows this section).

kilowatt A commonly used measurement of electrical power equal to 1000 watts and corresponding to the expenditure of 1000 joules per second.

The equation defining power as energy divided by time, introduced above, may be rewritten as follows:

$$\text{energy (in joules)} = \text{power (in watts)} \times \text{time (in seconds)}$$

This means that if you know how much power you are using and how long you use it, you can calculate the total amount of energy expended. The electric company calculates your electric bill in this way. The equation tells us that if you use 100 watts of power for 1 hour (by burning a light bulb, for example), you will have expended 100 watt-hours (Wh) of energy, or one tenth of a kilowatt-hour (kWh). This is the number that appears on your electricity bill.

Terms and units related to work, energy, and power are summarized in Table 5–1.

Table 5–1 • **Important Quantities**

Quantity	Definition	Units
Force	mass × acceleration	newtons
Work	force × distance	joules
Energy	ability to do work	joules
Energy	power × time	joules
Power	work/time = energy/time	watts

Science in the Making

James Watt and the Horsepower

The *horsepower*, a unit of power with a colorful history, was devised by James Watt so that he could sell his steam engines. Watt knew that the main use of his engines was in mines, where owners traditionally used horses to drive pumps that removed water. The easiest way to promote his new engines was to tell the mining engineers how many horses each engine would replace. Consequently, he did a series of experiments to determine how much energy a horse could generate over a given amount of time. He found that, on average, a healthy horse can do 550 ft-lb of work every second over an average working day. Watt defined this unit to be the horsepower, and rated his engines accordingly. We still use the unit (the engines of many cars and trucks are rated in horsepower), although these days we seldom build engines to replace horses. ●

Example 5–3: Paying the Piper

A typical CD system uses 250 W of electric power per hour. If you play your system for 3 h in an evening, how much energy do you use? If energy costs $0.08/kWh, how much do you owe the electric company?

▶ **Reasoning and Solution:** You can solve this problem by remembering the equation that relates energy used to power:

$$\text{energy} = \text{power} \times \text{time}$$
$$= 250\ \text{W} \times 3\ \text{h}$$
$$= 750\ \text{Wh}$$

and, since 750 watts equals 0.75 kilowatt, the cost will be:

$$\text{energy} = 0.75\ \text{kWh}$$

$$0.08/\text{kWh} \times 0.75\ \text{kWh} = \$0.06\ \blacktriangle$$

Example 5–4: Power Lifting

In Example 5–2, we talked about an athlete lifting a 400-lb weight. Suppose that, grunting and groaning, he carries out the job in 3 seconds. We know that he did 1200 ft-lb of work. How much power is he expending?

▶ **Reasoning and Solution:**   Power is the work done divided by the time it takes to do it. Remember that in this problem we are using the English system of units. Thus

$$P \text{ (in ft-lb/s)} = \frac{W \text{ (in foot-pounds)}}{t \text{ (in seconds)}}$$

$$= \frac{1200 \text{ ft-lb}}{3\text{s}}$$

$$= 400 \text{ ft-lb/s}$$

But 1 horsepower is 550 ft-lb/s, so

$$P \text{ (in hp)} = \frac{400 \text{ ft-lb/s}}{550 \text{ ft-lb/s/hp}}$$

$$= \frac{400}{550} \text{ hp}$$

$$= 0.72 \text{ hp}$$

In other words, for this very short period, the trained athlete is developing as much power as a small hand drill. In general, human beings can produce about $\frac{1}{4}$ hp over extended periods of time—a good deal less than a horse. ▲

TYPES OF ENERGY

kinetic energy   The type of energy associated with moving objects: the energy of motion. Kinetic energy is equal to the mass of a moving object times the square of that object's velocity, multiplied by $\frac{1}{2}$.

potential energy   The energy a system possesses if it is capable of doing work, but is not doing work now. Types of potential energy include magnetic, elastic, electrical, and chemical.

Energy—the ability to do work—appears in many different kinds of physical systems, which give rise to many different kinds of energy. Identifying these varieties posed a great challenge to science in the nineteenth century. While the division of energy into different types is somewhat arbitrary, it will prove convenient to recognize two very broad categories: **Kinetic energy** is energy associated with moving objects, while stored or **potential energy** is energy waiting to be released (see Table 5–2).

Table 5–2   •   Some Kinds of Energy

Potential Energy	Kinetic Energy
Gravitational	Moving objects
Chemical	Heat
Elastic	Sound and other waves
Electromagnetic	
Mass	

Kinetic Energy

Think about a cannonball flying through the air. When the iron ball hits a wooden target, it exerts a force on the fibers in the wood, splintering and pushing them apart, thereby creating a hole. The cannonball in flight clearly has the *ability* to do work because it is in motion. This energy of motion is kinetic energy.

You can find countless examples of kinetic energy in nature. A fish moving through water, a bird flying, and a predator catching its prey all have kinetic energy. A speeding car, a flying frisbee, a falling leaf, and a running child all have kinetic energy also.

Our intuition tells us that two factors govern the amount of an object's kinetic energy. First, heavier objects have more energy than light ones: a bowling ball traveling 10 m/s (a very fast sprint) carries a lot more kinetic energy than a golf ball traveling at the same velocity. In fact, kinetic energy is proportional to mass—double the mass and the kinetic energy is doubled, too.

Second, the faster something is moving, the greater the force it is capable of exerting. A high-speed collision causes much more damage than a fender bender in a parking lot. It turns out that an object's kinetic energy increases as the square of its velocity. A car moving 40 mph has four times as much kinetic energy as one moving 20 mph, while at 60 mph a car carries nine times as much kinetic energy as at 20 mph. Thus a modest increase in speed can cause a large increase in kinetic energy.

These ideas, presented as equations, are as follows:

▶ **In words:**

Kinetic energy equals the mass of a moving object times the square of that object's velocity, multiplied by the constant $\frac{1}{2}$.

▶ **In equation form:**

kinetic energy (in joules) $= \frac{1}{2} \times$ [mass (in kg)] $\times$ [velocity (in m/s)]^2

▶ **In symbols:**

$$KE = \tfrac{1}{2} mv^2$$

When a baseball breaks a window, it demonstrates the existence of kinetic energy.

Example 5–5: Bowled Over?

What is the kinetic energy of a 4-kg (about 9-lb) bowling ball traveling down a bowling lane at 10 m/s (22 mph)? Compare this energy to that of a 250-g (half-pound) baseball traveling 50 m/s (110 mph). Which object would hurt more if it hit you (i.e., which object has the greater kinetic energy)?

▶ **Reasoning and Solution:** We have to substitute numbers into the equation for kinetic energy. For the 4-kg bowling ball traveling at 10 m/s,

$$
\begin{aligned}
\text{Kinetic energy (in joules)} &= \tfrac{1}{2} \times \text{[mass (in kg)]} \\
&\quad \times \text{[velocity (in m/s)]}^2 \\
&= \tfrac{1}{2} \times 4 \text{ kg} \times (10 \text{ m/s})^2 \\
&= \tfrac{1}{2} \times 4 \text{ kg} \times 100 \text{ m}^2/\text{s}^2 \\
&= 200 \text{ kg-m}^2/\text{s}^2 \\
&= 200 \text{ J}
\end{aligned}
$$

For the 250-g baseball traveling at 50 m/s:

$$
\begin{aligned}
\text{Kinetic energy (in joules)} &= \tfrac{1}{2} \times \text{mass (in kg)} \\
&\quad \times \text{[velocity (in m/s)]}^2 \\
&= \tfrac{1}{2} \times 250 \text{ g} \times (50 \text{ m/s})^2
\end{aligned}
$$

A gram is a thousandth
of a kilogram, so
250 g = 0.25 kg:

$$= \tfrac{1}{2} \times 0.25 \text{ kg} \times 2500 \text{ m}^2/\text{s}^2$$
$$= 312.5 \text{ kg-m}^2/\text{s}^2$$
$$= 312.5 \text{ J}$$

These results are summarized in Table 5–3. Even though a bowling ball is much more massive than a baseball, a hard-hit baseball carries more kinetic energy than a typical bowling ball because of its high velocity. ▲

Table 5–3 • Kinetic Energy

	Mass	Velocity	Energy
Bowling ball	4 kg	10 m/s	200 J
Baseball	0.25 kg	50 m/s	312.5 J

Potential Energy

Almost every mountain range in the country has a "balancing rock," a boulder precariously perched on top of a hill so that it looks as if a little push would send it tumbling down the slope (Figure 5–4). If the balancing rock were to fall, it would acquire kinetic energy, and it would do "work" on anything it smashed. This means that even though the balancing rock does not work while it is motionless, it still has the potential to do work. The boulder possesses energy just by virtue of having the potential of falling.

Energy that could result in the exertion of a force over distance, but is not doing so now, is called potential energy. In the case of the balancing rock, it is called *gravitational potential energy*, because it is the force of gravity that would cause the rock to move and exert its own force.

An object that has been lifted above the surface of the Earth possesses an amount of gravitational potential energy exactly equal to the total amount of work required to lift it from the ground to its present position.

▶ **In words:**

The gravitational potential energy of any object equals its weight (the force of gravity exerted by the object) times its height above the ground.

▶ **In equation form:**

gravitational potential energy (in joules)
= mass (in kg) × g (in m/s²) × height (in m)

where g is the acceleration due to gravity at the Earth's surface (see Chapter 3).

▶ **In symbols:**

$PE = mgh$

In Example 5–1, we saw that if you lift a 10-kilogram suitcase 1 meter into the air, it holds 98 joules of potential energy. This is the amount of

Figure 5–4
The influential nineteenth-century American painter Thomas Cole produced a series of five large works called "The Course of Empire." A human civilization in the foreground is seen to pass from infancy to opulence to destruction, while in the background of each canvas a distinctive large boulder sits precariously perched on a cliff edge. The great potential energy of the boulder is obvious to the viewer; Cole's point was that the most transient of natural phenomena outlasts the most permanent works of humans. (By permission of the New York Historical Society.)

work that would be done if the suitcase were allowed to fall, and it is the amount of gravitational potential energy stored in the elevated suitcase.

We encounter many kinds of potential energy in our daily lives in addition to gravitational. *Chemical potential energy,* which is associated with the energy stored in *chemical bonds* between atoms (see Chapter 11), is concentrated in the gasoline that moves your car and the flashlight batteries that power your radio. All animals depend on the chemical potential energy of food, and all living things rely on molecules that store chemical energy for future use.

Wall outlets in your home and at work provide a means to tap into *electrical potential energy,* waiting to turn a fan or drive a vacuum cleaner. A tightly coiled spring, a flexed muscle, and a stretched rubber band contain *elastic potential energy,* while a refrigerator magnet carries *magnetic potential energy.* In every case, energy is stored, ready to do work (Figure 5–5).

Heat, or Thermal Energy

Two centuries ago, scientists understood the behavior of kinetic and potential energy, but the nature of heat was far more elusive. We now know that atoms and molecules—the minute particles that make up all matter—move around and vibrate; therefore, particles possess kinetic energy. The tiny forces that they exert will be experienced only by other atoms and molecules, but that small scale doesn't make the force any less real. If molecules in a material begin to move more rapidly, they have more kinetic energy and are capable of exerting greater forces on each other in collisions. If you touch an object whose molecules are

Figure 5–5
Potential energy comes in many forms. (*a*) An explosive contains chemical potential energy. (*b*) A rock ready to fall possesses gravitational potential energy. (*c*) A drawn bow has elastic potential energy. All are examples of potential energy about to be converted into other forms.

(*a*)

(*b*)

(*c*)

moving fast, the collisions of those molecules with molecules in your hand will exert greater force, and you will perceive the object to be hot. What we normally call **heat,** therefore, is simply **thermal energy,** which is the kinetic energy of atoms and molecules.

Heat energy is measured in a unit called the **calorie,** which is defined to be the amount of heat required to raise the temperature of 1 gram of water 1 degree Celsius. The large Calorie (with a capital C) is the unit used to measure stored energy in food, and is equal to 1000 calories (with a lower-case c), or 1 kilocalorie. In the English system, heat energy is measured in the British thermal unit, or *BTU*, which is defined to be the amount of energy needed to raise the temperature of 1 pound of water 1 degree Fahrenheit.

heat (thermal energy) A measure of the quantity of atomic kinetic energy contained in every object.

calorie A common unit of energy defined as the amount of heat required to raise 1 gram of room-temperature water 1 degree Celsius in temperature.

Science in the Making

The Nature of Heat

What is heat? How would you apply the scientific method to determine its origins? That was the problem facing scientists 200 years ago.

They found that, in many respects, heat behaves like a fluid. It flows from place to place, and seems to spread out evenly like water that has been spilled on the floor. Some objects soak up heat faster than others, and many materials seem to swell up when heated, just like waterlogged wood. Thus, in the eighteenth century, after years of observations and experiments, many physicists mistakenly accepted the theory that heat is an invisible fluid, which they called "caloric." Supporters of the caloric theory of heat claimed that the best fuels, such as coal, were saturated with caloric, and they thought ice was virtually devoid of the substance.

The caloric theory of heat eventually failed as new observations failed to bear out the theory's predictions. In particular, the practical experience of machinists just did not bear out the idea of heat as a fluid. If heat is a fluid, then each object must contain a fixed quantity of that substance. But Benjamin Thompson (later Count Rumford, 1753–1814), an eighteenth-century American who spent some time as a cannon maker, discovered that the amount of heat generated in cannon boring had nothing at all to do with the quantity of brass to be drilled. Sharp tools, he found, cut brass quickly with minimum heat generation, while dull tools made slow progress and produced prodigious amounts of heat. The amount of heat that could be generated seemed to be boundless. As long as the cannon was turned, it produced heat.

Thompson proposed an alternative hypothesis. He suggested that heat is nothing more than a consequence of the mechanical energy of friction, instead of an invisible fluid. He proved his point by immersing an entire cannon-boring machine in water, turning it on, and watching the heat generated turn the water to steam. British chemist and popular science lecturer, Sir Humphry Davy (1778–1829), further dramatized Thompson's point when he generated heat by rubbing two pieces of ice together on a cold London day.

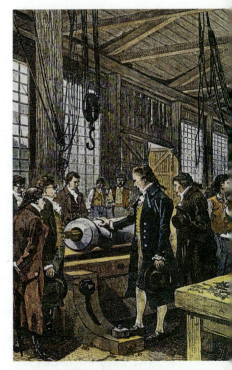

Benjamin Thompson, Count Rumford, in a Bavarian cannon foundry in 1798, calling attention to the transformation of mechanical energy into heat.

Figure 5–6
Joule's experiment demonstrated that heat is another form of energy by showing that the kinetic energy of a paddle wheel is transferred to heat energy of the agitated water.

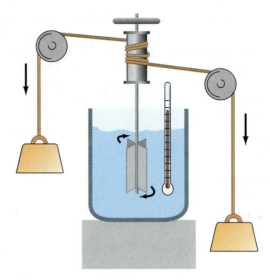

The work of Thompson, Davy, and others inspired English researcher James Prescott Joule to devise a special experiment to test the predictions of the rival theories. As shown in Figure 5–6, Joule's apparatus employed a weight that was attached to a rope and lifted up. The rope turned a paddle wheel immersed in a tub of water. The weight had gravitational potential energy and, as it fell, that energy was converted into kinetic energy of the rotating paddle. The paddle wheel's kinetic energy, in turn, was transferred to kinetic energy of water molecules. As Joule suspected, the activity caused the water to heat up. Heat, he declared, is just another form of energy; he even worked out how much heat was the equivalent of other kinds of energy. In modern units, Joule found that

$$1 \text{ calorie} = 4.186 \text{ J} \quad \bullet$$

Wave Energy

Anyone who has watched surf battering a seashore knows that waves carry energy. In the case of water waves, large amounts of water are in rapid motion, and therefore possess kinetic energy. We see this energy being released when waves hit the shore.

Waves possess a form of kinetic energy.

Other kinds of waves possess energy as well. For example, when a **sound wave** is generated, molecules in the air are set in motion and the energy of the sound wave is associated with the kinetic energy of those molecules. In Chapter 8, we will discuss waves associated with electromagnetic radiation such as the radiant energy (light) that streams from the Sun. This sort of wave stores its energy in changing electrical and magnetic fields.

sound wave Sound is a longitudinal wave that is created by a vibrating object and transmitted only through the motion of molecules in a solid, gas, or liquid. The energy of the sound wave is associated with the kinetic energy of those molecules.

Mass as Energy

Certain atoms, such as uranium, spontaneously release energy as they disintegrate, the phenomenon known as *radioactivity*. The discovery of radioactivity led to the realization in the early twentieth century that mass is also a form of energy. This principle is the focus of Chapter 16, but the main idea is summarized in Albert Einstein's most famous equation:

▶ **In words:**

Every object contains potential energy equivalent to the product of its mass times a constant, which is the speed of light squared.

▶ **In equation form:**

energy (in joules) = mass (in kg) × [speed of light (in m/s)]^2

▶ **In symbols:**

$E = mc^2$

where c is the symbol for the speed of light, a constant equal to 300,000,000 meters per second (3×10^8 m/s).

This famous equation, which has achieved the rank of a cultural icon, tells us that it is possible to convert mass into energy and to convert energy into mass. (Note: This equation does *not* mean the mass has to be traveling at the speed of light; the mass can be at rest or moving at any velocity.) Furthermore, because the speed of light is so great, the amount of energy stored in even a tiny amount of mass is enormous.

Example 5–6: The Energy of a Grape

According to Einstein's equation, how much potential energy is contained in the mass of a 1–g grape sitting on your desk?

▶ **Reasoning and Solution:** This problem can be solved by direct substitution of the mass, 1 g, into Einstein's equation. Remember that 1 g is a thousandth of a kilogram, and the speed of light is a constant, 3×10^8 m/s.

energy (in joules) = mass (in kg) × [speed of light (in m/s)]^2
$$= 0.001 \text{ kg} \times (3 \times 10^8 \text{ m/s})^2$$
$$= 0.001 \times 9 \times 10^{16} \text{ kg-m}^2/\text{s}^2$$
$$= 9 \times 10^{13} \text{ J}$$

Figure 5–7
A pole vaulter uses the energy of food to power his muscles. Kinetic energy of running is converted into elastic potential energy of the bent pole. That elastic potential energy is converted to kinetic energy, giving the skilled vaulter sufficient gravitational potential energy to clear the bar.

The energy contained in one gram of mass is prodigious—almost 10 trillion joules or 25 million kwh. The average American family uses about 1000 kwh of electricity per month, so a single grape (if we had the means to convert its mass entirely to electrical energy) could satisfy your home's energy needs for the next 10,000 years! ▲

In practical terms, Einstein's equation pointed the way for using mass to generate electricity in nuclear power plants. This technology means a few pounds of nuclear fuel is enough to power an entire city for a year.

THE INTERCHANGEABILITY OF ENERGY

Throughout this chapter, you continue to see how energy is changed from one form to another in your everyday life. Plants absorb light streaming from the Sun and convert that radiant energy into the stored chemical energy of cells and plant tissues. You eat those plants and convert that chemical energy into the energy of your muscles. The energy of motion can in turn be converted into gravitational potential energy when you climb a flight of stairs, into elastic potential energy when you stretch a rubber band, or into heat when you rub your hands together (see, for example, Figure 5–7).

Energy changes form thousands of times every day. When you strike a match, the kinetic energy of the match head is transformed into heat energy as it rubs against the striking surface, and this heat triggers a chemical reaction that produces a flame. The energy in the flame comes from stored chemical energy in the match head. Similarly, hydroelectric power plants convert the gravitational potential energy of dammed-up water into electrical energy, which in turn can be transformed to heat energy or light energy in your home. When you drive your car, the chemical energy stored in gasoline converts into kinetic energy of the moving car, and eventually into heat energy that may radiate into space. The lesson from these examples is clear:

> **The many different forms of energy are interchangeable.**

Bungee jumping provides a dramatic illustration of how energy can be transformed from one form to another (see Figure 5–8). Bungee jumpers climb to a high bridge or the top of a construction crane, where elastic cord is attached to their ankles. They then launch themselves into space and fall toward the ground until the cords stretch, slow them down, and stop their fall.

The bungee jumper uses the chemical potential energy generated from food to walk up to the launching platform. This work, done against gravity, provides the jumper with gravitational potential energy. During the long descent, the gravitational potential energy diminishes, while the jumper's kinetic energy simultaneously increases. As the cords begin to stretch, the jumper slows down, and kinetic energy is gradually converted to stored elastic potential energy in the cords. Eventually, the gravitational potential energy that the jumper had at the beginning is completely transferred to the stretched elastic cords, which then rebound,

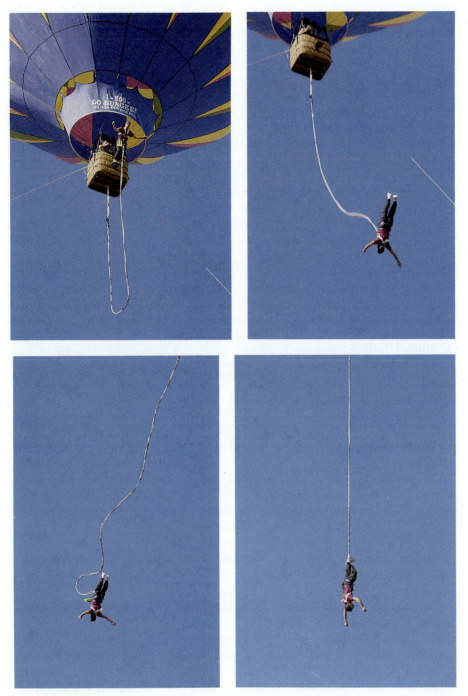

Figure 5–8
Bungee jumping provides a dramatic example of energy changing from one form to another. Can you identify points at which the jumper has maximum kinetic energy? Maximum gravitational potential energy? Maximum elastic potential energy? Does the jumper ever have a combination of all three? What other kinds of energy are involved?

converting some of that stored elastic energy back into kinetic energy and gravitational potential energy. All the time, some of the energy is also converted to heat—heat in the stressed cord, frictional heat on the jumper's ankles, and heat in the air as it is pushed aside.

Every form of energy on our list can be converted to every other form of energy. This is one of the most fundamental properties of the universe in which we live.

THE FIRST LAW OF THERMODYNAMICS: ENERGY IS CONSERVED

conservation law Any statement that says that a quantity in nature does not change.

system A part of the universe under study and separated from its surroundings by a real or imaginary boundary.

first law of thermodynamics The law of the conservation of energy. In an isolated system the total amount of energy, including heat energy, is conserved.

Scientists always look for constants in their efforts to describe a changing universe. Is the total number of atoms or electrons constant in the universe? Is the total amount of electrical charge fixed? Any statement that says that a quantity in nature does not change—that it is conserved—is called a **conservation law.**

The concept of a **system** is key in understanding how laws of nature are studied. You can think of a system as an imaginary box into which you put some matter and some energy that you'd like to study. Scientists might want to study systems containing a pan of water, a forest, or the entire planet Earth. Doctors examine your nervous *system*, astronomers explore the solar *system*, and biologists observe a variety of eco*systems*. In each case, our investigation of nature is simplified by focusing on one small part of the universe.

If the object being studied can exchange matter and energy with its surroundings, such as water that is heated on a stove and gradually evaporates, then we have an *open system*. An open system is like an open box where you can take things out and put things back in. If matter and energy do not freely exchange with their surroundings, as in a tightly shut box, then the system is said to be *closed* or *isolated*. The Earth and its primary source of energy, the Sun, together make a system that may be thought of for some purposes as closed (although energy is constantly being radiated into space).

The most important conservation law in the sciences is the conservation of energy, which applies to all closed systems. This law is also called the **first law of thermodynamics.** *Thermodynamics*—literally, the study of the movement of heat—is a term used for the science of heat, energy, and work. The law can be stated as follows:

> **In a closed system, the total amount of energy, including heat energy, is conserved.**

This law tells us that, although the kind of energy in a given system can change, the total amount cannot. For example, when that bungee jumper hurls herself into space, the total amount of gravitational energy she had at the beginning of the fall is still present in her and her surroundings at any time during the fall. When she's moving, some of the gravitational potential energy is converted into kinetic energy, some into elastic potential energy, and some into heat. At each point during the fall, however, the sum of kinetic, elastic potential, gravitational potential, and heat energies has to be the same as the gravitational energy at the beginning.

Energy is something like an economy with an absolutely fixed amount of money. You can earn it, store it in a bank or under your pillow, and spend it here and there when you want to. But the total amount of money doesn't change just because it passes through your hands. Likewise, in any physical situation you can shuffle energy from one place to another. You could take it out of the account labeled "kinetic" and put into the account labeled "potential"; you could spread it around into accounts labeled "chemical potential," "elastic potential," "heat energy" and so on. Whatever you do, the first law of thermodynamics tells us that you can never have more energy than you started with.

Example 5–7:   People as Light Bulbs

Every day you take in stored chemical energy when you eat. Then you use it in some way and return it to the environment as waste heat. As noted earlier, the energy content of your food is rated in Calories (with a capital C), equal to 1000 calories. Suppose you take in 2000 Calories per day and are not gaining or losing weight. What is your rate of power output?

▶ **Reasoning:**   Virtually all 2000 Calories that you take in during a day must leave you as heat, so this is the total amount of energy you are giving off in a day. Recall that a unit of power is the watt, or joule/second. Therefore, to solve this problem, we need to (1) find out how many joules there are in 2000 Calories, and (2) find out how many seconds there are in a day. Once we have done this, simple division will give us the answer to the problem.

▶ **Solution:**   Since 1 calorie equals 4.186 joules, it follows that

$$1 \text{ Cal} = 1000 \text{ cal} \\ \times 4.186 \text{ J/cal} \\ = 4186 \text{ J}$$

and your daily food intake corresponds to

$$2000 \text{ Cal} \times 4186 \text{ J/cal} = 8.4 \times 10^6$$

The number of seconds per day is

$$24 \text{ h/day} \times 60 \text{ min/h} \\ \times 60 \text{ s/min} = 86{,}400 \text{ s} \\ = 8.64 \times 10^4 \text{ s}$$

Therefore your power output is

$$\text{power} = \frac{\text{energy expended}}{\text{time}}$$
$$= \frac{8.4 \times 10^6 \text{ J}}{8.64 \times 10^4 \text{ s}}$$
$$= \frac{8.4}{8.64} \times 10^2 \text{ J/s}$$
$$= 97.2 \text{ W}$$

Each person in the audience is radiating about as much energy as a 100-watt light bulb. This is why the room often seems warm at the end of a concert.

In other words, a human being who uses 2000 Calories of food per day is radiating about as much energy into the environment as a 100-watt light bulb. This heat output explains why an auditorium gets so warm by the end of a concert. ▲

Science in the Making

Lord Kelvin and the Age of the Earth

The first law of thermodynamics provided physicists with a powerful tool for describing and analyzing their universe. Every isolated system, the law tells us, has a fixed amount of energy. Naturally, one of the first systems that scientists considered was the Earth and Sun.

British physicist William Thomson (1824–1907), knighted as Lord Kelvin (Figure 5–9), asked a simple question: How much energy could be stored in the Earth? And, given the present rate that energy radiates out into space, how old might the Earth be? This question had profound implications for philosophers and theologians who had their own ideas about the Earth's relative antiquity. Some Biblical scholars believed that the Earth could be no more than a few thousand years old. Most geologists, on the other hand, saw evidence in layered rocks to suggest an Earth at least hundreds of millions of years old, while biologists also required vast amounts of time to account for the gradual evolution of life on Earth. Who was correct?

Kelvin believed, as did most of his contemporaries, that the Earth formed from a cloud of dust and rock (see Chapter 17). He thought that the Earth was a hot, molten ball early in its history because impacts of large objects during its formation must have converted huge amounts of gravitational potential energy into heat. He used new developments in mathematics to calculate how long it would take for the hot Earth to cool to its present temperature. He assumed that there were no sources of energy inside the Earth, and found that the age of the Earth had to be less than about 100 million years. He soundly rejected the geologists' and biologists' claims of an older Earth as being incompatible with the laws of physics.

Seldom have scientists come to such a bitter impasse. Two competing theories about the age of the Earth, each supported by seemingly sound observations, were at odds. Physicists' calculations seemed unassailable, yet the observations of biologists and geologists in the field were equally meticulous. What could possibly resolve the disagreement? Had the scientific method failed? The solution to the dilemma came from a totally unexpected source. In the 1890s, scientists discovered that rocks hold a previously unknown source of energy, radioactivity (see Chapter 14) in which heat is generated by the conversion of mass. Lord Kelvin's rigorous age calculations were in error only because he and his contemporaries were unaware of this critical component of the Earth's energy budget.

The Earth, we now know, gains approximately half of its internal heat from the energy of radioactive decay. Revised calculations suggest an age of several billions of years, in conformity with geological and biological observations. ●

Figure 5–9
William Thomson, Lord Kelvin (1824–1907)

Science by the Numbers

Diet and Calories

The first law of thermodynamics has a great deal to say about the American obsession with weight and diet. Human beings take in energy with their food, energy we usually measure in Calories. When a certain amount of energy is taken in, the first law says that only one of two things can happen to it: it can be converted into work and waste heat, or it can be stored. If we take in more energy than we expend, the excess is stored in fat. If, on the other hand, we take in less than we expend, energy must be removed from storage to meet the deficit, and the amount of body fat decreases.

Here are a couple of rough rules you can use to calculate Calories in your diet:

1. Under most circumstances, normal body maintenance uses up about 15 Calories per day for each pound of body weight.

2. You have to take in about 3500 Calories to gain a pound of fat.

Suppose you weigh 150 lb. To keep your weight constant, you have to take in

$$150 \text{ lb} \times 15 \text{ Cal/lb/day} = 2250 \text{ Cal/day}$$

If you wanted to lose one pound (3500 Calories) in a week (7 days), you would have to reduce your daily Calorie intake by

$$\frac{3500 \text{ Cal}}{7 \text{ days}} = 500 \text{ Cal/day}$$

Another way of saying this is that you would have to reduce your calorie intake to 1750 Calories, a reduction equivalent to skipping dessert every day.

Alternatively, the first law says you can increase your energy use through exercise. Roughly speaking, to burn off 500 Calories you'd have to run 5 miles, bike 15 miles, or swim for an hour.

While it is a lot easier to refrain from eating than to burn off the weight by exercise, a combination of dieting and exercise may be the most effective approach to controlling weight. In fact, most researchers now say that the main benefit of exercise in weight control has to do with its ability to help people control their appetites. ●

THE GREAT CHAIN OF ENERGY, REVISITED

We opened this chapter by asking you to think about the epic chain of events that led up to the last time you used energy by driving a car or turning on a light bulb. We're now ready to look at those events in more detail. Hundreds of millions of years ago, energy was generated by converting mass in the core of the Sun; this process is called nuclear fusion (see Chapter 14). Through collisions of atoms and other processes, this energy eventually worked its way to the surface of the Sun. From

Energy from the Sun is stored in the molecules in plants and can be converted, with time, heat, and pressure, into coal.

there, light energy streamed outward into space in the form of energy waves—ordinary sunlight. Most of that radiant energy left the solar system and was radiated into outer space, but a portion reached the Earth. Some of that radiant energy was reflected by clouds or particles in the atmosphere, so that it, too, spread throughout the vastness of space. However, a fraction of the original energy from the Sun came down to the surface of the Earth where it was absorbed. That absorbed energy played a vital role in processes at the Earth's surface—it evaporated surface water to form rain clouds, powered ocean currents, and caused the winds to blow, for example.

Some of that energy was absorbed by plants and used to drive the chemical reactions of life. Those biological processes created clusters of atoms that formed the fiber of trees and shrubs in a tropical forest and stored the sunlight as chemical potential energy. As time went by, the plants and shrubs died, were buried under the ooze of the tropical swamp, and began a long stay within the Earth. The organic matter in the plants was gradually transformed by the Earth's interior pressure and temperatures into coal. Throughout this process, the original energy from the sunlight remained, passing through an energy chain from living plant material to buried organic matter to coal.

Today, miners bring that coal from its underground reservoir to the surface of the Earth and ship it to a power plant, where it is burned to make electricity. When you turn on the light, you tap into that energy. Most of the electrical energy is used to produce the heat of the glowing light filament, but a small fraction of the energy is converted directly into light. That light energy is then converted into heat and absorbed by surrounding objects.

But that's not the end of the story. When you turn on the light, the energy that was stored in the sunlight is eventually converted entirely into heat. Some of this heat raises the temperature of the light bulb and its socket, some warms the walls of your room slightly, some of it goes into motion of the air that is heated when it absorbs light. Wherever it went, however, the result is the same. Because you turn on your light, the Earth is a little bit warmer than it would be had you not done so. The net effect of this warmth is that the Earth will radiate heat into space.

The flow of energy never ceases. There is no point at which we can say "Here is where energy is created," or "Here is where it is destroyed." Energy flows through the Earth and every other system in the universe, constantly changing form, constantly shifting from one type to another, but always remaining the same in total quantity.

Environment

Fossil Fuels

All life is rich in the element carbon, which plays a key role in almost all chemicals that make up our cells. These carbon-based substances are formed using the Sun's energy, directly through photosynthesis or indirectly through food. These substances are rich in chemical potential energy. When living things die, they may collect in layers on the land or at the bottom of ponds, lakes, or the oceans. Over time, as the layers are buried, the Earth's temperature and pressure may alter the chemicals of life into deposits of fossil fuels.

Fossil fuels, carbon-rich deposits of ancient life that burn with a hot flame, have been the most important energy source during the past century and a half. *Coal, oil (petroleum),* and *natural gas,* the most common fossil fuels, are consumed in prodigious quantities around the world (Figure 5–10). Fossil fuels now account for 90% of all energy consumed by industrial nations. In the United States alone, approximately 1 billion tons of coal and 2.5 billion barrels of petroleum are used every year.

Figure 5–10
Sources of energy for industry. Note that most of our energy comes from fossil fuel.

Coal seam in Colstrip, Montana.

About 90% of the energy used by industrialized nations comes from fossil fuels such as petroleum.

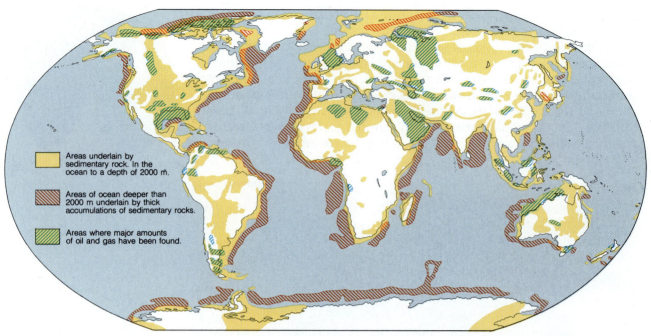

Figure 5–11
This map shows areas in which large accumulations of oil and gas have been found; all these areas are underlain by a type of rock formed from layers of sediment.

Legend:
- Areas underlain by sedimentary rock. In the ocean to a depth of 2000 m.
- Areas of ocean deeper than 2000 m underlain by thick accumulations of sedimentary rocks.
- Areas where major amounts of oil and gas have been found.

Some geologists estimate that it takes tens of millions of years of gradual burial under layers of sediments, combined with the transforming effects of temperature and pressure, to form a coal or petroleum deposit. Coal forms from layer upon layer of plants that thrived in vast ancient swamps, while petroleum represents primarily the organic matter once contained in plankton (microscopic organisms that float near the ocean's surface). While these natural processes continue today, the rate of coal and petroleum formation in the Earth is only a small fraction of the fossil fuels being consumed. For this reason, fossil fuels are classified as *nonrenewable resources*. (See Figure 5-11.)

One consequence of this situation is clear. Humans cannot continue to rely on fossil fuels forever. Reserves of high-grade crude oil and the cleanest-burning varieties of coal may last less than 100 years. Less efficient forms of fossil fuels, including lower grades of coal and *oil shale,* in which petroleum is dispersed through solid rock, could be depleted within a few centuries.

In addition, when fossil fuels burn, they may pollute the air and contribute to environmental problems such as acid rain and the greenhouse effect (see Chapter 19). ●

Technology

A Better Way to Burn

Many engineers and scientists are working to provide new ways of supplying energy sources for the future. Some of this work, as we shall see in Chapter 13, focuses on finding new ways to tap other natural sources, such as the Sun for solar energy. Other work is devoted to finding new ways of using familiar sources, such as coal.

The *fluidized bed combustor* is an example of that sort of technology. All coal-burning power plants convert water to steam, which drives electric generators. The fluidized bed combustor does the job more efficiently than conventional power plants.

As shown in Figure 5–12, this device is a high-tech "furnace" in which coal is first ground into a powder, added to a much larger amount of an inert material such as sand or limestone, and then supported on a cushion of blowing hot air as it burns. During this process, the fine particles float around like a liquid, as opposed to a pile of solids. The floating, burning assembly is called a fluidized bed. Water boils to steam in pipes that pass through the fluidized bed.

This method of burning coal has many advantages over the standard boiler. For one thing, since the inert sand or limestone in the bed isn't actually burning, once the bed is heated up the coal is burned much more efficiently than in a normal boiler. In addition, the pipes carrying the water that is to be heated can be put right down in the bed, so that a lot more of the heat actually gets to the water.

In addition to being more efficient, the fluidized bed combustor is also less polluting than a normal boiler. For example, some of the sulfur impurities in the coal combine with limestone in the bed to form a solid slag, which doesn't go out the smokestack and into the air. Furthermore, the increased efficiency of burning means that the combustor can operate at a lower temperature than a normal boiler, so that fewer nitrogen molecules in the air will be combined with oxygen as the coal burns. Thus this method of burning coal contributes much less to acid rain (see Chapter 19) than standard methods.

This modern coal-burning furnace is one example of a large number of advanced technologies that scientists and engineers are developing to deal with the pollution problem. While it may not cure the problem entirely, this technology is one step toward making our lives a little better in the future. ●

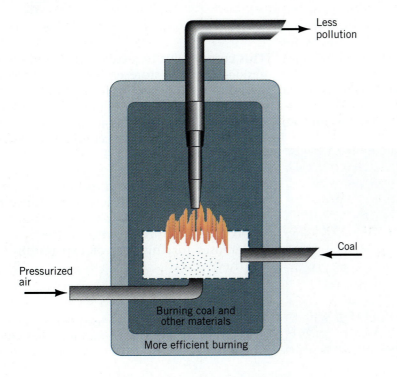

Less pollution

Coal

Pressurized air

Burning coal and other materials

More efficient burning

Figure 5–12
Many modern coal-burning furnaces employ this efficient, less-polluting "fluidized bed" design. Air jets suspend coal particles, which burn efficiently and facilitate efficient heat transfer from the burning fuel. This technology allows more useful energy to be extracted from the coal and reduces pollution from sulfur impurities in the coal.

THINKING MORE ABOUT
ENERGY

Alternative Energy Sources and the Storage Problem

One of the great problems facing the industrialized world today is reducing our dependence on fossil fuels. Solar energy is often mentioned as a "clean" and "natural" substitute for fossil fuels, but it isn't always available. Since the Sun doesn't always shine, we need some means of collecting the energy when it's available and storing it so that it can be used when it's needed.

Let's consider what would be involved in storing energy to heat a home in New England during the winter. First, solar collectors are placed on a roof so they can gather energy from the Sun during the day. One of the more popular means of storage is a large tank of water. As the Sun shines, the temperature of the water is raised and heat energy is stored. This energy is used to heat the house when sunlight isn't available. How big would such a tank have to be to store one day's worth of heat so that we could keep the house warm on a cloudy winter day?

A typical house in the northern United States might require 750,000 BTU to remain comfortable for 24 hours in the winter. (In the United States, calculations involving heating and cooling are usually done in English units). A solar collector might be capable of raising the temperature of a large tank of water to 100°F, and delivering useful heat to the house until the water temperature falls to 80°F.

How big a tank of water would the house need to have in the basement? It requires 1 BTU to raise the energy of 1 pound of water 1 degree, and we can get that energy back as the water cools. In our hypothetical storage system, the amount of heat we can extract from 1 pound of water is

$$\text{energy (in BTU/lb)} = (100°F - 80°F) \\ \times (1 \text{ BTU/°F})/\text{lb} \\ = 20 \text{ BTU/lb}$$

In other words, it takes 20 BTU of energy from sunlight to raise the temperature to 100°F, and we can get that 20 BTU back when the water cools to 80°F.

To store a day's worth of energy, therefore, would require

$$\text{amount of water (in lb)} \\ = \frac{(750,000 \text{ BTU})}{(20 \text{ BTU/lb})} \\ = 37,500 \text{ lb}$$

A cubic foot of water weighs about 64 lb, so the amount of space this amount of water would occupy is

$$\text{space (in cubic feet)} \\ = \frac{37,500 \text{ lb}}{64 \text{ lb/ft}^3} \\ = \frac{37,500}{64} \text{ ft}^3 \\ = 586 \text{ ft}^3$$

This corresponds to a cube a little over 8 feet on a side. Thus, to store enough energy to keep a house warm for three or four cloudy days would require a tank of water big enough to fill a fairly large room. Similar storage problems are associated with most alternate energy sources such as the wind and the tides.

Would you be willing to pay more for a house that derived its heat from the Sun? Should there be legislation requiring these sorts of heating systems? Should there be low-cost government loans or tax breaks for people who use them? Are there better uses for this sort of expenditure? What are they?

Energy from the Sun can be used directly by homeowners, but it must be stored for use when the Sun isn't shining.

▶ Summary

Work, measured in *joules* (or foot-pounds), is defined as a force applied over a distance. Every time you move an object, you are doing work. Every action of our lives requires *energy* (also measured in joules), which is the ability to do work. *Power*, measured in *watts* or *kilowatts*, indicates the rate at which energy is expended.

Energy comes in several varieties. *Kinetic energy* is the energy associated with moving objects such as cars or cannonballs. *Potential energy*, on the other hand, is stored energy, ready for use, such as the chemical energy of coal, the elastic energy of a coiled spring, the gravitational energy of dammed-up water, or the electrical energy in a wall socket. *Thermal energy* or *heat* is the form of kinetic energy associated with vibrating atoms and molecules. Energy can also take the form of waves, such as *sound waves* or light waves. Early in the twentieth century, it was discovered that mass is also a form of energy. Energy constantly shifts from one form to another, and all of these kinds of energy are interchangeable.

The most fundamental notion about energy is expressed in the *first law of thermodynamics*: Energy is *conserved*; the total amount of energy in an isolated *system* never changes. Different kinds of energy can shift back and forth, but the sum of all energy is constant.

▶ Key Equations

work (in joules)
$$= \text{force (in newtons)} \times \text{distance (in meters)}$$

$$\text{power (in watts)} = \frac{\text{energy (in joules)}}{\text{time (in seconds)}}$$

energy (in joules) = power (in watts) × time (in seconds)

kinetic energy (in joules)
$$= \tfrac{1}{2} \times [\text{mass (in kg)}] \times [\text{velocity (in m/s)}]^2$$

gravitational potential energy (in joules)
$$= g \times \text{mass (in kg)} \times \text{height (in m)}$$

energy associated with mass (in joules)
$$= \text{mass (in kg)} \times [\text{speed of light (in m/s)}]^2$$

▶ Constant

$c = 3 \times 10^8$ m/s = speed of light

▶ Review Questions

1. What is the scientific definition of work? How does it differ from ordinary English usage?

2. Is the kilowatt-hour a unit of energy or power? How about kilowatt?

3. What is the difference between the watt and the horsepower?

4. What is the difference between the joule and the kilowatt-hour? Who uses which unit?

5. What is the difference between energy and power?

6. List some different kinds of energy. Explain how they differ from each other.

7. Find something in your classroom or dorm room that possesses gravitational potential energy.

8. What is the relation between the perception that something is hot and the motion of the molecules in it?

9. Describe the historical process by which the notion of heat as a form of energy was developed.

10. Can waves carry energy? If so, give an example.

11. What does it mean to say that different forms of energy are interchangeable?

12. What does it mean to say that energy is conserved?

13. How did the discovery that mass is a form of energy resolve the debate over the age of the Earth?

14. Explain what it means to say "Energy flows through the Earth."

▶ Fill in the Blanks

Complete the following paragraph with words and phrases from the list of key terms.

conservation law	potential energy
energy (measured in joules)	power (measured in watts or kilowatts)
first law of thermodynamics	thermal energy (heat)
kinetic energy	work (measured in joules)

When a force is exerted over a distance, _____ is done. When an object is capable of exerting a force over a distance, we say that it possesses _____. _____ measures the rate at which energy is expended.

Energy comes in many forms. Energy associated with motion is called _____. A drawn bow or a rock sitting high on a hill are examples of _____. Energy associated with the motions of atoms or molecules is called _____.

The _____ tells us that the total amount of energy in a closed system does not change, though it can change from one form to another. The type of natural law that says a quantity does not change is called a _____.

▶ Discussion Questions

1. How does the discovery of heat as a form of energy illustrate the scientific method?

2. You use energy to heat your home. What ultimately happens to the energy that you pay for in your heating bill?

3. Think about your activities today. Pick one of them and identify the chain of energy that led to it. Where will the energy in that chain eventually wind up?

4. What happens on a hot summer day when the energy demand on your local power plant exceeds its energy output?

5. What kinds of energy are present in the following systems:

    a. water behind a dam

    b. a swinging pendulum

    c. an apple on an apple tree

    d. a uranium atom deep in the Earth

6. Identify four sources of energy around us that are constantly being renewed. What sources of energy do we use that are not constantly renewed?

7. Describe how you might convert the elastic potential energy of a coiled spring into heat energy. How might you convert heat energy into gravitational potential energy?

8. Plants and animals are still dying and falling to the ocean bottom today. Why, then, do we not classify fossil fuels as renewable resources?

9. Some people say that you lose more Calories by eating celery than you gain. How could that be?

▶ Problems

1. How much work against gravity do you do when you climb a flight of stairs 3-m high? Compare this work to that done by a 60-watt light bulb in an hour. How many flights of stairs would you have to climb to equal the work of the light bulb?

2. Would you rather be hit by a 1-kg mass traveling 10 m/s, or a 2-kg mass traveling 5 m/s?

3. Compared to a car moving at 10 mph, how much kinetic energy does that same car have when it moves at 20 mph? At 30 mph? At 60 mph? What do these numbers suggest to you about the difficulty of stopping a car as its speed increases?

4. According to Einstein's famous equation, $E = mc^2$, how much energy is contained in a pound of feathers? A pound of lead? (*Hint:* You will first need to convert pounds into kilograms.)

5. A small air compressor operates on a 1.5-hp electric motor for 8 hr a day. How much energy is consumed by the motor daily? If electricity costs 10 cents a kwh, how much does it cost to run the compressor each day? (*Hint:* 1 hp equals about 750 W.)

6. Joules and kilowatt-hours are both units of energy. How many joules are equal to one kilowatt-hour?

7. Zak, helping his mother rearrange the furniture in their living room, moves a 50-kg sofa 6 m with a constant force of 20 N.

    a. What is the work done by Zak on the sofa?

    b. What is the average acceleration of the sofa?

8. Georgie was pulling her brother (20 kg) in a 10-kg sled with a constant force of 25 N for one block (100 m).

    a. What is the work done by Georgie?

    b. How long would a 100-W light bulb have to glow to produce the same amount of energy exerted by Georgie?

9. A weight lifter can lift a 25-lb weight 1.5 ft during a bicep curl. If he does 50 repetitions of this lift, what is the total work done in ft-lb and joules?

10. The stair stepper is a novel exercise machine that attempts to reproduce the work done against gravity by walking up stairs. With each step, Brad (60 kg) will climb a distance of 0.2 m with this machine. If Brad exercises for 15 minutes a day with a stair stepper at a frequency of 60 steps per minute, what is the total work that he does?

11. Calculate the amount of energy produced in joules by a 100-W light bulb lit for 2.5 hr.

12. Normally the energy expended during a brisk walk is 3.5 Calories per minute. How long (in minutes) do you have to walk in order to use up the Calories contained in a candy bar (approximately 280 Calories)?

13. How long (in minutes) do you have to walk to produce the same amount of energy as a 100-W light bulb that is lit for one hour?

14. You throw a softball (250 g) straight up into the air, it reaches a maximum altitude of 15 m, and then it returns to you. (Assume that the ball departed and returned at ground level.)

    a. What is the gravitational potential energy (in joules) of the softball at its highest position?

    b. What is the kinetic energy of the softball as soon as it leaves your hand? (Assume no energy loss by the softball while it is in the air.)

    c. What is the kinetic energy of the softball when it returns to your hand?

    d. From the kinetic energy, calculate the velocity of the ball as it left your hand.

15. Sleeping will normally consume 1.3 Calories of energy per minute for a typical 150-lb person. How many Calories are expended during a good night's sleep of 8 hr?

16. Above the Earth's atmosphere we receive 1.35 × 10^3 W/m² of energy from the Sun (solar constant). If you can convert 100% of this energy into usable electricity, how large a collecting area of solar cells is necessary to produce a 1-gigawatt power plant?

17. You leave your 75-W portable color TV on for 6 hours during the day and evening, and you do not pay attention to the cost of this electricity. If the dorm (or your parents) charged you for your electricity use and the cost was \$0.10 per kWh, what would be your monthly (30 days) bill?

18. While skiing in Jackson, Wyoming, your friend (85 kg) starts his descent down the bunny run, 25 m above the bottom of the run. Assume he starts at rest and converts all of his gravitational potential energy into kinetic energy.

  a. What would be your friend's kinetic energy at the bottom of the bunny run?

  b. What would be his final velocity?

  c. Is this speed "reasonable"?

19. Lora (50 kg) is an expert skier. If she starts at 3 m/s at the top of a run, 35 m above the bottom, what is her final speed if she converts all her gravitational potential energy into kinetic energy? What is her final kinetic energy at the bottom of the ski run?

20. In order to lose 1 lb per week, you have to reduce your daily intake by 500 Calories per day.

  a. How long would a 60-W light bulb have to burn in order to produce this much energy?

  b. How long would you have to walk per day (3.5 Calories per minute) to produce this amount of energy?

  c. How many stairs (approximately 1 ft in height) would you have to climb to produce this amount of energy?

21. In the Sun, 1 g of hydrogen consumed in nuclear fusion reactions produce 0.0071 g of material that is converted into energy.

  a. How much energy does this produce in joules?

  b. How high could you raise the Mt. Palomar 5-m telescope (4.5×10^5 kg) with this energy?

  c. If you can convert 1 gm of hydrogen into energy every second through nuclear fusion, the energy produced would be equivalent to how many 1-gigawatt power plants?

22. In our storage problem (see Thinking More About Energy), instead of using water, we now want to use ethylene glycol (Prestone) as our heat-storage medium. Ethylene glycol has a heat capacity of 0.53 BTU/lb/°F. One cubic foot of ethylene glycol weighs 51.2 lb. Calculate the volume of ethylene glycol needed to store the same amount of heat as the water in our storage exercise.

▶ Investigations

1. From your most recent electric bill, find the cost of one kilowatt-hour in your area.

  a. Look at the back of your CD player or an appliance and find the power rating in watts. How much does it cost for you to operate the device for 1 hour?

  b. If you leave a 100-W light bulb on all the time, how much will you pay in a year of electric bills?

  c. If you had to pay \$10.00 for a high-efficiency bulb that provides the same light as the 100-W bulb with only 10 W of power, how much would you save per year of electric bills, assuming you use the light 5 hr/day? Would it be worth your while to buy the energy-efficient bulb if the ordinary bulb cost \$1.00?

2. What kind of fuel is used at your local power plant? What are the implications of the first law of thermodynamics regarding our use of fossil fuels? Our use of solar energy?

3. Keep track of the Calorie content of what you eat for a few days. If you kept eating at this rate, what would your weight eventually become?

4. Check your household's electric bills for the past year and calculate your total electric consumption for the year.

  a. How many 10-kg weights would you have to raise 100 m to produce a gravitational potential energy equal to this consumption?

  b. How much mass is equal to this consumption ($E = mc^2$)?

5. Investigate the history of the controversy between Lord Kelvin and his contemporaries regarding the age of the Earth. When did the debate begin? How long did it last? What kinds of evidence did biologists, geologists, and physicists use to support their differing calculations of the Earth's age?

6. In this chapter we introduced several energy units: joule, foot-pound, kilowatt-hour, BTU, calorie, and Calorie. Many other energy units are used, including erg, electron-volt, and reciprocal centimeter. Find a table of energy units and their conversation factors. Arrange these units in order of smallest to largest. Why do people use so many different energy units? Who uses which ones and why?

7. At one point, the Clinton-Gore administration advocated the imposition of a "BTU tax" on energy. What is this tax? Where does the name BTU tax come from? How might it be implemented? What are the pros and cons? If new revenues were essential to prevent the budget deficit from growing, how would you vote on legislation to impose a BTU tax? Why?

▶ Additional Reading

Burchfield, Joe D. *Lord Kelvin and the Age of the Earth*. Chicago: University of Chicago Press, 1990.

Cardwell, D.S.L. *From Watt to Clausius: The Rise of Thermodynamics in the Early Industrial Age*. Ithaca, N.Y.: Cornell University Press, 1971.

Fowler, John M. *Energy and the Environment*. 2d ed. New York: McGraw-Hill, 1984.

6

TEMPERATURE, HEAT, AND THE SECOND LAW OF THERMO- DYNAMICS

ENERGY ALWAYS GOES FROM A MORE USEFUL TO A
LESS USEFUL FORM.

The Cafeteria

Next time you're in the cafeteria, as you go through the food line,
think about how each course is prepared. Then try to imagine these
processes in reverse. You can peel a piece of fruit, but you can't put it
back together. It's easy to scramble eggs, but impossible to unscramble
them. You can cook vegetables, but there's no way to uncook them.
And once popcorn is popped, it can't be unpopped. But why is this so?
Nothing in Newton's laws of motion or the law of gravity suggests
that events work only in one way. Nothing that we have learned
about energy suggests that nature works in only one direction.

At the cafeteria you've probably noticed that foods and drinks that
are very hot get cooler, while those that are very cold get warmer. A
glass of ice water gradually gets warmer, while a plate of hot pasta gets
cooler. Ice cream gradually melts, while hot fudge sauce hardens.
These everyday events are so familiar that we don't give them a second
thought, yet underlying the popping of popcorn and the cooling of a
hot drink is one of nature's most subtle and fascinating laws—the
second law of thermodynamics.

NATURE'S DIRECTION

Molecules move from the perfume bottle out into a room, but, once out of the bottle, they are unlikely to find their way back in.

Think about the many events that occur every day that have only one direction. A drop of perfume quickly pervades an entire room with scent, but you'd be hard pressed to collect all those perfume molecules into a single drop again. Your dorm room seems to get messy in the course of the week all by itself, but it takes time and effort to clean it up. And everyone and everything gets older—there is no turning back the clock.

The first law of thermodynamics—conservation of energy—in no way forbids events to progress in the "wrong" direction. In fact, it takes exactly as much energy to unscramble an egg as it does to scramble it. The energy of a room with perfume molecules dispersed throughout is the same as the energy of the room with those molecules tightly bottled. And the energy that went into strewing things about your room is exactly the same energy it takes to put everything back again. Yet there seems to be some natural tendency for things to become less orderly with time.

This directionality in nature can be traced, ultimately, to the behavior of atoms and molecules in materials. If you hold a glass of water in your hand, for example, its atoms are moving at more or less the same average velocity. If you introduce one more atom into this collection—an atom that is moving much faster than any of its neighbors—you will have a situation in which many atoms move slowly and one atom moves fast. Over the course of time, the fast atom will collide again and again with the others. With each collision, the fast atom will probably lose some of its energy, much as a fast-moving billiard ball slows down after it collides with a couple of other billiard balls. Over a period of time, the fast-moving atom will probably share its energy with all the other atoms. Consequently, every atom in the collection will be moving slightly faster than those in the original collection, and there will be no single atom moving much faster than the others.

It is extremely unlikely that the atoms would ever arrange themselves in such a way that one atom moves very fast in a collection of very slow ones. In the language of physicists, the original *state* with only one fast

Figure 6–1
Many events work in only one direction. (*a*) As a pool game begins, the cue ball contains all the kinetic energy, but this energy soon becomes distributed evenly through all the balls. (*b*) A building collapses, transforming the highly ordered structure into a disordered pile of rubble. (*c*) A flower, once cut, begins to wilt and decay.

(*a*)

atom is highly improbable. Over the course of time, any unlikely initial state will evolve into a more probable state—a situation like the one in which all the atoms have approximately the same energy.

The tendency of all systems to change from improbable to more probable states accounts for the directionality that we see in the universe. There is no reason from the point of view of the first law of thermodynamics that improbable situations can't occur. Fifteen slow-moving billiard balls could transfer their kinetic energy to produce one fast-moving ball. The fact that this situation doesn't occur in nature is an important clue as to how things work at the atomic level (Figure 6–1).

Scientists in the nineteenth century discovered the underlying reasons for nature's directionality by studying heat, which is a consequence of the motion of atoms and molecules. In order to understand these discoveries, as summarized in the second law of thermodynamics, let's consider some properties of heat.

COMING TO TERMS WITH HEAT

Atoms never sit still. They are always moving, and in the process, they distribute the kinetic energy of moving atoms, or heat. If you have ever tried to warm a house during a cold winter day, you have experienced this phenomenon. If you turn off the furnace, the heat in the house gradually leaks out, and the house begins to cool. The only way to stay warm is to keep adding heat.

Like a furnace, our bodies constantly produce heat, and they distribute that energy to maintain our core body temperature close to 98.6°F (37°C). Both your furnace and your body produce heat on the inside; heat that will inevitably flow to the outside. You simply use that heat "on the fly," as it were.

In order to understand the nature of heat and its movement, we need to define three closely related terms: heat, temperature, and heat capacity.

(b)   (c)

Heat and Temperature

In everyday conversation, we often use the words temperature and heat interchangeably, but to scientists the words have somewhat different meanings. Heat is a form of energy that flows from a hot object to a colder object. Any object that is hot stores and can transfer a fixed and measurable amount of this energy. A gallon of boiling water contains more heat energy than a pint of boiling water. Heat, therefore, is related to the quantity of atomic kinetic energy contained in every object (see Chapter 5).

Temperature, on the other hand, is a relative term and reflects how vigorously atoms in a substance are moving and colliding. Two objects are at the same **temperature** if no heat energy flows spontaneously from one to the other when they are in contact. A gallon of boiling water, therefore, is at the same temperature as a pint of boiling water. The difference in temperature between two objects is one of the factors that determines how quickly heat energy will be transferred between those two objects: the larger the temperature difference, the more rapidly heat energy will be transferred.

Temperature scales provide a convenient way to compare the temperatures of two objects. Many different temperature scales have been proposed; all scales and temperature units are arbitrary, but every scale requires two easily reproduced temperatures for calibration. The freezing and boiling points of pure water are commonly used standards today in the Fahrenheit scale (32° and 212° for freezing and boiling, respectively) and the Celsius scale (0° and 100° for freezing and boiling, respectively). The Kelvin temperature scale also uses 100-degree increments between freezing and boiling water, but it defines 0 Kelvin as **absolute zero,** which is the coldest attainable temperature; this is the temperature at which it is impossible to extract any heat energy at all from atoms or molecules. The temperature of absolute zero is approximately −273°C, or −459°F. It turns out, therefore, that freezing and boiling occur at about 273 and 373 Kelvin, respectively (Figure 6–2).

To convert from Celsius to Fahrenheit degrees (something American travelers have to do in most parts of the world), we use the following formula:

$$°F = (1.8 \times °C) + 32$$

The 1.8 in this formula reflects the fact that the Fahrenheit degree is smaller than the Celsius, while the 32 reflects the fact that water freezes at 32°F but 0°C.

temperature A quantity that reflects how vigorously atoms are moving and colliding in a material.

absolute zero The temperature, 0 Kelvin, at which no energy can be obtained from atoms; the coldest attainable temperature, equal to −273.16°C or −459.67°F.

(*a*)

(*b*)

(*a*) Buckled train tracks illustrate how metals expand when heated. (*b*) An expansion joint in a bridge.

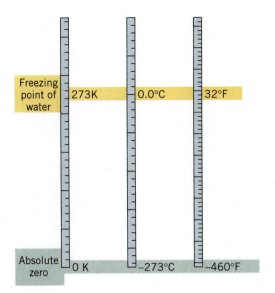

Heat Capacity

Heat capacity is a measure of the ability of a material to absorb heat energy, and is defined as the constant of proportionality between an amount of heat and the change in temperature that this heat produces in the material. *Specific heat* is the quantity of heat required to raise the temperature of a unit mass of that material by 1°C. Water displays the largest heat capacity of any familiar substance; by definition, 1 calorie is required to raise the temperature of 1 gram of water by 1°C. By contrast, you know that metals heat up quickly in a fire, so a small amount of heat energy can cause a significant increase in a metal's temperature.

Think about the last time you boiled water in a copper-bottomed pot. It doesn't take long to raise the temperature of a copper pot to above the boiling point of water because copper, like most other metals, can't hold much heat energy. In fact, 1 calorie will raise the temperature of 1 gram of copper by about 10°C. But water is a different matter; it must absorb 10 times more energy per gram than copper to raise its temperature. Thus, even at the highest stove setting, it can take several minutes to boil a pot of water. This ability of water to store heat energy plays a critical role in the Earth's climate, which is moderated by the relatively steady temperature of the oceans.

heat capacity   A measure of the ability of a material to absorb heat energy, defined as the constant of proportionality between an amount of heat and the change in temperature that this heat produces in the material.

Environment

The Heat Capacity of the Land and Oceans

The Earth's climate is affected by many things, but one important ingredient is the heat stored in the oceans. During warm weather, sunlight falling on the oceans raises the water's temperature. Since water has a high heat capacity, it takes a long time for it to warm up and a long time for it to cool off. The rocks and other materials found on land, however, have a lower heat capacity. They heat up and cool off much more quickly than water.

An eagle flying on rising warm air.

Several regular features of the Earth's climate are a direct result of the high heat capacity of the water in the Earth's oceans. For example, the lands surrounding the Indian Ocean experience a seasonal wind pattern known as the *monsoon*. During summer, the land on the Indian subcontinent heats up quickly while the ocean warms more slowly. The warm air over the land rises, drawing cooler air in from the east. These winds, after traveling over the ocean, carry a great deal of moisture. When they rise over the heated land, they produce clouds and regular rainfall (the rainy season).

In winter, on the other hand, the ocean cools more slowly than the land. Consequently, in winter, air over the ocean is warmer than that over the land, and the wind pattern reverses. Warm air rises over the ocean, drawing in cooler air from over the land. The rains stop and a six-month dry season begins. Thus the weather experienced over a large area of our planet can be traced directly to the high heat capacity of water. ●

Developing Your Intuition

At the Beach

If you spend time along a beach, you will probably notice that a breeze is usually blowing out to sea in the morning, but that the breeze calms down during the day, and then blows in toward the land by late afternoon. Think about why this should happen.

At night, the land cools off. As a result, in the morning air over the ocean is warmer and it will rise, drawing cooler air from the land to replace it. As the day passes, the land warms up, eventually becoming warmer than the sea. Then warm air rises over the land, and relatively cooler air from the ocean moves in to replace it. The midday calm occurs when the two temperatures are about the same. ●

HEAT TRANSFER

Heat will not stay in one place. You can't confine heat forever; you can only slow down its movement. In fact, scientists and engineers have spent many decades attempting to improve thermal insulation by studying the phenomenon known as **heat transfer**—the process by which heat moves from one place to another. Heat is transferred by three everyday mechanisms: conduction, convection, and radiation.

heat transfer The process by which heat moves from one place to another, through the mechanisms of conduction, convection, or radiation.

Conduction

Have you ever reached for a pan on a hot stove, only to have your fingers burned when you grasped the metal handle? If so, you have experienced **conduction,** which is the movement of heat by atomic collisions.

Conduction occurs because of the action of individual atoms or molecules that are linked together by chemical bonds. If a piece of metal such as a pot is heated at one end, the atoms at that end begin to move faster. When they vibrate and collide with atoms farther away from the heat source, they are likely to transfer energy to those atoms, so that those

conduction The movement of heat by collisions between vibrating atoms or groups of atoms (molecules); one of three mechanisms by which heat moves.

Figure 6–3
Steelworkers test a molten sample from the furnace. Why does the worker at the left wear thick insulated gloves to hold the cool end of the rod? Heat transfers from the hot end to the cool end by the process of conduction.

molecules will begin moving faster as well. A chain of collisions occurs, with atoms progressively farther and farther away from the heat source moving faster (see Figure 6–3).

To an outside observer, it appears that heat somehow flows like a liquid from the heat source through the metal pot into the handle. There's nothing particularly mysterious about this process. Heat conduction is a result of collisions between vibrating atoms or molecules. When a fast-moving object collides with a slow-moving one, the fast object usually slows down and the slow object usually speeds up.

When we pay our home heating bills, we are in large measure paying for the conduction of heat. The process works like this: Air inside the house in winter is kept warmer than the air outside, so that the molecules of the air inside are moving faster than those in the air outside. When those molecules collide with materials in the wall (a window pane, for example), they impart some of their heat energy to the molecules in the wall. At that point, conduction takes over and the heat is transferred to the outside of the wall. There, the heat energy is transferred outdoors by convection and radiation, processes that we will describe in a moment. In essence, your house becomes a kind of conduit carrying heat from the interior to the outside.

One way of slowing the flow of heat out of a house is to add insulation to the walls, or to use thermal glass for the windows. Both of these materials are effective because of their low **thermal conductivity,** which is their ability to transfer heat energy from one molecule to the next by conduction. Have you ever noticed that a piece of wood at room temperature feels "normal," while a piece of metal at the same temperature feels cold to the touch? The wood and metal are initially at exactly the same temperature, but the metal feels cold because it is a good *heat conductor*; it moves heat rapidly away from your skin, which is generally warmer than air temperature. The wood, on the other hand, is a good *heat insulator*; it impedes the flow of heat and so it feels warmer than the metal. You wouldn't want to live in a house made entirely of metal, and we usually use relatively good insulators such as wood or masonry as primary building materials. The insulation in homes, especially in newer ones, is designed to have especially low thermal conductiv-

thermal conductivity The ability of a material to transfer heat energy from one molecule to the next by conduction. When thermal conductivity is low, as in wood or fiberglass insulation, the transfer of heat is slowed down.

ity, so that heat transfer is slowed down (but never completely stopped). Thus, when you use special insulated window panes or put certain kinds of insulation in your walls, you make it more difficult for heat to flow to the outside, and thereby allow yourself to use heat more efficiently.

Convection

Let's look carefully at a pot of boiling water on the stove. On the surface of the water you will see a rolling, churning motion as the water moves and mixes. If you put your hand above the water, you feel heat. Heat has been transferred from the water at the bottom of the pot to the top by **convection,** which is the transfer of heat by the bulk motion of a liquid or gas, as shown in Figure 6–4. Convection may be *forced*—for example, when a fan circulates cool air—or it may be driven by gravity, as in the case of boiling water.

Water near the bottom of the pan expands as it is heated by the flames. Therefore, it weighs less per unit volume than the colder water immediately above it. A situation like this, with colder, denser water above and warmer, less dense water below, is unstable. The denser fluid tends to descend and displace the less dense fluid, which in turn begins to rise. Consequently, the warm water from the bottom rises to the top as shown, while the cool water from the top sinks to the bottom. In convection, masses of water move in bulk and carry the fast-moving molecules with them. Heat is transferred by the actual physical motion of these masses of water.

convection   The transfer of heat by the physical motion of masses of fluid: dense, cooler fluids (liquids and gases) descend in bulk and displace rising warmer fluids, which are less dense; one of three mechanisms by which heat moves.

Figure 6–4
Convection. Heat is transferred by the bulk motion of the water.

Convection is a continuous, cyclic process as long as heat is added to the water. As cool water from the top of a pot arrives at the bottom, it begins to be heated by the burner. In the same way, as hot water gets to the top, its heat is sent off into the air. The water on the top cools and contracts, while the water on the bottom heats and expands. The original situation is repeated continuously, with the less dense fluid on the bottom always rising and the more dense fluid on the top always sinking. This transfer of fluids results in a kind of a rolling motion, which you see when you look at the surface of boiling water. Each of these regions of rising and sinking water is called a **convection cell.**

convection cell   A region in a fluid in which heat is continuously being transferred by a bulk motion of heated fluid from a heat source to the surface of the fluid, where heat is released. The cooled fluid then sinks and the cycle repeats.

The areas of clear water, which seem to be bubbling, are the places where warm water is rising. The places where old bubbles (and scum, if the pot is dirty) tend to collect—the places that look rather stagnant—are where cool water is sinking. Heat is carried from the burner through the convection of the water, and is eventually transferred to the atmosphere.

Convection is a very efficient way of transferring heat. If you carefully study water in a pot on a stove, you will notice that for a while the surface of the water simply gets hotter and hotter. During this period, the heat transfers to the surface by ordinary conduction through the water molecules. Once the water contains more heat than can be transferred by conduction alone, the water starts to move. First, little bubbles begin to form, then they become fully defined convection cells. At this point, convection takes over and transfers the heat through the water.

Convection is a very common process in nature. From the small-scale circulation of cold water in a glass of ice tea, to air rising above a radiator or toaster, to large-scale motions of the Earth's atmosphere and ocean's currents, convection is at work. You may even have seen convection cells in operation in large urban areas. When you're in the parking lot of a shopping mall on a hot summer day, you can probably see the air shimmer. What you are seeing is air, heated by the hot asphalt, rising upward. Some place farther away, perhaps out in the countryside, cooler air is falling. The shopping center with all its concrete is called a "heat island" and is the source of upward rising air. It is the hot part of a convection cell, while the rest of the convection cell is the downward flowing air elsewhere.

You may have noticed that the temperature in big cities is usually a few degrees warmer than in the outlying suburbs. Cities help create their own weather through the creation of convection cells. Rainfall is typically higher in cities than in the surrounding atmosphere precisely because the warmer cities develop convection cells that draw in cool moist air from surrounding areas.

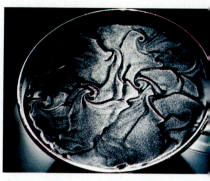

Convection in a coffee cup.

Technology

Home Insulation

Today's home builders take heat convection and conduction very seriously. An energy-efficient dwelling has to hold onto its heat in winter and remain cool in summer. A variety of high-tech materials provide effective solutions to this insulation problem.

Fiberglass, the most widely used insulation, is made of loosely intertwined strands of glass. It works by taking advantage of the fact that motionless air is an outstanding insulator. By trapping air into pockets, fiberglass minimizes the opportunities for conduction and convection of heat out of your home (see Figure 6–5). Solid glass is a rather poor heat insulator, but it takes a long time for heat to move along a

Figure 6–5
The interweaving of fibers in an ordinary piece of fiberglass, magnified in this photomicrograph, reduces heat transfer by convection and conduction.

thin, twisted glass fiber, and even longer for heat to transfer across the occasional contact points between pairs of crossed fibers. A clothlike mat of fiberglass, furthermore, disrupts air flow and prevents heat transfer by convection. A thick, continuous layer of fiberglass in your walls and ceiling thus acts as an ideal barrier to the flow of heat.

If our houses were constructed with solid walls, then fiberglass would serve all our needs, but windows pose a special problem. Have you ever sat near an old window on a cold winter day? Old-style single-pane windows conduct heat rapidly, as you can feel by putting your hand on such a window in winter. But how do we let light in without letting heat out? One solution is double-paned windows with sealed, airtight spaces between the panes that greatly restrict heat conduction. In addition, builders employ a variety of caulking and foam insulation to seal any possible leaks around window and doors. As a result, modern homes can be almost completely airtight (although some air has to be let in to prevent the house from becoming too stuffy). ●

Animal Insulation: Fur and Feathers

Houses aren't the only place where insulation can be seen in the natural world. Two kinds of animals, birds and mammals, maintain constant body temperatures despite the temperature of their surroundings, and both have evolved methods to control the flow of heat into and out of their bodies. Part of these strategies involve their natural insulating materials, such as furs, feathers, and fat. Most of the time, an animal's body is warmer than the environment, so insulation usually works to keep heat in.

Whales, walruses, and seals are examples of animals that have thick layers of fat to insulate them from the cold ocean waters in which they swim. Fat is a poor conductor of heat, and plays much the same role in their bodies as the fiberglass insulation in your attic.

Feathers also provide insulation; in fact, most biologists believe that feathers evolved first to help birds maintain their body temperature, and were only later adapted for flight. Feathers are made of light, hollow tubes connected to each other by an array of small interlocking spikes. They have some insulating properties, but their main effect comes from the fact that they trap air next to the body, and stationary air is an excellent insulator.

Birds often react to extreme cold by contracting muscles in their skin so that the feathers fluff out. This has the effect of increasing the thickness (and hence the insulating power) of the layer of trapped air. (Incidentally, birds need insulation more than we do because their normal body temperature is 41°C, or 106°F.)

Hair (or fur) is actually made up of dead cells similar to those in the outer layer of the skin. Like feathers, hair serves as an insulator and traps a layer of air near the body. In some animals (such as polar bears) the insulating power of the hair is increased by the fact that each hair contains tiny bubbles of trapped air. The reflection of light from these bubbles makes polar bear fur appear white—the strands of hair are actually translucent.

Animals use fur, feathers, and fat for insulation.

Hair grows from follicles in the skin, and small muscles allow animals to make their hair stand up to increase its insulating power. Human beings have lost much of their body hair as well as the ability to make most of it stand up. Our mammalian nature is revealed, however, in the phenomenon of "goose bumps," which is the attempt by muscles in the skin to make the nonexistent hair stand up.

Radiation

Everyone has experienced coming in on a cold day and finding a fire in the fireplace or an electric heater glowing red hot. The normal reaction is to walk up to the source of heat, hold out your hands, and feel the warmth on your skin.

radiation The transfer of heat by electromagnetic radiation; the only one of the three mechanisms of heat transfer that does not require atoms or molecules to facilitate the transfer process.

How did the heat energy get from the fire to your hands? It couldn't have done so by conduction—it's too hard to carry heat through the air that way. It couldn't have been convection either, because you don't feel a hot breeze. The air in the room is almost stationary.

What you experience is the third kind of heat transfer—**radiation,** or the transfer of heat by *infrared radiation,* which is a form of wave energy that we will discuss in Chapter 8. A fire, an electric heater, and the Sun all radiate heat energy in this form. This radiation travels like light from the source of heat to your hand, where it is absorbed and converted into kinetic energy of molecules. You perceive heat because of the energy that the infrared radiation carries to your hand (see Figure 6–6).

Every object in the universe radiates energy. Under normal circumstances, as an object gives off radiation to its surroundings, it also receives radiation from the surroundings. Thus a kind of equilibrium is set up, and there is no net loss of energy because the object is at the same temperature as its surroundings. If, however, the object is at a higher temperature than its surroundings, it will radiate more energy than it receives. Your body, for example, constantly radiates energy into its cooler surroundings. This energy can be detected easily at night with infrared goggles. You will continue to radiate this energy as long as your body processes the food that keeps you alive.

Radiation, such as *infrared energy,* is the only kind of energy that can travel through the emptiness of space. Conduction requires atoms or molecules that can vibrate and collide with each other. Convection requires atoms or molecules of liquid or gas in bulk, so that they can move. But radiation doesn't require anything in the environment to facilitate it; radiation can even travel through a vacuum. The energy that falls on the Earth in the form of sunlight—almost all of the energy that sustains life on Earth—travels through 93 million miles of intervening empty space in the form of radiation.

In the real world, all three types of heat transfer—conduction, convection, and radiation—are constantly occurring. Think about the heat generated by your body: heat is conducted through your bones and teeth, convects as your blood circulates (an example of forced convection), and radiates from your skin into the cooler surroundings, where it is eventually absorbed (and, in turn, produces slightly higher temperatures in the surrounding air). The same three phenomena operate in a pot of boiling water. Heat moves through the metal bottom and sides of the pot by

Figure 6–6
A fire transfers most of its energy by infrared radiation.

conduction, through the boiling water by convection, and from the sides and surface of the water by radiation. In fact, everywhere in the natural world, heat is constantly being transferred by these three mechanisms.

THE SECOND LAW OF THERMODYNAMICS

Throughout the universe, the behavior of energy is regular and predictable. According to the first law of thermodynamics, the total amount of energy is constant, though it may change from one form to another over and over again. Energy in the form of heat can flow from one place to another by conduction, convection, and radiation. But, as we pointed out earlier, energy flows in one direction. Hot things tend to cool off; cold things tend to warm up; and an egg, once broken, can never be reassembled. These examples illustrate the concept of the second law of thermodynamics—one of the most fascinating and powerful ideas in science.

The **second law of thermodynamics** places restrictions on the ways heat and other forms of energy can be transferred and used to do work. We will explore three different statements of this law:

1. Heat will not flow spontaneously from a cold to a hot body.
2. You cannot construct an engine that does nothing but convert heat to useful work.
3. Every isolated system becomes more disordered with time.

Although these three statements appear very different, they are actually equivalent. Given any one statement, you can derive either of the others as a consequence. Given the statement that heat flows from hot to cold objects, for example, a physicist can produce a set of mathematical steps that would show that no engine can convert heat to work with 100% efficiency. In this sense, the three statements of the law all say the same thing.

1. Heat Will Not Flow Spontaneously from a Cold to a Hot Body

The first statement of the second law of thermodynamics relates to the relative temperature of two objects. If you take an ice cream cone outside on a hot summer afternoon, it will melt. Heat will flow from the warm atmosphere to the cold ice cream cone and cause its temperature to rise above the melting point. By the same token, if you take a cup of hot chocolate outside on a cold day, it will cool as heat energy flows from the cup into the surroundings.

From the point of view of energy conservation, there is no reason why things should work this way. Energy would be conserved if heat stayed put, or even if heat flowed from an ice cream cone into the warm atmosphere, making the ice cream cone colder than it was at the beginning. Our everyday experience (and many experimental confirmations) convince us that our universe does not work this way. In our universe heat flows in one direction only—from hot to cold. It does not go the

second law of thermodynamics Any one of three equivalent statements: (1) Heat will not flow spontaneously from a colder to a hotter body; (2) It is impossible to construct a machine that does nothing but convert heat into useful work; (3) The entropy of an isolated system always increases.

The second law of thermodynamics says that heat will always flow from a warm body (the air) into a cold one (the ice cream cone).

other way spontaneously. This everyday observation may seem trivial, but in this statement is hidden all the mystery of those changes that make the future different from the past: the directionality in the universe.

If you think about energy at the molecular level, the explanation of this version of the second law is easy to understand. If two objects collide, and one of them is moving faster than the other, chances are that the slower object will speed up and the faster object will slow down. It's possible, but very unlikely, that events will go the other way. Thus, as we saw in the discussion of heat conduction, faster-moving molecules tend to share their energy with slower-moving ones. On the macroscopic scale, this process is seen as heat flowing from warm regions to cooler ones by conduction.

For the second law to be violated, the molecules in a substance would have to collide in such a way that slower-moving molecules slowed down even more, giving up their energy to faster molecules so they could go even faster. Our experience tells us that this doesn't happen. The second law takes this experience and makes it into a general law of nature.

The second law does *not* state that it is impossible for heat to flow from a cold to a hot body. Indeed, you know that in every refrigerator that's what happens. When a refrigerator is operating, heat energy is removed from the colder inside to the warmer outside. This fact can be verified by putting your hand under the refrigerator and feeling the warm air. The second law merely states that this action cannot take place *spontaneously*, of its own accord. If you wish to cool something in this way, you must supply energy. In fact, an alternative statement of the second law of thermodynamics could be: A refrigerator won't work unless it's plugged in.

The second law doesn't tell you that you can't make ice cubes, only that you can't make ice cubes without expending energy. Paying the electric bill, of course, is another piece of our everyday experience.

2. You Cannot Construct an Engine That Does Nothing But Convert Heat to Useful Work

The second statement of the second law of thermodynamics restricts the way we can use energy. Recall that energy is defined as the ability to do work. This second statement of the second law tells us that whenever energy is transformed from heat to another type—from heat to an electrical current, for example—some of that heat must be dumped into the environment and is unavailable to do work. The energy is neither lost nor destroyed, but it can't be used to make electricity to play your radio or gasoline to drive your car.

efficiency The amount of work you get from an engine, divided by the amount of energy you put in; a quantification of the loss of useful energy.

Scientists and engineers use the term efficiency to quantify the loss of useful energy. **Efficiency** is the amount of work you get from an engine, divided by the amount of energy you put into it.

In Chapter 5, we discovered that heat and other forms of energy are interchangeable and the total amount of energy is conserved. According to the *first* law of thermodynamics, there is no reason why energy in the form of heat could not be converted to electrical energy with 100% efficiency. But the *second* law of thermodynamics tells us that such a process is not possible. The flow of energy has a direction. Another way of stating this law is to say that energy always goes from a more useful to a less useful form.

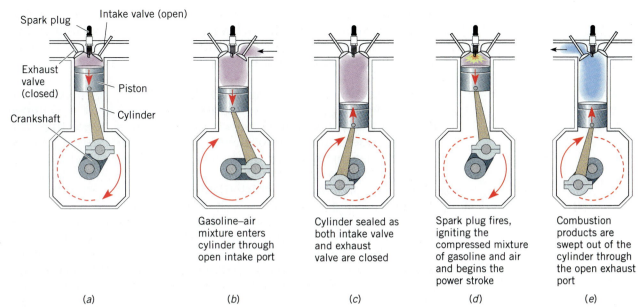

Spark plug
Intake valve (open)
Exhaust valve (closed)
Piston
Crankshaft
Cylinder

Gasoline–air mixture enters cylinder through open intake port

Cylinder sealed as both intake valve and exhaust valve are closed

Spark plug fires, igniting the compressed mixture of gasoline and air and begins the power stroke

Combustion products are swept out of the cylinder through the open exhaust port

(*a*)   (*b*)   (*c*)   (*d*)   (*e*)

Figure 6–7
The cycle of an automobile engine's piston. (*a*) The beginning of the intake stroke. (*b*) The middle of the intake stroke as a gasoline-air mixture enters the cylinder. (*c*) The beginning of the compression stroke. (*d*) At the beginning of the power stroke the spark plug fires, igniting the compressed mixture of gasoline and air. (*e*) At the beginning of the exhaust stroke combustion products are swept out. Each cycle involves two complete rotations of the crankshaft.

Your car engine provides a familiar example of this rule of nature. In the engine, a mixture of gasoline and air creates a very high-temperature, high-pressure gas that pushes down on a piston. The motion of this piston is converted into rotational motion of a series of machine parts that eventually turn the car's wheels. Some of the energy would go off as waste heat because of friction, but, in point of fact, the second law of thermodynamics would restrict our use of the energy even if friction did not exist, and even if every machine in the world was perfectly designed.

Look closely at the various stages of an engine's operation (Figure 6–7). Why couldn't heat energy in the exploding air-gas mixture be converted with 100% efficiency into energy of motion of the engine's piston? The reason is that you can't just think about the downward motion of a piston—what engineers call the *power stroke*—in the operation of your car's engine. If that was all the engine did, then the engine in every car could turn over only once. The problem is that, once you have the piston pushed all the way down, and once you have extracted all the useful work you're going to extract from the air-gas mixture, you still have to return the piston to the top of the cylinder so that the cycle can be repeated. (In actuality, the pistons in modern cars go up, down, and up again before they get back to the point where they can return energy to the system.) In order to reset the engine to its original position so that more useful work can be done, some heat has to be dumped into the environment.

Ignore for a moment the fact that a real engine is more complicated than the one we are discussing. Suppose that all you have to do is lift the piston up after the work has been done. The cylinder is full of air and, consequently, when you lift the piston up, the air will be compressed and heated. In order to return the engine to the precise state it was in before the explosion, the heat from this compressed air has to be taken away. In practice, it is expelled into the atmosphere as exhaust.

Physicists call the exploding hot gas-air mixture a *high-temperature reservoir*, and the ambient atmosphere into which this hot gas is dumped a *low-temperature reservoir*. The second law of thermodynamics says that any engine operating between two temperatures must dump some energy in the form of heat into the low-temperature reservoir. You can see how this works in the gasoline engine when the hot air produced in resetting the piston must be expelled.

The consequence of this situation is that some of the energy stored in the gasoline can be used to run the car, but some must be dumped into the low-temperature reservoir of the atmosphere. Once that heat energy has gone into the atmosphere, it can no longer be used to run the engine. That energy simply dissipates and is no longer available. Thus this version of the second law tells us that any real engine operating in the world, even an engine in which there is no friction, must waste some of the energy that goes into it.

This version of the second law explains why petroleum reserves and coal deposits play such an important role in the world economy. They are sources of *high-grade energy* that can be used to produce very high-temperature reservoirs. If these fossil fuels are burned to produce a high-temperature reservoir and generate electricity, a large portion of energy is wasted.

Although the second law applies to engines that work in cycles, it does not apply to many other uses of energy. No engine is involved if you burn natural gas to heat your home or use solar energy to heat water, for example. Consequently, these limits needn't apply. In other words, burning fossil fuels or employing solar energy to supply heat directly can be considerably more efficient than using it to generate electricity.

Science by the Numbers

Efficiency

The second law of thermodynamics can be used to calculate the maximum possible efficiency of an engine. Let's say that the high-temperature reservoir is at a temperature T_{hot} and the low-temperature reservoir is at a temperature T_{cold} (where all temperatures are measured in the Kelvin scale). The maximum theoretical efficiency—that is, the percentage of energy available to do useful work—of any engine in the real world can be calculated as follows:

▶ **In words:**

Efficiency is obtained by comparing the temperature difference between the high-temperature and low-temperature reservoirs, with the temperature of the high-temperature reservoir.

▶ **In equation form:**

$$\text{efficiency (in percent)} = \frac{(\text{temperature}_{hot} - \text{temperature}_{cold})}{\text{temperature}_{hot}} \times 100\%$$

▶ **In symbols:**

$$\text{efficiency (\%)} = \frac{T_{hot} - T_{cold}}{T_{hot}} \times 100\%$$

Any loss of energy due to friction in pulleys or gears or wheels in a real machine will make the actual efficiency less than this theoretical maximum. This maximum in the second law is *not* due to friction. This maximum is actually a very stringent constraint on real engines.

Consider the efficiency of a normal coal-fired generating plant. The temperature of the high-energy steam—the high-temperature reservoir—is about 500 K, while the temperature of the air into which waste heat must be dumped—the low-temperature reservoir—is around room temperature, or 300 K.

The maximum possible efficiency of such a plant is given by the second law to be

$$\text{efficiency (in percent)} = \frac{T_{hot} - T_{cold}}{T_{hot}} \times 100\%$$

$$= \frac{500\,\text{K} - 300\,\text{K}}{500\,\text{K}} \times 100\%$$

$$= 40.0\%$$

In other words, more than half of the energy produced in a typical coal-burning power plant must be dumped into the atmosphere as waste heat. This fundamental limit is independent of the engineers' ability to design the plant to operate efficiently. In fact, engineers have succeeded in making most generating plants operate within a few percent of the efficiency allowed by the second law of thermodynamics. The maximum possible efficiency of an engine is shown in Figure 6–8. ●

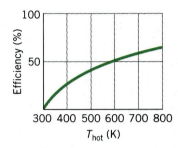

Figure 6–8
The maximum possible efficiency of an engine where the low-temperature reservoir is the outside air at a temperature of 300 K. Why does the efficiency never surpass 100%?

3. Every Isolated System Becomes More Disordered With Time

The third statement of the second law of thermodynamics is in many ways the most profound. It tells us something about the order of the universe itself. The idea of increasing disorder is perhaps the most familiar way of looking at the second law. It can be stated in colloquial form such as "the world is going to hell in a handbasket," or "things always seem to get worse." For physicists, however, this statement of the second law has a very precise meaning.

To understand what this statement means, you have to understand exactly what a physicist means by the terms "order" and "disorder" (Figure 6–9). An ordered system is one in which a number of objects, be they atoms or automobiles, are positioned in a completely regular and predictable pattern. For example, atoms in a perfect crystal or automobiles in a perfect line form highly ordered systems. A disordered system, on the other hand, contains objects that are randomly situated, without any obvious pattern. Atoms in a gas or automobiles after a multi-car pileup on a freeway are good examples of disordered systems.

A more formal definition of order and disorder requires thinking about the number of different ways a system can be arranged. Consider a garage containing exactly three spaces. Suppose you were told that there were exactly three identical cars parked in that garage. In this situation, you would know exactly what state the parking lot was in—each of the spaces would be filled by one of those three cars. This system is fully *ordered*.

Figure 6–9
Highly ordered, regular patterns of objects are less likely to occur than disordered, irregular patterns.

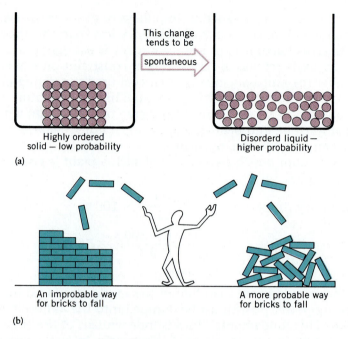

Highly ordered
solid — low probability

Disorderd liquid —
higher probability

(a)

An improbable way
for bricks to fall

A more probable way
for bricks to fall

(b)

Now, however, suppose you were told that there was only one car in the garage with three spaces. In this case, you couldn't be certain about the state of the garage. The car could be in the first parking space, the second, or the third. Thus the car could be in any of three different states. In such a situation, we would say that the garage was in a state of higher *disorder*, or, to use the technical term, higher entropy.

The disorder of a system, then, has a one-to-one correspondence with the number of states in which the system could adopt; or, equivalently, with our level of ignorance of the exact state of the system. The term entropy is used to describe this concept. **Entropy** is a measure of the disorder in a physical system. In terms of entropy, the statement of the second law reads

> **The entropy of an isolated system remains constant or increases.**

In other words, any system left to itself will change in the direction of the most disordered state. Without careful chemical controls, atoms and molecules tend to become more intermixed; without careful driving, automobiles also tend to become more disordered.

The example we gave at the beginning of the chapter, with the one fast atom in a collection of slow atoms, clearly shows how such a process works. In the most likely situation, when all of the atoms are in the same low energy states, the entropy is maximized. A much less probable situation occurs when one of the atoms is in a high-velocity state; the entropy is lower because one atom is very different from the others. Another way of saying this is that systems tend to avoid states of high improbability.

The concept of probability explains a number of paradoxes that puzzled scientists around the turn of the century. For example, it is possible

entropy The thermodynamic quantity that describes the degree of randomness of a system. The greater the disorder or randomness, the higher the statistical probability of the state, and the higher the entropy.

that all the air molecules in the room in which you are sitting could suddenly rush over to one side of the room, leaving you in a perfect vacuum. You don't worry about this happening because you know that such an event is highly unlikely. In fact, the probability is so low as to make it extremely unlikely you would see it even if you waited the entire lifetime of the universe.

While systems tend to become more disordered, the second law does not require that *every* system must approach a state of lower order. Think about water, a substance of high disorder because water molecules are arranged at random. If you put water into a freezer, it will become an ice cube, a much more ordered state in which water molecules have formed a regular crystal structure. You have caused a system to evolve to a state of higher order. How can this ordering be reconciled with the third statement of the second law?

The answer to this paradox is that the third statement refers only to systems that are isolated. The freezer in which you make the ice cubes is not an isolated system because it has a power cord that is plugged into the wall and, ultimately, connected to a generating plant. The isolated system in this case is the freezer plus the generating plant. The second law of thermodynamics says that in this particular isolated system, the total entropy must increase. However, it does not say that the entropy has to increase in all the subparts of the system.

In this example, one part of the system (the ice cube) becomes more ordered, while another part of the system (the generating plant, its burning fuel, and the surrounding air) becomes more disordered. All that the second law requires is that the amount of disorder at the generating plant be greater than the amount of order at the ice cube. As long as this requirement is met, the second law is not violated. In fact, in this particular example, the disorder at the generating plant greatly exceeds any possible order that could take place inside the freezer.

Science in the Making

The Heat Death of the Universe

The nineteenth-century discovery of the second law of thermodynamics was a gloomy event. The prevailing philosophy of the time was that life, society, and the universe in general were on a never-ending upward spiral of progress. Darwin's 1859 publication of *On the Origin of Species*, which proposed that more complex forms of life could evolve from less complex forms, reinforced this particular notion of an ever-improving world. In this optimistic climate, the discovery that the energy in the universe was being steadily and irrevocably degraded was difficult for nineteenth-century scientists and philosophers to accept.

In fact, they felt that the second law inevitably meant that all the energy in the universe would eventually be degraded into waste heat, and that everything in the universe would eventually be at the same temperature. They called this the "heat death" of the universe, and they saw it as the ultimate end of the laws of thermodynamics.

This notion even affected the literature and music of the time. For example, here is an excerpt from Algernon Swinburne's "The Garden of Proserpine":

> From too much love of living
> From hope and death set free
> We thank with weak thanksgiving
> Whatever gods there be
> That no man lives forever
> That dead men rise up never
> That even the weariest river
> Flows somewhere safe to sea.

Today, we have a rather different view of how the universe will end—a view that is discussed in detail in Chapter 22. Suffice to say, however, that Swinburne and his colleagues may have been premature in their gloom. ●

CONSEQUENCES OF THE SECOND LAW

The Arrow of Time

We live in a world of four dimensions. Three of these dimensions define space and have no obvious directionality. You can go east or west, north or south, and up or down in our universe. But the fourth dimension, time, behaves differently. Time has direction; we can never revisit the past, nor can we foresee the future.

Thomas Cole's "The Voyage of Life" is an artistic depiction of the one way direction of time.

Take one of your favorite home movies or just about any videotape and play it in reverse. Chances are that before too long you'll see something silly—something that couldn't possibly happen, that will make you laugh. Springboard divers fly out of the water and land completely dry on the diving board. From a complete stop, golf balls fly off toward the tee. Ocean spray coalesces into smooth waves that recede from shore. Most physical laws, such as Newton's laws of motion or the first law of thermodynamics, say nothing about time. The motions predicted by Newton and the conservation of energy are independent of time—they work just as well if you play a video forward or backward.

But the second law of thermodynamics is different. It takes into account a *sequence* of events. For example, heat flows from hot to cold; concentrated fuels burn to produce waste heat; the disorder of isolated systems never spontaneously decreases; we all must get older. We have established a direction to time. We experience the passage of events as dictated by the second law. Scientists cannot answer the deeper philosophical question of why we perceive the arrow of time in only one direction, but, through the second law, they can describe the effects of that directionality.

Built-in Limitations of the Universe

The second law of thermodynamics has both practical and philosophical consequences. It poses severe limits on the way that human beings can manipulate nature, and on the way that nature itself operates. It tells us that some things cannot happen in our universe.

At the practical level, it tells us that if we continue to generate electricity by burning fossil fuels or by nuclear fission, we are using up a good deal of the energy that is locked in those concentrated nonrenewable resources. These limitations are not a question of sloppy engineering or poor design—they're simply built into the laws of nature. If you could design an engine or other device that would extract energy from coal and oil with higher efficiency than the second law limits, then you could also design a refrigerator that worked when it wasn't plugged in.

At the philosophical level, the second law tells us that nature has a built-in hierarchy of more useful and less useful forms of energy. The lowest or least useful state of energy is the reservoir into which all energy eventually gets dumped. Once the energy is in that lowest energy reservoir, it can no longer be used to do work. For the Earth, energy passes through the region that supports life, the *biosphere*, but is eventually radiated into the black void of space.

THINKING MORE ABOUT
THE SECOND LAW

Evolution and Entropy

Entropy is such an important concept, so fundamental to the working of the universe, that it appears in many discussions that seem to have little to do with heat transfer. Economists, for example, use it to discuss disordered aspects of the economy, and communications engineers routinely use it to discuss the information (i.e., order) in transmissions. But perhaps the most interesting place where it crops up is in the intense debate about the origin of life on Earth.

Creationists, who believe that life appeared as a result of a single miraculous creation a few thousand years ago, point out that life is a highly ordered system in which trillions of atoms and molecules must occur in exactly the right sequence. They argue that life could not possibly have arisen spontaneously without violating the second law. How, they ask, could a natural system go spontaneously from a disordered state of non-life to the ordered state of life?

This argument fails to take into account the fact that the Earth is not itself an isolated system. The energy that drives living systems is sunlight, so that the "isolated system" that the second law speaks of is the Earth *plus the Sun*. To make the evolution of life consistent with the second law, the order observed in living things must be offset by a greater amount of disorder in the Sun. As with the earlier example of the ice cube, this requirement is easily met by the Sun and the Earth taken together.

Science cannot yet describe in detail how life arose, and some aspects of the process may never be known for sure. The scientific method, for example, is not an appropriate way to answer the question of whether God was involved in the origin of life (see Chapter 1). However, science can help explain how the development of life, as a natural process, is consistent with the universal laws of thermodynamics.

► Summary

All objects in the universe are at a *temperature* above *absolute zero*; thus they hold some heat energy, the kinetic energy of moving atoms. *Heat capacity* defines how much heat energy a substance can hold.

There are three modes of *heat transfer* between two objects that are at different temperatures. *Conduction* involves the transfer of heat energy through the collision of individual atoms and molecules; *thermal conductivity* is a measure of how easily this energy transfer occurs. *Convection* involves the motion of a mass of fluid in a *convection cell*, in which warmer atoms are physically transported from one place to another. Heat can also be transferred by *radiation*, which is *infrared energy* or other forms of light that travel across a room or across the vastness of space until they are absorbed.

The first law of thermodynamics promises that the total amount of energy never changes, no matter how you shift it from one form to another. The *second law of thermodynamics* places restrictions on how energy can be shifted. Three different but equivalent statements of the second law underscore these restrictions:

1. Heat will not flow spontaneously from a colder to a hotter body.
2. It is impossible to construct a machine that does nothing but convert heat into useful work. No engine can operate with 100% *efficiency*.
3. The *entropy* of an isolated system always increases.

► Key Equations

$$\text{efficiency (in percent)} = \frac{T_{hot} - T_{cold}}{T_{hot}} \times 100\%$$

Temperature conversions:
$$°F = (1.8 \times °C) + 32$$
$$°C = \frac{°F - 32}{1.8}$$
$$°C = K - 273$$

► Review Questions

1. Identify the three ways that heat can be transferred and give examples of each.

2. What kind of heat transfer depends only on collisions between individual atoms and molecules?

3. What kind of heat transfer depends on the bulk motion of large numbers of molecules?

4. By what process can heat be transferred across the vacuum of empty space?

5. What is the difference between temperature and heat?

6. What is heat capacity? Is it the same for all materials?

7. Describe three common temperature scales. What fixed points are used to calibrate them?

8. State the second law of thermodynamics in three different ways.

9. What is the high-temperature reservoir in your car's engine? What is the low-temperature reservoir?

10. What is entropy?

11. To what does the phrase "heat death of the universe" refer?

12. How does hair help keep humans and other mammals warm?

▶ Fill in the Blanks

Complete the following paragraph using words and phrases from the list.

absolute zero
conduction
convection
convection cell
efficiency
entropy

heat capacity
heat transfer
radiation
the second law of thermo-
    dynamics
temperature

_____ is the quantity that reflects how rapidly the atoms or molecules in a substance are moving, and not the total amount of energy in an object. The lowest possible temperature, at which no further energy can be extracted from a system, is called _____. The amount of heat energy a substance must absorb to raise its temperature one degree is called its _____, and when two objects are at different temperatures, it is possible for the process of _____ to take place between them. Heat can move from one place to another by _____, a process involving collisions between individual atoms. It can also be transferred by _____, which involves continuous motion of materials in a _____. Finally, heat can be transferred by _____, the only process by which energy can travel through empty space. The _____ explains the direction of time we observe in the universe. It tells us that heat always flows from hot to cold bodies, that the _____ of an engine (defined as useful work out divided by energy in) has a definite limit, and that the amount of disorder, or _____, in a closed system always increases.

▶ Discussion Questions

1. Why don't all the atoms in the room you're sitting in move to one side, leaving you in a vacuum?

2. Why are there big cooling stacks around nuclear reactors and coal-fired generating plants?

3. Identify three examples of the second law of thermodynamics in action that have occurred since you woke up this morning.

4. Imagine lying on a hot beach on a sunny summer day. In what different ways is heat transferred to your body? In each case what was the original source of the heat energy?

5. "Cogeneration" is a term used to describe systems in which waste heat from electrical generating plants is used to heat nearby homes with efficiencies much greater than 50%. Does cogeneration violate the second law? Why or why not?

6. Why is a perpetual-motion machine impossible?

7. Seawater is full of moving molecules that possess kinetic energy. Could we extract this energy from seawater? Why or why not?

8. Describe the kinds of heat transfer that occur while you are cooking a meal. Where does the heat energy come from? Where does it end up?

9. Why do some animals roll up into a ball when they are cold?

10. Why are feather beds warm, and why is goose down considered the best filling for a parka?

11. Why do human beings wear clothes? Compare our behavior in this regard to other warm-blooded animals.

12. In your own words, define temperature.

13. How is absolute zero related to your definition of temperature?

14. What is the difference between heat capacity and heat transfer?

15. Why is it impossible to have an engine that has 100% efficiency?

16. Outline the three major modes of heat transfer. In this outline, state if a medium is necessary, compare the motion of molecules in this medium, and relate the heat transfer to the presence or absence of a temperature difference in the medium.

17. When ice freezes, water goes from a state of more disorder to one with more order. Does this violate the second law of thermodynamics? Explain.

18. In the equation defining efficiency, why must you always use the Kelvin temperature scale? (*Hint*: Consider the sign of each temperature, and also dividing by zero.)

19. The average specific heat of the human body is 83% of the heat capacity of water. This is higher than most other solids, liquids, or gases. Why do you think the heat capacity of the human body is closest to water?

20. Human beings must lose heat so their internal temperatures do not increase substantially above 37°C. One mechanism to lose heat is sweating. Explain why this is an efficient mechanism to lose energy from the body.

▶ Problems

1. Calculate the maximum possible efficiency of a power plant that burns natural gas at a temperature of 600 K, with low-temperature surroundings at 300 K. How much more efficient would the plant be if it were built in the Arctic, where the low-temperature reservoir is at 250 K? Why don't we build all power plants in the Arctic?

2. The Ocean Thermal Electric Conversion system (OTEC) is an example of a high-tech electrical generator. It takes advantage of the fact that in the tropics, deep ocean water is at a temperature of 4°C, while the surface is at a temperature around 25°C. The idea is to find a material that boils between these temperatures. The material in the fluid form is brought up through a large pipe from the depths, and the expansion associated with its boiling is used to drive an electrical turbine. The gas is then pumped back to the depths, where it condenses back into a liquid and the whole process repeats.

    a. What is the maximum efficiency with which OTEC can produce electricity? (*Hint:* Remember to convert all temperatures to the Kelvin scale.)

    b. Why do you suppose engineers are willing to pursue the scheme, given your answer in *a*?

    c. What is the ultimate source of the energy generated by OTEC?

3. Convert the following Fahrenheit temperatures to Kelvin.

    a. 120°F

    b. −40°F

    c. 11,500°F

    d. −456°F

4. Convert the following Celsius temperatures to Fahrenheit.

    a. 300°C

    b. −180°C

    c. 6000°C

    d. 40°C

5. Convert the following Kelvin temperatures to Celsius.

    a. 80 K

    b. 300 K

    c. 6000 K

    d. 545 K

6. At what temperature are the Celsius and Fahrenheit values the same?

7. A pound of crushed ice, initially at 0°C, was placed on the kitchen table and melted.

    a. How much energy is needed to melt the ice?

    b. Where does the ice get this energy?

    c. If the same amount of energy calculated in part *a* was used to heat water at 0°C to 50°C, how much water could be heated?

8. The heat capacity of the average human body is 83% of the heat capacity of water. How much energy would it take to raise the temperature of a 60-kg human being by 1°F? By 1°C? By 1 Kelvin?

9. During a 1-hr jog, Patricia (55 kg) will increase her metabolism by 330 W. Her body will lose heat at a rate of 310 W.

    a. What is the total energy generated by Patricia during this jog in 1 hour?

    b. How much heat is lost by Patricia in 1 hour?

    c. What is the increase in internal temperature of Patricia after her 1-hour jog?

10. The Golden Gate Bridge is 1280 meters long and made out of steel beams. If the temperature of San Francisco varies from 40°F to 80°F throughout the year, at what temperature will the bridge be the shortest? The longest?

11. A 1.25-kg steel ball is dropped from the top of a building, 40 feet high, and lands directly on the pavement. If all of the energy is retained by the ball to increase its temperature, what is this temperature increase?

12. Many solar homes use concrete and stone blocks to store excess heat (see Chapter 5). The heat capacity of concrete and stone is 31.7 BTU/ft³°F, and that of water is 62.4 BTU/ft³°F. Carlos uses approximately 800 ft³ of concrete and stone for heat storage in his solar home, where the average temperature difference between day and night in the solar room is 100°F and 75°F. How much additional energy in the form of heat can Carlos's concrete block store?

13. Generally, the human body loses heat by evaporation from exhalation and through the skin. The amount of water loss is about 0.65 kg per day.

    a. What is the total amount of heat lost per day by evaporation?

    b. What is the rate of heat loss in watts?

    c. What percent of total daily energy consumption for a human being is this?

▶ Investigations

1. Research the daily high and low temperatures for the past week in a nearby city and one of its surrounding towns. On average, what is the difference in temperature? What causes this difference?

2. What kind of insulation is installed in your home? What could you do to improve your home's insulation?

3. Get an aluminum cup, a coffee mug, and a plastic drinking cup. Simply by feeling these objects, can you guess their relative thermal conductivities? What experiment might you perform to test your guess?

4. Visit a building supply store and look at the doors and windows they sell. What steps have manufacturers taken to reduce heat flow from homes?

5. Play a home movie or videotape backward. How many violations of the second law of thermodynamics can you spot?

▶ Additional Reading

Atkins, P. W. *The Second Law*. New York: Scientific American Library, 1984.

7 ELECTRICITY AND MAGNETISM

ELECTRICITY AND MAGNETISM ARE TWO DIFFERENT
ASPECTS OF ONE FORCE—THE ELECTROMAGNETIC
FORCE.

Late for Work at the Copy Center

Imagine waking up on a cold, dry winter day and realizing you're late for work. You rush to throw on your clothes and comb your hair with rapid, vigorous strokes. When you look in the mirror, you notice that the folds of your shirt are sticking together and your hair is standing on end. At work, you rush to photocopy an important report, only to find the copies sticking to each other, slowing your efforts. Static electricity is at work.

Imagine walking home from work on a hot, humid summer day. The sky is dark and you hear the distant rumble of thunder. Suddenly, a jagged lightning bolt slices the horizon. You hurry to get home before the thunderstorm. Static electricity has, quite literally, struck again.

The phenomenon that causes your clothes to stick together and your hair to stand on end in the cold, dry winter is the same phenomenon that causes lightning in the hot, humid summer. This force is called electromagnetism.

NATURE'S OTHER FORCES

According to Newton's laws of motion, nothing happens without a force; however, Newton's law of gravity cannot explain many everyday occurrences. How does a refrigerator magnet cling to metal, thereby defying gravity? How does a compass needle swing around to the north? How can static cling wrinkle your shirt? How can lightning shatter an old tree? All of these phenomena involve a force that is different from gravity.

Newton may not have known about these forces, but he did give us a method for studying them. First observe natural phenomena and learn how they behave, then organize those observations into a series of natural laws, and finally use those laws to predict future behavior of the physical world. This process is known as the scientific method, which you learned about in Chapter 1.

In particular, Newton's first law of motion (see Chapter 4) will be useful in our investigation of nature's other forces. According to this law, whenever we see a change in the state of motion of any material object, we know that a force has acted to produce that change. Thus, whenever we see such a change and can rule out the action of known forces such as gravity, we can conclude that the change must have been caused by an unknown force. We shall use this line of reasoning to show that both the electric and magnetic forces exist in the natural world.

Our understanding of the phenomena associated with static electricity and magnetism began in the eighteenth century with a group of scientists in Europe and North America who called themselves "electricians." These researchers were fascinated by the many curious phenomena associated with nature's unknown forces. Their thoughts were not focused on practical applications, nor could they have imagined how their work would help transform the world.

When the girl touched the electrically charged sphere, her hair became electrically charged as well. Individual hairs repel each other and thus "stand on end."

STATIC ELECTRICITY

Many phenomena related to electricity have been known since ancient times. The Greeks knew that if they rubbed a piece of amber (hardened tree resin) with cat's fur and then touched other objects with the amber, those other objects would repel each other. The same thing would happen, they found, if they rubbed a piece of glass with silk: objects that the glass touched repelled each other. On the other hand, if objects that had been touched with the amber were brought near objects touched with the glass, they were attracted toward each other. Objects that behave in this way are said to possess **electric charge,** or to be charged.

The force that moves objects toward and away from each other in these simple demonstrations was given the name of **electricity** (from the Greek word for amber). In these simple experiments, it was shown that the electric charge doesn't move once it has been placed on an object. This force is thus called **static electricity** or the *electric force.*

The electric force is different from gravity, in that gravity can never be repulsive. When a gravitational force acts between two objects, it always pulls them together. The electric force, on the other hand, can attract some objects toward each other and push other objects away from

electric charge An excess or deficit of electrons on an object.

electricity A force, more powerful than gravity, that moves objects both toward and away from each other, depending upon their charge.

static electricity A manifestation of the electric force, caused by the transfer of electrons between objects: often observed as lightning, or as sparks produced when walking across a wool rug on a cold day.

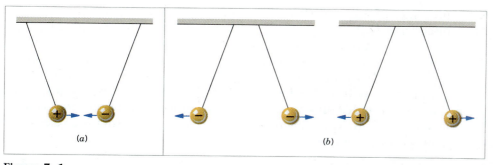

Figure 7–1
(*a*) If charges on the two objects are opposite, they will attract each other.
(*b*) If charges on the two objects are the same, they will repel each other.

each other. Furthermore, for most everyday objects the electric force between protons and electrons is vastly more powerful than gravity. A simple comb charged with static electricity easily lifts a piece of paper against the gravitational pull of the entire Earth. (Try this on a cold day: run a comb through your hair and bring it close to a small piece of tissue paper.)

Today we understand that the properties of the electrical force arise from the existence of two kinds of electric charge (Figure 7–1). We say that objects touched with the same source, whether it's amber or glass, have the same kind of electrical charge and are repelled from each other. On the other hand, one object touched with amber will have a different kind of electrical charge than a second object touched with glass. These objects are then attracted to each other.

Science in the Making

Benjamin Franklin and Electric Charge

The most famous North American "electrician" was Benjamin Franklin (1706–1790), one of the pioneers of electrical science, as well as a central figure in the founding of the United States of America (Figure 7–2). Franklin began his electrical experiments in 1746 with a study of electricity generated by friction. Most scientists of the time thought that electrical effects resulted from the interaction of two different "electrical fluids." Franklin, however, became convinced that all electrical phenomena could be explained as the result of a *single* electrical fluid being transferred from one object to another. He characterized these objects as having an excess or a deficiency of this fluid; to describe each of these conditions, he said that one was "negative" and the other one "positive." There was no particular significance to this choice of words—he could have chosen red and green, up and down, or any other combination to indicate the two states.

In June 1752, Franklin performed his famous (and extremely dangerous) experiment, in which he flew a kite in a thunderstorm. A rather mild lightning stroke probably hit his kite and passed along the wet string to produce sparks and an electric shock—proof that light-

Figure 7–2
Benjamin Franklin engaged in his famous (and potentially lethal) kite experiment.

ning was a form of electricity. Franklin followed up his experimental discoveries with the invention of the lightning rod, a simple device designed to reduce the incidence of fires. The lightning rod is a metal rod with one end in the ground and the other sticking up above the roof of a building. This rod carries electrical charge into the earth, diverting it away from the building. Lightning rods caught on quickly in the wooden cities of North America and Europe, and prevented countless deadly fires. The descendants of Franklin's lightning rods are in widespread use today. ●

(a)

(b)

(a) Since Franklin's discovery of the electrical nature of lightning, rods like this one have protected buildings around the world.
(b) Lightning hitting a car. Note the flow of charge to the ground from the front wheel. As a result, the person inside is safe.

The Movement of Electrons

We now understand that there are two kinds of electric charge—positive and negative. In modern language, we say that all objects are made up of minute building blocks called atoms, and all atoms are made up of still smaller particles that have electric charge. As you will see in Chapter 9, in every atom, negatively charged *electrons* move around a heavy, positively charged *nucleus* near the center, much as planets move around the Sun. The electrons and the nucleus have opposite electric charges, so an attractive force exists between them. This force plays roughly the same role that gravity plays in keeping the solar system together. Most atoms are electrically neutral, because equal numbers of positive protons and negative electrons cancel each other's charge.

Electrons, particularly electrons further from the nucleus, tend to be rather loosely bound to their nucleus. These electrons may be knocked loose in collisions and, once they're knocked loose, they can move around freely through the material.

When electrons are pulled out of a material—for example, by friction—they no longer cancel the positive charges in the nucleus. The result is a net excess of positive charge in the object. When this occurs, the object as a whole acquires a **positive electric charge.** Similarly, an object acquires a **negative electric charge** when extra electrons are pushed into it. This is what happens to a comb when you run it through your hair on a dry day. Electrons are transferred from your hair into the comb, so it acquires a negative charge. Your hair loses electrons,

positive electric charge   A deficiency of electrons on an object.

negative electric charge   An excess of electrons on an object.

so individual strands becomes positively charged. As Benjamin Franklin discovered more than 250 years ago, the processes of acquiring both positive and negative charge results from the flow of electrons—the particles that constitute Franklin's single electric "fluid."

During a thunderstorm, the same kind of electron transfer occurs on a much larger scale, as wind and rain disrupt the normal distribution of electrons in clouds. When a charged cloud passes over a tall tree or tower, the violent electric discharge, called lightning, results from the attraction of the positive charges on the ground and negative charges in the cloud (although in the case of lightning, both the positive and negative charges move).

Although historical investigations of electric charge tended to concentrate on somewhat artificial experiments, we now know that electrically charged particles play an important role in many natural systems. Virtually all of the atoms in the Sun, for example, have lost electrons due to violent collisions with other atoms at extreme temperatures. The Sun's positively charged atoms are thus surrounded by a sea of unattached electrons (though the Sun is neutral overall). In the cells of all advanced life forms, including human beings, charged atoms are routinely moved into and out of cells to maintain the processes of life. In fact, as you read these words, charged potassium and sodium atoms are moving across the membranes of cells in your optic nerve to carry signals to your brain.

Coulomb's Law

The phenomenon of electricity remained something of a mild curiosity until the mid-eighteenth century, when scientists began applying the scientific method to investigate it. One of the first things that had to be done was to propose a precise statement about the nature of electric force. French scientist Charles Augustin de Coulomb (1736–1806) is renowned for his work in this area. During the 1780s, at the same time the United States Constitution was being written by Benjamin Franklin and others, Coulomb devised a series of experiments in which he passed different amounts of electric charge onto objects and then measured the force between them (see Figure 7–3). After repeated careful measurements, he discovered that the electrical force was in some ways very similar to the gravitational force that Isaac Newton had discovered a century earlier.

Charles Augustine de Coulomb (1736–1806)

Coulomb summarized his discoveries in a simple relationship known as **Coulomb's Law,** which states:

▶ **In words:**

> The force between any two electrically charged objects is proportional to the product of their charges divided by the square of the distance between them.

▶ **In equation form:**

$$\text{force (in newtons)} = k \times \frac{(\text{1st charge} \times \text{2nd charge})}{\text{distance}^2}$$

Figure 7–3
Under normal circum-
stances, there are as many
positive as negative charges
in a material (*a*). When a
negatively charged rod is
brought near (*b*), the negatively charged particles (electrons) are pushed away,
leaving a material with
more positive than negative
charges (*c*).

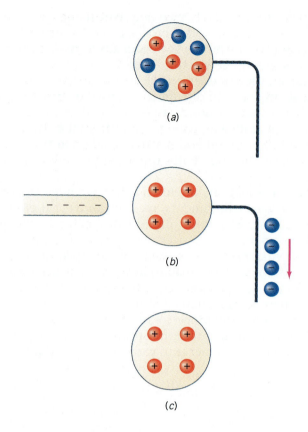

(*a*)

(*b*)

(*c*)

▶ **In symbols:**

$$F = k \times \frac{(q_1 \times q_2)}{d^2}$$

where distance *d* is measured in meters, charge *q* is measured in a unit
called the coulomb (see below), and *k* is the *Coulomb constant*. The Coulomb constant plays the same role in electricity that the gravitational
constant, *G*, plays in gravity. Like *G*, *k* is a number (9.00×10^9 newton-
meter^2/coulomb^2 in one common system of units) that can be determined
experimentally, and which turns out to be the same for all charges and
all separations of those charges anywhere in the universe.

Coulomb observed that if two electrically charged objects are moved
farther and farther away from each other, the force between them gets
smaller and smaller, just like gravity (see Figure 7–4). In fact, if the distance
between two objects is doubled, the force decreases by a factor of four;
this is the 1/distance^2, or inverse square, relationship that we saw in the
law of universal gravitation in Chapter 3. The farther you are from a
charge, the less you are affected by it. Coulomb also discovered that
the size of the force depends on the product of the charges of the two
objects—double the charge on one object and the force doubles; double
the charge on both objects and the force increases by a factor of four.
Coulomb's law summarizes a large number of experiments done on stationary electrical charges. It can also be thought of as the summary of
the behavior of static electricity.

Figure 7–4
The electrical force obeys
an inverse square law (see
Chapter 2).

In order to be able to draw this conclusion, scientists had to define a unit of electrical charge. It is now called the *coulomb* (abbreviated C) after the scientist who did so much of this important work. A coulomb is approximately equal to the charge on 6.3×10^{18} electrons—a very large number, indeed. When this many electrons have been moved onto an object or removed from it, that object is said to have one coulomb of electrical charge.

Though you might not realize it, you see Coulomb's law in action every day in the behavior of ordinary water. Water molecules, which are composed of two hydrogen atoms linked to one oxygen atom, have a negatively charged end and a positively charged end. This phenomenon is known as *polarity*. When you sprinkle ordinary salt crystals—the compound sodium chloride—into water, positively charged sodium atoms and negatively charged chlorine atoms are pulled apart by electrical forces. The salt dissolves.

The behavior of salt contrasts with that of plastic, which does not dissolve in water. Since plastics are not made of particles that have negatively and positively charged ends, water cannot exert an electrical force and pull them apart.

Science by the Numbers

Two Forces Compared

In Chapter 9 we will examine compelling evidence that the atoms that compose all the materials in our physical surroundings have a definite internal structure. Recall that atoms are made up of negatively charged electrons and a positively charged nucleus. When electrons circle in orbits around a positively charged nucleus, two different forces act within an atom: the force of gravity and the electrical force. What is the relative strength of these two forces?

The simplest atom is hydrogen, in which a single electron circles a single positively charged particle known as a proton (see Chapter 9). The masses of the electron and proton are 9×10^{-31} kg and 1.7×10^{-27} kg, respectively. The charge on the proton is 1.6×10^{-19} C, and the charge on the electron has the same magnitude, but is negative. A typical separation of these two particles in an atom is 10^{-10} m.

Given these numbers, we can calculate the values of the electrical and gravitational attractions between the two particles. ●

Gravity.   The force of gravity between the two particles will be

force of gravity (in newtons)

$$= G \times \frac{\text{mass}_1 \times \text{mass}_2 \text{ (in kg)}}{[\text{distance (in m)}]^2}$$

$$= (6.7 \times 10^{-11} \text{ m}^3/\text{kg-s}^2) \times \frac{(1.7 \times 10^{-27} \text{ kg}) \times (9 \times 10^{-31} \text{ kg})}{(10^{-10} \text{ m})^2}$$

$$= 1.0 \times 10^{-47} \text{ N}$$

Electricity. The electrical force is given by Coulomb's law to be

force of static electricity (in newtons)

$$= k \times \frac{\text{charge}_1 \times \text{charge}_2 \text{ (in C)}}{[\text{distance (in m)}]^2}$$

$$= (9 \times 10^9 \text{ N-m}^2/\text{C}^2) \times \frac{(1.7 \times 10^{-19} \text{ C}) \times (1.7 \times 10^{-19} \text{ C})}{(10^{-10} \text{ m})^2}$$

$$= 2.6 \times 10^{-8} \text{ N}$$

Ratio of Electrical and Gravitational Forces. The ratio of these forces is given by:

$$\frac{\text{electrical force}}{\text{gravitational force}} = \frac{2.6 \times 10^{-8} \text{ N}}{1.0 \times 10^{-47} \text{ N}} = 2.6 \times 10^{39}$$

From this calculation we can see that the electric force in the atom is almost 40 orders of magnitude larger than the gravitational force. For this reason, in our discussion of the atom in subsequent chapters, we will ignore the effects of gravity completely. ●

Developing Your Intuition

Lightning Strikes

People who have been on the ground near a place where lightning strikes often report that the hair on the back of their necks stood up just before the bolt hit. Is this just a folk tale, or could there be a physical basis for it?

Just before a lightning strike, a cloud with a large negative charge must be overhead (Figure 7–5). The negative charges exert a force on electrons in everything beneath, including those in your hair. These electrons are repelled by the Coulomb force and are pushed away, leaving behind positive charges. The individual hairs, now charged, will repel each other, and the "standing up" is the result of their moving away from each other. ●

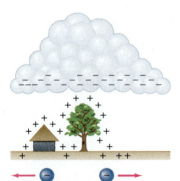

Figure 7–5
Before a lightning strike, negative charges in the lower layers of clouds drive electrons away from objects on the ground, leaving them with a positive charge.

The Electric Field

Suppose an electrically charged object such as a piece of lint is sitting at a point (Figure 7–6). If you brought a second charged object to a spot near that piece of lint, the second object would feel a force. If you then moved that second object to another spot near the lint, it would still feel a force, but that force would, in general, be of a different size and would point in a different direction than it did at the first spot. In fact, the second charged object would feel a force at every point in space around the piece of lint.

Figure 7–6 illustrates, in two dimensions, the distribution of these forces in the space around the charged lint particle. The arrow at each point around the lint represents the force that would be felt by another charged object, if that object were brought to the point in space where the arrow originates. The collection of all the arrows that represent these

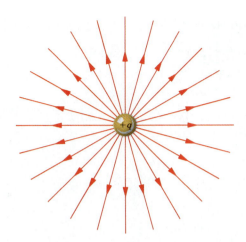

Figure 7–6
An electric field sur-
rounding a positive charge,
+q, may be represented by
lines of force radiating outward. Any charged object
that approaches +q, experiences a greater and greater
electrical force the closer it
gets. Positively charged objects will be repelled, while
negatively charged objects
will be attracted.

forces is called the **electric field** of the original charged object. Every charged object is surrounded by such a field. Notice that the electric field is defined as the force that *would* be experienced by another charge if that charge were located at a particular point, so that the field is present even if no other charge is in the region.

Technically, the electric field at a point is defined to be the force that would be experienced by a positive 1–coulomb charge if it were brought up to that point. The field is usually drawn so that the direction of the arrow corresponds to the direction of that force, and the length of the arrow to the magnitude of the force.

electric field The force that would be exerted on a positive charge if it were brought to a position near a charged object. Every charged object is sur-rounded by an electric field.

MAGNETISM

Just as electrical phenomena were known to the ancient philosophers, so too were the phenomena we call magnetism. The first known magnets were naturally occurring iron minerals. If one of these minerals (a com-mon one is magnetite or "lodestone") is placed near a piece of iron, the iron will be attracted to it. If you have seen experiments in which magnets attracted nails, which jumped up and hung from them, you have seen magnetism in action.

Magnetism is another force in nature—a force that acts differently from electricity or gravity. Electrical attraction doesn't make the nails move, nor is it gravity that causes the nails to jump up. The simple experiment of picking up a nail with a magnet illustrates the **magnetic force,** which can be identified and described using the scientific method.

magnetic force The force ex-erted by magnets on each other.

While electricity remained a curiosity until well into the nineteenth century, magnetism was put to practical use many centuries ago. The *compass*, invented in China and used to navigate the oceans by Europeans during the age of exploration, is the first magnetic device on record. A sliver of lodestone, left free to rotate, will line itself up in a north-south direction.

In the late sixteenth century, English scientist William Gilbert (1544–1603) conducted the first serious study of magnets. Though revered in his day as a doctor (he was physician to both Queen Elizabeth I and King James I), his lasting fame came from his discovery that every magnet can be characterized by what he called **poles.** If you take a piece of naturally

poles The two opposite ends of a magnet, named north and south, which repel a like mag-netic pole and attract an unlike magnetic pole.

William Gilbert (1544–1603) illustrates the properties of magnets.

occurring magnet and let it rotate, one end points north and the other end points south. These poles have been given the labels *north* and *south*, based on the direction that a magnetized compass needle will point.

In the course of his research, Gilbert discovered many important properties of magnets. He learned to magnetize iron and steel rods by stroking them with a lodestone. He discovered that hammered iron becomes magnetic and found that iron's magnetism can be destroyed when heated. He realized that the Earth, itself, is a giant magnet, a fact that, as we shall see, explains how a compass works. Gilbert also documented many of the most basic aspects of the magnetic force. He found that if two magnets are brought near each other so that the north poles are close together, a repulsive force develops between the magnets and they move apart. The same thing happens if two south poles are brought together. If, however, the north pole of one magnet is brought near the south pole of another magnet, the resulting force is attractive. In this respect, magnetism seems to mimic the eighteenth-century studies of static electricity. William Gilbert's results can be summarized in two simple statements.

> **Every magnet has at least two poles.**
> **Like magnetic poles repel each other; unlike poles attract.**

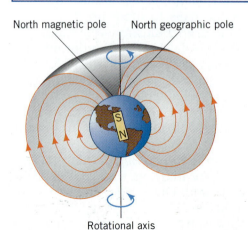

Figure 7–7
A compass needle and the Earth. Any magnet will twist because of the forces between its poles and those of the Earth.

Figure 7–8
A magnetic field. Small magnets placed near a large one orient themselves along the lines of the magnetic field, as shown.

Once you know that a magnet has two poles, then you can understand how a compass behaves. The Earth itself is a giant magnet, with one pole in Canada and the other pole in Antarctica. If a magnetized iron compass needle is allowed to rotate freely, one of its poles will be attracted and twist around toward Canada in the north, and the other end will point to Antarctica in the south (see Figure 7–7).

Just as the electric force can be represented in terms of an electric field, so too can the magnetic force be represented in terms of a **magnetic field.** If a small compass needle is brought near a magnet as shown in Figure 7–8, the forces exerted by the magnet will twist the needle around. The direction of the needle will, in general, be different at different locations around the magnet.

magnetic field   A collection of lines that map out the direction that compass needles would point in the vicinity of a magnet.

Magnetic Navigation

Humans aren't the only navigators who make use of the Earth's magnetic field. The fact that the Earth is surrounded by a magnetic field with north and south poles is used by many living things to guide their movements. This fact was established by scientists at the Massachusetts Institute of Technology in 1975, when they were studying a single-celled bacterium that lives in the ooze at the bottom of nearby swamps. They found about 20 tiny pieces of an iron ore called *magnetite* (the material often found in natural magnets) in their bodies. These pieces were strung out in a line, in effect forming a microscopic compass needle.

It turns out that the Earth's magnetic field lines dip into the surface in the Northern Hemisphere, and they rise up out of the surface in the Southern Hemisphere. This feature gives the Massachusetts bacterium a built-in magnetic "up" and "down" indicator. Its internal magnets are used to navigate down into the nutrient-rich ooze at the bottom of the pond. Related bacteria in the Southern Hemisphere, similarly, follow the field lines in the opposite magnetic direction to get to the bottom of their ponds.

Since 1975, similar internal magnets have been discovered in many animals. Some migratory birds, for example, use internal magnets as one of several cues to guide them on flights thousands of miles in length. In the case of the Australian silvereye, the bird can apparently "see" the magnetic field of the Earth, through a process involving modification of molecules that are normally involved in color vision.

Grains of iron minerals in this bacterium allow it to tell up from down.

A ship's compass tells the pilot the direction of the north magnetic pole.

The Dipole Field

The magnetic field shown in Figure 7–9 plays a very important role in nature. We call this field, which arises whenever a magnet has both a north and a south magnetic pole, the *dipole field,* because the bar magnet has two poles. All magnets found in nature have both a north and a south pole; one is never found without the other. Even if you take an ordinary bar magnet and cut it in two, you don't get a north and a south pole in isolation, but rather two small magnets each with a north and a

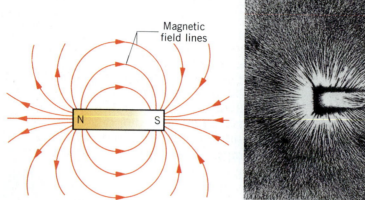
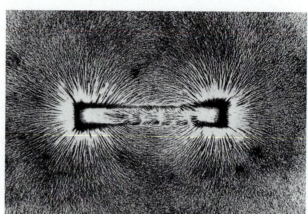

Figure 7–9
(*a*) A bar magnet and its magnetic dipole field. (*b*) Iron filings placed near a bar magnet align themselves along the field.

south pole (Figure 7–10). If you took each of those halves and cut them in half, you would continue to get smaller and smaller dipole magnets. In fact, it is a general rule that

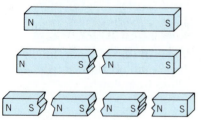

Figure 7–10
If you break a dipole magnet in two, you get two smaller dipole magnets, not an isolated north or south pole.

> **There are no isolated magnetic poles in nature.**

According to physicists, a single isolated north or south magnetic pole would be called a *magnetic monopole.* Although physicists have conducted extensive searches for monopoles, no experiment has yet found unequivocal evidence for their existence. In the next section, we'll see why.

CONNECTIONS BETWEEN ELECTRICITY AND MAGNETISM

In our everyday experience, static electricity and magnetism seem to be two unrelated phenomena. Yet scientists in the nineteenth century, probing deeper into the electric and magnetic forces, discovered remarkable connections between the two. The discovery transformed every aspect of technology.

Science in the Making

Luigi Galvani and Life's Electric Force

Scientists of the late eighteenth century discovered remarkable links between life and electricity. Of all the phenomena in nature, none fascinated these scientists more than life, the mysterious force that allowed animals to move and grow. An old doctrine called "vitalism" held that there was a special force found only in living things, and not found in the rest of nature. Luigi Galvani (1737–1798), an Italian physician and anatomist, added fuel to the debate about the nature of life with a series of classic experiments demonstrating the effects of electricity on living things.

Luigi Galvani (1737–1798)

His most famous investigations employed an electric spark to induce convulsive twitching in amputated frogs' legs—a phenomenon similar to a person's involuntary reaction to a jolt of electricity. Later, he was able to produce a similar effect simply by poking a frog's leg simultaneously with one fork of copper and one of iron. In modern language, we would say that the electric charge and the presence of the two metals in the salty fluid in the frog's leg caused a flow of electrical charge in the frog's nerves, a process that caused contractions of the muscles. Galvani, however, argued that his experiments showed that there was a vital force in living systems—something he called "animal electricity"—that made them different from inanimate matter. This idea gained some acceptance among the scientific community, but provoked a long debate between Galvani and Italian physicist Allesandro Volta (1745–1827). Volta argued that Galvani's effects had to do with chemical reactions between the metals and the salty fluids of the frog's legs.

In retrospect, we can see that both of these scientists were partly correct. Muscle contractions are indeed initiated by electrical signals, even if there is no such thing as animal electricity, and electrical charges can be induced to flow by the influence of chemical reactions.

The idea behind the legend of Frankenstein may have been suggested by early experiments on animal electricity.

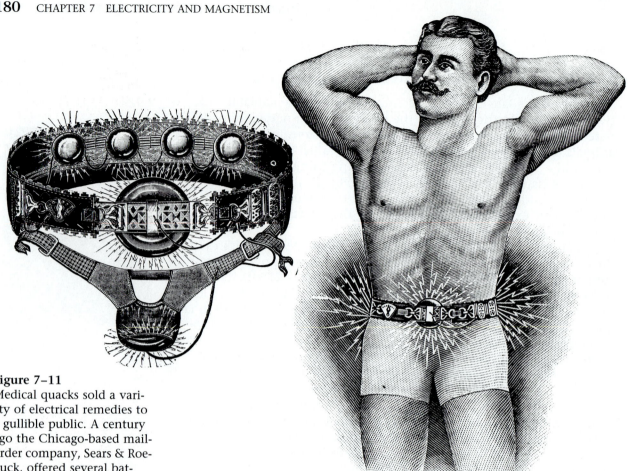

Figure 7–11
Medical quacks sold a variety of electrical remedies to a gullible public. A century ago the Chicago-based mail-order company, Sears & Roebuck, offered several battery-operated therapeutic devices, including this bizarre electrical belt, guaranteed to "restore manly vigor."

The controversy that surrounded Galvani's experiments had many surprising effects. On the practical side, Volta's work on chemical reactions led to the invention of the battery. Indirectly, it led to our modern understanding of electricity. The notion of animal electricity also proved a great boon to medical quacks and con men; for centuries, various kinds of electrical devices were palmed off on the public as cures for almost every known disease (Figure 7–11).

Finally, in a bizarre epilogue to Galvani's research, other researchers used batteries to study the effects of electric currents on human cadavers. In one famous public demonstration, a corpse was made to sit up and kick its legs by electrical stimulation. Such experiments inspired Mary Shelley's famous novel, *Frankenstein*. ●

Batteries and Electric Current

Although we frequently encounter static electricity, most of our contact with electricity comes from moving charges. In your home, for example, negatively charged electrons move through wires and run all of your electrical appliances. A flow of charged particles is called an **electric current.**

electric current A flow of charged particles, measured in amperes.

Until the work of Allessandro Volta, scientists could not produce persistent electrical currents in their laboratories and, therefore, knew little about them. As a result of his investigations into Galvani's work,

Volta developed the first **battery**—a device that converts the stored chemical energy in the battery materials into kinetic energy of electrons running through an outside wire.

The first battery was rather crude, but today you use batteries for everything from running your Discman to starting your car. In fact, the car battery is a reliable and beautifully engineered device that routinely performs for years before it needs replacing. It is made of alternating plates of two kinds of material, lead and lead oxide, immersed in a bath of sulfuric acid. When the battery is being discharged, the lead plate interacts with the acid, producing lead sulfate (the white crud that collects around the posts of old batteries) and some free electrons. These electrons run through an external wire to the other plates, where they interact with the lead oxide and sulfuric acid to form lead sulfate. The electrons running through the outside wire start your car.

When the battery is completely run down, or *discharged*, it consists of plates of lead sulfate immersed in water, a situation from which no energy can be obtained. Running a current backward through the battery, however, runs all the chemical reactions in reverse and restores the original configuration. This process is known as *recharging* the battery. Once this is done, the whole cycle can proceed again. In your car, the generator constantly recharges the battery whenever the engine is running, so that it is always ready to use.

> **battery** A device that converts stored chemical energy into kinetic energy of charged particles (usually electrons) running through an outside wire.

MAGNETIC EFFECTS FROM ELECTRICITY

In the spring of 1820, a strange thing happened during a physics lecture in Denmark. The lecturer, a professor by the name of Hans Christian Oersted (1777–1851), was using a battery to demonstrate some properties of electricity. By chance, he noticed that whenever he connected the battery (so that an electric current began to flow), a compass needle on a nearby table began to twitch and move. When he disconnected the battery, the needle went back to pointing north. This accidental discovery led the way to one of the most profound insights in the history of science. Oersted had discovered nothing less than the fact that electricity and magnetism—two forces that seem as different from each other as night and day—are, in fact, closely related.

In subsequent studies, Oersted and his colleagues established that whenever an electrical charge flows through a wire, a magnetic field will occur around that wire. A compass brought near the wire will twist around until it points along the direction of that field. This leads to an important experimental finding in electricity and magnetism:

Hans Christian Oersted
(1770–1851)

> **Magnetic fields can be created by the motion of electrical charges.**

Like all fundamental discoveries, the discovery of this law of nature has important practical consequences. Perhaps most significantly, it led to the development of the **electromagnet:** a device composed of a coil of wire that produces a magnetic field whenever an electrical charge runs through the wire. Almost every electrical appliance in modern technology uses this device.

> **electromagnet** A device that produces a magnetic field from a moving electrical charge.

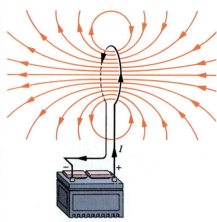

Figure 7–12
A schematic drawing of an electromagnet reveals the principal components: a loop of wire and a source of electric current. When a current flows through the wire loop, a magnetic field is created around it.

The Electromagnet

Electromagnets work on a very simple principle, as illustrated in Figure 7–12. If an electric current flows in a coil of wire, then a magnetic field will be created around the wire, just as Oersted discovered in 1820. That magnetic field will have the shape of a dipole magnetic field as shown in Figure 7–9.

In other words, we can create an electromagnet that is the equivalent of a piece of magnetite or magnetized piece of iron, simply by running electrical current around a loop of wire. The stronger the current (that is, the more electrical charge we push through the wire), the stronger the magnetic field will be. But unlike a piece of magnetite or a bar magnet, an electromagnet can be turned on and off. To differentiate between these two sorts of magnets, we often refer to magnets made from materials such as iron as permanent magnets.

The electromagnet can be used in many practical ways, including buzzers, switches, and electric motors. In each of these devices, a piece of iron is placed near a magnet. When the current flows in the loops of wire, the iron is pulled toward the magnet. In some cases, the electromagnet can be used to complete an electrical circuit by pulling a switch closed. As soon as the current is turned off in the electromagnet, a spring pushes the iron back, and the current in the larger circuit also shuts off.

Electromagnetic switches are found in many common household appliances. For example, your home is probably heated by a furnace that is linked to a thermostat on the wall in your living room or hallway. You set the thermostat to a specific temperature. If the temperature in the room falls below the temperature you set, the thermostat responds by using an electromagnet to close a switch, allowing a small current to flow. While the current is flowing in the electromagnet, the switch is closed and the furnace operates, heating the rooms in your house. As soon as the temperature reaches the level you have set, however, the thermostat stops the current that flows to the electromagnet, the switch opens, and the furnace shuts off. In this way, you can adjust the temperature of your house without having to watch it constantly and run to the basement every time you need to turn on the furnace.

Electromagnets, which can be turned on and off, are ideal for moving scrap iron at a junkyard.

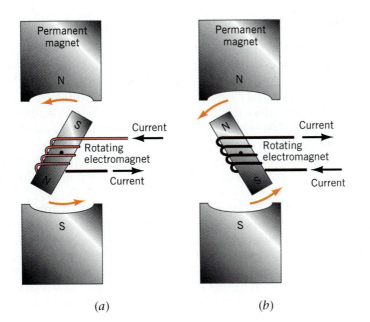

(a) (b)

Figure 7–13
An electric motor. The simplest motors work by placing an electromagnet that can rotate between two permanent magnets. (*a*) When the current is turned on, the north and south poles of the electromagnet are attracted to the south and north poles of the permanent magnets. (*b*) As the electromagnet rotates, the current direction is switched, causing the electromagnet to continue rotating.

Technology

The Electric Motor

All **electric motors**, which play a key role in modern society, incorporate electromagnets. They all operate by supplying a current to an electromagnet to make it move and generate mechanical power. The motor employs a permanent magnet, as shown in Figure 7–13, and a rotating loop of wire inside the poles of those magnets. The current in the rotating loop adjusts so that when it is oriented as shown in Figure 7–13*a*, the south pole of the electromagnet lies near the north pole of the permanent magnet, and the north pole of the electromagnet lies near the south pole of the permanent magnet. The attractive force between north and south poles causes the loop of wire to spin. As soon as the loop of wire gets to the position shown in Figure 7–13*b*, the current reverses so that the south pole of the electromagnet lies just past the south pole of the permanent magnet, and the north pole of the electromagnet lies just past the north pole of the permanent magnet. The repulsive force between these like magnetic poles perpetuates the rotation. Alternating the current in the loop keeps a continuous rotational force on the wire, and keeps the wire turning. This force is what operates an electric hand drill, an electric fan, and the motors in an electric razor and hair dryer.

The simple diagram described above contains all the essential features of an electric motor; however, real electric motors are much more complex than the one we have shown. Typically, they have three or more different electromagnets and at least three permanent magnets, and the alternation of the current direction is somewhat more complicated than we have indicated. By artfully juxtaposing electromagnets and permanent magnets, inventors have produced an astonishing variety of electric motors: fixed-speed for your CD player, variable-speed for your food processor, reversible motors for power drills and screwdrivers, and specialized motors for many precision industrial uses. ●

electric motor A device that operates by supplying current to an electromagnet to make the magnet move and generate mechanical power. The motor employs permanent magnets and rotating loops of wire inside the poles of these magnets.

Why Magnetic Monopoles Don't Exist

Earlier we stated that there are no isolated magnetic poles in nature. Electromagnets provide a basis for understanding this law. Every magnetic field that we know in the universe results from the motion of electrical charges. If you think of an electron moving around a nucleus, as described earlier, you'll realize that the moving single electron constitutes a tiny "current." This current is identical to the circle of wire in an electromagnet. The only difference is that the current in the atomic loop consists of a single electron going around and around, while the current in the wire consists of many electrons moving around and around in a much larger loop.

The magnetism in magnetite and other permanent magnets can be traced ultimately to the cumulative effects of countless tiny current loops made by electrons going around in atoms (see Chapter 9). This fact explains why ordinary magnets can never be broken down into magnetic monopoles. If you break a magnet down to one last individual atom, you still have a dipole field because of the atomic-scale current loop. If you try to break the atom down further, the dipole field will disappear and there will be no magnetism except that associated with the particles themselves. Thus magnetism in nature is ultimately related to the arrangement and motion of electrical charges, rather than to anything intrinsic to matter itself.

ELECTRICAL EFFECTS FROM MAGNETISM

Now that you know magnetic effects arise from electricity, it should come as no surprise that the opposite is true—that electrical effects arise from

Figure 7–14
Electromagnetic induction. (*a*) When a current flows in the circuit on the left, a current is observed to flow in the circuit on the right *even though there is no battery or power source in that circuit;* (*b*) moving a magnet into the region of a coil of wire causes a current to flow in the circuit, even in the absence of a battery or other source of power.

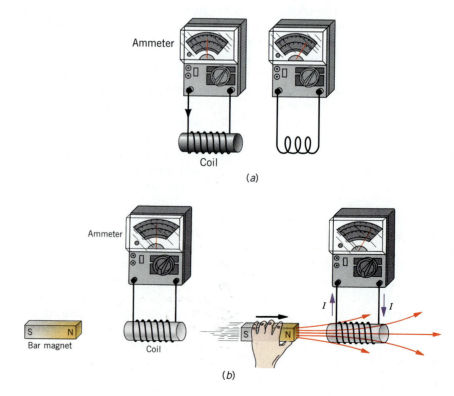

Ammeter

Coil

(a)

Ammeter

Bar magnet

Coil

S   N

I   I

(b)

magnetism. The British physicist Michael Faraday (1791–1867) is usually credited with this discovery.

Faraday's key experiment took place on August 29, 1831, when he placed two coils of wire side by side in his laboratory. He used a battery to pass an electric current through one of the coils of wire, and he watched what happened to the other coil. Even though the second coil of wire was not connected to a battery, a strong electrical current was produced in that coil. Faraday reasoned that the loop of wire through which a changing current was running produced a changing magnetic field in the neighborhood of the second loop. This changing magnetic field, in turn, produced a current in the second loop through a process called *electromagnetic induction* (Figure 7–14). An identical effect was observed when Faraday waved a permanent magnet in the vicinity of his wire coil; he produced an electric current without a battery.

The law summarizing Michael Faraday's research is

> **Electric fields and electric currents can be produced by changing magnetic fields.**

Figure 7–15 illustrates the **electric generator,** or dynamo, a vital tool of modern technology that demonstrates this effect. Place a loop of electrical wire with no batteries or other power source between the north and south poles of a strong horseshoe magnet. As long as the wire does not move, no current flows in it, but as soon as we begin to rotate the loop, a current flows in the wire. This current flows despite the fact that there is no battery or other power source in the wire.

electric generator A source of energy producing an alternating current in an electric circuit through the use of electromagnetic induction, which may be located in a power plant many miles from the energy's ultimate use, or under the hood of a car.

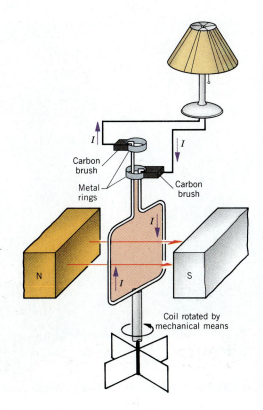

Figure 7–15
An electric generator. As long as the loop of wire rotates, there is a changing magnetic field near the loop and a current flows in the wire.

Any rotation changes the orientation of the magnetic field. When the magnetic field changes, a current flows in the loop. If we spin the loop continuously, a current flows continuously. The current flows in one direction for half of the rotation, then flows in the opposite direction for the other half of the rotation. This extremely important device, the electrical generator, followed immediately from Faraday's discovery of electromagnetic induction.

In an electrical generator, some source of energy—water passing over a dam, steam produced by a nuclear reactor or coal-burning furnace, or wind-driven propeller blades, for example—turns a shaft. In your car, the energy to turn the coils in a magnetic field comes from the gasoline that is burned in the motor. In every generator, the rotating shaft links to coils of wire that spin in a magnetic field. Because of the rotation, electrical current flows in the wire and can be tapped off onto external lines. Almost all the electricity used in the United States is generated in this way.

You may have already noticed a curious fact about electric motors and generators. In an electric motor, electrical energy is converted into the kinetic energy of the spinning shaft, while in a generator, the kinetic energy of the spinning shaft is converted into electrical energy. Motors and generators are thus, in a sense, exact opposites in the world of electromagnetism.

Because the current in the generating coils flows first one way and then the other, it will do the same thing in the wires in your home. This kind of current, used in household appliances, is called **alternating current,** or **AC,** because the direction keeps alternating. In contrast, chemical reactions in a battery cause electrons to flow in one direction only and produce what is called **direct current,** or **DC.**

alternating current (AC) A type of electric current, commonly used in household appliances and cars, in which charges alternate their direction of motion.

direct current (DC) A type of electric current in which the electrons flow in one direction only; for example, in the chemical reaction of a battery.

Electricity is generated when steam or water power turns a turbine, which contains coils of wire and magnets.

Science in the Making

Michael Faraday

Michael Faraday, one of the most honored scientists of the nineteenth century, did not come easily to his profession. The son of a black-smith, he received only a rudimentary education as a member of a small Christian sect. Faraday was apprenticed at the age of 14 to a London book merchant, and he became a voracious reader as well as a skilled bookbinder. Chancing upon the *Encyclopedia Britannica*, he was especially drawn to scientific articles, and he determined then and there to make science his life.

The young Faraday pursued his scientific career in style. He attended a series of public lectures at the Royal Institution by London's most famous scientist, Humphrey Davy, a world leader in physical and chemical research. Then, in a bold and flamboyant move, Faraday transcribed his lecture notes into beautiful script, bound the manuscript in the finest tooled leather, and presented the volume to Davy as his calling card. Michael Faraday soon found himself working as Davy's laboratory assistant.

After a decade of work with Davy, Faraday had developed into a creative scientist in his own right. He discovered many new chemical compounds, including the liquid benzene, and enjoyed great success with his own popular lectures for the general public at the Royal Institution in London. But his most lasting claim to fame was a series of classic experiments through which he discovered a central idea that helped link electricity and magnetism. ●

Michael Faraday (1791–1867)

Maxwell's Equations

We now understand the basic facts that govern the behavior of electrical and magnetic phenomena in our world. Electricity and magnetism are not distinct phenomena at all, but simply different manifestations of one underlying fundamental entity, **electromagnetism.** In the 1860s, Scottish physicist James Clerk Maxwell (1831–1879) realized that four very different statements about electricity and magnetism, which we have already talked about, constitute a single, coherent description of electromagnetism. He also recognized that the law that described the production of magnetic fields by electrical phenomena was incomplete. Maxwell was able to insert the missing piece.

electromagnetism A term used to refer to the unified nature of electricity and magnetism.

In the end, his four statements have come to be known as *Maxwell's equations*, in part because he was the first to realize their true import, and because he was the first to manipulate the mathematics to make important predictions, which we will discuss in detail in the next chapter. In summary, the four fundamental laws of electricity and magnetism known as Maxwell's equations are

1. Coulomb's law—like charges repel, unlike attract.
2. There are no magnetic monopoles in nature.
3. Magnetic phenomena can be produced by electrical effects.
4. Electrical phenomena can be produced by magnetic effects.

ELECTRIC CIRCUITS

electric circuit An unbroken path of material that carries electricity and consists of three parts: a source of energy, a closed path, and a device to use the energy.

We often come into contact with electrical phenomena through electric circuits in our homes and cars. An **electric circuit** is an unbroken path of wire and other material that carries electricity. For example, the electric light that you are using to read this book is part of an electrical circuit that begins at a power plant with an electric generator, many miles away. That electricity moves through power lines into your town and is distributed around your community on overhead wires. Once it reaches your home, copper wires in the walls transfer electricity to the circuit containing the light. These wires first connect to a circuit breaker, which breaks the circuit in the case of an *overload*, then to a switch, and finally to the light bulb. When you turn the switch on, you complete an unbroken path that runs from the generating plant to the light and back to the plant. When electricity flows through the light, the current heats up the filament until it glows. When you turn the switch off, it's like raising a drawbridge; the current can no longer flow into this part of the circuit and none reaches the light. The room becomes dark.

Every circuit consists of three parts: a source of energy, a closed path (usually made of metal wire through which the current can flow), and a device that uses electrical energy, such as a motor or light bulb (Figure 7–16).

This, the world's largest battery, is used to store energy in the electrical power system in California.

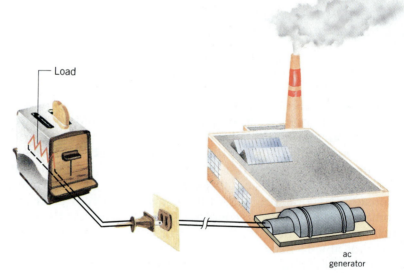

Figure 7–16
Your toaster completes part of an electric circuit that extends all the way back to your local power generating plant.

The energy source of the circuit is usually a battery or an electrical generator. Batteries, which are ideal for powering small, portable electric circuits, rely on chemical energy to push a direct current of electrons through wires. Electrical generators, which may be located hundreds of miles from where the electricity is finally used, produce alternating current, as we have seen. Your car carries both a battery and a generator under the hood.

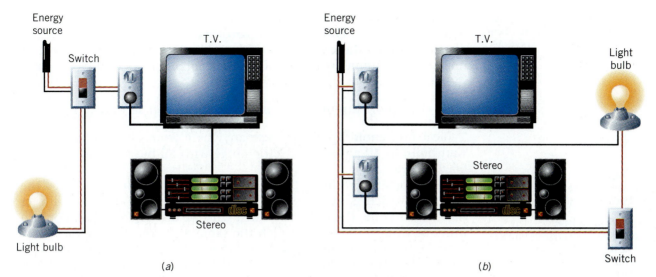

Energy source | Switch | T.V. | Stereo | Light bulb
(a)

Energy source | T.V. | Stereo | Light bulb | Switch
(b)

Figure 7–17
(a) A series circuit. (b) A parallel circuit.

Series and Parallel Circuits

There are two ways that you can imagine hooking elements of a circuit together. One way, as shown in Figure 7–17a, would be to connect the elements so that the electrical current flows through one element, then through the next, and so on. This is called a *series circuit* and is characterized by the fact that the same current flows through each element. If one element of a series circuit is defective (if a single light bulb burns out, for example) the current cannot flow and no other element will receive the current it needs to function.

Alternatively, we can arrange the circuit as shown in Figure 7–17b. This is a *parallel circuit*, and each element has its own loop of wire to carry current. Thus, if one element is defective, the others will continue to receive current and function. All elements in a parallel circuit have the same voltage across them but can, in general, have different currents flowing through them.

Amperage, Voltage, and Resistance

The way electrical circuits work, with electrons flowing through a wire, can be compared to water flowing through a pipe (Table 7–1). In the case of the water, we use two quantities to characterize the flow: the amount of water that passes a point each second, and the pressure behind that water. The measurements we use to describe the flow of electrons in a circuit are similar.

Table 7–1 • Terms Related to Electric Circuits

Term	Definition	Unit	Plumbing Analog
Voltage	Electrical pressure	volt	Water pressure
Resistance	Resistance to electron flow	ohm	Pipe diameter
Current	Flow rate of electrons	amp	Flow rate
Power	Current × voltage	watt	Rate of work done by moving water

ampere (amp) A unit of measurement for the amount of current (number of charges) flowing in a wire or elsewhere per unit time.

The amount of current, or the number of electrons, that flows in a wire is measured in a unit called the **ampere** or **amp,** named after French physicist Andre-Marie Ampere (1775–1836). One amp corresponds to a flow of one coulomb (the unit of electric charge) per second past a point in the wire.

1 amp of current = 1 coulomb of charge per second

Typical household appliances use anywhere from about 1 amp (100-watt bulb) to 40 amps (an electric range with all burners and the oven blazing away).

voltage The pressure produced by the energy source in a circuit, measured in volts.

The pressure produced by the energy source in a circuit is called the **voltage,** which is measured in *volts* (abbreviated V) and named after Alessandro Volta, the Italian scientist who we discussed earlier. You can think of voltage in circuits much the same way you think about water pressure in your plumbing system. More volts in a circuit mean more "oomph" to the current, just as more water pressure increases the water flow. Typically, a new flashlight battery produces 1.5 volts, a fully charged car battery about 15 volts (even though they are called 12-volt batteries), and ordinary household circuits operate on either 115 or 230 volts.

electrical resistance The quantity, measured in ohms, that represents how hard it is to push electrons through a material. High-resistance wires are used when electron energy is to be converted into heat energy. Low-resistance wires are used when energy is to be transmitted from one place to another with minimum loss.

ohm A unit of measurement for the electrical resistance of a wire.

It is harder to push electrons through some wires than others. The quantity that measures how hard it is to push electrons through wires is called **electrical resistance,** and it is measured in a unit called the **ohm.** The higher the resistance, the more of the electron energy is converted into heat. Ordinary copper wire, for example, has a low resistance, which explains why we use it to carry electricity around our homes. Toasters and space heaters, on the other hand, employ high-resistance wires so that they will glow bright red and give off large amounts of heat when current flows through them. Transmission lines, in which as much energy as possible must get from one end of the line to the other, use very thick low-resistance wires.

As electrons move through a wire, they collide with atoms. As a result of these collisions, some of the kinetic energy of the electrons is transferred to kinetic energy of the atoms; this process heats up the wire. In fact, electrons collide so often with atoms that their progress through a wire is rather slow. You can walk faster than a single electron can move through the copper wires in your home.

Ohm's Law

The relation between the resistance in a circuit, the current that flows, and the voltage is called *Ohm's law*, after the German scientist Georg Ohm (1787–1854). It states

▶ **In words:**

The current in circuits is directly proportional to the voltage and inversely proportional to the resistance. The higher the electrical "pressure," the higher the flow of charge. The higher the resistance to flow, the smaller the flow.

▶ **In equation form:**

voltage (in volts) = current (in amps) × resistance (in ohms)

▶ **In symbols:**

$$V = I \times R$$

In most situations, the voltage in a circuit is determined by the source of electricity. For example, most household outlets operate at 115 volts, while the battery in your calculator may be 1.5 volts. The amount of current drawn from the source will depend on the resistance of the appliance; the higher the resistance, the less charge will flow (see Figure 7–18).

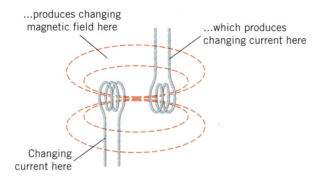

...produces changing magnetic field here

...which produces changing current here

Changing current here

Figure 7–18
In an electrical transformer, changes in current in one loop affect the current that flows in another.

Developing Your Intuition

The Transformer

When electricity is moved from one place to another, it is often necessary to change the voltage. For alternating current, this task is done with a device known as a *transformer*. A simplified transformer consists of two loops of wire located close together (Figure 7–19). Alternating current flows in the right-hand loop, and alternating current at a different voltage flows out the other. Given that the wires don't touch, how can there be any current at all in the second loop?

Think about it this way: The current in the right-hand loop will create a magnetic field as discussed above. This field will extend through the second loop. Since the current in the first loop is always changing, the magnetic field associated with it will always be changing as well. Thus at the position of the second loop, there is a constantly changing magnetic field. But according to Faraday's results, this means that a current has to flow in that loop. Can you see that the current in the second loop will also be alternating?

It turns out that if the number of turns of wire in the two loops are different, then the current in the second loop will have a different voltage *and* a different amperage from the current in the first, subject to the condition that the power (volts × amps) is the same in both. Why does the power have to be the same? ●

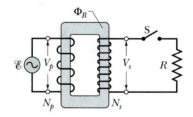

Primary   Secondary

Figure 7–19
A sketch of a simplified transformer, consisting of two coils of wire wound on an iron core. The current flowing in the left-hand coil induces a current in the right-hand coil. The iron ring helps concentrate the magnetic field.

Example 7–1: Your Toaster

If you plug a toaster into a wall socket, it probably draws 10 amps of current. What is the resistance of the device?

▶ **Reasoning and Solution:** This problem can be solved by remembering the equation that relates current, voltage, and resistance. We can rewrite Ohm's law as $R = V/I$. Since we know both voltage (115 V) and current in this case, we can calculate the resistance:

$$R = \frac{V}{I}$$

$$= \frac{115 \text{ volts}}{10 \text{ amps}}$$

$$= 11.5 \text{ ohms} \ \blacktriangle$$

Power Consumption

The *load* in any electrical circuit is the device where useful work gets done. The filament of a light bulb, the heating element in your hair dryer, or an electromagnetic coil of wire in a washer motor are typical loads in household circuits. The power used by the load depends both upon how much current flows through it and the voltage. The greater the current and the higher the voltage, the more power is used. The following equation allows us to calculate the amount of electrical power used:

▶ **In words:**

The power consumed by an electrical appliance is equal to the product of the current and the voltage.

▶ **In equation form:**

power (in watts) = current (in amps) × voltage (in volts)

▶ **In symbols:**

$$P = I \times V$$

This equation tells us that both the current and the voltage have to be high for a device to consume high levels of electrical power.

Example 7–2: Starting Your Car

In the early days of automobiles, most vehicles were started by a hand crank, which might have required 100 watts of power, which could be generated by an adult. Today, when you turn on the ignition of your automobile, your 15-volt car battery must turn a 400-amp starter motor. How much power is required to start your car?

▶ **Reasoning and Solution:** This problem can be solved by applying the equation for electrical power:

power (in watts) = current (in amps) × voltage (in volts)

$$= 400 \text{ amps} \times 15 \text{ V}$$

$$= 6000 \text{ W}$$

$$= 6 \text{ kw}$$

Modern, high-compression automobile engines require much more starting power than could be generated by one person. ▲

Example 7–3: The Power of Sound

In Example 5–3, we talked about a 250–watt compact disc system. Assuming that this system is plugged into a normal household outlet rated at 115 volts, how much current will flow through the stereo at full power?

▶ **Reasoning and Solution:** We know the power (250 watts) and voltage (115 volts) of the system. The power consumed by the CD player can be calculated from the equation for electrical power:

$$\text{power (in watts)} = \text{current (in amps)} \times \text{voltage (in volts)}$$

We can manipulate this equation to find the current:

$$\text{current} = \frac{\text{power}}{\text{voltage}}$$

$$= \frac{250\ \text{W}}{115\ \text{V}}$$

$$= 2.17\ \text{amps}$$

This amount is slightly more current than flows through two 100-watt light bulbs. ▲

Environment

Electric Cars

Air pollution generated by automobiles in cities may be significant because most cars operate by burning gasoline, a fossil fuel. In some areas, such as the large cities on the eastern seaboard of North America

Electric cars run on the stored chemical energy in batteries.

and in the Los Angeles Basin, governments have enacted or are considering enacting laws requiring, in effect, that a certain fraction of the cars in their jurisdiction run on electrical energy.

An electric car operates on batteries that can be recharged at night, when (presumably) the car is not in use. Although electric cars do contribute somewhat to pollution (the electricity has to be generated somewhere), they are estimated to produce only about half the pollution per passenger mile produced by a car using gasoline. In addition, since the energy for electric cars is produced in a central location, pollution may be easier to monitor and control than exhausts from thousands of cars scattered around a city.

So why aren't more electric cars on the road? The answer is that most available batteries can't hold enough energy to carry the car more than 40 to 50 miles between charges. This distance is equivalent to that driven by many commuters who live an hour from work. Let's do some numbers to see what the problem is.

A typical battery is rated in amp-hours—that is, it is supposed to deliver a certain current for a certain length of time before it runs down. Suppose a standard 15–volt car battery is rated at 25 A—2 h (i.e., it will deliver 25 amps of current for 2 hours). How much energy is actually stored in it?

We know that the power expenditure for the battery is

$$
\begin{aligned}
\text{power} &= \text{volts} \times \text{amps} \\
&= 15\ \text{V} \times 25\ \text{A} \\
&= 375\ \text{W} \\
&= 0.375\ \text{kW}
\end{aligned}
$$

But from Chapter 5 we know that

$$
\begin{aligned}
\text{energy} &= \text{power expended} \times \text{time} \\
&\quad \text{during which it is expended} \\
&= 0.375\ \text{kW} \times 2\ \text{h} \\
&= 0.75\ \text{kWh}
\end{aligned}
$$

So a fully charged battery carries about three quarters of a kilowatt-hour of energy—enough to run a stereo for about three hours. This isn't very much energy. This isn't a problem in the normal operation of a gasoline-powered car, because the battery is expected to give only a short burst of current to start the car, after which it is recharged. In an electric car, however, it *is* a problem, because the heavy battery has to be carried along with the rest of the car and its passengers. In fact, it takes about 800 pounds of ordinary lead-acid batteries to store as much energy as there is in a gallon of gasoline, and it's hard to imagine designing a car that would carry much more of a battery load. This means that if you want an electric car to be able to go 100 miles between charges, it has to be designed to get the equivalent of 100 miles per gallon.

Current research on electric cars, therefore, is focusing on (1) developing batteries that store more energy for less weight, and (2) improving the efficiency of the car to lower its energy requirements. ●

THINKING MORE ABOUT
ELECTROMAGNETISM

Basic Research

It's hard to imagine modern American society without electricity. We use it to move goods, to communicate with each other, to light and warm our homes, and to manufacture the necessities and amenities of life. Chances are that nothing within view as you read these pages would be as it is if it weren't for electricity.

Yet the people who gave us this marvelous gift were not, by and large, primarily concerned with developing better lamps or clocks or modes of transportation. In terms of the categories we introduced in Chapter 1, they were doing basic research. Galvani and Volta, for example, were drawn to the study of electricity by their studies of frog muscles that contracted by jolts of electrical charge. Volta's first battery was built to duplicate the organs found in electric fish. Scientific discoveries, even those that bring enormous practical benefits to humanity, may derive from unusual and unexpected sources.

What does this tell you about the problem of allocating government research funding? Can you imagine trying to justify funding Galvani's experiments on frogs' legs to a government panel on the grounds that it would lead to something useful? Would a federal research grant designed to produce better lighting systems have produced the battery (and, eventually, the electric light), or would it more likely have led to an improvement in the oil lamp? How much funding do you think should go to offbeat areas (on the chance that they may produce a large payoff) as compared to projects that have a good chance of producing small but immediate improvements in the quality of life?

While you're thinking about these issues, you might want to keep in mind Michael Faraday's response to a question. When asked by a political leader what good his electric motor was, he is supposed to have answered, "What good is it? Why, Mr. Prime Minister, someday you will be able to tax it!"

▶ Summary

Electricity and *magnetism* are phenomena that involve forces quite different from the universal gravitational force that Newton described in the seventeenth century. Nevertheless, Newton's laws of motion provided scientists of the eighteenth and nineteenth centuries a way to describe and quantify a range of intriguing electromagnetic behavior.

The phenomenon of *static electricity*, which includes lightning, static cling, and the small sparks produced by walking across a wool rug on a cold winter day, are caused by the transfer of electrons between objects. An excess of electrons imparts a *negative electric charge*, while a deficiency causes an object to have a *positive electric charge*. Objects with like charges experience a repulsive force, while objects with opposite charges attract each other. These observations were quantified in *Coulomb's law*, which states that the magnitude of electrostatic force between any two objects is proportional to the charges of the two objects, and inversely proportional to the square of the distance between them.

Other scientists, investigating the phenomenon of magnetism, observed that every *magnet* has a *north and south pole*, and that magnets exert forces on each other. No matter how many times a magnet is divided, each of its pieces will have two poles: there are no isolated magnetic poles. Like magnetic poles repel each other, while opposite poles attract. A compass is a needle-shaped magnet that is designed to point at the poles of the Earth's magnetic field.

Both electric and magnetic forces can be described in terms of force *fields*—imaginary lines that reveal the directions of forces that would be experienced in the vicinity of electrically charged or magnetic objects. Scientists in the nineteenth century discovered that the seemingly unrelated phenomena of electricity and magnetism were actually two aspects of a single *electromagnetic force*.

Hans Oersted found that an *electric current* passing through a coil of wire produces a magnetic field. The *electromagnet* and *electric motor* were direct results of his work. Michael Faraday discovered the opposite effect when he induced an electric current by placing a wire coil near a magnetic field, thus designing the first *electric generator*, which produced an *alternating current (AC)*. *Batteries*, on the other hand, develop a *direct current (DC)*. All electric currents (measured in *amperes*) are characterized by an electric "push" or *voltage* (measured in *volts*) and *electrical resistance* (measured in *ohms*). A closed path through which electricity flows is called an *electric circuit*.

James Clerk Maxwell realized that the many independent observations about electricity and magnetism, taken together, constitute a complete description of *electromagnetism*.

▶ Key Equations

electrostatic force (in newtons)
$$= k \times \frac{[\text{1st charge} \times \text{2nd charge (in coulombs)}]}{\text{distance}^2}$$

1 coulomb = the charge on 6.3×10^{18} electrons

electric power (in watts)
$$= \text{current (in amps)} \times \text{voltage (in volts)}$$

1 ampere of current = 1 coulomb of charge per second

voltage (in volts)
$$= \text{current (in amps)} \times \text{resistance (in ohms)}$$

▶ Universal Electrostatic Constant

$k = 9.00 \times 10^9$ newton-meter^2/coulomb^2

▶ Review Questions

1. How do you know that there is such a thing as an electrical force?

2. How does the electrical force differ from gravity in its manifestations?

3. What would a modern scientist say that Franklin's "single fluid" was?

4. How can the movement of negative charges such as the electron produce a material that has a positive charge?

5. What is an electric field?

6. How do you know that there is such a thing as a magnetic force?

7. What was the first practical application of magnetism?

8. What is a dipole field?

9. Under what circumstances can electric charges produce a magnetic field?

10. Describe the working and function of an electromagnet.

11. How does an electromagnet differ from a permanent magnet?

12. How are electromagnets used in an electric motor?

13. How can a single atom produce a magnetic field?

14. Under what circumstances can a magnetic field produce an electric current?

15. How does an electric generator work?

16. What are Maxwell's equations?

17. What does the volt measure? Where do you run across this term in your everyday experience?

18. What does the ampere measure? Where do you run across this term in your everyday experience?

19. What does the ohm measure? Where do you run across this term in your everyday experience?

20. What is the relation between the power an appliance consumes, the voltage across it, and the current through it?

▶ Fill in the Blanks

Complete the following paragraphs with words and phrases from the list.

alternating current (AC)	electricity
ampere	electromagnet
Coulomb's law	magnetic field
direct current (DC)	magnetic force
electric circuit	negative charge
electric current	ohms
electric field	poles (north and south)
electric generator	positive charge
electric motor	static electricity
electric charge	volts
electrical resistance	

Sheets of paper stick together when they come out of a copying machine because of _____. Objects that respond to the electrical force are said to possess _____. In the atom, the electron carries a _____, while the nucleus carries a _____. _____ describes the electric force, and the _____ is defined to be the force that would be felt by a unit positive charge at any point in space.

When a compass needle points north, it demonstrates the existence of a _____. Every magnet has two _____, and the _____ maps out the force that would be felt by a small magnet at any point in space.

A changing magnetic field can produce an _____, or flow of charge, in a wire, a phenomenon known as _____. This effect operates in the _____, a device that converts mechanical energy into electrical current. Likewise, moving charges can produce a magnetic field, an effect that is put to practical use both in the _____ and the _____. Generators produce _____, while batteries produce _____.

An _____ is an unbroken path of conducting material. Current in an electric circuit is measured in _____, while the electrical "pressure" is measured in units called _____. The amount of energy that moving charges give up to heat in moving through a material is a function of that material's _____, which is measured in _____.

▶ Discussion Questions

1. Your car uses a 12-volt battery, but do you need all of the energy to start the car? (*Hint:* What happens if you leave the lights on for half an hour—can you still start it?)

2. If you took an electric motor and turned it by hand, what do you think would happen in the coils of wire?

3. What would you say was the hypothesis in Faraday's classic experiment on electromagnetic induction?

4. How did Benjamin Franklin know that lightning had an electrical nature?

5. Why can't a magnetic monopole exist?

6. Identify five things in your room that would not be possible without discoveries in electromagnetism.

7. How does the first law of thermodynamics apply to electrical circuits?

8. How does the second law of thermodynamics apply to electric circuits?

9. How is the electrical force similar to the gravitational force?

10. Think about the definition of an electrical current.
   a. If excess electrons are moving from right to left across this page in a line, do you have a current?
   b. If excess protons are moving from right to left across this page in the same line, do you have a current?
   c. Is the direction of the current in situation *a* the same as the direction of the current in situation *b*?

11. List Maxwell's four equations. Retell each equation in your own words.

12. Using Maxwell's equations, outline the steps for an electrical generator to work.

13. When you place a strong magnet near your television set, the picture gets out of focus. What is happening to the TV's electron beam?

▶ Problems

1. Three light bulbs in their sockets are connected to a battery, as illustrated in Figure 7–20.

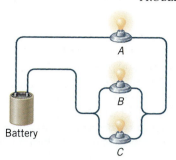

Figure 7–20

   a. If bulb *A* is unscrewed, will bulbs *B* and *C* remain lit? Why?
   b. If bulb *B* is unscrewed, will bulbs *A* and *C* remain lit? Why?
   c. If bulb *C* is unscrewed, will bulbs *A* and *B* remain lit? Why

2. In a simple DC circuit for a flashlight, a 1.5-V battery powers the light bulb, which has a resistance of 2 ohms.
   a. What is the current drawn by the flashlight?
   b. What is the power generated by the flashlight?

3. In a typical household 100-W bulb, the current drawn is about 1 amp. What is the resistance of a 100-W light bulb?

4. An electric range is on a 220-V circuit and draws up to 40 amps of current when all the burners and oven are on.
   a. What is the maximum total power generated by the electric range?
   b. What is the total resistance of the burners and oven?

5. A single electron has a charge of 1.60×10^{-19} C. How many electrons will it take to produce one ampere of current if they all pass a specific point in 1 second?

6. Assume that in interstellar space the distance between two electrons is about 0.1 cm.
   a. Is the electric force between the two electrons repulsive or attractive?
   b. Calculate the electric force between these two electrons.

7. A typical 1.5-V alkaline D battery is rated at 3.5 amp-hours.
   a. What is the power that can be expended by the battery?
   b. What is the total energy stored by this battery?

8. A standard flashlight will have two 1.5-V D batteries arranged so that they add in voltage, with a 2-W bulb. These batteries are rated at 3.5 amp-hours.
   a. What is the total power that can be expended by the batteries?
   b. What is the total energy stored in the flashlight batteries?

c. How long can the flashlight keep the bulb lit at its rated power of 2 W?

9. What is the total power in watts for a 10-hp motor?

10. Rafael and Becky decided to impress their physical science professor with a simple experiment investigating the resistance of a 100-W light bulb. They used a VARIAC (which provides a variable voltage to a circuit) in a series circuit with one 100-W light bulb. They measured the current and voltage, which are listed in the table:

Voltage (volts)	Current (amps)
120	0.81
100	0.72
80	0.62
60	0.51
40	0.40
20	0.23

a. Calculate the resistance of the light bulb for each voltage setting.

b. Does the resistance of the 100-W bulb increase or decrease with the voltage? With current?

c. From your personal experience, can you predict if the temperature of the filament of the light bulb increases or decreases with voltage?

d. Using this information, speculate on the reason(s) the resistance of the light bulb changes.

11. Based on their electric charges and separations, which of the following atomic bonds is strongest?

a. A +1 sodium atom separated by 2.0 distance units from a −1 chlorine atom in table salt.

b. A +1 hydrogen atom separated by 1.0 distance units from a −2 oxygen atom.

c. A +4 silicon atom separated by 1.5 distance units from a −2 oxygen atom.

(*Hint:* You are interested only in the *relative* strengths, which depend only on the relative charges and distances.)

12. When a car battery runs down, it is recharged by running current through it backward. Typically, you might run 5 amps at 12 volts for an hour. How much energy does it take to recharge a battery?

13. Most household circuits have fuses or circuit breakers that open a switch when the current in the circuit exceeds 15 amps. Would the lights go off when you plug in an air conditioner (1 kw), a TV (250 W), and two 100-W light bulbs? Why?

14. Energy-efficient appliances are important in today's economy. Suppose that a light bulb gave as much light as a 100-W bulb, but consumed only 20 W, while costing $2.00 more. If electricity costs $0.08/kwh, how long will the bulb have to operate to make up the difference in price?

15. An energy-efficient air conditioner draws 7 amps in a standard 115-V circuit. It costs $40 more than a standard air conditioner that draws 12 amps. If electricity costs $0.08/kwh, how long would you have to run the efficient air conditioner to recoup the difference in price?

▶ Investigations

1. Make an inventory of all your electrical appliances. How many watts does each use?

2. Most household circuits have fuses or circuit breakers that open a switch when the current in the circuit exceeds 15 amps. How many of the appliances in (1) could you run on the same circuit without overloading it?

3. Read Jules Verne's *The Mysterious Island*. Do you think it is possible for castaways to build an electrical generator as described in the book?

4. Examine carefully your most recent electric bill. How much power did you use? How much did it cost? Is there a discount for electricity used at off-peak hours? Examine your living place for all the locations where you use electricity. Plan a strategy for reducing your electric bill by 10% next month. You can reduce consumption by turning off lights and appliances when not in use, installing lower-wattage bulbs, or using electricity during low-rate times.

5. Find out where your electric power is generated. Does your local utility buy additional power from some other place? What kind of fuel or energy is used to power or drive the turbines? Are there special pollution controls that restrict the use of certain kinds of fuels at your local power plant?

6. How many kilowatts of electrical power does a typical commercial power plant generate? How much electricity does the United States use each year? Is this amount going up or down?

7. Take an old electric razor or other small motor-driven appliance and dissect the motor. How many permanent magnets are inside? How many separate coils of wire?

8. Identify the major electrical circuit components in your automobile. Which require the greatest power?

9. How do electric eels generate electrical shocks?

10. How does an electroencephalogram (EEG) work? How does it differ from an electrocardiogram (EKG)?

11. Many kinds of living things, from bacteria to vertebrates, incorporate small magnetic particles. Investigate the ways living things use magnetism.

12. Read the novel *Frankenstein* (or see the classic 1931 movie with Boris Karloff and Colin Clive). Discuss the ideas about the nature of life that are implicit in the story. Does it represent a realistic picture of scientific research? Why or why not?

▶ Additional Reading

Everitt, C. W. F. *James Clerk Maxwell: Physicist and Natural Philosopher*. New York: Scribners, 1976.

Franklin, Benjamin. *The Autobiography of Benjamin Franklin*. Boston: St. Martins Press, 1993.

MacDonald, D. K. C. *Faraday, Maxwell, and Kelvin*. New York: Doubleday, 1964.

Pera, Marcello. *The Ambiguous Frog: The Galvani-Volta Controversy on Animal Electricity*. Princeton: Princeton University Press, 1992.

8

WAVES AND ELECTRO- MAGNETIC RADIATION

WHENEVER AN ELECTRICALLY CHARGED OBJECT IS ACCELERATED, IT PRODUCES ELECTROMAGNETIC RADIATION—WAVES OF ENERGY THAT TRAVEL AT THE SPEED OF LIGHT.

A Day at the Beach

Few experiences are more relaxing than a day at the beach. The sight of waves washing ashore, the sound of good music, and the feel of the Sun's rays beating down help us forget about the pressure of exams and term papers. What might surprise you is that underlying all of those familiar experiences at the beach is the phenomenon of waves.

Waves are all around us. Waves of water travel across the surface of the ocean and come crashing against the land. Waves of sound travel through the air when we listen to music. Some parts of the United States suffer from earthquakes—mighty seismic waves of rock and soil. All of these waves must move through matter.

But the most remarkable waves of all are electromagnetic waves. These waves fill every corner of our environment every moment of our lives, and they travel through an absolute vacuum at the speed of light. Sunlight, which warms you at the beach and provides virtually all of the energy necessary for life on Earth, is transmitted through space by electromagnetic waves. The radio waves that carry your favorite music, the microwaves that heat your dinner, and the X-rays your dentist uses to check for cavities are also types of electromagnetic waves—all invisible forms of energy that travel at light speed.

THE NATURE OF WAVES

Waves are fascinating phenomena, at once familiar and yet somewhat odd. Scientists study waves by observing their distinctive behavior, particularly their unique ability to transfer energy without transferring mass. In this section, we will discuss different properties and types of waves.

Energy Transfer by Waves

Energy travels in two forms: the particle and the wave. Suppose you have a domino sitting on a table and you want to knock it over. This process requires transferring energy from you to the domino. One way to knock over the domino would be to take another domino and throw it. If you did this, the muscles in your arms would impart kinetic energy to the moving domino, which, in turn, would impart enough of that energy to the standing domino to knock it over (Figure 8–1a). In such a case, the energy is transferred by the motion of a solid piece of matter.

Suppose, however, that you lined up a row of standing dominoes. If you knocked over the first domino, it would then knock over the second, which in turn would knock over the third, and so on (Figure 8–1b). Eventually, the falling chain of dominoes would hit the last one and you would have achieved the same result—energy has been transferred from you to knock over the domino. In the case of the row of dominoes, however, no single object traveled from you to the most distant domino. You started a wave of falling dominoes, and that wave knocked over the final one. A **wave**, then, carries energy from place to place without requiring matter to travel across the intervening distance.

wave A traveling disturbance that carries energy from one place to another without requiring matter to travel across the intervening distance.

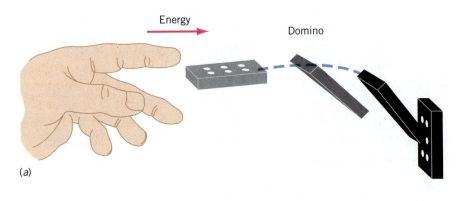

(a)

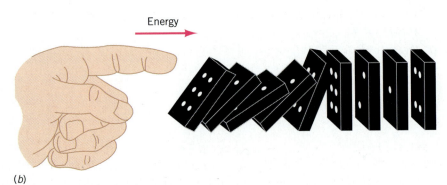

(b)

Figure 8–1
You can use dominoes to knock over another domino in two different ways: (a) you can throw a domino, or (b) you can trigger a wave of dominoes.

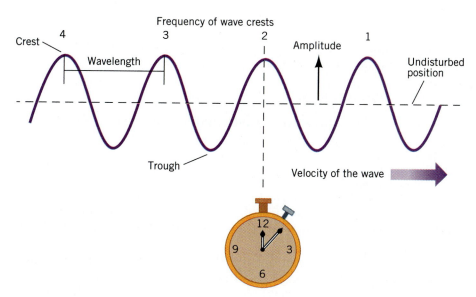

Figure 8–2
A cross-section of a wave reveals the characteristics of wavelength, velocity, and amplitude. Successive wave crests are numbered 1, 2, 3, and 4. An observer at the position of the clock records the number of crests that pass by in a second. This is the frequency, which is measured in cycles per second, or hertz.

The Properties of Waves

Think about a familiar example of waves. You are standing on the banks of a quiet pond on a crisp autumn afternoon. There's no breeze and the pond in front of you is still and smooth. You pick up a rock and toss it into the middle of the pond. As soon as the rock hits the water, a series of ripples moves outward from the point of impact. In cross-section, the ripples look like waves, as shown in Figure 8–2. You can use four measurements to characterize the ripples.

1. **Wavelength** is the distance between crests, or the highest points of adjacent waves. On a pond, the wavelength might be only a centimeter or two, while ocean waves may be tens or hundreds of meters between the crests.

2. **Frequency** is the number of wave crests that go by a given point every second. A wave that sends one crest by every second (completing one *cycle*) is said to have a frequency of one cycle per second, or one **hertz** (abbreviated 1 Hz). Small ripples on a pond might have a frequency of several Hz, while large ocean waves might arrive only once every few seconds.

3. *Velocity* is the speed and direction of the wave crest itself. Water waves typically travel a few meters per second, about the speed of walking or jogging, while sound waves travel about 340 meters (1100 feet) per second.

4. **Amplitude** is the height of the wave crest above the undisturbed surface, such as the level of the water.

wavelength   The distance between crests, the highest points of adjacent waves.

frequency   The number of wave crests that go by a given point every second. A wave completing one cycle (sending one crest by a point every second) has a frequency of 1 hertz (Hz).

hertz (Hz)   The unit of measurement for the frequency of waves; one wave cycle per second.

amplitude   The height of a wave crest above the undisturbed level of the medium.

The Relationship Between Wavelength, Frequency, and Velocity

A simple relationship exists among three fundamental wave measurements: wavelength, frequency, and velocity. If we know any two of the three measurements, we can calculate the third using a simple equation:

▶ **In words:**

The velocity of a wave is equal to the length of each wave times the number of waves that pass by each second.

▶ **In equation form:**

wave velocity (in m/s) = wavelength (in m) × frequency (in Hz)

▶ **In symbols:**

$$v = \lambda \times \omega,$$

where λ (Greek lambda) and ω (Greek omega) are the common symbols for wavelength and frequency. This simple equation holds for all kinds of waves.

To understand this relationship, think about waves on the water. Suppose you are sitting on a small sailboat, as shown in Figure 8–3, watching a series of wave crests passing by. You can count the number of wave crests going by every second (the frequency) and measure the distance between the crests (the wavelength). From these two numbers, the speed of the wave can be calculated. If, for example, one wave arrives every 2 seconds, or one-half wave per second (the frequency), and the wave crests are 6 meters apart (the wavelength), then the waves must be traveling 6 meters every 2 seconds, which is a velocity of 3 meters per second. You might look out across the water and see a particularly large wave crest that will arrive at the boat after four intervening smaller waves. You could then predict that the big wave is 30 meters away (five times the wavelength of 6 meters), and that it will arrive in 10 seconds (five divided by the frequency of one-half wave per second). That kind of information can be very helpful if you are plotting the best course for an America's Cup yacht race or estimating the path of potentially destructive ocean waves.

Figure 8–3
Waves passing a sailboat reveal how wavelength, velocity, and frequency are related. If you know the distance between wave crests (the wavelength) and the number of crests that pass each second (the frequency), then you can calculate the wave's velocity.

Example 8–1:   Seismic Waves

A pile driver pounds a metal I-beam two times per second, driving it into rock. Each shock sends a wave traveling 6 km/s through the rock. This kind of sound wave is called a **seismic wave,** and it plays a major role in how we measure the interior of the Earth (see Chapter 18). What is the wavelength of these seismic waves?

▶ **Reasoning and Solution:** The relationship among wavelength, frequency, and velocity applies to all waves. We know that

wave velocity (in m/s) = wavelength (in m) × frequency (in Hz).

In this case, the velocity is 6 km/s, which equals 6000 m/s, and the frequency is 2 times per second, which equals 2 Hz. We can rearrange the equation to solve for wavelength:

$$\text{wavelength (in m)} = \frac{\text{velocity (in m/s)}}{\text{frequency (in Hz)}}$$

$$= \frac{6000 \text{ m/s}}{2 \text{ Hz}}$$

$$= 3000 \text{ m}$$

These seismic waves, therefore, are quite long—3 kilometers, or about 1.8 miles. Nevertheless, they're much shorter than distances in the Earth, and thus travel many times their wavelength before they die out. ▲

seismic wave The form through which an earthquake's energy is transmitted. Some earthquakes cause the Earth to rise and fall like the surface of the ocean.

The Two Kinds of Waves: Transverse and Longitudinal

Imagine that a chip of bark or a piece of grass is lying on the surface of the pond when you throw a rock. When the ripples go by, that floating object and the water around it move up and down, but they do not move to a different spot. At the same time, however, the wave crest moves in a direction parallel to the surface of the water. This means that *the motion of the wave is different from the motion of the medium on which the wave moves.* This kind of wave, where the motion moves perpendicular to the direction of the wave, is called a **transverse wave.**

You can observe (and participate in) this phenomenon if you ever go to a sporting event in a crowded stadium where fans "do the wave." Each individual simply stands up and sits down, but the visual effect is of a giant sweeping motion around the entire stadium. In this way, waves can move great distances, even though individual pieces of the transmitting medium hardly move at all.

Not all waves are transverse waves like those on the surface of the water. We used the example of a pond simply because it is so familiar and easy to visualize. Sound is another form of wave, which moves through the air. When you talk, for example, your vocal cords move air molecules back and forth. The vibrations of these air molecules set the adjacent molecules in motion, which set the next set of molecules in motion and so forth. A circular wave moves out from your mouth. This wave looks very much like the ripples on a pond. The only difference is that, in the air, the wave crest that is moving out is not a raised portion of a water surface, but a denser region of air molecules. Sound is a *pressure*

transverse wave A kind of wave in which the motion of the wave is perpendicular to the motion of the medium on which the wave moves; pond ripples.

Figure 8–4
Transverse (*a*) and longitudinal (*b*) waves differ in the motion of the wave relative to the motion of individual particles.

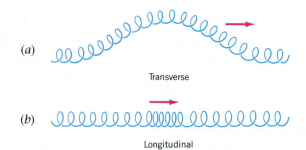

(*a*)

Transverse

(*b*)

Longitudinal

longitudinal wave A kind of wave in which the motion of the medium is in the same direction as the wave movement; pressure wave or sound wave.

wave or **longitudinal wave**. As a longitudinal wave of sound moves through the air, gas molecules vibrate forward and back *in the same direction as the wave*. This motion is very different from the transverse wave of a ripple in water, where the water molecules move *perpendicular to the direction of the waves* (see Figure 8–4). Note that in either longitudinal or transverse waves, the energy always moves in the direction of the wave.

Developing Your Intuition

The Wave

Let's look again at the wave that fans in a large stadium do during ball games. Watching from a distance you see a steady motion around and around the entire stadium. In this transverse wave, the individual people are the "particles" that move up and down, while the wave moves round and round. But think for a moment. Could fans in a stadium ever produce a longitudinal "wave," in which the particles move in the same direction as the wave?

In a longitudinal wave, the fans (the transmitting medium in this case) have to move in the same direction as the wave. One way to accomplish this would be to have everyone lean quickly to the right as the wave crest arrives, then slow straighten up again. Another (rather unlikely) way it might occur is if one entire column of seats, top to bottom in the stadium, were empty. If fans next to the empty column shifted seats to the right, rapidly in succession, a different kind of wave would travel around the stadium. It might not look as impressive as the usual wave, but it would be longitudinal. ●

When a crowd in a stadium does the "wave," they are demonstrating the properties of transverse waves.

Different size instruments in a New Orleans jazz ensemble play in different ranges. The large tuba on the left plays low notes, while the smaller trumpet at center plays in a higher range.

Science by the Numbers

The Sound of Music

The speed of sound in air is more or less constant for all kinds of sound. The way we perceive a sound wave, therefore, depends on its other properties such as wavelength, frequency, and amplitude. For example, we perceive the amplitude of a sound wave as loudness—the bigger the amplitude, the louder the sound. Similarly, we hear high-frequency sound waves (sound with short wavelengths) as high pitches, while we perceive low-frequency sound waves (with long wavelength) as low sounds.

You can see one such contrast when you watch a symphony orchestra. The highest notes are played by small instruments such as a piccolo and violin, while the lowest notes are played by the massive tuba and double basses. Each of a large pipe organ's thousands of pipes produces a single note. An open-ended organ pipe encloses a column of air in which a sound wave can travel back and forth, down the length of the pipe over and over again. The column is just long enough for one-half wavelength to be completed. The number of waves completing this circuit every second—the frequency— determines the pitch that you hear (Figure 8–5).

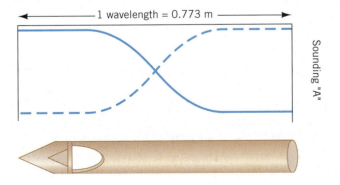

1 wavelength = 0.773 m

Sounding "A"

Figure 8–5
An organ pipe produces a single note. Air in the pipe vibrates and produces a sound wave with a wavelength exactly equal to twice the length of the pipe.

The note that we hear as "middle A," the pitch to which most orchestras tune, has a frequency of 440 Hz. Sound travels through air at about 340 m/s. An organ pipe open at both ends produces a note with a wavelength twice as long as the pipe. The length of an organ pipe air column that plays middle A, therefore, is given by half the wavelength in the equation

$$\text{wavelength (in m)} = \frac{\text{velocity (in m/s)}}{\text{frequency (in Hz)}}$$

$$= \frac{340 \text{ m/s}}{440 \text{ Hz}}$$

$$= 0.773 \text{ m (about 2 ft)}$$

The length of the organ pipe is half the wavelength:

$$\text{organ pipe length (in m)} = \frac{\text{wavelength (in m)}}{2}$$

$$= \frac{0.773 \text{ m}}{2}$$

$$= 0.387 \text{ m (about 15 in.)}$$

Notes in the middle range on the pipe organ are thus produced by pipes that are about a half-meter long. ●

Example 8–2:   The Limits of Human Hearing

The human ear can hear sounds at frequencies from about 20 to 20,000 Hz. What are the longest and shortest wavelengths you can hear? What are the longest or shortest open-ended organ pipes you are likely to see?

▶ **Reasoning and Solution:**   Each organ pipe has a fixed length and produces just one note. We have to calculate the wavelength needed for both the lowest and highest frequency.

The lowest audible note, at 20 Hz, would require a wavelength:

$$\text{wavelength (in m)} = \frac{\text{velocity (in m/s)}}{\text{frequency (in Hz)}}$$

$$= \frac{340 \text{ m/s}}{20 \text{ Hz}}$$

$$= 17 \text{ m (about 50 ft long)}$$

Similarly, the highest audible note, at 20,000 Hz, is produced by a wavelength:

$$\text{wavelength (in m)} = \frac{\text{velocity (in m/s)}}{\text{frequency (in Hz)}}$$

$$= \frac{340 \text{ m/s}}{20,000 \text{ Hz}}$$

$$= 0.017 \text{ m}$$
$$\text{(about two thirds of an inch)}$$

Open organ pipes producing these notes would be about half the wavelengths: approximately 8.5 and 0.009 m, respectively. Most large pipe organs have pipes ranging from about 8 m to less than 0.05 m in length. Next time you have the chance, visit a church or auditorium with a large pipe organ and look at the variety of pipes. Not only are there many different lengths, but there are also many distinctive shapes, each sounding like a different instrument.

At the extremes of human hearing, pitches take on a strange quality. The highest pitched notes have a piercing quality, whereas at the lowest frequencies we don't so much hear as feel the notes. Listen for the highest and lowest notes the next time you hear a large pipe organ playing. ▲

Use of Sound by Animals

Humans use sound to communicate, as do many other animals. But some animals have refined the use of sound to a much higher level. In 1793, the Italian physiologist Lazzaro Spallanzani did some experiments that established that bats use sound to locate their prey. He blinded bats that lived in the cathedral tower in Pavia, and then turned them loose. Weeks later, those bats had fresh insects in their stomachs, proving that they did not use their sight to locate food. Similar experiments with bats that were deafened, however, showed that they could neither fly nor locate insects.

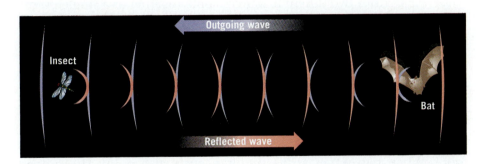

Figure 8–6
A bat emits a sound wave, which reflects off its target. By sensing the time it takes the sound to go out and back, the bat can tell how far away the target is.

Today, we understand that bats navigate by emitting high-pitched sound waves. They then listen for the echo of those waves off of other objects. By measuring the time it takes for a pulse of sound waves to go out, be reflected, and come back, the bat can determine the distance to surrounding objects, particularly the flying insects that constitute its diet (Figure 8–6). Typically, a bat can detect the presence of an insect up to 10 meters away. In addition, the bat can use the Doppler effect (see below) to tell whether the target is moving.

A bat navigates by emitting high-pitched sounds and listening for their echoes.

Some animals (elephants, for example) routinely use sound in the range of 20 to 40 Hz to communicate with each other over long distances. The mating call of the female elephant, for example, is experienced as a vibration by humans, but attracts bull elephants from many miles away.

Whales, dolphins, and porpoises use sound echoes as a navigation tool in the ocean, much as bats do in air. Sometimes, however, the sounds that they emit are in the audible range for humans. Perhaps the most famous sophisticated use of sound by animals is the songs of the humpback whales, which have appeared on innumerable commercial recordings. It's not clear what the functions of these songs are. It appears, however, that all of the whales in a wide area of ocean (the South Atlantic, for example) sing approximately the same song, although some individual whales may leave out parts. The songs change every year, but the whales in a given area seem to change their songs together.

Interference

interference   When waves from two different sources come together at a single point, they interfere with each other, and the observed wave amplitude is the sum of the amplitudes of the interfering waves.

Waves from different sources may overlap and affect each other in the phenomenon called **interference.** Interference describes what happens when waves from two different sources come together at a single point: each wave interferes with the other, and the observed wave amplitude is the sum of the amplitudes of the interfering waves.

Consider the common situation shown in Figure 8–7. Suppose you and a friend each threw rocks into a pond at two separate points, labeled *A* and *B* in the figure. The waves from each of these two points travel outward and will eventually meet at a point like the one labeled *C*. What will happen when the two waves come together?

Imagine that every part of each wave carries with it instructions for the water surface, such as "move down three feet," or "move up one foot." When two waves arrive simultaneously at a point, the surface responds to both sets of instructions. If one wave says to move down three feet and the other to move up one foot, the result will be that the water surface will move down two feet.

Figure 8–7
Two waves originating from different points create an interference pattern. Bright regions correspond to constructive interference, while dark regions correspond to destructive interferences.

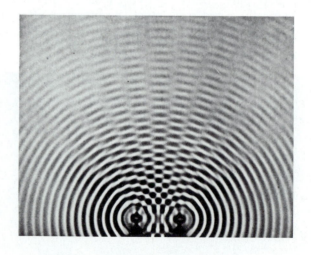

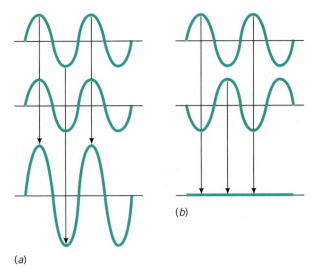

(a)

(b)

Figure 8–8
Cross-sections of interfering waves illustrate the phenomena of constructive (*a*) and destructive (*b*) interference.

Each point on the surface of the water, then, moves a different distance up or down, depending on the instructions that are brought to it by the waves from point *A* and point *B*. One possible situation is shown in Figure 8–8*a*. Two waves, each carrying the command "go up one foot," arrive at a point together. The two waves act together to lift the water surface to the highest possible height it can have. This is called **constructive interference** or reinforcement. On the other hand, you could have a situation like the one shown in Figure 8–8*b*, where the two waves arrive at the point in such a way that one is giving the instruction to go up one foot and the other to go down one foot. In this case, the water surface will not move at all. This situation is called **destructive interference** or cancellation. Waves can interfere anywhere between these two extremes.

Sound waves provide a familiar example of destructive interference. Occasionally, an auditorium is designed in such a way that almost no sound can be heard in certain seats. This unfortunate situation results when two waves—for example, one directly from the stage and one bouncing off the ceiling or a wall—arrive at those seats in just such a way as to cause partial or total destructive interference. One of the main goals of acoustical design of auditoriums, a field that relies on complex computer modeling of sound interference patterns, is to avoid such problems.

Interference can also be seen at service stations on hot afternoons. If you look carefully at the oil slicks on the pavement, you often see a kind of iridescence, a rainbow play of colors on the dark oil surface. In this case, two light waves are interfering with each other. One wave is the sunlight that bounces off the top of the oil film; the other wave goes through the oil film and bounces off the bottom. These two waves may interfere constructively or destructively. As we shall see later, different wavelengths of light correspond to different colors. Light from different parts of the oil slick, then, produces different colors and creates the rainbow display.

constructive interference A situation in which two waves act together to reinforce or maximize the wave height at the point of intersection.

destructive interference A situation in which two waves intersect in a way that decreases or cancels out the wave height at the point of intersection.

Developing Your Intuition

Iridescence

When you clean the windows of your car on a warm sunny day, you often see a strange phenomenon. As you wipe the squeegee along the glass, an iridescent sheen of colors appears. The colors seem to move around and then fade. What's going on?

What you are seeing is another effect of constructive interference. The squeegee creates a thin film of water on the windshield. Two light waves, one reflecting from the top of the film of water and the other reflecting from the bottom, come together at your eye to produce colors as described above. But as the Sun evaporates the water, the thickness of the film changes, so that the wavelength corresponding to constructive interference changes as well. You see this as a change in color. When the film evaporates completely, the interference (and hence the display of colors) disappears. ●

Light that reflects from the front and back surfaces of soap film interferes to produce a pattern of different colors.

THE ELECTROMAGNETIC WAVE

Whenever you have had your teeth X-rayed, cooked a meal in a microwave oven, watched television, or listened to the radio, you have had firsthand experience with the phenomenon of electromagnetic waves.

Waves are characterized by a mathematical expression called a *wave equation*, which describes the movement of the medium for every wave, whether it's a water wave moving through a liquid, a sound wave in air, or a seismic wave causing an earthquake. (If you've ever taken a course in trigonometry, you will remember a solution to one such equation, called the sine function.) Physicists have learned that whenever a phenomenon can be described by an equation of this type, a corresponding wave should be seen in nature.

Soon after Maxwell wrote down the four equations that describe electricity and magnetism (see Chapter 7), he realized that some rather straightforward mathematical manipulation led to yet another equation—an equation that describes waves (see Figure 8–9). The waves that Maxwell predicted were rather strange, and we'll describe their anatomy in more detail later. The significance of his discovery was that these waves transferred energy not through matter, but rather through electric and magnetic fields. Based on the equations, whenever an electric charge is accelerated, for example, one of these waves is emitted. Maxwell called this phenomenon **electromagnetic waves** or **electromagnetic radiation.** Electromagnetic waves consist of electric and magnetic fields fluctuating together. They move by themselves, through the vacuum of space, without any medium.

Maxwell's equations also predicted exactly how fast the waves could move. The wave velocity depends only on known constants such as the universal electrostatic constant in Coulomb's equation (see Chapter 7). These numbers are known from experiments by Coulomb and other researchers. When Maxwell put the numbers into his expression for the velocity of his new waves, he found a very surprising answer. The predicted velocity of the wave turned out to be 300,000 km/sec (186,000 mi/s).

Figure 8–9
A sine function, one type of solution to the wave equation.

electromagnetic wave A form of radiant energy that reacts with matter by being transmitted, absorbed, or scattered. A self-propagating wave made up of electric and magnetic fields fluctuating together. A wave created when electrical charges accelerate, but requiring no medium for transfer; electromagnetic radiation.

If you just had an "aha" moment, you can imagine how Maxwell felt. This number is the speed of light, which means that the waves described by his equation are actually waves of light. Maxwell's equations reveal that the speed of light depends only on measurable quantities, such as the magnitude of forces between electrical charges or between magnets. This fact will become very important when we talk about relativity in Chapter 16.

Maxwell's result was astonishing. For centuries scientists had puzzled over the origin and nature of light. Newton and others had discovered natural laws that describe the connections between forces and motion, as well as the behavior of matter and energy. But light remained an enigma. How did radiation from the Sun travel to the Earth? What caused the light produced by a candle? Scholars had been hard pressed to advance beyond the poetry of the third verse of the Old Testament: "And God said, 'Let there be light': and there was light."

There is no obvious reason why static cling, refrigerator magnets, or the workings of an electric generator should be connected in any way to the behavior of visible light. Yet Maxwell discovered the connection: **light** and other forms of electromagnetic radiation are a kind of wave that is generated whenever electrical charges are accelerated.

light A form of electromagnetic wave to which the human eye is sensitive. Light travels at a constant speed in a vacuum and needs no medium for transfer.

The Anatomy of the Electromagnetic Wave

A typical electromagnetic wave, shown in Figure 8–10, consists of electric and magnetic fields arranged at right angles to each other and perpendicular to the direction the wave is moving. To understand how the waves work, go back to Maxwell's equations that describe how a changing magnetic field produces an electric field, and vice versa (Chapter 7). At the point labeled A in Figure 8–10, the electric and magnetic fields associated with the wave are at maximum strength, but these fields are changing slowly, so both fields decrease in strength. At point B, on the other hand, the fields are at minimum strength, but they are changing rapidly and so begin to increase. Thus, at point A, the magnetic and electric fields begin to die out, while at point B just the reverse happens. In this way, the electromagnetic wave leapfrogs through space, bouncing its energy back and forth between electric and magnetic fields as it goes.

Once you understand that the electromagnetic wave has this kind of ping-pong arrangement between electricity and magnetism, you can understand one of the most puzzling aspects of this relationship—the fact that it can travel through a vacuum. Every other wave discussed up to this point is easy to visualize because the wave moves on a medium. We know that the motion of the wave is not the same as the motion of

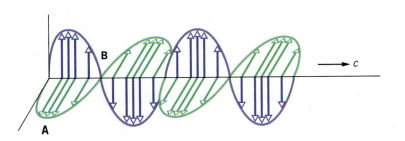

Figure 8–10
A diagram of an electromagnetic wave shows the relationship of the electric field, the magnetic field, and the direction that the wave is moving. A and B indicate points of maximum and minimum field strength.

the medium, but in every other way, the medium is there to give the wave solid support. Thus sound waves move through air, ocean waves move through water, and seismic waves move through rock. The electromagnetic wave is different. It is a wave that has no medium whatsoever, but that simply keeps itself going through its own internal mechanisms.

Remember that every wave carries energy. In the case of electromagnetic waves, energy is in the form of *radiation*, or *radiant energy*. This radiation is created when electrical charges accelerate, but once electromagnetic waves start moving, they operate according to their own mechanisms and no longer depend on the source that emitted them.

Science in the Making

The Ether

When Maxwell first proposed his idea of electromagnetic radiation he was not prepared to deal with a wave that required no medium. Earlier scientists who had studied light, including such luminaries as Isaac Newton, assumed that light must travel through a hypothetical substance called "ether" that permeates all space. Ether, they thought, served as the medium for light, and so Maxwell assumed that ether provided the medium for electromagnetic waves as well. Maxwell envisioned the ether to be something like a transparent tenuous Jell-O that filled all of space. An accelerating charge shook the Jell-O at one point, and after that the electromagnetic waves moved outward at the speed of light.

The idea of ether in one form or another goes back a long time, all the way to the Greeks; for most of recorded history, people filled the vacuum of space with this imaginary substance. It wasn't until 1887 that two U.S. physicists, Albert A. Michelson (1852–1931) and Edward W. Morley (1838–1923), working at what is now Case Western Reserve University in Cleveland, Ohio, did a series of experiments that demonstrated that the ether could not be detected. The failure to detect ether was interpreted by many scientists to mean that it did not exist.

The concept of the experiments was simple. They reasoned that if an ether really existed, then the motion of the Earth around the Sun and the Sun around the center of our part of the universe would produce an apparent ether "wind" at the surface of the Earth. They used very sensitive instruments to search for interference effects in light waves—effects that would result from the deflection of light by the ether wind. When their experiment showed no such deflection, they concluded that the ether could not exist.

In 1907, Albert Michelson became the first U.S. scientist to win a Nobel prize, an honor that recognized his pioneering experimental studies of light. ●

Albert Michelson

Light

Once Maxwell understood the nature of light, his equations allowed him to draw several important conclusions. For one thing, since the velocity of the electromagnetic waves depends entirely on the nature of interactions between electrical charges, it cannot depend on the properties of the wave itself. This means that all electromagnetic waves, regardless of their wavelength or frequency, have to move at exactly the same velocity. This velocity is the **speed of light,** which is so important in science that we give it a special letter, *c*. The speed of electromagnetic waves in a vacuum is one of the fundamental constants of nature.

For electromagnetic waves, then, the relation between velocity, wavelength, and frequency takes on a distinctive form:

$$\text{wavelength} \times \text{frequency} = c$$
$$= 300{,}000 \text{ km/s} \ (= 186{,}000 \text{ mi/s})$$

In other words, if you know the wavelength of an electromagnetic wave, you can calculate its frequency and vice versa.

The Energy of Electromagnetic Waves

Think about how you might produce an electromagnetic wave with a simple comb. Electromagnetic waves are generated any time a charged object is accelerated. Imagine combing your hair on a dry winter day when the comb picks up a static charge. Each time you move the comb back and forth, an electromagnetic wave traveling 300,000 kilometers per second is sent out from the comb.

If you wave the electrically charged comb up and down slowly, once every second, you create electromagnetic radiation, but you're not putting much energy into it. You produce a low-frequency, low-energy wave with a wavelength of about 300,000 kilometers. That electromagnetic wave moves outward 300,000 kilometers in a second, so that distance will be the separation between wave crests.

If, on the other hand, you could vibrate the comb vigorously—say at 300,000 times per second—you would produce a high-energy, high-frequency wave with a 1-kilometer wavelength. By putting more energy into accelerating the electrical charge, you have more energy in the electromagnetic wave.

Visible light, the first example of an electromagnetic wave known to humans, bears out this particular kind of reasoning. A glowing ember has a dull red color, corresponding to relatively low energy. Hotter, more energetic fires, on the other hand, show a progression of more energetic colors, from the yellow of a candle flame, to the blue-white flame of a blowtorch. These colors are merely different frequencies, and therefore different energies, of light; higher frequencies correspond to higher energies. High-frequency light corresponds to a blue color, low-frequency to a red color.

Red light has a wavelength corresponding to the distance across about 7000 atoms, or about 700 nanometers (a nanometer is 10^{-9} meter, or about 40 billionths of an inch, and is abbreviated nm). Red light is the longest wavelength that the eye can see, and is the least energetic of the visible electromagnetic waves. Violet light has a shorter wavelength, corresponding to the distance across about 4000 atoms, or about 400 nm, and is the most energetic of the visible electromagnetic waves. All of the other colors, such as green, yellow, and so on, have wavelengths and energies between those of red and violet.

speed of light (*c*) The velocity at which all electromagnetic waves travel in a vacuum, regardless of their wavelength or frequency; equal to 300,000 kilometers per second (about 186,000 miles per second).

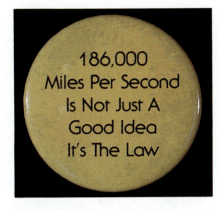

186,000 Miles Per Second Is Not Just A Good Idea It's The Law

Example 8–3: Figuring Out Frequency

The wavelength of red light is about 700 nm, or 7×10^{-7} m. What is the frequency of a red light wave?

▶ **Reasoning and Solution:** Again, recall the relationship of wavelength, frequency, and velocity. We know that for all electromagnetic waves

$$\text{wavelength} \times \text{frequency} = 300{,}000 \text{ km/s}$$
$$= 3 \times 10^8 \text{ m/s}$$

We want to determine frequency, so we have to rearrange this equation:

$$\text{frequency} = \frac{(3 \times 10^8 \text{ m/s})}{\text{wavelenth}}$$

This means that for red light with a wavelength of 7×10^{-7} m:

$$\text{frequency} = \frac{3 \times 10^8 \text{ m/s}}{7 \times 10^{-7} \text{ m}}$$
$$= 0.43 \times 10^{15} \text{ Hz}$$
$$= 4.3 \times 10^{14} \text{ Hz}$$

Remember, a hertz equals 1 cycle per second.

In order to generate red light by vibrating a charged comb, you would have to wiggle it at a frequency of more than four hundred trillion (400,000,000,000,000) times per second. ▲

The Doppler Effect

Once waves have been generated, their motion is independent of the source. It doesn't matter what kind of charged object accelerates to produce an electromagnetic wave: once produced, all such waves behave exactly the same way. This statement has an important consequence that was discovered in 1842 by the Austrian physicist Christian Johann Doppler (1803–1853), and is now called the Doppler effect in his honor. The **Doppler effect** describes the way the frequency of a wave appears to change if there is relative motion between the wave source and the observer.

Doppler effect The change in frequency or wavelength of a wave detected by an observer because the source of the wave is moving.

Let's take sound as an example. In Figure 8–11a, we show the way a sound wave looks when the source is stationary relative to a listener, such as when you listen to your radio. In this case everything sounds "normal."

If the source of sound is moving relative to the listener, like the racing fire engine illustrated in Figure 8–11b, a different situation occurs. Each cycle a pulse of sound moves away from its source and travels out in a sphere *centered on the spot where the source was when that particular crest was emitted.* By the time the source is ready to emit other crests, it will have moved. Thus the second sound-wave sphere emitted will be centered at the new location. In a similar way, the source will emit sound waves centered farther and farther to the right, producing a characteristic pattern as shown.

To an observer standing in front of the source, the crests appear to be crowded together. In other words, to this observer, the frequency of the wave appears to be higher than it would be to an observer moving with the source. In the case of a sound wave, this means that the sound will be higher pitched than it would have been had the source been

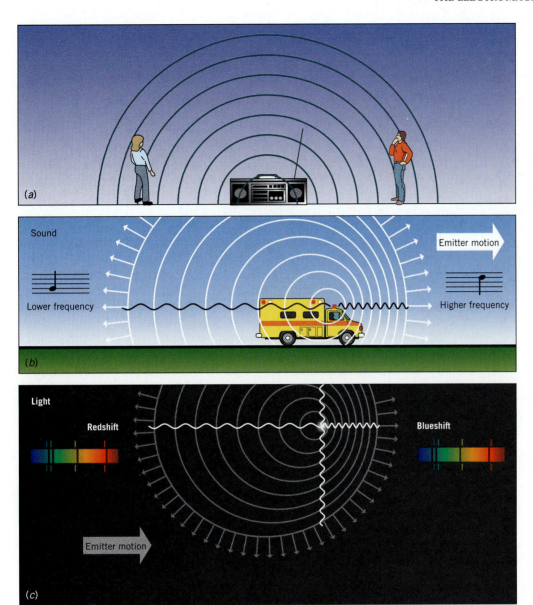

Figure 8–11
The Doppler effect occurs whenever a source of waves is moving relative to the observer of the waves. (*a*) If the source is not moving, everyone hears the same frequency. (*b*) Sound waves seem to increase or decrease in pitch, depending on whether the sound is approaching or receding. (*c*) The Doppler shift for light waves causes a blueshift for approaching sources, and a redshift for receding sources.

stationary. On the other hand, if the observer were standing in back of the source, the distance between crests would be stretched out, and the frequency and the pitch of the resulting sound would be lower.

You have probably heard the Doppler effect. Think of standing on a highway while cars go by at high speeds. The engine noise appears to be very high pitched as the car approaches you and then suddenly drops in pitch as the car passes you. This effect is particularly striking at automobile races where cars are moving at very high velocity. This change of pitch was, in fact, the first example of the Doppler effect to be studied. Scientists hired a band of trumpeters to sit on an open railroad car and blast a single long, loud note as the train whizzed by at a carefully controlled speed. Musicians on the ground determined the pitch they heard as the train approached and as it receded, and they compared those pitches to the actual note the musicians were playing.

The same sort of crowding together and stretching out of crests can occur for any wave, including light. If you are standing in the path of a source of light that is moving toward you, the light you see will be of higher frequency, and hence will look bluer than it would ordinarily. (Remember, blue light has a higher frequency than red light.) We say the light is **blueshifted.**

If, on the other hand, you are standing in back of the moving source, the distance between crests will be stretched out, and it will look to you as if the light has a lower frequency. We say that it is **redshifted.** In Chapter 22, we will see that the redshifting of light from distant moving sources is one of the main clues that we have about the structure of the universe.

The Doppler effect also has practical applications. Police radar units send out a pulse of electromagnetic waves that is absorbed by the metal in your car, then reemitted. The waves that come back will be Doppler shifted, and by comparing the frequency of the wave that went out and the wave that comes back, the speed of your car can be deduced. A similar technique is used by meteorologists, who use Doppler radar to measure wind speed and direction during the approach of potentially damaging storms (see Chapter 19).

Transmission, Absorption, and Scattering

The only way that we can learn about electromagnetic radiation is to observe its interaction with matter. For example, our eyes interact with light and send nerve impulses to our brain. These impulses are interpreted as what we "see."

When an electromagnetic wave hits matter, it interacts in one of three ways:

1. *Transmission.* Electromagnetic waves often pass right through matter in a process called **transmission.** Light is transmitted through your window, for example, and radio waves are transmitted through the air. During transmission, light may be modified in two related ways: it may bend, and its speed may be reduced. Light bends when passing through a transparent material by a process called **refraction** (Figure 8–12a). We take advantage of the phenomenon of refraction in eyeglasses and other optical devices, in which lenses focus light. During its transit through matter, light generally slows down. In air, this reduction in speed is negligible, but in some clear solids such as diamond, the speed of light is almost halved, to less than 200,000 kilometers per second.

2. *Absorption.* Matter may soak up an electromagnetic wave and its energy in the process of **absorption.** The energy of absorbed electromagnetic radiation is converted into some other form of energy,

blueshifted The result of the Doppler effect on light waves, when the source of light moves toward the observer: light wave crests bunch up and have a higher frequency.

redshifted The result of the Doppler effect on light waves, when the source of light moves away from the observer: light wave crests are farther apart and have a lower frequency.

transmission A process by which light energy passes through matter.

refraction A process by which light slows down or changes direction as it passes through matter.

absorption A process by which light energy is converted into some other form, usually heat energy.

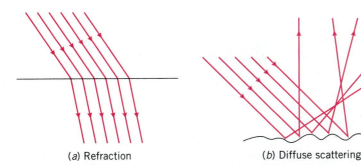

(a) Refraction   (b) Diffuse scattering   (c) Reflection

Figure 8–12
When light interacts with matter, it may be (a) refracted, (b) scattered diffusely, or (c) reflected.

usually heat. Black and dark colors, for example, absorb visible light: you've probably noticed how hot black pavement can become on a sunny day.

3. *Scattering.* Alternatively, electromagnetic waves may be absorbed and rapidly reemitted in the process of **diffuse scattering** (Figure 8–12b). Most white materials, such as a wall or piece of paper, scatter light in all directions. White objects such as clouds and snow, which scatter light from the Sun back into space, play a major role in controlling the Earth's climate (see Chapter 19). Mirrors, on the other hand, scatter light at the same angle as the original wave in the process of **reflection** (Figure 8–12c).

All electromagnetic waves are detectable in some way. For each of them to be useful, researchers must find appropriate materials to transmit, absorb, and scatter the waves. For each wavelength there must be instruments that produce the waves, and others that detect their presence. The optical instrument we call the human eye detects only a narrow range of electromagnetic waves, but scientists have devised an extraordinary array of transmitters and detectors to produce and measure electromagnetic radiation that we can't see. We'll learn about some of these transmitters and detectors in the following section.

diffuse scattering   A process by which light waves are absorbed and reemitted in all directions by a medium such as clouds or snow.

reflection   A process by which light waves are scattered at the same angle as the original wave; for example, from the surface of a mirror.

THE ELECTROMAGNETIC SPECTRUM

A profound puzzle accompanied Maxwell's original discovery that light is an electromagnetic wave. Waves can be of almost any length. Water waves on the ocean, for example, range from tiny ripples to globe-spanning tides. Yet visible light spans an extremely narrow range of wavelengths—only about 3.9 to 7.1×10^{-9} meter.

According to the equations that Maxwell derived, electromagnetic waves could exist at any wavelength (and, consequently, any frequency) whatsoever. The only constraint is that the wavelength times the frequency must be equal to the speed of light. Yet when Maxwell looked into the universe, all he could see was the narrow band we call visible light. It was as if a splendid symphony, ranging from the deep bass of the tuba to the sharp shrill of the piccolo, was poised to play, but you could only hear a couple of notes. There had to be more out there.

In such a situation, it was natural to wonder what had happened to the rest of the waves. Scientists looked at Maxwell's equations, looked at nature, and realized that something was missing. The equations predicted

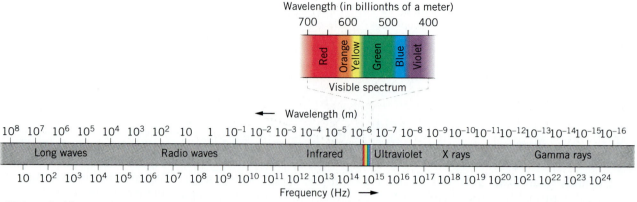

Figure 8–13
The electromagnetic spectrum includes all kinds of waves that travel at the speed of light, including radio, microwave, infrared, visible light, ultraviolet, X-rays, and gamma rays. Note that sound waves, water waves, seismic waves, and other kinds of waves that require matter to move travel much slower than light speed.

electromagnetic spectrum The entire array of waves, varying in frequency and wavelength, but all resulting from an accelerating electric charge; includes radio waves, microwaves, infrared, visible light, ultraviolet, X-rays, gamma rays, and others.

that there ought to be more kinds of electromagnetic waves than visible light. There should be other electromagnetic waves performing the waltz between electricity and magnetism, but with frequencies and wavelengths different from those of visible light.

These as yet unseen waves would have exactly the same structure as the one shown in Figure 8–10, scientists thought, but they could have either longer or shorter wavelengths than visible light depending on the acceleration of the electric charge that created them. These waves would move at the speed of light, and would be exactly the same as visible light except for the differences in the wavelength and frequency.

Between 1885 and 1889, German physicist Heinrich Rudolf Hertz (1857–1894), after whom the unit of frequency is named, performed the first experiments that confirmed these predictions. He identified what we now know as radio waves. Subsequently, all manner of electromagnetic waves have been discovered, from those with wavelengths longer than the radius of the Earth to those with wavelengths shorter than the size of the atom's nucleus. They include radio waves, microwaves, infrared, visible light, ultraviolet, X-rays, and gamma rays. This entire symphony of waves is called the **electromagnetic spectrum** (see Figure 8–13). Remember that every one of these waves, no matter what its wavelength or frequency, is the result of an accelerating electric charge. Furthermore, the product of the wavelength and frequency is a constant, the speed of light.

Radio Waves

radio wave Part of the electromagnetic spectrum that ranges from the longest waves—wavelengths longer than the Earth—to waves a few meters long.

The **radio wave** portion of the electromagnetic spectrum ranges from the longest waves—those whose wavelength is longer than the size of the Earth—to waves a few meters long. The corresponding frequencies, from roughly a *kilohertz* (1000 cycles per second, abbreviated kHz) to several hundred *megahertz* (1 million cycles per second, abbreviated MHz), are familiar to you as the numbers on your radio dial. From a practical point of view, the most important fact about radio waves is that, like light, they can penetrate long distances through the atmosphere. This makes radio waves very useful in communication systems.

Have you ever been driving at night and picked up a radio signal from a station a thousand miles away? If so, you have experienced firsthand the ability of radio waves to travel long distances through the atmosphere. Astronomers take advantage of this "radio window" in the atmosphere, which allows Earth-based telescopes to monitor radio waves emitted by objects in the sky.

A typical radio wave used for communication can be produced by pushing electrons back and forth rapidly in a tall metal antenna with a radio amplifier. This acceleration of electrons produces outgoing radio waves, just as throwing a rock in a pond produces outgoing ripples. A fraction of a second later, when these waves encounter another piece of metal (for example, the antenna in your radio or TV set), the electric fields in the waves accelerate electrons in that metal, so that its electrons move back and forth. The electronics in your radio receiver detect these electron motions and convert them into a sound or a picture.

Most construction materials are at least partially transparent to radio waves. Thus you can listen to the radio even in the basement of most buildings. In long tunnels or in deep valleys, however, absorption of radio waves by many feet of rock and soil may limit reception.

Technology

AM and FM Radio Transmission

Radio waves carry signals in two ways: amplitude modulation (AM) and frequency modulation (FM). Broadcasters send out their programs at only one narrow range of frequency—a situation very different from music and speech, which use a wide range of frequencies. Thus radio stations cannot simply transform a range of sound wave frequencies into a similar range of radio wave frequencies. Instead, the information to be transmitted must be impressed, in some way, on the narrow frequency range of your station's radio waves.

This problem is similar to one you might experience if you had to send a message across a lake with a flashlight at night. You could adopt two strategies. You could send a coded message by turning the flashlight on and off, thus varying the brightness (the amplitude) of the light. Alternatively, you could change the color (the frequency) of the light by passing blue or red filters in front of the beam.

Radio stations also adopt these two strategies (Figure 8–14). All stations begin with a *carrier wave* of fixed frequency. AM radio stations typically broadcast at frequencies between about 530 and 1600 kHz (thousands of hertz), while the carrier frequencies of FM radio stations range from about 88 to 110 MHz (millions of hertz). The AM frequencies are partially reflected off the layers of the atmosphere, which is why AM radio stations can be heard over greater distances.

The process called **amplitude modulation,** or **AM,** depends on varying the strength (or amplitude) of the radio's carrier wave according to the sound signal to be transmitted. Thus the shape of the sound wave is impressed on the radio's carrier wave signal. This signal is absorbed by your radio antenna, and the electronics in the radio are designed so that the original sound signal is recovered and used to run the speakers. The original sound signal is what you hear when you turn on your radio.

amplitude modulation (AM) A process that a radio station uses to vary the amplitude of a radio wave signal being transmitted, usually between 530 and 1600 kilohertz. After transmission, the signal is converted to sound by the radio receiver.

Figure 8–14
AM (amplitude modulation) and FM (frequency modulation) transmission differ in the way that a sound wave (*a*) is superimposed on a carrier wave of constant amplitude and frequency (*b*). The frequency of the carrier wave can be varied, or modulated, to carry information.

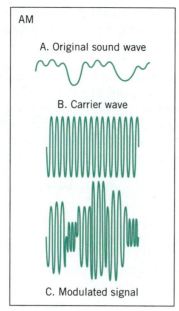

(*a*) Amplitude modulation

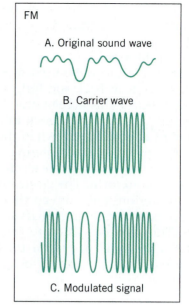

(*b*) Frequency modulation

Alternatively, you can vary the frequency of the radio's wave slightly to convey the signal you want to transmit, a process called **frequency modulation,** or **FM,** as shown in Figure 8–14*b*. A radio that receives this particular signal will unscramble the changes in frequency and convert them into electrical signals, which are amplified and converted to sound by the speakers, so that you can hear the original signal. TV broadcasts, which use carrier frequencies much higher than FM radio, typically send the picture on an AM signal, and the sound on an FM signal at a slightly different frequency. ●

frequency modulation (FM) A process by which information is transmitted by varying the frequency of a signal. After being transmitted, the signal may be converted to sound by circuits in the receiver.

Example 8–4:   The FM Band

You turn on your car radio and hear an announcer give the call letters of the station and add "at one-oh-three point five on your FM dial." What is the wavelength of the radiation being absorbed by your car's antenna?

▶ **Reasoning and Solution:**   The announcer is saying that the carrier frequency of the station is 103,500,000 Hz, or 103.5 MHz. Since we know that radio waves travel at the speed of light (300,000,000 m/s), we can find the wavelength as follows:

$$\text{wavelength} = \frac{\text{velocity}}{\text{frequency}}$$

$$= \frac{300,000,000 \text{ m/s}}{103,500,000/\text{s}}$$

$$= 2.899 \text{ m}$$

In other words, the wave that is accelerating the electrons in your car's antenna as you listen to the announcer is about 3 m (about 9 ft) long. ▲

A satellite dish is a lens that focuses microwaves beamed down from a transmitter in orbit above the Earth.

Microwaves

Microwaves are electromagnetic waves whose wavelengths range from about 1 meter (a few feet) to 1 millimeter (a thousandth of a meter, or about a twentieth of an inch). There are no absolute boundaries to the different kinds of electromagnetic waves, and the wavelengths of microwaves and radio waves overlap. Longer wavelengths of microwaves travel easily through the atmosphere, and they are not reflected off atmospheric layers like the radio part of the spectrum. Therefore, microwaves are used extensively for line-of-sight communications. Most satellites broadcast signals to the Earth in microwave channels, so they are commonly used for long-distance telephone calls and television broadcasts. The dish-shaped satellite antennas that you see at private homes and businesses are designed primarily to receive and use microwave transmissions, as are the large cone-shaped receivers attached to the microwave relay towers that are situated on many hills or tall buildings.

The distinctive transmission and absorption properties of microwaves make them ideal for use in aircraft radar. Solid objects, especially those made of metal, reflect most of the microwaves that hit them. By sending out timed pulses of microwaves and listening for the echo, you can judge the direction, distance (from the time it takes the wave to travel out and back), and speed (from the Doppler effect) of a flying object. Modern military radar is so sensitive that it can detect a single fly at a distance of a mile. To counteract this sensitivity, aircraft designers have developed planes with "stealth" technology, which combines microwave-absorbing materials, angled shapes that reduce the apparent cross section of the plane, and electronic jamming to avoid detection.

microwave Electromagnetic waves, with wavelengths ranging from approximately 1 meter to 1 millimeter, which are used extensively for line-of-sight communications and cooking.

Technology

Microwave Ovens

The same kind of waves that are used for phone calls, television broadcasts, and radar can be used to cook your dinner in an ordinary micro-

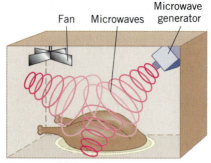

Figure 8–15
Every microwave oven contains a device that generates microwaves by accelerating electrons, and walls that scatter the microwaves until they are absorbed, usually by water molecules in the material being heated.

infrared radiation Wavelengths of electromagnetic radiation that extend from a millimeter to a micron; felt as heat radiation.

wave oven. The oven works as shown in Figure 8–15. A special electronic device in the oven accelerates electrons rapidly and produces the microwave radiation, which is a form of energy. These microwaves are guided into the main cavity of the oven, which is composed of material that scatters microwaves. Thus the wave energy remains inside the box until it is absorbed.

Some microwaves are absorbed quickly by water molecules. This means that the energy that was used to create microwaves is carried by those waves to food inside the oven, where it is absorbed by water and converted into heat. This absorption of microwave energy results in a very rapid rise in temperature and thus rapid cooking. Microwaveable paper, glass, and other materials that do not contain water molecules are not heated by microwaves. Despite the different applications, there is no fundamental difference between the microwaves that are used for cooking and those that are used for communication. ●

Infrared Radiation

Infrared radiation includes wavelengths of electromagnetic radiation that extend from a millimeter (a thousandth of a meter) down to about a micrometer (a millionth of a meter). Our skin, which absorbs infrared radiation, provides a crude kind of detector. You feel infrared radiation when you put your hands near a warm fire or the cooking element of an electric stove. Infrared waves are what we feel as heat radiation.

Warm objects emit infrared radiation, and this fact has been used extensively in both civilian and military technology. Infrared detectors are used to guide air-to-air missiles to the exhaust of jet engines in enemy aircraft, and infrared detectors are often used to "see" human beings and other warm objects at night. Similarly, many insects such as mosquitoes and moths and other nocturnal animals (including opossums and some snakes) have developed sensitivity to infrared radiation, and they can thus "see" in the dark.

Infrared detection is also used to find heat leaks in homes and buildings (Figure 8–16). If you take a photograph of a house at night using film that is sensitive to infrared radiation, places where heat is leaking out will show up as bright spots on the film. This information can be used to deal with the heat loss and thus conserve energy. In a similar way, Earth scientists often monitor volcanoes with infrared detectors. The appearance of a new "hot spot" may signal an impending eruption several days in advance.

Figure 8–16
This "false-color" image is coded so that white is hottest, followed by red, pink, blue, and black.

Figure 8–17
A glass prism separates white light into the visible spectrum.

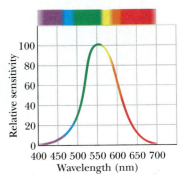

Figure 8–18
The relative sensitivity of the human eye differs for different wavelengths. Our perception peaks at wavelengths that we perceive as yellow, though the colors we see have no special physical significance.

Visible Light

All of the colors of the rainbow are contained in **visible light,** whose wavelengths range from red light at about 700 nm down to violet light at about 400 nm (Figure 8–17). The visible energy that we detect with our eyes is a very small part of the total electromagnetic spectrum (see Figure 8–13).

Our eyes distinguish several different *colors*, but these portions of the electromagnetic spectrum have no special significance except in our perceptions (Figure 8–18). In fact, the distinct colors that we see, such as red, yellow, green, and blue, represent very different-sized slices of the spectrum. The red and green portions of the spectrum are rather broad, spanning more than 50 nanometers of frequencies; we thus perceive many different wavelengths as red or blue. In contrast, the yellow part of the spectrum is quite narrow, encompassing only wavelengths from about 570 to 590 nanometers.

Why should the colors we perceive as yellow cover so restricted a range of wavelengths? The answer may be related to the fact that the Sun's light is especially intense in this part of the spectrum. Some biologists suggest that our eyes evolved to be especially sensitive to these wavelengths, to take maximum advantage of the Sun's light. Our eyes are ideally adapted for the light produced by our Sun during daylight hours. Animals such as owls and cats, who hunt at night, have eyes that are more sensitive to infrared wavelengths; this radiation makes warm living things stand out against a cooler background.

visible light Electromagnetic waves with a wavelength that can be interpreted by receptors in the brain: wavelengths range from 700 nanometers for red light to 400 nanometers for violet light.

A variety of chemical reactions produce light energy. (*a*) Fire. (*b*) Light sticks. (*c*) A firefly.

(*a*)

(*b*)

(*c*)

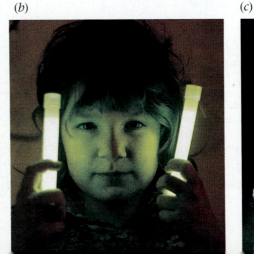

(a)

Two examples of fluorescence: (*a*) Fluorescent crayons under ultraviolet lights. (*b*) As they fluoresce, two different minerals show up as different colors under ultraviolet light. Willemite appears green and calcite appears red.

(b)

Ultraviolet Radiation

ultraviolet radiation   High-frequency wavelengths, shorter than visible light, ranging from 400 nanometers to 100 nanometers.

At wavelengths shorter than visible light, we begin to find waves of high frequency and therefore high energy and potential danger. The wavelengths of **ultraviolet radiation** range from 400 nanometers down to about 100 nm (the size of a thousand atoms placed end-to-end) in length. The energy contained in longer ultraviolet waves can cause a chemical change in skin pigments, the phenomenon known as tanning. This lower-energy portion of the ultraviolet is not particularly harmful by itself.

Shorter-wavelength ultraviolet radiation, on the other hand, carries more energy. It produces enough energy to damage skin cells, causing sunburn and skin cancer in human beings. The wave's energy is absorbed by cells and can cause extensive damage to DNA, the molecule that carries chemical instructions in our bodies. This damage may lead to errors in the way the body's chemicals are produced, called *mutations*. The ability

During the restoration of the tomb of Nefertiti in Egypt, special paper was applied to the surface to stop disintegration. Under ultraviolet light, the adhesive on the paper fluoresces and appears blue.

of ultraviolet radiation to kill living cells is used by hospitals to sterilize equipment and kill unwanted bacteria that might cause infections.

The Sun produces intense ultraviolet radiation in both the longer and shorter wavelengths. Our atmosphere absorbs much of the harmful short wavelengths and thus shields living things. Nevertheless, if you spend much time outdoors under a bright Sun, you should protect exposed skin with a sunblocking chemical, which is transparent (colorless) to visible light, but prevents harmful ultraviolet rays from reaching your skin.

The energy contained in both long and short ultraviolet wavelengths can be absorbed by atoms, which in special materials may subsequently emit a portion of that absorbed energy as visible light. Remember, both visible light and ultraviolet light are forms of electromagnetic radiation, but visible light has longer wavelengths, and therefore less energy. This phenomenon, called **fluorescence,** provides the so-called "black light" effects so popular in stage shows and night clubs. Fluorescence is also used by the postal service, where they commonly tag postage stamps with invisible fluorescent ink. This tagging enables automated machines with ultraviolet lights to locate and cancel stamps with great efficiency.

fluorescence A phenomenon in which energy contained in ultraviolet wavelenths is absorbed by the atoms in some materials and partially emitted as visible light.

X-rays High-frequency and high-energy electromagnetic waves that range in wavelength from 100 nanometers to 0.1 nanometers, used in medicine and industry.

X-rays

X-rays are electromagnetic waves that range in wavelength from about 0.1 to 100 nanometers (a billionth of an inch). These high-frequency (and thus high-energy) waves can penetrate several centimeters into most solid matter, but are absorbed to different degrees by all kinds of materials. This fact allows X-rays to be used extensively in medicine to form visual images of bones and organs inside the body (Figure 8–19). Bones and teeth absorb X-rays much more efficiently than skin or muscle, so a detailed picture of inner structures emerges. X-rays are also extensively used in industry to inspect for defects in welds and manufactured parts.

The X-ray machine in your doctor's or dentist's office is something like a giant light bulb with a glass vacuum tube. At one end of the tube, a thin wire of the metal called a *filament* is heated to very high temperature by an electrical current, just like in an incandescent light bulb. At the other end of the tube is a polished metal plate. X-rays are produced by applying an extremely high voltage, which is negative on the filament and positive on the metal plate, so electrons stream off the filament and smash into the metal plate at high velocity. The sudden deceleration of the negatively charged electrons releases a flood of high-energy electromagnetic radiation—the X-rays that travel from the machine to you at the speed of light.

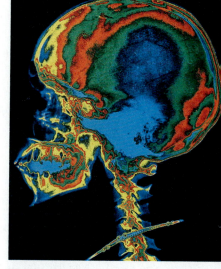

Figure 8–19
False color X-ray of the skull of a young woman wearing a necklace.

Gamma Rays

The highest energies in the electromagnetic spectrum are called **gamma rays.** Their wavelengths range from slightly less than the size of an atom (about 0.1 nm, or 10^{-10} m) to the size of a nucleus, which is less than a trillionth of a meter. Gamma rays are normally emitted only in very high-energy nuclear and particle reactions (see Chapter 15), and they are not as common in nature as the other kinds of radiation that we have discussed.

gamma ray The highest-energy wave of the electromagnetic spectrum, with wavelengths between the size of an atom and the size of a nucleus, less than one trillionth of a meter; normally emitted in very high-energy nuclear particle reactions.

Gamma rays have many uses in medicine. Some types of medical diagnosis involve giving the patient a radioactive chemical that emits gamma rays. If that chemical concentrates at places where bone is actively healing, for example, then doctors can monitor the healing by locating the places where gamma rays are emitted. The gamma ray detectors used in this specialized form of nuclear medicine are both large (to capture the energetic waves) and expensive. Doctors also use gamma rays to treat cancer in humans. In these treatments, high-energy gamma rays are directed at tumors or malignancies that cannot be removed surgically. If the gamma ray energy is absorbed in those tissues, the tissues will die and the patient will have a better chance to live.

Gamma rays are also used extensively in astronomy because many of the interesting processes going on in our universe involve bursts of very high energy and, hence, the emission of gamma rays.

Environment

Is ELF Radiation Dangerous?

We usually think of electromagnetic radiation in terms of relatively high-frequency waves, like radio waves and microwaves at millions or billions of hertz, but Maxwell's equations tell us that *any* accelerated charge will emit waves of electromagnetic radiation, not just those that are shaken rapidly. In particular, electromagnetic radiation is generated when electrons move back and forth in wires at tens or hundreds of cycles per second to produce the alternating current in household wiring and overhead power lines. Every object in which AC (alternating current) electric power flows, from power lines to toasters, is a source of this weak, extremely low-frequency radiation or ELF radiation.

For more than a century, human beings in industrialized countries have lived in a sea of weak ELF radiation, but, until recently, no questions were raised about whether that radiation might have an effect on human health. In the late 1980s, however, a series of books and magazine articles created a minor sensation by claiming evidence that exposure to ELF radiation causes some forms of cancer, most notably childhood leukemia.

Most scientists downplayed these claims. They pointed out that the electric fields due to power lines at the location of a cell are a thousand times smaller than those due to natural causes (such as electrical activity in nearby cells). They pointed out that age-corrected cancer rates in the United States (with the exception of lung cancer, which is caused primarily by smoking) have remained constant or dropped over the last 50 years, though exposure to ELF radiation has increased enormously. They also questioned the validity of studies claiming correlations with cancer, arguing that when a more detailed analysis of results was conducted, the correlation disappeared.

Activists who suspected a lurking danger in power lines argued that the effect they were trying to measure was very small and, therefore, easy to lose in the mass of carcinogens to which Americans are ex-

Power lines produce ELF radiation. There is a debate about whether this type of radiation can affect living things.

posed. Opponents counter that numerous health scares in the past have proved to be utterly without basis, and predict that ELF will play out the same way. They point out that original studies of problems such as the link between smoking and lung cancer provoked unambiguous answers from the start.

This angry debate is typical of encounters at the border between science and public health. Preliminary data indicate a possible health risk in the environment, but do not prove that the risk is real. Settling the issue by further study takes years, and in the meantime, people have to make life-style decisions. In addition, as in the case of ELF radiation, the cost of removing the risk is often very high.

Suppose you were a scientist who had inconclusive evidence that some common food such as bread or a familiar kind of fruit could be harmful. What responsibility would you have to make your results known to the general public? If you stress the uncertainty of your results and no one listens, should you make sensational (perhaps unsupported) claims to get people's attention? ●

THINKING MORE ABOUT
THE ELECTROMAGNETIC SPECTRUM

Real Estate

Since Hertz's discovery of radio waves more than a century ago, the electromagnetic spectrum has been used extensively for communication. We have discussed AM and FM signals and how they work, but what happens if two stations send out signals at the same frequency? A receiver will pick up both transmissions and, in general, will not be able to interpret either. Consequently, someone (usually an agency of the government) has to assign frequencies and ensure that only a single signal is being broadcast on each one. In the United States, the Federal Radio Commission (now the Federal Communication Commission, or FCC) was established in 1927 to do this job.

Think about the electromagnetic spectrum from radio to microwave; then, consider it as valuable publicly owned real estate. People who want to rent that space include radio and TV stations, satellite broadcasters, navigators, cellular phone companies, and various civilian band broadcasters from police and fire departments to forest rangers and game wardens. The FCC monitors the electromagnetic real estate to ensure that it is distributed so that everyone can use it without interfering with each other. In Figure 8–20, we show the current radio spectrum allocations.

Different kinds of uses require different spreads of frequencies. Each TV channel, for example, gets about 6 MHz (megahertz, or million hertz), from Channel 2 (50 to 60 MHz) to channel 69 (800 to 806 MHz). Radio stations require a narrower band; a typical AM radio station might be assigned 10 KHz (kilohertz), so a station whose nominal frequency is 100 KHz would be permitted to broadcast from 95 to 105 KHz.

In the beginning, the FCC assigned frequencies free of charge. Any broadcaster who met a set of standards could apply for and receive the right to use a particular frequency in a particular area. In July 1994, however, this policy was changed, and the FCC began holding auctions to sell off its valuable real estate. Both national and local licenses were sold, netting the Treasury some $833 million. Analysts expect proceeds of over $10 billion from such sales during the next few years.

Do you think that people and businesses should have to pay for the right to use the electromagnetic spectrum? Should the government require public service as well as money from companies that use the spectrum?

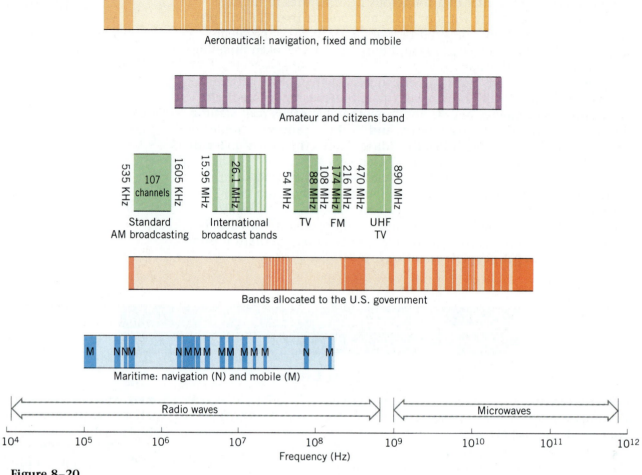

Figure 8–20
Radio spectrum allocations.

▶ Summary

Waves provide a way to transfer energy from one place to another through a medium without matter travelling across the intervening distance. Every wave can be characterized by four characteristics: *wavelength*, *velocity*, *amplitude*, and *frequency* (measured in cycles per second, or *hertz*). *Transverse waves*, such as swells on the ocean, occur when the medium moves perpendicular to the direction of the waves. *Longitudinal waves*, such as sound, occur when the medium moves in the same direction as the wave.

Two waves can interact with each other, causing constructive or destructive *interference*. The observed frequency of a wave depends on the relative motion of the wave's source and the observer—a phenomenon known as the *Doppler effect*.

The motion of every wave can be described by a characteristic wave equation. James Clerk Maxwell, by

manipulating his equations that describe electricity and magnetism, predicted the existence of *electromagnetic waves* or *electromagnetic radiation*. He found that alternating electric and magnetic fields can travel through a vacuum at the *speed of light*. This discovery solved one of the oldest mysteries of science, the nature of *light*. While visible light was the only kind of electromagnetic radiation known to Maxwell, he predicted the existence of other kinds with longer and shorter wavelengths. Soon thereafter, a complete *electromagnetic spectrum* of waves, including *radio waves*, *microwaves*, *infrared radiation*, *visible light*, *ultraviolet radiation*, *X-rays*, and *gamma rays*, was recognized.

Electromagnetic radiation can interact with matter in three ways: it can be *transmitted*, *absorbed*, or *scattered*. We use these properties in countless ways every day—radio and TV, heating and lighting, microwave ovens, tanning salons, medical X-rays, and more. Much of science and technology during the past one hundred years has been an effort to find new and better ways to produce, manipulate, and detect this extraordinary range of electromagnetic radiation.

▶ Key Equations

wave velocity (in m/s)
$$= \text{wavelength (in m)} \times \text{frequency (in Hz)}$$

$$\text{Wavelength (in m)} = \frac{\text{Velocity (in m/s)}}{\text{Frequency (in Hz)}}$$

1 hertz = 1 cycle/second

For light:
$$\text{wavelength (in m)} \times \text{frequency (in Hz)} = c$$

▶ Constant: Speed of Light

$$c = 300{,}000 \text{ km/s} = 3 \times 10^8 \text{ m/s}$$

▶ Review Questions

1. Identify the two different forms in which energy can be transferred.

2. What are three characteristics that can be used to describe a wave? How are these three things related?

3. What are some everyday examples of waves?

4. Describe the difference between a longitudinal wave and a transverse wave.

5. What happens when two different waves overlap?

6. What causes an electromagnetic wave?

7. What features do all electromagnetic waves have in common?

8. Why did Maxwell think there were kinds of electromagnetic radiation other than visible light?

9. What was the first kind of electromagnetic radiation other than visible light to be discovered? Who discovered it? How was he honored by other scientists?

10. What three things can happen when electromagnetic radiation hits a piece of matter?

11. What are some uses of microwaves?

12. Give one important difference between the effects of long and short wavelengths of ultraviolet light.

13. What are some uses of gamma rays?

14. What kinds of electromagnetic radiation can you detect with your body?

15. What are similarities and differences between water waves and light waves?

16. Describe the Doppler effect.

17. Identify an object that absorbs radio waves.

18. Identify an object that absorbs microwaves.

19. Identify an object that absorbs visible light.

20. Explain how sunlight can damage your skin.

21. Describe seismic waves. How are they used?

22. What kind of wave is produced at a ball game as people stand and sit.

23. What properties are exhibited by particles, but not by waves?

24. How do animals such as bats perceive their prey?

25. Describe the phenomenon of interference.

26. What is the difference between constructive and destructive interference?

▶ Fill in the Blanks

Complete the following paragraphs with words and phrases from the list.

constructive interference	infrared radiation
destructive interference	interference
Doppler effect	light
electromagnetic spectrum	microwaves
	radio waves
electromagnetic wave, or electromagnetic radiation	the speed of light, c
	ultraviolet radiation
	visible light
frequency	wave
gamma rays	wavelength
hertz	X-rays

   Energy can be transmitted from one point to another by a particle or a _____. A wave is characterized by its _____, the distance between crests; its _____, the number of crests passing a point each second; and its velocity. A wave for which one crest passes each second has a _____ of one _____. _____ is a phenomenon that occurs when two waves come together at a point. If the waves reinforce each other, we say that there is _____ _____; if they cancel each other out, we say there is _____ _____. Waves emitted by a moving source exhibit the _____, which changes the apparent frequency depending on whether the source is moving toward or away from the observer.

   Maxwell's equations predicted the existence of a particular kind of wave, the _____, which moves at a speed of 300,000 km/s, also known as _____. The many different wavelength of this radiation are referred to as the _____, which includes (in increasing order of frequency) _____, _____, _____, _____, _____, _____, and _____.

▶ Discussion Questions

1. Describe the major differences between a sound wave and a radio wave.

2. If a tree falls in a forest, what kinds of waves are created? Where did the energy that produced those waves come from?

3. Propose an experiment to test whether lead is transparent to radio waves.

4. Imagine that you are looking into a store window on a sunny day. You see yourself reflected in the glass as you look at the items for sale. Describe the different ways that the sunlight is interacting with matter.

5. How did Michelson and Morley demonstrate that ether does not exist? What was their hypothesis?

6. What do you think causes the beautiful iridescent sheen of a butterfly wing?

7. Why do people wear light-colored clothing in summer and dark-colored clothing in winter?

8. Why might a car radio emit static when you are driving through an area with lots of tall buildings?

9. What characteristic of X-rays makes them well suited for medical diagnosis? Are X-rays fundamentally different from other kinds of electromagnetic radiation?

10. Retell the principle of interference in your own words.

11. The "wave" is a common activity in large football stadiums. Is the "wave" an example of a transverse or longitudinal wave? Explain.

12. In the elementary grades, some teachers will illustrate a wave with the following activity. The students will stand in a straight line side-by side, shoulder-to-shoulder, with their hands at their sides. When the first student is tapped on the shoulder by the teacher, that student then touches the next student at her side, who in turn will then touch the next student down the line until the final student is touched.

   a. This is an example of what type of wave? Explain.

   b. If the pupils were holding hands, would the time for the wave to reach the last student be longer or shorter than the first example? Why?

   c. If the students were standing at arms' length, instead of side-by-side, would the time it takes for the wave to reach the last student be shorter or longer than the first example? Why?

   d. Using this information, can you relate the speed of a longitudinal wave to the density of the medium?

   e. Instead of touching the next student, the students will jump up off the floor, and the next pupil cannot jump until the previous student touches the floor. In this case, which type of wave do you have? Explain.

13. List radio waves, microwaves, radar, infrared rays, yellow light, red light, green light, ultraviolet light, and X-rays according to:

   a. increasing wavelength

   b. increasing frequency

   c. speed (assume that the waves are in vacuum)

14. Why was "ether" introduced to explain electromagnetic waves?

15. Do you agree that water waves and light waves have similar properties? Explain your response.

16. Using a diagram or graph, outline a redshift with a stationary observer.

17. While surfing at the Hawaiian Pipeline, Aaron noticed that about every 25 minutes the waves were at their maximum height. Of what principle is this an example? Justify your answer.

▶ Problems

1. An organ pipe is 3.1 m long. What is the frequency of the sound it produces? Extra credit: To what pitch does that frequency correspond?

2. An ocean liner experiences broad waves, called swells, with a frequency of one every 10 s (0.1 Hz) and a wavelength of 220 ft. Assume the waves are moving due east. If the liner maintains a speed of 15 mph, will it have a smoother trip going east or west? Why?

3. Radio and TV transmissions are being emitted into space, so "Star Trek" episodes are streaming out into space. The nearest star is 9.5×10^{17} m away. If life exists on a planet near this star, how long will they have to wait for the next episode?

4. Classify the following electromagnetic waves into the human colors:

   a. 650 nm

   b. 7.5×10^{14} Hz

   c. 580 nm

   d. 0.45 micrometers

5. You are pushing your little sister on a swing and in 1.5 minutes you make 45 pushes. What is the frequency (in hertz) of your swing-pushing efforts?

6. Andrea was watching her brother in the ocean and noticed that the waves were coming in on the beach at a frequency of 0.33 Hz. How many waves hit the beach in 15 s?

7. Andrea asked her brother to take a 6-ft floating raft out on the water. Using this raft as a measuring tool, she estimated that the wavelengths of these particular ocean waves were about 9 ft. How fast are these surface ocean waves if the frequency remains 0.33 Hz?

8. An octave is a note that has twice the frequency of the corresponding note.

   a. What frequency is the first higher octave of middle A?

   b. What frequency is the first lower octave of middle A?

   c. What is the wavelength of the second octave (higher) of middle A?

9. While hiking, a common practice to calculate how far away lightning may be is to count the number of

seconds it takes for you hear thunder after you have seen the lightning stroke from a cloud. If it takes 5.5 s for thunder to reach you, how far away is the cloud? (Use 340 m/s for the velocity of sound.)

10. Jose was studying astronomy this semester and decided to check the relationship between the redshift of a certain spectral line (measured in microns) and the velocity of several galaxies (measured in km/s). Jose's data are presented below.

Velocity of Galaxy (km/s)	Change in Spectral Line (microns)
3,000	0.010
4,500	0.015
9,000	0.030
18,000	0.060
27,000	0.090
54,000	0.180
60,000	0.200

a. Describe and identify in words the pattern that you and Jose observe.

b. Using a graph, plot the data that will best illustrate this pattern.

c. Express this trend or pattern in an equation with words.

d. Express this trend or pattern in an equation with symbols.

11. What is the length of the FM wave transmitted from FM station "97.7 (MHz) on your dial?"

12. Rosa and Jon were asked by their physical science teacher to determine the speed of sound. While walking to their dormitories after class, Jon yelled out in frustration, which Rosa and Jon heard momentarily as an echo. The echo bounced off a building that was 300 ft away. Jon and Rosa realized that they could not measure the time for a clap to return, so they had a brilliant idea. Jon yelled and then started to clap as soon as he heard the echo, and he then continued this synchronized clapping so Rosa could measure the frequency. Rosa counted 56 of Jon's claps in one-half minute.

a. What is the frequency of Jon's clapping?

b. What is the speed of sound as determined by Rosa and Jon?

c. How does this compare with the speed of sound at room temperature (about 340 m/s)?

13. Rosa and Jon decided to test their results, so they walked an additional 100 ft away from the wall. Using their calculated sound speed in problem 12, answer the following,

a. Predict their new clapping frequency. Will it be greater than, equal to, or less than the frequency determined in problem 12?

b. What is their new frequency of clapping?

c. How many claps will Jon have to make in half a minute?

▶ Investigations

1. Visit a local hospital and see how many types of electromagnetic radiation are used on a regular basis. From radio waves to gamma waves, how are the distinctive characteristics of absorption and transmission for each segment of the spectrum used at the facility?

2. What frequencies of electromagnetic radiation are used for communications by the emergency response groups (police, fire, ambulance) in your community? What are the corresponding wavelengths of these signals?

3. Examine a microwave oven or, better yet, obtain an old broken microwave oven that you can dissect. Locate the source of microwaves. Which materials in the oven transmit microwaves? Which ones scatter microwaves? Do you think any of the components absorb microwaves? Why?

4. In large metropolitan areas, a license to broadcast electromagnetic waves at an AM frequency may change hands for millions of dollars.

a. Why is electromagnetic "real estate" so valuable? Investigate how frequencies are divided up and who regulates the process. Should individuals or corporations be allowed to "own" portions of the spectrum, or to buy and sell pieces of it?

b. Currently, the only portions of the electromagnetic spectrum that are regulated by national and international law are the longer wavelengths, including radio and microwave. Why are the shorter wavelengths, including infrared, visible light, ultraviolet, and X-ray wavelengths, not similarly regulated?

5. Whales communicate over distances of thousands of miles. Investigate how they do this. What kind of wave is used? What is its speed? Its frequency? Its wavelength? Listen to a recording of humpback whale songs.

6. Find out how sonar works. Compare it to the use of sound by bats. Discuss the analogy between the defenses of moths and the use of radar detectors by motorists. (Perhaps the bats would like to outlaw these kinds of detectors the way some states do.)

▶ Additional Reading

Lewis, Thomas. *Empire of the Air: The Men Who Made Radio*. New York: Edward Burlingmane Books, 1991.

Sobel, Michael I. *Light*. Chicago: University of Chicago Press, 1989.

9 THE ATOM

ALL OF THE MATTER AROUND US IS MADE OF ATOMS—
THE CHEMICAL BUILDING BLOCKS OF OUR WORLD.

Breathing Lessons

Imagine being exposed to a substance so corrosive that it reacts violently with almost anything it touches. Upon contact, shiny metals become pitted and stained, while many everyday objects burst into flame or explode when exposed to the slightest spark. To the best of our knowledge, this dangerous substance exists on only one planet, Earth. The substance is oxygen.

Oxygen is a chemical element, one of about a hundred basic building blocks that make up everything around us. Oxygen, for example, makes up about 20% of the air around you. Unlike most other elements, however, oxygen is starved for electrons. Oxygen gas reacts chemically with many other elements, taking their electrons and, in the process, releasing energy. Once an oxygen atom seizes negative electrons from other atoms, however, it becomes electrically charged and may become locked into a new chemical substance. Glass, pottery, and almost all the rocks at the Earth's surface are more than half oxygen atoms, tightly bonded into durable solids. Even the water you drink is almost 90 percent oxygen by weight.

With every breath you take, oxygen from the air enters your body and sustains life by participating in chemical reactions that release the energy you use to grow, move, and think. Without a very corrosive element like oxygen, these reactions couldn't take place. Our very lives depend on the fact that oxygen is nasty stuff.

Figure 9–1
Repeatedly dividing a bar of gold, just like cutting the paper repeatedly, produces smaller and smaller groups of atoms, until you come to a single gold atom. Dividing that atom into two parts produces fragments that no longer have the properties of gold.

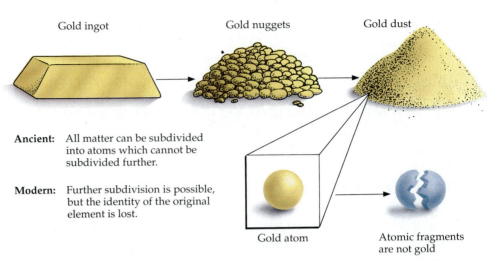

Gold ingot   Gold nuggets   Gold dust

Ancient: All matter can be subdivided into atoms which cannot be subdivided further.

Modern: Further subdivision is possible, but the identity of the original element is lost.

Gold atom   Atomic fragments are not gold

THE SMALLEST PIECES

Think about the paper in this book. You could take a page from the book and cut it in half. Then you could cut the half in half, and then cut half of the half of the half, and so on (Figure 9–1). If paper is really smooth and continuous, there would be no end to this process—no smallest piece of paper that couldn't be cut further. But is the paper really made that way?

The Greek Atom

In about 530 B.C., a group of Greek philosophers, the most famous of whom was a man named Democritus, gave this question some serious thought. Democritus argued (purely on philosophical grounds) that if you took the world's sharpest knife and started slicing chunks of matter, you would eventually come to a smallest piece that could not be divided further. He called this smallest piece the "atom," which translates roughly as "that which cannot be divided." He argued that all material was formed from these atoms, and that the atoms themselves are eternal and unchanging, but that the relationship between the atoms is constantly shifting.

These Greek philosophers did not really engage in science—their reasoning had none of the interplay between observation and hypothesis that characterizes the scientific method. It wasn't, in fact, until the beginning of the nineteenth century that the modern theory of atoms was born.

John Dalton (1766–1844)

The Beginning of an Atomic Theory

The modern atomic theory is generally attributed to English meteorologist John Dalton (1766–1844). In 1808, Dalton published a book called *New System of Chemical Philosophy*, in which he argued that the new knowledge being gained by chemists about materials provided evidence, in and of itself, that matter was composed of atoms. Chemists knew that most materials can be broken down into simpler chemicals. If you burn wood, for example, you get carbon dioxide, water, and all sorts of materials in the ash. If you use an electric current to break down water, you get two gases, hydrogen and oxygen. In addition, no matter how much of a material you break down, it always breaks down in the same proportion. Water, for example, always yields one part hydrogen to eight parts oxygen by weight (that is, the oxygen always weighs eight times more than the hydrogen).

On the other hand, a few materials could not be broken down into other things. Chemists could burn wood to get charcoal (pure carbon), for example; but, try as they might, they couldn't break the carbon down any further. Those materials that could not be broken down further were called chemical **elements.**

The hypothesis that we now call *atomism* was very simple. Dalton suggested that for each chemical element there was a corresponding species of indivisible objects called **atoms.** He borrowed the name from the Greeks, though his concept of the atom was more developed than the Greek view. Atoms that are linked together in a definite proportion constitute a chemical *compound*; a **molecule** is the smallest unit of a compound that retains the distinct characteristics of that compound. You can think of a molecule as a collection of atoms bound together. Molecules make up most of the different kinds of material we see around us. Water, for example, forms from one oxygen atom and two hydrogen atoms, giving the familiar composition designated H_2O in chemical notation (see Chapter 11).

element Materials made from a single type of atom. Elements cannot be broken down any further by chemical means.

atom Fundamental building blocks for all matter. A neutral particle having one nucleus; the smallest representative sample of an element.

molecule A cluster of atoms that can be isolated; the basic constituent of many different kinds of material.

Figure 9–2
Atoms may be envisioned as solid balls that stack together to form crystals, like fruit at the supermarket. Atomic models are often drawn with spheres, though we now know that atoms are not solid objects.

In Dalton's view, atoms were truly indivisible. He thought of them as little spheres (Figure 9–2). Thus in Dalton's world, indivisible atoms provide the fundamental building blocks of all matter. This picture was not only intellectually attractive, but it also conformed to much of what chemists had discovered by Dalton's time. It explained why chemical elements couldn't be broken down; they were made from a single type of atom that was, by definition, indivisible. It also explained the regular proportions of elements in compounds, such as the one-to-eight ratio of hydrogen to oxygen in water. Because of these successes, scientists rapidly accepted the atomic picture, which became a part of the way we think about the world. Even though we know today that atoms are made up of even smaller parts, we still use the old name "atom" for historical reasons.

Discovering Chemical Elements

Describing and isolating chemical elements provided a major challenge for chemists of the nineteenth century. In 1800, fewer than 30 elements had been isolated—not enough to establish any systematic trends in their chemical behavior. In the early 1800s, a new electrical process called *electrolysis*, facilitated by Volta's invention of the battery (see Chapter 7), allowed many new elements to be separated by means of electric current. As a result, more than two dozen more elements were discovered in the first half of the nineteenth century.

These discoveries led to Dmitri Mendeleev's conception of the first periodic table of the elements in 1869 (see Chapter 1). Recall that Mendeleev's original table listed several dozen elements on the basis of their atomic weights (in rows from the upper left) and by groups with distinctive chemical properties (in columns). This regular arrangement of chemical elements and their properties pointed the way to even more new elements. What Mendeleev did not realize, however, was that his table revealed much about the underlying structure of atoms and their electrons.

In the early 1800s, the list of known chemical elements was rapidly expanding, but contained only a few dozen entries. Today, the periodic table lists 110 elements, of which about 90 occur in nature and the rest have been produced artificially (the 110th, for example, was made in a German laboratory in October, 1994). Most of the materials we encounter in everyday life are not elements, but compounds of two or more elements bound together. Table salt, plastics, stainless steel, paint, window glass, and soap are all made from a combination of elements. Nevertheless, we do have daily contact with a few chemical elements, including the following:

- *Helium:* A light gas that has many uses besides filling party balloons and blimps. In liquid form, helium is used to maintain special equipment in laboratories and hospitals at low temperatures.
- *Carbon:* Pencil lead, charcoal, and diamonds are all examples of pure carbon. The differences between these materials has to do with the

Lighter-than-air craft, like this blimp, float because they are filled with helium.

(*a*) The elements gold and silver. (*b*) The elements mercury, copper, and carbon.

(*a*)   (*b*)

way the atoms of carbon are linked together, as we discuss in Chapter 11.

- *Aluminum:* A lightweight metal used for many purposes. The dull white surface of this metal is actually a combination of aluminum and oxygen, but if you scratch the surface, the shiny material underneath is pure elemental aluminum.

- *Copper:* The reddish metal of which pennies and pots are made. Copper wire also provides a cheap and efficient conductor of electricity.

- *Gold:* A soft, yellow, dense, and highly valued metal. For thousands of years, this element has been coveted as a symbol of wealth. Today it coats critical electrical contacts in spacecraft and other sophisticated electronics.

Although we know of more than 90 different elements in nature, living systems are constructed almost entirely from just half a dozen. Most of the atoms in your body are either hydrogen, carbon, oxygen, or nitrogen (with smaller but important roles played by phosphorus and sulfur). The atoms that make up our bodies are a small subset of the collection of atoms in nature.

THE STRUCTURE OF THE ATOM

Dalton's idea of the atom as a single indivisible entity was not destined to last. In 1897, English physicist Joseph John Thomson (1856–1940) unambiguously identified a particle called the **electron,** an object that has a negative electrical charge, and is much smaller and lighter than even the smallest atom known. Since there was no place for a particle like the electron to come from other than inside the atom, Thompson's discovery provided incontrovertible evidence for what people had suspected for a long time—that atoms are not the fundamental building block of matter, but rather are made up of things that are smaller and more fundamental still.

electron   Tiny, negatively charged particles that circle in orbits around a positively charged nucleus of an atom.

Ernest Rutherford and Hans Wilhelm Geiger.

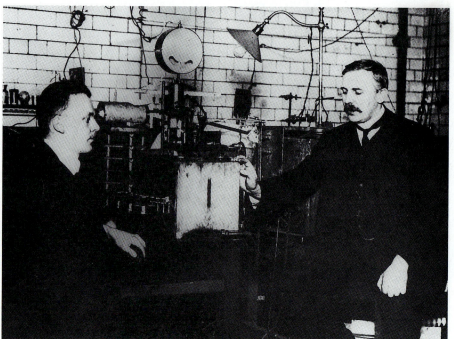

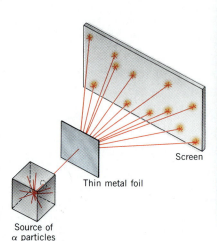

Figure 9–3
In Rutherford's experiment, a beam of radioactive particles was scattered by atomic nuclei in a piece of gold foil. A lead shield protected researchers from the radiation.

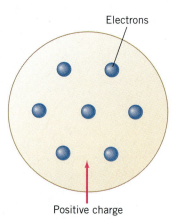

Figure 9–4
In the "raisin bun" atom, the negatively charged electrons were thought of as being distributed throughout a thin, positively charged material.

The Atomic Nucleus

The most important discovery about the structure of the atom was made by New Zealand-born physicist Ernest Rutherford (1871–1937) and his co-workers in Manchester, England, in 1911. The basic idea of the experiment is sketched in Figure 9–3. The experiment started with a piece of *radioactive* material, which is matter that sends out energetic particles (see Chapter 14). For our purposes, you can think of radioactive materials as sources of tiny subatomic "bullets." The particular material Rutherford used produced bullets that scientists had named *alpha particles*, which are thousands of times heavier than electrons. By arranging the apparatus as shown, Rutherford produced a stream of these subatomic bullets moving toward the right in the figure. In front of this stream, he placed a thin foil of gold.

This experiment was designed to measure the structure of the atom. At the time, people believed that the small, negatively charged electrons were distributed around the entire atom, more or less like raisins in a bun (Figure 9–4). What Rutherford was trying to do was to shoot bullets into the "bun" and see what happened. He expected that electrical forces would deflect a few of the bullets by a small angle. What the experiment actually revealed, however, was little short of astonishing to Rutherford and his colleagues. Most of the subatomic bullets either passed right through the gold foil unaffected or were scattered through very small angles, as expected. This result meant that most of the heavy alpha particles passed through spaces between gold atoms, and that those that hit the gold atoms were only moderately deflected by the relatively light electrons embedded in the diffuse positive charge.

One alpha particle in a thousand, however, was scattered through a large angle; some particles even bounced straight back (see Figure

9–3). After almost two years of puzzling over these unexpected results, Rutherford concluded that a large part of each atom's mass is located in a very small, compact object at the center—what he called the **nucleus.** In other words, about 999 times out of 1000, the alpha particles either missed the atom completely or went through the low-density material in the outer reaches of the atom. About 1 time out of 1000, however, the alpha particle hit the nucleus and was bounced through a large angle.

You can think of the Rutherford experiment in this way: If the atom were a large ball of mist or vapor taller than a skyscraper, and the nucleus were a bowling ball at the center of that sphere of mist, then most bullets shot at the atom would go right through. Only those that hit the bowling ball would be bounced through large angles. In this analogy, the bowling ball plays the part of the nucleus, while the mist is the domain of the electrons.

As a result of Rutherford's work, a new picture of the atom emerged—the one that is most familiar to us. Rutherford described a small, dense, positively charged nucleus sitting at the atom's center, while light, negatively charged electrons circle it, like planets orbiting the Sun. Indeed, Rutherford's model has become an icon of the modern age; stylized images of electrons orbiting a nucleus adorn everyday objects from postage stamps to bathroom cleaners (Figure 9–5). Nevertheless, Rutherford's model is not an accurate one.

In the next few decades, physicists came to realize that the nucleus itself is made up primarily of two different kinds of particles (see Chapter 14). One of these carries a positive charge, and is called a *proton.* The other, whose existence was not confirmed until 1932, carries no electrical charge and is called a *neutron.*

For each positively charged proton in the nucleus of the atom, there is normally one negatively charged electron moving around it. Thus the electrical charges of the electrons and protons cancel out, and every atom is normally electrically neutral. In some cases, atoms may either lose or gain electrons. In these cases, they acquire an electrical charge and are called **ions.**

Table 9–1 lists some of the important definitions related to atoms.

nucleus The very small, compact object at the center of an atom; made up primarily of protons and neutrons.

Figure 9–5
Many businesses and organizations have adopted a highly stylized atomic model as their logo.

ion An atom that has an electrical charge, from either the loss or gain of one or more electrons.

Table 9–1 • Important Terms Related to Atoms

Element:	A chemical substance that cannot be broken down further by chemical reactions.
Atom:	The smallest particle that retains its chemical identity.
Molecule:	The smallest unit of a compound that retains the distinctive characteristics of that compound.
Electron:	An atomic particle with negative charge and low mass.
Nucleus:	The small, massive central part of an atom.
Proton:	Positively charged nuclear particle.
Neutron:	Electrically neutral nuclear particle.
Ion:	An electrically charged atom.

Why the Rutherford Atom Couldn't Work

The picture of the atom that Rutherford developed is logical and appealing, particularly because it recalls to us the familiar stately orbits of our solar system. However, we have already learned enough about the behavior of nature to know that the atom described in the text could not possibly exist in nature. Why do we say this?

We learned in Chapter 4 that an object travelling in a circular orbit is constantly being accelerated; it is not in uniform motion because it is continually changing direction. Furthermore, we learned in Chapter 8 that any accelerated electric charge must give off electromagnetic radiation, as a consequence of Maxwell's equations. Thus, if an atom was of the type Rutherford described, the electrons moving in their orbits would constantly be giving off energy in the form of electromagnetic radiation. This energy, according to the first law of thermodynamics, would have to come from somewhere (remember conservation of energy!), so the electrons would gradually spiral in toward the nucleus as they converted their kinetic energy to electromagnetic radiation. Eventually, the electrons would have to fall into the nucleus and the atom would cease to exist in the form we know it.

In fact, the life expectancy of the Rutherford atom turns out to be less than a second. Given the fact that many atoms have survived billions of years since almost the beginning of the universe, this calculation poses a serious problem for the simple orbital model of the atom.

WHEN MATTER MEETS LIGHT

Almost from its inception, the Rutherford model of the atom encountered difficulties. Some of the problems involved its violations of fundamental physical laws as described above, while others were more mundane; the Rutherford atom simply did not explain all the known behavior of atoms. The first decades of the twentieth century, then, was a period of tremendous ferment in the sciences as people scrambled to find a new way of describing the nature of atoms.

The Bohr Atom

In 1913, Niels Bohr (1885–1962), a young Danish physicist working in England, produced the first model of the atom that addressed the kinds of objections encountered by Rutherford's model. The **Bohr atom** is very strange: it does not match well with our intuition about the way things "ought to be" in the real world. About the only thing to be said in the Bohr atom's favor is that it worked.

The Bohr atom began with a hunch. The young scientist was deeply immersed in studying the way that atoms interact with light and other forms of electromagnetic radiation. He realized that one way of explaining what he saw in the laboratory was that electrons circling the nucleus, *unlike planets circling a sun*, could not maintain their orbits at just any distance from the center. He suggested that there were only certain orbits—he called them *allowed orbits*—located at specified distances from the center of the atom, in which an electron could exist for long periods

Niels Bohr (1885–1962) with Aage Bohr, one of his five sons. Both won Nobel Prizes in physics.

Bohr atom   A model of the atom, developed by Niels Bohr in 1913, in which electrons exist only in allowable orbits, located at fixed distances from the center of an atom. In these orbits, the electrons maintain fixed energy for long periods of time, without giving off radiation.

of time *without giving off radiation.* In fact, in Bohr's picture the atom looked like the one in Figure 9–6. The idea was that the electron could exist at a distance r_1 from the nucleus, or at a distance r_2 from the nucleus, or at a distance r_3 from the nucleus, and so on. As long as the electron remained at one of those distances, its energy was fixed. The electron could not ever, at any time, be in orbit any place between these allowed distances.

One way of thinking about the Bohr atom is to imagine what happens when you climb a flight of steps. You can stand on the first step or you can stand on the second step. It's very hard, however, to imagine what it would be like to stand somewhere between two steps. In just the same way, an electron can be in the first allowed orbit, or in the second allowed orbit, but it can't be in between.

Each orbit in the Bohr atom corresponds to a different *energy level* of an electron. Each time you change steps in your home, your gravitational potential energy changes. Similarly, each time an electron changes orbits, its energy level also changes. This situation is illustrated schematically in Figure 9–7.

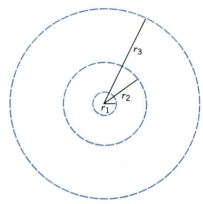

Figure 9–6
A schematic diagram of the Bohr atom reveals three of the many possible locations for electrons.

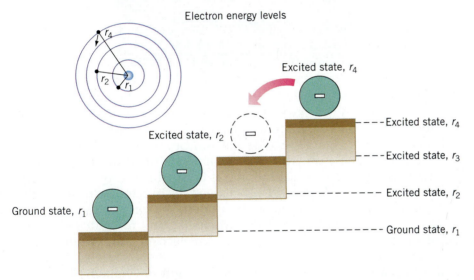

Figure 9–7
Stairs provide an analog to energy changes associated with electrons in the Bohr atom.

An electron in the atom can be in any one of a number of allowed orbits, and each allowed orbit is at a different energy. You can see this by noting that you would have to exert a force over a distance to move an electron from one allowed orbit to another to overcome the electrical attraction to the nucleus just as your muscles would have to exert a force to get you up a flight of stairs. The allowed energy levels of an atom, then, occur as a series of steps as shown in the figure. The electron in the lowest energy level is called the *ground state,* while all energy levels above the ground state are called *excited states.*

Photons: Particles of Light

The major feature of the Bohr atom is that an electron in an orbit of higher energy can move down into an available orbit of lower energy.

Figure 9–8
Electrons may jump between energy levels and in the process emit or absorb energy in the form of a photon.

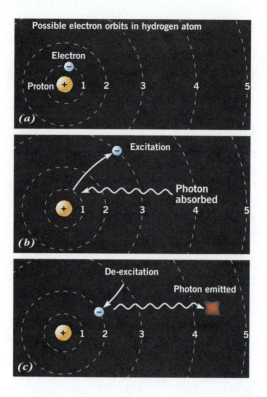

This process is analogous to that by which a ball at the top of a flight of stairs can bounce down the flight of stairs under the influence of gravity.

Just for the sake of argument, assume that an electron is in an excited state, as shown in Figure 9–8. The electron can move to a lower state, as shown, but if it does, something must happen to the extra energy. Energy can't just disappear. This realization was Bohr's great insight. The energy that's left over when the electrically charged electron moves from a higher state to a lower one is emitted by the atom in the form of a single packet of electromagnetic radiation. This particle-like unit of light was called a **photon** by Einstein in 1905. Every time an electron jumps from a higher to a lower energy level, a photon moves away at the speed of light. (We will come back to the perplexing nature of light in Chapter 10, once we have learned more about the behavior of atoms.)

The interaction of atoms and electromagnetic radiation provides the most compelling evidence for the Bohr atom. If electrons are in excited states, and if they make transitions to lower states, then photons are emitted. If we look at a group of atoms in which these transitions are occurring, we will see light. Thus, when you look at the flame of a fire or the glowing heating coil of an electric stove in your kitchen, you are actually seeing photons that were emitted by electrons jumping between allowed states in that material's atoms.

Not only does the Bohr atom give us a picture of how matter emits radiation, it also provides an explanation for how it absorbs radiation. Start with an electron in a low-energy state, perhaps its ground state. If a photon arrives that has just the right amount of energy so that it can raise the electron to a higher energy, the photon can be absorbed and the electron will be pushed up to an excited state (Figure 9–8b). Light absorption by an atom is thus the mirror image of light emission.

photon A particle-like unit of electromagnetic radiation, emitted or absorbed by an atom when an electrically charged electron changes state; the form of a single packet of electromagnetic radiation.

The interaction of matter and radiation encompasses two key ideas. First, when an electron moves from one allowed orbit to another, it cannot ever, at any time, be at any place in between. This rule is built into the definition of an allowed orbit. This means that the electron must somehow disappear from its original orbit and reappear in its final orbit *without ever having to traverse any of the positions in between.* This process, called a **quantum leap** or **quantum jump,** cannot be visualized, but it is something that seems to be fundamental in nature.

Second, if an electron is in an excited state, it can, in principle, get back down to the ground state in a number of different ways. Look at Figure 9–9. An electron in an upper energy level can move to the ground state by making one large jump and emitting a single photon with large energy (Figure 9–9*a*). Alternatively, it can move to the ground state by making two smaller jumps (Figure 9–9*b*). Each of these jumps will emit photons of somewhat less energy. The energies emitted in the two different jumps will generally be different from each other, but the sum of the two energies will equal that of the single large jump. We would expect, if we had a large collection of atoms of this kind, that some electrons would make the large leap while others make the two smaller ones. Thus when we look at a collection of these atoms, we would see three different energies of photons.

quantum leap A process by which an electron changes orbit without ever traversing any of the positions between the original and final orbits; also known as quantum jump.

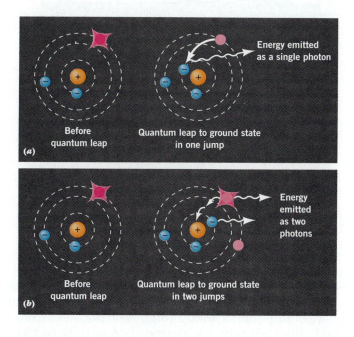

Before quantum leap

Quantum leap to ground state in one jump

Energy emitted as a single photon

(a)

Before quantum leap

Quantum leap to ground state in two jumps

Energy emitted as two photons

(b)

Figure 9–9
An electron can jump from a higher to lower energy level in a single quantum leap (*a*), or by multiple quantum leaps (*b*).

In Chapter 8 we saw that the energy of electromagnetic radiation is related to its frequency. In the special case of visible light, for example, the colors of the light we perceive represent photons of different energies. Thus, if we look at a collection of atoms of the type shown in Figure 9–9, we would see three different frequencies, one corresponding to each of the quantum leaps in the picture.

Another aspect of the Bohr atom that needs to be examined is how electrons get to the excited states in the first place. We know that it requires energy to lift an electron from the ground state to any excited state, and this energy has to come from somewhere. One possibility, as noted above, is that the atom will absorb a photon of just the right frequency to raise the electron to a higher energy level. Electrons can also absorb energy in other ways. If the material is heated, for example, atoms will move fast, gain kinetic energy, and undergo energetic collisions. In these collisions, an atom can absorb energy that moves an electron to a higher orbit. When you heat an object, countless electrons jump to excited states and then cascade down again, causing the object to glow and radiate energy.

Science by the Numbers

All Possible Quantum Leaps

How many different frequencies of photons might you see coming from an atom that has one electron and four allowed energy levels? To answer this question, we can list all possible quantum jumps by numbering the levels 1, 2, 3, and 4 (Figure 9–10).

Figure 9–10
A one-electron atom with four available energy levels can emit photons of six different frequencies, corresponding to the six possible quantum jumps.

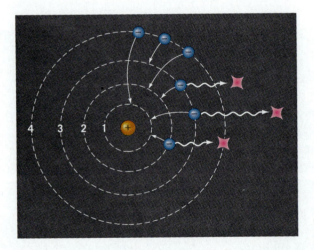

The electron in level 4, the highest state, can make one transition to level 1, the ground state, yielding us one photon of high energy. Alternatively, it could jump to level 3 (one more photon), from which point it could either jump to the ground state directly (yet another photon) or make two successive jumps to get to the ground state (two more photons). Finally, the atom could jump to level 2, the first excited state (one additional photon) and from there to the ground state. All intermediate quantum jumps are included in the jumps just described for an electron in the highest energy state. If you add up all of these possible transitions, you will find that this atom gives off photons of six different energies. ●

An Intuitive Leap

As we pointed out above, Bohr first proposed his model of the atom based on an intuition guided by experiments and ideas about the behavior of things in the subatomic world. In some ways the Bohr model was completely unlike anything we experience in the macroscopic world. Indeed, to some scientists of the time the model at first seemed a little bit "crazy." It took two decades for scientists to develop a theory called *quantum mechanics* that showed why electrons can have only certain allowed energies and not in between. We will discuss this justification for the Bohr atom in Chapter 10, but it should be remembered that the justification occurred long after the initial hypothesis. The Bohr atom was accepted by physicists because it worked; it explained what they saw in nature and allowed them to make predictions about the behavior of real matter.

How could Bohr have come up with such a strange picture of the atom, in which electrons magically disappear from one orbit and instantaneously reappear in another? He was undoubtedly guided by some of the early work by other scientists that would eventually lead to the theory of quantum mechanics. In the end, however, this is an unsatisfactory explanation. Many other people at the time studied the interactions of atoms and light, just as Bohr did, but only Bohr was able to make the leap of intuition to his description of the atom. This insight, like Newton's realization that gravity might extend to the orbit of the Moon, remains one of the great intuitive achievements of the human mind.

SEEING THE LIGHT

Whenever energy is added to a system with many atoms in it, electrons in some atoms are pushed to excited states. As time goes by, these electrons will make quantum leaps down to the ground state, giving off photons as they do. If some of those photons are in the range of visible light, the source will appear to glow.

Why Things Aren't Gray

You may not realize it, but you have looked at glowing collections of atoms all your life. Common mercury vapor street lamps contain bulbs filled with mercury gas. When the gas is heated, electrons are moved up

Sodium vapor street lamps give off a yellow light characteristic of a particular quantum jump in the sodium atom.

to excited states. When they jump down they emit photons that give the lamp a bluish-white color. In the same way, at freeway interchanges you occasionally see intense yellow street lights that use bulbs filled with sodium atoms. When sodium atoms are excited, the most frequently emitted photons lie in the yellow range, so the lamps look yellow.

Yet another place where you can see photons emitted directly by single quantum leaps is in Day-Glo colors, which are the vivid colors often used in sports clothing and for other purposes. From these examples, you can draw two conclusions:

1. Quantum leaps are very much in evidence in your everyday life.
2. Different atoms give off different characteristic photons, resulting in different characteristic colors.

The elements (a) sodium, (b) potassium, and (c) lithium impart distinctive colors to a flame.

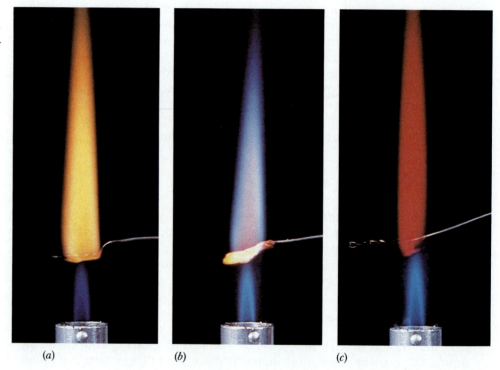

(a)   (b)   (c)

The second of these two facts is extremely important for scientists. If you think about the structure of an atom, the idea that different atoms have different characteristic photons shouldn't be too surprising. Electron energy levels depend on the electrical attraction between the nucleus and the electrons, just as the orbits of the planets depend on the gravitational attraction between the planets and the Sun. Different nuclei have different numbers of protons, so electrons circling them are in different orbits. In fact, the arrangement of the allowed energy levels is different in each of the hundred or so different chemical elements. Since the energy and frequency of the photons emitted by an atom depend on the differences in energy between the allowed levels, each chemical element gives off a distinct set of characteristic photons.

Spectroscopy

You can think of the collection of characteristic photons emitted by each chemical element as a kind of "fingerprint"—something that is distinctive for that chemical element and none other. This feature opens up a very interesting possibility. The total collection of photons emitted by a given atom is called its **spectrum,** which is the characteristic "fingerprint" that can be used to identify chemical elements, even when they are very difficult to identify by any other means.

Spectroscopy, which is the study of the characteristic radiation emitted by atoms or molecules, has become a standard tool that is used in almost every branch of science. Astronomers use emission spectra to find the chemical composition of distant stars, and they study absorption lines to determine the chemical composition of interstellar dust and the atmospheres of the outer planets. Spectroscopic analysis is also used to search for impurities on production lines, to monitor pollution in the atmosphere, and to identify small traces of unknown materials in courts of law.

spectrum The characteristic signal from the total collection of photons emitted by a given atom that can be used to identify the chemical elements in a material; the atomic fingerprint.

spectroscopy The study of emission and absorption spectra of materials to discover the chemical makeup of that material; a standard tool used in almost every branch of science.

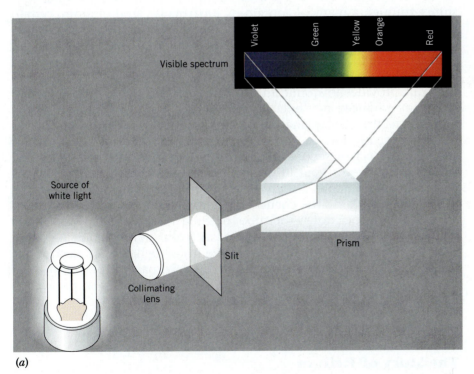

(a)

Figure 9–11
A glass prism spreads out the colors of the visible spectrum, thereby allowing measurements of the intensities of different wavelengths of light.

In practice, the identification process works like this: Light from the atoms is spread out by being passed through a prism (Figure 9–11). Each possible quantum jump corresponds to light at a specific wavelength, so each type of atom produces a set of *emission lines* to form an *emission spectrum*, as shown in Figure 9–12. This spectrum is what we have called the atomic fingerprint.

The Bohr picture suggests that if an atom gives off light of a specific wavelength and energy, then it will also absorb light at that wavelength. The emission and absorption processes involve quantum jumps between

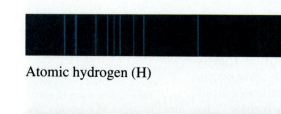

Atomic hydrogen (H)

Sodium (Na)

Neon (Ne)

Figure 9–12
Line spectra, shown here for hydrogen, sodium, and neon, provide distinctive fingerprints for elements and compounds.

the same two energy levels but in different directions. Thus, if white light shines through a material containing a particular kind of atom, certain wavelengths of light will be absorbed. Observing that light on the other side of the material, you will see certain colors missing. The dark areas corresponding to the absorbed wavelengths are called *absorption lines*, and the complete set of lines is called the *absorption spectrum*. The absorption spectrum provides an atomic fingerprint just as diagnostic as the emission spectrum produced by ''hot'' atoms. Indeed, one early triumph of the Bohr model was its explanation of the spectral lines of hydrogen.

Spectroscopy has countless applications in scientific research. For example, in a classic set of experiments in the early 1940s, scientists used spectroscopy to work out in detail how chemical reactions are assisted by *enzymes* (see Chapter 12). In these experiments, a fluid containing the materials undergoing the chemical reactions was allowed to flow down a tube. The farther down the tube the fluid was, the farther along the reaction was. By measuring spectra at different points along the tube, scientists were able to follow the change in the behavior of electrons as the chemical reactions went along. In this way, part of the enormously complex problem of understanding the chemistry of life was unraveled.

Science in the Making

The Story of Helium

You have probably had some experience with helium, perhaps when you used it to inflate party balloons for your little brothers and sisters. Helium gas is a fascinating material, not only for its properties (it's so light that it floats in air), but because of the history of its discovery.

The word helium refers to *helios*, the Greek word for Sun, because helium was discovered in spectral lines in the Sun in 1868 by English scientist Joseph Norman Lockyer (1836–1920). Helium is very rare in the atmosphere of the Earth, and before Lockyer's discovery, scientists were not even aware of its existence. Following the discovery, there was a period of about 30 years when astronomers accepted the fact that the element helium existed in the Sun, but they were unable to find it on the Earth.

The detection of helium in the Sun but not on Earth led to a very interesting problem. Could it be that there were chemical elements in the Sun that simply did not exist on our own planet? If so, our ability to describe the rest of the universe would be limited. If we can't isolate and study an element in our laboratories, then we can never really be sure that we understand its properties.

The existence of helium on Earth wasn't confirmed until 1895, when Lockyer identified its spectrum in a sample of radioactive material. ●

Technology

The Laser

The Bohr atom provides an excellent way to understand the workings of the laser, one of the most important devices in modern science. The word **laser** is an acronym for *l*ight *a*mplification by *s*timulated *e*mission of *r*adiation. At the core of every laser is a collection of atoms—a crystal of ruby, perhaps, or a gas enclosed in a glass tube. The term "stimulated emission" refers to a process that goes on when light interacts with these atoms. If an electron is in an excited state, as shown in Figure 9–13, and one photon of just the right energy passes nearby, the electron may be "stimulated" to make the jump to a lower energy state, thus releasing a second photon. By "just the right energy" for the first photon, we mean a photon whose energy corresponds to the energy gap between two electron energy levels in the atom.

laser An instrument that uses a collection of atoms, energy, and mirrors to emit photons that have wave crests in exact alignment. The instrument name is the acronym for *l*ight *a*mplification by *s*timulated *e*mission of *r*adiation.

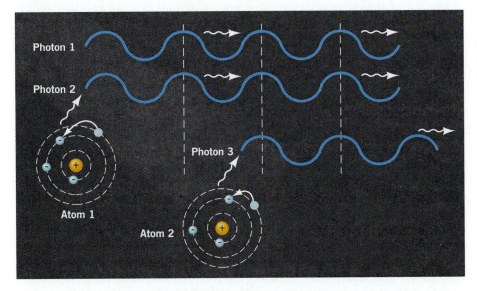

Photon 1
Photon 2
Photon 3
Atom 1
Atom 2

Figure 9–13
Lasers produce a beam of light when one photon stimulates the emission of other photons.

Furthermore, the stimulated atom emits photons in a special way. Remember that light is a form of electromagnetic radiation that can be described as a wave. In a laser, the crests of all the emitted photon waves line up exactly with the crests of the first photon, and the signal

Figure 9–14
The action of a laser. Electrons in the laser's atoms are continuously "pumped" into an excited state by an electric current, and the beam of photons is released when the electrons return to their ground state.

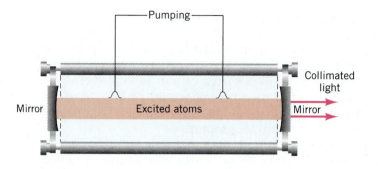

is enhanced by *constructive interference* (see Chapter 8). Thus the effect of stimulated emission is such that, whereas you had one photon at the beginning of the process, you have two superimposed photons at the end.

Now suppose that you have a collection of atoms where most of the electrons are at the excited state, as shown in Figure 9–14. If a single photon of the correct frequency enters this system from the left and moves to the right, it will pass the first atom and stimulate the emission of a second photon. You will then have two photons moving to the right. As these photons encounter other atoms, they, too, stimulate emission so that very quickly you have four photons. Light amplification in a laser occurs very quickly, cascading so that soon there is a flood of photons moving to the right through the collection of atoms. Energy added to the system from outside continuously returns atoms to their excited states—a process called *pumping*—so that more and more photons can be produced.

In a laser, the collection of excited atoms is bounded on two sides by mirrors, so that photons moving to the right hit the mirror, are reflected, and make another pass through the material, stimulating even more emission of photons as they go. If a photon happens to be lined up exactly perpendicular to the mirrors at the end of the laser, it will continue bouncing back and forth forever. If its direction is off by even a small angle, however, it will eventually bounce out through the sides of the laser and be lost. Thus only those photons that are exactly aligned will wind up bouncing back and forth between the mirrors, constantly amplifying the signal. Aligned photons will traverse the laser millions of times, building up an enormous cascade of photons in the system.

The mirrors are designed to be partially reflective; perhaps 95% of the photons that hit the mirror are reflected back into the laser, and 5% are allowed to leak out. Those photons that leak out form the familiar laser beam, which is made up of intense, coherent light.

Laser beams have found thousands of uses in science and industry since their development in the 1960s. Low-power lasers are ideal for optical scanners, such as the ones in your supermarket checkout line, and they make ideal light pointers for lectures and slide shows. The fact that the beam of light travels in a straight line makes the laser invaluable in surveying over long distances; for example, modern subway tunnels are routinely surveyed by using lasers to provide a straight line underground. Lasers are also used to detect movement of seismic faults in order to predict earthquakes (see Chapter 18). In this case, a

(a)

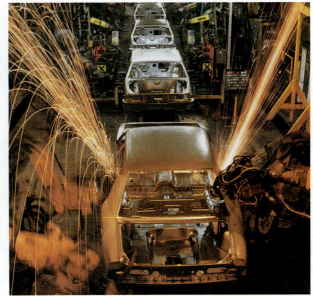

(b)

Lasers are now used extensively in scientific research, industry, and medicine. (*a*) A laser beam facilitates delicate eye surgery. (*b*) Robot arms use lasers to complete precise welding on an automobile assembly line.

laser is directed across a fault, so that small motions of the ground are easily measured.

Lasers come in many sizes and shapes. Finely-focused laser beams have revolutionized delicate procedures such as eye surgery. Much more powerful lasers that transfer large amounts of energy are often used as cutting tools in factories, as well as implements for performing surgery. The military has embraced laser technology, both in targeting and range finders, and in designs for futuristic energy-beam weapons. Scientists use lasers to make extremely precise and detailed measurements of the properties and structures of atoms and molecules. Indeed, almost all modern studies of the atom depend in one way or another on the laser. ●

Environment

Detecting Pollution by Using Lasers and Spectroscopy

Our knowledge of the inner workings of the atom has proved to be invaluable in efforts to control things such as air pollution. The first step in controlling pollution is to measure how many unwanted materials are in the air and to identify what they are. One way of doing this is to fly airplanes to collect air samples and then analyze them in a laboratory. This technique is expensive, however, and it does not give immediate information about air quality. It only tells how the air quality was at a specific time and place in the past.

During the 1980s, scientists increasingly turned to the use of spectroscopy to monitor air quality. For example, if you want to know on a minute-by-minute basis what kinds of molecules are being emitted from a smokestack, it is possible to set up a spectroscope in the smokestack itself. The device shines light (either visible, infrared, or ultraviolet) through the smoke as it moves up, and the light is collected by a

Lasers are used to measure pollution and monitor environmental quality.

spectrometer on the other side of the stack. By looking at which frequencies of light have been absorbed, you can identify various atoms and molecules in the smoke; and by looking at how much light is absorbed at each frequency, you can tell how much of each atom or molecule is present. This use of absorption spectroscopy gives the plant manager instant knowledge of what is being emitted into the atmosphere.

Lasers also add an important dimension to the monitoring of air quality, especially through the use of a technique called *lidar* (*li*ght *de*tection *a*nd *r*anging). Lidar works exactly like radar (see Chapter 8), except that light waves rather than radio waves are sent out. In its simplest form, laser pulses are sent out to be reflected from particles in the air, and the amount of reflected light is used to measure the density of these particles.

More sophisticated uses of lasers involve processes in which two frequencies of light are sent out. One frequency is adjusted so that it corresponds exactly to a quantum leap in a particular atom or molecule, while the other does not correspond to any quantum leap in that particular atom or molecule. The difference in reflected light between the two frequencies then tells us how much of that particular species is in the air. In this way, we can find out about molecules such as sulfur dioxide and nitrogen dioxide, which contribute to acid rain, as well as ozone and various hydrocarbons, which have a direct negative effect on human health.

Laser techniques are so sensitive that scientists are developing instruments that can detect pollution from individual cars as they go by on a highway. In this way, they hope to develop a measure of automobile emissions that is both more realistic and easier to administer than the current system of requiring periodic emissions tests in garages. ●

THE PERIODIC TABLE OF THE ELEMENTS

Dmitri Mendeleev discovered the regularity or *periodicity* in properties of the known chemical elements (see Chapter 1), but he was not aware of

Figure 9–15 The periodic table of the elements. Elements include the main table with groups IA–VIIIA/0 and the lanthanide and actinide series below.

	IA (1)	IIA (2)	IIIB (3)	IVB (4)	VB (5)	VIB (6)	VIIB (7)	VIII (8)	VIII (9)	VIII (10)	IB (11)	IIB (12)	IIIA (13)	IVA (14)	VA (15)	VIA (16)	VIIA (17)	0 (18)
1	1 H 1.00794																	2 He 4.00260
2	3 Li 6.941	4 Be 9.01218											5 B 10.81	6 C 12.011	7 N 14.00674	8 O 15.9994	9 F 18.99840	10 Ne 20.1797
3	11 Na 22.98977	12 Mg 24.3050											13 Al 26.98154	14 Si 28.0855	15 P 30.97376	16 S 32.066	17 Cl 35.4527	18 Ar 39.948
4	19 K 39.0983	20 Ca 40.078	21 Sc 44.9559	22 Ti 47.88	23 V 50.9415	24 Cr 51.9961	25 Mn 54.9380	26 Fe 55.847	27 Co 58.93320	28 Ni 58.69	29 Cu 63.546	30 Zn 65.39	31 Ga 69.723	32 Ge 72.61	33 As 74.92159	34 Se 78.96	35 Br 79.904	36 Kr 83.80
5	37 Rb 85.4678	38 Sr 87.62	39 Y 88.90585	40 Zr 91.224	41 Nb 92.90638	42 Mo 95.94	43 Tc 98.9072	44 Ru 101.07	45 Rh 102.90550	46 Pd 106.42	47 Ag 107.8682	48 Cd 112.411	49 In 114.82	50 Sn 118.710	51 Sb 121.75	52 Te 127.60	53 I 126.90447	54 Xe 131.29
6	55 Cs 132.90543	56 Ba 137.327	57 *La 138.9055	72 Hf 178.49	73 Ta 180.9479	74 W 183.85	75 Re 186.207	76 Os 190.2	77 Ir 192.22	78 Pt 195.08	79 Au 196.96654	80 Hg 200.59	81 Tl 204.3833	82 Pb 207.2	83 Bi 208.98037	84 Po 208.9824	85 At 209.9871	86 Rn 222.0176
7	87 Fr 223.0197	88 Ra 226.0254	89 †Ac 227.0278	104 Unq 261.11	105 Unp 262.114	106 Unh 263.118	107 Uns 262.12	108 Uno (265)	109 Une (266)	110								

*	58 Ce 140.115	59 Pr 140.90765	60 Nd 144.24	61 Pm 144.9127	62 Sm 150.36	63 Eu 151.965	64 Gd 157.25	65 Tb 158.92534	66 Dy 162.50	67 Ho 164.93032	68 Er 167.26	69 Tm 168.93421	70 Yb 173.04	71 Lu 174.967
†	90 Th 232.0381	91 Pa 231.0359	92 U 238.0289	93 Np 237.0482	94 Pu 244.0642	95 Am 243.0614	96 Cm 247.07003	97 Bk 247.0703	98 Cf 251.0796	99 Es 252.083	100 Fm 257.0951	101 Md 258.10	102 No 259.1009	103 Lr 260.105

Atomic number — 1 H 1.00794 — Atomic weight

Noble gases

the underlying atomic structure that was reflected in his table. Today, each element is assigned a number, called the *atomic number*, that corresponds to the number of protons in the atom. That number, in turn, dictates the number of electrons in orbit around the nucleus of an uncharged atom. If you arrange the elements as shown in Figure 9–15, with elements getting progressively heavier in general as you read from left to right and top to bottom as in a book, then elements in the same vertical column have very similar chemical properties. Mendeleev called this discovery the **periodic table of the elements,** and you may have seen versions of it in your high-school science classroom.

Periodic Chemical Properties

The most striking characteristic of the periodic table is the similarity of elements in any given column. In the far left-hand column of the table, for example, are highly reactive elements called *alkali metals* (lithium, sodium, potassium, etc.). Each of these soft, silvery elements forms compounds called *salts* by combining in a one-to-one ratio with any of the elements in the seventh column (fluorine, chlorine, bromine, etc.). Water dissolves these compounds, which include common table salt—sodium chloride.

The second column, including beryllium, magnesium, and calcium, is called the *alkaline earth metals*, and they too display similar chemical properties. All of these elements, for example, combine with oxygen in a one-to-one ratio to form colorless, solid compounds with very high melting temperatures.

Elements in the far right-hand column (helium, neon, argon, and so on), by contrast, are all colorless, odorless gases that are almost impossible

Figure 9–15 The periodic table of the elements. The weights of the elements increase from left to right. Each vertical column group elements with similar chemical properties.

periodic table of the elements An organizational system, first developed by Dmitri Mendeleev in 1869, now listing more than 110 elements by atomic weight (in rows) and by chemical properties (in columns). The pattern of elements in the periodic table reflects the arrangement of electrons in their orbits.

to coax into any kind of chemical reaction. These *noble gases* find applications when ordinary gases are too reactive. Helium is used to fill blimps, because it is the next lighter-than-air gas after the explosive hydrogen. Argon fills incandescent light bulbs, because the hot filament would burn up in air.

In the late nineteenth century, scientists knew that the periodic table "worked." It organized the 63 elements known at that time, and it successfully predicted their properties. Their faith in the periodic table was buttressed by the fact that when Mendeleev first wrote it down, there were holes in the table: places where he predicted elements should go, but for which no element was known. The ensuing search for the missing kinds of atoms eventually produced the elements we now call scandium (in 1876) and germanium (in 1886). But chemists had no idea *why* the periodic table worked.

Why the Periodic Table Works

With the advent of Bohr's atomic model and its modern descendants, we finally have an understanding of why the periodic table works. We now realize that the pattern of elements in the periodic table reflects the arrangement of electrons in the atom.

The atom is largely empty space. When two atoms come near enough to each other to undergo a chemical reaction, such as when a carbon atom and an oxygen atom combine in a burning piece of coal, the outermost electrons interact with each other first. You will see in Chapter 11 that these outermost electrons govern the chemical properties of materials. In fact, we have to understand the behavior of these electrons in order to understand the periodic table.

Pauli exclusion principle A statement that says no two electrons may occupy the same state at the same time.

To do this, we need to know one more curious fact about electrons. Electrons are particles that obey what is called the **Pauli exclusion principle,** which says that no two electrons can occupy the same state at the same time. One analogy is to compare electrons to cars in a parking lot; each car takes up one space, and once a space is filled, no other car can go there. Electrons behave the same way; once an electron fills a particular niche in the atom, no other electron can occupy the same niche. A parking lot can be full long before all the space in the lot is taken up with cars; the driveways and space between cars must remain empty. So too can a given Bohr orbit be filled with electrons long before all the available space in the orbit is taken.

electron shell A specific orbit in an atom that can be filled with a predetermined number of electrons.

It turns out that there are only two spaces that an electron can fill in the lowest, innermost Bohr atom orbit, which is called the first **electron shell.** One of these spaces corresponds to a situation in which the electron "spins" clockwise on its axis, and the other to a situation in which it "spins" counterclockwise on its axis. When we start to catalog all possible chemical elements in the periodic table, we have element one (hydrogen) with a single electron in the lowest orbit, and element two (helium) with two electrons in that same orbit. At this point the first shell is said to be *filled* or *closed*. After this, if we want to add one more electron, it has to go into the second electron shell. This situation explains why only hydrogen and helium appear in the first row in the periodic table (Figure 9–15).

Adding a third electron yields an atom with two electrons in the innermost shell, and a single electron in the second electron shell. This element is lithium, the one just below hydrogen in the periodic table. Lithium and hydrogen both have a single electron in their outermost electron shell, so they have very similar chemical properties.

The second electron shell has room for 8 electrons, a fact reflected in the eight elements of the periodic table's second row, from lithium with 3 electrons to neon with 10. Neon appears directly under helium, and we expect these two gases to have similar chemical properties because both have completely filled outer electron shells.

Thus a simple counting of the states available to electrons in the lowest two orbits explains why the first row in the periodic table has two elements in it, and the second row eight. By similar (but somewhat more complicated) arguments, you can show that the Pauli exclusion principle requires that the next row of the periodic table have 8 elements, the next 18, and so on. Thus, with an understanding of the structure of the atom, the seemingly mysterious regularity that Mendeleev found among the chemical elements becomes an example of the basic laws of nature at work.

To facilitate thinking about these electron arrangements, chemists often use a simple arrangement of circles and dots to represent an atom's electrons (Figure 9–16). The distribution of electrons into shells around

Figure 9–16
Hydrogen is in the first row of the periodic table. The lowest shell holds only two electrons, so the third element, lithium, has one electron in its outermost shell, and will therefore have chemical properties similar to hydrogen. Other common atoms with their electron configurations are also represented in this figure.

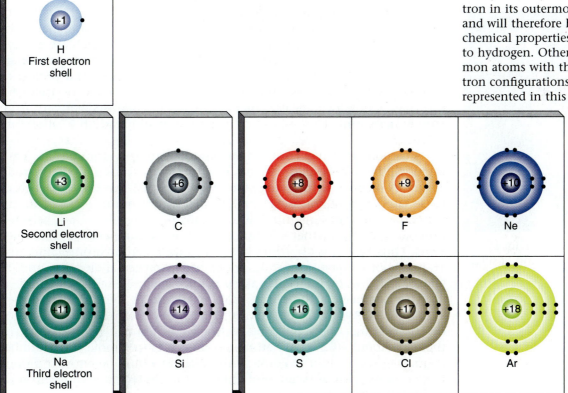

an atom provides one of the most important concepts in chemistry. As two atoms approach each other, their electrons may rearrange themselves to form a bond between the atoms (see Chapter 11).

Developing Your Intuition

The Heaviest Elements

Uranium is the heaviest naturally occuring element, but as you can see from the periodic table (Figure 9–15), scientists have succeeded in producing heavier elements. Suppose they keep making heavier artificial elements. What would you expect the chemical properties of elements 117, 118, and 119 to be?

Counting along the columns of the periodic table, you can see that the last row will end with element 118. Element 118, therefore, will have a filled outer shell. Like helium and neon, it will not enter into chemical reactions or form chemical bonds, and will therefore probably be a gas. Element 117, with one electron needed to have a filled outer shell, and element 119, with one electron more than a filled outer shell, will have chemical properties similar to chlorine and sodium, respectively. The two can be expected to react to produce a heavy version of table salt. ●

Counting Atoms

One mole of carbon, of lead, of water, and of sulfur.

The periodic table arranges elements in order of increasing number of protons, which corresponds closely to the order of increasing mass. The mass of atoms is generally given in *atomic mass units*, which are defined as one twelfth of the mass of a carbon atom with six protons and six neutrons. Think of this unit, then, as roughly the mass of a proton or neutron. An atomic mass unit is so small that it is not very useful for everyday laboratory measurements. How do chemists convert from atomic mass units to the kind of mass you would read on a laboratory balance?

A carbon atom has 12 atomic mass units, an iron atom has 56. Thus there will be just as many atoms in 12 grams of carbon as there are in 56 grams of iron. In the language of the chemist, we say that each of these samples contains one *gram molecular weight*, or one *mole*, of atoms. A mole is the very large number of atoms or molecules contained in a number of grams of that substance equal to its number of atomic mass units. Thus 12 grams of carbon has one mole of carbon atoms. Similarly, 16 grams of oxygen atoms, 32 grams of O_2 oxygen molecules, and 18 grams of H_2O water molecules all constitute one mole of those materials.

The number of atoms in a mole is known as Avogadro's number, after Amadeo Avogadro, the eighteenth-century Italian scientist who first determined that there must be such a fixed number. It turns out that there are about 6.02×10^{23} atoms or molecules in a mole. This number is huge—if you had a mole of pennies, there would be enough money to give every man, woman, and child in the United States a grand total of 24 trillion dollars!

THINKING MORE ABOUT
ATOMS

Are Atoms Real?

Throughout the nineteenth century, physicists and chemists came to realize that matter really behaved *as if* it were made up of atoms. The question remained, however, whether the atoms were real or simply useful fictions. After all, when you look at the paper on this page you don't see atoms. In a sense, this debate mirrors the current argument over whether even smaller components of matter called *quarks* (see Chapter 15) are real or not.

By the end of the nineteenth century, however, an important series of observations was made that argued that atoms are real. If you look at very small objects such as grains of pollen suspended in a liquid, they seem to jitter around. Over long periods of time, each grain tends to wander in a random fashion, changing directions all the time (Figure 9–17). This phenomenon is called *Brownian motion*. It is clear from Newton's laws of motion that something in the water has to be causing the tiny movements by exerting a force, but what force could it be?

In 1905, Albert Einstein demonstrated that individual collisions with large numbers of water molecules were driving the motion of the pollen grains. Einstein realized that a small object suspended in liquid would be constantly bombarded by moving atoms. At any given moment, there will, purely by chance, be more atoms hitting on one side than the other. The object will be pushed toward the side with fewer collisions. A moment later, however, more atoms will be hitting another surface, and the object will change direction. Einstein argued that, over time, atomic collisions would produce precisely the sort of erratic motion that you see in the microscope.

Einstein used the mathematics of statistics to make a number of predictions about how fast and how far the suspended grains would move based on the hypothesis that the motion was due to collisions with real atoms. French physicist Jean Baptiste Perrin (1870–1942) tested the theory and published the results of his careful experiments on Brownian motion in 1909. His measurements agreed with Einstein's calculations. With this work, the question of the reality of atoms was laid to rest for good; hypothetical constructs cannot exert forces.

Until the late twentieth century, all evidence for atoms was indirect though compelling; no one had ever "seen" an atom. In the late 1980s, however, scientists devised instruments called scanning probe microscopes, with which it is possible to "photograph" atoms and thus provide direct visual evidence of their reality (Figure 9–18).

Figure 9–18
This electronic image representing individual atoms was taken with an instrument called a scanning probe microscope. The "mountains" correspond to individual xenon atoms that were placed on a "plain" of nickel. This image reveals the first time an atomic structure had been "hand built."

Figure 9–17
The motion of small bits of material as they are bombarded by atoms, called Brownian motion, was a major piece of evidence for the reality of atoms.

When in the chain of historical events would you have been willing to believe that atoms are real? When Dalton explained the existence of elements? When Einstein explained Brownian motion? When you were shown a picture like the one in Figure 9–18? Never? What does it take to make something "real"? And, finally, does it make a difference to science whether atoms are "real" or not?

▶ Summary

All the solids, liquids, and gases around us are composed of about 100 different chemical *elements*. *Atoms*, the building blocks of our chemical world, combine in groups of two or more called *molecules*. For thousands of years, atoms were discussed purely as hypothetical objects, but studies of *Brownian motion* early in the twentieth century, and recent imaging of individual atoms in new kinds of microscopes, have confirmed the existence of these tiny particles.

Each atom contains a massive central *nucleus* made from positively charged protons and electrically neutral neutrons. Around the nucleus move *electrons*; these are swift, negatively-charged particles that have only a small fraction of the mass of protons and neutrons. Early models of this kind of atom treated electrons like planets orbiting around the Sun. Those models were flawed, however, because each electron, constantly accelerating, would have to continuously emit electromagnetic radiation. Bohr proposed an alternate model in which electrons rest in various energy levels, much as you stand on different levels of a flight of stairs.

Electrons in the *Bohr atom* can shift to a higher energy level by absorbing the energy of heat or light. They can also drop into a lower energy level, and in the process release a *photon*—an individual electromagnetic wave. These tiny changes in electron energy level are called *quantum leaps* or *quantum jumps*. *Spectroscopy*, the study of the light emitted or absorbed by atoms (the atom's *spectrum*), reveals the nature of each atom's electron energy levels.

Each of an atom's energy levels, or *electron shells*, holds only a limited number of electrons. The innermost electron shell will accommodate only 2 electrons, the second and third shells contain up to 8 electrons, while the fourth and fifth shells hold 18 electrons. This electronic structure is reflected in the organization of the *periodic table of elements*, which lists all the elements in rows corresponding to each shell, and in columns corresponding to elements with similar numbers of outer electrons and thus similar chemical behavior.

▶ Review Questions

1. What experiment first implied the existence of atoms?

2. What three particles make up every atom? What are the major differences among these particles?

3. In what ways does the nucleus differ from electrons?

4. What is a molecule? a compound?

5. Why did some alpha particles bounce backward off atoms in Rutherford's experiment?

6. Describe the main features of the Bohr atom.

7. What is a quantum leap? What can cause an electron to undergo such a change?

8. What is the major difference in the appearance of an emission spectrum and an absorption spectrum?

9. Cite three examples of everyday objects with vivid emission spectra.

10. How can scientists on Earth determine the elements that occur in stars?

11. Describe the workings of a laser.

12. In what ways are all the elements in a given column of the periodic table similar?

13. Name a chemical element that is considered corrosive and dangerous.

14. Explain Dalton's view of the atom.

15. Why couldn't Rutherford's atom work?

16. Describe the ground state and the excited states of an electron.

17. What is a photon? Under what circumstances are photons emitted?

18. What is spectroscopy?

19. What is a mole?

20. How were scientists convinced that atoms really exist?

21. What is an electron shell?

▶ Fill in the Blanks

Complete the following paragraphs with words and phrases from the list.

atoms	Pauli exclusion principle
Bohr atom	periodic table of the ele-
chemical element	ments
electrons	photon
electron shell	quantum leap, or
laser	quantum jump
molecules	spectrum
nucleus	spectroscopy

Some materials can be broken down by ordinary chemical reactions. Those that cannot are called _____, and to each of these there corresponds one type of _____. Atoms are composed of negatively charged _____ in orbit around a positively charged _____. Combinations of atoms are called _____.

In the _____, electrons can exist only in certain allowed orbits. Electrons move between orbits by means of a _____, a process that can result in the emission of a bundle of electromagnetic radiation known as a _____. Every atom and molecule emits a specific set of electromagnetic waves, and this collection is known as the _____ of the atom or molecule. The science of _____ is devoted to unscrambling these "atomic fingerprints." The _____ is a device that produces a coherent beam of photons as a result of quantum leaps in atoms.

The _____ is a chart of all the chemical elements in which atoms get heavier, in general, as we read from left to right, and atoms in the same column have similar chemical properties. Because of the _____, two electrons cannot exist in the same state, so only a certain number of electrons can be fit into each _____. This rule explains the structure of the periodic table.

▶ Discussion Questions

1. In what ways are atoms like the solar system of planets orbiting around the Sun? In what ways are they different?

2. The leaves of a tree are bright green. What do you think a leaf's absorption spectrum might look like?

3. Rutherford's experiment involved firing nucleus-sized "bullets" at atoms of gold. He found that one atom in a thousand bounced backward. What might Rutherford have seen if atoms were completely uniform in mass? What if electrons were more massive than alpha particles? (*Hint:* What happens when a bowling ball collides with a ping-pong ball?)

4. Based on your knowledge of Newton's laws of motion, the laws of thermodynamics, and the nature of electromagnetic radiation, explain why the Rutherford model of the atom couldn't work.

5. People sometimes use the phrase "quantum leap" to signify a major breakthrough or stunning advance. Knowing what you do about the Bohr atom, is that an appropriate meaning? Why or why not?

6. When you shine invisible ultraviolet light ("black light") on certain objects, they fluoresce with brilliant colors. How does this behavior relate to the Bohr atom?

7. Why do different lasers have different-colored beams?

8. What does it mean to say the periodic table was useful because it "worked"? How does this relate to

the scientific method? Describe an imaginary discovery that might have invalidated the periodic table.

9. If you were told that fluorine is an extremely reactive element—that is, it combines readily with other elements—what other elements could you guess were also extremely reactive? Why?

10. Space probes often carry compact spectrometers among their scientific hardware. What kind of spectroscopy might scientists use to determine the surface composition of cold, outer planets that orbit the Sun? How might they use spectroscopy to determine the atmospheric composition of these planets?

11. What are the differences and similarities between the concept of the Greek atom and today's concept of the atom?

12. Outline or diagram Rutherford's experiment *from the point of view* of an alpha particle bullet. In this diagram be sure to include one bullet that passed through unimpeded, slightly impeded, or completely "bounced" back.

13. Consider yourself an electron bound to a nucleus of a Bohr atom. Describe in a step-by-step outline the processes that you will undertake for the following conditions:

    a. A photon passes by with the exact energy it takes for you to reach the next highest level.

    b. A photon passes by with sufficient energy to take you past the next highest level, but not to the second highest level.

    c. You decide to drop to the next lowest energy level.

14. Consider yourself a photon that passes close to an atom in a laboratory.

    a. What is your speed, as measured by a person in the lab?

    b. If you were "absorbed" by the atom, what statement can you make about the energy level structure of that atom?

15. Consider yourself a photon in a laser that is "tuned" to the right energy. Describe a typical interaction with the atoms in the laser.

16. The word *maser* means *m*icrowave *a*mplification by the *s*timulated *e*mission of *r*adiation, and applies to molecules. Explain a maser using your definition of laser.

17. Explain the Pauli exclusion principle in your own words.

18. Why is the Pauli exclusion principle important in the structure of the elements?

19. Why is electrolysis significant in the discovery of chemical elements?

20. If the atom is almost all empty space, why can't you walk through walls?

▶ Problems

1. If a one-electron atom has five different available energy levels, how many emission lines might appear in that atom's spectrum?

2. How many electrons do the following ions have? (Refer to the periodic table in Figure 9–15.)

    a. +1 of sodium (Na)

    b. −2 of oxygen (O)

    c. −3 of iron (Fe)

3. How many grams comprise a mole of the following substances? (Refer to the periodic table in Figure 9–15.)

    a. the element gold (symbol Au)

    b. the element neon (symbol Ne)

    c. carbon dioxide (composed of one carbon atom and two oxygen atoms, or CO_2)

    d. sodium chloride (composed of one atom of sodium and one atom of chlorine, or NaCl)

▶ Investigations

1. Simple hand-held spectroscopes are available in many science labs. Look at the spectra of different kinds of light bulbs: an incandescent bulb, a fluorescent bulb, a halogen bulb, and any other kinds available to you. What differences do you observe in their spectra? Why? Place pieces of transparent materials between a strong light source and the spectrometer. Does the spectrum change? Why?

2. Why do colors look different when viewed indoors, under fluorescent light, and outdoors in sunlight? How might you devise an experiment to quantify these differences?

3. Look around you and make a short catalog of 15 or 20 objects within a few feet of where you are sitting. You might observe shiny metal jewelry, soft cotton fabric, colorful plastic, transparent glass, fizzy soda, crunchy pretzels, and so on. How might you determine what elements these objects are made of? Why are these objects so different from each other?

4. Which of the chemical elements do you encounter in more or less pure form in your daily experience? What are the properties of these elements and how are they used? Check off the positions of these elements on the periodic table. Do you see any patterns? Are there any groups of elements that you do not encounter?

5. If you had never been told about atoms in science class, what observations might convince you of their existence? What might be an alternative hypothesis to the atomic theory of matter?

▶ Additional Reading

Perrin, Jean. *Atoms*. Woodbridge, Connecticut: Oxbow Press, 1990 (translation of the 1913 French edition).

von Beyer, Hans Christian. *Taming the Atom*. New York: Random House, 1992.

10 QUANTUM MECHANICS

AT THE SUBATOMIC SCALE EVERYTHING IS
QUANTIZED. ANY MEASUREMENT AT THAT SCALE
SIGNIFICANTLY ALTERS THE OBJECT BEING MEASURED.

Take Me Out to the Ball Game

Imagine yourself at a big league ball game under the lights of a great stadium. Cheering fans fill the stands, roving vendors sell their food and drink, and the pitcher and batter play out their classic duel. The pitcher stares the batter down, winds up, and hurls a fastball. But the batter is ready and pounces on the pitch. The ball leaps off the bat with a sharp crack. And then all the lights go out.

Where is the ball? You can't see it, but given a videotape replay of the bat hitting the ball, you could determine the exact position and velocity of the ball at the time the lights went out. You could, at least in principle, calculate the subsequent path of the ball and predict where it would eventually land. We know how objects like baseballs behave, even if we can't see them.

The behavior of particles at the submicroscopic scale of electrons is different. Early in this century, scientists realized that you can never know the exact position and the exact velocity of an electron simultaneously. For the first time, the scientific community was faced with the fact that there are some things we cannot know about our physical universe. Despite these limits, we take advantage of electron properties in thousands of applications, from VCRs to microwave ovens. Our present understanding of the electron results from advances in quantum mechanics, the branch of science devoted to the study of the behavior of the atom and its constituents. Given the importance of fax machines, video games, electronic fuel injection systems, and the myriad other devices that depend on the microelectronics industry, it's very difficult to imagine modern civilization operating without this particular branch of science.

THE WORLD OF THE VERY SMALL

In Chapter 9, we saw that when an electron moves between energy levels and emits a photon, it is said to make a *quantum leap*. The word quantum comes from a Latin word related to "bundle," while mechanics, as we discussed in Chapter 3, is the study of the motion of material objects. **Quantum mechanics,** then, is the branch of science that is devoted to the study of the motion of objects that come in small bundles, or *quanta*. We have already discussed how the atom contains little bundles of matter called electrons, which move around another bundle of matter called the nucleus. This matter is said to be **quantized.**

Electrical charges are also quantized. Every electron has a charge of exactly -1, while every proton has a $+1$ charge. We have seen that photons emitted by an atom can have only certain values of energy, so that energy levels and energy emitted are also quantized. In fact, inside the atom, *everything* comes in quantized bundles.

This behavior at the atomic level may be hard to imagine since our everyday world isn't like this at all. Although we have been told since childhood that the objects around us are made up of atoms, for all intents and purposes we experience matter as if it were smooth, continuous, and infinitely divisible. Indeed, for almost anything we want to do in the physical world, matter behaves as if it is smooth and continuous.

The quantum world is foreign to our senses. All of the intuition that we have built up about the way the world operates—all of the "gut feelings" we have about the universe—comes from our experiences with large-scale objects made up of apparently continuous material. If it should turn out (as it does) that the world of the quantum does not match our intuition, we should not be surprised. We have never dealt with this kind of world, so we have no particular reason, based on observations or experience, to believe that it should behave one way or the other.

MEASUREMENT AND OBSERVATION IN THE QUANTUM WORLD

Every measurement in the physical world incorporates three essential components:

1. A sample—a piece of matter to study.
2. A source of energy—light or heat or kinetic energy that interacts with the sample.
3. A detector to observe and measure that interaction.

When you look at a piece of matter such as this book, you can see it because light (a source of energy) bounces from the book (a sample) to your eye (a detector). When you examine a piece of fruit at the grocery store, you apply energy by squeezing it to detect if it feels too ripe.

Many different professions use sophisticated devices to make measurements. Air traffic controllers reflect microwaves off airplanes to determine their positions. Oceanographers bounce sound waves off deep-ocean sediments to map the sea floor. Dentists pass X-rays through pa-

quantum mechanics The branch of science that is devoted to the study of the motion of objects that come in small bundles, or quanta.

quantized Whenever energy or other property of a system can have only certain definite values, and nothing in between those values, it is said to be quantized.

The radar antenna sends out microwaves that interact with flying airplanes, are reflected, and detected on their return. This allows air traffic controllers to keep track of where airplanes are in the sky.

tients' teeth and gums to look for cavities. We assume that such interactions of matter and energy do not change the objects being measured in any appreciable way. Microwaves don't alter an airplane's flight path, nor do sound waves noticeably disturb the topography of the ocean's bottom. And while prolonged exposure to X-rays can be harmful, the dentist's brief exploratory X-ray photograph has no obvious immediate effects on the tooth. Our experience tells us that we can measure something large enough to be seen without a microscope without altering that object, because the energy of the probe is much less than the energy of the object.

The situation is rather different in the quantum world. If you want to "see" an electron, you still have to bounce something energetic off it so that the information can be carried to your detectors. But nothing at your disposal can interact with the electron without simultaneously affecting it. You can bounce a photon off it, but in the process the electron's energy will change. You can bounce another particle off it, but the electron will recoil as a pool ball does when it is struck by another pool ball. When the energy of the probe is too close to the energy of the thing being measured, the object being measured will be altered. Therefore, the electron cannot fail to be altered by the interaction.

Measurements in the world of quantum mechanics are like sending a car down a tunnel and listening for a crash. The car won't be the same after the collision as it was before.

The act of measurement in the quantum world poses a dilemma analogous to trying to discover if there is a car in a tunnel when the only means of finding out is to send another car into the tunnel and listen for a crash. With this technique you can certainly discover whether the first car is there—you can probably even find out where it is by measuring the time it takes the probe car to crash. You *cannot*, however, assume that the first car is the same after the interaction as it was before. In the same way, nothing in the quantum world can be the same as it was before after the interaction associated with a measurement.

In principle, this argument applies to any interaction, whether it involves photons and electrons, or photons and bowling balls. As we demonstrate in the Science by the Numbers section later in this chapter, the effects of an interaction for objects like baseballs are so tiny that they can simply be ignored. In the case of interactions at the atomic level, the effects are apparent. This fundamental difference between the quantum and macroscopic worlds is what makes quantum mechanics quite different from the classical mechanics of Isaac Newton. The consequences of small-scale interactions make the quantum world different than the world with which we are familiar, because of the disruptive nature of its interactions.

THE HEISENBERG UNCERTAINTY PRINCIPLE

In 1927, a young German physicist, Werner Heisenberg (1901–1976), put the idea of limitations on quantum-scale measurements into precise mathematical form. His work, which was one of the first results to come from the new science of quantum mechanics, is called the Heisenberg **uncertainty principle.** The central concept of the uncertainty principle is this:

uncertainty principle The idea quantified by Werner Heisenberg in 1927, that at a quantum scale, the precise location and velocity of an object can never be known at the same time, because quantum-scale measurement affects the object being measured. Specifically, "The error or uncertainty in the measurement of an object's position, times the error or uncertainty in that object's velocity, must be greater than a constant, h, divided by the object's mass."

Werner Heisenberg (1901–1976)

> **At the quantum scale, any measurement significantly alters the object being measured.**

Suppose, for example, you have a particle such as an electron in an atom and want to know where it is *and* how fast it's moving. The uncertainty principle tells us that it is impossible to measure both the position and the velocity with infinite accuracy at the same time.

The reason for the uncertainty principle is that every measurement changes the object being measured. Just as the car in the tunnel changed after its first measurement was taken, so too will the quantum object change. The result is that the more precisely you measure one property, such as the position of an object, the less precise your knowledge becomes of a property such as velocity.

The uncertainty principle doesn't say that we cannot know a particle's location with great precision. It is possible, at least in principle, for the uncertainty in position to be zero, which would mean that we know the exact location of a quantum particle. In this case, however, the uncertainty in the velocity has to be infinite. Thus, at the point in time when we know exactly where the particle is, we have no idea whatsoever how fast it is moving. By the same token, if we know exactly how fast the quantum particle is moving, we cannot know where it is.

In practice, every quantum measurement involves trade-offs. We accept some imprecision in the location of the particle, or some imprecision regarding the velocity, playing the two off of each other to get the best solution to whatever problem we're working on. Since we cannot have precise knowledge of both characteristics at the same time, we can settle for knowing one characteristic accurately.

Heisenberg put his notion into a mathematical relationship, which is a complete and exact statement of the uncertainty principle.

▶ **In words:**

> The error or uncertainty in the measurement of an object's position, times the error or uncertainty in that object's velocity, must be greater than a constant (called Planck's constant after German physicist Max Planck, 1858–1947) divided by the object's mass.

▶ **In equation form:**

$$\text{(uncertainty in position)} \times \text{(uncertainty in velocity)} > \frac{h}{\text{mass}}$$

> where h is *Planck's constant* (see below).

▶ **In symbols:**

$$\Delta x \times \Delta v > \frac{h}{m}$$

This equation is simply a precise, shorthand way of saying that you can never know both the position and velocity of an object with perfect accuracy. The difference between the macroscopic world and the world inside the atom hangs on the question of the numerical value of h/m, the numbers on the right side of Heisenberg's equation. In SI units (see Appendix A), Planck's constant, h, has a value of 6.63×10^{-34} joule-seconds.

The important point about the Heisenberg relationship is not the exact value of the number, h/m, on the right side, but the fact that the number is greater than zero. For example, if you make a very precise measurement of a particle's location, then the uncertainty in position, Δx, must be small. In this situation, it follows that the uncertainty in velocity, Δv, must be large. We can use the uncertainty principle to calculate the uncertainty in velocity for a given uncertainty in position, and vice versa.

Science by the Numbers

Uncertainty in the Newtonian World

To understand why we do not have to be concerned about the uncertainty principle in our everyday lives, we can calculate the uncertainty in measurements in two separate situations: macroscopic objects such as automobiles, and submicroscopic objects such as electrons (see Figure 10–1).

1. Small Uncertainties with Large Objects.   A moving automobile with a mass of 1000 kg is located in an intersection that is 5 m across. How precisely can you know how fast the car is traveling?

We can solve this problem by noting that if the car is somewhere in an intersection 5 m across, then the uncertainty in position of the car is about equal to 5 m. Because, we know the car's mass and uncertainty in position, and we can calculate the uncertainty in velocity from the uncertainty equation:

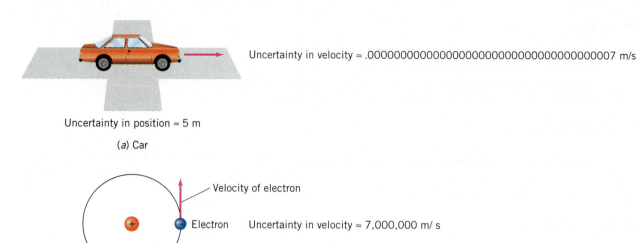

Uncertainty in velocity ≈ .0000000000000000000000000000000000007 m/s

Uncertainty in position ≈ 5 m

(a) Car

Velocity of electron

Electron   Uncertainty in velocity ≈ 7,000,000 m/ s

Uncertainty in position ≈ 10^{-10} m

(b) Atom

Figure 10–1
The uncertainty in velocity of a car in an intersection affects only the 37th decimal place. For an electron in an atom, however, the uncertainty is more than 2% of the speed of light.

First, we must re-arrange this equation to solve for uncertainty in velocity:

$$\text{(uncertainty in position)} \times \text{(uncertainty in velocity)} > \frac{h}{\text{mass}}$$

$$\text{uncertainty in velocity} > \frac{h/\text{mass}}{\text{uncertainty in position}}$$

$$> \frac{6.63 \times 10^{-34}\,\text{J-s}/1000\,\text{kg}}{5\,\text{m}}$$

$$> \frac{6.63 \times 10^{-37}}{5}\,\text{J-s/kg-m}$$

$$> 1.33 \times 10^{-37}\,\text{J-s/kg-m}.$$

Recall the definition of a joule:

$$1\ \text{joule} = 1\ \text{kg-m}^2/\text{s}^2$$

Substituting this definition into the unit, J-s/kg-m:

$$\text{J-s/kg-m} = \frac{[(\text{kg-m}^2/\text{s}^2)\text{s}]}{(\text{kg-m})}$$

$$= \text{m/s}$$

The solution is thus in units of velocity, m/s, as expected. The uncertainty in the velocity of the automobile is greater than 1.33×10^{-37} m/s. This uncertainty is extremely small. Theoretically, we could know the velocity of the car to an accuracy of 37 decimal places! In practice, however, we have no method of measuring velocities with present or foreseeable human technology to an accuracy remotely approaching this. The uncertainty is, for all practical purposes, indistinguishable from zero. Therefore, for objects with significant mass like automobiles, the effects of the uncertainty principle are totally negligible. The equation confirms our experience that Newtonian mechanics works perfectly well in dealing with everyday objects.

2. Large Uncertainties with Small Objects. Contrast the above example with the uncertainty in velocity of an electron located within an atom, an area about 10^{-10} m on a side. To what accuracy can we measure the velocity of that electron?

The mass of an electron is 9.11×10^{-31} kg. If we take the uncertainty in position to be 10^{-10} m, then according to the uncertainty principle:

$$\text{(uncertainty in velocity)} > \frac{h/\text{mass}}{\text{uncertainty in position}}$$

$$> \frac{[(6.63 \times 10^{-34}\,\text{J-s})/(9.11 \times 10^{-31}\,\text{kg})]}{(10^{-10}\,\text{m})}$$

$$> 7.3 \times 10^{6}\,\text{m/s}.$$

This uncertainty is very large indeed. If we know that an electron is somewhere in an atom, then we cannot know its velocity to any better accuracy than this.

For ordinary sized objects like cars and bowling balls, whose mass is measured in kilograms, the number on the right side of the uncertainty relation is so small that we can treat it as being zero. Only when the masses get very small, as they do for particles like the electron, does the number on the right get big enough to make a practical difference. ●

Developing Your Intuition

Absolute Zero

In the nineteenth century, people used to define absolute zero (see Chapter 6) as the temperature at which all atomic motion stopped. Why did quantum mechanics force a change in the definition of absolute zero?

The reason has to do with the uncertainty principle. If an atom stops moving completely, then there is no uncertainty in its velocity, which must be zero. But if we know the velocity, then we cannot tell where the atom is. It could be anywhere in the universe. If the atom is located in a laboratory apparatus, then there will be an uncertainty in our knowledge of the velocity. The velocity could be non-zero, in other words, without our being able to tell. Such movement would violate the definition of absolute zero.

In fact, sophisticated calculations in quantum mechanics predict that even at absolute zero, atoms will have some velocity. This phenomenon is called "zero point motion" or "zero point energy," and its existence can actually be confirmed in many different kinds of experiments. ●

PROBABILITIES

The uncertainty principle has consequences that go far beyond simple statements about measurement. In the quantum world we must radically change the way that we describe events. Consider an everyday example in which the uncertainties are much larger (but easier to picture) than those associated with Heisenberg's equation. Think of the example of a batter hitting a ball during a night baseball game, with the stadium lights going out at the moment of impact. Where will the ball be in 5 seconds?

Figure 10–2
The position of a "quantum baseball" cannot be determined precisely. Instead, you can predict only the probabilities of the ball being various distances from home plate, as discussed in the text. The most likely location is at the peak of the curve, but the ball could be anywhere else.

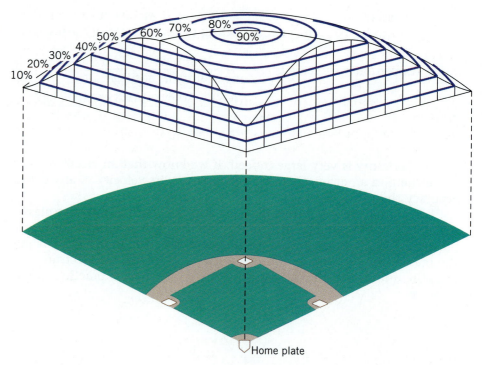

If you were an outfielder, this would be more than a philosophical question. You need to know where to go to make your catch. In a Newtonian world, you would have no problem in doing this. If you knew the position and velocity of the ball at the instant the lights went out, some simple calculations would tell you exactly where the ball will be at any time in the future.

If you were a quantum outfielder in an atom-sized ball field, on the other hand, you would have a much harder time of it. You wouldn't know both the position and velocity of the quantum ball when the lights go out—at best, you would be able to put some bounds on them. You may, for example, be able to say something like "It's somewhere inside this 3–foot circle and traveling between 30 and 70 feet per second." This means that when you try to guess where it will be in 5 seconds, you can't do so with any precision. If you were thinking in Newtonian terms, you would have to say that the ball could be 147 feet from the plate (if it were traveling 30 feet per second and located at the back of the 3-foot circle), 353 ft from the plate (if it were traveling 70 feet per second and located at the front of the circle), or anyplace in between. You could predict the likelihood, or **probability,** that the ball would be anywhere in the outfield, and could present these probabilities on a graph like the one shown in Figure 10–2.

probability   The likelihood that an event will occur or that an object will be in one state or another; how nature is described in the subatomic world.

This example shows that the uncertainty principle requires a description of quantum-scale events in terms of probabilities. Just like the baseball in our example of the darkened stadium, there must be uncertainties in the positions and velocities for all quantum objects when we first observe them. Hence there will be uncertainties at the end of our observations—uncertainties that can be dealt with by reporting probabilities.

Probabilities are extremely important. They require us to think of quantum events in a different way from ordinary events in our lives. In particular, probabilities require us to rethink concepts such as regularity, predictability, and causality at the quantum level.

For technical reasons, physicists find it useful to talk about the *wave function* of an object, which is defined to be the square root of the probability of finding the object at a point. Thus the graph in Figure 10–2, which records the probability of finding the baseball at a given distance from the plate, shows the square of the baseball's wave function. When we describe the motion of any particle in the quantum world, that description will not be in terms of the Newtonian position-and-velocity, but in terms of a wave function, or a collection of probabilities.

WAVE-PARTICLE DUALITY

Quantum mechanics is sometimes called **wave mechanics,** because it turns out that quantum objects sometimes act like particles, and sometimes like waves. This dichotomy is known as the problem of "wave-particle duality," and it has puzzled some of the best minds in science. To understand it, consider how particles and waves behave in our macroscopic world.

wave mechanics Another term for quantum mechanics indicating the dual (wave and particle) nature of quantum objects.

The Double-Slit Test

Recall from Chapter 8 that energy travels either as a wave or as a particle. Particles transfer energy through collisions, while waves transfer energy through collective motion of the medium in which they travel or through electromagnetic fields. Every phenomenon in our everyday world can be neatly divided into particles or waves, and many experiments can be performed to determine whether something is a particle or a wave. The most famous of these experiments is called the *double-slit apparatus*, which consists of a barrier that has two slits in it (Figure 10–3). If particles such as baseballs are thrown from the left side, a few will make it through the slits, but most will bounce off. If you were standing on the right side of the barrier, you would expect to see the baseballs coming through more or less in the two openings shown, accumulating in two piles behind the barrier (Figure 10–3a).

If, however, waves of water were coming from the left side, you would expect to see the results of constructive and destructive interference (see Chapter 8). Rather than the two piles of baseballs, we would see perhaps half a dozen regions of high waves beyond the barrier, interspersed with regions of still water (Figure 10–3b).

Now, let's use the same arrangement to see whether light behaves as a particle or a wave. In Chapter 9, we learned that light is emitted in quanta of energy, called *photons*. On the one hand, photons behave like particles in the sense that they can be localized in space. You can set up experiments in which a photon is emitted at one point, then received somewhere else after an appropriate lapse of time, just as a baseball is "emitted" by a pitcher and "received" later by a catcher (Figure 10–3a). If, on the other hand, you shine light—a flood of photons—on the two-slit apparatus, you will observe an interference pattern on the right (Figure 10–3b). In that experiment, photons act like waves. The question that

Figure 10–3
(*a*) The two-slit experiment may be used to determine whether something is a wave or a particle. A stream of particles striking the barrier will accumulate in the two regions directly behind the slits. (*b*) When waves converge on two narrow slits, however, constructive and destructive interference results in a series of peaks.

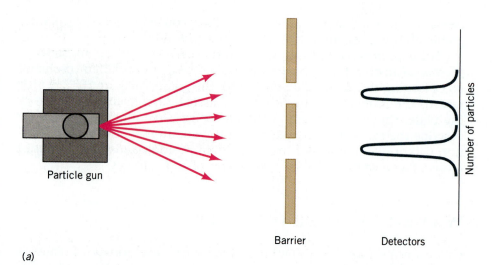

(*a*)

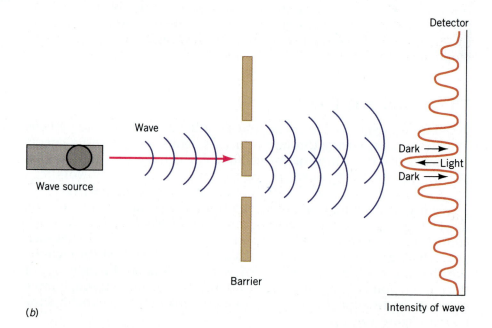

(*b*)

remains, however, is how can photons sometimes act like waves and sometimes like particles?

You can make the experiment more complicated by setting up the apparatus so that only one photon at a time comes through the slits. If you do this, you find that each photon arrives at a specific point at the screen; this is the behavior you would expect of a particle. If you allow photons to accumulate over long periods of time, however, they will arrange themselves into the interference pattern characteristic of a wave (Figure 10–4).

You could do a similar series of experiments with any quantum object, such as electrons or even atoms. They all exhibit the properties of both particles and waves, depending on what sort of experiment is done. If you do an experiment that tests the particle properties of these things,

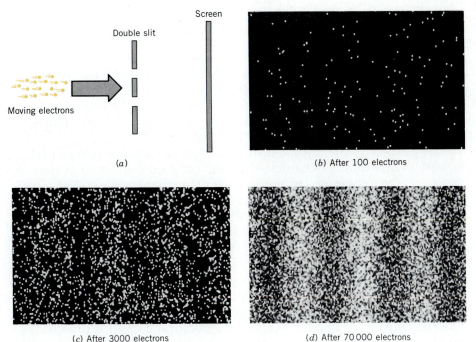

Screen

Double slit

Moving electrons

(a)

(b) After 100 electrons

(c) After 3000 electrons

(d) After 70 000 electrons

Figure 10–4
When 100 electrons or photons (light particles) pass through a two-slit apparatus one at a time (a), they cause 100 single spots on a photographic film (b). As the number of subatomic particles increases (c and d), a wavelike interference pattern emerges. The bright spots are places where constructive interference occurs, and the dark spots correspond to destructive interference.

they look like particles. If you do an experiment to test their wave properties, they look like waves. Whether you see quantum objects as particles or waves depends on the experiment that you do.

Some experimenters have gone so far as to try to "trick" quantum particles such as electrons into revealing their true identity. Such experiments use modern electronics to decide whether a particle- or wave-type experiment is being done *after* the quantum object is already on its way into the apparatus. Scientists who do these experiments find that the quantum object seems to "know" what experiment is being done, because the particle experiments always turn up particle properties, while the wave experiments always turn up wave properties.

At the quantum level, the objects that we talk about are neither particles nor waves in the classical sense. In fact, we can't really visualize them at all, because we have never encountered anything like them. They are a third kind of object, neither particle nor wave, but exhibiting the properties of both. If you persist in thinking about them as if they were baseballs or surf coming into a beach, you will quickly lose yourself in confusion. To understand how foreign this concept is, consider someone who has seen only the colors red and green in her entire life. If she has decided that everything in the world has to be either red or green, she will be totally confused by seeing another color, such as blue. What she has to realize is that the problem is not in nature, but in her assumption that everything has to be either red or green.

In the same way, the problem of wave-particle duality arises because we assume that everything has to be either a wave or a particle. If we allow ourselves the possibility that quantum objects are things that we have never encountered before, we can better accept that these objects might have unencountered properties. If we can accept that possibility, then we can accept that such objects cannot be visualized.

Technology

The Photoelectric Effect

When photons strike some materials, their energy can be absorbed by electrons, which are shaken loose from their atoms. If the material in question is in the form of a thin sheet, then when light strikes one side, electrons are observed coming out of the other. This phenomenon is called the *photoelectric effect* (see Figure 10–5).

One aspect of the photoelectric effect played a major role in the history of quantum mechanics. The time between the arrival of the light and the appearance of the electrons is extremely short—far too short to be explained by the relatively gentle action of a wave nudging the electrons loose. In fact, it was Albert Einstein who pointed out that the explanation of this rapid response depended on the particle-like nature of the photon. He argued that the interaction between the light and the electron is something like the collision between two billiard balls, with one ball shooting out instantly after the collision. It was this work that led to our modern concept of the photon, which was the basis of Einstein's Nobel Prize in 1921.

The conversion of light energy into electrical current is used in many familiar devices. In your camera, for example, a photoelectric device measures the amount of light available. If your camera is automatic, the amount of electric current produced by the light is used to determine how wide to open the lens and what the shutter speed should be. In telephone systems that use *fiber optics*—fibers that act like pipes for visible light—light signals striking sophisticated semiconductor devices (see Chapter 13) shake loose electrons to form a current that, ultimately, drives the diaphragm in your telephone and produces the sound that you hear. In medical CAT scans (see Technology section following), photoelectric devices convert X-ray photons into electrical currents whose stength can be used to produce a picture of a patient's internal organs. As all of these examples show, an understanding of the way that objects interact in the quantum world can have enormous practical consequences. ●

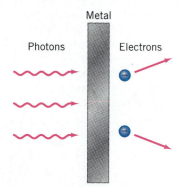

Figure 10–5
When light falls on a thin sheet of metal, electrons are knocked loose and come out of the metal. This process is called the photoelectric effect.

The light meter in a camera uses the photoelectric effect to measure the amount of light falling on the film.

The CAT Scan

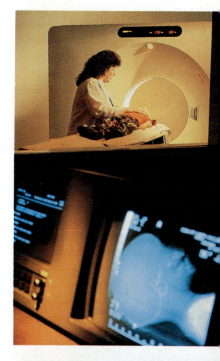

Ordinary X-ray photographs depend on the differences in density (and therefore in the ability to absorb X-rays) of bone, skin, and other materials in the body. In these photographs, the X-rays make one pass-through, in one direction only, to produce the pictures that you have seen in your doctor's or dentist's office. This technique cannot produce a three-dimensional image of the interior of the body, nor can it produce sharp images of an organ whose density is similar to that of its surroundings. These shortcomings are overcome by a different X-ray technique, known as *computerized axial tomography*, or the *CAT scan*.

The easiest way to visualize a CAT scan is to think of dividing the body into imaginary slices perpendicular to the backbone, with each slice being a millimeter or so in width. The material in each slice is probed by successive short bursts of X-rays, lasting only a few milliseconds each, that cross the slice in different directions. Each part of the body is thus traversed by many different X-ray bursts, each of which contains the same number of photons. A photoelectric device (see above) measures the number of X-rays that pass through the body (i.e., those not absorbed by material along their path).

Once all the data on a given slice have been obtained, a computer works out what the density of each point of the body would have to be and produces a detailed cross-section of the body along that particular slice. A complete picture of the body (or a specific part of it) can then be built up by looking at successive slices. ●

WAVE-PARTICLE DUALITY AND THE BOHR ATOM

In Chapter 9 we saw that an atom's electrons are restricted to a few energy levels. Treating electrons as waves helps explain why. Every quantum object displays a simple relationship between its speed (when we think of it as a particle) and its wavelength (when we think of it as a wave). It turns out that for electrons, protons, and other quantum objects, a faster speed always corresponds to a shorter, more energetic wavelength (which corresponds to a higher frequency).

If you think of an electron as a particle, then you can treat its motion around an atom's nucleus in the same Newtonian way you treat the motion of the Earth in orbit around the Sun. That is, for any given distance from the nucleus, the electron must have a precise velocity to stay in a stable orbit. Provided it is moving at this velocity, it will stay in that orbit, just as the Earth stays in a stable orbit around the Sun. If it is any faster, it must adopt a higher orbit; if it is any slower, it will move closer to the nucleus.

If we choose to think about the electron as a wave, a different set of criteria can be used to decide how to put the electron into its orbit. A wave on a straight string (on a guitar, for example) vibrates only at certain

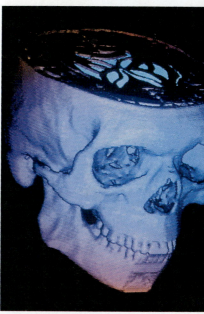

A boy having CAT scan with the video monitor in the foreground; A CAT scan of a human skull and brain is shown at the bottom.

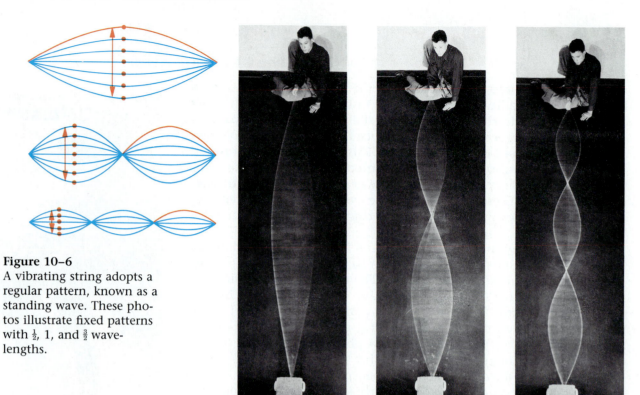

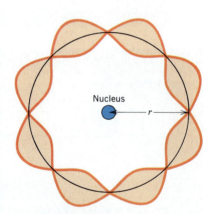

Figure 10–6
A vibrating string adopts a regular pattern, known as a standing wave. These photos illustrate fixed patterns with $\frac{1}{2}$, 1, and $\frac{3}{2}$ wavelengths.

Figure 10–7
An electron in orbit about an atom adopts a standing wave like a vibrating string. This illustration shows a standing wave with four wavelengths fitting into the orbit's circumference.

frequencies; the frequencies depend on the length of the string (Figure 10–6). These frequencies correspond to $\frac{1}{2}$ wavelength, 1 wavelength, $\frac{3}{2}$ wavelengths, and so on, on the string in the figure. Now imagine bending the guitar string around into a circular orbit. In this case, you will only be able to fit certain waves in the orbit, as shown in Figure 10–7.

Consider the different situations for particles and waves just discussed. Are there any orbits for which the wave and particle descriptions are consistent? In other words, are there any possible orbits in which the velocity needed to keep the electron from falling in (when we think of it as a particle) gives an electron wavelength that just fits on the orbit (when we think of it as a wave)? The answer turns out to be yes—a few such orbits can be found. They turn out to be the "allowed orbits" of the Bohr atom we talked about in Chapter 9. It seems, then, that the only orbits that are allowed in the atom are those for which it makes no difference whether we think of the electron as a particle or a wave. In a sense, then, the wave-particle duality exists in our minds, and not in nature.

QUANTUM WEIRDNESS

The fact that quantum objects behave so differently from objects in our day-to-day lives causes many people to worry that nature has somehow become "weird" at the subatomic level. The description of particles in terms of a wave defies our common sense. Situations in which a photon or electron seems to "know" how an apparatus will be arranged before the arranging is done seems wrong and unnatural. Many people, scientists and nonscientists alike, find the conclusions of quantum mechanics to be quite unsettling.

Albert Einstein argued that the reason quantum mechanics was so strange was that it was not complete. He felt that there must be some way of getting at the missing information about quantum objects—in effect, some way of measuring them without disturbing them. This way of looking at quantum mechanics eventually came to be called the *hidden variable theory*. It holds that the variables we see, like position and velocity, are not adequate to describe nature at the atomic scale. If only we could get all the information about quantum objects, Einstein believed, then we would be able to describe the quantum-scale universe exactly, just as we do classical objects. The weirdness would simply disappear. In effect, Einstein argued that we end up having to describe nature in terms of probabilities simply because we don't know enough about the quantum world.

During the 1960s, a Scottish physicist named John Bell thought about this problem and realized that experiments could be devised that would have different results if quantum particles "really" had to be described in terms of wave functions, or if they "really" were something like little baseballs but we just didn't know all the things we needed to know to describe them. A typical experiment might involve an atom in an excited state emitting photons back-to-back, one to the left and one to the right. What Bell showed was that there are certain measurements you can make on these photons that will be different depending on how you describe them.

Although Bell's actual predictions are beyond this text, you can get some of the flavor of his work from the following (unrealistic) example. Bell's predictions might say that if quantum mechanics is correct, and if the photon is always described as a set of probabilities rather than as a "real" particle, then every time you find the photon moving to the right spinning clockwise, you would find the photon on the left spinning counterclockwise. If, on the other hand, the photon is "really" an ordinary particle whose true nature we simply have failed to discern, then every time you find the photon on the right spinning clockwise you will find the one moving to the left doing the same. (We emphasize that this is not what actually happens in the experiment, but it gives you a sense of the logic of the argument.) By doing an experiment of this type, it is possible to get more precise information about the actual photon.

During the late 1970s, researchers in Europe and the United States, using tremendous technological ingenuity, succeeded in doing a whole series of experiments of this type. The results were incontrovertible. The predictions of quantum mechanics were borne out, and those of the hidden variable theories were not.

The success of quantum mechanical predictions means that we have to live with the description of the subatomic world in terms of probabilities. The quantum world is what it is, and there's no way around it. This conclusion has led philosophers to think a great deal about what quantum mechanics tells us about the world. Some have argued, for example, that the advent of quantum mechanics means that ordinary causality no longer holds; we can no longer predict how a physical system will evolve in time. For these philosophers, the comforting clockwork Newtonian world has vanished forever.

Einstein and Bohr sitting together.

Most scientists (the authors included) do not agree with this view. The successful applications of quantum mechanics in our electronic age are proof enough that there is a satisfactory way to describe an atomic-scale system. Newtonian notions such as position and velocity aren't appropriate for the quantum world, which must be described from the beginning in terms of waves and probabilities. Thus quantum mechanics becomes a way of predicting how subatomic objects change in time. If you know the behavior of an electron now, you can use quantum mechanics to predict the behavior of that electron in the future. This approach to the atomic world is similar to the application of Newton's laws of motion in the macroscopic world. If you know the state of nature now, you can predict the state of nature in the future. The only difference is that in the quantum world, the state of the system is a probability.

In the view of most scientists, quantum mechanics is a marvelous tool that allows us to do all sorts of experiments and build extraordinarily useful pieces of equipment. The fact that we can't visualize the quantum world in terms that are familiar to us seems a small price to pay for all the benefits we receive.

Science in the Making

A Famous Interchange

Many people are disturbed by the fact that nature must be described in terms of probabilities at the subatomic level. Many scientists were also disturbed when quantum mechanics was first developed in the early twentieth century. Even Albert Einstein, one of the founders of quantum mechanics, could not accept what it was telling us about the world. He spent a good part of the last half of his life trying to refute it. His most famous statement from this period was "I cannot believe that God plays dice with the universe."

Confronted one too many times with this aphorism, Einstein's life-long friend and colleague Neils Bohr is supposed to have replied "Albert, stop telling God what to do." ●

THINKING MORE ABOUT
QUANTUM MECHANICS

Uncertainty and Human Beings

The ultimate Newtonian view of the universe was the concept of the Divine Calculator (see Chapter 4). This mythical being, given the position and velocity of every particle in the universe, could produce an exact prediction of every future state of those particles. The difficulty with this concept is that if the future of the universe is laid out with clockwork precision, it allows no room for human action. No one can make a choice about what he or she will do, because that choice has been predetermined.

Quantum mechanics allows for human choice. Heisenberg tells us that, although we might be able to predict the future if we knew the position and velocity of every particle exactly, we can never actually get those two numbers simultaneously. The Divine Calculator in a quantum world, therefore, is doomed to wait forever for the input data with which to start the calculation.

One area where quantum mechanics is starting to play a somewhat unexpected role regards an old philosophical argument about the connection between the mind and the brain. The brain is a physical object, an incredibly complex organ that processes information in the form of nerve impulses. The problem: what is the connection between the physical reality of the brain—the atoms and structures that compose it—and the consciousness that we all experience?

Many scientists and philosophers have argued that the brain or mind is really nothing more than physical structures. They have traditionally run into a problem, however, because if the brain is purely a physical object, its future states should be predictable. But what accounts for human creativity, and human diversity, if this is the case?

Recently, scientists (most notably Roger Penrose of Cambridge University) have argued that quantum mechanics can introduce a kind of unpredictability, an unprogrammability, that corresponds more closely with our perceptions of our own minds. How could quantum mechanics do this? Is there anything about the probabilities of quantum objects that might make it difficult (or even impossible) to make precise predictions of the future state of the brain?

▶ Summary

Matter and energy at the atomic scale come in discrete packets called quanta. The rules of *quantum mechanics*—the laws that allow us to describe and predict events in the quantum world—are different from Newton's laws of motion.

At the quantum scale, unlike our everyday experience, any measurement of the position or velocity of a particle causes the object to change in unpredictable ways. The mere act of making a measurement alters the object being measured. Werner Heisenberg quantified this situation in the Heisenberg *uncertainty principle*, which states that the uncertainty in the position of a particle multiplied times the uncertainty in its velocity must be greater than a small positive number. Unlike the Newtonian world, you can never know the exact position *and* velocity of a quantum particle.

These uncertainties preclude us from describing atomic-scale particles in the classical way, with precise position and velocity throughout time. Instead, quantum descriptions are given in terms of *probabilities* that an object will be in one state or another. Furthermore, quantum objects are not simply particles or waves, a dichotomy familiar to us in the macroscopic world. They represent something completely different from our experience, incorporating properties of both particles and waves.

▶ Key Equation

(uncertainty in position) × (uncertainty in velocity)
$$> \frac{h}{\text{mass}}$$

▶ Review Questions

1. What does "quantum mechanics" mean?

2. Give three examples of properties that are quantized at the scale of an electron.

3. What are the three essential parts of every physical measurement?

4. In what way is a measurement at the quantum scale of an electron different from a measurement at large scales of everyday objects?

5. State Heisenberg's uncertainty principle.

6. Under what circumstances can you know the position of an electron with great accuracy?

7. Why is quantum mechanics sometimes called "wave mechanics"?

8. What is "wave-particle duality"?

9. How has the acceptance of the uncertainty principle changed the definition of absolute zero?

10. Explain the quantum mechanical definition of absolute zero.

11. How can the double-slit experiment be used to define the wave-particle duality of nature?

12. Why are probabilities used in our descriptions of an electron?

13. Describe a wave function. How is it related to probabilities?

14. How is the photoelectric effect a demonstration of the particle description of nature?

15. Why is the double-slit experiment a demonstration of the wave description of nature?

16. Using the wave description of nature, carefully construct (on a qualitative basis), one of the allowed electron orbits of the Bohr atom, using a diagram, outline, or words.

▶ Fill in the Blanks

Complete the following paragraph with words and phrases from the list.

probability   wave function
quantum mechanics   wave-particle duality
uncertainty principle

_____ is the study of the motion of objects in the atomic world, where everything comes in bundles. The fact that measurements in this world cannot be made without disturbing the object being measured is quantified in the _____. As a consequence, quantum mechanical predictions are often made in terms of _____, rather than in Newtonian terms. The _____ of a particle is related to the probability that a particle will be present at a particular point, and the ability of quantum objects to appear as both waves and particles is called _____.

▶ Discussion Questions

1. Identify the sample, source of energy, and detector in the following experiments:

    a. Measuring the speed of a tennis ball.

    b. Observing bacteria in a microscope.

    c. Testing the electrical resistance of a piece of wire.

    d. Determining the bending strength of an iron bar.

2. Which properties of electrons are particle-like? Which are wavelike?

3. What role does probability play in description of subatomic events?

4. Sketch a possible probability diagram for the final resting position of a golf ball on a driving range. Assume that the golf tee is the starting point and that an average drive is 250 yds. (*Hint:* There is no one correct answer; different golfers will have different probability diagrams.)

5. In Chapter 4 we discussed the fact that chaotic systems are, for all practical purposes, unpredictable. How does this sort of unpredictability differ from that associated with quantum mechanics?

6. Present an argument in terms of the wave nature of the electron that shows that electrons in Bohr orbits cannot emit radiation and spiral in toward the nucleus, as they might be expected to do on the basis of Maxwell's equations. (*Hint:* See "Why the Rutherford Atom Couldn't Work," in Chapter 9.)

7. How does light behave like a particle? How does it behave like a wave?

8. If you threw baseballs through a large two-slit apparatus, would you produce a diffraction pattern? Why or why not?

▶ Problems

1. A ball (mass 0.1 kg) is thrown with a speed between 20.0 and 20.1 m/s. How accurately can we determine its position?

2. In Science By the Numbers, we converted the unit, J-s/kg-m, to the unit of velocity (m/s) without comment. Demonstrate the equivalence of these two units.

3. An atom of iron (mass 10^{-25} kg) travels at a speed between 20.0 and 20.1 m/s. How accurately can we determine its position? Is the accuracy in either of situations attainable? How does it compare to the size of anatom? Of a nucleus?

4. In the "Science By the Numbers" example you could have been solving for the uncertainty in the position, for which the unit, J-s/(kg-m/s), would give you a unit of meters. Show this unit conversion.

5. In a fictitious model of the Bohr atom, the wavelength of the first allowable state is 6.28 ($2 \times \pi$) cm.

    a. If the first allowable orbit includes only one complete wavelength, what is the radius of that orbit? (*Hint:* The circumference of a circle equals $2 \times \pi \times$ radius.)

    b. If the second allowable orbit contains two complete wave lengths, what is the radius of the second allowable orbit?

    c. Draw a qualitative wave function or probability function for each orbit as a function of radial distance from the nucleus.

6. In our fictitious model in problem 5, can an orbit be allowed at a distance of 3.5 cm?

7. A second fictitious model of the Bohr atom is constructed with the one allowable orbit located at 4 cm from the nucleus. If this orbit contained five complete wavelengths, what is this wavelength?

▶ Investigations

1. Look up the doctrine of predestination in an encyclopedia. Does it have a logical connection to the notion of the Divine Calculator? Which came first historically?

2. Werner Heisenberg was a central, and ultimately controversial, figure in German science of the 1930s and 1940s. Read a biography of Heisenberg. Discuss how his early work in quantum mechanics influenced his prominent scientific role in Nazi Germany.

3. What changes in artistic movements were taking place during the period 1900 (just before the discoveries of quantum mechanics) and the mid-twentieth century? Are there any connections between the artistic and scientific movements?

4. Some people interpret the Heisenberg uncertainty principle to mean that you can never really know anything for certain. Would you agree or disagree?

▶ **Additional Reading**

Cassidy, David. *The Life and Times of Werner Heisenberg.* New York: W. H. Freeman, 1992.

Casti, John L. *Paradigms Lost.* New York: William Morrow, 1989.

Mermin, David. "Is the Moon Really There When Nobody Looks?" *Physics Today*, April 1985, pp 38–47.

Penrose, Roger. *The Emperor's New Mind.* New York: Oxford University Press, 1992.

Trefil, James. "Quantum Physics Works." *Smithsonian Magazine*, August 1987.

11 ATOMS IN COMBINATION: THE CHEMICAL BOND

ATOMS BIND TOGETHER BY THE REARRANGEMENT
OF ELECTRONS.

Throwing Things Away

Think about all the things you've thrown away during the past month. Every day you toss out aluminum cans, plastic wrappers, glass jars, food scraps, and many types of paper. From time to time you also discard used batteries, disposable razors, dirty motor oil, worn-out shoes and clothes, and maybe even old tires or broken furniture. What happens to all that stuff after it becomes trash?

More and more, communities try to recycle much of their waste. Plastic, glass, aluminum, and newspaper, for example, can be collected and turned into new products and packaging. Used motor oil can be re-processed and tires retreaded. But most of the things you throw out must be put into landfills, where, we hope, they will eventually break down into soil.

The situation that faces our society is more than a little ironic. Everything with which you have contact each day, everything you throw away, is made from collections of atoms bonded together. We want these products and their packaging to last and keep looking new while they're in the store and for as long as we use them. But as soon as we throw them out, we would like our disposable materials to disappear. One way to achieve this end is to engineer *biodegradable* paper, plastics, fabrics, and other materials that are designed to break apart when thrown away.

But what holds materials together in the first place? Why do certain atoms, when brought near each other, develop an affinity and stick together? How do the clusters of atoms that play such an important role in our lives retain their identity? And how can we design new materials that will fall apart when their useful lives are over? The answers lie in the nature of the chemical bond.

285

ELECTRON SHELLS AND CHEMICAL BONDING

Think about how two atoms might interact. You know that the atom is mostly empty space, with a tiny, dense nucleus surrounded by swift electrons. If two atoms approach each other, what happens?

The State of Lowest Energy

When two atoms meet, their outer electrons encounter each other first. Whatever holds two atoms together thus has to do primarily with those outer electrons, which are called *valence electrons*. Chemical bonding often involves exchanging or sharing valence electrons; the number of an atom's electrons that are exchanged is called its *valence*. According to chemists, the valence represents the combining power of a given atom.

The most stable arrangement of electrons—that is, the electron configuration of lowest energy—is a completely filled outer shell. It is a fact of chemical life that different electron shells hold different numbers of electrons, which gives rise to the structure of the periodic table of the elements (see Chapter 9). A glance at the periodic table tells us that atoms with a total of 2, 10, 18, or 36 electrons (the atoms that appear in the extreme right-hand column) have completely filled shells. Atoms with this many electrons in their outermost orbits are *inert gases*, which do not combine readily with other materials. Helium, neon, argon, and the other noble gases, with atomic numbers 2, 10, and 18, have completely filled electron shells; these are the only common elements ordinarily found as isolated atoms. The atoms of all other elements are found in combinations of two or more atoms.

Every object in nature, including every atom, tries to reach a state of lowest energy. Atoms that do not have filled electron shells are more likely to react with other atoms to produce a state of lower energy. You are familiar with this kind of process in many natural systems. If you put a ball on top of a hill, for example, it will tend to roll down to the bottom, creating a system of lower gravitational potential energy (see Chapter 5). Similarly, a compass needle tends to align itself spontaneously with the Earth's magnetic field, thereby lowering its magnetic potential energy (see Chapter 7). In exactly the same way, when two or more atoms come together, the electrons tend to rearrange themselves to minimize the chemical potential energy of the entire system. This situation may require that they exchange or share electrons, and, as often as not, that process involves creating an outermost shell with 2 or 8 electrons in it.

chemical bond The name given to the attraction resulting from the redistribution of electrons that leads to a more stable configuration between two or more atoms—particularly by filling the outer electron shells—that holds the two atoms together. The principal kinds of chemical bonds are ionic, covalent, and metallic.

The **chemical bond** is the name we give to the attraction that results from the redistribution of electrons that leads to a more stable configuration between two or more atoms. Most atoms adopt one of three strategies to achieve a filled shell and bind themselves together: they give away electrons, accept electrons, or share electrons.

If the bond formation takes place spontaneously, without outside intervention, energy will be released in the reaction. The burning of wood or paper (once their temperature has been raised high enough) is a good example of this sort of process, and the heat you feel when you put your hands toward a fire derives ultimately from the chemical potential energy that is given off as electrons are reshuffled. Alternatively, atoms may be

When compounds form spontaneously, energy is released, as in this explosion.

changed into new configurations by adding energy to systems; for example, by heating them. Much of industrial chemistry, from the smelting of iron to the synthesis of plastics, operates on this principle, as do such everyday processes as baking a cake or frying an egg.

Compounds and Mixtures

Sprinkle some grains of table salt into your hand and look at them closely. Then do the same thing for a small pile of sand or soil. Do you notice a difference? The salt grains all look the same, while sand and soil reveal a variety of materials mixed together.

Table salt is an example of a *pure substance*, in which all the atoms or molecules are the same and occur in a fixed proportion. (Recall from Chapter 9 that a *molecule* of a material is the smallest collection of atoms that retains the chemical identity of that material.) A pure substance may form from a single element, such as gold or oxygen, or it may be a *compound*, which forms from two or more different elements that are chemically bonded. A compound may be represented by a *chemical formula*, in which the relative amounts of different elements are indicated by symbols for the elements followed by subscripts. Water, for example, forms molecules with one oxygen atom and two hydrogen atoms, and thus has the chemical formula H_2O.

Sand and soil, on the other hand, are both examples of a *mixture*, in which two or more different substances occur together but are not chemically bonded and may occur in different proportions. Other familiar mixtures are seawater (a mixture of water and salt), the atmosphere (a mixture of oxygen, nitrogen, and a few trace gases), and most rocks.

Ordinary table salt is a pure substance, made up almost entirely of crystals of sodium chloride. A handful of dirt, on the other hand, is a mixture.

TYPES OF CHEMICAL BONDS

Atoms link together by three principal kinds of chemical bonds, known as ionic, metallic, and covalent, all of which involve redistributing electrons between atoms. In addition, three types of attractions, called polarization, van der Waals interactions, and hydrogen bonding, result from shifts of electrons within their atoms or groups of atoms. Each type of bonding or attraction corresponds to a different way of rearranging electrons, and each produces distinctive properties in the materials it forms.

Ionic Bonds

ionic bond A chemical bond in which the electrostatic force between two oppositely charged ions holds the atoms in place; often formed as one atom gives up an electron while another receives it, lowering chemical potential energy when atom shells are filled.

We have seen that atoms with filled outer electron shells are particularly stable. By the same token, atoms with only one electron or one missing electron in the outer shell are particularly reactive; in effect, they are "anxious" to fill their outer shells. Such atoms form **ionic bonds:** chemical bonds in which the electrical force between two oppositely charged ions holds the atoms together.

Ionic bonds often form as one atom gives up an electron, while another receives it. Sodium (a soft, silvery white metal), for example, has 11 electrons in an electrically neutral atom—2 in the lowest shell, 8 in the next, and 1 electron in its outer shell. Sodium's best bonding strategy, therefore, is to lose one electron. The 17th element, chlorine (a yellow-green toxic gas), on the other hand, is one electron shy of a filled shell. Highly corrosive chlorine gas will react with almost anything that can give it an extra electron. When you place sodium in contact with chlorine gas, the result is predictable: in a fiery reaction, each sodium atom donates its extra electron to a chlorine atom (see Figure 11–1).

Figure 11–1
Sodium, a highly reactive element, readily transfers its single valence electron to chlorine, which is one electron shy of a filled shell. The result is the ionic compound sodium chloride—ordinary table salt. In these diagrams, electrons are represented as dots in shells. The inner shell can contain no more than two electrons, while the second and third shells are limited to eight electrons.

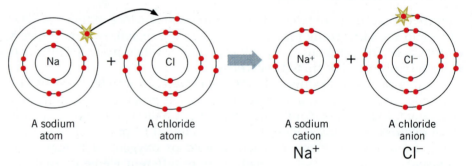

A sodium atom	A chloride atom	A sodium cation	A chloride anion
		Na⁺	Cl⁻

In the process of this vigorous electron exchange, atoms of sodium and chlorine become electrically charged—they become ions. Neutral sodium has 11 positive charges in its nucleus, balanced by 11 negative electrons in orbit. By losing an electron, sodium becomes an ion with one unit of positive charge, shown as Na^+ in Figure 11–1. Similarly, neutral chlorine has 17 protons and 17 electrons. The addition of an extra negative electron creates a chloride ion with one unit of negative charge, shown as Cl^- in the figure. The mutual electrical attraction of positive sodium and negative chloride ions is what forms the ionic bonds between sodium and chlorine. The resulting compound, NaCl, is called sodium chloride or common table salt, which has properties totally different from either sodium or chlorine.

Under normal circumstances, sodium and chloride ions will lock together into a *crystal*, a regular arrangement of atoms such as the one shown in Figure 11–2. Alternating sodium and chloride ions form an elegant repeating structure in which each Na^+ is surrounded by six Cl^-, and vice versa.

Ionic bonds may involve more than a single electron transfer. The 12th element in the periodic table, magnesium, for example, donates two electrons to oxygen, which has eight electrons. In the resulting compound, MgO (magnesium oxide), both atoms have stable filled shells of 10 electrons, and the ions, Mg^{2+} and O^{2-}, form a strong ionic bond. Ionic bonds involving the negative oxygen ion O^{2-} and positive ions such as calcium (Ca^{2+}), magnesium (Mg^{2+}), and iron (Fe^{2+}), are found in many common objects: in most rocks and minerals, in ceramics such as china, in glass, and in bones and eggshells.

Ionic bonds in these compounds can be very strong, but only in certain ways. You can picture how this works by thinking about Tinkertoys. Tinkertoy structures can be quite strong; once assembled, it is difficult to break one apart by just pushing in the directions of the sticks. But Tinkertoy bonds break easily when the sticks are twisted or snapped. In the same way, ionic bonds hold atoms together, but if the atoms should for some reason become displaced, the bond can't hold them very well. As a consequence, ionic-bonded materials such as rock, glass, or eggshells are usually quite brittle. These materials are strong in the sense that you can pile a lot of weight on them; however, once they shatter—or, in the case of ionic bonds, once they have been broken—they can't be put back together again.

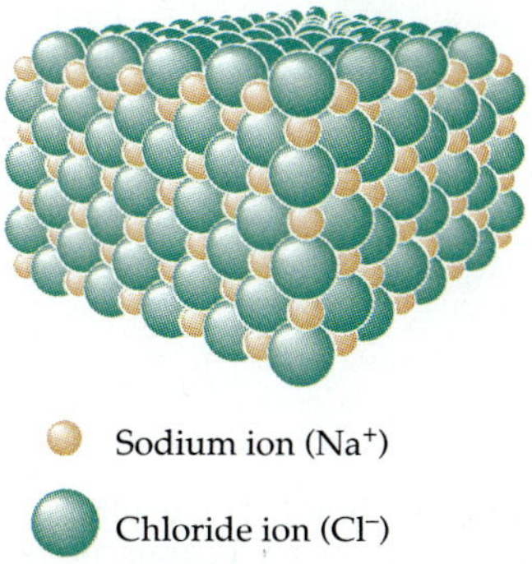

⬤ Sodium ion (Na^+)

⬤ Chloride ion (Cl^-)

Figure 11–2
The atomic structure of a sodium chloride crystal consists of a regular pattern of alternating sodium and chloride ions.

Example 11–1: Ionic Bonding of Three Atoms

Calcium chloride, which plays an important role as an industrial desiccant (a substance that speeds drying), is an ionic-bonded compound with one atom of calcium to two atoms of chlorine ($CaCl_2$). How are the electrons arranged in this compound?

▶ **Reasoning:** Calcium and chlorine are elements 20 and 17, respectively, in the periodic table. Calcium, therefore, has 18 electrons (2 + 8 + 8) in inner shells and 2 valence electrons. Chlorine has 10 electrons (2 + 8) in its inner shells and 7 electrons in the outer one, meaning it is 1 electron short of a filled outer shell (see Figure 11–3).

Figure 11–3
Calcium and chlorine neutral-atom electron configurations.

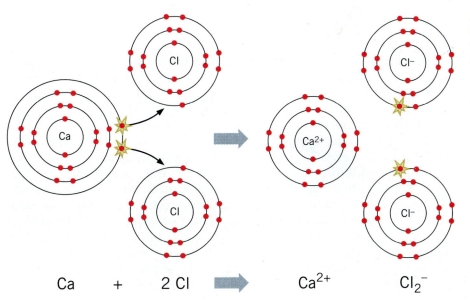

$$Ca \quad + \quad 2\ Cl \quad \Rightarrow \quad Ca^{2+} \quad Cl_2^-$$

▶ **Solution:**   Calcium has two electrons to give, and chlorine seeks one electron, to achieve stable filled outer shells. Therefore, calcium gives one electron to each of two chlorine atoms, and the resulting Ca^{2+} ion attracts two Cl^- ions to form calcium chloride, $CaCl_2$. ▲

Metallic Bonds

Atoms in an ionic bond transfer electrons from one atom to another. Atoms in a metal also give up electrons, but they use a very different bonding strategy. In the **metallic bond,** electrons are redistributed so that they are *shared* by all the atoms as a whole.

Sodium metal, for example, is made up entirely of individual sodium atoms. All of these atoms begin with 11 electrons, but they release 1 to achieve the more stable 10–electron configuration. The extra electrons move away from their parent atoms to move around the metal, forming a kind of sea of negative charge. In the negative electron sea, the positive sodium ions adopt a regular crystal structure, as shown in Figure 11–4.

You can think of a metallic bond as one in which each atom shares its outer electron with all the other atoms in the system. Picture the free electrons as a kind of loose glue in which the metal atoms are placed. In fact, the idea of a metal as being a collection of marbles (the ions) in a sea of stiff, gluelike liquid provides a useful analogy.

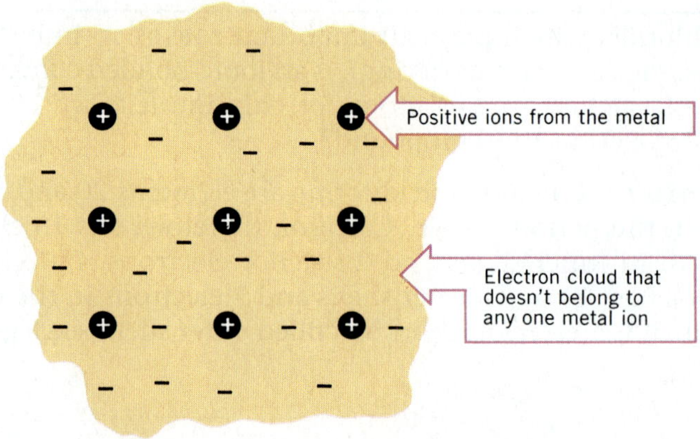

metallic bond   A chemical bond in which electrons are redistributed so that they are shared by all the atoms in the material as a whole.

Figure 11–4
Metallic bonding, in which a bond is created by the sharing of electrons between individual metal atoms.

Positive ions from the metal

Electron cloud that doesn't belong to any one metal ion

metal   Characterized by a shiny luster and ability to conduct electricity, an element or combination of elements in which the sharing of a few electrons results in a more stable electron arrangement.

Metals, characterized by their shiny luster and ability to conduct electricity, are formed by almost any element or combination of elements in which the sharing of electrons results in a more stable electron arrangement.

The great majority of chemical elements are known to occur in the metallic state. Some elements, including aluminum, iron, copper, and gold, find widespread applications. Other elements solidify into a metal at extreme conditions. For example, elements that we normally think of as gases, such as hydrogen or oxygen, become metals at very high pressure. In addition, two or more elements can mix to form a metal *alloy*, such as brass (a compound of copper and zinc) or bronze (a compound of copper and tin). Modern specialty steel alloys often contain more than half a dozen different elements in carefully controlled proportions.

The special nature of the metallic bond explains many of the distinctive properties we observe in metals. For example, if you attempt to deform a metal by pushing on the "marble-and-glue" bonding system, atoms will gradually rearrange themselves and come to some new configuration. It is very hard to shatter a metallic bond just by pushing or twisting, because the atoms are able to rearrange themselves. Thus, when you hammer on a piece of metal, you leave indentations but you do not break it, in sharp contrast to what happens when you hammer on a ceramic plate. This quality of metals, called *malleability*, follows directly from the nature of the metallic bond.

In Chapter 13, we will examine more closely the electrical properties of materials held together by the metallic bond. We will see that this particular kind of bond produces materials through which electrical current can flow.

Covalent Bonds

In ionic bonds, one atom donates electrons to another. In the metallic bond, some electrons wander freely throughout the material. In between these two situation is the **covalent bond,** in which well-defined clusters of neighboring atoms share pairs of electrons. These strongly bonded groups of atoms may consist of anywhere from two to many millions of atoms.

covalent bond   A chemical bond in which neighboring molecules share electrons in a strongly bonded group of at least two atoms.

The simplest covalently bonded molecules contain two atoms of the same element, such as the diatomic (two-atom) gases hydrogen (H_2), nitrogen (N_2), and oxygen (O_2). In the case of hydrogen, for example, each atom has a single electron, which is a relatively unstable situation. Two hydrogen atoms can pool their electrons, however, to create a much more stable two-electron arrangement. The two hydrogen atoms must remain close to each other for this sharing to continue, so a chemical bond is formed, as shown in Figure 11–5. Similarly, two oxygen atoms, each with eight electrons, share pairs of electrons.

Hydrogen, oxygen, nitrogen, and other covalently bonded molecules have lower chemical potential energy than isolated atoms because electrons are shared. This lower chemical potential energy means these molecules are less likely to react chemically than the isolated atoms.

Figure 11–5
Two hydrogen atoms become an H_2 molecule by sharing each of their electrons in a covalent bond. This bonding may be represented schematically in a dot diagram (*a*), or by the merging of two atoms with their electron clouds (*b*).

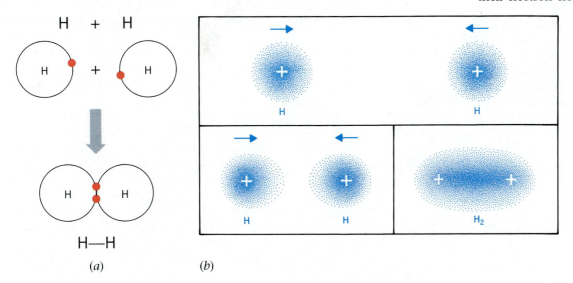

(*a*)   (*b*)

Carbon and the Covalent Bond

The most fascinating of all covalently bonded elements is carbon, the basis of all life on Earth. Carbon, with two electrons in its inner shell and four in its outer shell, presents a classic case of a half-filled shell. When carbon atoms approach each other, therefore, a real question arises as to whether they should accept or donate four electrons to achieve a more stable arrangement. You could imagine, for example, a situation where some carbon atoms gave four electrons to their neighbors, while other carbon atoms accepted four electrons, to create a compound with extremely strong ionic bonds between C^{4+} and C^{4-}. Alternatively, carbon might become a metal in which every atom releases four electrons into an extremely dense electron sea. But neither of these possibilities happens, because they each require quite a bit of energy to pull those electrons loose.

In fact, the strategy that lowers the energy of the carbon-carbon system the most is for the carbon atoms to share their outer electrons. Once bonds between carbon atoms have formed, the atoms have to stay close to each other for the sharing to continue. Thus bonds are generated here, just as it was in the case of hydrogen and oxygen. The case of carbon is unusual, however, because the shape of its electron shells allows a single carbon atom to form covalent bonds with up to four other atoms by sharing one of its four valence electrons with each. A *single bond* (shown as C—C) is formed when each atom shares one of its electrons, while a *double bond* (shown as C=C results when each atom shares two electrons with the other atom (see Figure 11–6).

By forming bonds among several adjacent carbon atoms, you can make rings, long chains, branching structures, planes, and three-dimensional frameworks of carbon in almost any imaginable shape. There is virtually no limit to the complexity of molecules you can build from such carbon-carbon bonding (see Figure 11–7). So important is the study of carbon-based molecules that chemists have given it a special name, *organic chemistry*.

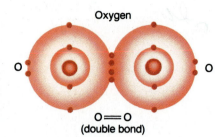

Oxygen

O=O
(double bond)

Figure 11–6
An example of a double bond resulting from the sharing of two pairs of electrons by oxygen atoms.

Figure 11–7
Carbon-based molecules may adapt almost any shape. The molecules may consist of long, straight chains of carbon atoms that lead to fibrous materials such as nylon (*a*), or they may incorporate complex rings and branching arrangements that form lumpy molecules such as cholesterol (*b*).

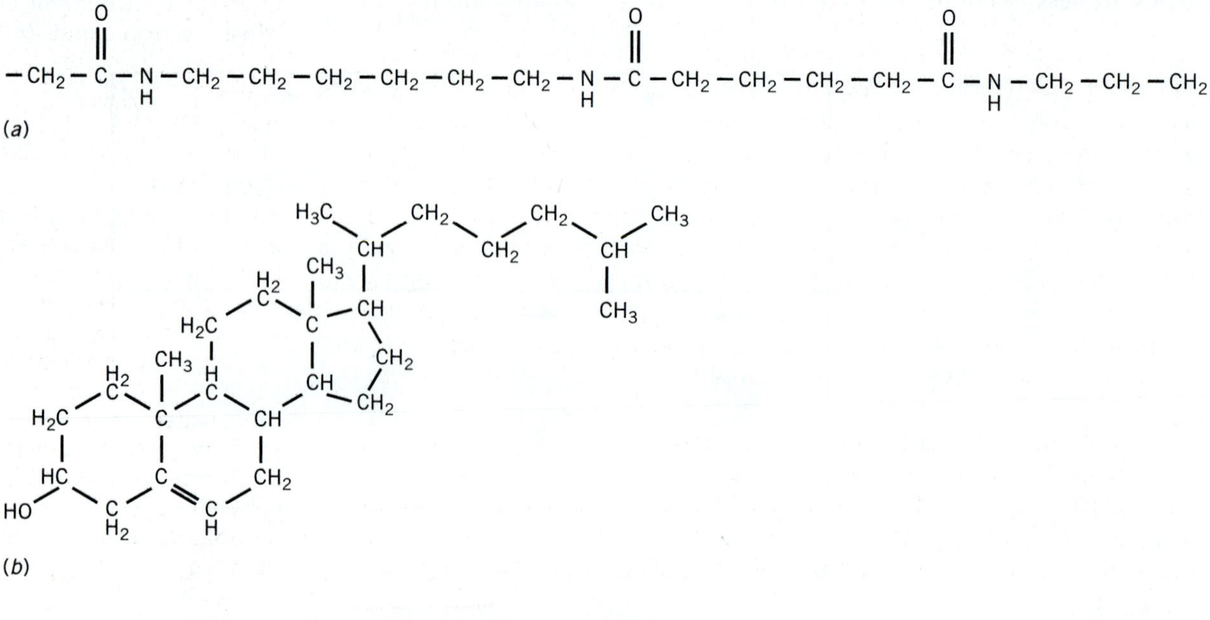

(a)

(b)

In Chapter 12, we discuss organic chemistry further, as we examine how all the molecules in your body and in every other living thing are held together, at least in part, by covalent bonds in carbon chains. Covalent bonds also drive much of the chemistry in the cells of your body and play a role in holding together the DNA molecules that carry your genetic code. This does not mean that covalent bonds appear only in living materials, however. The transistors in the microchip that runs your computer are made from covalently bonded silicon, an element that has four electrons in its outer shell just like carbon.

Like ionic and metallic bonds, covalent bonds can vary in strength. They hold together soft human tissue, but they also hold together the diamond, the hardest substance known.

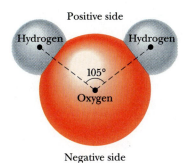

Figure 11–8
The water molecule and its polarity.

Polarization and Hydrogen Bonds

Ionic, metallic, and covalent bonds form strong links between individual atoms, but molecules also experience forces that hold them to each other. In many molecules, the electrical forces are such that, although the molecule by itself is electrically neutral, one part of the molecule has more positive or negative charge than another. In water, for example, the electrons tend to spend more time around the oxygen atom than around the hydrogen atoms. This uneven electron distribution has the effect of making the oxygen side of the water molecule more negatively charged, and the two "Mickey Mouse ears" of the hydrogen atom more positively charged (see Figure 11–8). Atom clusters of this type, with a positive and negative end, are called *polar molecules*.

The electrons of an atom brought near a polar molecule such as water will tend to be pushed away from the negative side and shifted toward the positive side. Thus the side of an atom facing the negative end of a polar molecule will become slightly positive itself. This subtle electron shift, called *polarization*, in turn, will give rise to an attractive electrical force between the negative end of the polar molecule and the positive side of the approaching atom, even though the atom and molecule in this scheme may both be electrically neutral.

A process related to the forces of polarization leads to the **hydrogen bond,** a weak interaction that may form after a hydrogen atom links to another atom (oxygen or nitrogen, for example) by a covalent or ionic bond. Because of the kind of rearrangement of electrical charge described above, the hydrogen will become polarized and develop a slight positive charge, which attracts another atom to it. You can think of the hydrogen atom as a kind of bridge in this situation, causing a redistribution of electrons that, in turn, holds the larger atoms or molecules together. Individual hydrogen bonds are weak, but in many molecules they occur repeatedly, and therefore play a major role in determining the molecule's shape and function. Note that while most hydrogen bonds require hydrogen atoms, not all hydrogen atoms are involved in hydrogen bonds (see Figure 11–9).

Hydrogen bonds are common in virtually all biological substances, from everyday materials such as wood, plastics, silk, and candle wax, to the complex structures of every cell in your body. Hydrogen bonds in every living thing link the two sides of the DNA double helix together (see Chapter 12), although the sides themselves are held together by

hydrogen bond   A bond that may form when a polarized hydrogen atom links to another atom by a covalent or ionic bond.

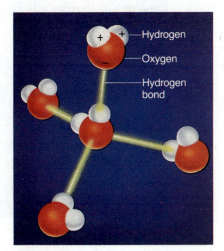

Figure 11–9
The formation of the hydrogen bond between water molecules. Slightly positively charged hydrogen atoms of one water molecule align next to slightly negative portions of another water molecule. These weak interactions therefore result from the attraction between oppositely charegd portions of adjacent molecules.

covalent bonds. Ordinary egg white is made from molecules whose shape is determined by hydrogen bonds; when you heat the material—for example, when you fry an egg—hydrogen bonds are broken and the molecules rearrange themselves, so that instead of a clear liquid you have a white, gelatinous solid.

The Properties of Water

Have you ever watched a water bug dancing across the surface of a pond, or beads of water forming on a smooth surface? Have you ever wondered how crystals of salt or sugar seem to disappear in a glass of water? These and other familiar properties of water and other liquids result from attractive forces of polarization and hydrogen bonding.

Many of water's distinctive properties result from strong attractive forces between adjacent water molecules. Molecules on the surface of water tend to be pulled back into the body of the liquid by this force. This means that there is a net force at the surface, called *surface tension*, that keeps molecules from leaving the surface, and causes the surface to pull together. You can see this effect when water beads up on the freshly waxed surface of a car.

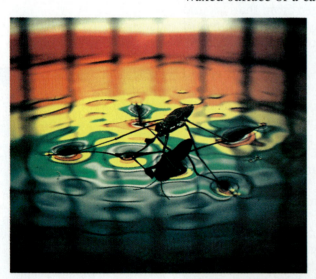

The water bug is supported by the water's surface tension.

Polar water molecules attract each other. Water thus tends to form spherical beads and droplets, which have the minimum surface area per volume.

Underground, the ability of water to dissolve limestone produces enormous beautiful underground caverns, such as Lehman Cave in New York.

Another way to think about surface tension is in terms of energy. If you want to take a molecule in the surface and move it outward, you have to exert a force to overcome the attraction of the rest of the molecules, and you have to exert that force over a distance. To change the shape of the surface, in other words, you have to do work. The state of lowest energy is the surface that requires the least work to form, and this is why liquids tend to pull together to make their surface area as small as possible.

Water has a relatively high surface tension. This high tension allows small bugs to walk on water without breaking through. Their weight is simply not big enough to do the work necessary to break through the surface.

Water can *dissolve* many materials. This is a consequence of forces due to polarization. Water is made up of strongly polar molecules, which exert forces that tend to pull ions such as Na^+ and Cl^- (the elements of table salt), or polar molecules such as sugar out of crystals. The resulting *solution* is a random mixture of the original water molecules and the dissolved ions or molecules. The Earth's oceans are complex solutions with dozens of different *solutes*, or dissolved elements and compounds, that have been added gradually over hundreds of millions of years (see Chapter 19).

There are limits, however, to the amount of a substance that water can dissolve. A solution that holds a maximum of solute for a given temperature is said to be *saturated*. If a solution becomes too concentrated, the opposite effect to dissolving occurs. This process, called *precipitation*, occurs when a solid (or *solute*) forms and falls to the bottom of the liquid solution. The next time you are at the ocean, scoop up a little salt water and, as the water evaporates, watch as tiny salt crystals precipitate on your palm.

Water's ability to dissolve and transport molecules is critical to all living things. In fact, most of your body's weight is this solvent, water.

Environment

Is Your Drinking Water "Pure"?

The next time you pour yourself a glass of water, think about what you are drinking. Is it pure H_2O? The answer is no. Water, one of the best naturally occurring solvents, is never pure, no matter what bottled-water advertisers might lead you to believe.

All water comes from sources that have been in contact with rocks and soils, which dissolve to add calcium, magnesium, iron, and other ions to the water. Water also invariably contains dissolved gases, including nitrogen, oxygen, and carbon dioxide, as well as trace amounts of many other substances. Drinking water usually contains between 100 and 200 parts per million dissolved substances.

In some regions, dissolved chemicals can cause severe problems for homeowners. Well water in regions underlain by limestone may be so concentrated in calcium and magnesium compounds that new limestone-like deposits form in household plumbing systems, just like limestone formations in caves. In these regions of "hard" water, water must be "softened" by chemically extracting these inconvenient minerals. An even more serious problem faces owners of aging homes and apartment buildings with old lead pipes. Tap water that circulates through these plumbing systems can dissolve dangerous amounts of lead, which is a poisonous heavy metal that gradually builds up in the human body and constitutes a major health problem in American cities. There is even concern about pipes joined together with lead-based solder. A non-lead solder is now being used.

Another concern is the increasing contamination of water supplies by human activities. Fertilizers and other agricultural chemicals, sewage and detergents from homes, drainage from landfills and chemical dumps, and industrial byproducts enter our water supply in every part of North America. Federal, state, and local governments are spending billions of dollars annually in an effort to control these sources of water pollution. ●

Developing Your Intuition

Rain on Distant Worlds

Rain on Earth is composed of the polar molecule water, which evaporates off the surface, condenses in clouds, and falls back to the surface. But imagine a cold, distant planet with an atmosphere composed in part of nonpolar methane (CH_4), which is liquid between about -161 and $-182°C$. What would methane rain look like on that planet?

The key difference between the behavior of rain on the two planets lies in the polarity of water molecules versus methane molecules. In humid air, polar water molecules exert forces on each other so that they stick together until they become heavy enough to fall as rain drops. In a methane-rich atmosphere, on the other hand, nonpolar methane molecules do not have the same tendency to form large droplets, so rain as we know it on Earth couldn't form. It is possible that on days when the planet's atmosphere was saturated with methane vapor, a thick methane fog would settle over the alien landscape.

Methane lakes, rivers, and oceans on the planet's surface might appear from a distance to be similar to bodies of water on the Earth's surface, but a closer inspection would reveal important differences. Methane is a poor solvent, so the distant oceans would not be "salty." The low surface tension of liquid methane, furthermore, would mean that any alien bugs would sink to the bottom, unlike bugs that walk on water on Earth. ●

van der Waals Forces

The hydrogen bond exists because atoms or molecules can become polarized; their electrons can shift to one side or another, thereby creating localized electric fields. In the molecules we've discussed so far, that electrical field arises from charges that are more or less permanently locked into polar molecules in a fixed or static arrangement. Another attractive force, called the **van der Waals force,** results from the polarization of electrically neutral atoms or molecules, but in this case, the atoms and molecules are not themselves polar.

van der Waals force A net attractive force resulting from the polarization of electrically neutral atoms or molecules.

When atoms or molecules are brought near each other, there is an electrical force exerted on every part of one atom or molecule by all the parts of the other atoms or molecules. For example, an electron in one atom will be repelled by the electrons of but attracted to the nucleus of an adjacent atom. The net result of these forces exerted on the electron may be a temporary shift of the electron to another part of the atom. The same thing happens to every electron in any nearby atom or molecule, and the net result is that every electron is constantly shifting because of the presence of others.

Once the charges in neighboring atoms have shifted, there will be a complex set of forces acting. A given atom will feel an attractive force because of the attraction between its (relocated) electrons and its neighbor's nucleus, but will feel repulsive forces from the neighbor's electrons. In some cases, the attractive forces prevail, and a weak attractive force develops between the two atoms, even though both atoms are electrically neutral and no charges are exchanged between them. This weak force that binds two atoms or molecules together is called van der Waals attraction (named after Dutch physicist Johannes van der Waals, 1837–1923).

If you take a piece of clay and rub it between your fingers, you will notice that your fingers pick up a slick coating of the material, even though the clay breaks apart easily. The reason for this behavior is that the clay is made up of sheets of atoms. Within each sheet of clay, atoms are held together by strong ionic bonds. One sheet is held to another, however, by comparatively weak van der Waals forces. Compare this situation to the way a stack of photocopying paper will stick together on a dry day. Each sheet of paper is relatively strong, but the stack of paper is stuck together by much weaker electrostatic forces. It is easy to pull the stack apart, but very difficult to rip the stack in two. When you break apart clay in your fingers, therefore, you are breaking weak van der Waals forces between layers, but preserving the stronger bonds that hold each layer together. The clay stains on your hands are thin sheets of atoms, held together by ionic and covalent bonds, and the crumbling you feel is the breaking of the van der Waals forces.

Many other examples of van der Waals forces are common, everyday occurrences (see Figure 11-10). When you rub talcum powder on your

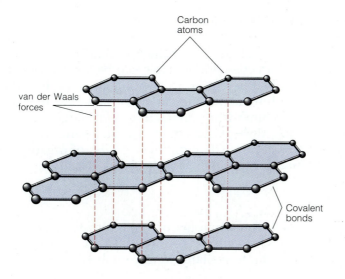

Carbon atoms

van der Waals forces

Covalent bonds

Figure 11–10
Graphite, a form of carbon that serves as the "lead" in your pencil, contains layers of carbon atoms strongly linked to each other by covalent bonds (represented by solid lines). These layers are held to each other by much weaker van der Waals forces (represented by dashed lines).

body, for example, you use a pure white material not unlike the clay discussed above; that is, material in which strongly bonded sheets are held together by van der Waals forces. Similarly, when you write with a "lead" pencil, van der Waals bonded layers of graphite, a form of carbon, are transferred from pencil to paper. As you draw the pencil across the paper, you break the van der Waals forces and leave behind sheets of graphite in a dark streak across the paper. Van der Waals forces also hold together molecules in many liquids and soft solids, from candle wax and soap to Vaseline and other petroleum products.

Technology

Superslick Coatings

Imagine a material so slippery and so unattractive to other substances that nothing would stick to it. It could take the form of a surface coating that couldn't be soiled, stained, or covered with graffiti, or airplane wings to which ice could not adhere. Recent discoveries by scientists at Dow Chemical Company may soon lead to commercial production of such a material.

Two objects stick together because of atomic interactions. The best glues, for example, form strong covalent bonds with a wide variety of materials. What electron configuration, therefore, must any nonstick surface have? What materials in nature refuse to bond to any other?

Inert gas atoms almost never form a chemical bond or experience attractive forces; thus they are, in effect, "nonstick" atoms. A nonadhesive coating must look to any other atom or molecule exactly like a collection of inert gas atoms, with completely filled outer electron shells. The new Dow material, which is water-based and can be brushed on just like paint, forms a hard, clear surface layer with carbon and fluorine atoms that are bonded in such a way that they have completely filled outer shells. The resulting coating is impervious to water and common solvents, while Magic Marker ink, paints, and dyes simply bead up and roll off the surface. ●

STATES OF MATTER

So far, we have been talking about the way that the limited number of chemical elements in the periodic table can be locked together to form materials with many different properties. But chemical bonds alone cannot explain everything about the variety of materials in our world. All of our experience comes from interactions with large groups of atoms. Depending on how these groups of atoms are organized, they may take on many different forms. These different modes of organization, called the **states of matter,** include gases, plasmas, liquids, and solids.

states of matter   Different modes of organization of atoms or molecules, which result in properties of gases, plasmas, liquids, or solids.

gas   Any collection of atoms or molecules that expands to take the shape of and fill the volume available in its container.

Gases

A **gas** is any collection of atoms or molecules that have little or no attractive force between them. As a result, a gas expands to take the shape and fill the volume available in its container. Most common gases, including those that form our atmosphere, are invisible, but the force of a gust of wind is proof that matter is involved. The individual particles that comprise a gas may be isolated atoms such as helium or neon, or small molecules such as nitrogen (N_2), oxygen (O_2), or carbon dioxide (CO_2).

If we could magnify an ordinary gas a billion times, we would see these particles randomly flying about, bouncing off each other and anything else with which they collided. The gas pressure that inflates a basketball or tire is a consequence of these countless collisions. If we pump in more air, or heat the tire, the number of collisions, and thus the pressure, will go up.

Plasma

At the extreme temperatures characteristic of the Sun and other stars, high-energy collisions between atoms may strip electrons off, creating a new state of matter called a **plasma,** in which positive nuclei move about in a sea of electrons. Such a collection of electrically charged objects is something like a gas, but it displays unusual properties not seen in other states of matter. Plasmas, for example, are efficient conductors of electricity. Furthermore, though they are gaslike, plasmas can be confined in a strong magnetic field or "magnetic bottle."

Plasmas are to us the least familiar state of matter, yet more than 99.9% of all the visible mass in the universe exists in this form. Not only are most stars composed of a dense hydrogen- and helium-rich plasma mixture, but several planets, including the Earth, have regions of thin plasma (the *ionosphere*) in their outer atmospheres.

Some gradations exist between gas and plasma. Partially ionized gases in neon lights or fluorescent light bulbs, for example, have a small fraction of their electrons in a free state. While not a complete plasma, these ionized gases do conduct electricity.

plasma A state of matter existing under extreme temperatures in which electrons are stripped from their atoms during high-energy collisions, forming an electron sea surrounding positive nuclei.

Liquids

Any collection of atoms or molecules that has no fixed shape but maintains a fixed volume is called a **liquid.** Other than water and biological fluids, there are few naturally occurring liquids on Earth. Water, by far the most abundant liquid on the Earth's surface, is a major cause of geological change (see Chapter 19), while water-based solutions are essential to all life.

liquid Any collection of atoms or molecules that has no fixed shape but maintains a fixed volume.

Most of the visible matter in the universe, including the Sun, is in the form of plasma.

Figure 11–11
Individual salt crystals reveal a characteristic cube shape.

solid All materials that possess a fixed shape and volume, with chemical bonds that are both sufficiently strong and directional to preserve a large-scale external form.

crystal A group of atoms that occur in a regularly repeating sequence. Crystal structure is described by first determinining the size and shape of the repeating boxlike group of atoms, and then recording the exact type and position of every atom that appears in the box.

glass A solid with predictable local environments for most atoms, but no long-range order to the atomic structure. Compared to a crystal, glass lacks the repeating unit of atoms.

plastic A solid composed of intertwined chains of molecules called polymers, with an ability to be molded or formed into virtually any desired shape: clear film, dense casting, strong fiber, colorful molding.

At the molecular level, liquids behave something like a container full of sand grains. The grains fill whatever volume they are poured into, freely flowing over each other without ever taking on a fixed shape. Unlike sand grains, however, liquid atoms or molecules experience attractive forces that hold the liquid together. At the surface of the liquid, these attractive forces link adjacent atoms or molecules, which might otherwise escape into the atmosphere. All liquids, however, release some atoms or molecules into a gas phase. This phenomenon is reflected in daily weather reports of *relative humidity*, a number that indicates the percentage of water molecules in the atmosphere, compared to the maximum amount of water that air of that temperature could hold. (At higher temperatures, the air can hold more water vapor.)

Solids

Solids include all materials that possess a more-or-less fixed shape and volume. In all solids, the chemical bonds are both sufficiently strong and directional to preserve a large-scale external form. At the atomic scale, however, solids adopt several quite different kinds of structures.

In **crystals,** groups of atoms occur in a regularly repeating sequence, with the same atom or atoms appearing over and over again in a predictable way. A crystal structure can be described by first determining the size and shape of the tiny boxlike unit that repeats, and then recording the exact type and position of every atom that appears in the box. In common salt (Figure 11–11), for example, the box is a tiny cube less than a billionth of a meter on a side, and each box contains sodium atoms at the cube center and corners, and chlorine atoms at the center of every side. The regular atomic structure of crystals often leads to large single crystals with beautiful flat faces.

Common single crystals include grains of sand and salt, computer chips, and the gemstones found in jewelry. Most crystalline solids, however, are composed of numerous interlocking crystal grains. Two important groups of these *polycrystalline* materials are metals and ceramics. Metals and alloys are characterized by metallic bonds, and are employed primarily for their strength and durability, distinctive luster, and ability to conduct electricity. *Ceramics* include a wide variety of hard, durable solids, including bricks, concrete, pottery, porcelain, and numerous synthetic abrasives, as well as teeth and bones and most rocks and minerals.

Amorphous materials, in contrast to crystals, are solids with no regular atomic structures. Two common types of amorphous materials are glass and plastics. **Glass** has predictable local environments for most atoms, but no long-range order to the atomic structure (see Figure 11–12). In most common window and bottle glasses, for example, the most abundant elements, silicon and oxygen, form a strong three-dimensional framework. Every silicon atom is surrounded by a fixed number of oxygen atoms, and most oxygens are linked to two silicons. If you were placed on any atom in a piece of glass, chances are you could predict the kind of adjacent atoms. Nevertheless, glass has no regularly stacked boxes of structure. Travel more than two or three atom diameters from any starting point, and there is no way that you could predict whether you'd find a silicon or an oxygen atom.

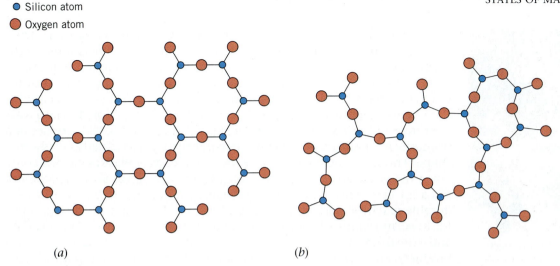

● Silicon atom
● Oxygen atom

(a)   (b)

Figure 11–12
The arrangement of atoms in a crystal (*a*) is regular and predictable over distances of thousands of atoms. The arrangement of atoms in a glass (*b*) is regular only on a local scale, and irregular over a distance of three or four atoms.

Plastics form another important class of solid materials. Along with numerous biological molecules such as animal hair, plant cellulose, cotton, and spider's webs, plastics are examples of **polymers,** which are solids composed of extremely long and large molecules that form from

polymer A material composed of extremely long and large molecules that are formed from numerous smaller molecules, like links in a chain, with predictable repeating sequences of atoms along the chain.

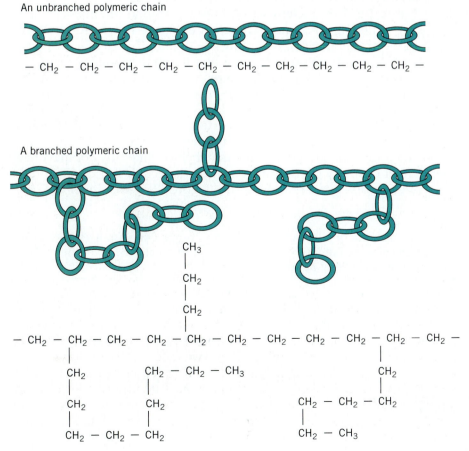

An unbranched polymeric chain

$$- CH_2 - CH_2 - CH_2 - CH_2 - CH_2 - CH_2 - CH_2 - CH_2 - CH_2 - CH_2 -$$

A branched polymeric chain

Figure 11–13
Branching and unbranching polymers form from small molecules, just as long chains can form from individual links.

Most plastics can be recycled, thus saving petroleum (the raw material for all plastics) and reducing the volume of solid waste.

numerous smaller molecules, like links forming a chain. The atomic structure of these materials, therefore, is essentially one-dimensional, with predictable repeating sequences of atoms along the polymer chain (see Figure 11–13). Plastics consist of complex intertwined polymer strands, much like the strands of fiberglass insulation. When heated, these strands slide across each other to adopt new shapes. When cooled, the plastic fiber mass solidifies into whatever shape is available. Plastics are defined primarily by their ability to be molded or formed into virtually any desired shape. Though almost unknown a few decades ago, plastics have become our most versatile commercial material, providing for an extraordinary range of uses: clear films for lightweight packaging, dense castings for durable machine parts, thin strong fibers for clothing, colorful moldings for toys, and many other uses. While most plastics are used in packaging and construction, they also serve as paints, inks, glues, sealants, foam products, and insulation. And new kinds of tough, resilient plastics have revolutionized many sports, with products such as high-quality bowling and golf balls, durable football and ice hockey helmets, not to mention a host of completely new products from Frisbees to in-line skates.

Technology

Liquid Crystals and Your Hand Calculator

Almost every material on Earth is easily classified as a solid, a liquid, or a gas, but in the 1970s scientists synthesized an odd intermediate state of matter, called *liquid crystals*. These materials are now used in many kinds of electronic devices, including the digital display of your pocket calculator. The distinction between a liquid and a crystal is one of atomic-scale order: atoms are disordered in a liquid, but ordered in a crystal. But what happens in the case of a liquid formed from very long molecules? Like a box of uncooked spaghetti, in which the individual pieces are mobile but well oriented, these molecules may adopt a very ordered arrangement even in the liquid form.

Liquid crystal displays are used in most pocket calculators.

If the molecules are polar, they may behave like tiny compass needles. Under ordinary circumstances, these molecules will occur in random orientations like a normal liquid (Figure 11–14a). Under the in-

Figure 11–14
Schematic diagram of a liquid crystal. Under normal circumstances, the elongated molecules are randomly oriented (*a*), but in an electric field, the molecules align in an orderly pattern (*b*).

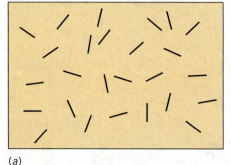

(*a*)

(*b*)

fluence of an electric field, however, the molecules may adopt a partially regular structure, in which the long molecules line up side-by-side (Figure 11–14*b*). This change in structure may even change some of the liquid's physical properties, such as its color or light-reflecting ability. This phenomenon is now widely used in liquid crystal displays in watches and computers, in which electric impulses align molecules in select regions of the screen to provide the rapidly-changing visual display.

Are liquid crystals found in nature? Every cell membrane is composed of a double layer of elongated molecules, called *lipids* (see Chapter 12). Many scientists now believe that these "lipid bilayers" originated in the primitive ocean as molecules similar to today's liquid crystals. ●

Science in the Making

The Discovery of Nylon

Nature's success in making strong, flexible fibers inspired scientists to try the same thing. American chemist Wallace Carothers (1896–1937) began thinking about polymer formation when he was a graduate student in the 1920s. At the time, no one was sure how natural fibers formed, or what kinds of chemical bonds were involved. Carothers wanted to find out.

The chemical company, Du Pont, took a gamble by naming Carothers head of their new "fundamental research" group in 1928. No pressure was placed on him to produce commercial results, but within a few years his team had developed the synthetic rubber neoprene, and by the mid-1930s they had learned to make a variety of extraordinary polymers, including nylon, the first human-made fiber. Carothers also demonstrated conclusively that polymers in nylon are covalently bonded chains of small molecules, each with six carbon atoms.

Du Pont made a fortune from the production of nylon and related synthetic fibers. Nylon had many advantages over natural fibers. It could be melted and squeezed out of spinnerets to form strands of al-

Wallace Carothers (1896–1937)

Nylon stockings, woven from synthetic polymers; created a sensation when they were first marketed in the United States.

change of state Transition between the solid, liquid, and gas states caused by changes in temperature and pressure. The processes involved are freezing and melting (for solids and liquids), boiling and condensation (for liquids and gases), and sublimation (for solids and gases).

A glass of iced tea demonstrates the coexistence of different phases of water.

most any desired size: threads, rope, surgical sutures, tennis racket strings, and paintbrush bristles, for example. These fibers could be made smooth and straight like fishing line, or rough and wrinkled like wool, to vary the texture of fabrics. Nylon fibers could also be kinked with heat, to provide permanent folds and pleats in clothing. The melted polymer could even be injected into molds to form durable parts such as tubing or zipper teeth. And nylon was cheap, formed, as Du Pont advertised, from "air, water, and coal."

Wallace Carothers did not live to see the impact of his extraordinary discoveries. Suffering from increasingly severe bouts of depression and convinced that he was a failure as a scientist, Carothers took his own life in 1937, just a year before the commercial introduction of nylon. ●

Changes of State

Place a tray of liquid water in the freezer and it will turn to solid ice. Heat a pot of water on the stove and it will boil away to a gas. These daily phenomena are examples of **changes of state**—transitions among the solid, liquid, and gaseous states. *Freezing* and *melting* involve changes between liquids and solids, while *boiling* and *condensation* are changes involving liquids and gases. In addition, some solids may transform directly to the gaseous state by a process called *sublimation*.

Temperature induces these transitions by changing the speed at which molecules vibrate. An increase in the temperature of ice to 0°C (32°F) or above, for example, causes molecular vibrations to increase to the point that individual molecules jiggle loose and the crystal structure starts to break apart. As a result, a liquid forms. At 100°C (212°F) and above, individual water molecules move fast enough to break free of the liquid surface and form a gas. As a gas, these molecules remain as H_2O and do not break apart. These transformations require a great deal of energy, because a great many interactions between molecules must be altered to change from a solid to a liquid, or from a liquid to a gas. Thus a pot of water may reach boiling temperature fairly quickly, but it takes a long time to disrupt all of the attractive interactions between water molecules and boil the water away. By the same token, a glass of ice water will remain at 0°C for a long time, even on a warm day, until enough energy has been absorbed by the ice to disrupt the attractive interactions that hold molecules in their crystal form. Only after the last bit of ice is gone can the water temperature begin to rise.

In the past few decades, scientists have discovered that changes in pressure also induce changes of state. Many common solids and liquids become gaseous at low or vacuum pressures, while all known gases and liquids become solid at modest pressures equivalent to a few tens of thousands of atmospheres. Researchers have found that at even higher pressures, equivalent to several million times the pressure of the Earth's atmospheres, many nonmetallic elements and salts become metallic.

Every change of state is accompanied by a transfer of energy, as can be seen in our daily lives. We can't make ice cubes on a summer day without plugging in a refrigerator, for example. Nor can we boil water without turning on a burner. In fact, water is a good example for illustrating the energy transfers that occur during any change of state. It requires

80 calories to melt 1 gram of ice (a quantity called the latent *heat of fusion*). While the ice is melting, it remains at a temperature of 0°C. It requires 100 calories to raise the temperature of the resulting melted ice to 100°C (1 calorie per gram), and then 580 calories to convert that water into steam (the latent *heat of vaporization*).

Developing Your Intuition

Fire and Ice

What happens if you throw an ice cube into a fire? How is energy transferred from one form to another?

First, the ice cube absorbs heat from the fire until it reaches a temperature of 0°C. It then remains at that temperature until it has absorbed 80 calories for each gram of ice. When this has happened, it will have been converted into liquid water and will continue to absorb heat from the fire. One calorie is required to raise the temperature by 1 degree, so 100 calories per gram will raise the temperature to 100°C. At this point, the water will boil, remaining at a temperature of 100°C until all of it has been converted to steam. At this point, the steam will simply escape into the atmosphere. ●

Example 11–2: A Nice Cool Drink

You have 300 g of soda in a cup at a temperature of 30°C (about room temperature). How much ice at 0°C do you have to add to cool the soda to 5°C? Assume the soda's heat capacity (the heat energy required to raise the temperature 1 degree) is the same as that of water.

▶ **Reasoning:** First we have to figure out how many calories have to be removed from the soda to lower its temperature by 25°C. Then we have to figure out how much ice this amount of heat would melt. All sodas are water-based solutions, with dissolved carbon dioxide (the fizz) and a variety of flavorings. For simplicity, however, we will assume that a soda behaves like pure water.

▶ **Solution:** By definition, 1 cal is the heat energy required to change the temperature of 1 g of water by 1°C. The heat energy, E, required to lower the temperature of 300 g of water by 25°C is thus

$$E = 300 \text{ g} \times 1 \text{ cal/g°C} \times 25°C$$
$$= 300 \times 1 \times 25 \text{ cal}$$
$$= 7500 \text{ cal}$$

Ice requires 80 cal/g to melt, so the amount of ice needed to cool 300 g of soda is:

$$\text{grams of ice} = \frac{7500 \text{ cal}}{80 \text{ cal/g}}$$
$$= 93.75 \text{ g}$$

In other words, you would have to use about a 3–to-1 ratio of soda to ice to cool your drink. ▲

THINKING MORE ABOUT
CHEMICAL BONDING

Biodegradeable Plastics

We opened this chapter with a description of some of the things we throw away into landfills. We would like our trash to decompose into small molecules that will return safely to the soil. But many common plastics seem almost indestructible, lasting for many decades without breaking down. Plastic bottles, utensils, packing materials, and other everyday items just keep piling up. How can we minimize this growing problem? Recycling provides one obvious way to maximize our use of resources, but we can't recycle every bit of plastic. In some cases, the development of *biodegradeable plastics* may prove to be an ideal alternative.

Biodegradeable materials are substances that bacteria and other organisms can eat. For a plastic to be edible, it must possess small groups of atoms that can be attacked chemically and digested. Electrons in the plastic have to be able to shift

and regroup, breaking old bonds and forming new ones. In traditional tough plastics, including Teflon and polyurethane, all surface electrons are tightly held in covalent bonds, and these materials may persist for long periods of time. (Indeed, some scientists have noted that the late twentieth century may well be identified by archeologists many thousands of years from now as the "Teflon Age," because of the indestructibility of its many plastic items.

Tough plastics can be altered to facilitate breakdown under bacterial action. Typically, the long polymer backbone is modified by adding side groups of atoms that are polar. These side groups, such as OH^- or a double-bonded oxygen, have active electrons that can easily form new bonds during the digestive process. As bacteria attack the polar side groups, the polymer chain breaks down into smaller pieces and the plastic decomposes.

▶ Summary

Atoms link together by *chemical bonds*, which form when electrons are rearranged, which lowers the potential energy of the electron system, particularly by the filling of outer electron shells. *Ionic bonds* lower chemical potential energy by transferring one or more electrons to create atoms with filled shells. The positive and negative ions created in the process bond together through electrostatic forces. In *metals*, on the other hand, isolated electrons in the outermost shell wander freely throughout the material and create *metallic bonds*. *Covalent bonds* occur when adjacent atoms or molecules share a pair of bonding electrons. *Hydrogen bonding* and *van der Waals forces* are special cases involving distortion of electron distributions to create electrical polarity—regions of slightly positive and negative charge that can bind together.

Atoms combine to form several different *states of matter*. *Gases* are composed of atoms or molecules that can expand to fill any available volume. *Plasmas* are ionized gases in which electrons have been stripped off the atoms. *Liquids* have a fixed volume but no fixed shape. *Solids* have fixed volume and shape. Solids include *crystals*, with a regular and repeating atomic structure; *glasses*, with a nonrepeating structure; and *plastics*, which are composed of intertwined chains of molecules

called *polymers*. The various states of matter can undergo *changes of state*, such as freezing, melting, or boiling, with changes in temperature or pressure.

▶ Review Questions

1. What is a valence electron?

2. What is a filled shell? How many electrons do the first few shells hold?

3. Describe the ionic bond. Give an example of a material that results from such a bond.

4. Describe the metallic bond. What properties of metal follow from the properties of the bond?

5. What is an alloy? Give an example.

6. Describe the covalent bond. Give an example of a material that results from such a bond.

7. Describe the hydrogen bond. Explain how it affects the properties of water.

8. Describe the van der Waals bond. Give an example of a material in which it is important.

9. What is a gas? Give an example.

10. What is a liquid? Give an example.

11. What is a solid? Give an example.

12. What is a plasma? Where are plasmas found in the solar system?

13. How is a glass constructed from its atoms?

14. How is a plastic constructed from its atoms?

15. Define freezing and melting, boiling and condensation, and sublimation.

16. What are the three types of attractions that result from shifts in electrons within groups of atoms?

17. What is a crystal and how is it formed?

18. Describe the characteristics of a liquid crystal.

19. What is the basis of organic chemistry?

20. What is the difference between a compound and a mixture?

21. What is the difference between a single bond and a double bond in carbon compounds?

22. What are some effects of surface tension?

23. How is solution related to precipitation? What happens in each case?

24. What is the valence of an atom?

25. The numbers 2, 10, 18, 36, 54, and 86 are ''magic numbers'' for atoms. Why?

▶ Fill in the Blanks

Complete the following paragraph using key words and phrases from the list.

changes of state	liquid
chemical bonds	metallic bond
covalent bond	plasma
crystals	plastics
gases	polymer
glass	solid
hydrogen bonds	states of matter
ionic bond	van der Waals forces

_____ form when electrons shift around so that attractive forces develop between atoms. If one atom gives up an electron while another accepts it, an _____ is formed. If electrons are shared by all atoms in a material, a _____ bond is formed, while the sharing of electrons between neighboring atoms produces a _____ bond. _____ and _____ are examples of attractions produced by internal shifting of charges within electrically neutral atoms or molecules.

Matter appears in many _____. A material that has no fixed volume or shape is called a _____. A gaslike material in which some electrons have been torn from the nuclei is called a _____. A material that has a fixed volume but not a fixed shape is a _____, while a material with fixed volume and shape is called a _____. There are many kinds of solids, including _____, made up of regularly repeating structures; _____, in which local structures are regular but there is no large-scale pattern; and _____, consisting of long-chain molecules made from linking sim-

pler molecules together. These long chains are called _____.

Freezing, boiling, condensation, and sublimation are all examples of _____, in which the atomic constituents of a material do not change, but the forces between them do.

▶ Discussion Questions

1. Identify objects around you that use the different kinds of chemical bonding. Which objects combine two or more kinds?

2. Classify the solid objects around you as ceramics, metals, glass, and plastic.

3. During icy winter conditions, we often spread salt on sidewalks, driveways, and streets. What occurs when salt comes in contact with ice? What physical reaction is involved? How do you think the melting point of salt water compares to that of pure water?

4. Identify three examples of changes of state (other than in water) that occur in your everyday experience.

5. What are some common liquids, other than water or water-based solutions, in your everyday experience? Make an educated guess whether the atoms or molecules of these liquids are polar or nonpolar.

6. Atoms, like most objects in nature, try to reach the state of lowest energy. How can you relate this principle to

    a. chemical bonds?

    b. valence electrons?

    c. ''closed'' electron shells?

7. List three strategies that atoms use to achieve their lowest energy states during chemical bonding.

8. When paper or wood burns, heat is given off. With respect to the energy of the wood or paper molecules, what is occurring?

9. When an egg is fried, are the egg molecules in a higher or lower energy state compared to their energy state before they were fried? Why?

10. With regard to the ''movement'' of electrons in an atom, what is the common theme about electrons' movement that produces the three principle chemical bonds and three attractions?

11. From the perspective of metallic bonding, explain the property of malleability and pliability of metals.

12. Electrical conductors are materials that have an ability to transfer electrical currents. From the perspective of metallic bonding, why are metals usually good conductors?

13. Diatomic gases are examples of the simplest covalently bonded substances and are generally less reactive chemically. When two oxygen atoms pool their electrons to form diatomic oxygen, what happens to the total chemical potential energy of the two oxygen atoms, and why does this help explain the lower state of reactivity of diatomic oxygen?

14. What is the difference between single and double covalent bonds? Why are these bonds so prevalent in the element carbon? In what other element(s) would you suspect these bonds to be prevalent?

15. How is the surface tension of water related to the polarization of the water molecule?

16. You can easily dissolve sugar or salt in water.
  a. How is this an example of the polarization of the water molecule?
  b. Outline or diagram what is occurring at the molecular level when salt dissolves in water.

17. What is occurring when water is "softened"?

18. You can write with a graphite "lead" pencil. How is this related to the principles behind van der Waals forces?

19. Using the "magnified" description of a gas from the text, predict how the pressure of a gas is related to the following:
  a. An increase in the temperature of the gas (i.e., increase in the average speed of the particles).
  b. An increase in the number of particles in the gas.

20. Classify a crystal, a glass, and a plastic according to the range and direction of their repeating atomic structure.

21. Describe the changes of state for a simple water molecule that goes from a solid to liquid to a gas from the perspective of the "forces" that it experiences from its neighboring molecules.

22. Describe the changes of state for a simple water molecule that goes from a solid to liquid to a gas from the perspective of its average kinetic energy.

▶ Problems

1. Demonstrate how electrons might rearrange to form ionic bonds between a silicon atom and two oxygen atoms.

2. Demonstrate how electrons might rearrange to form ionic bonds between an aluminum atom and three fluorine atoms.

3. How much energy is required to completely boil away 500 g of 25°C tap water?

4. How much energy is required to make ice cubes out of 500 grams of 25°C tap water?

5. To form ionic bonds, some atoms have the ability either to donate or lose electrons depending on their valence electrons, and thus form ions. Indicate if the following elements, designated by their atomic numbers, would most likely participate in ionic, covalent, or no bonding, and why.

Atomic Number	Element Name	Type of Bonding	Reason
1			
2			
6			
7			
8			
10			
11			
13			
14			
16			
17			
18			
19			
20			
26			
29			
47			
79			
82			
86			

6. The maximum amount of water vapor that air can contain at 25°C is 23 g/m³ of air. If the amount of water vapor measured at this temperature is 4.5 g/m³, what is the relative humidity of the air?

7. The relative humidity of the air at a temperature of 10°C is 45%. If the maximum capacity of water vapor for this temperature is 10 g/m³, how much water vapor is there in the air?

8. The relative humidity is 65% and the amount of water vapor in the air is measured at 7.3 g/m³. What is the maximum capacity of water vapor in the air?

9. The dew point is the temperature at which the air would be saturated with water vapor or 100% relative humidity. In today's weather report, the dew point is 5°C (capacity for water vapor: 5 g/m³) and the temperature is 25°C. What is today's relative humidity?

FOR THE FOLLOWING PROBLEMS, USE THE FOLLOWING LATENT HEATS:
  Freezing and melting of ice = 80 cal/g
  Evaporating and condensing of water = 540 cal/g
  Sublimation (change from a solid to a gas) of dry ice = 137 cal/g
  Heat capacity of water = 1 cal/g per °C
  Heat capacity of ice = 0.5 cal/g per °C

10. How much energy does it take to raise the temperature of 150 g of water from 10°C to 57°C?

11. How much energy does it take to melt 120 g of ice at a temperature of 0°C?

12. There are 2400 cal of energy available to raise the temperature of 40 g of ice from −80°C to −20°C. Is there enough energy to raise the temperature of all the ice?

13. Twenty grams of ice at −15°C are placed into an insulated cup that contains 1000 g of water initially at 0°C. What are the final temperature and composition (amount of ice and water) of the cup?

14. The average human body loses approximately 600 g of water every day through perspiration and breath evaporation. How much energy per day does the average human being use in doing this?

15. At the end of an energetic workout, Priscilla has lost 0.45 kg of water through perspiration and evaporation. If all of this loss is a result of her one-hour exercise activity:

　　a. How much energy did Priscilla use?

　　b. What was her rate of energy use in watts?

16. Every morning your mother fixes you a piping hot cup (0.25 liters) of tea from boiling water. By the time the tea is ready for you, it is at a temperature of 90°C, which is still too hot for you. You add two ice cubes (20 g each) to the tea. You let the tea melt all of the ice. What is the final temperature of your tea?

▶ Investigations

1. Look around your home and school and list the variety of plastic objects. What strategy might you develop to recycle plastics? Note the numbers surrounded by a triangle on many disposable plastics. What do the numbers mean, and how are they used?

2. Do a historical research project on the kinds of everyday materials in the United States 200 years ago. What materials were available for the construction of buildings, furniture, and transportation devices? What modern technologies would be difficult or impossible with just those materials?

3. Investigate the production of aluminum. What are natural sources of aluminum, and where is it mined? How is it processed, and how does chemical bonding play a role? What are the advantages of recycling aluminum cans?

▶ Additional Reading

Snyder, Carl. *The Extraordinary Chemistry of Ordinary Things.* New York: John Wiley and Sons, 1995.

Yergin, Daniel. *The Prize: The Epic Quest for Money, Oil, and Power.* New York: Simon and Schuster, 1990.

12 CHEMICAL REACTIONS

CHEMICAL REACTIONS, WHICH REARRANGE ATOMS
AND MOLECULES, ARE DRIVEN BY ENERGY.

Waking Up

The alarm clock begins to buzz, waking you from a deep sleep. You groan and yawn, get out of bed, and stumble to the bathroom. You wash your face, brush your teeth, and begin to feel a bit more alert.

In the kitchen you make some coffee, cook yourself an egg, and take a vitamin. Soon you're ready to drive to school, but on the way you stop for gasoline. Just an ordinary morning—and already you've participated in dozens of different chemical reactions.

Atoms and molecules are constantly rearranged in our surroundings. Soap and toothpaste react with impurities, which can then be washed away. Water dissolves many common chemicals, including sugar, vitamins, and the collection of molecules that makes coffee. Eggs cook. Gasoline burns. And in your body, countless thousands of chemical reactions take place every minute in every cell.

What causes these chemical changes? What principles underlie this extraordinary range of phenomena?

CHEMICAL REACTIONS AND ENERGY

chemical reaction The process by which atoms or smaller molecules come together to form large molecules, or by which larger molecules are broken down into smaller ones; involves the rearrangement of atoms in elements and compounds, as well as the rearrangement of electrons to form chemical bonds.

The process by which a group of atoms or molecules in one substance is rearranged into a new substance is called a **chemical reaction.** Chemical reactions are an integral part of our daily lives. When we take a bite of food, light a match, wash our hands, or drive our cars, we initiate chemical reactions.

All chemical reactions involve the rearrangement of atoms in elements and compounds, as well as the rearrangement of electrons to form chemical bonds. Thus, while countless millions of chemical reactions are known, many of the most familiar ones can be grouped into a few broad categories based on how atoms and electrons are distributed.

Chemical Reactions and Bonding

Before we discuss the way that atoms can combine to form molecules, let's consider why these reactions occur in the first place. The fundamental reason has to do with energy.

Consider, for example, one of the electrons in the neutral sodium atom shown in Figure 11–1. This electron has mass and it is moving around the nucleus, so it must have kinetic energy (see Chapter 5). In addition, the electron possesses potential energy because it has a negative electrical charge and is a certain distance from the positively charged nucleus. You can convince yourself of this in several ways. For example, you can notice that if the electron were at rest, rather than moving in its orbit, it would fall in toward the nucleus, just as the Earth would fall in toward the Sun in the analogous situation. Thus, just by virtue of its position, the electron is capable of doing work. This is the way potential energy was defined in Chapter 5. Alternatively, you could imagine starting with the electron close to the nucleus and ask what you would have to do to bring it to the position of the orbit. First, you would have to exert a force (to overcome the attraction of the electron to the nucleus), then you would have to exert that force over a distance. Thus, just as lifting this book up against the force of gravity gives it gravitational potential energy, lifting the electron up against the pull of the nucleus gives it electrical potential energy.

Finally, the electron has an additional component of potential energy because of the electrical repulsion between it and all of the other electrons in the atom. This is analogous to the small contribution to the Earth's potential energy from the other planets. The sum of these three energies—the kinetic energy associated with orbital motion, the potential energy associated with the nucleus, and the potential energy associated with the other electrons—is the total energy of the single electron.

The sum of the energies of all of the electrons is the total energy of the atom. For an isolated sodium atom, the total energy is the sum of the energy of the 11 electrons; for the chlorine atom, the total energy is the sum of the energies of the 17 electrons. The total energy of the sodium and chlorine together is the sum of the individual energies of the two atoms.

Once the ionic bond is formed between sodium and chlorine, the force on each individual electron is different than it was before. For one thing, the number of electrons in each atom has changed, and for another, the atoms are no longer isolated, so that electrons in the sodium can exert forces on the electrons (and the nucleus) in the chlorine and vice versa. Consequently, the paths of all the electrons will shift slightly due to the formation of the bond. This means that each electron will find itself in a slightly different position with regard to the nucleus, will be moving at a slightly different speed, and will experience a slightly different set of forces than it did before. The total energy of each electron will be different after the bond forms, the total energy of each atom will be different, and the total energy of the sodium plus chlorine atoms will be different. This process is illustrated for the simple case of two hydrogen atoms in Figure 11–5.

Exothermic and Endothermic Reactions

Whenever two or more atoms come together to form chemical bonds, the total energy of the system changes. These changes can occur in either of two ways: the final energy is less than the initial energy, or the final energy is greater than the initial energy.

The reaction that produces sodium chloride from sodium and chlorine is an example of the first kind of reaction. The total energy of all the electrons in the system is less after the atoms have come together than it was before. The difference in the energies is given off during the reaction in the form of heat, light, and sound (there is an explosion). A chemical reaction in which the final energy of the electrons is less than the initial energy, and which therefore gives off energy in some form, is said to be *exothermic*.

Many examples of exothermic reactions occur every day. The energy that moves your car is a result of the chemical combination of gasoline and oxygen in the car's engine. If your kitchen stove runs on natural gas, then the energy used to cook your food comes from a chemical reaction of natural gas and oxygen. The chemical reactions in the batteries that run your Discman also produce energy, although in this case, some of the energy is in the form of kinetic energy of electrons (electricity) in a wire.

If the final energy of the electrons in a reaction is greater than the initial energy, then you have to supply energy to make the chemical reaction occur. Such reactions are said to be *endothermic*. The chemical reactions that go on when you are cooking (frying an egg, for example, or baking a cake) are of this type. You can put the ingredients of a cake together and let them sit for as long as you like, but nothing happens until you turn on the oven and supply energy in the form of heat. When the energy is available, electrons can move around and rearrange their chemical bonds. The result is a cake, where before there was only a mixture of flour, sugar, and other materials.

Chemical reactions can be compared to a ball lying on the ground. If the ball happens to be at the top of a hill, it will lower its potential energy by rolling down the hill, giving up the excess energy in the form of frictional heat. If the ball is at the bottom of the hill, energy needs to be supplied in order to get it to the top.

In just the same way, exothermic reactions correspond to systems that "roll down the hill," going to a state of lower energy and giving off excess energy in some form. Endothermic reactions, on the other hand,

The combustion of butane, a compound of carbon and hydrogen, produces heat in an exothermic reaction.

Metallic sodium and water react violently in an exothermic reaction that sends out a shower of sparks.

Energy must be added to a set of ingredients to drive the endothermic reactions that take place during baking.

have to be "pushed up the hill," and hence absorb energy from their surroundings.

Although we have discussed the energies involved in chemical reactions in terms of two atoms coming together to form a single molecule, the arguments we've used are equally applicable to reactions where two or more molecules or atoms are produced in the final state, in which more than two atoms or molecules come together to start the reaction, or in which a large molecule breaks up into smaller ones. The important question is whether the sum of the energies of all the electrons and of the nucleus in the beginning is greater or less than the sum of the energies of all the electrons at the end.

COMMON CHEMICAL REACTIONS

Many millions of chemical reactions take place in the world around us. Some occur naturally, and some occur as the result of human design or intervention. In your everyday world, however, you are likely to see a few types of chemical reactions over and over again. Let's examine a few of these common reactions in more detail.

Oxidation and Reduction

oxidation   A chemical reaction in which an atom such as oxygen accepts electrons while combining with other elements; for example, rusting of iron metal into iron oxide, and animal respiration.

Perhaps the most distinctive chemical feature of our planet is the abundance in the atmosphere of the highly reactive gas, oxygen. This trait leads to many of the most familiar chemical reactions in our lives. **Oxidation** includes any chemical reaction in which oxygen combines with another element. Rusting, in which iron metal combines with oxygen to form a reddish iron oxide, is a common consequence of gradual oxidation. Burning, or *combustion*, is a much more rapid oxidation in which oxygen combines with carbon-rich materials to produce carbon dioxide and other often-polluting by-products. In all oxidation reactions, the combining element loses electrons to oxygen.

Hydrocarbons, which are chemical compounds of carbon and hydrogen, provide the most common fuels for combustion, with only carbon dioxide and water (hydrogen oxide) as products. In words, this reaction is

Iron exposed to air will gradually oxidize, producing iron oxide or rust.

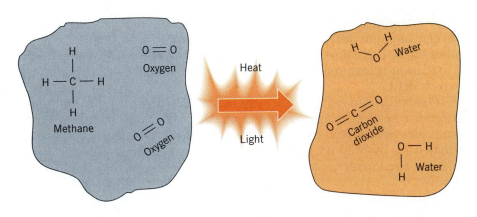

Figure 12–1
In pictorial form, an oxidation reaction involves the transfer of electrons to oxygen atoms. When natural gas (CH_4) burns, it combines with two oxygen molecules (O_2) to form a molecule of carbon dioxide (CO_2) and two molecules of water (H_2O).

$$\text{hydrocarbon} + \text{oxygen } (O_2) \longrightarrow \text{carbon dioxide} + \text{water} + \text{energy}$$

where the symbol $\longrightarrow$ should be read "reacts to form." If you heat or cook with natural gas, then you are using an oxidation reaction to generate energy in your home. The term *natural gas* refers to a compound that chemists call methane. A methane molecule consists of a single carbon atom covalently bonded to four hydrogen atoms, as shown on the left in Figure 12–1. The oxidation reaction involved in the burning of methane is written

$$CH_4 + 2O_2 \longrightarrow CO_2 + 2H_2O$$

The opposite of oxidation is **reduction,** a chemical reaction in which oxygen is removed from a compound. In the process, the reduced material gains electrons. Thousands of years before scientists discovered oxygen, primitive metalworkers had learned how to reduce metal ores by smelting. In the iron-smelting process, which was used at hundreds of North American furnaces 250 years ago, iron makers heated a mixture of ore (iron oxide) and lime (calcium oxide) in an extremely hot charcoal fire. The lime lowered the melting temperature of the entire mixture, which then reacted to produce iron metal and carbon dioxide.

reduction A chemical reaction in which electrons are transferred from an atom to other elements, resulting in a gain in electrons for the material being reduced.

Iron ore was reduced to iron metal in a blast furnace, as shown in this historic illustration from about 1875.

When a tank of concentrated nitric acid ruptured in a Denver, Colorado, switching yard in 1983, sodium carbonate was pumped by a highway snow blower to neutralize the acid.

acid Any material that when put into water produces positively charged hydrogen ions (i.e., protons) in the solution; for example, lemon juice and hydrochloric acid.

base A class of corrosive material that when put into water produces negatively charged hydroxide ions; usually tastes bitter and feels slippery.

A material that has been oxidized, therefore, has lost electrons to oxygen or some other atom. A material that has been reduced, on the other hand, has gained electrons. You may find the mnemonic OIL RIG ("oxidation is loss, reduction is gain") to be a useful way of keeping the two straight.

Oxidation and reduction are essential to life, and define the principal difference between plants and animals. All animals take in carbon-based molecules as food and oxidize this food in their cells. Carbon dioxide is released as a by-product. Plants, on the other hand, take in carbon dioxide and use the energy in sunlight to reduce it, releasing oxygen as a by-product.

Acid-Base Reactions

Acids are common substances, used by people for thousands of years. The word has even entered our everyday vocabulary; we may refer to someone with an "acid" wit when we mean something sharp and corrosive. Acids corrode (eat away) many metals and have a sour taste. For our purposes, we can make a technical definition of an acid as follows: an **acid** is any material that, when put into water, produces positively charged hydrogen ions (i.e., protons) in the solution. Lemon juice, orange juice, and vinegar are examples of common weak acids, while sulfuric acid (used in car batteries) and hydrochloric acid (used in industrial cleaning) are strong acids.

Bases are another class of corrosive materials. They taste bitter and generally feel slippery between your fingers. For our purposes, we can define a **base** as any material that, when put into water, produces negatively charged ions with one hydrogen atom and one oxygen atom, or OH^-. This ion, called the hydroxide ion, has an extra electron and thus a charge of -1. Most antacids (e.g., milk of magnesia) are weak bases, while many cleaning fluids and common drain cleaners are examples of stronger bases.

Although the common definitions of acids and bases involve taste and feel, you shouldn't try these tests yourself. Many acids and bases are *extremely* dangerous and corrosive; for example, battery acid (sulfuric acid, H_2SO_4) and lye (sodium hydroxide, NaOH).

When acids and bases are brought together in the same solution, the H^+ and the OH^- ions come together to form water, and we say that the substances "neutralize" each other. If mixed in the correct proportions, there is no acid or base left. For example, if we mix hydrochloric acid

Some common acids.

Some common bases.

and lye, we will see the following reaction

$$HCl + NaOH \longrightarrow H_2O + NaCl$$

As a water molecule forms, a positively charged hydrogen ion and a negatively charged hydroxide ion are "removed from solution" in the sense that they form a water molecule. The other ions, sodium and chlorine, remain in solution and make the water salty. In fact, all molecules formed by the neutralization of an acid and a base are called *salts*.

If you take an antacid when you have an upset stomach, you initiate a neutralization reaction in your body. Ordinary over-the-counter antacids contain bases such as aluminum hydroxide [$Al(OH)_3$] or sodium bicarbonate ($NaHCO_3$), which react with some of the stomach's natural acid. These products do not neutralize all the stomach's hydrochloric acid, only enough to alleviate the symptoms from an excess of acid.

The definition of acids and bases, in terms of whether they add positively charged protons or negatively charged hydroxide groups to water, leads to a simple way of measuring the strength of a solution. Although you might not think so at first glance, pure distilled water always contains some protons and hydroxide groups. A small number of water molecules are always being broken up, at the same time that elsewhere in the liquid protons and hydroxide groups come together to form new molecules of water. In fact, in pure water there are almost exactly 10^{-7} moles of positively charged particles per liter. Acids contain more positive charges than this, while bases contain fewer.

This fact is used to set the scale for measuring acids and bases. Pure water is defined to have a *pH* ("power of Hydrogen") of 7. An acid solution that has more positive charges (a concentration, for example, of 10^{-6} moles per liter) would have a lower pH (a pH of 6 in this example). A base that has a lower concentration of positive charges—10^{-10} moles per liter, for example—would have a higher pH (a pH of 10 in this example). Some common pH values are given in Figure 12–2.

Litmus paper turns blue when dipped in an acid and red when dipped in a base.

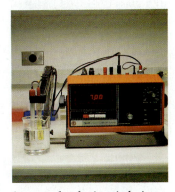

A neutral solution is being tested by a digital pH meter.

Figure 12–2
The strength of acids and bases is measured on the pH scale, with a pH of 7 indicating pure water, lower pH indicating a base, and higher pH indicating an acid.

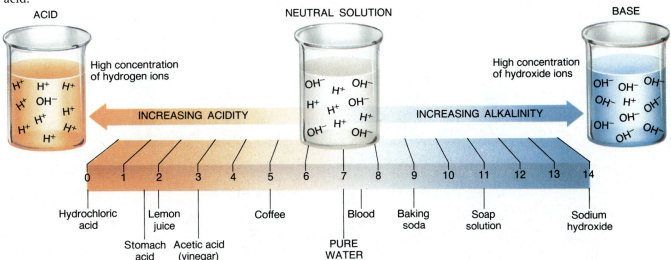
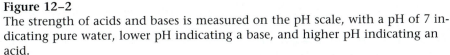

ACID

High concentration of hydrogen ions

NEUTRAL SOLUTION

BASE

High concentration of hydroxide ions

INCREASING ACIDITY

INCREASING ALKALINITY

0 1 2 3 4 5 6 7 8 9 10 11 12 13 14

Hydrochloric acid

Lemon juice

Stomach acid

Acetic acid (vinegar)

Coffee

Blood

PURE WATER

Baking soda

Soap solution

Sodium hydroxide

Fibers of the polymer nylon can be pulled from liquids in a glass dish. The same process is used to make nylon commercially.

polymerization A reaction that includes all chemical reactions that form long strands of polymer fibers by linking small molecules.

Polymerization and Depolymerization

The molecular building blocks of most common biological structures are small, consisting of at most a few dozen atoms (see the section "Modular Chemistry" below). Yet most biological molecules are huge, with up to millions of atoms in a single unit. How can these small building blocks yield the large structures characteristic of living things? The answer lies in a process known as polymerization.

A *polymer* is a large molecule that is made by linking smaller, simpler molecules together repeatedly to build up a complex structure. The word comes from the Greek *poly-* (many) and *meros* (parts). In spiders' webs, growing hair, clotting blood, and thousands of other materials, living systems have mastered the art of combining small molecules into long chains.

Polymerization is a process that includes all chemical reactions that form long strands from small molecules. Synthetic polymers usually begin in liquid form, with relatively small molecules that freely move past their neighbors. The common plastic polyethylene, for example, begins as a liquid called ethylene with molecules containing just eight atoms (two carbons and six hydrogens), while the liquid that makes nylon contains molecules with six atoms of carbon, eight of hydrogen, one of nitrogen, and two of oxygen—a combination that chemists write as $C_6H_8NO_2$. Polymers form when the ends of these molecules begin to link up.

In the case of nylon, the polymer forms by a *condensation reaction* in

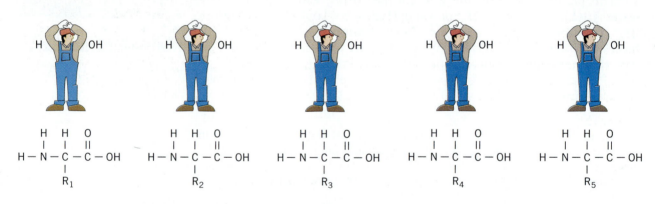

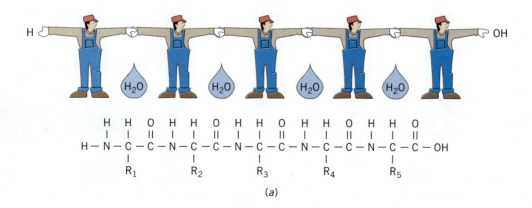

(a)

which each new polymer bond forms by the release of a water molecule (Figure 12–3a). Polyethylene, on the other hand, forms by *addition polymerization*, in which the basic building blocks are simply joined end to end (Figure 12–3b).

Polymers play a key role in our lives. The useful properties of these varied materials are related to the shapes of the molecules and the way they come together to form materials (Figure 12–4). In polyethylene, for example, long-chain molecules wrap themselves together into something like a plate of hairy spaghetti. It's hard for water molecules to penetrate into this material, so polyethylene is widely used in packaging. The "plastic" on prepackaged fruit or meat at your supermarket may be made from polyethylene.

A closely related polymer is polyvinyl chloride (often sold as PVC), whose basic building block, vinyl chloride, is just an ethylene molecule in which one of the four hydrogens has been replaced by a chlorine atom. Because the chlorine atom is bigger than hydrogen, the molecules of the polymer are lumpy, and cannot pack too closely together. Commercial PVC, widely used in water and sewer pipes, contains other kinds of molecules that move between the polymers, lubricating the system and making the resulting material highly flexible. Your credit cards are also probably made from this material. Other common polymers include polypropylene (artificial turf), polystyrene ("foam" cups and packaging), and Teflon (nonstick cookware).

Most garbage bags are made from polyethylene, a common plastic.

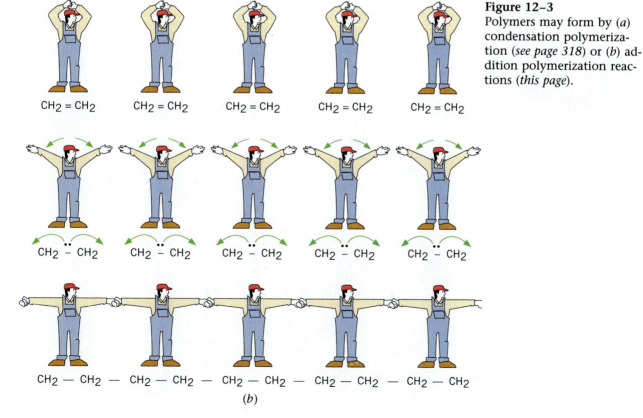

Figure 12–3
Polymers may form by (*a*) condensation polymerization (*see page 318*) or (*b*) addition polymerization reactions (*this page*).

$CH_2 = CH_2$ $CH_2 = CH_2$ $CH_2 = CH_2$ $CH_2 = CH_2$ $CH_2 = CH_2$

$CH_2 - CH_2$ $CH_2 - CH_2$ $CH_2 - CH_2$ $CH_2 - CH_2$ $CH_2 - CH_2$

$CH_2 - CH_2 - CH_2 - CH_2 - CH_2 - CH_2 - CH_2 - CH_2 - CH_2 - CH_2$

(*b*)

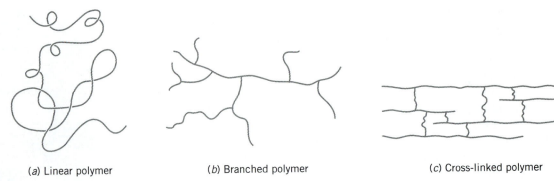

(*a*) Linear polymer   (*b*) Branched polymer   (*c*) Cross-linked polymer

Figure 12–4
Polymers can come in many forms: (*a*) a twisted chain, (*b*) a branched chain, and (*c*) chains that are linked together.

Many polymers are extremely long-lasting and difficult to break down—a situation that provides us with long-lasting plastic products, but one that also presents a growing problem in an age of diminishing landfills. Nevertheless, most polymers are not permanently stable. Given time, they will decompose into smaller molecules. This breakdown of a polymer into short segments is called *depolymerization*.

Perhaps the most familiar depolymerization reactions occur in your kitchen. Polymers cause the toughness of uncooked meat and the stringiness of many raw vegetables. We cook our food, in part, to break down these polymers. Chemicals such as meat tenderizers and marinades can also contribute to depolymerization and can improve the texture of some foods.

Not all depolymerization is desirable, however. Museum curators are painfully aware of the breakdown process, which affects leather, paper, textiles, and other historic artifacts made of organic materials. Storage in an environment of low temperature, low humidity, and an inert atmosphere (preferably without oxygen) may slow the depolymerization process, but there is no known way to repolymerize old brittle objects.

Styrofoam, made of the polymer polystyrene, is often used as insulation in buildings.

Fresh Kills landfill on New York's Staten Island covers almost five square miles and is a major repository of New York City's trash.

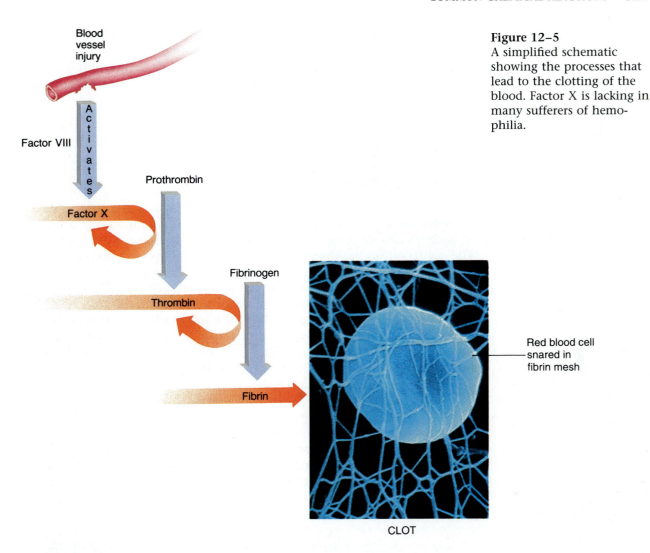

Figure 12–5
A simplified schematic showing the processes that lead to the clotting of the blood. Factor X is lacking in many sufferers of hemophilia.

CLOT

The Polymerization of Blood

Whenever you get a cut that bleeds, your blood begins a remarkable and complex sequence of chemical reactions called *clotting* (Figure 12–5). Normal blood, a liquid crowded with cells and chemicals that distribute nutrients and energy throughout your body, flows freely through the circulatory system. When that system is breached and blood escapes, however, the damaged cells release a chemical called prothrombin.

Prothrombin, itself, is inactive, but other blood chemicals convert it into the active chemical thrombin. The thrombin breaks apart other normally stable chemicals that are always present in blood, and thus produces small molecules that immediately begin to polymerize. The new polymer, called fibrin, congeals quickly and forms a tough fiber net that traps blood cells and seals the break in a matter of minutes.

Clotting chemistry differs depending on the nature of the injury and the presence of foreign matter in the wound. Biologists have discovered more than a dozen separate chemical reactions that may occur during the process. A number of diseases and afflictions may occur if some part of this complex chemical system is not functioning properly. Hemophiliacs, who suffer from a disease that causes them to bleed continuously from small cuts, lack one of the key clotting chemicals. Some lethal snake venoms, on the other hand, work by inducing clotting in a circulatory system.

Science by the Numbers

Balancing Chemical Equations

When chemical reactions occur, atoms are rearranged and new molecules are formed. In this process, however, the total number of atoms of each species must be the same after the reaction as it was before. The conservation requirement, that no atoms be created or destroyed, dictates that every chemical reaction must be "balanced."

Your car battery, for example, contains plates of lead (Pb) and lead dioxide (PbO_2) immersed in a solution of sulfuric acid (H_2SO_4). After the battery has been discharged, both the lead and the lead dioxide have been converted into lead sulfate ($PbSO_4$) and the sulfuric acid solution has been replaced by water (H_2O). If you wrote a chemical reaction that included all these substances, you would get

$$Pb + PbO_2 + H_2SO_4 \longrightarrow PbSO_4 + H_2O$$

This equation, however, is not balanced. There are two lead atoms on the left side, for example, and only one lead atom on the right. To balance the equation and to make sure that there are as many atoms at the end as at the beginning, you can write

$$Pb + PbO_2 + 2H_2SO_4 \longrightarrow 2PbSO_4 + 2H_2O$$

As you can see, there are now two atoms of lead on each side of the equation, two SO_4 groups, four hydrogens, and two oxygens. This balanced equation represents the reshuffling of atoms that goes on when your battery discharges. It tells us that two molecules of sulfuric acid will be used for each molecule of lead, and that two molecules of water and two of lead sulfate will be produced at the end.

Every chemical reaction, no matter how complex, must be balanced in just this way. ●

Environment

Life Cycle Costs

Every year, chemists around the world develop thousands of new materials and introduce them to the market. Some of these materials do a particular job better than those they replace, some do jobs that have never been done before, and some do jobs more cheaply. All of them, however, share one characteristic. When the product is worn out or replaced, it will have to be disposed of in a way that is not harmful to the environment. Until very recently, engineers and planners had given little thought to this problem.

Think about the battery in your car, for example. The purchase price covered the cost of mining the lead in its plates, pumping the oil that was made into its plastic case, assembling the final product, and so on. When that battery runs down, all of the materials that make up the battery have to be dealt with responsibly. If the battery is thrown into a ditch somewhere, for example, the lead may eventually wind up in nearby streams and wells.

One way of dealing with this sort of problem is to recycle materials—in the case of the battery, pull the lead plates out of the battery, reprocess them, and then use them again. But even in the best system, some materials can't be recycled, either because they have become contaminated with other chemicals or because we don't have technologies capable of recycling the material. These materials have to be disposed of in a way that isolates them from the environment. The question then becomes, "Who pays?"

In the United States, the tradition has been that the person who does the dumping—in effect, the last user—must be responsible for the disposal. In some European countries, however, a new concept called "life cycle costing" is being introduced. This notion is built around the proposition that once a manufacturer uses a material, he or she owns it forever, and is responsible for its disposal. The manufacturer's price of a product such as a new car has to include the cost of dealing with that car if it is worn out or abandoned.

Life cycle costing increases the immediate price of commodities, contributing to inflation in the process. What do you think the proper trade-off is in this situation? How much extra cost should be imposed up front compared to eventual costs of disposal? ●

When we are through driving our cars, we still must dispose of the materials in them.

MODULAR CHEMISTRY: BUILDING THE MOLECULES OF LIFE

Next time you're outside, look closely at a tree. The trunk and limbs divide over and over again—a branching formation that is mirrored by the hidden root system. The tree has countless almost identical leaves on every limb, as well as myriad seeds in their season. The complex structure of the tree is made by repeating the same simple building blocks over and over again.

Buildings in your city show the same pattern. They feature stacks of identical bricks, row after row of identical windows, and numerous identical shingles, slates, or other roofing materials. The sidewalk is made of slab after slab of concrete, while street lamps and signs also repeat over and over again.

Indeed, almost any complex structure designed by nature or humans is modular—composed of a few simple pieces that combine to form larger objects. The chemicals of life are no different. A few very simple molecules combine to create the wonderful complexity of life around us.

(a)

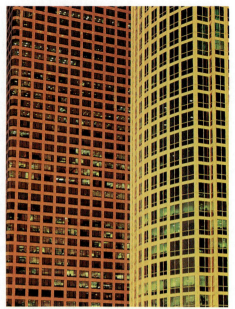

(b)

(a) Repeating patterns of trunks, branches, twigs, and leaves can be seen in any forest.

(b) Modern architecture relies on the repetition of simple forms.

(c) Nature's architecture displays repeating patterns, as in this elegant chambered nautilus.

(c)

Building Molecules: The Hydrocarbons

As an example of how a wide variety of materials can be made by putting the same molecular building blocks together in different ways, let's start with the methane molecule shown in Figure 12–1 and build a family of molecules known to chemists as *alkanes*. Alkanes are flammable materials (often either gases or liquids) that burn readily and are used as fuels. Most of the components of the gasoline in your car, for example, are members of this family. Alkanes are one example of *hydrocarbons*, molecules made completely from hydrogen and carbon atoms.

You can think of the methane as being composed of a carbon and

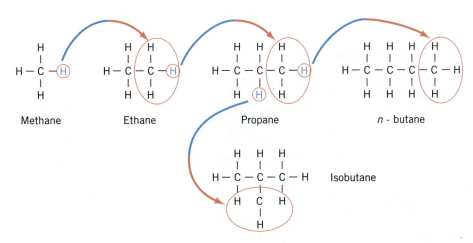

Figure 12–6
Building up the alkane series by adding methyl groups to methane. The first three members of the group are methane, ethane, and propane.

three hydrogens (what chemists call the methyl group) and a fourth hydrogen. We begin by noticing that we can replace the hydrogen in the methane by another methyl group to form a molecule with two carbons in it. This larger molecule is ethane, a volatile, flammable gas (see Figure 12–6).

You can keep going. Adding a third methyl group produces propane, a three-carbon chain. Propane is widely used as a fuel for portable stoves—you may have used it on your last camping trip. The next step is to add another methyl group to form a four-carbon chain, a molecule called butane. But there is an ambiguity here. We could, as shown, add the new group at the end of the chain so that all four carbon atoms form a straight line. This process would give us a form of butane known as n-butane (or "normal" butane). However, we could equally well add the methyl group to the interior carbon atom. In this case, the molecule would be known as isobutane. Isobutane and n-butane have exactly the same numbers of carbon and hydrogen atoms, but are actually quite different materials (to give just one example, the former boils at $-11.6°C$ while the latter remains a liquid up to $-0.5°C$). Molecules that contain the same number and types of atoms but have different structures are said to be *isomers*.

As we continue the building process, moving to molecules with five carbons (pentane), six carbons, (hexane), seven carbons (heptane), eight carbons (octane) and beyond, the number of different ways to assemble the atoms grows very fast. Octane, for example, may form 18 different isomers. Some of these variants have long chains, others are branched. As we shall see, these structural differences play an important role in a molecule's usefulness as an automotive fuel.

All other things being equal, the carbon chain length affects whether the alkane is a solid or liquid: the longer the chain, the higher the temperature at which that material can remain a solid. If carbon chains are straight, then alkanes with a half dozen or so carbons are liquid, but those with more than 10 are soft solids. Good quality paraffin candle wax, which melts only near a hot flame, for example, is composed primarily of chains with 20 to 30 carbon atoms. The presence of branches in the chain, however, makes it more difficult for the molecules to pack together efficiently. One consequence of branching, therefore, is that the melting points are generally lowered compared to those of straight alkanes.

Technology

Refining Petroleum

Deep underground are vast lakes of a thick, black liquid called *petroleum*, derived from many kinds of transformed molecules of former life. Petroleum is an extremely complex mixture of organic chemicals, as much as 98% molecules of hydrogen and carbon (mostly in the form of hydrocarbons), with about 2% of other elements. Engineers must separate this mixture into much purer fractions through the process of *distillation*.

Hydrocarbons with different numbers and arrangements of carbon atoms and different degrees of saturation have very different boiling temperatures. The key to distillation, therefore, is to successively boil off and collect different kinds of molecules, according to their boiling points. The most *volatile* hydrocarbon—the one with lowest boiling point—is simple methane (CH_4), or natural gas. At the opposite extreme are very long-chain hydrocarbons with dozens of carbons, as in the molecules that comprise hard waxes, asphalt, and tar.

Petroleum products.

Modern chemical plants bristle with tall distillation columns, in which petroleum products are purified.

Modern chemical plants bristle with tall cylindrical towers that distill petroleum. Engineers pump crude oil into a tower, which is heated from below to create a temperature gradient up the tower (see Figure 12–7). At various levels of the tower, useful petroleum products, such as gasoline or heating oil, are recovered and sent to other parts of the chemical plant for further refining and processing.

The gasoline you buy at a service station is usually rated in "octanes." The octane rating of a gasoline is based on its ability to withstand high compression in a cylinder without igniting. A fuel mixture that ignites while the piston is still moving up and compressing the gas in the cylinder will cause the engine to knock, a highly undesirable quality. In general, the more highly branched a molecule is, the better it is able to reduce knocking.

A particular isomer of octane that has five carbons in a row and the others on the sides turns out to have very good anti-knock proper-

Fractionating tower

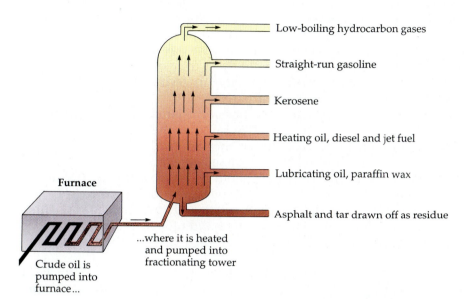

Low-boiling hydrocarbon gases

Straight-run gasoline

Kerosene

Heating oil, diesel and jet fuel

Lubricating oil, paraffin wax

Asphalt and tar drawn off as residue

Furnace

...where it is heated
and pumped into
fractionating tower

Crude oil is
pumped into
furnace...

Figure 12–7
A distillation column in a
chemical plant separates hydrocarbons into fractions
useful as gases, gasoline, kerosene, heating oil, lubricating oils and paraffin, asphalt, and tar.

The "octane" rating of a gasoline depends on how well
it can withstand compres-
sion in your car's cylinders.

ties. This isomer is called isooctane, and an octane rating of 100 for
any fuel mixture means that it is as good as pure isooctane. At the opposite extreme, n-heptane is an isomer with seven carbons in a
straight chain that produces constant knocking. An octane rating of
zero means the fuel mixture is as bad as pure n-heptane. Thus the octane rating of a fuel is simply a statement that it performs as well as a
particular mixture of isooctane and n-heptane. A fuel rated at 95 octane, for example, performs as well as a mixture of 95% isooctane and
5% n-heptane. ●

Organic Molecules

Wood. Leather. Hair. Cotton. Skin. All of these materials originated in
living systems on our planet. And like all other materials found in living
things, they share four basic chemical characteristics.

**1. Most molecules in living systems are based on the chemistry
of carbon.**   In Chapter 11, we saw that carbon atoms possess the ability
to form long chains, branches, and rings. In fact, chemists usually refer
to molecules containing carbon as **organic molecules,** whether or not
they are or were part of a living system. The branch of science devoted
to the study of such carbon-based molecules and their reactions is called
organic chemistry.

organic molecules   Molecules
that contain carbon, whether
or not they come from living
systems.

2. Life's molecules form from very few different elements.   In
terms of the percentages of atoms, just four elements—hydrogen, oxygen,
carbon, and nitrogen—comprise 99.4% of our bodies. Phosphorus and
calcium represent 0.2% of our atoms, while all the other elements make
up the remaining 0.4%.

Large structures, be they buildings or molecules, can be put together from simple modules.

3. The molecules of life are modular—composed of simple building blocks. You might imagine that large and complicated molecules could be put together in two contrasting ways. One way would be to build each one from scratch, so that no piece of one molecule would be part of another. A second very different way would be to make the molecules modular; that is, to build them from a succession of simpler, widely available parts so that each large molecule differs from another only in the arrangement of those parts.

Nature, for the most part, has chosen the latter strategy in designing the molecules of living systems. We say that the molecules of life are built from simple molecules, or that they are modular.

In a sense, the choice between these two extreme modes of making molecules boils down to one of economics. It would take a great deal of work, time, and money to custom design every door and window and other component of your home. You might end up with a better-designed structure, but at a very high price. By building your home with widely available parts, you save money and still end up with a satisfactory dwelling.

As it happens, almost all of the large molecules that are crucial in forming and operating living systems are built in a modular manner. Though life's molecules come in an extraordinary variety of shapes and functions, they are made from collections of just a few smaller molecules. This modularity does not mean that the final products are simple, just as there's nothing particularly simple about a house. It simply means that if we wish to understand how large molecules behave, we first have to talk about the simple pieces from which they are built.

4. Shape determines the behavior of organic molecules—in other words, molecular geometry controls the chemistry of life. The connection between geometry and the behavior of organic molecules can be understood if you remember one important thing about chemical bonds. All chemical bonds are formed by the shifting of electrons among specific pairs or groups of atoms. This bonding property is particularly true of atoms that tend to form ionic, covalent, and hydrogen bonds (see Chapter 11).

A very large and complex molecule may have millions of atoms arranged in a complicated shape. If this large molecule is to take part in chemical reactions—if it is to bind to another molecule, for example—then that binding must take place through the actions of the outer-shell electrons of atoms near the outsides of the two molecules. Specific atoms in each molecule must be able to get near enough to each other so that their electrons can form a bond. Consequently, the geometrical shape of the molecule plays a crucial role, because it determines whether atoms that can form bonds in each molecule will be able to get close enough together for the bonds actually to form.

In principle, an infinite number of molecules could be constructed according to these four rules. In fact, when we examine natural systems we find that only four general classes of molecules—proteins, carbohydrates, nucleic acids, and lipids—govern most of life's main chemical functions. Keep in mind that all of these molecules conform to the four rules: they are carbon-based, they are built from just a few elements, they are modular in structure, and their behavior depends on their shapes.

Science in the Making

The Synthesis of Urea

In the early 1800s, scientists were not convinced that molecules in living systems are formed according to the same chemical rules that govern the chemical behavior of nonliving systems. Such molecules, after all, had not been produced in the laboratory. In 1828, a German chemist by the name of Friedrich Wohler (1800–1882) performed a series of experiments that were crucial in establishing the ordinariness of organic molecules.

Like many scientists at that time, Wohler had a breadth of knowledge that is unheard of today. Before he began his career as a chemistry teacher, for example, he became a medical doctor and qualified in the specialty of gynecology. He was also interested in the practical aspects of chemistry, and collected minerals from the time he was a child. He described his crucial experiments this way: "I found that whenever one tried to combine cyanic acid [a common laboratory chemical] and ammonia, a white crystalline solid appeared that behaved like neither cyanic acid nor ammonia."

After extensive testing, Wohler found that the white crystals were identical to urea, a substance routinely found in the kidneys and (as the name implies) in urine, where it plays the role of removing nitrogen wastes from the body. In other words, the appearance of that "white crystalline solid" showed that it is possible to take ordinary chemicals off the shelf and produce a substance found in living systems. He had demonstrated that organic molecules may be formed by the same chemical processes as other materials.

With humor some might find uncharacteristic of academicians, Wohler announced his findings in a letter as follows: "I can no longer, as it were, hold back my chemical urine: and I have to let out that I can make urea without needing a kidney, whether of man or dog." ●

Friedrich Wohler (1800–1882)

Proteins: The Workhorses of Life

The molecules we call proteins play many key roles in living systems. Some proteins form building materials from which large structures are formed. Your hair, your fingernails, the tendons that hold your muscles in place, and much of the connective tissue that hold your body together, for example, are made primarily of protein molecules. Proteins also serve to regulate the movement of materials across cell walls, and thus control what goes into and out of each cell in your body. In addition, proteins serve as *enzymes*, which are molecules that control the rate of complex chemical reactions in living things.

We are all familiar with protein as a vital component of our diet. We must regularly take in proteins to supply our bodies with building materials to effect repairs and growth.

As with all complex biological molecules, proteins are modular; they are made up of strings of a few basic building blocks called **amino acids.** A typical amino acid molecule is sketched in Figure 12–8*a*. All amino acids incorporate a characteristic backbone of atoms. One end terminates in a *carboxyl group* (COOH), which is a combination of carbon, oxygen,

amino acid   The building block of protein, incorporating a carboxyl group (COOH) at one end, an amino group (NH₂) at the other end, and a side group (which varies from one amino acid to the next).

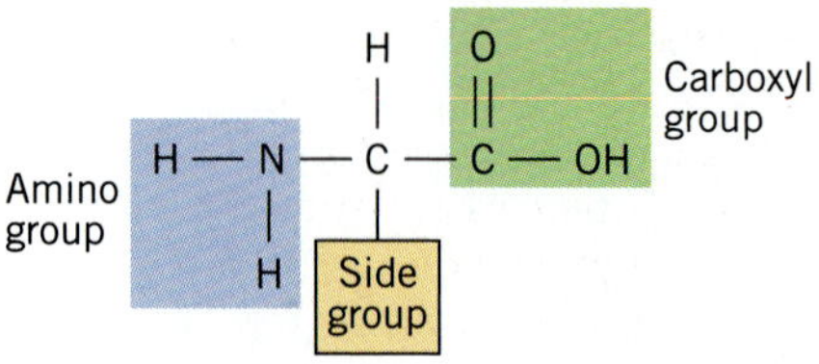

(a)

Figure 12–8
(a) An amino acid, showing the amino and carboxyl groups and the side group. (b) Several amino acids. The side group varies from one type of amino acid to another, and gives a particular amino acid the chemical properties that distinguish it from any other.

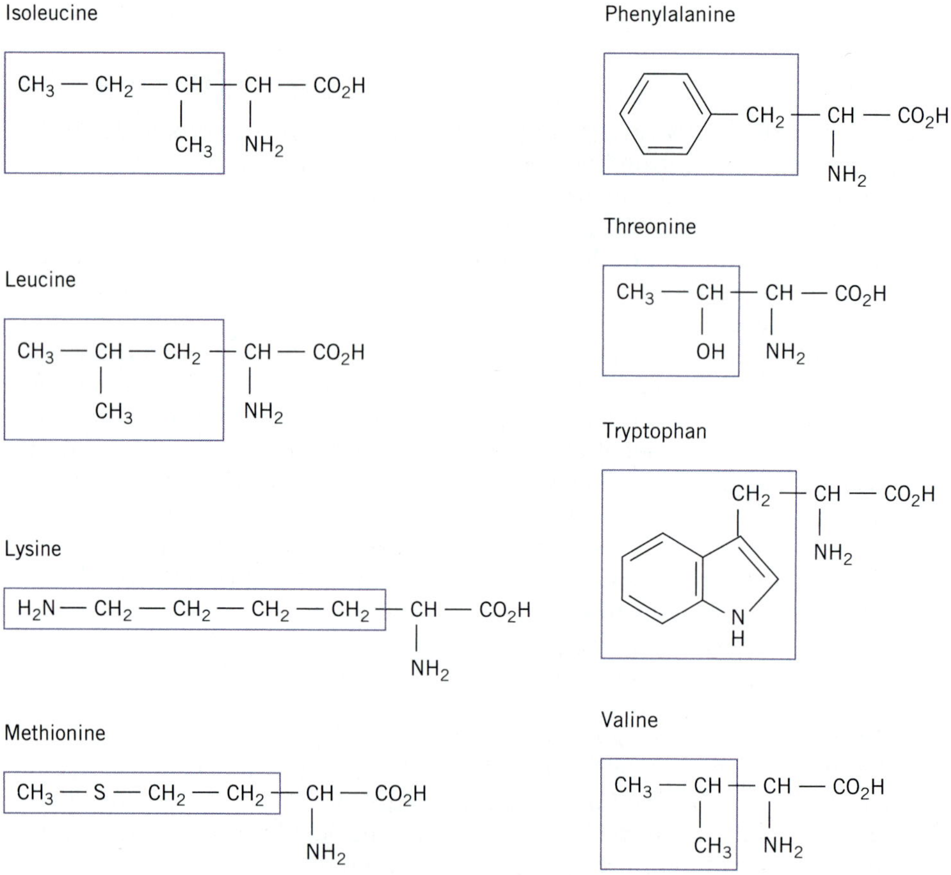

Isoleucine

Leucine

Lysine

Methionine

Phenylalanine

Threonine

Tryptophan

Valine

(b)

and hydrogen; on the other end is an *amino group* (NH_2), which is a nitrogen bonded to two hydrogen atoms (this group gives the molecules their name). Between these two ends a carbon atom completes the backbone.

Branching off the central carbon atom is another atom or cluster of atoms—the side group—that makes amino acids unique and interesting. Hundreds of different amino acids can be made in the laboratory, each with its characteristic side group, but only 21 appear in living systems. A few common amino acids are sketched in Figure 12–8b to give you a sense of the kind of diversity that is possible within this basic structure.

Two amino acids can bond together in a very simple way. The hydrogen (H) from one end of one amino acid will connect to the hydroxyl (OH) from the end of another amino acid to form a molecule of water (H_2O). This water molecule moves off (you can think of this process as "squeezing out" the water), leaving the two amino acids hooked together in what is called a *peptide bond*. This process is identical to a reaction known as *condensation polymerization*, which is often used to manufacture plastics and other polymers (see Chapter 11). Chemists often refer to a bonded chain of amino acids as a *polypeptide*.

Once two amino acids have joined together with a peptide bond, more amino acids can be hooked onto either end by the same process to form a long string of amino acids. A **protein** is a large molecule

protein A molecule which can consist of hundreds of amino acids and thousands of atoms formed in a chain structure. Proteins function as enzymes and direct the cell's chemistry.

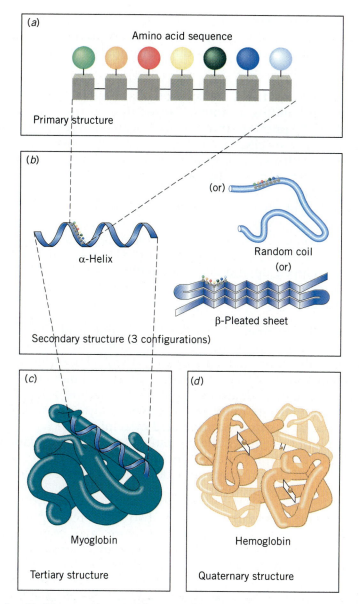

Figure 12–9
The structure of a protein can be described in four steps. Primary structure (*a*) is the sequence of amino acids. Secondary structure (*b*) is the way the sequence kinks or bends. Tertiary structure (*c*) is the shape of the completely folded protein. Quaternary structure (*d*) is the clustering of several chains to form the active structure, in this case hemoglobin, which carries oxygen in your bloodstream.

formed by linking amino acids together in this way. There are many different amino acids to choose from, and different proteins correspond to a different ordering (as well as the different total number) of the amino acids in the string.

Proteins are complex molecules, sometimes consisting of many hundreds of amino acids and thousands of atoms. One of the great triumphs of modern science has been to determine the exact atomic structures of many of these large molecules. To accomplish this feat, biochemists first have to isolate quantities of pure protein, and then form delicate single crystals in which the large molecules line up in a regular array. These crystals are examined by X-ray crystallography, which reveals the distribution of atoms in space.

A protein structure is usually described in four stages, each representing an increasing order of complexity (see Figure 12–9):

1. Primary Structure.   The exact order of amino acids that go into a given protein is called its *primary structure*. Every protein has a distinctive primary structure; that is, it has a different number and order of amino acids along its string.

2. Secondary Structure.   Formation of a string of amino acids is just the first step in building a protein. Depending on the arrangements of amino acids in the primary structure, hydrogen bonds can form that give the final protein a specific shape. Some proteins, for example, take the form of a long helix or long spring. Others bend or fold into a zig-zag shape. These shapes formed by the string of amino acids are called its *secondary structure*.

When you cook an egg, you can see the effect of secondary structure. The proteins in egg white are wrapped up into tiny spheres scattered throughout the fluid. This is why uncooked egg white is transparent. When you cook the egg white, you break the hydrogen bonds that keep the protein wrapped up and allow the molecules to unfold. The tough mat they form when they interlock gives the cooked egg white its characteristic texture.

3. Tertiary Structure.   As parts of the amino acid chain fold back on themselves, atoms in the side groups can come into contact with each other. As a result, additional cross-linking chemical bonds form between side groups in amino acids in different parts of the chain. One common link occurs between sulfur atoms in different side groups. The distinctive shape of human insulin, for example, arises because of bonds that form between sulfur atoms in the amino acid cysteine.

As a result of these links, the protein will twist around, kink up, and fold itself into a complex shape, much as a string will fold itself into a complex shape if it's dropped on a table. This complex folding is the *tertiary structure* of the protein.

4. Quaternary Structure.   Finally, two or more long chains, each with its own secondary and tertiary structure, may come together to form a single larger unit. This joining of separate chains determines the *quaternary structure* of the protein.

The prediction of the exact shape that will be assumed by a given sequence of amino acids remains one of the great goals of modern biochemistry—a goal that we are still far from reaching. But whether or not we can predict the ultimate shape of a protein, the fact remains that each different sequence of amino acids will produce a large molecule with a different three-dimensional shape. This fact is critical to the roles that proteins play in living systems.

Proteins as Enzymes

A *catalyst* is a molecule that facilitates a reaction between two other molecules, but which is not itself consumed or altered in that overall reaction. Because of the presence of the catalyst, the chemical reaction takes place at a much faster rate than it otherwise would. An *enzyme* is a catalyst in a living system and is made of protein. Enzymes play a role in chemical reactions similar to a broker or an agent in a business deal.

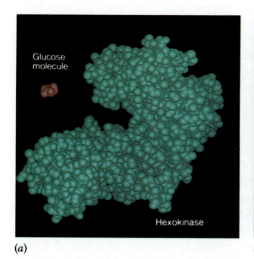

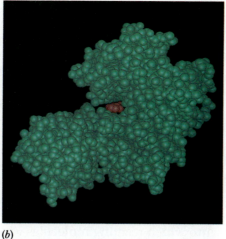

(a)   (b)

An enzyme in action: A glucose molecule approaches the enzyme hexokinase (*a*). When the two molecules combine, the shape and function of the enzyme change (*b*).

The role of the broker is to bring together a buyer and seller, but brokers themselves do not buy or sell. The buyer and seller might eventually find each other without the help of the broker, but the deal progresses much quicker if the broker is there. In the same way, a molecule that plays the role of an enzyme may bring together two other molecules and facilitate their forming a bond, or it may tear a molecule apart, without itself being included in the chemical reaction. Because of the enzyme, the reaction takes place relatively quickly.

Enzymes illustrate the primary importance of geometrical shape in determining how chemical reactions take place among large molecules. You can visualize the workings of an enzyme in Figure 12–10. Each large molecule has places on it—atoms or small groups of atoms—where chemical bonding can take place. Think of these locations as sticky spots somewhere on a large, convoluted molecular shape. In order for two or more molecules to interact, their respective sticky spots have to come into contact. More precisely, the atoms whose electrons will eventually

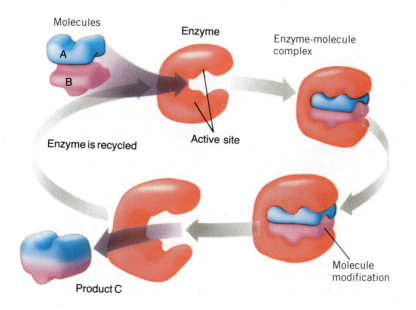

Figure 12–10
An enzyme in action joins two molecules (designated *A* and *B*) and produces a product (*C*). The product is released and the enzyme is free to repeat the process.

form the bonds must be brought close enough together for the electrons to interact. This proximity is often unlikely to happen at random.

Think of each molecule as a large pile of string that is dumped in the corner of a room, and think of the spots that could form chemical bonds as patches of glue located at odd places along the string. If you took two such pieces of string and tossed them together into the corner, the chances are very slight that any of the sticky spots would come near each other. If, on the other hand, you picked up the string and arranged it so that the sticky spots were next to each other, you could make sure that they formed a bond. In this "chemical reaction," you are playing the role of the enzyme. You cause bonds to form that probably wouldn't form without you, but you are not yourself involved in the bonding process.

Enzymes perform an analogous function in organic reactions. Typically, an enzyme is a large molecule that has particular spots on its surface into which reacting molecules will fit precisely. The enzyme first attracts one of the molecules and then the other. In some enzymes, a pair of specific molecules attach themselves in only one way, so it is guaranteed that their "sticky spots" will be near each other. The chemical bonds that hold them together will form. Once the bonds have formed, the overall shape of the resulting composite molecule becomes different from that of either of the two molecules. Consequently, the new large molecule no longer fits into the appropriate grooves and valleys of the enzyme, and it will spontaneously break free and wander off by itself. This separation leaves the enzyme free to mediate the same reaction again and again, each time with two new molecules. Another kind of enzyme, like those in your stomach, performs the opposite function; they break apart large molecules into smaller units, over and over again.

If you think about the way an enzyme works, you will realize why molecular shape is so important. The hills and valleys on the surface of an enzyme serve as resting places for the molecules that interact on the enzyme's surface. Versatile protein molecules, which adopt many different shapes and can therefore provide resting places for many different kinds of interacting molecules, are ideally suited for functioning in this way. For precisely this reason, proteins participate in most of the chemical reactions in living organisms.

Technology

Designer Drugs

Many of the drugs we take produce their effect because of the shape of their molecules. They may, for example, alter our body chemistry by blocking the action of enzymes. You can understand how such a drug might work by looking at Figure 12–9. The efficacy of the enzyme depends on whether the shape of its surface matches the shape of the molecules involved in the reaction. A drug molecule that attaches itself to one of those crucial sites on the enzyme would block that site, preventing one of the molecules involved in the original reaction from occupying it. As a result, the enzyme would not be able to facilitate the reaction as it normally does, and the chemical balance of the cell would be changed. When you take an aspirin, for example, you block the action of an enzyme that facilitates the production of molecules called prostaglandins. These molecules, among other things, affect the transmission of nerve signals.

Throughout most of the history of medicine, the search for new drugs has been carried out in a somewhat random way. If you think of the place where the drug is supposed to bind as a lock, and the drug molecule as a key that will fit the lock, then the traditional search for new medicines would correspond to trying one key after another until you find one that fits. It is not unusual, for example, for drug companies to test 10,000 different molecules to produce one effective drug.

As our understanding of organic molecules and their shapes has increased, however, scientists are increasingly able to consider producing a molecule with the right shape from scratch. In terms of our lock and key analogy, this process would correspond to studying the lock and then making a key that fits it. Products made in this way have been nicknamed "designer drugs." One such drug (called captopril) has been in use since 1975. This drug blocks the action of an enzyme that produces molecules that contribute to hypertension, and so is used to control that condition. Designer drugs for treating psoriasis, glaucoma, AIDS, and some forms of cancer and arthritis are in advanced stages of testing and may be on the market soon. Given the rate of progress in understanding the molecular basis of life, it seems clear that there will be many more designer drugs in our future. ●

carbohydrate A class of modular molecules, which are made of clusters of sugar molecules (carbon, hydrogen, and oxygen), and which form part of the solid structure of plants and play a central role in how living things are supplied with energy.

sugars The simplest of the carbohydrates: common sugars contain five, six, or seven carbon atoms arranged in a ring-like structure.

Carbohydrates

Carbohydrates, the second important class of modular molecules found in all living things, are made of carbon, hydrogen, and oxygen. They play a central role in the way that living things acquire and use energy, and form many of the solid structures of living things. You use carbohydrates every day in many of the foods you eat, the fuels you burn, the clothes you wear, and even in the paper of this book. They are truly a diverse group of molecules.

The simplest carbohydrates are **sugars,** which are molecules that usually contain five, six, or seven carbon atoms arranged in a ringlike structure. *Glucose,* an important sugar in the energy cycle of living things, is sketched in Figure 12–11. Glucose figures prominently in the energy metabolism of every living cell. The "burning" of glucose in our cells supplies the energy that we use to move and grow.

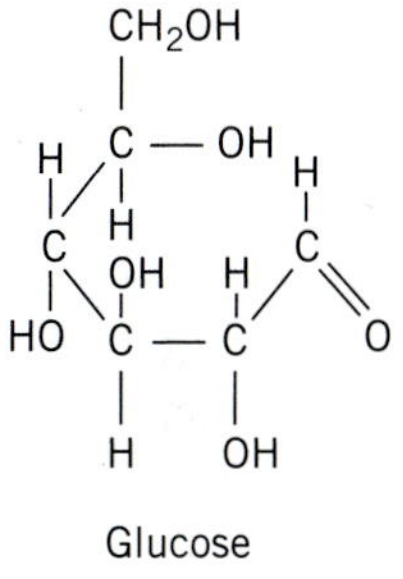

The general chemical formula for sugar is $C_nH_{2n}O_n$, or $C_n(H_2O)_n$. In this formula, n stands for a small integer. Glucose, for example, with $n = 6$, has the formula $C_6H_{12}O_6$. As often happens with organic molecules, other forms of the molecule have the same chemical composition, but have the components arranged differently. In Figure 12–12, for example, we show the sugar fructose. As the name implies, this sugar is commonly found in fruit. It has the same number of carbon, hydrogen, and oxygen molecules as glucose, but the atoms are arranged slightly differently, and this different arrangement gives fructose a different chemical behavior.

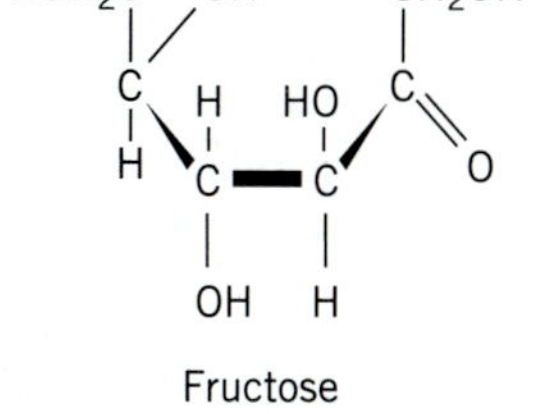

Chemists call individual sugar molecules *monosaccharides,* meaning "one sugar." (The same root word is used when we describe an overly sentimental story as "saccharin.") The carbohydrates that we eat, however, are usually formed from two or more sugar molecules strung together in a covalently bonded chain to form a sugar polymer. Ordinary table

Glucose

Figure 12–11
The structure of glucose.

Fructose

Figure 12–12
The structure of fructose. It has the same number and kinds of atoms as glucose, but in a different arrangement.

The cellulose of cabbage and the starch of potatoes are different polymers of the sugar glucose.

sugar, for example, is made from two sugars, glucose and fructose, linked together by covalent bonds.

When many sugars are strung together in a chain, the resulting molecules are called *polysaccharides* ("many sugars"). The two most familiar polysaccharides are starch and cellulose. Both of these kinds of molecules are made from long chains of glucose molecules. They differ from each other only in the details of the way the glucose molecules bind to each other.

Starches, a common component of the human diet, are a large family of molecules in which the glucose constituents link together at certain points along the ring. They are found in many plants, such as potatoes and corn. Humans break down starch molecules with an enzyme in the digestive system, thus releasing individual glucose molecules, which provide the fundamental fuel used by cells.

Cellulose, a long stringy polymer that provides the main structural element in plants, including stems, leaves, and the trunks of trees, is also formed from glucose molecules. Because the glucose molecules are linked in a different way, however, human beings cannot digest cellulose; we simply do not manufacture an enzyme that can separate individual glucose molecules from the cellulose polymer. Consequently, humans don't go out and graze on the lawn at lunch time, though people on diets often eat cellulose-rich celery and other "roughage" or "fiber." On the other hand, cellulose can be digested and broken down by many single-celled organisms such as bacteria. Cows, for example, have organisms in their stomachs that perform this function for them, as do wood-eating termites.

The wood fibers in the paper of this page are made from glucose molecules bonded into cellulose—basically the same chemical as in the stalk of a celery stick. The same glucose molecules, bonded in a different way, form the flour in the spaghetti you ate the last time you had an Italian dinner. The modular construction of organic molecules is responsible for such an amazing diversity.

Nucleic Acids

The discovery of the nature and function of nucleic acids has fundamentally transformed the study of biological systems in the past three decades. **Nucleic acids,** so called because they were originally found in the nucleus of cells, include *DNA* and *RNA*, the molecules that govern both the inheritance of physical traits by offspring and the basic chemical operations of the cell.

Proteins (chains of amino acids) and carbohydrates (clusters of sugar molecules) can form large structures from a single kind of building block. Nucleic acids, on the other hand, are assembled from subunits that are themselves made from three different kinds of smaller molecules. The assemblage of three molecules is called a *nucleotide*, and the nucleic acids are formed when nucleotides are strung together in a long chain.

The first of the smaller molecules that go into an individual nucleotide is a sugar; in DNA it's *deoxyribose*, which is the sugar that gives DNA its complicated name, **deoxyribonucleic acid** (see Figure 12–13). The second small molecule of the nucleotide is the *phosphate group*, which consists of one phosphorous atom surrounded by a tetrahedron of four

nucleic acid   A molecule originally found in the nucleus of cells that carries and interprets the genetic code; includes DNA and RNA.

deoxyribonucleic acid (DNA) A strand of nucleotides with alternating phosphate and sugar molecules in a long chain, and with base molecules adenine, guanine, cytosine, and thiamine at the side. The nucleotide strand bonds with a second nucleotide strand to make a molecule with a ladderlike double helix shape. DNA stores the genetic information in a cell.

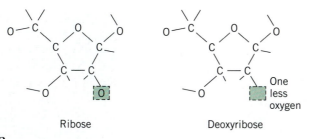

Figure 12–13
Ribose and deoxyribose. Deoxyribose has one "missing" oxygen. The hydrogen atoms are omitted from this figure to help clarify the difference between these two sugar molecules, but the bonds to the missing atoms are shown for completeness.

atoms of oxygen. Finally, each nucleotide incorporates one of four different kinds of molecules that are called *bases*. The four different base molecules are often abbreviated by a single letter—*A* for adenine, *G* for guanine, *C* for cytosine, and *T* for thymine.

Each nucleotide combines the three basic building blocks—a sugar, a phosphate, and a base (see Figure 12–14). These three molecules bond together with the sugar molecule in the middle. You can think of a nucleotide as something like a prefabricated wall in a house. Both DNA and RNA are made by putting nucleotides together in a specific way.

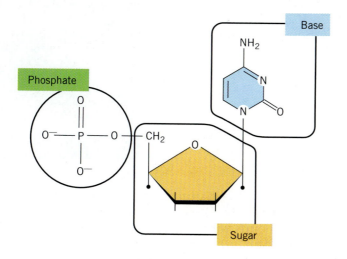

Figure 12–14
A nucleotide, formed by a sugar, base, and phosphate group.

DNA Structure

We can start putting DNA together by assembling a long strand of nucleotides. In this strand, the alternating phosphate and sugar molecules form a long chain, and the base molecules hang off the side (see Figure 12–15). This chain looks like a ladder that has been sawn in half vertically through the rungs. DNA consists of two such strands of nucleotides joined together to form a complete "ladder." The bases sticking out to the side provide the natural points for joining the two single strands. As you can see from

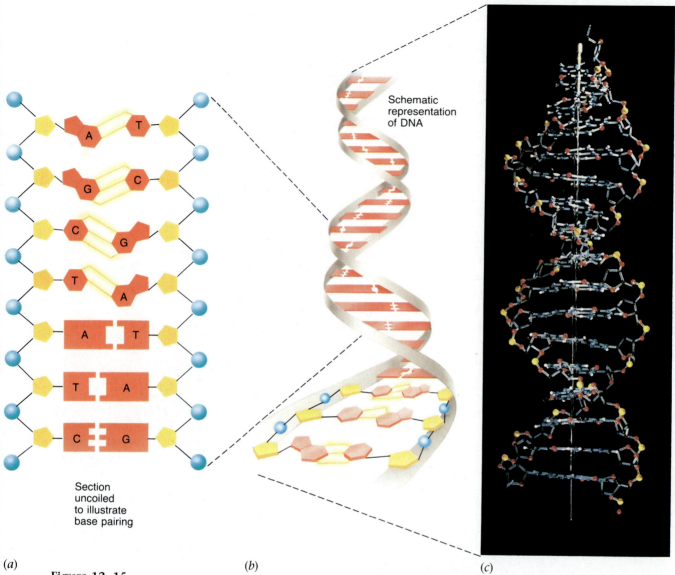

(a)   *(b)*   *(c)*

Figure 12–15
The structure of DNA illustrated schematically uncoiled (*a*), as a double helix (*b*), and in an atomic model (*c*).

Section
uncoiled
to illustrate
base pairing

Schematic
representation
of DNA

Figure 12–16, however, the distinctive shapes of the four bases ensure that only certain pairs of bases can form hydrogen bonds. Adenine, for example, can form bonds with thymine, but not with any of the other bases or with itself. Similarly, cytosine can form a bond with guanine, but not with itself, thymine, or adenine.

As a consequence, there are only four possible "rungs" that can exist in a DNA "ladder." They are:

$$\begin{array}{c} AT \\ TA \\ CG \\ GC \end{array}$$

base pair   One of four possible bonding combinations of the bases adenine, thymine, guanine, and cytosine on the DNA molecule: AT, TA, GC, and CG.

With the bonding of these **base pairs,** the complete DNA molecule is formed into a ladderlike double strand. It turns out that, because of the

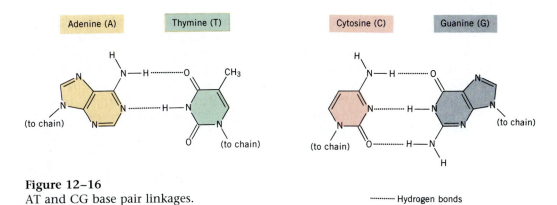

Figure 12–16
AT and CG base pair linkages.

·········· Hydrogen bonds

details of the shape of the bases, each rung is twisted slightly with respect to the one before it. The net result is that this ladder comes to resemble a spiral staircase, as shown in Figure 12–17—a helical shape that gives DNA its common nickname, "the **double helix.**"

The principal function of DNA is to carry all the coded information necessary to build an organism. The molecule accomplishes this amazing task with its four-letter "alphabet"—A, C, G, and T. Any message, no matter how complex, can be transmitted by an appropriate arrangement of these four molecular "letters." Furthermore, by splitting the DNA ladder down the middle and assembling new sides (see Figure 12–18), the critical DNA message can be copied and passed on to future generations.

double helix The twisted double strand of nucleotides that forms the structure of the DNA molecule.

Figure 12–17
James Watson

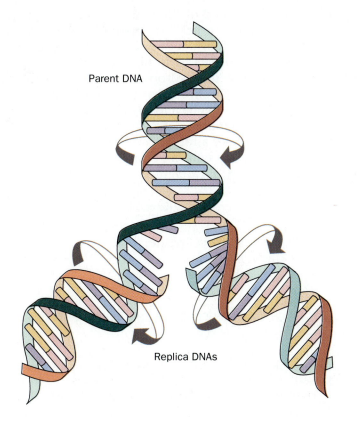

Parent DNA

Replica DNAs

Figure 12–18
A DNA double helix may be split, thus exposing bases on both strands. Two new identical double helices may then be formed from the original.

The code works this way: Each set of three bases on certain stretches of DNA serves as the starting point for a long chemical process whose end effect is the addition of a particular amino acid to a protein. In this way, the information contained in succeeding triplets of bases along the DNA codes for the primary structure of a protein. The protein, as we discussed earlier, acts as an enzyme that governs one chemical reaction in the cell. Thus the information in the DNA governs the chemistry of every living cell on Earth.

Lipids

lipid An organic molecule that is insoluble in water.

The final important set of organic molecules that make up living things is something of a grab-bag category called lipids. **Lipids** include a variety of water-insoluble molecules, such as fats in food, waxes in candles, greases for lubrication, and a wide variety of oils. If you think of drops of oil or bits of fat floating around on top of a pot of soup, you have a pretty good picture of what large clumps of lipid molecules are like.

At the molecular level, lipids play two important roles in living things. First, they form the cell walls that separate living materials from their environment, as well as separating one part of the cell from another. Like proteins, carbohydrates, and nucleic acids, many lipid molecules can come together to form large modular structures in every cell. Of special importance are *phospholipids*, which are long and thin, with a carbon backbone, as shown in Figure 12–19. In phospholipids, one end of the molecule has oxygen atoms that tend to be negatively charged, so that the end of the molecule is attracted to water (we say it is "water loving," or hydrophilic). The other end of the molecule, however, is repelled by water (we say it is "water fearing," or hydrophobic). When placed in water, these molecules typically adopt a double-layered structure like the one shown in Figure 12–20. The hydrophobic ends of the molecule line up facing each other, while the hydrophilic ends face to the outside. In this way, water is kept away from the hydrophobic ends and is kept near the hydrophilic ends. A double-layered structure of molecules like this functions very well as a membrane of a cell. It is flexible and can change its shape, but it also provides a tough barrier.

Lipids are also extremely efficient storehouses of energy. In the human body, for example, excess weight is usually carried in the form of

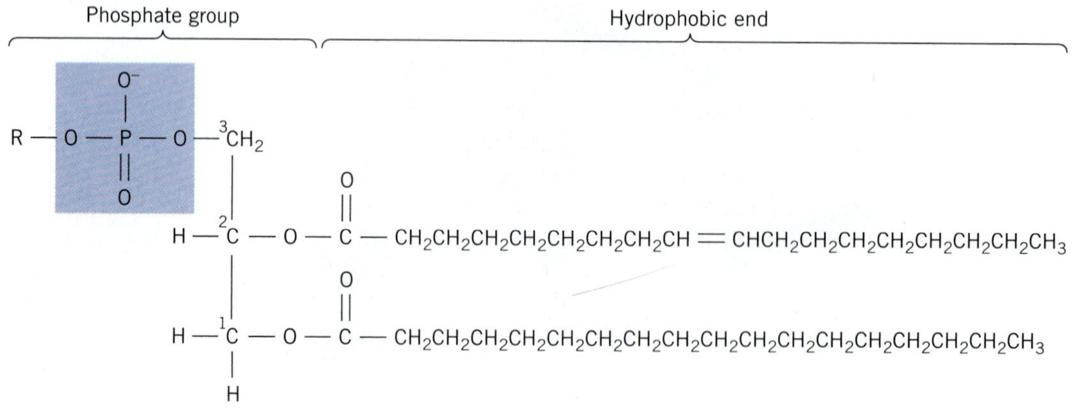

Figure 12–19
A phospholipid molecule, showing the negatively charged phosphate group at one end, and ordinary hydrocarbon chains at the other. The end with the phosphate group is attracted to water, and the hydrocarbon end is repelled by it. Different collections of molecules in the group labeled *R* correspond to different kinds of phospholipids.

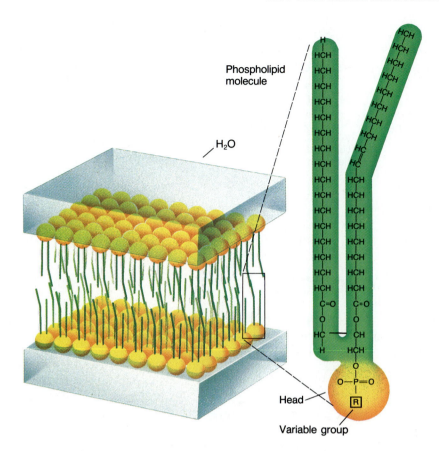

Phospholipid molecule

H₂O

Head

Variable group

Figure 12–20
The structure of a lipid bilayer. The hydrophobic ends of the molecules face each other, while the hydrophilic ends are in the surrounding water.

fat, which is different than the lipids in cell membranes. A typical gram of fat, for example, will contain twice as much chemical energy as a gram of either protein or carbohydrate.

Saturated and Unsaturated Fats

Every carbon atom in a lipid chain forms exactly four bonds to neighboring atoms (see Chapter 11). In a straight chain, each carbon bonds to two adjacent carbons along the chain and two hydrogens on the sides. Carbons of this type are *saturated*, which means they are bonded to four other atoms.

In some lipids, adjacent carbon atoms will have only three neighbors—two carbon atoms and one hydrogen atom. A kinked "double bond" will thus form between the two carbons (see Figure 12–21). A chain with one double bond is *monounsaturated*, while two or more double bonds yields a *polyunsaturated* lipid.

Saturated fats in our diet provide the raw materials from which our bodies can synthesize *cholesterol*, an essential component of all cell membranes. However, high levels of cholesterol in the blood can lead to fatty deposits that clog arteries. For this reason, many food producers emphasize their use of cholesterol-free ingredients that are rich in polyunsaturated fats. Food advertisers often suggest that unsaturated foods, in contrast to high cholesterol foods, are "good for you."

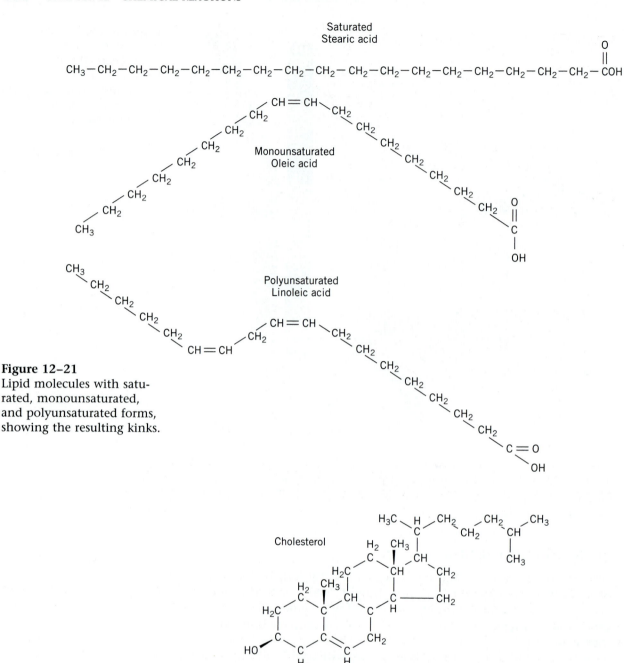

Figure 12–21
Lipid molecules with saturated, monounsaturated, and polyunsaturated forms, showing the resulting kinks.

Be warned, however. Many of these polyunsaturated products require further processing. This process, called *hydrogenation*, gives some foods a pleasing texture and consistency. Popular chocolate candies, for example, are made from fats and oils that soften near body temperature. Hydrogenation—the addition of hydrogen atoms into the carbon chains— eliminates carbon-carbon double bonds. Two hydrogen atoms bond to adjacent doubly-bonded carbon atoms and thus partially saturate the lipid chains. Many popular cooking oils that begin with highly unsaturated lipids are "partially hydrogenated for freshness and consistency" at the manufacturing plant, a process that reverses the positive aspects of the unsaturated bonds.

The partial hydrogenation of liquid vegetable oil (*a*) converts it to a solid product (*b*).

(*a*)   (*b*)

CH₂—O₂C—(CH₂)₇—CH=CH—(CH₂)₇—CH₃

CH —O₂C—(CH₂)₇—CH=CH—(CH₂)₇—CH₃ + 3H₂ →(catalyst)→

CH₂—O₂C—(CH₂)₇—CH=CH—(CH₂)₇—CH₃

Three *monounsaturated* fatty acid side chains

CH₂—O₂C—(CH₂)₁₆—CH₃

CH —O₂C—(CH₂)₁₆—CH₃

CH₂—O₂C—(CH₂)₁₆—CH₃

Three fully saturated fatty acid side chains

Figure 12–22
Catalytic hydrogenation.

THINKING MORE ABOUT THE MOLECULES OF LIFE

Dietary Fads

Humans have long believed that the body's well-being depends on the foods we eat. This belief is bolstered by the understanding that the cell's basic structures are built from molecules brought in through the digestive system. This understanding, coupled with the current preoccupation with health and fitness in the United States, leads occasionally to fads in which one food or another is touted as a new cure-all. It's hard to get enough information to analyze a fad while it's in full swing, but studies of fads after the fact can teach us a lot about them. The rise and fall of oat bran is a particularly enlightening case.

In the mid-1980s, people began to understand that high levels of cholesterol in the blood were correlated to the incidence of heart disease. Studies available at the time indicated that the inclusion of fiber in the diet, particularly oat bran, helped lower blood cholesterol levels. Oat bran

became a fad food, and for a time it was almost impossible for stores to keep it in stock.

Then, in 1990, newspaper headlines blared that a study in the prestigious *New England Journal of Medicine* had shown that oat bran did not, in fact, lower cholesterol levels. The oat bran industry, running at $54 million a year, collapsed. Processing plants closed and thousands of people lost their jobs.

Was this a reasonable response to the *New England Journal* paper? Let's look at the study that was reported and try to find out. The study took 20 people, all healthy hospital employees with ages ranging from 23 to 49 and low cholesterol levels, and tested them for 6 weeks on two diets: one with high-fiber oat bran, and the other low in fiber. The result? The mean cholesterol levels of the subjects was 172 ± 28 milligrams per deciliter on the low-fiber diet, and 172 ± 25 on oat bran. (Physicians usually start to worry when a person's cholesterol level gets to the neighborhood of 200). This inconclusive result, based on 20 healthy people, provided the basis for the headlines.

Does this study tell you anything about what would happen if someone with high cholesterol went on an oat bran diet? How representative of the entire population are 20 hospital employees in Boston? Given the spread of cholesterol levels in the group, could the actual levels have gone down (or up) without the researchers being able to detect it?

▶ Summary

Chemical bonds form during *chemical reactions*, which may involve the synthesis or decomposition of chemical elements or compounds. Reactions in which atoms combine with oxygen and lose electrons are called *oxidation* reactions, and they include rusting, burning, and our body's respiration. In the opposite reaction, called *reduction*, a material's atoms lose oxygen and gain electrons. Examples include the smelting of many metal ores and plant photosynthesis.

An *acid* is any material that, when put into water, produces positively charged hydrogen ions (i.e., H^+ or protons) in the solution. A *base* is any material that, when put into water, produces negatively charged hydroxide (OH^-) ions. When acids and bases are brought together in the same solution, the H^+ and the OH^- ions come together to form water, and we say that the substances neutralize each other.

All life depends on *polymerization* reactions, in which small molecules link together to form long polymer fibers such as synthetic polyesters, vinyl, cellophane, and other plastics; and natural hair, silk, plant fiber, and skin. *Hydrocarbons*, which are widely used as fuels, are chainlike molecules of carbon and hydrogen atoms. High temperatures and certain chemicals can cause the breakdown of polymers—depolymerization—which is often a key objective in cooking.

Organic molecules share several characteristics: they are carbon based, they are usually formed from only three or four different kinds of atoms, they are generally modular structures, and their chemical function is largely determined by their geometrical shape. The four main types of biological molecules are proteins, carbohydrates, nucleic acids, and lipids.

Proteins form from chains of *amino acids* to make many of the body's physical structures, such as hair and muscle. Proteins in cells also function as *enzymes*, which are molecules that increase the rate of reaction between other molecules, but are themselves unaffected by the reaction.

Carbohydrates are modular molecules built from *sugars*, which are relatively simple molecules built from carbon, oxygen, and hydrogen. Carbohydrates provide an essential source of energy for all animals, and they provide much of the solid structure in the cellulose of plants.

The *nucleic acid* known as *DNA* (*deoxyribonucleic acid*) is a double-stranded polymer consisting of a chain of small molecules called nucleotides. Each nucleotide, in turn, incorporates three smaller molecules—a sugar (deoxyribose), a phosphate group (PO_4), and a base (one of four different molecules, each with a distinctive shape). The entire molecule is twisted into a *double helix* shape.

Lipids, including fats and oils, are molecules that will not dissolve in water. If the carbon atoms form only single bonds, the lipid is said to be saturated; molecules in which adjacent carbon atoms form double bonds are said to be unsaturated. All cell membranes are constructed from bilayers of lipid molecules.

▶ Review Questions

1. What is a chemical reaction? Give three examples.

2. Why do chemical reactions occur?

3. What is an exothermic reaction? Give an example.

4. What is an endothermic reaction? Give an example.

5. What is an oxidation reaction? Give an example.

6. What is a reduction reaction? Give an example.

7. What is a solution? Give an example.

8. What is a precipitation reaction? Give an example.

9. What is a polymerization reaction? Give an example.

10. What is a depolymerization reaction? Give an example.

11. What kind of reaction takes place when blood clots?

12. What is an organic molecule?

13. What is an enzyme? How does it work?

14. What is an amino acid? What groups of atoms are common to all amino acids?

15. How is a protein constructed from amino acids?

16. What is the primary structure of a protein?

17. What chemical role do proteins play in living things?

18. What is a carbohydrate?

19. What is the difference between saturated and unsaturated fats? Give examples of each.

20. What is the difference between cellulose and starch at the molecular level? Give examples of each substance.

21. What is a nucleotide?

22. Name the four bases that occur in DNA.

23. Describe the construction of the double helix structure of DNA.

24. What roles do lipids play in the human body?

25. What is a salt? How can it form?

26. What is combustion and how does it occur?

27. Give three examples of alkanes. Indicate how they are used.

28. What are the secondary, tertiary, and quaternary structures of a protein?

▶ Fill in the Blanks

Complete the following paragraphs with words and phrases from the list.

acids	lipids
amino acids	nucleic acid
bases	oxidation
carbohydrates	pH
chemical reactions	polymerization
DNA	proteins
double helix	reduction
enzymes	sugar
hydrocarbons	

Atoms and molecules can combine in _____. If an atom or molecule loses electrons, it is said to have undergone _____, whereas if it gains electrons, it has undergone _____. Materials that contribute free protons when in solution are called _____, and those that contribute OH^- ions are called _____. The strength of these substances is measured on the _____ scale.

The linking together of simple molecules to form long chains is called _____. _____ are molecules made from carbon and hydrogen. They are often used for fuels.

_____ are small molecules that include an amine group. _____ are made from amino acids. They often serve as _____: that is, they facilitate the reactions of other molecules but are not themselves affected by the reaction. _____, often used as an energy source in the body, are made from carbon, hydrogen, and oxygen atoms in specific proportions. Starches and cellulose are made from _____ molecules.

The genetic information in living things is carried by _____, a molecule which is an example of a _____. It has the shape of a _____. A fourth class of important molecules in living things are _____, which, among other functions, form cell membranes.

▶ Discussion Questions

1. Based on your experience today, list as many everyday chemical reactions as you can. Try to classify these reactions according to their type (i.e., polymerization, oxidation, etc.)

2. The chemistry of the planet Mars is quite similar to Earth except that there is almost no water. Which common chemical reactions that occur on Earth would you not expect to see on Mars?

3. Cooks often tenderize meat by soaking it in a liquid such as lemon juice or vinegar for several hours. What chemical reaction do you think is taking place in the meat? How is this reaction analogous to heating in an oven?

4. Marathon runners are often advised to "carbo load" (i.e., eat a lot of carbohydrates) before a race. Why do you suppose they are given this advice?

5. Given what you know about the structure of protein molecules, why might a single mistake in the sequence of amino acids make a significant difference in the shape of a protein?

6. Explain why rust occurs. What process is taking place?

7. Discuss how reduction comes into play in everyday life.

8. With regard to position and energy of the individual atoms and their electrons, what occurs during a chemical reaction?

9. List the major changes in an atom that occur during chemical bond formation.

10. Outline or diagram all the changes in the total energy of the system of atoms or compounds for an exothermic reaction; for an endothermic reaction. You may use the sodium-chlorine system that is discussed in your text as a guideline.

11. List some obvious signs that a chemical reaction took place for both endothermic and exothermic reactions (e.g., an explosion took place).

12. Classify a reaction in terms of the following changes in the total energy of a system of electrons.

    a. Initial energy of the system is greater than the final energy of the system.

    b. Initial energy of the system is equal to the final energy of the system.

    c. Initial energy of the system is less than the final energy of the system.

13. What are the principal characteristics of oxidation and reduction reactions?

14. How is the natural cycle of the transfer of oxygen and carbon dioxide between plants and animals an example of oxidation and reduction?

15. Use an outline or diagram to classify neutral,

acidic, and alkaline (basic) solutions. Use pure water as a reference.

16. List the procedures for blood clotting that illustrate the principal ideas behind polymerization. How can you explain the problem hemophiliacs have in producing adequate blood clots?

17. Why is it possible to have many isomers for a given hydrocarbon?

18. List the four basic characteristics of organic molecules that form the basis of organic chemistry.

19. Explain the process of forming a polypeptide bond.

20. Why is the geometric shape of a protein important in its function as an enzyme?

▶ Problems

1. For sugars, the general chemical formula is $C_nH_{2n}O_n$, or $C_n(H_2O)_n$. Show that these two chemical formulas are equivalent.

2. Using the number of carbon atoms as an indicator, list the name and number of hydrogen atoms for the simple alkane hydrocarbon sequence (C_nH_{2n+2}). Start at one carbon atom and end at 10.

3. Balance the following equations by filling in the brackets indicated by [   ].

a. $2H_2 + O_2 \longrightarrow [\quad]H_2O$
b. $[\quad]NO + O_2 \longrightarrow 2NO_2$
c. $Fe_2O_3 + [\quad]CO \longrightarrow 2Fe + 3CO_2$
d. $[\quad]KClO_3 \longrightarrow 2KCl + [\quad]O_2$
e. $CH_4 + [\quad]O_2 \longrightarrow [\quad]CO_2 + [\quad]H_2O$
f. $2Al + [\quad]Cl_2 \longrightarrow [\quad]AlCl_3$
g. $[\quad]Al + [\quad]HCl \longrightarrow [\quad]AlCl_3 + 3H_2$

4. Balance the following equations.

a. $Fe_2O_3 + S \longrightarrow Fe + SO_2$
b. $NaNO_3 + H_2SO_4 \longrightarrow Na_2SO_4 + HNO_3$
c. $Mg(OH)_2 + H_3PO_4 \longrightarrow Mg_3(PO_4)_2 + H_2O$

5. List the pH readings for the following concentrations of positively charged hydrogen ions. Indicate if they are acids or bases

a. $10^{-7.5}$ moles per liter
b. $10^{-6.5}$ moles per liter
c. 10^{-15} moles per liter
d. 10^{-2} moles per liter

6. The pH readings for some common household substances are listed. Give the molecular concentration of positively charged hydrogen ions for each substance.

Substance	pH Value
Vinegar	3.0
Blood	7.4
Tomatoes	4.3
Milk of magnesia	10.5
Lemon juice	2.3
Household ammonia	12.0

7. A fuel is rated at 90 octane. What mixture of isooctane and n-heptane has the same rating?.

8. Joel wants a fuel for his power mower that has 25% n-heptane. What octane rating is this fuel? Assume that the fuel is made of n-heptane and isooctane only.

▶ Investigations

1. Investigate why is it that materials in landfills don't break down into their constituent chemical parts.

2. Read *The Double Helix* by James Watson, co-discoverer with Francis Crick of the DNA structure. What data did they use to unravel the structure? What were the key steps in solving the double helix structure? Who helped them in their research?

3. Make a detailed record of one week's intake of food. Consult nutrition charts and determine the percentage by weight of protein, carbohydrate, fat, and other substances that you consumed. What percentage of the fat consumed was saturated? What changes in your diet could reduce the total percentage of fat consumed, and lower the ratio of saturated to unsaturated fat?

▶ Additional Reading

Crick, Francis. *What Mad Pursuit.* New York: Basic Books, 1988.

Fruton, Joseph. *A Skeptical Biochemist.* Cambridge, Mass.: Harvard University Press, 1992.

Snyder, Carl. *The Extraordinary Chemistry of Ordinary Things.* New York: John Wiley and Sons, 1995.

Watson, James. *The Double Helix.* New York: Athaneum, 1985.

13 PROPERTIES OF MATERIALS

A MATERIAL'S PROPERTIES RESULT FROM ITS ATOMS
AND THE ARRANGEMENTS OF CHEMICAL BONDS
HOLDING THOSE ATOMS TOGETHER.

James Trefil and His Volkswagen

A
RANDOM
WALK

When I was a kid, I had to sit around and listen while the grown-ups reminisced about what a great car the Model T had been. It was dependable, I was told, and, most importantly, no matter what went wrong it could be fixed by anyone with a wrench and a pair of pliers. Needless to say, I didn't give much credibility to all this.

Later, as a student, I acquired the first of a long string of Volkswagen Beetles. Now let me tell you, my friends, *that* was a sweet car! There were never any problems with the cooling system (the radiator, hoses, and water pump in your car), for the simple reason that there wasn't one—the engine was cooled by air. And almost any repair could be made by someone with reasonable mechanical ability and a set of tools. I spent many happy hours during graduate school under that car, adjusting this or that.

But I never work on my cars any more. When I look under the hood, all I see is a complex array of computers and microchips—nothing a person can get a wrench around. Yet the car I drive today, provided everything is working, is much more user friendly than my old Volkswagen, and infinitely easier to operate than my parent's Model T. I can start it just by turning a key, for example, rather than getting in front of the car and turning a crank. The flow of gasoline to the cylinders is regulated by a small onboard computer, rather than by a mechanical carburetor, as it was in the VW Beetle.

This story about cars turns out to be a pretty good allegory for the way in which the science of materials has developed in the twentieth century. In the beginning, industry turned out big, relatively simple things that were easy to understand and work with, such as iron wheels for railroads, steel springs for car suspensions, wooden chairs and tables for the home. Today, industry turns out items that perform the same jobs more efficiently, but they are made from new kinds of materials such as plastics, composites, and semiconductors. Instead of manipulating large chunks of material, we now control the way atoms fit together. Like the modern car, modern materials do their job, but when they stop working well, they require more than a simple craftsperson with simple tools to repair them.

So while the materials we use are becoming more efficient and easier to use, it becomes increasingly difficult for us to understand what those materials are. I might have been able to fix my Volkswagen myself, but there is no way I can look under the hood and alter the configuration of atoms in my modern car's microchip. In a sense, the improved performance of modern materials has been bought at the price of our ability to understand them. To a large extent, the emphasis of modern materials science has shifted away from manipulating large blocks of stuff, which are readily available to our senses, to manipulating atoms in ever more complex ways.

MATERIALS AND THE MODERN WORLD

The materials people use, perhaps more than any other single feature of a culture, define a society's technical sophistication. We speak of primitive groups as Stone Age societies, and recognize Iron Age and Bronze Age cultures as progressively more advanced. Given that perspective, how would we define our culture today?

Take a moment and look around your room. How many different kinds of materials do you see? The lights and windows are made of glass, a brittle, transparent material. The walls may be made out of gypsum, a chalklike mineral that has been compressed in a machine and placed between sheets of heavy paper. Your chair is probably made out of metal and plastic, possibly along with wood and woven fabric.

Many of these materials would have been familiar to Americans 200 years ago, when almost everything was made from less than a dozen common substances: wood, stone, pottery, glass, animal skin, natural fibers, and a few basic metals such as iron and copper. But thanks to the discoveries of chemists, the number of everyday materials has increased a thousandfold in the past two centuries. Steel transformed the nineteenth-century world with railroads and skyscrapers, while aluminum provided a lightweight metal for hundreds of applications. The development of rubber, synthetic fibers, and a vast array of other plastics affected every kind of human activity from industry to sports. Brilliant new pigments enlivened art and fashion, while new medications cured many ailments and prolonged lives. And, in our electronic age, the discovery of semiconductor and superconductor materials has changed life in the United States in ways that our eighteenth-century ancestors could not have imagined.

How can scientists devise so many different materials? They succeed, in part, because materials display so many different physical properties: color, smell, hardness, luster, flexibility, heat capacity, solubility in water, texture, melting point, strength—the list goes on and on. Each new material holds the promise of doing some job better or cheaper than any other. We rely on superhard abrasives and fine lubricants, tenacious glues

A typical room is filled with high-tech materials: synthetic fibers, specialized glass, colorful plastics, metal alloys.

and efficient solvents, flexible fibers and rigid plastics—a million products for a million uses.

Based on our understanding of atoms and their chemical bonding (see Chapters 10 and 11), we now realize that the properties of every material depend on three essential features:

1. The kind of atoms that make it up.
2. The way those atoms are arranged.
3. The way the atoms are bonded together.

In this chapter, we will look at different properties of materials and see how they relate to their atomic architecture. We will examine the strength of materials—how well they resist outside forces. We will look at the ability of materials to conduct electricity, and whether or not they are magnetic. And, finally, we will describe perhaps the most important new materials in modern society—the semiconductor and the microchip—to see how the ability to arrange atoms into new materials can lead to dramatic changes in human society.

THE STRENGTHS OF MATERIALS

Have you ever carried a heavy load of groceries in a plastic bag? You can cram a bag full of bottles and cans and lift it by its thin handles without fear of breakage. How can something as light, flexible, and inexpensive as a piece of plastic be so strong?

Strength is the ability of a solid to resist changes in shape. Strength is one of the most immediately obvious material properties, and it bears a direct relationship to chemical bonding. A strong material must be made from strong chemical bonds. By the same token, a weak material, like a defective chain, must have weak links between atoms. While no type of bond or attraction is universally stronger than the other kinds,

strength The ability of a solid to resist changes in shape; directly related to chemical bonding.

The strength of materials is vital to many activities. Bone and muscle provide a weightlifter's strength.

van der Waals forces (see Chapter 11) are generally the weakest on an individual basis. You will probably be able to pull a material with Van der Waals forces apart with your hands. You experience this softness whenever you use baby powder, graphite lubricant, or soap.

By contrast, the materials we normally think of as strong and durable, such as rocks, glass, and ceramics, are usually held together primarily by ionic bonds. Next time you see a building under construction, look at the way beams and girders link diagonally to form a rigid framework. Chemical bonds in strong materials do the same thing. A three-dimensional network of ionic bonds in these materials holds them together like a framework of steel girders.

The strongest materials we know, however, incorporate long chains and clusters of carbon atoms held together by covalent bonds. The extraordinary strength of natural spider webs, synthetic Kevlar (used to make bulletproof vests), diamonds, and plastic shopping bags all stem from the strength of covalent carbon bonds.

The Nature of Strength

Every material is held together by the bonds between its atoms. When an outside force is applied to the material, the atoms shift around, the bonds stretch and compress, and a force is generated inside the material to oppose the force imposed from the outside. The strength of a material is thus related to the size of the force it can generate when it is pushed.

Material strength is not a single property, because there are different ways of placing an object under stress. Scientists and engineers recognize three very different kinds of strength when characterizing materials:

1. A material's ability to withstand crushing, known as *compressive strength*.
2. Its ability to withstand pulling apart, known as *tensile strength*.
3. Its ability to withstand twisting, known as *shear strength*.

Your everyday experience will convince you that these three properties are often quite independent. A loose stack of bricks, for example, can withstanding crushing—you can pile tons of weight on it without having the stack collapse. But the stack of bricks has little resistance against a push from the side. Indeed, it can be toppled by a child. A rope, on the other hand, is extremely strong when pulled, but has little strength under twisting or crushing.

Materials will break when their elastic limit is exceeded.

(a)

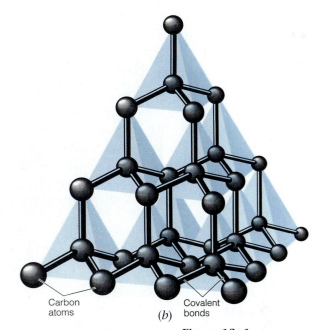

Carbon atoms   (b)   Covalent bonds

Figure 13–1
The girder framework of a skyscraper (a) and the diamond crystal structure (b) are both strong because of numerous strong cross-connections. In a building the connections are steel girders, while in diamond they are carbon-carbon covalent bonds.

The point at which a material stops resisting external forces and begins to deform permanently is called its *elastic limit*. We see examples of this phenomenon every day. When you break an egg, crush an aluminum can, snap a rubber band, or crease a piece of paper, you exceed an elastic limit and permanently change the object. When the materials in your body exceed their elastic limit, the consequences can be catastrophic. Our bones may break if put under too much stress, while arteries under too high pressure may rupture in an aneurysm.

The strengths of a material reflect its arrangements of chemical bonds. Think about how you might design a structure out of Tinkertoys that would be strong, regardless of how it is squeezed, twisted, or pulled apart. The strongest arrangement would have a lot of short sticks with triangular patterns. Diamond is exceptionally strong under all three kinds of stress because of its three-dimensional framework of strong carbon-carbon bonds (see Figure 13–1). Glass, ceramics, and most rocks also feature rigid frameworks of chemical bonds; thus they are also relatively strong. However, many plastics like the one in your shopping bag have strong bonds in only one direction (a common feature of *polymers*, as discussed in Chapter 12) and are thus strong when stretched, but have little strength when twisted or crushed. On the other hand, materials with layered atomic structures, like a stack of paper, are generally strong when squeezed but quite weak under other stresses.

Thus the strength of a material depends on the kind of atoms it contains, the way the atoms are arranged, and the kind of chemical bonds that hold the atoms together.

Composite Materials

If you have ever driven by a building site or a place where a highway was under repair, you almost certainly saw people using a class of materials

In composite materials, such as plywood and reinforced concrete, one material's weakness is compensated by the other's strength.

composite material   A combination of two or more substances in which the strength of one of the constituents is used to offset the weakness of another, resulting in a new material whose strengths are greater than any of its components; for example, plywood and reinforced concrete.

that were made by combining two or more substances into a **composite material.** In these kinds of materials, the strength of one of the constituents is used to offset the weakness of another, resulting in a material whose strengths are greater than any of its components.

One of the most common examples of a composite material is plywood, which is a glued composite of thin wood layers with alternating grain direction. Thin sheets of wood tend to break easily along the grain, making them inadequate for many structural uses. In plywood, however, the weakness of one sheet along the grain is compensated by the strength of the neighboring sheets. Not only is plywood much stronger than a solid board of the same dimension, but it can also be produced from much smaller trees by slicing layers off a rotating log, like removing paper from a roll.

Reinforced concrete provides another common composite material in which steel rods (with great tensile strength) are embedded in a concrete mass (with great compressive strength). A similar strategy is used in fiberglass, formed from a cemented mat of glass fibers, and in new

Safety glass is a composite material in which a thin sheet of plastic is sandwiched between two sheets of glass.

carbon fiber composites that provide extraordinarily strong and light-weight structural materials for industry and sports applications.

The modern automobile features a wide variety of composite materials. Windshields are layered to resist shattering and to reduce sharp edges in a collision. Tires are intricately formed from rubber and steel belting for strength and durability. Car upholstery commonly mingles natural and artificial fibers, and dashboards often employ complexly laminated surfaces. The bodies of many cars are formed from a fiberglass or other lightweight molded composite. And, as we shall see, all of a modern automobile's electronics, from radio to ignition, depend on semiconductor composites of extraordinary complexity.

ELECTRICAL PROPERTIES OF MATERIALS

Of all the properties of materials, none are more critical to our world than those that control the flow of electricity. Look around you and tally up the number of electrical devices nearby. Chances are your list will quickly grow to several dozen. Almost every aspect of our technological civilization depends on electricity, so scientists have devoted a good deal of attention to materials that are useful in electrical systems. (See Chapter 7 for a review of electricity and magnetism.) If the job at hand is to send electrical energy from a power plant to a distant city, for example, we need a material that will carry the electrical energy without much loss. If, on the other hand, the job is to put a covering over a wall switch so that we will not be endangered by electricity when we turn on the light, we want a material that will not conduct electricity at all. In other words, a large number of different kinds of materials contribute to any electrical circuit.

Conductors

Any material capable of carrying electrical current—that is, any material through which electrons can flow freely—is called a **conductor** (see Figure 13–2). Metals are the most common conductors, but there are many others. Tap water, for example, has ions of hydrogen (H^+) and hydroxide (OH^-), as well as dissolved ions of various kinds of minerals in it. These ions carry a charge, and they, too, are free to move if they become part of an electrical circuit. As a result, handling electrical appliances when you are standing on a wet surface or sitting in a bathtub is extremely dangerous. The electricity can be conducted through the water, and then through your body.

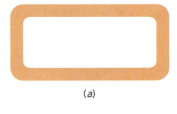
(a)

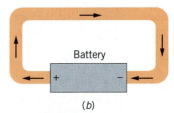
Battery
(b)

Figure 13–2
Electrons in a loop of copper wire (a) don't have any systematic motion. Put a battery in the loop (b) and electrons will flow. This behavior is characteristic of electrical conductors.

conductor   A material capable of carrying an electric current; any material through which electrons can flow freely.

Common electric conductors include metal wire, pots, and pans.

We can find out if a material will conduct electricity by making it part of an electrical circuit and seeing if a current will flow through it. You already know that metals such as copper conduct electricity; in fact, they almost certainly carry the electricity through wires in the walls of the building in which you are now sitting.

The arrangement of a material's electrons determines its ability to conduct electricity. In the case of metals, you will recall, some electrons are bonded fairly loosely and shared by all the atoms. If such a material is made part of an electrical circuit, those electrons are free to move in response. For example, if you connect the poles of a battery across a metal wire, electrons will flow toward the positive pole of the battery and away from the negative pole.

The motion of electrons in electrical circuits is seldom smooth. Under normal circumstances, electrons moving through a metal will collide continuously with the much heavier ions in that metal. In each of those collisions, electrons lose some of the energy they have received from the battery, and that energy is converted to the faster vibration of ions. We perceive these vibrations as heat. As we saw in Chapter 7, when we discussed Ohm's Law, the property by which materials drain the energy away from a current is called *electrical resistance*. Even very good conductors have some electrical resistance. The inverse of electrical resistance is **electrical conductivity,** or the ease with which a material allows electrons to flow. Resistance and conductivity are thus two different ways of describing the same property.

electrical conductivity A number that measures the ease with which a material allows electrons to flow. The inverse of electrical resistance.

A cut away of a cable reveals colorful plastic insulation surrounding metal wires.

Electrical equipment like this plastic electrical outlet and cord is often covered by insulating material to protect the user from shocks.

Insulators

Many materials incorporate chemical bonds in which few electrons are free to move in response to an electric field. In rocks, ceramics, and many organic materials such as wood and hair, for example, the electrons are bound tightly to one or more atoms by ionic or covalent bonds (see Chapter 11). It takes considerable energy to pry electrons loose from those atoms; in fact, the amount of energy would be much greater than the energy supplied by a battery or an electrical outlet. These materials will not conduct electricity (unless subjected to an extremely high voltage, which can pull the electrons loose). If they are made part of an electrical circuit, no electricity will flow through them. We call these materials **insulators.**

The primary use of insulators such as glass or plastic in electrical circuits is to channel the flow of electrons, and to keep people from touching wires that are carrying current. The shields on light switches and household power outlets, as well as the shields and casings for most car batteries, for example, are often made from plastic, a reasonably good insulating material that has the added advantages of low cost and flexibility. Similarly, electrical workers use protective rubber boots and gloves when working on dangerous power lines. In the case of high-power lines, glass or ceramic components are preferred because of their strength and superior insulating properties.

Semiconductors

Many materials in nature are neither good conductors nor perfect insulators. We call such materials **semiconductors.** A semiconductor will carry electricity, but will not carry it very well. Typically, the resistance of silicon, a common semiconductor, is a million times higher than the resistance of a conductor such as copper. Nevertheless, silicon is not an insulator, because some of its electrons do flow in an electric circuit. In Figure 13–3, we sketch a silicon crystal as it appears under normal circumstances to see why it behaves this way.

In this crystal, all the electrons are taken up in the covalent bonds that hold each silicon atom to its neighbor. At absolute zero, in fact, no electrons would be free to move around the material at all. (Recall from Chapter 6 that absolute zero is the lowest possible temperature—the temperature at which no heat energy can be removed from an object.) At normal temperatures, however, the silicon atoms vibrate, and a few of the electrons that go into making the covalent bonds are shaken loose—think of them as picking up a little of the vibrational energy of the atoms. These *conduction electrons* are free to move around the crystal. If the silicon is made part of an electrical circuit, a modest number of conduction electrons are free to move through the solid.

When a conduction electron is shaken loose, it leaves behind a defect in the silicon crystal because of the absence of an electron. This missing electron is called a *hole*. Just as electrons can move around in response to electrical charges, so too can holes (see Figure 13–4).

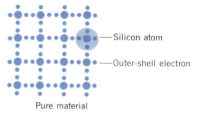

Figure 13–3
A normal silicon crystal displays a regular pattern of silicon atoms. Some of its electrons are shaken loose by atomic vibrations, and these electrons are free to move around and conduct electricity.

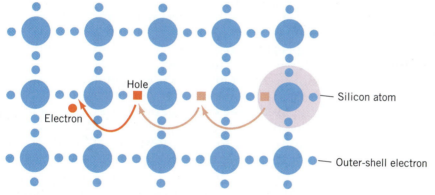

Figure 13–4
A hole in a semiconductor is produced when an electron is missing. Holes can move, just like electrons. As an electron moves to fill a hole, it creates another hole where it used to be.

Figure 13–5
As cars in a traffic jam move slowly forward, "holes" in traffic can be described as moving backward.

The motion of holes in semiconductors is something like what you see in heavy traffic on a crowded expressway. A space will open up between two cars, after which one car moves up to fill the space, then another car moves up to fill that space, and another car moves up to fill that space, and so on (see Figure 13–5). You could describe this sequence of events as the successive motion of a number of cars. But you could just as easily (in fact, from a mathematical point of view, *more* easily) say that the space between cars—the hole—has moved backwards down the line. In the same way, you can either describe the effects of the successive jumping of electrons from one atom to another, or you can talk about the hole moving through the material. Although relatively few semiconducting materials are found in nature, they have played an enormous role in the microelectronics industry, as we shall see later.

Superconductors

superconductivity The property of some materials to exhibit the complete absence of any electrical resistance, usually when cooled to within a few degrees of absolute zero.

Some materials cooled to within a few degrees of absolute zero exhibit a property known as **superconductivity**—the complete absence of any electrical resistance. Below some very cold critical temperature, electrons in these materials are able to move without surrendering any of their energy to the atoms. This phenomenon, discovered in Holland in 1911, was not understood until the 1950s. Today, superconducting technologies provide the basis for a billion-dollar annual industry worldwide. The principal reason for this success is that once a material becomes superconducting and is kept cool, current will flow in it forever. This behavior means that if you take a loop of superconducting wire and hook it up to a battery to get the current flowing, the current will continue to flow even if you close the loop of wire and take the battery away.

In Chapter 7, we learned that current flowing in a loop creates a magnetic field. If we make an electromagnet out of superconducting material and keep it cold, the magnetic field will be maintained at no energy cost except for the refrigeration. Indeed, superconductors provide strong magnetic fields much more cheaply than any conventional copperwire electromagnet because they don't heat up from electrical resistance. Superconducting magnets are used extensively in many applications where very high magnetic fields are essential. Some examples include particle accelerators (see Chapter 15), and magnetic resonance imaging systems for medical diagnosis. Perhaps, eventually, they will be used in everyday transportation.

How is it that a superconducting material can allow electrons to pass through without losing energy? The answer, in at least some cases, has to do with the kind of electron-ion interactions that occur. At very low temperatures, heavy ions in a material don't vibrate very much, and can be thought of as being fixed more or less in one place. As a fast-moving electron passes between two positive ions, the ions will be attracted toward the electron and will start to move slowly toward it. By the time the ions have responded, however, the electron is long gone. Nevertheless, when the ions move close together, they create a region in the material with a more positive electrical charge than normal. This region attracts a second electron and pulls it in. Thus the two electrons can move through the superconducting material something like the way two bike racers move down a track, with one running interference for the other.

At the very low temperatures at which a material becomes superconducting, electrons hook up in pairs, and then the pairs start to interlock like links of a complex, tangled matrix. While individual electrons are very light, the whole collection of interlocked electrons in a superconductor is quite massive. If one electron encounters an ion, it can't easily be deflected. In fact, to change the velocity of any electron—something you would have to do to get energy from it—you would have to change the velocity of *all* the electrons. Since this can't be done, no energy is given up in such collisions, and electrons simply move through the material together. If the temperature is raised, though, the ions will vibrate more vigorously and will no longer be able to form the delicate "minuet" required to produce the electron pairs. Thus, above a critical temperature, superconductivity breaks down.

Science in the Making

High-Temperature Superconductors

Until the mid-1980s, all superconducting materials had to be cooled in liquid helium, an expensive refrigerant that boils at a few degrees above absolute zero, because none of these materials was capable of sustaining superconductivity above about 20 K (about 250°C *below* zero). Acting on a hunch, scientists Karl Alex Müller and George Bednorz of IBM's Zurich, Switzerland, research laboratory began a search for new superconductors. Traditional superconductors are metals and alloys, but Bednorz and Müller decided instead to focus on oxides—chemical compounds that contain oxygen like most rocks and ceramics. It was an odd choice: oxides usually make the best electrical insulators, but occasionally one will conduct electrons.

Working with little encouragement from their peers and no formal authorization from their employers, they spent many months mixing chemicals, baking them in an oven, and testing for superconductivity, but with no success. All they seemed to be doing was wasting their time and IBM's resources, but they continued doggedly.

The breakthrough came on January 27, 1986, when a small black wafer of baked chemicals was found to become superconducting at greater than 30° above absolute zero—a temperature that shattered the old record and ushered in the era of "high-temperature" (though still extremely cold) superconductors. Their compound of copper, oxygen, and other elements seemed to defy all conventional wisdom, and it began a frantic race to study and improve the novel material.

Today, oxide materials closely related to those first described by Bednorz and Müller have been found to superconduct at temperatures

Karl Müller and George Bednorz won the Nobel Prize for their discovery of a new group of superconductors.

(a)

(b)

Figure 13–6
(a) A magnet floats "magically" above a black disk made from the new "high-temperature" superconductor. The clouds in the background form above the cold liquid-nitrogen refrigerant. (b) Japanese engineers have designed this magnetically levitated train that uses superconducting magnets to float above the tracks.

as high as 160° above absolute zero (see Figure 13–6). It may soon be possible to make commercially useful electrical devices out of these materials. Such a prospect would lower the costs of using superconducting magnets, because it is much cheaper to keep materials at 160 K than near absolute zero.

Perhaps equally important, high-temperature superconductors have taken superconductivity from the domain of a few low-temperature laboratories and brought it into classrooms around the world. As a new generation of scientists grows up with this new kind of superconductor, exciting new ideas and inventions are sure to be found. Within the next generation we may have frictionless electric motors that rotate a million revolutions per minute while floating above a superconducting magnet, superconducting electrical storage facilities that reduce our energy bills, and magnetically levitated trains that travel at jet speeds between cities. ●

MAGNETIC PROPERTIES OF MATERIALS

The magnets that lie at the heart of most electrical motors and generators, though critical to almost everything we do, are not usually evident in our daily activities. Similarly, we are usually unaware of the magnets that make our stereo speakers, telephones, and other audio systems work. Even refrigerator magnets and compass needles are so common that we take them for granted. But why do some common materials such as iron display strong magnetism, while other substances seem to be unaffected by magnetic fields?

In Chapter 7, we learned that one of the fundamental laws of nature is that every magnetic field is due, ultimately, to the presence of electric currents. In particular, the movement of electrons around atoms can be thought of as a small electrical current, so each electron in an atom acts like a little electromagnet. Thus an atom with many electrons can be thought of as being composed of many small electromagnets, each with a different strength and pointing in a different direction. The total magnetic field of the atom arises by adding together the magnetic fields of all the tiny electron electromagnets.

It turns out that many atoms have magnetic fields that closely approximate the dipole type (originally shown in Figure 7–5), so that each atom in the material can be thought of as a tiny dipole magnet (Figure 13–7). If this is the case, it is not hard to see how a solid material like a piece of lodestone could have a magnetic field. It is somewhat harder to understand why most materials do *not* have magnetic fields.

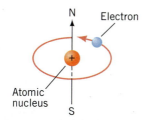

Figure 13–7
An electron in orbit is an electric current, and it produces a dipole magnetic field.

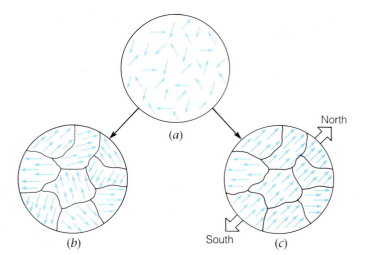

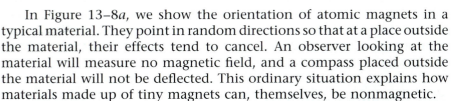

(a)

(b)   South   (c)

Figure 13–8
Different magnetic behavior in materials. (*a*) Nonmagnetic materials have random orientations of atomic spins. (*b*) Ferromagnetic materials with randomly oriented domains are not magnetic. (*c*) A permanent magnet has uniformly oriented atomic spins.

In Figure 13–8*a*, we show the orientation of atomic magnets in a typical material. They point in random directions so that at a place outside the material, their effects tend to cancel. An observer looking at the material will measure no magnetic field, and a compass placed outside the material will not be deflected. This ordinary situation explains how materials made up of tiny magnets can, themselves, be nonmagnetic.

A few materials in nature, including iron, cobalt, and nickel metals, do not show a random scrambling of atomic magnets. This effect, called **ferromagnetism** (from the Latin word for iron, *ferrum*), has to do with the details of the sizes of atoms and the separations between them. In a material such as iron, for example, we often find a situation like the one shown in Figure 13–8*b*. Forces between atoms in these materials tend to favor situations in which neighboring atoms line up with each other into small magnetic *domains*, which are regions a few thousand atoms in width. Thus, in a normal piece of iron, atoms within a specific domain will all be lined up pointing in the same direction, but the orientation of domains is random. Again, someone standing outside this material will not measure a magnetic field, because the many tiny magnetic fields due to different domains will cancel each other. For every domain with

ferromagnetism The property of a few materials in nature, such as iron, cobalt, and nickel metals, in which the individual atomic magnets are arranged in a nonrandom manner, lined up with each other into small magnetic domains to produce an external magnetic field.

a north pole facing one direction, another domain will have a north pole facing more or less in the opposite direction.

In special cases, as when a ferromagnetic material such as iron cools from a very high temperature in the presence of a strong magnetic field, all of the neighboring domains may line up and reinforce each other. Only when most of the magnetic domains line up (as shown in Figure 13–8c) do we find a material that exhibits an external magnet field—the arrangement that occurs in "permanent" (as opposed to electro-) magnets.

The ultimate cause of the magnetic field is the movement of individual electrons around individual atoms. These motions must be aligned in the material in order to form a permanent magnet. This alignment requires that magnetic domains be created within the material, and lined up parallel to each other.

MICROCHIPS AND THE INFORMATION REVOLUTION

Every material has dozens of different physical properties that can be measured in a laboratory. We have already seen how strength, electrical conductivity, and magnetism result from the properties of individual atoms, and how those atoms bond together. We could continue in this vein for many more chapters, examining optical properties, elastic properties, thermal properties, and so on. But such a treatment would overlook a key idea about materials: new materials often lead to new technologies that change society.

Of all the countless new materials discovered in the last century, none has transformed our lives more than silicon-based semiconductors. From personal computers to auto ignitions, portable radios to sophisticated military weaponry, microelectronics are a hallmark of our age. Indeed, semiconductors have fundamentally changed the way we manipulate a society's most precious resource—information.

The key to this revolution is our ability to fashion complex crystals atom-by-atom from silicon, a material that is produced from ordinary beach sand.

Doped Semiconductors

The element silicon, by itself, is not a very useful substance in electrical circuits. What makes silicon useful, and what has driven our modern microelectronic technology, is a process known as doping. **Doping** is the addition of a minor impurity to an element or compound. The idea behind silicon doping is simple. When silicon is melted before being made into circuit elements, a small amount of some other material is added to it. One common additive, for example, is phosphorus, which is an element that has five electrons in its outermost shell, as opposed to the four of silicon.

When the silicon crystallizes to form the structure shown in Figure 13–9, the phosphorus is incorporated into the crystalline structure. Of the five outer electrons in each phosphorus atom, however, only four are needed to make bonds in the crystal. The fifth electron is not locked

doping The addition of a minor impurity to a semiconductor.

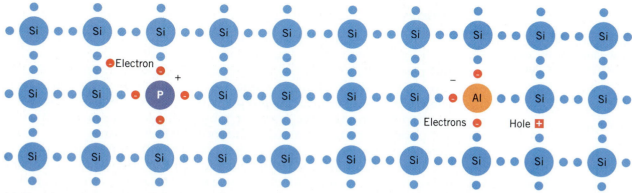

(a) Phosphorus-doped silicon *n*-type semiconductor

(b) Aluminum-doped silicon *p*-type semiconductor

Figure 13–9
(*a*) Phosphorus-doped silicon *n*-type semiconductors and (*b*) aluminum-doped silicon *p*-type semiconductors are formed from silicon crystals with a few impurity atoms.

in at all. In this situation, it does not take long for the extra electron to be shaken loose and wander off into the body of the crystal. This action has two important consequences:

1. Conduction electrons move through the lattice.
2. The phosphorus ions have a positive charge.

A semiconductor prepared in this way is said to be an *n-type semiconductor*, because the moving charge is a negative electron.

Alternatively, silicon can be doped with an element such as aluminum, which has only three electrons in its outer shell. In this case, when the aluminum is taken into the crystal structure, there will be a missing electron, or a hole, which is capable of carrying electricity. The hole need not stay with the aluminum, of course, but is free to move around within the semiconductor as described earlier. Once it does so, the aluminum atom, which has now acquired an extra electron, will have a negative charge. This type of semiconductor is called *p-type*, because a hole, representing a missing negative electron, acts as a positive charge would act.

A technician wears protective clothing to maintain the ultraclean environment required in a semiconductor factory.

Diodes

diode   An electronic device that allows electric current to flow in only one direction.

You can understand the basic workings of a microchip by conducting an experiment in your mind. Imagine taking a piece of *n*-type semiconductor and placing it against a piece of *p*-type semiconductor. As soon as the two types of material are in contact, electrons diffuse from the *n*-type semiconductor over into the *p*-type, while holes diffuse back the other way. Over a period of time, however, most of the electrons in the semiconductor boundary region actually fall into the holes (provided that there were equal numbers of each), so that eventually both free electrons and free holes disappear near the *n-p* contact. On one side of the boundary between the two semiconductors, there would be a region where negative aluminum ions—ions locked into the crystal structure by the doping process—had acquired an extra electron. Conversely, on the other side of the boundary, you would see an array of positive phosphorus ions, each of which had lost an electron, but which were nonetheless locked into the crystal structure.

Figure 13–10
The semiconductor diode consists of a negatively-charged *n* region attached to a positively-charged *p* region. Electrons in this diode can easily flow from the negative to the positive region, but will not flow the opposite way. The result is a one-way valve for electrons.

A semiconducting device like this, which is formed from one *p* and one *n* region, is called a **diode** (see Figure 13–10). Once a diode is constructed, a permanent electrical field tends to push electrons across the boundary in only one direction—from the *n* side to the *p* side. As electrons are pushed "with the grain" in the diode—from negative to positive—the current flows through normally. When the current is reversed, however, the electrons are blocked from going through by the presence of the built-in electric field. Thus the diode acts like a one-way gate, allowing the electrical current to move through in only one direction.

The semiconductor diode has many uses in technology. One use, for example, can be found in almost any electronic device that is plugged into a wall outlet. As we saw in Chapter 7, electricity is sent to homes in the form of alternating current, or AC. It turns out, however, that most home electronics such as televisions and stereos require direct current, or DC. A semiconductor diode can be used to convert that alternating current into direct current by blocking off half of it. In fact, if you examine the insides of almost any electronic gear, the power cord leads directly to a diode and other components that convert AC into DC, as shown in Figure 13–11.

Figure 13–11
A diode converts alternating current to direct current in many electronic devices. Half of the alternating current passes through the diode, but the other half is blocked.

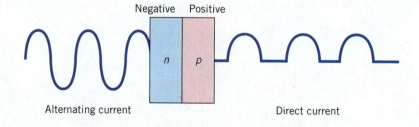

Negative   Positive

Alternating current   Direct current

Environment

The Promise of Solar Energy

Many of the most serious environmental problems we encounter in the world today arise from the extensive use of fossil fuels (see Chapters 5 and 19). The use of coal and petroleum fuels not only adds to problems of air pollution and acid rain, but also inevitably adds carbon dioxide to the atmosphere. Furthermore, the supply of these fuels is limited, and someday we will run out of these resources. Because of these circumstances, a good deal of thought has been given to replacing fossil fuels with solar energy, an energy source that is, for all intents and purposes, inexhaustible. Now that you understand the working of semiconductors, you can understand one such scheme. This is a plan in which an appreciable fraction of the world's electricity would be generated by devices called *photovoltaic cells*.

A photovoltaic cell is nothing more than a large semiconductor diode (see Figure 13–12). A thin layer of *n*-type material overlays a thicker layer of *p*-type material. Sunlight striking the top *n*-type layer shakes electrons loose from the crystal structure. (In other words, solar energy is converted into electrical energy.) These electrons are then accelerated through the *n-p* boundary and pushed out into an electrical circuit to light a light or run a motor. Thus, while the Sun is shining, the photovoltaic cell acts in the same way as a battery. It provides a constant push for electrons and moves them through an external circuit. If large numbers of photovoltaic cells are put together, they can generate enormous amounts of current.

Photovoltaic cells enjoy many uses today. Your hand calculator, for example, may very well have a photovoltaic cell in it that recharges the batteries. Photovoltaic cells are also used in regions where it is hard to bring in traditional electricity, such as water pumping stations in remote sites, or electrical generating facilities in wilderness areas of the national parks.

Solar energy suffers from the serious drawback that the Sun doesn't shine all the time. If solar energy is to be our sole source of energy, some means must be found to store the energy. Alternatively, a region's electrical energy network can be designed so that a certain base

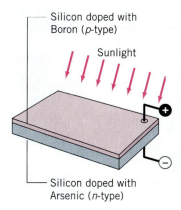

Silicon doped with Boron (*p*-type)

Sunlight

Silicon doped with Arsenic (*n*-type)

Figure 13–12
A solar energy panel is just a semiconducting diode in which sunlight shakes electrons loose and then accelerates them across the junction.

Some pocket calculators can be recharged by the Sun's energy.

The Sun's energy is converted to electricity by photovoltaic panels at a southern California generating plant.

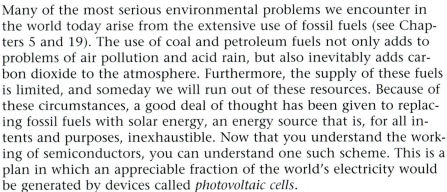

The first transistor, invented in 1947, was a bulky device. Now we can put thousands of these on a chip the size of a postage stamp.

transistor A device that sandwiches *p* and *n* semiconductors in an arrangement that can amplify an electrical current running through it; a device that played an essential role in the development of modern electronics.

level of generation is handled by conventional coal or nuclear plants, with solar plants coming into use during summer days when the demand for air-conditioning reaches its peak. ●

The Transistor

The device that drives the entire information age, and perhaps more than any other has been responsible for the transformation of our modern society, is the transistor. Invented just two days before Christmas, 1947, by Bell Laboratory scientists John Bardeen, Walter Brattain, and William Shockley, the early **transistor** was simply a three-layer sandwich of *n*- and *p*-type semiconductors.

In *pnp* transistors, two *p*-type semiconductors form the "bread" of a sandwich, while the *n*-type semiconductor is the "meat." Another kind of transistor uses the *npn* configuration (see Figure 13–13). Three electrical leads connect to each of the three semiconductor regions of the transistor. An electrical current goes into the region called the *emitter*, the thin slice of semiconductor in the middle is called the *base*, and the third semiconductor region is the *collector*.

Thus the transistor contains two built-in electrical fields, one at each *n-p* junction. The idea behind the transistor is that a small amount of electrical charge running into or out of the base can regulate these electrical fields, in effect opening and closing the gates of the transistor. One way to think of the transistor is to make an analogy to a water pipe. The electrical current that flows from emitter to collector is like water that flows through the pipe, and the base is like a valve in the pipe. A small amount of energy applied to turning the valve can have an enormous effect on the flow of the water. In the same way, a small amount of charge run onto the base can have an enormous effect on the current that runs through the transistor.

In your tape deck, for example, small electrical currents are created when the magnetized tape is run past the tape heads. This small current can be fed into the base of a transistor, and can thus be impressed upon the much larger current that is flowing from the emitter to the collector; the large current will rise and fall in step with the small current. A device that takes a small current and converts it into a large one is called an *amplifier* (see Figure 13–14). The amplifier in your tape deck takes the small signal created by the tape itself and converts it into the much larger current that runs the speakers.

As important as the transistor's amplifying properties are, probably its most important use has been as a switch. If you run enough negative charge onto the base, it can repel any electrons that are trying to get through. Thus moving the electrical charge onto the base will shut off the flow of current through the transistor, while removing the electrical charge will turn it on again. The transistor operates as an electron switch in much the same way as the faucet on your sink acts as a water switch. The fact that a transistor can be operated as a switch means that it can be used to manipulate and store data as discussed shortly in the section titled "Information." Thus the transistor serves as the basic working element of the computer—arguably the most important device developed in the twentieth century.

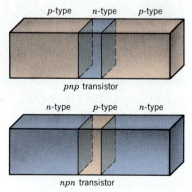

p-type   *n*-type   *p*-type

pnp transistor

n-type   *p*-type   *n*-type

npn transistor

Figure 13–13
A *pnp* transistor and an *npn* transistor.

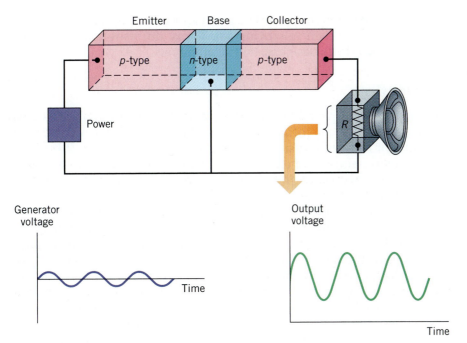

Generator voltage

Time

Output voltage

Time

Figure 13–14
A transistor acting as an amplifier. A small amount of energy, supplied by a power source such as the output from a tape reader can be fed into the base of a transistor. The current is then amplified.

Microchips

Individual diodes and transistors play a vital role in modern electronics, but these simple devices have been largely replaced by much more complex arrays of *p*- and *n*-type semiconductors, called **microchips** (see Figure 13–15). Microchips may incorporate hundreds or thousands of transistors in one *integrated circuit*, specially designed to perform a specific function. An integrated circuit microchip lies at the heart of your pocket calculator or microwave oven control, for example. Similarly, arrays of integrated circuits store and manipulate data in your personal computer, and they regulate the ignition in all modern automobiles.

The first transistors were bulky, about the size of a golf ball, but today a single microchip can integrate hundreds of thousands of these devices. California's Silicon Valley has become a well-known center for the design and manufacture of these tiny integrated circuits.

Producing hundreds of thousands of transistors on a single silicon chip requires exquisite control of atoms. One way to accomplish this is to put a thin wafer of silicon into a large, heated vacuum chamber. Around the edges of the chamber is an array of small ovens, each of which holds a different element, such as aluminum or phosphorus. The side ovens are heated in carefully controlled sequence and opened to allow small amounts of other elements, known as dopants, to be vaporized and enter the chamber along with silicon.

To make a *p*-type semiconductor, for example, you could mix a small amount of aluminum with the silicon in the chamber and let it deposit onto the silicon plate at the bottom. Typically, the silicon chip is partially masked so that the *p*-type semiconductor is deposited only in designated parts of the chip. The vapor is then cleared from the chamber, a new mask is put on, and another layer of material is laid down. In this way,

microchip A complex array of *p*- and *n*-type semiconductors, which may incorporate hundreds or thousands of transistors in one integrated circuit.

Figure 13–15
A microchip incorporates many complex circuits built into a single piece of silicon. For comparison, the eye of an ordinary sewing needle is shown in the background.

a complex three-dimensional structure can be built up at a microscopic scale. In the end, each microchip has many different transistors in it, connected exactly as designed by the engineers.

Information

The single most important use of semiconducting devices is in the storage and manipulation of information. In fact, the modern revolution in information technology, such as the development of arrays of interconnected computers, global telecommunications networks, vast databanks of personal statistics, digital recording, and the credit card, is a direct consequence of materials science.

While it may appear strange to say so, almost all the things we normally consider as conveying information, such as the printed or spoken word, pictures, or music, can be analyzed in terms of their information content and manipulated by the microchips we've just discussed. The term "information," like many words, has a precise meaning when it is used in the sciences. That meaning is somewhat different from colloquial usage. In its scientific context, information is measured in a unit that is called the *binary digit*, or **bit.**

bit Binary digit: a unit of measurement for information equal to "yes-no" or "on-off."

A bit contains the two possible answers to any simple question: yes or no, on or off, up or down. A single transistor being used as a switch, for example, can convey one bit of information, either on or off. Any form of communication, including writing, Morse code, or the genetic code in your DNA molecules (see Chapter 12), contains a certain number of bits of information. As we shall see shortly, the computer is simply a device that stores and manipulates this kind of information.

Let's begin by asking a very basic question. What is the information content of a single letter of the alphabet? The answer is simply the minimum number of questions with "yes or no" answers that are required to identify a letter of the alphabet unambiguously. The following five questions are ones you might ask to specify the letter E.

1. Is it in the first half of the alphabet? (yes)
2. Is it in the first six letters of the alphabet? (yes)
3. Is it in the first three letters of the alphabet? (no)
4. Is it D? (no)
5. Is it E? (yes)

From this simple example, you can see that five questions need to be answered. In other words, you need five bits of information to specify a single letter of the alphabet. (You might be lucky and guess the answer in fewer questions, but five questions are always enough to pinpoint any one of the 26 letters.) We would say, then, that the information content of a single letter of the alphabet is five bits.

Mathematically, five bits of information represent $2 \times 2 \times 2 \times 2 \times 2 = 32$ different things that can be specified. Thirty-two items is not enough, however, to handle all the letters and digits from 0 to 9, as well as capitalization, punctuation marks, and other symbols. Six bits of information are capable of specifying $2 \times 2 \times 2 \times 2 \times 2 \times 2 = 64$ different things, and you could argue that all the things you'd need to specify on a printed page could be included in those 64. Thus the information content of a single ordinary printed symbol is six bits. If you were using switches to store the information on a normal printed page, you would need six of them lined up in a row to specify each symbol.

The average word is six letters long, so the information content of a typical word is

6 letters × 6 bits/letter = 36 bits

The average printed page of a novel contains about 500 words, corresponding to an information content of almost 20,000 bits. A 200–page book in this scheme thus contains about 4 million bits, or 4 megabits, of information (remember, the prefix "mega" stands for one million).

Historically, switches in computers were lumped together in groups of eight. Such a group is capable of storing eight bits of information, or one **byte.** In terms of this unit, a 200–page book would contain 500,000 bytes, or half a megabyte.

byte In a computer, a group of eight switches storing eight bits of information; the basic information unit of most modern computers.

Science by the Numbers

Is a Picture Really Worth a Thousand Words?

Pictures and sounds can be analyzed in terms of information content, just like words. Your television screen, for example, works by splitting the picture into small units called *pixels*. In North America, the picture is divided into 525 segments on the horizontal and vertical axes, giving a total of about 275,000 pixels for one picture on the TV screen. Your eye integrates these dots into a smooth picture. Every color can be thought of as a combination of the three colors (red, green, and blue), and it is usual to specify the intensity of each of these three colors by a number that requires 10 bits of information to be recorded (in practice, this means that the intensity of each color is specified on a scale of about 1 to 1000.) Thus each pixel requires 30 bits to define its color. Thus the total information content of a picture on a TV screen is

275,000 pixels × 30 bits = about 8 million bits.

In other words, it requires about 8 megabits, or 1 megabyte, to specify a single frame of a TV picture. We should also mention that a TV picture typically changes 30 times a second, so that the total flow of information on the TV screen may exceed 200 million bits per second.

It would appear, then, that a picture is not only worth a thousand words, but much more. In fact, if a word contains 36 bits of information, then the picture will be worth:

$$\frac{8,000,000 \text{ bits/picture}}{36 \text{ bits/word}} = 220,000 \text{ words/picture.}$$

The old saying, if anything, underestimates the truth! ●

A TV picture contains a large amount of information.

Developing Your Intuition

The Information Content of a Civilization

Scientists occasionally have to think in terms of very large quantities of information. For example, you can imagine a situation in which we make contact with an extraterrestrial civilization and wish to tell it about the human race. How many bits of information would we have to send to do so?

Although we haven't had much experience communicating with civilizations through space, we have a lot of experience communicating with other civilizations through time. Most of what we know about the ancient Greek and Egyptian civilizations, for example, comes from studies of books and works of art that have survived to the present time. In a similar way, you could argue that everything that we might want to convey to an extraterrestrial could be contained in a large library—one with 10 million volumes.

If a book contains 4 megabits of information, as we argue in the text, then 10 million books would contain 40 trillion bits of information. Throw in a few pictures and films, and you could argue that the content of human civilization could be conveyed in about 50 trillion (5×10^{13}) bits of information.

Modern high-speed data links typically transmit about 10 million (10^7) bits per second. At that rate it would take

$$\frac{5 \times 10^{13} \text{ bits}}{10^7 \text{ bits/s}} = 5 \times 10^6 \text{ s}$$

or a little less than 60 days to transmit the entire library. ●

Computers

computer A machine that stores and manipulates information.

A **computer** is a machine that stores and manipulates information. The information is stored in transistors, with groups of transistors acting as switches that carry information. In principle, if you had a machine that had a few million transistors in it, you could store the text for this entire book. In practice, however, computers do not normally work in this way. They have a *central processing unit (CPU)* in which transistors store and manipulate relatively small amounts of information at any one time. When the information is ready to be stored—for example, when you have finished working on a text in a word processor, or writing a program to perform a calculation—it is removed from the CPU and stored elsewhere. It might, for example, be stored in the form of magnetically oriented particles on a *floppy disc* or a *hard drive*. In these cases, a bit of information is no longer a switch that is on or off, but a bit of magnetic material that has been oriented either "north pole up" or "north pole down."

A typical computer can store hundreds of millions of bits (hundreds of megabytes) in magnetic form, and can hold additional tens of millions of bits of information in its CPU. The ability to store information in this way is extremely important in modern society. As just one example, think about the last time you made an airline reservation. You called in, and your travel agent consulted the airline's computer. Stored within that

In less than a quarter century, the computer has evolved from a specialized research aid to an essential tool for business and education.

computer are the flights, the seating assignments, the ticket arrangements, and often the addresses and phone numbers of every passenger who will be flying on the particular day when you want to fly. This information is all stored in strings of bits in the computer. When you change your reservation, make a new one, or perform some other manipulation, the information is taken out of storage, brought to the central processing unit, manipulated by changing the exact sequence of bits, and then put back into storage. This process—the storage and manipulation of vast amounts of data—forms the very fabric of our modern society.

Technology

The Computer and the Brain

When computers first came into public awareness in the 1960s, there was a general sense that we were building a machine that would, in some way, duplicate the human brain. Concepts such as *artificial intelligence*, or the development of machines that can think for themselves, were sold (some would say oversold) on the basis of the idea that they would soon be able to perform all those functions that we normally think of as being distinctly human. In fact, this has not happened. The reason has to do with the difference between the basic unit of the computer, which is the transistor, and the basic unit of the brain, which is the nerve cell.

The transmission of electrical signals among the brain's cells is fundamentally different than in normal electrical circuits (see Chapter 7). This difference in signal transmission alone, however, does not make a brain so different from a computer. A computer normally performs a sequential series of operations—that is, a group of transistors takes two numbers, adds them together, feeds the answer to another group of transistors that performs another manipulation, and so on. Some computers are now being designed to have some parallel capacity. These machines will perform addition and other operations simultaneously, rather than sequentially. As they stand now, however, the natural configuration for computers is to have each transistor hooked to a few others.

A nerve cell in the brain, however, operates in quite a different manner. A typical nerve cell may have 1000 projections, called *dendrites*. Each dendrite on a nerve cell in the brain is connected to a different neighboring nerve cell. Thus each of the trillions of cells in the brain is connected to a thousand other cells. The behavior of nerve cells depends on a complex integration of all the signals that come into that cell from a thousand other cells.

This complex arrangement means that the brain is a system that is highly interconnected—more interconnected than any other system known in nature. In fact, if the brain has 50 billion cells and each cell has a thousand connections, there will be on the order of 25,000,000,000,000 connections between brain cells. Building a computer of this size and level of connectedness is presently beyond the capability of technology. ●

THINKING MORE ABOUT MATERIALS

Thinking Machines

One of the questions that often intrigues people when they think about complex computers is whether or not a computer can be built that would, in some way, mimic or replace the human brain. Computers today can add faster and remember more than any single human being, but these specific abilities by themselves do not seem to be crucial in developing a machine that thinks. The real question is whether or not a machine will be designed that is, by general consensus, regarded as "alive" or "conscious."

The British mathematician Alan Turing (1912–1954) proposed a test that was designed to address this question. The "Turing test" operates this way: A group of human beings in a room interacts with something through some kind of computer terminal. The individuals might, for example, type questions into a keyboard, and read answers on a screen. Alternatively, they could talk into a microphone and hear answers played back to them in some kind of voice synthesizer. These people are allowed to ask the hidden "thing" any questions they like, and, at the end, they have to decide whether or not they are talking to a machine or to a human being. If they can't tell the difference, the machine is said to have passed the Turing test.

As of this date, no machine has passed this test, although there have been occasional contests in Silicon Valley in which computers have taken part. But what if a machine did actually pass? Would that mean we have invented a truly intelligent machine?

The philosopher John Searle at the University of California at Berkeley has recently challenged the whole idea of the Turing test as a way of telling if a machine can think by proposing a paradox he calls the "Chinese Room." It works like this: An English-speaking person sits in a room and receives typed questions from a Chinese-speaking person in the adjacent room. The English-speaking person does not understand written Chinese characters, but has a large manual of instructions. The manual might say, for example, that if a certain group of Chinese characters is received, then

Data, from *Star Trek, the Next Generation*, is an android. Computers that think like people have been a staple of science fiction stories, but could they really be built?

a second group of Chinese characters should be sent out. The English-speaking person could, at least in principle, pass the Turing test if the instructions were sufficiently detailed and complex. However, the English speaker has no idea of what he or she is doing with the information that comes in or goes out. Thus, argues Searle, the mere fact that a machine passes the Turing test tells you nothing about whether it is aware of what it is doing.

Do you think a machine that can pass the Turing test must be aware of itself? Do you see any way around Searle's argument for the Chinese room? What moral and ethical problems might arise if human beings could indeed make a machine that everyone agreed was aware and conscious?

► Summary

All materials, from building supplies and fabrics to electronic components and food, have properties that arise from the kind of constituent atoms of which they are made, and the ways those atoms are bonded together. The high *strength* of materials such as stone and synthetic fibers depends on interconnected networks of ionic or covalent bonds, whereas many soft and pliable materials such as soap and graphite incorporate weaker van der Waals forces. *Composite materials*, such as plywood, fiberglass, and reinforced concrete, merge the special strengths of two or more materials.

The electrical properties of materials also depend on the kinds of constituent atoms and the bonds they form. For example, electrical resistance—a material's resistance to the flow of an electric current—depends on the mobility of its electrons. *Electrical conductivity* measures the ease with which a material allows electrons to flow. Metals, which are characterized by loosely bonded outer electrons, make excellent *conductors*, while most materials with tightly held electrons in ionic and covalent bonds are good electrical *insulators*. Materials such as silicon that conduct electricity, but not very well, are called *semiconductors*. At very low temperatures, some compounds lose all resistance to electricity and become *superconductors*.

Magnetic properties also arise from the collective behavior of atoms. While most materials are nonmagnetic, *ferromagnets* have domains in which electron spins are aligned parallel to each other.

New materials play important roles in modern technology. Semiconductors, in particular, are vital to the modern electronics industry. Semiconductor material, usually silicon, is modified by *doping* with small amounts of another element. Phosphorous doping, for example, adds a few extra negative electrons to produce an *n-type* semiconductor, while aluminum doping provides positive holes in *p-type* semiconductors. Devices formed by juxtaposing *n-* and *p*-type semiconductors act as switches and valves for electricity. A *diode* joins single pieces of *n-* and *p*-type material, for example, to act as a one-way valve for current flow. *Transistors*, which incorporate a *pnp* or *npn* semiconductor sandwich, act as amplifiers or switches for current. *Microchips* can combine up to thousands of *n* and *p* regions in a single integrated circuit that performs complex operations.

Semiconductor technology has revolutionized the storage and use of information. All kinds of information can be reduced to a series of simple yes-no questions, or *bits*. Eight-bit words, called *bytes*, are the basic information unit of most modern *computers*.

► Review Questions

1. Define three different kinds of strengths. Give examples of materials that are strong in each of these modes.

2. What kinds of bonds are in the strongest materials we know? Why are these bonds so strong?

3. What three atomic factors determine the strength of a material?

4. Diamonds and graphite are both made from carbon atoms. Why are diamonds so much stronger?

5. What is a composite material? Give an example.

6. What is an electrical insulator? Give an example.

7. What is an electrical conductor? Give an example.

8. What is a superconductor? What advantages are gained when they are used to make a magnet?

9. What is a semiconductor? Give an example.

10. Explain how holes can move in a semiconductor.

11. How do electrons behave in the atoms of a ferromagnet? How might a natural ferromagnet form?

12. If the magnetic fields of individual atoms in a material are arranged randomly, is the material magnetic? Why or why not?

13. How can you create a permanent magnet?

14. What is a semiconductor diode?

15. How do diodes convert AC into DC?

16. What is a transistor?

17. What are the base, emitter, and collector of a transistor?

18. What is an integrated circuit? How might one be made?

19. What is a bit of information? A byte of information?

20. What is a computer?

21. How many connections are there between cells in the brain? How does this compare to the connections in a large computer?

22. What is the Turing test?

► Fill in the Blanks

Complete the following paragraph with words and phrases from the list.

bit	electrical conductivity
byte	ferromagnet
composite materials	insulators
computer	microchip
conductors	semiconductors
diode	superconductors
domains	transistor
doping	

In _____, the strengths of one material compensate for the weakness of the other. Materials that allow electrical current to flow are called _____, while those that do not are called _____. _____ measures the ease with which electrons flow through a material. _____ conduct electricity somewhat, but not as well as metals, while _____ exhibit no electrical resistance at all at low temperatures. A _____ is a material in which the magnetic dipoles of neighboring atoms line up over regions called _____, and then these regions themselves line up.

The modern electronic industry is based on the process of _____, in which atoms other than silicon are added to crystals. A _____ is made of two semiconductors, a _____ of three. A _____ is a device with thousands of semiconducting devices built into it.

All information can be represented in _____, the answer to a simple yes-no question. Eight of these units are called a _____. A _____ is a machine that can process information in digital form.

▶ Discussion Questions

1. How do the principles of physics and chemistry both come into play when developing new materials?

2. From the point of view of atomic architecture, how would you design a material that has great strength when pulled apart? Great strength when twisted?

3. How can you explain electrical resistance in terms of the second law of thermodynamics?

4. Why do workers often wear rubber gloves when dealing with electricity?

5. Was the research conducted by Bednorz and Müller on superconductivity an example of "big science" or "little science?"

6. Now that the cold war between the United States and the former Soviet Union is over, the United States government recently announced plans to share its hitherto secret research on strong yet lightweight materials with U.S. automakers. Do you agree with this plan? What benefits and risks do you see?

7. Identify five different materials that were invented in the twentieth century and that you use in your daily life.

8. Identify 10 objects in your home that use semiconductors. What other kinds of materials with special electrical properties are found in all of these objects?

9. How would your life change if you had none of your electrical appliances?

10. Why does a magnet become demagnetized when you repeatedly hit it with a hammer? What other ways can you destroy a permanent magnet? Why?

11. Shortly after the discovery of high-temperature superconductivity, many newspapers and TV shows ran features on how these new materials would change society. In what ways might superconductivity change society? Historically, what other new materials have caused significant changes in human societies?

12. List the three essential features of materials that determine its properties.

13. Classify the three strength categories of materials (weak, moderately strong, and very strong) according to the type of chemical bonds found in them and the relative strengths of these chemical bonds. In this classification, include information on the relative strength of these bonds in one, two, or three dimensions (or directions.)

14. What is the principal idea relating the compressive, tensile, and twisting strength of a material with the chemical bonds of a material?

15. What is the principal idea relating the compressive, tensile, and twisting strengths of a material and the size of the force this material can generate when it is pulled, squeezed, or twisted? Are these three strengths related or independent properties of a material?

16. How is the elastic limit related to the three strengths of a material?

17. What is the main idea behind the development of composite materials?

18. Metals are generally good conductors of electricity. Why is this the case? Can you relate this generality to the type of chemical bonding that occurs in metals? Explain.

19. Rocks, ceramics, wood, and hairs are generally good insulators. Why is this the case? Can you relate this generality to the type of chemical bonding that is present in these materials? Explain.

20. From the point of view of an electron moving among other electrons and atoms in a material, describe your motion in an insulator. In a conductor. In a superconductor. In a semiconductor.

21. N-and p-type semiconductors are determined by a doped element.
   a. What is the origin of the names *n-* and *p-* type?
   b. How is the doped material related to the semiconductor type?

22. How is the CPU related to the functions of a computer?

▶ Problems

1. There is an effort in the world today to convert television into so-called high definition TV (HDTV). In HDTV, the picture is split up into as many as 1100 by 1100 (as opposed to 525 by 525) pixels. What is the information content of a HDTV picture? What is the information content that must be transmitted each second in a HDTV broadcast?

2. Construct a set of yes-no questions to specify any letter of the alphabet or digit from zero to nine.

3. Construct a set of yes-no questions to specify any of the 50 states in the United States.

4. Estimate the total amount of information contained in the printed words in this chapter.

5. Calculate the number of bytes in one word.

6. While visiting a scientific museum, Reneda noticed that one of the first modern personal computers had 64 kilobytes of memory. She wanted to compare that to one of today's modern computers that has 250 megabytes of memory. What is the ratio of memory storage in bytes, and in bits, between the two computers?

7. A typical 3.5-inch floppy disk can store 1.44 megabytes of information in the high-density mode, and 720 kilobytes in the normal mode.

  a. What is the ratio of information storage between the two disk modes?

  b. How many words (1 word = 36 bits) can each disk mode store?

8. Julio writes a 20-page paper for his extra credit history grade. Each page has an average of 26 lines, with 12 words per line.

  a. How many bits of information did Julio generate in this paper?

  b. Can Julio store this paper on a normal-mode 3.5-inch floppy disk?

9. A modem is a device that transfers electronic data from a computer to telephone lines. Two fairly common modems are rated at 2400 baud and 9600 baud (1 baud equals 1 bit per second). How long would it take to send Julio's 20-page paper (see Problem 8) over the telephone lines using these two modems?

10. Janice asked her mother to help her choose a modem for her computer. Being on a limited budget, Janice could only afford a 2400 baud modem. One of Janice's needs would be to transfer financial documents over the telephone lines every night. These documents are equivalent to a 250-page book. How long would it take Janice to transfer all of this data over the telephone lines using the 2400 baud modem? Is this practical for Janice?

11. What is the rate of information transfer in bytes per second for a color television? For a black-and-white television? Assume a 525×525 pixel television.

12. If a typical dendrite in a brain cell has about 1000 projections on it, and every 20 projections is capable of one bit of information, how many bits of information could the typical dendrite store?

▶ Investigations

1. Read about materials available to scientists and engineers at the time of the American Revolution. What modern technologies would have been impossible with those materials? If you traveled back to the 1770s, what new technologies could you introduce using only those materials?

2. What new materials were introduced during the period 1930 to 1945? How were those materials used during World War II, and what impact did they have on the outcome?

3. Imagine that you are a science fiction writer. Concoct a description of a new material with unique (but plausible) properties and describe how that material might change a society.

4. Research the status of magnetically levitated trains, such as the one now operating in Japan. How does it operate? How fast might it go? How soon might such a train operate in North America?

5. Read the book or watch the movie *The Man in the White Suit*. What unique material properties are described, and how is the new technology received by society?

6. Visit a sports equipment store. Learn about the new materials that are used in tennis rackets, football helmets, and sports clothing.

7. Write a short story in which a new material with unique properties plays a central role.

8. What kinds of materials do surgeons use to replace broken hip bones? What are the advantages of these materials?

9. Until recently, plastic surgeons used silicone-filled implants for breast enlargement and other procedures. Intensive research is now underway to understand the effects of silicone on the human body, due to numerous adverse reactions to these implants. Investigate the nature of silicone and this recent research. What is silicone composed of? What are the results of these investigations? Originally, why did scientists think silicone would be a safe substance? What did they neglect to take into account?

▶ Additional Reading

Adams, James L. *Flying Buttresses, Entropy, and O-Rings*. Cambridge, Mass.: Harvard University Press, 1991.

Gordon, J. E. *The New Science of Strong Materials: Or Why You Don't Fall through the Floor*. Princeton: Princeton University Press, 1976.

Gordon, J. E. *Structures, or Why Things Don't Fall Down*. New York: Plenum, 1978.

Hazen, Robert M. *The Breakthrough: The Race for the Superconductor*. New York: Summit Books, 1988.

Hazen, Robert M. *The New Alchemists: Breaking the Barriers of High Pressure*. New York: Times Books, 1994.

Penzias, Arno. *Ideas and Information*. New York: Norton, 1989.

Rossotti, Helen. *Why Things Aren't Grey*. Princeton, N.J.: Princeton University Press, 1987.

14 THE NUCLEUS OF THE ATOM

NUCLEAR ENERGY DEPENDS ON THE CONVERSION
OF MASS.

A Book Full of Energy

Right now you are sitting down reading a heavy book. How much energy are you holding in your hands? You could drop the book, releasing a little bit of gravitational potential energy (enough to hurt your toe, anyway). You could burn the book (we don't encourage this experiment) and generate heat energy by releasing chemical potential energy for a few minutes. But your book holds vastly more energy than that.

If you could tap all the energy stored in this book, you could generate all the electricity used by a very large city in a year. That's a lot of energy—it corresponds to the output of three large power plants operating continuously, burning tons of coal day and night. How can all that energy be stored in such a small container?

The answer: mass is itself a form of energy, as we saw in Chapter 5. We don't usually observe this principle in operation in our daily lives, but inside the nucleus of the atom, where matter is densely packed, the relationship between mass and energy is of the utmost importance.

EMPTY SPACE, EXPLOSIVE ENERGY

Imagine that you are holding a basketball. At the same time, 25 kilometers (about 15 miles) away, a few grains of sand whiz around. All of the vast intervening space—enough to house a fair-sized city—is absolutely empty. No rocks or soil, no grass or trees, not even any air or water fill the void. In some respects, that empty void resembles an atom, though on a much larger scale. The nucleus is the basketball, while the grains of sand represent the electrons (remember that electrons in an atom display characteristics of both particles and waves). The atom, with a diameter 100,000 times that of its nucleus, is almost all empty space. (Refer to Chapter 9 for a review of the structure of an atom.)

The previous chapters explored the properties of atoms in terms of their electrons. Chemical reactions, the way a material handles electricity, and even the shape and strength of objects all depend on the way that electrons in different atoms interact. In terms of our basketball analogy, all of the properties of the atoms that we have studied so far result from things that are taking place 25 kilometers from where the basketball-sized nucleus is sitting. The incredible emptiness of the atom is a key to understanding two important facts about the relation of the atom to its nucleus.

1. *What goes on in the nucleus of an atom has little to do with the atom's chemistry, and vice versa.* In other words, the chemical bonding of an atom's electrons has no effect at all on what happens to the nucleus. In most situations, you can regard the orbiting electrons and the central nucleus as two separate and independent systems. This is why physicists instead of chemists usually study the nucleus.

2. *The energies available in the nucleus are much greater than those available to electrons.* The particles inside the nucleus are tightly bound. It takes a great deal more energy to pull them out than it does to remove an electron from an atom.

The enormous amount of energy we can get from the nucleus follows from the equivalence of mass and energy (which we will discuss in more detail in Chapter 16). This relationship is defined in Einstein's most famous equation:

▶ **In words:**

Mass is a form of energy. When mass is converted into energy, the amount of energy produced is equal to the mass of the object multiplied by the square of the speed of light.

▶ **In equation form:**

Energy = mass × (speed of light)²

▶ **In symbols:**

$E = mc^2$

Remember from Chapter 8 that the constant c, the speed of light, is a very large number (3×10^8 meters per second), and that large number is squared in Einstein's equation to give an even larger number. Thus even very small amounts of mass are equivalent to very large amounts of energy, as shown in the calculation in the upcoming Science by the Numbers.

Einstein's equation tells us that a given amount of mass can be converted into a specific amount of energy in any form, and vice versa. Technically, this statement is true for any process involving energy. When hydrogen and oxygen combine to form water, for example, the mass of the water molecule is infinitesimally less than the sum of the masses of the original atoms. This lost mass has been converted to heat energy and dissipated. Similarly, when an archer draws a bow, the mass of the bow increases by an infinitesimal amount because of the increased elastic potential energy in the bent material.

The change in mass of objects in common events such as chemical reactions or a drawn bow is so small that it is customarily ignored. In nuclear reactions, however, we cannot ignore the mass effects. A nuclear power plant, for example, can transform fully 20% of the mass of a proton into energy in each reaction by a process we will soon discuss. Thus nuclear reactions can convert significant amounts of mass into energy, while chemical reactions, which involve only relatively small changes in electrical potential energy, involve only infinitesimal changes in mass. This difference explains why an atomic bomb, which derives its destructive force from nuclear reactions, is so much more powerful than conventional explosives such as dynamite and conventional weapons, which depend on chemical reactions in TNT and similar substances.

When a bow is drawn, its mass has increased by a tiny amount.

Science by the Numbers

Mass and Energy

On the average, each person in the United States uses energy at the rate of about 1 kilowatt-hour each hour. In effect, this means that each person uses an amount of energy equivalent to a toaster going full blast all the time. How much mass would have to be converted completely to energy to produce your year's supply of energy?

First, calculate the number of hours in a year:

24 hours/day × 365 days/year = 8760 hours/year.

The total energy used by one person annually is approximately 8760 kilowatt-hours.

In Appendix A, we find that 1 kilowatt-hour of energy is the same as 3.6 million joules, so every year each of us uses

$$(8760 \text{ kwh}) \times (3.6 \times 10^6 \text{ J/kwh}) = 8760 \times 3.6 \times 10^6 \text{ J}$$
$$= 3.15 \times 10^{10} \text{ J}$$

In order to calculate the mass that is equivalent to this large amount of energy, we need to use Einstein's equation, which we can rewrite as

$$\text{mass} = \frac{\text{energy}}{(\text{speed of light})^2}$$

Written in this form, the number we seek (the mass) is expressed in terms of two numbers we already know. The speed of light, c is 3×10^8 meter/second, so we find

$$\text{mass} = \frac{(3.15 \times 10^{10} \text{ J})}{(3 \times 10^8 \text{ m/s})^2}$$
$$= \frac{(3.15 \times 10^{10})}{(9 \times 10^{16})} \text{ J}/(\text{m}^2/\text{s}^2)$$
$$= 3.5 \times 10^{-7} \text{ J-s}^2/\text{m}^2$$
$$= 3.5 \times 10^{-7} \text{ kg}$$

In the last step, we have to remember that a joule is defined as a kilogram-meter^2/second^2, so the unit "joule-second^2/meter^2" in the answer above is exactly the same as kilograms (see Appendix A). Our year's energy budget could be satisfied by a mass that weighs less than a millionth of a kilogram, which is equivalent to the mass of a very fine sand grain—if you could unlock that energy from the nucleus! ●

Each grain of sand along a beach contains enough energy, in the form of its mass, to provide all the energy you use in a year.

THE ORGANIZATION OF THE NUCLEUS

Like the atom, the nucleus is made up of smaller pieces. The primary building blocks are the proton and the neutron, which are approximately equal in mass.

The **proton** (from the Greek for "the first one") has a positive electrical charge of +1 and was the first of the nuclear constituents to be discovered and identified. The number of protons determines the electrical charge of the nucleus. An atom in its electrically neutral state will have as many negative electrons in orbit as protons in the nucleus. Thus the number of protons in the nucleus determines the chemical identity of an atom.

proton One of two primary building blocks of the nucleus, with a positive electrical charge of +1 and a mass ($1.6726430 \times 10^{-24}$ g) approximately equal to that of the neutron.

When people began studying nuclei, they quickly found that the mass of a nucleus is significantly greater than the sum of the mass of its protons. In fact, for most atoms, the nucleus was found to be more than twice as heavy as its protons. What did this observation imply? Scientists realized that additional mass must be in the nucleus. We now realize that this extra mass is supplied by a particle with no electric charge called the **neutron** (for "the neutral one"). The neutron has about the same mass as the proton, so a nucleus with equal numbers of protons and neutrons will have twice the mass of the protons alone.

The mass of the proton (or the almost identical mass of the neutron) is about 2000 times the mass of the electron. Thus almost all of the mass of an atom is contained within the protons and neutrons in its nucleus. In other words, the electrons give an atom its size, but the nucleus gives an atom its mass.

neutron A type of subatomic particle, located in the nucleus of the atom, which carries no electrical charge, but has approximately the same mass as the proton; one of two primary building blocks of the nucleus.

Element Names and Atomic Numbers

The most important fact in describing any atom is the number of protons in the nucleus, known as the **atomic number.** This number defines the *element* in the *periodic table* (see Chapter 9). All atoms of gold (atomic number 79) have exactly 79 protons, for example. In fact, the name gold is simply a convenient shorthand for "atoms with 79 protons." Every element has its own atomic number: all hydrogen atoms have just 1 proton, carbon atoms must have 6 protons, iron atoms always have 26, and so on. The periodic table of elements, which we described in Chapter 9, can be thought of as a chart in which the number of protons in the atomic nucleus increases as we read from left to right.

atomic number The number of protons in the nucleus, which determines the nuclear charge and therefore the chemical identity of an atom.

Isotopes and the Mass Number

Two atoms with the same number of protons (i.e., atoms of the same element) may have different numbers of neutrons. Such atoms are said to be different **isotopes** of each other, and they have different masses. (The name isotope comes from the Greek, meaning "same place," because all isotopes of a given element appear in the same place on the periodic table.) The total number of protons and neutrons is called the *mass number*.

isotope Two nuclei are isotopes if they have the same number of protons but a different number of neutrons.

Every known element has several different isotopes, each with a different number of neutrons. All atoms of carbon, for example, have 6 protons, but the number of neutrons varies from atom to atom. The most common isotope of carbon, for example, has 6 neutrons, so it has a mass number of 12 (6 protons + 6 neutrons); it is usually written ^{12}C, carbon-12, or $^{12}_{6}C$, and is called "carbon twelve." (The subscript refers to the atomic number, the superscript to the mass number.) Other isotopes of the carbon nucleus, such as carbon-13 with 7 neutrons and carbon-14 with 8 neutrons, are heavier than carbon-12, but they have the same number of electrons as each other and, therefore, the same chemical identity. A neutral carbon atom that has 8 neutrons in its nucleus, for example, still must have 6 electrons in orbit to balance the required 6 protons, but it has a mass number of 14 (Figure 14–1) and is written as $^{14}_{6}C$.

Figure 14–1
Every chemical element has many isotopes. Here are three isotopes of carbon, which differ from each other only in the number of neutrons in the nucleus.

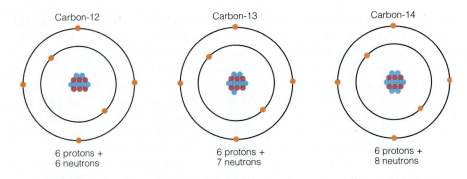

Carbon-12 — 6 protons + 6 neutrons

Carbon-13 — 6 protons + 7 neutrons

Carbon-14 — 6 protons + 8 neutrons

A list of the complete set of all the isotopes, containing every known combination of protons and neutrons, is often illustrated on a graph that plots number of protons versus number of neutrons (see Figure 14–2). Several features are evident from this graph. First, every chemical element has many known isotopes; in some cases, such as iron, there are dozens of them. Close to 2000 isotopes have been documented, compared to the hundred or so different elements. This plot also reveals that the number of protons is not generally the same as the number of neutrons. While many of the light elements, up to around calcium (with 20 protons), have nearly equal numbers of protons and neutrons, heavier elements tend to have more neutrons than protons. This fact plays a key role in the phenomenon of radioactivity, as we shall see.

Figure 14–2
A chart of the isotopes. Stable isotopes appear as black squares, while radioactive isotopes are open squares. Each of the approximately 2000 known isotopes has a different combination of protons (Z on the vertical scale) and neutrons (N on the horizontal scale). Isotopes of the light elements (toward the bottom left of the chart) have similar numbers of protons and neutrons and thus lie close to the diagonal $N = Z$ line at 45 degrees. Heavier isotopes (on the upper right part of the chart) tend to have more neutrons than protons and thus lie well below this line.

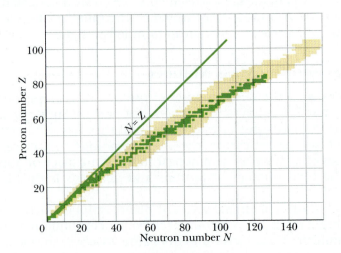

Example 14–1: Inside the Atom

We find an atom with 7 protons and 8 neutrons in its nucleus and 10 electrons in orbit.

a. What element is it?

b. What is its mass number?

c. What is its electrical charge?

d. How is it possible that the numbers of protons and electrons are different?

▶ **Reasoning and Solution:** We can find the first three answers by looking at the periodic table, but we will refer back to Chapter 9 and the discussion of stable electron orbits for the last answer.

a. The element name depends on the number of protons, which is 7. A glance at the periodic table reveals that element number 7 is nitrogen.

b. Next, we calculate the mass number, which is the sum of protons and neutrons: 7 + 8 = 15. This isotope is nitrogen-15.

c. The electrical charge equals the number of protons (positive charges) minus the number of electrons (negative charges): 7 − 10 = −3. The ion is thus N^{-3}.

d. The number of positive charges (7 protons) differs from the number of negative charges (10 electrons) because this atom is ionized. Atoms with 10 total electrons (8 in their outermost shell) are particularly stable (see Chapter 9), so nitrogen commonly occurs in nature as an ion with −3 charge. ▲

Example 14–2: A Heavy Element

How many protons, neutrons, and electrons are contained in the atom ^{238}U when it has a charge of +4?

▶ **Reasoning:** Once again we can look at the periodic table for the first two answers, but we will have to do a simple calculation for the last answer. Remember, the number of protons is the same as the atomic number; the number of neutrons is the mass minus the number of protons; and we compare the number of protons and the +4 charge to determine the number of electrons.

▶ **Solution:** From the periodic table, the element U, uranium, is element number 92, so it has 92 protons.

The number of neutrons is simply the mass number, 238, minus the number of protons: 238 − 92 = 146 neutrons.

The number of electrons is equal to the number of protons minus the charge on the ion, which in this case is +4. Thus there are 92 − 4 = 88 electrons in orbit in this case. ▲

Uranium, a source of nuclear power, is one of the densest elements known.

The Strong Force

In Chapter 7 we learned that one of the fundamental laws of electricity is that like charges repel each other. If you think about the structure of the nucleus for a moment, you will realize that the nucleus is made up of a large number of positively charged objects (protons) in close proximity to each other. Why doesn't the electrical repulsion between the protons push them apart and disrupt the nucleus completely?

The only way the nucleus can be stable is for there to be an attractive force capable of balancing or overcoming the electrical repulsion at the incredibly small scale of the nucleus. You can get some idea of the magnitude of this unknown force by thinking about the strength of the electrical repulsion between protons that it must overcome. If two protons in a nucleus were scaled up to the size of basketballs, and if their charges were raised proportionately, then even if you encased those two basketball-sized protons in a room-sized block of solid steel they would be blasted apart almost instantaneously. The electrostatic repulsion would be so strong that it would quite literally tear the steel apart as the protons moved away from each other.

Physicists in the twentieth century have made a major effort to understand the nature of the force that holds the nucleus together. Whatever this force is, it must be vastly stronger than gravity or electromagnetism, the only two forces we've discussed up to this point. For these reasons, it is called the **strong force.** The strong force must operate only over very short distances, which are characteristic of the size of the nucleus, because our everyday experience tells us that the strong force doesn't act on large objects. Both with respect to its magnitude and its range, then, the strong force is somehow confined to the nucleus. In this respect, the strong force is unlike both electricity and magnetism.

The strong force has another distinctive feature. If you weigh a dozen apples and a dozen oranges, their total weight is simply the sum of the individual pieces of fruit. But this is not true of protons and neutrons in the nucleus. The mass of the nucleus is always slightly less than the sum of the masses of the protons and neutrons. One way of thinking about the generation of the strong force that holds the nucleus together is to imagine that, when protons and neutrons come together, some of their mass is converted into the energy that binds them together. We know this must be true, because it requires energy to pull most nuclei apart. This so-called binding energy varies from one nucleus to another. The iron nucleus is the most strongly bound of all the nuclei. This fact will become important in Chapter 21 when we discuss the death of stars.

RADIOACTIVITY

The vast majority of atomic nuclei in objects around you—more than 99.99% of the atoms in our everyday surroundings—are stable. In all probability, the nuclei in those atoms will never change to the end of history. But some kinds of atomic nuclei are not stable. Uranium-238, the most common isotope of the rather common element uranium, for example, has 92 protons and 146 neutrons in its nucleus. If you were to put a block of uranium-238 on a table in front of you and watch it for a while, you would find that a few of the uranium nuclei in that block would disintegrate spontaneously. One moment all the uranium atoms in the block will be normal, and the next moment a few of those uranium atoms will disintegrate into smaller atoms and no uranium. At the same time, fast-moving particles will speed away from the uranium block into the surrounding environment.

This process of spontaneous change, and the associated emission of energetic particles, is called **radioactivity,** or **radioactive decay.** The emitted particles themselves are referred to as *radiation*. The term radiation used in this sense is somewhat different from the electromagnetic radiation that we introduced in Chapter 8. In this case, radiation refers to whatever emerges from the spontaneous decay of nuclei, either electromagnetic waves or actual particles with mass.

strong force The force responsible for holding the nucleus together; one of the four fundamental forces in nature. This force operates between particles in the nucleus over extremely short distances and also between quarks (see Chapter 15) to hold elementary particles together.

radioactivity The spontaneous release of energy by certain atoms, such as uranium, as these atoms disintegrate. The emission of one or more kinds of radiation from an isotope with unstable nuclei.

radioactive decay The process of spontaneous change of unstable isotopes.

A safety officer in protective clothing uses a Geiger counter to examine waste for radioactivity.

What's Radioactive?

Although almost all of the atoms around you are stable, a small fraction of these atoms are radioactive isotopes. Carbon, for example, is stable in its most common isotopes of carbon-12 and carbon-13. However, carbon-14, which constitutes about a millionth of the carbon atoms in the world, is radioactive. A few elements such as uranium, radium, and thorium have no stable isotopes at all. Even though most of our surroundings are composed of stable isotopes, a quick glance at the chart in Figure 14–2 reveals that most of the 2000 or so known isotopes (most of which are produced for brief periods in a laboratory) are unstable and undergo radioactive decay of one kind or another.

Science in the Making

Becquerel and Curie

The nature of radioactivity was discovered, more or less by accident, by Antoine Henri Becquerel (1852–1908) in 1896. Becquerel had obtained samples of radioactive minerals—natural compounds that incorporate significant amounts of uranium and other radioactive elements. By chance, he put some of these samples in a drawer of his desk along with an unexposed photographic plate and a metal key. When he developed the photographic plate some time later, the silhouette image of the key was clearly visible on it. From this photograph, he concluded that some as yet unknown form of radiation had traveled from the mineral sample to the plate. The key seemed to have absorbed the radiation and blocked it off, but the radiation that silhouetted the key and got through to the film delivered enough energy to the plate to cause the chemical reactions that normally go into photographic development. Becquerel knew that whatever had exposed the plate must have originated in the minerals, and must have traveled at least as far as the plate.

Becquerel's discovery was followed by an extraordinarily exciting time for chemists, who began an intensive effort to isolate and study the elements from which the radiation originated. The leader in the field we now call radiochemistry was also one of the best known women scientists of the modern era, Marie Sklodowska Curie (1867–1934). Born in Poland and married to the distinguished French scientist, Pierre Curie, she conducted her pioneering research in France, often under extremely difficult conditions because of the unwillingness of her colleagues to accept her. She worked with exotic uranium-bearing minerals from mines in Bohemia, and isolated minute quantities of previously unknown elements such as radium and polonium from tons of ore. One of her crowning achievements was the isolation of 22 milligrams of pure radium chloride, which became an international standard for measuring radiation levels. She also pioneered the use of X-rays for medical diagnosis during World War I. For her work, she became the first scientist to be awarded two Nobel Prizes, one in physics and one in chemistry. She also was one of the first scientists to die from prolonged exposure to radiation, whose harmful effects were not

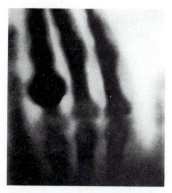

The first X-ray photograph of human bones was taken by Henri Becquerel, who exposed his wife's hand with her wedding ring.

The Curie family with Marie Sklodowska, Pierre, and their child Irene. Both parents received the Nobel Prize in Chemistry in 1911 for isolating radium and polonium. Irene received the 1935 Nobel Prize with her husband Frederic Joliot-Curie.

Pitchblende and yellow cake, which are about 75% uranium oxide, are principal ores for nuclear fuel.

alpha decay The loss by the nucleus of a large and massive particle composed of two protons and two neutrons.

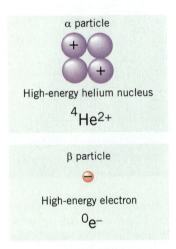

α particle

High-energy helium nucleus

$^{4}He^{2+}$

β particle

High-energy electron

$^{0}e^{-}$

γ radiation

High-energy electromagnetic radiation

Figure 14–3
The components of alpha, beta, and gamma radiation.

Figure 14–4
The Rutherford experiment led to the identification of the alpha particle, which is the same as a helium nucleus.

known at that time. Her fate, unfortunately, was shared by many of the pioneers in nuclear physics. ●

The Kinds of Radioactive Decay

Physicists who studied radioactive rocks and minerals discovered three different kinds of radioactive decay, each of which changes the nucleus in its own characteristic way. Each type plays an important role in modern science and technology. These three kinds of radioactivity were dubbed alpha, beta, and gamma radiation to emphasize that they were unknown and mysterious when first discovered. We retain the names today even though we now know what they are (Figure 14–3).

1. Alpha Decay. Some radioactive decays involve the emission of a relatively large and massive particle composed of two protons and two neutrons. Such a particle is exactly the same as the nucleus of a helium-4 atom. It is called an alpha particle, and the process by which it is emitted is called **alpha decay.** (An alpha particle is often represented in equations and diagrams by the Greek letter α.)

The nature of alpha decay was discovered by Ernest Rutherford (who is also credited with discovering the nucleus of the atom) in the first decade of the twentieth century. His simple and clever experiment, sketched in Figure 14–4, involved placing a small amount of radioactive material known to emit alpha particles in a sealed tube. After several months, careful chemical analysis revealed the presence of a small amount of helium in the tube, even though helium hadn't been present when the tube was sealed. Based on this observation, Rutherford concluded that alpha particles must be associated with the helium atom. Today we would say that Rutherford observed the emission of the helium nucleus in radioactive decay, followed by the acquisition of two electrons to form an atom of helium gas.

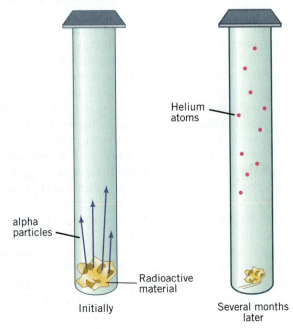

alpha particles

Radioactive material

Helium atoms

Initially

Several months later

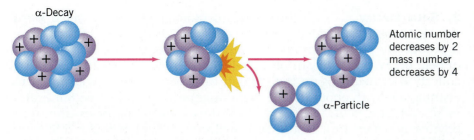

α-Decay

Atomic number
decreases by 2
mass number
decreases by 4

α-Particle

Figure 14–5
In alpha emission, a nucleus emits an alpha particle: two protons and two neutrons, the equivalent of a helium nucleus. The result is an entirely different chemical element.

Rutherford received the Nobel Prize in chemistry for his chemical studies and his work in sorting out radioactivity. He is one of the few people in the world who made his most important contributions to science—in this case the discovery of the nucleus—*after* he received the Nobel Prize.

When the nucleus emits an alpha particle, it loses two protons and two neutrons in the process. The resulting nucleus is referred to as the *daughter nucleus*, which will have two fewer protons than the original, called the *parent nucleus* (Figure 14–5). If the original nucleus is uranium-238 with 92 protons, for example, the daughter nucleus will have only 90 protons—a completely different chemical element called thorium. The mass number of the new atom will be 234, so alpha decay causes uranium-238 to transform to thorium-234. The thorium nucleus with 90 protons can accommodate only 90 electrons in its neutral state. This change means that soon after the decay, two of the original complement of electrons will wander away, leaving the daughter nucleus with its allotment of 90. The result of the alpha decay process, therefore, reduces the mass and changes the chemical identity of the decaying nucleus.

Radioactivity is nature's "philosopher's stone." The philosopher's stone, according to medieval alchemists, was supposed to turn lead into gold. The alchemists never found their stone because almost all of their work involved what we today would call chemical reactions; that is, they were trying to change one element into another by manipulating electrons. Given what we now know about atomic structure, we realize that they were approaching the problem from the wrong end. We know that to change one chemical element into another, you have to manipulate the nucleus. This is precisely what happens in the process of radioactivity.

When the alpha particle leaves the parent nucleus, it typically travels at very high speed (often at an appreciable fraction of the speed of light). This means it has a lot of kinetic energy. This energy, like all nuclear energy, comes from the conversion of mass: the mass of the daughter nucleus and the alpha particle, added together, is somewhat less than the mass of the parent uranium nucleus. If the alpha particle is emitted by an atom that is part of a solid body, then it will undergo a series of collisions as it moves from the parent nucleus into the wider world. In each collision, it will share some of its kinetic energy with other atoms. The net effect of the decay as far as the surrounding materials is concerned is that the kinetic energy of the alpha particle is eventually converted into heat, and the material warms up. Much of the Earth's interior heat comes from exactly this kind of energy transfer. As we shall see in Chapter 18, this heat is responsible for most of the major surface features of the Earth.

beta decay A kind of radioactive decay in which a neutron spontaneously transforms into a collection of particles that includes an electron and a proton.

2. Beta Decay. The second kind of radioactive decay, called **beta decay,** involves the emission of an electron. (Beta decay and the electron it produces are often denoted by the Greek letter β.) The simplest kind of beta decay that one can observe is for a single neutron. If you put a collection of neutrons on the table in front of you, they would start to disintegrate, with about half of them disappearing in the first 8 minutes or so. When a neutron decays, it produces several particles, the most obvious of which are a proton and an electron—both particles that carry an electrical charge and are therefore very easy to detect. This production of one positive and one negative particle from a neutral one does not change (i.e., it conserves) the total electrical charge of the system.

In the 1930s, when beta decay of the neutron was first seen in the laboratory, the experimental equipment available at the time easily detected and measured the energies of the electron and proton. Scientists looking carefully at beta decay were troubled to find that the process appeared to violate the law of conservation of energy, as well as some other important conservation laws in physics. When they added up the mass and kinetic energies of the electron and proton after the decay, they amounted to less than the mass tied up in the energy of the original neutron. They realized that if only one electron and one proton were given off, the conservation law of energy would be violated.

Rather than face this possibility, physicists at the time followed the lead of Wolfgang Pauli (see Chapter 9) and postulated that another, undetected particle had to be emitted in the decay—a particle that they could not detect at the time, but that carried away the missing energy and other properties. It wasn't until 1956 that physicists were able to detect this missing particle in the laboratory; it was referred to as the *neutrino* or "little neutral one." This particle has no electric charge, travels at the speed of light, and, if stopped, would have no mass. Today, at giant particle accelerators (see Chapter 15) neutrinos are routinely produced and used to probe matter in other experiments.

When beta decay takes place inside a nucleus, one of the neutrons in the nucleus is converted into a proton, an electron, and a neutrino. The lightweight electron and the neutrino speed out of that nucleus, while the proton remains. The electron that comes off in beta decay is *not* one of the electrons that was originally circling the nucleus in a Bohr orbit. The electrons emitted from the nucleus come out so fast that they are long gone from the atom before any of the electrons in orbit have time to react. The new atom has a net positive charge from the added proton, however, and eventually may acquire a stray electron from the environment.

The net effect of a beta decay is that the daughter nucleus has approximately the same mass as the parent (it has the same total number of protons and neutrons), but has one more proton and one less neutron. It is therefore a different element than it was before. Carbon-14, for example, undergoes beta decay to become an atom of nitrogen-14 (Figure 14–6). If you place a small pile of carbon-14 powder—it would look like black soot—in a sealed jar and come back in 10,000 years, most of the powder will have disappeared and the jar will be filled with colorless, odorless nitrogen gas. Beta decay, therefore, is a type of radioactivity in which the chemical identity of the atom is changed, but its mass is virtually the same before and after. (Remember, the electron and neutrino

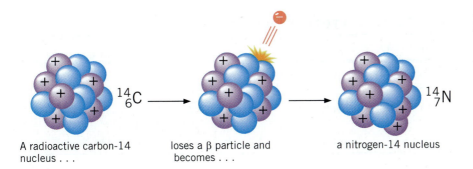

A radioactive carbon-14 nucleus . . .

loses a β particle and becomes . . .

a nitrogen-14 nucleus

Figure 14–6
When a radioactive carbon-14 nucleus emits a beta particle (a fast electron), it becomes a nitrogen-14 nucleus.

that are emitted are extremely lightweight and make almost no difference in the atom's total mass.)

What force in nature could cause an uncharged particle like the neutron to fly apart? The force is certainly not gravitational attraction between masses. It is not the electromagnetic force that causes oppositely charged particles to fly away from each other. And beta decay seems to be quite different from the strong force that holds protons together in the nucleus. In fact, beta decay is an example of the operation of the fourth fundamental force in nature, a nuclear interaction that is called the *weak force*. This force will be discussed in more detail in Chapter 15.

3. Gamma Radiation. The third kind of radioactivity, called **gamma radiation,** is different from alpha and beta decay. (Gamma decay and gamma radiation are often denoted by the Greek letter γ.) A "gamma ray" is simply a generic term for a very energetic photon, which is one unit of electromagnetic radiation (see Chapter 8). In Chapter 8, we saw that all electromagnetic radiation comes from the acceleration of charged particles: that is what happens in gamma radiation.

When an electron in an atom shifts from a higher energy level to a lower one, we know that a photon will be emitted, typically in the range of visible light. In just the same way, the particles in a nucleus can also shift between different energy levels. These shifts, or nuclear quantum leaps, involve energy differences that are thousands or millions of times greater than those of orbiting electrons. When protons in a nucleus undergo shifts from higher to lower energy levels, some of the emitted gamma radiation is in the range of the X-rays, while others are even more energetic.

Gamma rays are emitted from a nucleus any time the protons and neutrons inside reshuffle to a lower energy state. Neither the protons nor the neutrons change their identity as a result, so the daughter atom has the same mass, the same isotope number, and the same chemical identity as the parent. Nevertheless, highly energetic radiation is produced by this process.

gamma radiation A kind of radioactivity involving the emission of energetic electromagnetic radiation from the nucleus of an atom, with no change in the number of protons or neutrons in the atom.

Moving Down the Chart of the Isotopes

The three kinds of radioactivity, summarized in Table 14–1, affect the nuclei of isotopes. Gamma radiation changes the energy of the nucleus without altering the number of protons and neutrons, which means that the isotope doesn't change. Alpha and beta radiation, however, result in a new element by changing the numbers of protons and neutrons.

Table 14–1 • **Types of Radioactive Decay**

Type of Decay	Particle Emitted	Net Change
alpha	alpha particle	new element with two less protons, two less neutrons
beta	electron	new element with one more proton, one less neutron
gamma	photon	same element, less energy

Refer back to Figure 14–2 and look at how alpha and beta radiation are represented on the chart of the isotopes. Alpha decay removes two protons and two neutrons, so we shift diagonally down and to the left two spaces on the chart. Alpha decay thus provides a way to move from heavier to lighter elements.

Beta decay, on the other hand, results in one less neutron and one more proton. On the chart of the isotopes we move diagonally up to the left by one space. Beta decay provides a way to reduce the ratio of neutrons to protons, which is generally greater for heavier elements. Combinations of beta and alpha decays, therefore, provide a means to move from heavier radioactive isotopes to lighter stable isotopes.

Radiation and Health

Why is nuclear radiation so dangerous? One reason is that alpha, beta, and gamma radiation all carry a great deal of energy; in fact, they carry enough energy to damage the molecules that are essential to the workings of your cells. The most common type of damage is *ionization*, the stripping away of one or more of an atom's electrons. An ionized atom cannot bond normally, and any structures that depend on that atom—a cell wall or a piece of genetic material, for example—will be damaged. Prolonged exposure to ionizing radiation can so disrupt an organism's cells that it will die.

It takes a great deal of radiation to cause sickness or death. Only in unusual circumstances, such as the aftermath of the nuclear weapons used on the Japanese cities of Hiroshima and Nagasaki at the end of World War II, or the nuclear reactor accident at Chernobyl in the Ukraine in 1986, do people die shortly after exposure. More insidious are the long-term effects of exposure to low levels of radiation. While such exposure will not cause death in the short term, it may affect a body's ability to replace old cells and can increase the risk of cancer. Radiation may also damage the genetic material that carries the coded information required to reproduce. Birth defects are tragic consequences of nuclear radiation that may not appear until decades after exposure to radiation.

No one can escape high-energy radiation entirely. Radioactive decay of uranium and other elements in rocks and soils, as well as cosmic radiation from space, constantly bombard us. Radioactive atoms make up a small fraction of the material in our bones and tissues. While certain kinds of radiation, such as medical X-rays and radioactive isotopic tracers for diagnosis, have saved countless lives, it is important to minimize the risk of unnecessary exposure to radiation sources.

Students on spring break expose themselves to the harmful effects of ionizing radiation while sunbathing.

Environment

The Radioactive Environment

Human beings cannot sense radiation and it has only been in the last century that we have become aware of its existence. Thus it is easy for us to assume that the presence of radioactivity in our environment represents a new and unprecedented threat to our health. In fact, the human race (and every other living thing on our planet) evolved in a radioactive environment, whether we were aware of it or not.

Natural radioactivity comes from many sources. One is the presence of natural radioactive elements in the soil and rocks that surround us. Every ton of rock on the Earth's surface, for example, contains an average of about 2 grams of uranium. Furthermore, cosmic rays from the Sun and other stars are constantly raining down on the Earth. In addition to producing the same effects as ordinary radiation when they enter our body, they constantly produce showers of nuclear fragments, a large proportion of which are radioactive. Finally, a certain fraction of the chemical elements we incorporate into our bodies from the environment are radioactive, so that the human body is itself a source of radiation. In fact, it's not an exaggeration to say that *everything* is radioactive.

The modern unit for measuring radiation dose is called the *Seivert* (abbreviated Sv). The Sv is defined to be the amount of radiation that has the same biological effect on one kilogram of body tissue as the absorption of one joule of energy from X-rays or gamma rays. Normal radiation doses are measured in *milliseiverts*, (mSv), or 0.001 Sv. For reference, a standard chest X-ray delivers about 0.15 mSv of radiation.

In terms of this unit, every year the average person in the United States receives about 0.3 mSv from cosmic rays, 0.3 mSv from rocks and soil, 0.4 mSv from his or her own body, and 0.6 mSv from medical X-rays. We all receive an additional dose of radiation from indoor radon (see the next Environment section); the magnitude of this exposure is a subject of some dispute, but some scientists think it may be comparable to the above sources.

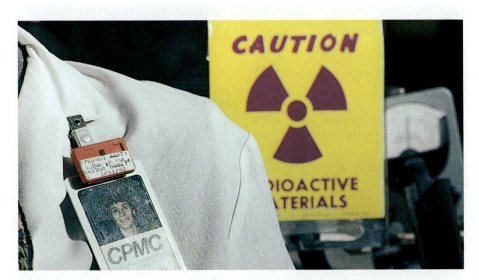

Badges like this are used to monitor radiation exposure. The radiation exposes special photographic film in the badge.

The fact that we live in a radioactive environment does not necessarily mean that we are in danger. Increased exposure to small amounts of radioactivity does not lead automatically to increased health risks. People in Denver, for example, receive twice the exposure from cosmic rays as people living at sea level because there is less atmosphere to shield them, yet cancer rates in Denver are no higher than elsewhere. Exposure to radiation is simply one of many risk factors in our lives, and compared to a hazard such as smoking cigarettes, it is a relatively minor one. ●

Developing Your Intuition

A Radiation-Free Environment

Suppose that you wanted to eliminate radiation completely from a particular area. Could you do it? While this may seem a strange question, many physics experiments are so sensitive that natural radiation would completely mask the effects scientists are trying to measure. So how would you design such an environment?

For starters, you'd have to get away from cosmic rays streaming down from space. This means that your environment would have to be deep underground, so that the overlying rocks would absorb the cosmic rays and their products. You would have to choose your site so that the amount of radioactive material in the surrounding rocks was low (although you could never make it zero). Then you would have to choose your construction materials carefully so that they, too, contained as small an amount of radioactive material as possible—a requirement not as easy as it sounds. Ordinary stainless steel, for example, contains measurable amounts of radioactive cobalt, and even concrete (like the rest of the Earth's soil) can contain products of the uranium decay chain. And, finally, the experiment would have to be shielded by lead from its surroundings, because even scientists themselves contain significant amounts of radioactive carbon-14.

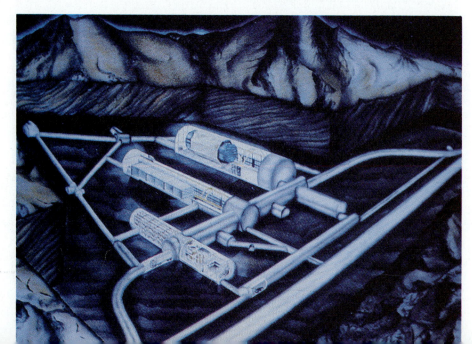

To shield their apparatus from background radiation, physicists often conduct experiments deep underground. This painting illustrates the Gran Sasso Laboratory in Italy, the largest underground laboratory in the world.

While a radioactive-free environment may not be achievable, a number of mines and deep tunnels around the world have been converted into physics laboratories with low background radiation for precisely the reasons outlined above. ●

Long-term storage of radioactive waste may be accomplished in deep underground tunnels, such as this one at the Hanford Nuclear Reservation in Washington.

Nuclear Tracers

Radioactive materials are useful in medicine and industry because they combine two key properties. First, these atoms emit radiation, so they can be followed and identified as they move from place to place. Second, radioactive isotopes are also chemical elements, so they take part in chemical reactions like ordinary atoms (recall that the chemistry of atoms is governed by their electrons, which operate independently of the nucleus). In other words, an isotope of a particular chemical that has a radioactive nucleus will undergo the same chemical reactions as the stable isotopes of the same element. If a radioactive isotope of iodine or phosphorus is injected into your bloodstream, for example, it will collect at the same places in your body as stable iodine or phosphorus, and physicians can thus monitor how your body is working.

Medical scientists can use this fact to study the functions of the human body and make diagnoses of diseases and abnormalities (see Figure 14–7). Iodine, for example, concentrates in the thyroid gland, so instruments that detect radiation from a radioactive isotope of iodine can monitor the workings of your thyroid by examining the decay products of the iodine that concentrates there. Similarly, phosphorus accumulates in bone; therefore, doctors can examine a healing bone fracture with radioactively enhanced phosphorus compounds injected into the bloodstream. Such radioactive materials, called *nuclear tracers*, are used extensively in medicine, in the earth sciences, in industry, and in all kinds of scientific research to follow the exact chemical progressions of different elements. Small amounts of radioactive material will produce measurable signals as they move through a system, allowing scientists and engineers to trace their pathways.

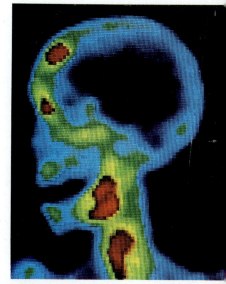

Figure 14–7
Radioactive tracers at work. The patient has been given a radioactive tracer that concentrates in the bone, and emits radiation that can be measured on a film. The dark spot in the front part of the skull indicates the presence of a bone cancer.

Half-Life

A single nucleus of an unstable isotope, left to itself, will decay spontaneously at some point in time. That is, the original nucleus will persist up until a specific time, then radioactive decay will occur, and from that point on you will detect only the fragments of the decay.

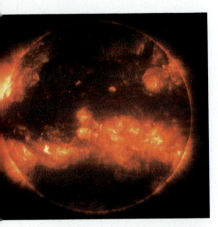

The Sun is a giant fusion reactor, in which hydrogen atoms fuse to produce helium and energy.

half-life The rate of radioactive decay measured by the time it takes for half of a collection of isotopes to decay into another element.

Watching a single nucleus undergo decay is like watching one kernel in a batch of popcorn. Each kernel will pop at a specific time, but all the kernels don't pop at the same time. Even though you can't predict when any one kernel will pop, you can predict how fast and for how long the popping will last. A collection of radioactive nuclei behaves in a similar fashion. Some nuclei decay almost as soon as you start watching; others persist for much longer times. The percentage of nuclei that decay in each second after you start watching remains more or less the same.

Physicists use the term **half-life** to describe the average time it takes for half of a batch of radioactive isotopes to undergo decay. If, for example, there are 100 nuclei at the beginning of your observation and it takes 20 minutes for 50 of them to undergo radioactive decay, then the half-life of that nucleus is 20 minutes. If you were to watch that sample for another 20 minutes, however, the nuclei would not all have decayed. You would have half of that sample or about 25 nuclei at the end, then at the end of another 20 minutes you would most likely have 12 or 13, and so on (Figure 14–8).

Figure 14–8
The number of radioactive nuclei left in a sample as the number of half-lives increases.

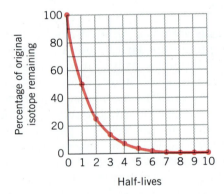

Half-lives

Radioactive decay is something like the popping of popcorn. You can predict, on average, how many decays or pops will occur, but you can't say for sure when a single atom or kernel will change.

Think about this process. Saying that a nucleus has a half life of an hour does *not* mean that all the nuclei will sit there for an hour, at which point they will decay. The nuclei, like the popcorn in our example, decay at different times. The half-life is simply an indication of the average time it takes for an individual nucleus to decay.

Radioactive nuclei display a wide range of half-lives. Some nuclei, such as uranium-227, are so unstable that they persist only a tiny fraction of a second. Others, such as uranium-238, have half-lives that range into the billions of years; these half-lives are comparable to the age of the Earth. Between these two extremes you can find a radioactive isotope that has almost any half-life you can imagine.

We do not yet understand enough about the nucleus to be able to predict half-lives. On the other hand, the half-life is a fairly easy number to measure and therefore can be determined for any nucleus. The fine print on most charts of the isotopes (expanded versions of Figure 14–1) usually includes the half-life for each radioactive isotope.

Radiometric Dating

The phenomenon of radioactive decay has provided scientists who study the Earth and human history with one of their most important methods of determining the age of materials. This remarkable technique, which depends on our knowledge of the half-life of radioactive materials, is called **radiometric dating.**

The best known radiometric dating scheme involves the isotope carbon-14. Every living thing takes in carbon during its lifetime. At this moment, your body is taking the carbon in your food and converting it to tissue, and the same thing is true of all other animals. Similarly, plants take in carbon dioxide from the air. Most of this carbon—about 99%—is in the form of carbon-12, while perhaps 1% is carbon-13. But a certain small percentage, no more than one carbon atom in every million, is in the form of carbon-14, a radioactive isotope of carbon with a half-life of about 5700 years.

The carbon-14 that resides in live tissues is constantly renewed in the same small proportion that is found in the general environment. When a living thing dies, however, it stops taking in carbon of any form. Therefore, when something or someone dies, the carbon-14 in the tissues is no longer replenished. Like a ticking clock, carbon-14 disappears atom by atom, to form an ever smaller percentage of the total carbon. It is this process that is used to determine the approximate age of a bone, a piece of wood, cloth, or other objects by carefully measuring the fraction of carbon-14 that remains. That information is then compared to the amount of carbon-14 that we know must have been in that material when it was alive. If the material happens to be a piece of wood taken out of an Egyptian tomb, for example, we have a pretty good estimate of how old the artifact is and, probably, when the tomb was built.

Carbon-14 dating often appears in the news when a reputedly ancient artifact is shown to be from more recent times. In a highly publicized experiment, the Shroud of Turin, a fascinating cloth artifact thought by many to be involved in the burial of Jesus, was shown by carbon-14 techniques to date from the twelfth century.

Carbon-14 dating has been instrumental in mapping out human history over the past several thousand years. When an object is more than about 70,000 years old, however, the amount of carbon-14 left in it is so small that this dating scheme cannot be used reliably. To date rocks and minerals that are millions of years old, scientists employ similar techniques, but they use radioactive isotopes that have much longer half-lives. Among the most widely used radiometric clocks in geology are those based on the decay of potassium-40 (half-life of 1.25 billion years), uranium-238 (half-life of 4.5 billion years), and rubidium-87 (half-life of 49 billion years). In these cases, scientists measure the total number of

radiometric dating A technique based on the radioactive half-lives of carbon-14 and other isotopes that is used to determine the age of materials.

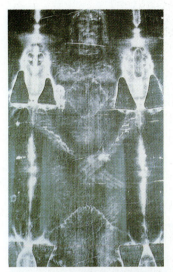

(a)

(b)

(a) The Shroud of Turin, with the ghostly image of a man, was dated by carbon-14 techniques to centuries after the death of Christ. (b) Radiometric dating was used to determine the age of this mummified human, discovered in glacial ice of the Alps.

Geologists collect rock samples for radiometric dating.

atoms of a given element, together with the relative percentage of a given isotope, to determine how many radioactive nuclei were present at the beginning. Most of the ages that we will discuss in the chapters on the Earth sciences and evolution are derived ultimately from these radiometric dating techniques.

Decay Chains

When a parent nucleus decays, the daughter nucleus will not necessarily be stable. In fact, in the great majority of cases, the daughter nucleus is as unstable as the parent. The original parent will decay into the daughter, the daughter will decay into a second daughter, on and on, perhaps for dozens of different radioactive decays. Even if you start with a pure collection of atoms of the same isotope of the same chemical element, nuclear decay will guarantee that eventually you'll have many different chemical species in the sample. A series of decays of this sort is called a *decay chain* (see Figure 14–9). The sequence of decays continues until a stable isotope appears. Given enough time, all of the atoms of the original element will eventually decay into that stable isotope.

The oldest human fossils are too ancient to be dated by carbon-14 methods. An alternate technique, called potassium-argon dating, is employed for dating the rocks in which these skulls, which are up to 3.7 million years old, were found.

To get a sense of a decay chain, consider the example we used at the beginning of this chapter—uranium-238. With a half-life of approximately 4.5 billion years, uranium-238 decays by alpha emission into thorium-234, another radioactive isotope. Thorium-234 undergoes beta decay (half-life 24.1 days) into protactinium-234 (half-life about 7 hours), which in turn undergoes beta decay to uranium-234. Thus, after three radioactive decays, we are back to uranium, albeit a lighter isotope with a 247,000–year half-life.

The rest of the uranium decay chain is shown in Figure 14–9. It follows a long path through eight different elements before it winds up as stable lead-208. Given enough time, all of the uranium-238 now in the Earth will eventually be converted into lead. Since the Earth is believed to be only about 4.5 billion years old, however, there's only been time

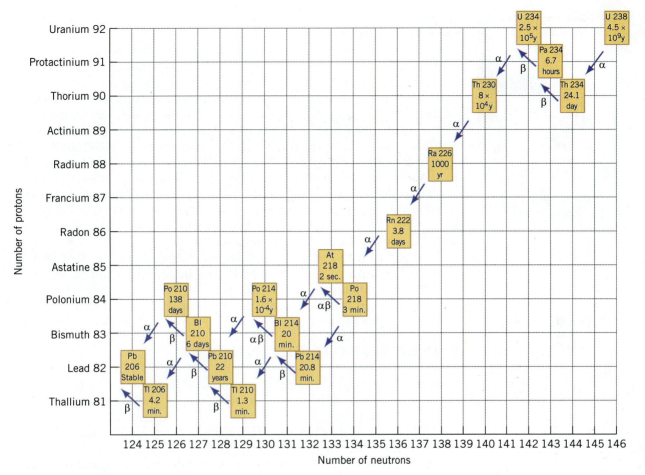

Figure 14–9

The uranium-238 decay chain. The nuclei in the chain decay by both alpha and beta emission until they reach lead-208, a stable isotope. Some isotopes may undergo either alpha or beta decay, as indicated by splits in the chain. Nevertheless, all paths lead eventually to lead-208 after 14 decay events.

for about half of the original uranium to decay, so at the moment (and for the foreseeable future) we can expect to have all the members of the uranium decay chain in existence on the Earth.

Environment

Indoor Radon

The uranium-238 decay chain is not an abstract concept, of interest only to theoretical physicists. In fact, the health concern over indoor radon pollution is a direct consequence of the uranium decay chain. Uranium is a fairly common element—about 2 grams out of every ton of rocks at the Earth's surface are uranium. The first steps in the uranium-238 decay chain produce thorium, radium, and other elements that remain sealed in ordinary rocks and soils. The principal health

Kits like these are used to test homes for the presence of indoor radon.

concern arises from the production of radon-222, about halfway along the path to stable lead.

Radon is a colorless, odorless, inert gas that does not bond chemically to its host rock. As radon is formed, it seeps out of its mineral host and moves into the atmosphere, where it undergoes alpha decay (half-life about 4 days) into polonium-218 and a dangerous sequence of short-lived, highly radioactive isotopes. Historically, radon atoms were quickly dispersed by winds and weather, and they posed no serious threat to human health. In our modern age of well-insulated, tightly-sealed buildings, however, radon gas can seep in and build up to alarming concentrations—occasionally, in poorly ventilated basements, hundreds of times that of normal levels. Exposure to such high radon levels might be dangerous because each radon atom is easily inhaled with the air we breathe and will undergo at least five more radioactive decay events in just a few days.

The solution to the radon problem is relatively simple. First, any basement or other sealed-off room should be tested for radon. Simple test kits are available at your local hardware store. If high levels of radon are detected, then the area's ventilation should be improved. ●

ENERGY FROM THE NUCLEUS

Most scientists who studied the nucleus and its decays were involved in basic research (see Chapter 1). In other words, they were interested in acquiring knowledge for its own sake. But, as frequently happens, knowledge pursued for its own sake is quickly turned to practical use. This certainly happened with the science of the nucleus.

As you now know, the atomic nucleus holds vast amounts of energy. One of the defining achievements of the twentieth century was developing an understanding of and learning to harness that energy. Two very different nuclear processes can be exploited in our search for energy. These processes are nuclear fission and nuclear fusion.

Nuclear Fission

fission A reaction that produces energy when heavy radioactive nuclei split apart into fragments that together have less mass than the original isotopes.

Fission means splitting; therefore, nuclear fission means the splitting of a nucleus. In most cases, energy is required to tear apart a nucleus. Some heavy isotopes, however, have nuclei that can be split apart into products that have less mass than the original. From such nuclei, energy can be obtained from the mass difference.

The most common nucleus from which energy is obtained by fission is uranium-235, an isotope of uranium that constitutes about 0.7% of the uranium in the world. If a neutron hits uranium-235, the nucleus splits into two roughly equal-sized large pieces and a number of smaller fragments (see Figure 14–10). Among these fragments will be two or three more neutrons. If these neutrons proceed to hit other uranium-235 nuclei, the process will be repeated, and a *chain reaction* will begin, with each split nucleus producing the neutrons that will cause more splittings. Using this basic process, large amounts of heat energy can be obtained from uranium.

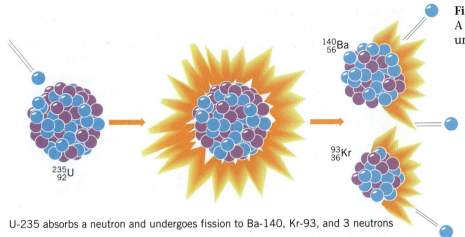

Figure 14–10
A typical fission reaction of uranium-235.

$^{140}_{56}$Ba

$^{93}_{36}$Kr

$^{235}_{92}$U

U-235 absorbs a neutron and undergoes fission to Ba-140, Kr-93, and 3 neutrons

The device that allows us to extract heat energy from nuclear fission is called the **nuclear reactor** (see Figure 14–11). The uranium in a reactor contains mostly uranium-238, but it has been *enriched* so that it contains much more uranium-235 than it would if it were found in nature. This uranium is stacked in long fuel rods, about the thickness of a lead pencil, surrounded by a metallic protector. Typical reactors will incorporate many thousands of fuel rods. Between the fuel rods is the **moderator**—a fluid, usually water, whose function is to slow down neutrons that leave the rods (fast neutrons are less likely to produce fission) (Figure 14–12).

A nuclear reactor works like this: a neutron strikes a uranium-235 nucleus in one fuel rod. The decay products, which include several fast-moving neutrons, leave that particular fuel rod, are slowed down by the moderator, and enter another fuel rod. Because they are moving slowly,

nuclear reactor   A device that controls fission reactions to produce energy when heavy radioactive nuclei split apart.

moderator   In a nuclear reactor, the fluid whose function is to slow down neutrons that leave the fuel rods.

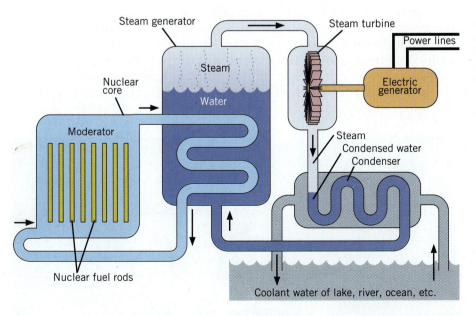

Steam generator

Steam turbine

Power lines

Steam

Nuclear
core

Water

Electric
generator

Moderator

Steam
Condensed water
Condenser

Nuclear fuel rods

Coolant water of lake, river, ocean, etc.

Figure 14–11
A nuclear reactor, shown here schematically, produces heat that converts water to steam. The steam powers a turbine, just as in a conventional coal-burning plant.

Figure 14–12
A map of locations of U.S. nuclear reactors.

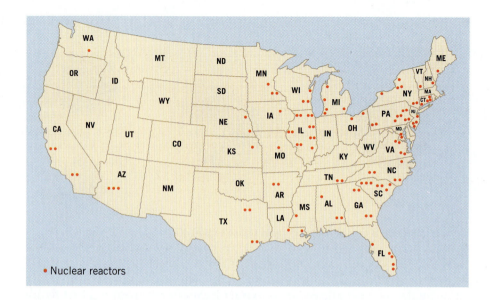

• Nuclear reactors

they are more likely to initiate fissions in that rod. A chain reaction in a reactor proceeds by steps in which neutrons move from one fuel rod to another. In the process the energy released by the conversion of small amounts of matter goes into heating the fuel rods and the water. The water is pumped to another location in the nuclear plant, where it is used to produce steam.

The steam runs a generator to produce electricity as described in Chapter 7 (see Figure 7–11). In fact, the only significant differences between a nuclear reactor and a coal-fired generating plant are the way in which they make steam, and the products left over after the energy is produced. In the former, the energy to produce steam comes from the conversion of mass in uranium nuclei; in the latter, it comes from the burning of coal.

Modern reactors are designed with numerous safety features, because nuclear reactors must keep a tremendous amount of nuclear potential energy under control, while confining dangerously radioactive material. The water that is in contact with the uranium, for example, is sealed in a self-contained system and does not touch the rest of the reactor. Another built-in safety feature is that nuclear reactors cannot function without the presence of the moderators. If there should be an accident in which the water was evaporated from the reactor vessel, the chain reaction

The nuclear reactor at Three Mile Island near Harrisburg, Pennsylvania, had to shut down after suffering a partial meltdown. Safety measures insured that no radioactive material was released to the environment.

would shut off. Thus a reactor cannot explode and it is *not* similar to an atomic bomb.

The most serious accident at a nuclear reactor could occur after the nuclear reactions are shut down, and the flow of water to the fuel rods is interrupted. When this happens, the enormous heat stored in the central part of the reactor could cause the fuel rods to melt. Such an event is called a *meltdown*. In 1979, a nuclear reactor at Three Mile Island in Pennsylvania suffered a partial meltdown, but released no radioactive material to the environment. In 1986, a less carefully designed and super-vised reactor at Chernobyl, Ukraine, underwent a meltdown accompanied by large releases of radioactivity.

Fusion

Fusion refers to a process in which two nuclei come together and fuse to form a third, larger nucleus. Under special circumstances, it is possible to push two nuclei together and make them fuse. When elements with low atomic numbers fuse, the mass of the final nucleus is less than the mass of its constituent parts. In these cases, it is possible to extract energy from the fusion reaction by converting the "missing" mass.

The most common and arguably the most important fusion reaction involves combining four hydrogen nuclei into a helium nucleus (see Figure 14–13). (Remember from Chapter 9 that the nucleus of the ordi-

fusion A process in which two nuclei come together to form a third, larger nucleus. When this reaction combines light elements to make heavier ones, the mass of the final nucleus is less than the mass of its constituent parts. The "missing" nuclear mass can be converted into energy.

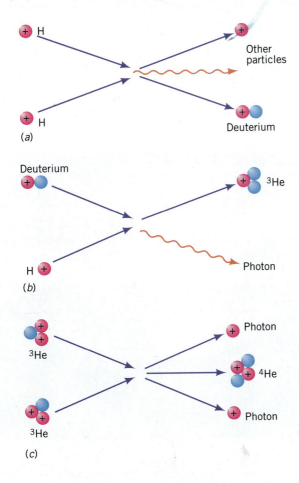

(a)

(b)

(c)

Figure 14–13
A fusion reaction releases energy as nuclei combine. Hydrogen nuclei enter into a multi-step process whose end product is a helium nucleus.

nary hydrogen atom is a single proton, with no neutron. Thus we use the terms "hydrogen nucleus" and "proton" interchangeably.) This reaction releases the energy of the Sun and other stars and thus is ultimately responsible for all life on Earth.

You cannot just put hydrogen in a container and expect it to form helium, however. Two positively charged protons must collide with tremendous force in order to overcome electrostatic repulsion and allow the strong force to prevail (remember, the strong force operates only over extremely short distances). In the Sun, high pressures and temperatures in the star's interior trigger the fusion reaction. The sunlight falling outside your window is part of a process in which 700 tons of hydrogen are converted into helium each second. The helium nucleus has a mass about half a percent less than the four hydrogen nuclei. The "missing" mass is converted to the energy that eventually radiates into space.

The energy you feel in sunlight comes ultimately from just this kind of fusion reaction in the Sun. Since the 1950s, there have been many attempts to harness nuclear fusion reactions to produce energy for human use. The problem has always been that it is very difficult to get protons to collide with enough energy to overcome the electrical repulsion between them and get close enough to initiate nuclear reaction in a controlled manner.

One promising but technically difficult method of accomplishing this is to confine protons in a very strong magnetic field while heating them with high-powered radio waves. Alternatively, powerful lasers can be used to heat pellets of liquid hydrogen. In this scheme, which is called *inertial confinement*, the rapid heating produces shock waves in the liquid. These shock waves raise the temperature and pressure to the point that fusion is initiated.

A number of expensive programs are now under way in the United States, Europe, and Japan, where researchers are investigating ways of producing commercially feasible nuclear fusion reactors, perhaps some-

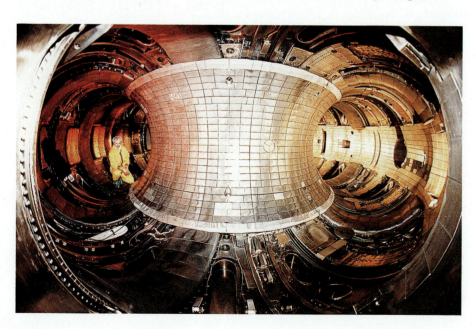

The Princeton Tokomak is a ring-shaped magnetic chamber. Hydrogen plasma is confined in the ring during attempts to achieve nuclear fusion reactions.

time in the twenty-first century. Whether they will be successful is not known. If they succeed, however, the energy crisis will be over forever, since there is enough hydrogen in the oceans to power fusion reactions virtually forever.

Technology

Nuclear Weapons

Whenever humans discover a new source of energy, that source quickly becomes a technology that is turned into weapons. Nuclear energy—the most concentrated energy source known—is no exception. Both fission and fusion bombs have been developed.

The earliest atomic bombs, developed in the United States as part of World War II's Manhattan Project, relied on fission reactions in concentrated uranium-235 or plutonium-239. The problem faced by the designers of the bomb was that a certain number of uranium-235 atoms are required to sustain a chain reaction to the point where large amounts of energy can be released. This amount, called the *critical mass*, is about 10 kilograms, which is equivalent to a chunk of uranium about the size of a grapefruit.

At the instant an atom bomb explodes, a critical mass of the radioactive isotope is compressed into a small volume by a conventional chemical explosive (see Figure 14–14). The normal chain reactions accelerate, as a flood of neutrons almost instantaneously splits all the available nuclear material and releases a destructive wave of heat,

Figure 14–14
In fission bombs, explosives force a critical mass of uranium-235 or plutonium together, which leads to a chain reaction and explosion.

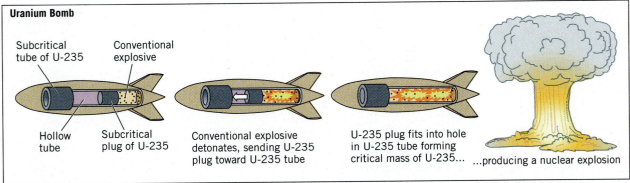

Uranium Bomb

Subcritical tube of U-235 · Conventional explosive

Hollow tube · Subcritical plug of U-235

Conventional explosive detonates, sending U-235 plug toward U-235 tube

U-235 plug fits into hole in U-235 tube forming critical mass of U-235...

...producing a nuclear explosion

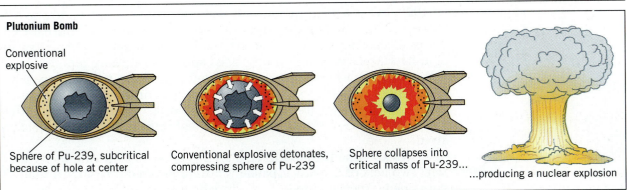

Plutonium Bomb

Conventional explosive

Sphere of Pu-239, subcritical because of hole at center

Conventional explosive detonates, compressing sphere of Pu-239

Sphere collapses into critical mass of Pu-239...

...producing a nuclear explosion

In the years following World War II, nuclear weapons tests were conducted on land. This test took place at the Bikini Atoll on July 26, 1946.

light, and other forms of radiation. The first atomic bombs carried the explosive power of many thousands of tons of chemical explosive.

Shortly after World War II, the United States and the Soviet Union developed the hydrogen bomb, which relies on nuclear fusion. In a hydrogen bomb, a compact hydrogen-rich compound is surrounded by atom bomb triggers. When the atomic fission bombs are set off, they trigger much more intense fusion reactions in the hydrogen. Unlike atom bombs, which are limited by the critical mass of nuclear explosive, hydrogen bombs have no intrinsic limit to their size. The largest nuclear warheads, which could fit in a small closet, carry payloads equivalent to many millions of tons (*megatons*) of conventional explosives. ●

Science in the Making

Cold Fusion

In 1990, two scientists at the University of Utah announced to the press that they had produced fusion reactions under very unusual circumstances. They claimed that they had experimental evidence that fusion had taken place in a simple glass apparatus at room temperature—an apparatus no more complicated than those normally used by high school students in carrying out their experiments. This announcement caused a flurry of excitement. Could it be, people wondered, that the energy crisis was over and a new age of unlimited resources had arrived?

The experiment that attracted all this attention was simplicity itself. Two bars of the metal palladium were lowered into a bath that contained a special kind of water in which some of the hydrogen atoms, which normally have a lone proton in the nucleus, are replaced by deuterium, the isotope of hydrogen whose nucleus contains one proton and one neutron. The fusion of two hydrogen atoms into deuterium is the first step in the fusion process. The experimenters claimed that in the presence of palladium, the deuterium in the water fused to form helium, releasing energy in the process. If their results turned out to be correct, the waters in the Earth's oceans could be "mined" for deuterium, and the human race would have acquired an energy source that was, for all practical purposes, inexhaustible.

Given such extraordinary claims, thousands of scientists rushed to reproduce the Utah experiment. Unfortunately, despite many scientist-years of research, no one was able to reproduce all the results originally claimed. Within about a month, the vast majority of scientists had concluded that the original experiment was flawed. Although the definitive explanation of what happened in the cold fusion experiment is yet to be written, it appears that measurements of both energy production and radioactivity (necessary to establish the presence of nuclear reactions) in the original experiment were simply incorrect.

Cold fusion is now cited as a classic example of how science is a self-correcting process. No result, no matter how compelling or satisfying, can stand without independent confirmation and testing. This incident also reveals why scientists are so reluctant to believe results that are announced directly to the press, without the standard (and exacting) process of peer review that goes into the publication of normal scientific results. ●

THINKING MORE ABOUT
THE NUCLEUS

Nuclear Waste

When power is generated in a nuclear reactor, many more nuclear changes take place than those associated with the chain reaction. The fast-moving debris from the fission of uranium-235 strike other nuclei in the system—both the ordinary uranium-238 that makes up most of the fuel rods, and the nuclei in the concrete and metal that make up the reactor. In these collisions, the original nuclei may undergo fission to become other elements, or absorb neutrons to become heavier isotopes. Many of these byproducts are themselves radioactive. The result is that even when the uranium-235 has all been used to generate energy, a lot of radioactive material remains in the reactor. This sort of material is called high-level nuclear waste. (The production of nuclear weapons accounts for another source of this kind of waste.)

The half-lives of some of the materials in the waste can run to hundreds of thousands of years. What should we do with the waste? In particular,

how can we dispose of this waste in a way that keeps it away from living things?

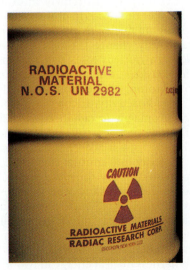

A container of radioactive material bearing the universal trident symbol for radioactivity.

Managing nuclear waste begins with storage. Power companies usually store used fuel rods at the reactor site for tens of years to allow the short-lived isotopes to decay. At the end of this period, long-lived isotopes that are left behind must be isolated from the environment. Some scientists have suggested that we incorporate these nuclei into stable minerals, either rocks or glass. The idea is that the electrons in these atoms will form the same kind of bonds that stable isotopes do, so that with the judicious choice of mineral, you can use the electrons to lock the radioactive nucleus into the mineral for long periods of time. The hope is that long-lived wastes can be sequestered from the environment until after they are no longer dangerous to human beings.

Plans now call for nuclear waste disposal by incorporating radioactive atoms into stable glasses, surrounding these glasses with successive layers of steel and concrete, and then burying them a kilometer or more under the surface of the Earth in stable rock formations. From our point of view, that seems to be the most reasonable available solution to this critical societal problem.

Nuclear waste leaves scientists and society with several difficult questions. What responsibility do we have to future generations to insure that the waste we bury stays where we put it? Should the existence of nuclear waste restrain us in our development of nuclear energy? Should we, as some scientists argue, keep nuclear waste materials at the surface and use them for things like tracers and fuel for reactors?

▶ Summary

The nucleus is a tiny collection of massive particles (by atomic standards), including positively charged *protons* and electrically neutral *neutrons*. The nucleus plays an independent role from the moving electrons that control chemical reactions, and the energies associated with nuclear reactions are much greater. The number of protons is the *atomic number*, which determines the nuclear charge, and therefore the type of element; each element in the periodic table has a different number of protons. The number of neutrons plus protons is the *mass number*, which determines the mass of the *isotope*. Nuclear particles are held together by the *strong force*, which operates only over extremely short distances.

While most of the atoms in objects around us have stable, unchanging nuclei, many isotopes are unstable and spontaneously change through *radioactive decay*. In *alpha decay*, a nucleus loses two protons and two neutrons; in *beta decay*, a neutron spontaneously transforms into a proton, an electron, and a neutrino. A third kind of radioactivity, involving the emission of energetic electromagnetic radiation, is called *gamma radiation*. The rate of radioactive decay is measured by the *half-life*—the time it takes for half of a collection of isotopes to decay into another element. Radioactive half-lives provide the key for *radiometric dating* techniques based on carbon-14 and other isotopes. Unstable isotopes are also used as radioactive tracers in medicine and other areas of science. Indoor *radon* pollution and nuclear waste are two societal problems that arise from the existence of radioactive decay.

We can harness nuclear energy in two ways. *Fission* reactions, as controlled in *nuclear reactors*, produce energy when heavy radioactive nuclei split apart into fragments that together weigh less than the original isotopes. *Fusion* reactions, on the other hand, combine light elements to make heavier ones, as in the conversion of hydrogen into a smaller mass of helium in the Sun. In each case the lost nuclear mass is converted into energy.

▶ Review Questions

1. Compare the size of a nucleus to the size of the atom.

2. Which is stronger—the binding of electrons into atoms, or the binding of protons and neutrons into nuclei?

3. What are some characteristics of protons?

4. Which are heavier—electrons or protons?

5. What are some characteristics of neutrons?

6. What fact about atomic nuclei suggests the existence of the neutron?

7. What is the atomic number?

8. What is an isotope?

9. Why do scientists think that a strong force exists?

10. How is the strong force different from gravity and electromagnetism?

11. What is radioactive decay?

12. Define the term "half-life."

13. What is the role of half-life in radiometric dating?

14. Explain how radioactivity can be used to date (*a*) a piece of wood, (*b*) a rock.

15. What is the range of half-lives of radioactive isotopes?

16. Describe the major achievement of Marie Curie.

17. What is alpha decay? How does it change the nucleus?

18. What is beta decay? How does it change the nucleus?

19. What is gamma radiation? How does it change the nucleus?

20. What does it mean to say, "Radioactivity is the philosopher's stone"?

21. How can we get energy from nuclear fission?

22. What is a chain reaction?

23. How does a nuclear reactor work?.

24. How can we get energy from nuclear fusion?

25. What is a critical mass?

26. What is a radioactive tracer?

27. Describe a situation in which a radioactive tracer might be used.

28. What is nuclear waste?

29. How is nuclear waste disposed of today? Why must this change?

▶ Fill in the Blanks

Complete the following paragraphs with words and phrases from the list.

alpha decay	isotopes
atomic number	nuclear reactor
beta decay	neutrons
fission	protons
fusion	radioactive decay
gamma radiation	radiometric dating
half-life	strong force

The nucleus of the atom is made up primarily of _____ and _____. The number of protons is called the _____, while nuclei with the same number of protons but different numbers of neutrons are called _____. The nucleus is held together by the _____. When nuclei disintegrate spontaneously, we say that they undergo _____.

There are three types of nuclear radiation. _____ consists of emission particles identical to the helium nucleus, _____ is emission of electrons, and _____ is the emission of high-energy photons. The time it takes for a sample of nuclei to lose half of its number through radiation is called its _____. _____ is a process of estimating age of materials by counting remaining radioactive nuclei.

Energy can be generated by splitting the nucleus, a process called _____, or by putting nuclei together, a process called _____. A _____ is a device for generating electricity from splitting nuclei.

▶ Discussion Questions

1. Would you say that Becquerel's discovery of radioactivity was good science or a lucky break? Explain your answer.

2. Why does radioactivity seem to be more common with heavier elements?

3. What was the hypothesis behind Rutherford's experiment on alpha decay? What did he prove?

4. What are the potential benefits and risks in using nuclear tracers in medical diagnosis.

5. "Critical mass" is a term that is widely used outside of nuclear science. What is its everyday meaning, and how does that relate to its scientific meaning?

6. Almost all of the atoms with which we come into daily contact have stable nuclei. Given that there are many more unstable isotopes than stable ones, how could this state of affairs have arisen? (*Hint*: What happens to unstable isotopes in nature?)

7. How does the principle of conservation of energy hold (*a*) in fission reactions, and (*b*) in fusion reactions?

8. Discuss the pros and cons of nuclear power.

9. Can nuclear radiation escape from nuclear power plants? If so, how?

10. How is the relative "emptiness" of space between a nucleus in an atom and its electrons related to the effect the nucleus has on the electrons? To the kinds of chemical bonds the atom form?

11. Classify the three fundamental particles in an atom according to their mass, size, position in the atom, charge, and the relative strength of the electrical, gravitational, and strong forces between each of the particles.

12. How is the number of protons, neutrons, and electrons related to the atom's identity, atomic mass, atomic number, and overall electrical charge?

13. Why is the number of protons in an atom significant in determining the atom's chemical properties?

14. How is the repulsive force(s) between protons related to the existence of the strong force?

15. Describe the "binding energy" of the nucleus of an atom.

16. How does the binding energy of an atom relate to the difference between the sum of the atomic weights of the number of protons and neutrons that make up the atom, and atomic weight of the nucleus of the atom?

17. What are the essential parts of an atom and the forces that are responsible for their presence in the atom?

18. How did Rutherford realize that alpha particles must be related to helium?

19. Is there a difference between a chain reaction and a decay chain?

20. How does nuclear fusion contrast with nuclear fission? Do they occur spontaneously, or do they require an initial dose of energy? If so, why?

▶ Problems

1. Use the periodic table to identify the element, atomic number, mass number, and electrical charge of the following combinations.

    a. 8 protons, 8 neutrons, 10 electrons.

    b. 8 protons, 9 neutrons, 10 electrons.

    c. 9 protons, 8 neutrons, 10 electrons.

    d. 8 protons, 8 neutrons, 9 electrons.

2. Use the periodic table to determine how many protons and neutrons are in each of the following atoms.

    a. ^{14}C

    b. ^{55}Fe

    c. ^{40}Ar

    d. ^{235}U

3. The average atomic weight of cobalt atoms (atomic number 27) is actually slightly greater than the average atomic weight of nickel atoms (atomic number 28). How could this situation arise?

4. Imagine that a collection of 1000 atoms of uranium-238 was sealed in a box at the formation of the Earth 4.5 billion years ago. Use the uranium-238 decay chain (Figure 14–9) to predict some of the things you would find if you opened the box today.

5. A particular isotope has a half-life of 5 days. A particular sample is known to have contained a million atoms when it was put together, but is now observed to have only about 100,000. Estimate how long ago the sample was assembled. Explain the relevance of this problem to the technique of radiometric dating.

6. Suppose that Ricardo wanted to demonstrate the meaning of a half-life to his father. He decided to make a graph and table, starting with 1000 red radioactive parent beads. Then for every half-life, he exchanged the appropriate number of parent beads with "daughter" beads. Help Ricardo out by constructing such a table, as indicated below.

Time (In half-lives)	Total Number of Beads	Number of Parents	Number of Daughters
0	1000	1000	0
1			
2			
3			
4			
5			
6			

7. The first four steps in the decay chain of ^{238}U are given below. Fill in the atomic mass, the atomic number, and the name of the isotope for these first four decays.

Decay Step	Element Uranium	Atomic Mass 238	Atomic Number 92
1. Alpha decay			
2. Beta decay			
3. Beta decay			
4. Alpha decay			

8. A typical fission reaction is a neutron capture of uranium-235, which produces uranium-236, which almost instantaneously splits into strontium-95, xenon-139, and two neutrons. For this nuclear fission reaction, write the atomic number and masses for each radioactive element and the neutrons.

9. Thorium-232 will undergo three beta decays and one alpha decay in a decay chain. Does the final daughter element depend on the order of the decay modes?

10. The EPA sets approximately 1.2 mSv as the maximum whole-body dose for a human being, exclusive of medical and natural radiation. How many standard chest X-rays is this?

11. Jill received the normal dose of X-rays for a year, 0.6 mSv, in standard chest X-rays. Due to a critical accident, she received an additional 0.9 mSv in X-rays. Calculate the total number of standard chest X-rays Jill received this year.

12. An ancient archeological site was being dated using the carbon-14 dating methods on a fireplace. Only 12½% of the original carbon-14 was detected. How old is this ancient site?

13. One atomic mass unit (amu) equals 1.66×10^{-27} kg. Calculate the equivalent of 1 amu in terms of energy in joules.

14. A common unit of energy in nuclear physics is the electron volt (eV), which is equal to 1.6×10^{-19} J. How much energy is 1 amu in electron volts?

15. When 1 gram of hydrogen is converted into helium through nuclear fusion, the difference between the masses of the original hydrogen and the final helium is 0.0072 g. How much energy is released when 1 gram of hydrogen undergoes nuclear fusion?

16. It has been said that the energy released by 1 gram of hydrogen in nuclear fusion is equivalent to 16 tons of oil. How much energy is released by 1 kg of oil?

17. What element is it? From the following information, name the element, its atomic mass, and its charge.

Element	Number of Protons	Number of Neutrons	Number of Electrons
A	1	2	2
B	6	8	5
C	13	14	13
D	27	33	29
E	54	77	51
F	82	148	87
G	90	142	90

▶ Investigations

1. Read a historical account of the Manhattan Project. What was the principal technical problem in obtaining the nuclear fuel? Why did chemistry play a major role? What techniques are now used to obtain nuclear fuel?

2. What is the current status of United States' progress toward developing a depository for nuclear waste? How do your representatives in Congress vote on matters relating to this issue?

3. What sorts of isotopes are used for diagnostics in your local hospital? Where are supplies of those radio-isotopes purchased? What are the half-lives of the isotopes, and how often are supplies replaced? What is the hospital's policy regarding the disposal of radioactive waste?

4. How much of the electricity in your area comes from nuclear reactors? What fuel do they use? Where are the used fuel rods taken when they are replaced? If the facility offers public tours, visit the reactor and observe the kinds of safety procedures that are used.

5. Obtain a radon test kit from your local hardware store and use it in the basement of two different buildings. How do the values compare? Is either at a dangerous level? If the values differ, what might be the reason?

6. Only about 90 elements occur naturally on Earth, but scientists are able to produce more elements in the laboratory. Investigate the discovery and characteristics of one of these human-made elements.

7. Read an account of the cold fusion episode. At what point in this history were conventional scientific procedures bypassed? Ultimately, do you think that the scientific method worked or failed?

8. Soon the United States government will take over the responsibility for the nuclear wastes of the 50 states. What options do we have for waste storage? Do you think they should all be stored in one place? Should we try to separate and use the radioactive isotopes? What are the factors—social, political, and economic—that will help determine what happens to this nuclear waste?

9. Helium is not naturally found in the Earth's atmosphere, but it is found in oil and gas wells. Investigate the origin of this helium gas.

▶ Additional Reading

Alvarez, Luis. *Adventures of a Physicist.* New York: Basic Books, 1987.

Pflaum, Rosalynd. *Grand Obsession: Madame Curie and Her World.* New York: Doubleday, 1989.

Rhodes, Richard. *The Making of the Atomic Bomb.* New York: Simon and Schuster, 1986.

The movie *Fat Man and Little Boy*

15 THE ULTIMATE STRUCTURE OF MATTER

ALL MATTER IS MADE OF QUARKS AND LEPTONS—THE
FUNDAMENTAL BUILDING BLOCKS OF THE UNIVERSE.

The Library

The next time you head over to the library, wander through the stacks
and think about the fundamental building blocks of the library. Your
first reaction might be to say that the fundamental building blocks are
books. After all, the stacks are filled with row after row, shelf after
shelf, of bound volumes. But a library is not just a collection of books;
the volumes are arranged in a particular order. You could also describe
the set of rules that dictates how books are arranged in libraries; for example, they might be organized according to the Dewey decimal system or the Library of Congress classification scheme. Thus a complete
description of a library at this most superficial level includes two
things: books as the fundamental building blocks, and rules about how
the books are organized.

If you look inside a book, you notice that various volumes are not
as different from each other as they might seem at first. They are all
made of an even more fundamental unit—the word. In this sense, you
can say that the word is the fundamental building block of the library;
and, as is the case for cataloged books, we require a set of rules, called
grammar, that tells us how to put words together to make books. Thus
the discovery of words and grammar takes you to a more basic level in
a probe of a library's reality.

You probably wouldn't be content very long with the notion of
the word as the fundamental building block, because it soon becomes
obvious that all of the thousands of words are different combinations
of a small number of a unit even more fundamental—letters. Only 26
letters provide the building blocks for all the thousands of words on
all the pages in all the books of the library. Furthermore, we need a set
of rules (spelling) that tells us how to put letters together into words.
The discovery of letters and spelling would constitute perhaps the ultimate basis of a library and its organization.

But, as we saw in Chapter 13, letters are a form of information that
can be stored and manipulated in terms of a binary code. Call these entries in this code the "dots and dashes" of the library. We could argue
that they, not the letters, are the most fundamental quantities in the library. And, just as we needed a rule to tell us how to use other "ele-

411

mentary" entities, we need a rule (call it a code) that tells us how to go from dots and dashes to letters.

The library, then, can be described in this way: A code tells us how to put the dots and dashes together into letters; then we use spelling to tell us how to put the letters together into words; then grammar tells us how to put the words together into books. Finally, we use the organizing rules to tell us how to put the books together into a library.

And this, as you shall see, is exactly the kind of process scientists use to describe the entire physical universe.

WHAT IS THE UNIVERSE MADE OF?

How many different kinds of materials do you see when you look up from this book? You may see a wall made of cinder blocks, a window made of glass, and a ceiling made from fiberglass panels. Outside the window you may see grass, trees, blue sky, and clouds. We encounter thousands of different kinds of materials every day. They all look different; what possible common ground could there be between a cinder block and a blade of grass?

For at least two millennia, people who have thought about the physical universe have asked similar questions. Is the universe just what we see, or is there some underlying structure, some basic stuff, from which it's all made? Herein for many researchers lies *the* scientific question.

Reductionism

The quest for the "ultimate building blocks" of the universe is referred to by philosophers as an example of *reductionism*. Reductionism is an attempt to reduce the complexity of nature by first looking for an underlying simplicity, and then trying to understand how that simplicity gives rise to the observed complexity. This pursuit perpetuates an old intellectual belief that the appearances of the world do not tell its true nature, but that its true nature can be discovered by combining the thinking process with experiment and observation.

The Greek philosopher Thales (640?–546 B.C.) suggested that all materials are made of water. This suggestion was based on the observation that in everyday experience water appears as a solid (ice), a liquid, and a gas (steam). Thus, alone among the common substances, water seemed to exhibit all the states of matter (see Chapter 11). To Thales, this observation suggested that water was in some sense fundamental. We no longer accept the idea that water is the fundamental constituent of matter, but we do believe that we can find other fundamental constituents.

The Building Blocks of Matter

The library analogy presents a simple and satisfying way of describing complex systems. Some would even argue that everything you could possibly want to know about the library is contained in the dots and

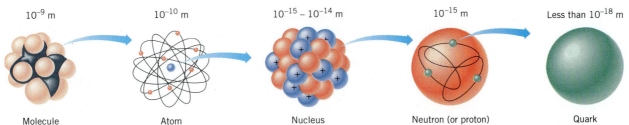

10^{-9} m 10^{-10} m $10^{-15} - 10^{-14}$ m 10^{-15} m   Less than 10^{-18} m

Molecule   Atom   Nucleus   Neutron (or proton)   Quark

Figure 15–1
The modern picture of the fundamental building blocks of the universe. Molecules are made from atoms, which contain nuclei, which are made from elementary particles, which in turn are made from quarks.

dashes and their organizing principles. In just the same way, scientists want to describe the complex universe by identifying the most fundamental building blocks and deduce the rules by which they are put together.

At first, it might seem that the most fundamental building block of the universe is the atom. You know that the myriad solids, liquids, and gases are made of just 100 or so different kinds of chemical elements. The complexity of materials that appears to the senses results from the many combinations of these relatively few kinds of atoms. The rules of chemistry tell us how atoms bind together to form all of the materials we see.

Early in this century, as was discussed in Chapters 9 and 14, scientists learned that atoms are not the most fundamental building blocks of matter. They are made up of smaller, more fundamental bits, including nuclei and electrons. These particles arrange themselves according to a set of rules, with massive neutrons and protons in the positively charged nucleus, and negatively charged electrons moving around the nucleus. Protons and neutrons together form nuclei, while the electrons move around the nucleus to form atoms. Electrons combine and interact to form all the known materials.

This picture of the universe, with protons, neutrons, and electrons as the only three fundamental building blocks, is very simple and appealing. But, just as letters, spelling, and grammar provide a false level of simplicity when analyzing the basic components of the library, the simple picture of a universe of protons, neutrons, and electrons hasn't stood up to more detailed experiments and observations. The nucleus contains much more than just protons and neutrons, although this fact did not become clear to physicists until the post-World War II era. To follow the reductionist perspective of the universe, we have to start thinking about what constitutes the nucleus. Scientists refer to the particles that make up the nucleus, together with particles like the electron, as *elementary particles*, to reflect the belief that they comprise the basic building blocks of the universe (Figure 15–1). The study of these particles and their properties is the domain of **high-energy physics** or **elementary-particle physics.**

elementary-particle physics or high-energy physics The study of particles that are believed to comprise the basic building blocks of the universe; for example, the particles that make up the nucleus and particles such as the electron.

DISCOVERING ELEMENTARY PARTICLES

Nowhere in nature is the equivalence of mass and energy more obvious than in the interactions of elementary particles. Imagine that you have a mechanism for accelerating a proton to very high velocities, approaching the speed of light. This mechanism might be astronomical in nature, or it might be a machine that accelerates particles. Once the proton has been accelerated, it has a very high kinetic energy. If this high-energy proton collides with a nucleus, the nucleus can be split apart. In this collision process, some of the kinetic energy of the original proton can be converted into mass according to the equation $E = mc^2$. When this happens, new kinds of particles that are not protons, electrons, or neutrons can be created.

Cosmic Rays

During the 1930s and 1940s, physicists used a natural source of high-energy particles to study the structure of matter. These particles, known as **cosmic rays,** are mostly protons that continuously rain down on the Earth's atmosphere after being emitted by stars in our galaxy and beyond.

cosmic rays Particles (mostly protons) that rain down continuously on the atmosphere of the Earth after being emitted by stars in our galaxy and in others.

Space is full of cosmic rays. When these rays hit the atmosphere, they collide with molecules of oxygen or nitrogen and produce sprays of very fast-moving secondary particles. Secondary particles, in turn, can collide further to produce even more particles as a cascade builds up in the atmosphere. It is not uncommon for a single incoming particle to produce billions of secondary particles by the time the cascade reaches the surface of the Earth. Indeed, on average, several rays pass through your body every minute of your life.

Physicists in the 1930s and 1940s set up their apparatus high on mountaintops so they could observe what happened when the fast-moving primary cosmic rays or the slightly slower-moving secondary particles collided with nuclei. A typical apparatus incorporated a gas-filled chamber several centimeters across (see Figure 15–2). A thin sheet of target material such as lead was located midway in the chamber. Cosmic rays occasionally collided with one of the nuclei in the piece of lead, producing a spray of secondary particles. By studying particles in that spray, physicists hoped to understand what was going on inside the nucleus.

By the early 1940s, when the international effort in physics research shut down temporarily because of World War II, physicists working with these cosmic ray experiments had discovered particles in the universe in addition to the proton, the neutron, and the electron. When the research effort started up again after the war, these discoveries multiplied, as more and more particles were found in the debris of nuclear collisions.

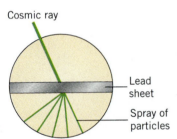

Figure 15–2
In a typical cosmic ray experiment, cosmic rays hit a lead nucleus, producing a spray of particles.

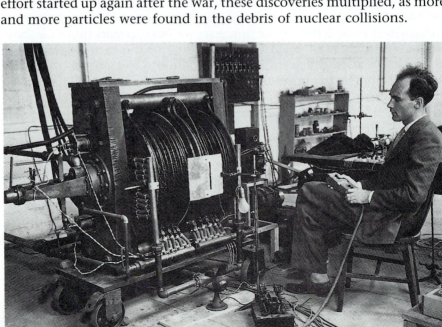

Physicist Carl Anderson, shown working with an electromagnet used in his cloud chamber.

A cosmic ray detector at the Gran Sasso Laboratory in the Italian Alps.

The net result of these discoveries was that the nucleus could no longer be thought of simply as a combination of protons and neutrons. Instead, researchers started to view the nucleus as a very dynamic place. All kinds of newly discovered elementary particles in addition to protons and neutrons were found there. These exotic elementary particles are created in the interactions inside the nucleus, and they give up their energy (and their very existence) in subsequent interactions to make other kinds of particles. This constant "dance" of the elementary particles inside the nucleus has been well documented since these early explorations.

Technology

Detecting Elementary Particles

If elementary particles are even smaller than an individual nucleus, how do we know they exist? The process of detecting elementary particles has been raised to a fine art over the years by experimental physicists. Nevertheless, the basic technique used in any detection process is the same: the particle in question interacts with matter in some way, and we measure the changes in matter that are the effects of that interaction.

If an elementary particle has an electrical charge, for example, it may tear electrons loose as it passes near an atom. Thus a charged elementary particle moving through material will leave a string of ions in its wake, much as a speedboat going across a lake leaves a trail of troubled water. Over the years, many techniques have been used to measure the presence of those ions. You can, for example, use photographic film on which the ions will show up as dark tracks when the film is developed. Alternatively, an apparatus can be set up in such a way that bubbles form in a liquid, or droplets of water appear in a gas,

False-color photograph of particles in a bubble-chamber—an apparatus where particle tracks are marked by bubbles in a liquid. In this photo, proton tracks are in red, electron tracks in blue.

The 4-story-high Fermilab CDF particle detector. Particle detectors have gone from being small, desk-top pieces of equipment to huge high-tech instruments like the one shown.

when an elementary particle passes through. In these detectors, a string of bubbles or droplets will mark the paths of the particles.

A more modern method of detecting charged particles is to allow the particles to pass though a grid of thin conducting wires. As a particle passes a wire, it exerts a force on the electrons in the metal, creating a small pulse of current. By measuring the time it takes for this pulse to arrive at the end of the wire, and putting together the information from many wires, a computer can reconstruct the path taken by a particle with high precision.

Uncharged particles such as neutrons are much more difficult to detect because they do not leave a string of ions in their path. Typically, we do not detect the passage of an uncharged particle directly, but wait until it collides with something. If that collision produces charged particles, then we can detect them by the techniques outlined above and can work backwards and deduce the property of the uncharged particle. ●

Particle Accelerators: The Essential Tool

particle accelerator A machine such as a synchrotron or linear accelerator that produces particles at near-light speeds for use in the study of the fundamental structure of matter.

Once upon a time, physicists had to sit around and wait for nature to supply high-energy particles in the form of cosmic rays, so that they could study the fundamental structure of matter. The arrival of cosmic rays could not be controlled, however, and it could be very time consuming waiting for one to hit. Physicists quickly realized that they had to build machines that could produce streams of "artificial cosmic rays." These machines, called **particle accelerators,** can be turned on and off at will by scientists. Particle accelerators take the place of the natural, but sporadic, production of cosmic rays in experiments. Early in the 1930s at the University of California at Berkeley, Ernest O. Lawrence began work on a new kind of accelerator called the *cyclotron,* an invention for which he won the 1939 Nobel Prize in physics.

This type of accelerator works by applying a large electrical force to groups of charged particles, usually electrons or protons. This action produces a large acceleration according to Newton's second law (force = mass × acceleration: see Chapter 4). The particles move in a circular path between the poles of a large magnet, so that they encounter the force over and over again. The particles are accelerated until they are moving at almost the speed of light. Once they have acquired this much kinetic energy, they are allowed to collide with other particles. Those collisions provide the interactions that scientists wish to study.

Lawrence's first cyclotron was no more than a dozen centimeters across and produced energies that were extremely small by today's standards. A modern particle accelerator, on the other hand, is a huge, high-tech structure capable of producing energies as high as all but the most energetic cosmic rays. Called a **synchrotron,** its main working part is a large ring of magnets that keep the accelerated particles moving in a circular track.

One aspect of Maxwell's equations that we didn't explore in detail in Chapter 7 is that magnetic fields exert a force on moving charged particles. That force tends to make charged particles move in a circular track. As a particle in a synchrotron moves around the circle, the fields of the large electromagnets are adjusted to keep its track within a small chamber (typically tens of centimeters on a side). The accelerated particles must travel in a near-perfect vacuum, to prevent unwanted collisions with stray atoms or molecules. This chamber, in turn, is bent into the large circle that marks the particles' orbit. Each time the particles come around to a certain point, an electric field boosts their energy. As the velocity increases, the field strength in the magnets is also increased to compensate, so that the particles continue around the circular track. Eventually, the particles reach the desired speed and they are brought out into an experimental area where they undergo collisions.

Ernest O. Lawrence posed in the 1930s with his invention, the cyclotron, which was the first particle accelerator.

synchrotron A particle accelerator in which magnetic fields are increased as particles become more energetic, keeping them moving on the same track.

Fermilab in Illinois (*a*) and Stanford Linear Accelerator Center (SLAC) in California (*b*) are two of the world's most powerful accelerators.

(*a*)

(*b*)

As the energy required to stay at the frontier of particle physics increases, so too does the size of accelerators. The highest-energy accelerator in the world today (a synchrotron) is at the Fermi National Accelerator Laboratory (Fermilab) outside of Chicago, Illinois. There, protons accelerate around a ring almost 2 kilometers (about 1 mile) in diameter.

linear accelerator A device for making high-velocity particles, which relies on a long, straight vacuum tube into which particles are injected to ride an electromagnetic wave down the tube.

The **linear accelerator** provides an alternative strategy for making high-velocity particles. This device relies on a long, straight vacuum tube into which electrons are injected. The electronics are arranged so that an electromagnetic wave travels down the tube, and the electrons ride this wave more or less the way a surfer rides a wave on the ocean. The largest linear accelerator in the world, at the Stanford Linear Accelerator Center in California, is about 3 kilometers (almost 2 miles) long.

Environment

Fermilab and the Restoration of the Tall-grass Prairie

We tend to think of scientific research institutions as monolithic sorts of places, devoted only to the pursuit of knowledge. Yet every once in a while something happens to change that view. The Fermi National Accelerator Laboratory, for example, is one of the main research centers in the world for the investigation of elementary particles. But it also occupies a 6800–acre site on the Illinois prairie, about 50 miles from downtown Chicago. Before construction started on the laboratory in the late 1960s, the site enclosed a small subdivision and a large expanse of typical midwestern farmland. The plowed fields had replaced the original prairie ecosystem that had occupied much of northern Illinois before farming had become a major industry. On Fermilab's site, only a relatively small portion of the land is actually occupied by the accelerator and its machinery. The rest is open space, meant to provide a buffer between the facility and surrounding residents.

At the same time that the accelerator at Fermilab was beginning to revolutionize our understanding of the world of elementary particles, a group of conservationists led by Robert Betz of Northeastern Illinois University realized that the open land at Fermilab presented an opportunity to reconstruct the tall-grass prairie ecosystem. With the enthusiastic approval of the laboratory directors, this group set out to rebuild the vanished prairie. Gathering seeds in old cemeteries and along railroad right-of-ways, they began by planting a few acres of land on the site. They knew that one element of the old prairie—regular autumn fires caused by lightning—was missing, so they proceeded to burn their small patches of prairie. Because the original plants sent their roots much deeper into the ground than weeds and imported varieties of plants, the periodic burning allowed the original ecosystem to establish itself. Today, large bands of volunteers gather to harvest seeds on hundreds of acres of tall grass prairie, and seeds from Fermilab are used to establish patches of prairie all over the Midwest.

Although Fermilab was not built to restore the original prairie ecosystems of the Midwest, that has been one of its unintended benefits. And it is quite easy to imagine that decades from now, when the accelerator has been superseded by other machines, the site of Fermilab will be established as a national ecological park. ●

The restored tall-grass prairie at Fermilab.

THE ELEMENTARY PARTICLE ZOO **419**

THE ELEMENTARY PARTICLE ZOO

At the beginning of the 1960s, when the first generation of modern particle accelerators began to produce copious results, scientists discovered that there must be a very large number of elementary particles inside the nucleus. At present, this list exceeds 200 and is still growing. These particles typically are named after Greek letters by their discoverers, and they come in a wide variety of types, summarized in Table 15–1. Following are some important classifications.

Table 15–1 • Summary of Elementary Particles

Type	Definition	Examples
leptons	do not take part in holding nucleus together	electron, neutrino, mu, tau
hadrons	participate in holding nucleus together	proton, neutron, roughly 200 others
antiparticles	particles with same mass, but opposite charge and other properties	positron

Leptons

Some elementary particles do not participate in the strong force that holds the nucleus together (see Chapter 14), and they are not part of the nuclear maelstrom. These non-nuclear particles are called **leptons,** for "weakly interacting ones." The two leptons that we have encountered so far are the *electron*, which is normally found in orbit around the nucleus rather than in the nucleus itself, and the *neutrino*, a light neutral particle that hardly interacts with matter at all. In the period since the 1940s, physicists have discovered a total of six kinds of leptons. Two of these particles are like the electron except that they are heavier (particles called the mu and the tau), and two others are kinds of neutrinos (one neutrino associated with each of the massive mu and tau particles). Thus the present classification of the elementary particles suggests the existence of six different kinds of leptons all told. If you keep in mind that the electron and the neutrino are typical leptons, you will have a pretty good idea of what they're like.

lepton A particle (such as the electron, muon, and neutrino) that participates in the weak, but not the strong, interaction.

Hadrons

All of the different kinds of particles that exist inside the nucleus are referred to collectively as **hadrons** (from the Greek for "strongly interacting ones"). The array of these particles is truly spectacular. Hadrons include particles that are stable like the proton, particles that undergo radioactive decay in a matter of minutes like the neutron (which undergoes beta decay), and still other particles that undergo radioactive decay in 10^{-24} seconds. The latter kind of particles do not live long enough even to travel across a single nucleus! Some hadrons carry an electrical charge, while others are neutral. But all of these particles are subject to the strong

hadron Particles, including the proton and neutron, that are made from quarks and are subject to the strong force.

force, and all participate in some way in holding the nucleus together, and thus in making the physical universe possible.

Antimatter

antimatter Particles that have the same mass as their matter twins, but with an opposite charge and opposite magnetic characteristics.

positron The positively charged antiparticle of the electron.

For every particle that we see in the universe, it is possible to produce a second particle with opposite properties. Every **antimatter** particle, or *antiparticle,* has the same mass as its matter twin, but opposite charge and opposite magnetic characteristics. The antiparticle of the electron, for example, is a positively charged particle known as the **positron.** It has the same mass as the electron, but a positive electrical charge. Antinuclei of antiprotons and antineutrons, orbited by positrons, could in principle form antiatoms. Indeed, some scientists have proposed the existence of entire galaxies, in every way like our own, but composed entirely of antimatter.

When a particle collides with its antiparticle, both masses are converted completely to energy in the most efficient and violent process known in the universe—a process called *annihilation*. The original particles disappear, and this means that energy appears as a spray of rapidly moving particles and electromagnetic radiation. This fact has long been adopted by science fiction writers in their descriptions of futuristic weapons and power sources.

Although antimatter is fairly rare in the universe, it is routinely observed as a by-product of high-energy collisions in particle accelerators around the world. High-energy protons or electrons strike nuclear targets, and the energy of the particles is converted to equal numbers of other particles and antiparticles. Thus the existence of antimatter is verified daily in laboratories.

Developing Your Intuition

The Starship *Enterprise*

In the television series "Star Trek," the starship *Enterprise* has an unusual design. The imaginary starship features two pods, one for matter and one for antimatter. Streams of particles are piped into the engi-

The fictional Starship *Enterprise* was supposed to be powered by matter-antimatter annihilation.

neering section, where they come together in the (imaginary) "di-lithium" crystal to move the ship to "warp drive." Why do you suppose the writers chose that particular means of propulsion?

Part of the reason was certainly the exotic nature of antimatter. But it also must have had to do with the extreme efficiency of the matter-antimatter annihilation process. Whatever happens in the di-lithium crystals, it must end with the complete conversion of matter into energy. Unlike nuclear fission, in which only a fraction of a percent of a uranium atom is converted into energy, in annihilation matter and antimatter are converted 100% to energy. In the more than quarter century that has elapsed since the original version of "Star Trek" burst on the scene, no one has managed to come up with a more efficient way of producing energy. So for now, we can expect the design of the good ship *Enterprise* to remain basically unaltered. ●

Science in the Making

The Discovery of Antimatter

In 1932, Carl Anderson, a young physicist at the California Institute of Technology, was performing a rather straightforward cosmic ray experiment of the type described earlier in this chapter. Cosmic rays entered a type of detector called a cloud chamber. In Anderson's cloud chamber, a cosmic ray particle would move through a moisture-laden gas, leaving behind a string of ions. By pulling out a piston at the bottom of the chamber, the gas pressure was lowered, and the liquid (usually alcohol) that had been in gaseous form was condensed into droplets. The ions acted as nuclei for the condensation of these droplets, so that the path of the particle was marked out by a string of droplets in the chamber.

The key innovation in Anderson's experiment was the positioning of the cloud chamber between the poles of a powerful magnet. This

Carl Anderson identified the positron, the antiparticle of the electron, from the distinctly curved path left in the cloud chamber. In Anderson's original photograph (*a*), the positron path curves up and to the left. In a more recent bubble chamber photograph (*b*), an electron (e−) and a positron (e+) curve opposite directions in a magnetic field.

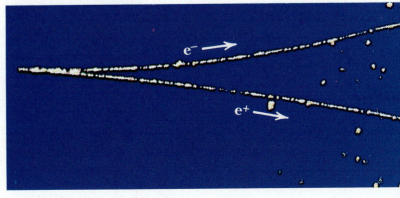

(*a*)   (*b*)

magnet caused electrically charged cosmic rays to move in curved tracks, with the amount of curving dependent on a particle's mass, speed, and charge. Furthermore, the tracks of positively and negatively charged particles curved in opposite directions under the influence of the magnetic field.

Soon after he turned his apparatus on, Anderson saw tracks of particles whose mass seemed to be identical to that of the electron, but whose tracks curved in the opposite direction from those of electrons being detected. This feature, he concluded, had to be the result of a "positive electron," a phrase he shortened to positron. Although no one realized it at the time, Anderson was, in fact, the first human being to see antimatter.

One amusing sequel to this discovery is that when Carl Anderson received the Nobel Prize for his work in 1936, he still had not been promoted to full professor at Caltech. ●

Positron Emission Tomography

The study of elementary particles at times seems quite abstract; you may feel that you are becoming lost in a fantasyland of strange names, strange concepts, and strange ideas. But new technologies based on elementary particles play a very important role in the real world. Positron emission tomography (PET) is one of these.

In this medical technique, glucose molecules (see Chapter 12) that carry an unstable isotope of oxygen (an isotope produced in nuclear reactions) are injected into a patient's bloodstream. These molecules are taken up by organs in the body, including the brain. They will go to the parts of the brain that need it—the parts that are using glucose at the time (see Figure 15–3).

This particular isotope of oxygen emits a positron, the antiparticle of an electron, which is relatively easy to detect from outside the body. A PET scan works like this: After the material is injected into the bloodstream, the patient is asked to do something, such as talk, read, do mathematical problems, or just relax. Each of these activities uses a different region of the brain. Scientists watching the emission of positrons can see

Figure 15–3
Positron emission tomography, commonly called the PET scan, reveals activity in the human brain. (*a*) A patient undergoing a PET scan. (*b*) A scan of a normal brain. The bright spots are places where large amounts of glucose (a simple sugar used by most cells for energy) are being used by the brain. (*c*) A scan of someone suffering from Alzheimer's disease.

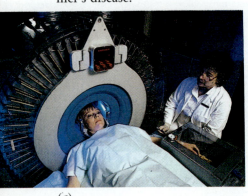

(*a*)

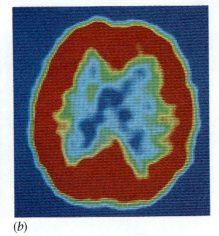

(*b*)

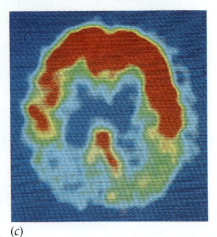

(*c*)

those regions of the brain quite literally "light up" as they are used. In this way, scientists can both study the normal working of the human brain without disturbing the patient, and can detect and study abnormalities that can perhaps be treated. Thus elementary particles have their role to play in the frontiers of medical research.

Quarks

When chemists understood that the chemical elements could be arranged in the periodic table, it wasn't long before they realized what caused this regularity. Different chemical elements were not "elementary," as Dalton had suggested, but were structures made up of things more elementary still. The same thing is true of the hundreds of "elementary" hadrons, or nuclear particles. They are not themselves elementary, but are made up of things even more elementary. These building blocks of matter are given the name **quark** (pronounced "quork"). First suggested in the late 1960s, quarks have come to be accepted by physicists as the truly fundamental building blocks of the hadrons. Even though they never have been (and probably cannot be) seen in the laboratory, the concept of quarks has brought order and predictability to the complex zoo of elementary particles.

Quarks are different from other elementary particles in a number of ways. It is now believed that quarks existed as free particles only briefly in the very first stages of the universe (see Chapter 22). Unlike any other known particle, they have fractional electrical charge; that is, they have electrical charges that are equal to $\pm\frac{1}{3}$ or $\pm\frac{2}{3}$ the charge on the electron. Scientists postulate that quarks make up all the hadrons, but once locked into these particles, no amount of experimental machination will ever pry them loose.

In spite of these distinctive properties, the quark picture of matter has proven very appealing to scientists. Instead of dealing with numerous hadrons, only six kinds of quarks (and six corresponding antiparticles, called *antiquarks*) occur in the universe. The quarks, like many things in elementary-particle physics, have been given fanciful names: up, down, strange, charm, top, and bottom (see Table 15–2). Scientists have now found evidence that supports the existence of all six of these elementary particles. The discovery of the sixth and final quark, the top quark, was made with front-page headlines and great fanfare in May, 1994. The

quark *(pronounced "quork")*
The truly fundamental building block of the hadrons: a particle that has fractional electrical charge and cannot exist alone in nature.

Table 15–2 • **Quark Properties**

Name of Quark	Symbol	Electrical Charge*
down	d	$-\frac{1}{3}$
up	u	$+\frac{2}{3}$
strange	s	$-\frac{1}{3}$
charm	c	$+\frac{2}{3}$
bottom	b	$-\frac{1}{3}$
top	t	$+\frac{2}{3}$

* Quarks with the same charge differ from each other in mass and other properties.

short scientific article describing this historic finding listed almost 450 coauthors, most of them from Fermilab, where the experiments were performed.

All of the hadrons that we know about—all of the hundreds of different kinds of particles that whiz around inside the nucleus—can be made from these six simple particles. (We have to stress, however, that leptons like the electron are *not* made from quarks.) The proton, for example, is the combination of two up quarks and one down quark, while the neutron is the combination of two down quarks and one up quark. In this scheme, the charge on the proton, equal to the sum of the charges on its three quarks, is

$$\tfrac{2}{3} + \tfrac{2}{3} + (-\tfrac{1}{3}) = +1$$

while the charge on the neutron is

$$\tfrac{2}{3} + (-\tfrac{1}{3}) + (-\tfrac{1}{3}) = 0$$

In the more exotic particles, pairs of quarks circle each other in orbit, like some impossible star system.

Quarks and Leptons: The Dots and Dashes of the Universe

The quark model gives us a picture of the universe that restores the kind of simplicity that was suggested by both Dalton's atoms and Rutherford's nucleus. All of the elementary particles in the nucleus are made from various combinations of the six kinds of quarks. These elementary particles are then put together to make the nuclei of atoms. The leptons, primarily the electrons, are stationed outside the nucleus to make the complete atoms, and different atoms interact with each other to produce what we see in the universe. In this scheme, the quarks and leptons are the dots and dashes of the universe; they are the basic stuff from which everything else is made. The fact that there are six leptons and six quarks provides a kind of symmetry that has not escaped the notice of physicists. This balance is built into almost all theories of elementary particles. The question of why nature should be arranged this way is still very much open.

Quark Confinement

It would be nice to be able to find quarks and study them in the laboratory; however, at present there is no generally accepted method for isolating a quark. In fact, particle theorists now believe that quarks cannot be isolated in the laboratory because they can never be pried loose from the particles in which they exist. In these theories, once a quark is taken up into a particle, it is "confined" in that particle forever.

To help you understand the notion of quark confinement, consider the following analogy. Suppose you cut a rubber band and were asked to isolate just one of its ends. Think of the rubber band as a particle and the very end of it as being the quark. You could grab hold of the rubber band and stretch it, perhaps even break it. You would then have two shorter rubber bands, but you would never have the end of a rubber band by itself. No matter how many times you broke the rubber band apart, you would get the same result. There just is no such thing as an "end" not attached to something else.

Elementary particles operate according to the same principle. You can hit them as hard as you like in an attempt to shake the quarks loose, but every time you start to pull a quark out, you've also supplied enough energy to the system to make more quarks and antiquarks; in fact, those new quarks and antiquarks will immediately be taken up into new particles.

What does it mean that the fundamental building blocks of the universe are things we can never isolate and study? Does that mean they aren't real or they don't exist? We know that elementary particles behave *as if* they were made up of quarks. This situation should remind you of the question of whether or not atoms really exist, which we discussed in Chapter 9.

THE FOUR FUNDAMENTAL FORCES

In our excursion into the library in the chapter opening, finding the dots and dashes wasn't enough to explain what we saw. We also needed to know the rules by which dots and dashes are converted into books. In the same way, if we are going to understand the fundamental nature of the universe, we have to understand not only the letters (quarks and leptons), but also the forces that arrange them and make them behave as they do.

One useful analogy is to think of the quarks and leptons as the bricks of the universe. The universe appears to be built of these two different kinds of bricks that are arranged in different ways to make everything we see. But you cannot build a house using bricks alone. There has to be something like mortar to hold the bricks together. The "mortar" of the universe—the things that hold the elementary particles together and organize the physical universe into the structures we know—are the forces. At the moment, we know of only four fundamental forces in nature. Two of these are *gravity* and *electricity*. These were known to nineteenth-century physicists and are key elements of our daily existence. They are forces with infinite range; that is, two objects such as stars and planets can exert these forces on each other even though they are very far apart.

The other two forces are less familiar to us, because they operate in the realm of the nucleus and the elementary particle. They have a range comparable to the size of the nucleus (or smaller), and hence play no role in our everyday experience. One of these forces, the *strong force*, holds the nucleus together. The other, the *weak force*, is responsible for processes such as beta decay (see Chapter 14) that tear nuclei and elementary particles apart.

Each of the four fundamental forces is different from the others in terms of its strength and range (see Table 15–3). It is important to understand the role of these four forces, because whenever anything happens in the universe, whenever an object changes its motion, it happens because one or more of these forces is acting.

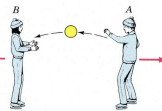

Figure 15–4
The exchange of a baseball between two skaters provides an analogy for the exchange of a gauge particle. The skater who throws the ball (*A*) recoils, and the other (*B*) recoils when the ball reaches her. Thus both skaters change direction, and, by Newton's first law, we say that a force acts between them.

gauge particle   A particle that produces a force when exchanged by two other particles; examples include photons, gravitons, gluons, and W and Z particles.

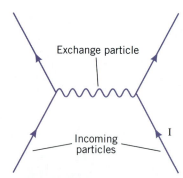

Figure 15–5
Exchange diagrams, introduced by physicist Richard Feynman, provide a model for particle interactions and the fundamental forces. Two incoming particles (such as two electrons) exchange a gauge particle (a photon) and are thus deflected by the force.

Table 15–3   •   **The Four Forces**

Force	Relative Strength*	Range	Gauge Particle
gravity	10^{-39}	infinite	graviton
electromagnetic	$\frac{1}{137}$	infinite	photon
strong force	1	10^{-13} cm	gluon
weak force	10^{-5}	10^{-15} cm	W and Z

* Relative to the strong force.

Force as an Exchange

As you have learned throughout this text, nothing happens without a force. We have talked about the gravitational force, the electromagnetic force, the strong force, and the weak force. Each has its own distinctive effects on nature. Now we'll examine more about how these forces work.

The modern understanding of forces may be thought of schematically, as in Figure 15–4. Every force between two particles corresponds to the exchange of a third kind of particle, called a **gauge particle,** for historical reasons. That is, a first particle (an electron, for example) interacts with a second particle (say another electron) by exchanging a gauge particle. The gauge particles produce the fundamental forces, like electricity, that hold everything together.

In Chapter 4, we used the analogy of someone standing on roller skates throwing baseballs to explain Newton's third law of motion. Suppose this same person was throwing baseballs, but another person standing on roller skates caught those baseballs some distance away. The person who threw the first baseball would recoil, as we discussed. The person who subsequently caught the baseball would also recoil. We could describe the situation this way: Two people stand still before anything happens. After some time, the two people are moving away from each other. From Newton's first law, we conclude that a repulsive force had acted between those two people. Yet it's very clear in this analogy that that repulsive force is intimately connected with (a physicist would say "mediated by") the exchange of the baseballs.

In the same way, scientists believe that every fundamental force is mediated by the exchange of some kind of gauge particle (see Figure 15–5). For example, the electrical force is mediated by the exchange of *photons.* That is, the magnet holding notes onto your refrigerator is exchanging huge numbers of photons with atoms inside the refrigerator metal to generate the magnetic force.

Similarly, the gravitational force is mediated by particles called *gravitons.* Right now you are exchanging large numbers of gravitons with the Earth, and it is this exchange that prevents you from floating up into space. The four fundamental forces and the gauge particles that are exchanged to generate each of them are listed in Table 15–3.

The two familiar forces of gravity and electromagnetism act over long distances because they are mediated by massless, uncharged particles (such as the photon). The weak interaction, on the other hand, has a short range because it is mediated by the exchange of very massive particles—the W and Z particles—that have masses about 80 times that of the proton. Like the photon, the W and Z are particles that can be seen in the laboratory. They were first discovered in 1983, and are now routinely produced at accelerators around the world.

The situation with the strong force is a bit more complicated. The force that holds quarks together is mediated by particles called *gluons* (they "glue" the hadrons together). These particles are supposed to be massless, like the photon, but like the quarks, they are confined to the interior of particles.

Unified Field Theories

Although a universe with six kinds of quarks, six kinds of leptons, and four kinds of forces may seem to be a relatively simple one, physicists have discovered an even greater underlying simplicity. The four fundamental forces turn out not to be as different from each other as their properties might at first suggest. The current thinking is that all four of these "fundamental" forces may be simply different aspects of a single underlying force. Theories in which the four forces are seen as different aspects of a single force go under the general name of **unified field theories.**

Richard Feynman (1918–1988), Nobel Prize–winning physicist.

unified field theory The general name for any theory in which fundamental forces are seen as different aspects of the same force.

Scientists believe that the four forces now appear to be different because we are observing them at a time when the universe has been around for a long time and is at a relatively low temperature. The situation is somewhat analogous to freezing water. When water freezes, it can adopt many apparently different forms. It can occur as powdered white snow, solid ice blocks, delicate hoar frost on tree branches, or a slippery layer on the sidewalk. You might interpret these forms of frozen H_2O as very different things, and in certain respects they are distinct; however, when subjected to heat, they all melt and take the form of water.

The first unified field theory in history was Isaac Newton's synthesis of Earth's gravity and the circular motions observed in the heavens. To medieval scientists, earthly and heavenly motions seemed as different as the strong and electromagnetic forces do to us. Nevertheless, they were unified in Newton's theory of universal gravitation. In a similar fashion, scientists today are working to unify the four fundamental forces, which look different only at the relatively low temperatures of our present existence. Scientists suspect that if matter is heated to trillions of degrees, the different forces are not really different at all.

Much of the research at the world's largest particle accelerators is designed to show that if the temperature can be raised high enough—that is, if enough energy can be pumped into an elementary particle—the underlying unity of the forces will become clear. At a few laboratories around the world, it is possible to take protons and antiprotons (or electrons and positrons), accelerate them to extremely high energies, and let them collide. (As noted above, such collisions involve the process of annihilation between particle and antiparticle). When these collisions occur, for a brief moment the temperature in the volume of space about the size of a proton is raised to temperatures that have not been seen in the universe since it was less than a second old. In the resulting maelstrom, particles are produced that can be accounted for only by assuming that the electromagnetic and weak forces become unified.

In 1983, experiments at the European Center for Nuclear Research and the Stanford Linear Accelerator Center demonstrated conclusively that this kind of unification does occur. When protons and antiprotons (at the former laboratory) or electrons and positrons (at the latter) were accelerated and allowed to collide head-on, W and Z particles were seen in the debris of the collisions. Not only were the reactions seen, but the properties of the resulting particles and their rates of production were exactly those predicted by the first unified field theories.

The expectations of today's physicists regarding the unification of forces are illustrated in a simple flow diagram (Figure 15–6). Scientists have already seen the unification of the electromagnetic and weak forces in their laboratories. The resulting force, which physicists call the *electroweak force*, will be studied in great detail by the next generation of particle accelerators. At much higher energies—energies that will probably never be attained in Earth-based laboratories—the strong force is expected to unify with the electroweak. The theories that make this prediction constitute the *standard model*—the best model we have today of the elementary particles. Physicists have accumulated a fair amount of experimental evidence supporting the standard model.

Figure 15–6
The four forces become unified at extremely high temperatures, equivalent to those at the beginning of the universe. At 10^{-43} second after the moment of creation, the universe had already cooled sufficiently that gravity separated from the other three forces. The strong force separated at 10^{-33} second, while the weak and electromagnetic forces separated at 10^{-10} second.

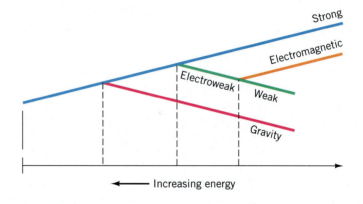

Finally, at still higher energies, physicists predict that the force of gravity will unify with the strong-electroweak force. No theory yet describes this unification successfully, and attempts to develop a viable model are very much a frontier issue.

The unification of the four forces has no obvious effect on our everyday world, since it requires such exotic high-energy conditions. Nevertheless, the unification of forces played a crucial role in the formation of the universe. Much of our understanding of what the universe is and how it came to be in its present state (Chapter 22) revolves around the various kinds of unification of the forces of nature.

Science in the Making

TOEs and TOOs

Unified theories that attempt to explain the existence of quarks, leptons, and the equivalence of all the forces in a single set of equations have been nicknamed *Theories Of Everything*, or TOEs. Some scientists—for example, Cambridge University physicist Stephen W. Hawking—believe that the discovery of a valid TOE will be the supreme scientific discovery. In his best-selling book, *A Brief History of Time*, he writes:

> If we do discover a complete theory, it should in time be understandable in broad principle by everyone, not just a few scientists. Then we shall all, philosophers, scientists, and just ordinary people, be able to take part in the discussion of the question of why it is that we and the universe exist. If we find the answer to that, it would be the ultimate triumph of human reason—for then we would know the mind of God.

Not all scientists agree with Hawking. Even the most elegant theory of everything is incomplete without additional knowledge of the organizing principles. It is doubtful, for example, that a physicist could start with a theory of everything and predict the existence of life. Thus we also need to find *Theories Of Organization*, or TOOs, that reveal how matter and energy are ordered into larger, complex systems. Such theories may or may not embody new general principles.

The relationship between TOEs and TOOs is something like that between computer hardware and software. A complete description of a computer system should include hardware diagrams that describe the machine down to every wire, screw, and microchip. But hardware by itself does nothing. Our description must also include listings and instructions for the computer program, which is what allows the computer to process information. ●

THINKING MORE ABOUT
ELEMENTARY PARTICLES

The Superconducting Supercollider

Until 1993, physicists were building what probably would have been the last of the great particle accelerators. The Superconducting Supercollider (SSC), begun in the rolling Texas prairie south of Dallas, had been hailed as the greatest high-tech engineering project in history. The SSC was designed basically as a synchrotron, but one of gigantic proportions. In a circular tunnel some 85 kilometers (about 50 miles) around, 100 meters (about 300 feet) underneath the surface, particles were to have circulated through a series of electromagnets made from superconducting wire (see Chapter 13). The goal was to achieve potentials of 80 trillion volts. While bunches of protons sped around through the system one way, antiprotons were to have circulated the other. When both groups of particle achieved their highest energy, they would have been allowed to collide head-on, so that all of their energy would have been dumped into a volume about the size of a proton, raising its temperature to unprecedented levels on Earth. In the resulting maelstrom, physicists

Had it been completed, the Superconducting Supercollider, here shown under construction, might have unraveled the mystery of why matter has mass.

hoped to see the details of the unification of weak and electromagnetic forces and, perhaps, to learn why particles have mass.

That dream is dead, defeated by a budget-conscious Congress, which felt it could not allocate the $10 billion required for the project. You might think that all scientists would have been enthusiastic about building the SSC, but many weren't. (The authors, for example, had completely different views of the desirability of building this machine.)

Proponents argued that maintaining the traditional U.S. leadership at the frontier of science required that the SSC be built here. They pointed out that the romance of finding the ultimate answers to questions about the nature of the universe would help attract young people into careers in science, and that the payoff for understanding the phenomenon of mass would have been even greater than that which followed from the nineteenth-century research into the nature of elec-

tricity. Like their nineteenth-century counterparts, they could not predict what the payoff from the SSC might have been.

Opponents argued that this enormous expenditure would result in a lowered support for other areas of science research and education, while the SSC would benefit only a few highly visible scientists. They supported more international cooperation with similar (though smaller) efforts in Europe. Also, they predicted that much more economic benefit would come from discoveries in small laboratories than from one more mega-experiment.

What is the best thing for governments to do when one field of science needs a large and expensive facility like the SSC? How much weight should be given to immediate payoffs, and how much to the long-term benefits of basic research? Is it essential that the United States be the leader in all areas of science?

▶ Summary

High-energy physics or *elementary-particle physics* deals with bits of matter that we cannot see, and forces and energies that we can barely imagine. Nevertheless, the study of the subatomic world holds the key to understanding the structure and organization of the universe.

All matter is made up of atoms, which are made up of even smaller particles—electrons and the nucleus—but these are not the most fundamental building blocks of the universe. Physicists originally studied the collisions between energetic *cosmic rays* and nuclei to learn about elementary particles. To help them with this quest, they now employ *particle accelerators*, including *synchrotrons* and *linear accelerators* to collide charged particles at near-light speeds. These scientists have discovered over 200 subatomic particles. One class of particles, the *leptons* (including the electron and neutrino), are not subject to the strong force and thus do not participate in holding the nucleus together. Nuclear particles called *hadrons* (including the proton and neutron), according to present theories, are made from *quarks*, which are odd particles that have fractional electrical charge and cannot exist alone in nature. Together, leptons and quarks are the most fundamental building blocks of matter that we know. Each of these particles has an *antimatter* particle, such as the *positron*, which is the positively charged *antiparticle* of the electron.

The four known forces—gravity, electromagnetism, the strong force, and the weak force—cause particle interactions that lead to all of the organized structures we see in the universe. Particle interactions are mediated by the exchange of *gauge particles*, a different gauge particle for each of the different forces. Two masses, for example, will exchange gravitons (the gauge particle of gravity) as they as are attracted to each other, and two charged particles will exchange photons, much the same way that two people will be "repelled" by each other if a mass is thrown from one to the other.

While the four known forces appear quite different to us, scientists believe that early in the universe, when temperatures were extremely high, the four forces were unified into a single force. At the forefront of modern physics research is the search for a *unified field theory* that describes this single force.

▶ Review Questions

1. What is reductionism?

2. How does this philosophy manifest itself in science?

3. What are fundamental building blocks of a library? Why is there more than one correct answer?

4. What is a synchrotron and how does it work?

5. How can we detect the presence of elementary particles?

6. Describe an early experiment that demonstrated the existence of elementary particles.

7. What is a lepton, and how do we know it exists?

8. Give two examples of a lepton.

9. What is a hadron, and how do we know it exists?

10. Give two examples of a hadron.

11. Describe the differences between a lepton and a hadron.

12. What is antimatter, and how do we know it exists?

13. What is the antiparticle of an electron? What are its properties?

14. Describe the process of annihilation.

15. What is a linear accelerator and how does it work?

16. What are cosmic rays and from where do they originate?

17. What is a quark? Is there any way to prove that it exists?

18. How many different kinds of quarks are there?

19. Describe how quarks and leptons are put together to make all the matter we see.

20. What are the properties of quarks?

21. Explain what it means for quarks to be confined.

22. What are the four fundamental forces? Which is the strongest? The weakest?

23. What role do gauge particles play in the universe?

24. What particle is exchanged to generate each of the four fundamental forces?

25. What is a unified field theory? Give an example.

26. In what sequence will the fundamental forces unify?

27. What is a PET scan? How does it employ elementary particles?

28. What was the SSC? Why was it discontinued?

29. What are gravitons and how do they work?

30. What is the difference between TOEs and TOOs?

▶ Fill in the Blanks

Complete the following paragraphs with words and phrases from the list.

antiparticle
cosmic rays
gauge particles
hadrons
high-energy physics, or elementary particle physics
leptons
linear accelerators
particle accelerators
positron
quarks
synchrotrons
unified field theories

_____ is the study of the particles that make up the atomic nucleus. Originally, such studies were done by allowing _____ to collide with nuclei, but currently _____ provide a steady, controllable beam of particles. The two types of accelerators are _____, in which particles travel in a circular path, and _____, in which they travel in a straight line.

The two kinds of elementary particles are _____, which can be found in the nucleus, and _____, like the electron, which cannot. For every particle, it is possible to make an _____ that is identical in mass but has opposite charge. The antiparticle of the electron is the _____.

Hadrons are made from _____, and forces between particles are due to the exchange of _____. Theories that see different forces as being aspects of the same fundamental force are called _____.

▶ Discussion Questions

1. Identify what might be considered the "fundamental units" and the rules of organization of (a) a grocery store, and (b) a parking garage. How many levels of organization can you identify? (Remember, not all questions have only one correct answer.)

2. Which particle-antiparticle interaction releases more energy: an electron-positron annihilation, or a proton-antiproton annihilation? How does the law of conservation of energy come into play?

3. What are the similarities in the modern argument over the reality of quarks and the nineteenth-century argument over the reality of atoms? What are the differences?

4. Why is it appealing for physicists to think that there are only six kinds of quarks, as opposed to more than a hundred hadrons?

5. What observations led Carl Anderson to conclude that he had discovered a particle with the same mass as the electron, but with a positive electrical charge?

6. What does it mean to say that all four fundamental forces were unified?

7. How might you detect the presence of a charged elementary particle?

8. How will we know when we have identified the truly fundamental building blocks?

9. Do you think the SSC should have been continued? If so, how can you justify such an expenditure? If not, how would you respond to an accusation that your position prevents scientific progress?

10. How does the unified field theory help physicists understand how the universe was formed?

11. Using a diagram or outline, show the various levels of the structure of matter that scientists have developed from large matter such as rocks, rain, human beings, etc., down to the basic quark level. For each stage, include a general statement of the defining rules that govern each level, and some indication of the dimensions for each level.

12. As stated in your text, elementary-particle physics is also known as high-energy physics. How is the adjective high-energy related to our understanding of fundamental elementary particles?

13. What are cosmic rays and why are they significant in the development of elementary-particle research?

14. Are there any differences in detecting charged particles such as protons and uncharged particles such as neutrons?

15. What do cosmic rays and particle accelerators have in common?

16. Leptons (weakly interacting) and hadrons (strongly interacting) were so named due to the force by which each participates with the nucleus. What force is this and how does it relate to the adjectives weak and strong?

17. How do quarks contrast with electrons with respect to electric charge?

18. Classify the four fundamental forces according to their increasing relative strengths and the range of their mediation.

19. From the perspective of gravitons, why do you remain on the surface of the Earth and not float off into space?

20. Classify the four gauge particles according to their mediating fundamental force, mass, charge, and the range over which they are mediated.

▶ Problems

1. What is the electric charge of an antiproton? An antineutron? Why?

2. A hadron called the lambda particle is made from two down quarks and one strange quark. What is the charge of the lambda particle?

3. A particle called the pi meson is made from an up quark and an anti-down quark. What is the charge of this particle?

4. A proton and an antiproton, both at rest with respect to each other, mutually annihilate into two gamma rays. How much energy is produced in this annihilation?

5. One naturally occurring reaction in a radioactive decay process is nitrogen-12 (atomic number 7) decaying into carbon-12 (atomic number 6) plus an unknown particle. What are the properties (atomic mass, number, and charge) of this unknown particle and how does it compare with the properties of an electron?

6. Some astronomers have theorized that galaxies (matter) and antigalaxies (antimatter) have existed. Each galaxy contains about 10^{11} masses of the Sun (the mass of the Sun is 1.9×10^{30} kg). Calculate the total energy released if a matter-antimatter galaxy pair annihilated.

7. Temperature can be related to energy via the equation $E = \frac{1}{2} kT$, where $k = 1.38 \times 10^{-23}$ J/K and the temperature T is in Kelvins. What is the difference in the temperatures generated between an electron-positron annihilation and a proton-antiproton annihilation?

▶ Investigations

1. How did your congressional representative and senators vote on SSC funding in Congress? Why did they vote that way? Why do you agree or disagree with their vote?

2. Discuss the SSC controversy (a) in terms of the big science versus little science dichotomy discussed in Chapter 1; and (b) in terms of the basic research versus applied research dichotomy.

3. Locate the nearest PET-scan facility and arrange a visit. Where do the physicians obtain the special form of glucose used in the procedure? What kind of educational training would you need to operate such a facility?

4. Watch an episode of "Star Trek" and discuss the use of matter and antimatter in the propulsion system of the *Enterprise*. Can you find any other uses of antimatter in science fiction stories?

▶ Additional Reading

Lederman, Leon. *The God Particle*. New York: Houghton-Mifflin, 1993.

Riordan, Michael. *The Hunting of the Quark*. New York: Simon and Schuster, 1987.

Trefil, James. *From Atoms to Quarks*. New York: Anchor, 1993.

Weinberg, Steven. *Dreams of a Final Theory*. New York: Pantheon, 1993.

16 ALBERT EINSTEIN AND THE THEORY OF RELATIVITY

ALL OBSERVERS, NO MATTER WHAT THEIR FRAME OF
REFERENCE, SEE THE SAME LAWS OF NATURE.

A
RANDOM
WALK

The Airport

You walk down the ramp, enter the plane, and get into your seat. It's
been a long day of classes, and maybe an exam or two. You doze off
for a while, then wake up with a start. For a moment, it appears to you
that the plane has started to move backward, away from the gate.
Then, as you become more alert, you realize that your plane isn't moving at all, but that another plane had pulled into the gate next to
yours.

During that brief moment between sleep and waking, you were
seeing the world through eyes unaffected by years of experience. You
were realizing that there is more than one way to view uniform motion. One way is to say that you are stationary and the other plane is
moving in relation to you. But you could also say that the other plane
is stationary and you are moving in relation to it.

Which point of view is right?

One of the great scientific discoveries of the early twentieth century, the theory of relativity, grew out of thinking about this sort of
question.

FRAMES OF REFERENCE

frame of reference The physical surroundings from which a person observes and measures the world.

A **frame of reference,** or *reference frame,* is the physical surroundings from which you observe and measure the world around you. If you read this book at your desk or in an easy chair, you experience the world from the reference frame of the room in which you are sitting, which is firmly attached to the solid Earth. If you read on a train or in a plane, your frame of reference is the vehicle that moves with respect to the Earth's surface. And you could imagine yourself in an accelerating spaceship in deep space, where your frame of reference would be different still. In each of these references frames you are what scientists call an "observer." An observer looks at the world from a particular frame of reference; some have a casual interest, while others pursue full-fledged laboratory investigations of phenomena. These investigations lead to a determination of natural laws.

For human beings who grow up on Earth, it is natural to think of the ground as a fixed, immovable frame of reference. After all, train or plane passengers don't think of themselves as stationary while the countryside zooms by. But, as we saw in the opening example, there are times when we lose this prejudice and see that the question of who is moving and who is standing still is largely one of definition.

From the point of view of an observer in a spaceship above the solar system, there is nothing "solid" about the ground you're standing on. The Earth is rotating on its axis and moving in an orbit around the Sun, while the Sun itself is performing a stately rotation around the Galaxy. Therefore, even though our reference frame tells us that the Earth is fixed in position, another reference frame would show a very different view of our rotating planet.

Descriptions in Different Reference Frames

A group of observers in different reference frames may provide very different accounts of the same event. To convince yourself of this idea, think about a simple experiment. While riding in a car, take a coin out of your pocket and flip it.

You know what will happen: the coin will go up in the air and fall straight back into your hand, just as it would if you flipped it while sitting in a chair in your room. But now ask yourself how a friend standing along the roadside, watching your car go by, would describe that flip of the coin?

To that observer, it would appear that the coin went up into the air, but by the time it came down again the car would have traveled some distance down the road. As far as your friend on the ground is concerned, the coin traveled in an arc (see Figure 16–1).

The perception of the coin's action differs between you and your friend. You, sitting in the car, say the coin went straight up and down, while someone on the ground says it travels in an arc. You are both correct in your respective frames of reference. The universe we live in possesses this general feature: different observers will describe the same event in different terms.

Does this mean that we live in a world where nothing is fixed, where everything depends on the frame of reference of the observer? Not necessarily. The possibility exists that even though different observers give

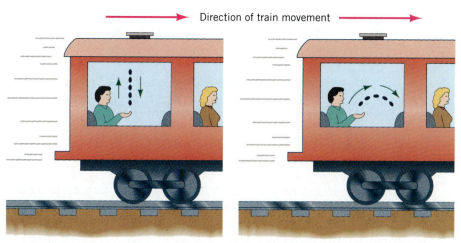

Direction of train movement

Apparent direction of coin fall

(a) Frame of reference: inside the train   (b) Frame of reference: outside the train

Figure 16–1
The path of a coin flipped in the air depends on the observer's frame of reference. (a) A rider in the car sees the coin go up and fall straight down. (b) An observer on the street sees the coin follow an arching path.

different descriptions of the same event, they will agree on the underlying laws that govern it. Even though the observers disagree on the path followed by the flipped coin, they may very well agree that motion in their frame is governed by Newton's laws of motion and the law of universal gravitation.

The Principle of Relativity

Albert Einstein's theories of relativity were a result of Einstein's realization that there was a fundamental contradiction between Newton's laws and Maxwell's equations. You can understand the problem by thinking about a simple example that will help illustrate this contradiction. Imagine you're on a moving railroad car and you throw a baseball. At what speed will the baseball travel according to a ground-based observer?

If you throw the ball forward at 40 kilometers per hour from a train traveling 100 kilometers per hour, the ball will appear to a ground-based observer to travel 140 kilometers per hour; 40 kilometers per hour (the speed of the ball) added to 100 kilometers per hour (the speed of the train). If, on the other hand, you throw the ball backward, the ground-based observer will see the ball moving at only 60 kilometers per hour—the train's 100 kilometers per hour minus the ball's 40 kilometers per hour. In our everyday world, we just add the two speeds to get the answer (Figure 16–2). This notion is reflected in Newton's laws.

Figure 16–2
An observer on the ground sees the baseball moving with a speed equal to the sum of the speed with which it's thrown and the speed of the train. If the ball is thrown in the same direction as the train (a) the speed seen from the ground is $u + v$, while if it's thrown backward (b), the speed is seen to be $u - v$.

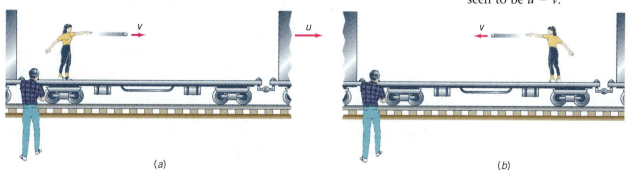

(a)   (b)

Figure 16–3
Both the observer on the ground and the observer on the train must see the speed of light as being c, regardless of the motion of the train.

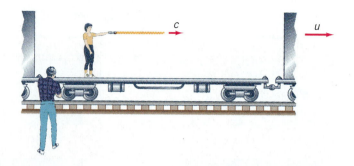

Suppose, however, that instead of throwing a ball you turned on a flashlight and measured the speed of its light. In Chapter 8, we saw that the speed of light is built into Maxwell's equations. If every observer is to see the same laws of nature, they all have to see the same speed of light. In other words, the ground observer would have to see light from the flashlight moving at 300,000 kilometers per second, and *not* 300,000 kilometers per second plus 100 kilometers per hour (Figure 16–3). In this case, velocities wouldn't add, as our intuition tells us they must.

Albert Einstein thought long and hard about this paradox, and he realized that it could be resolved in only one of three ways:

1. The laws of nature are not the same in all frames of reference (an idea Einstein was reluctant to accept on philosophical grounds).
2. Maxwell's equations could be wrong and the speed of light depends on the speed of the source emitting the light (in spite of abundant experimental evidence to the contrary).
3. Our intuitions about the addition of velocities could be wrong.

Einstein accepted the third of these possibilities, and set about to examine its consequences.

Einstein assumed that the laws of nature are the same in all frames of reference, an idea called the *principle of relativity*. This principle can be stated as follows:

> **Every observer must experience the same natural laws.**

theory of relativity An idea that the laws of nature are the same in all frames of reference and that every observer must experience the same natural laws.

This statement is the central assumption of Einstein's **theory of relativity.** Hidden beneath this seemingly simple statement lies a view of the universe that is both strange and wonderful. The extraordinary theoretical effort required to understand the consequences of this one simple assumption occupied Einstein during the first decades of the twentieth century.

Einstein's work becomes more accessible by recalling what Isaac Newton had demonstrated three centuries earlier: that all motion falls into one of two categories—uniform motion or acceleration (Chapter 3). Einstein, therefore, divided his theory of relativity into two parts, each dealing with one of these kinds of motion. The easier part, first published by Einstein in 1905, is called **special relativity** and deals with all frames of reference in uniform motion relative to one another—reference frames

special relativity The first of two parts of Einstein's theory of relativity, dealing with reference frames that do not accelerate.

that do not accelerate. It took Einstein another decade to complete his treatment of **general relativity,** which is mathematically a much more complex theory. General relativity applies to any reference frame whether or not these frames are accelerating relative to one another.

At first glance, the underlying principle of relativity seems obvious. The laws of nature *must* be the same everywhere—that's the only way in which scientists can explain how the universe behaves in an ordered way. But once you accept this central assumption of relativity, be prepared for some surprises. Relativity forces us to accept the fact that nature doesn't always behave as our intuition says it must. You may find it disturbing that nature sometimes violates our sense of the "way things should be." But you'll have little problem with relativity if you just accept the idea that the universe is what it is, and not necessarily what we think it should be.

Another way of saying this is to note that our intuitions about how the world works are built up from experience with things that are moving at modest speeds—a few hundred, or at most a few thousand, miles per hour. None of us has any experience with things moving near the speed of light, so when we start examining phenomena in that range, our intuitions don't necessarily apply. Strictly speaking, we shouldn't be surprised by anything we find.

general relativity   The second and more complex of two parts of Einstein's theory of relativity, which applies to all reference frames, whether or not those frames are accelerating relative to each other.

Albert Einstein (1879–1955) in the early 1900s, at his desk in the Bern Patent Office.

Relativity and the Speed of Light

As the example of the train and the flashlight shows, one of the most disturbing aspects of the principle of relativity has to do with our common notions of speed. According to the principle, any observer, no matter what his or her reference frame, should be able to confirm Maxwell's description of electricity and magnetism. Since the speed of light is built into these equations, it follows that

> The speed of light, c, is the same in all reference frames.

Although, strictly speaking, this statement is one of many consequences of the principle of relativity, so many of the surprising results of relativity follow from it that it is often accorded special status and given special attention in discussions of those theories.

Science in the Making

Einstein and the Streetcar

Newton and his apple have entered modern folklore as a paradigm of unexpected discovery. A less-well-known incident led Albert Einstein, then an obscure patent clerk in Bern, Switzerland, to relativity.

One day, while riding home in a streetcar, he happened to glance up at a clock on a church steeple (Figure 16–4). In his mind, he imagined the streetcar speeding up, moving faster and faster, until it was going at almost the speed of light. Einstein realized that if the streetcar were to accelerate to light speed, it would appear to someone on the

Figure 16–4
Albert Einstein, speeding away from a clock tower, imagined how different observers might view the passage of time. If Einstein were traveling at the speed of light, for example, the clock would appear to him to have stopped, even though his own pocket watch would still be ticking.

streetcar that the clock had stopped. The passenger would be like a surfer on a light wave crest—a crest that originated at 12 noon, for example—and the same image of the clock would stay with him.

On the other hand, a clock moving with him—his pocket watch, for example—would still tick away the seconds in its usual way. Perhaps, Einstein thought, *time as measured on a clock, just like motion, is relative to one's frame of reference.* ●

SPECIAL RELATIVITY

Special relativity deals with reference frames that do not accelerate, and thus is mathematically more accessible than general relativity. A few calculations can convince you how paradoxical time and distance can be in our universe when we travel near the speed of light.

Time Dilation

Think about how you measure time. The passage of time can be measured by any kind of regularly repeating phenomenon, such as a swinging pendulum, a beating heart, or an alternating electrical current. To understand the theory of relativity, it's easiest to think of an unusual kind of clock. Suppose, as in Figure 16–5, we had a flashbulb, a mirror, and a photon detector. The "tick-tock" of this clock would consist of the flashbulb going off, the light traveling to the mirror, bouncing down to the detector, and then triggering the next flash. By adjusting the distance between the light source and mirror, these pulses could correspond to any desired time interval. This unusual "light clock," therefore, serves the same function as any other clock; in fact, you could adjust it to be synchronized with anything from a grandfather's clock to a wristwatch.

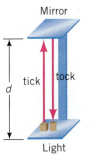

(*a*) Stationary light clock

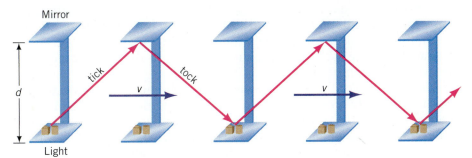

(*b*) Moving light clock

Figure 16–5
A light clock incorporates a flashing light and a mirror. A light pulse bounces off the mirror and returns to trigger the next pulse. Two light clocks, one stationary (*a*) and one moving (*b*), illustrate the phenomenon of time dilation. Light from the moving clock must travel farther, and so it appears to a stationary observer to tick more slowly.

Now imagine two identical light clocks: one is next to you on the ground, the other is whizzing by in a spaceship. Assume that the mirrors are adjusted so that both clocks would tick at the same rate if they were standing next to each other. How would the moving clock look to you?

Standing on the ground, you would see the ground-based clock ticking along as the light pulses bounce back and forth between the mirror and detector. When you looked at the moving clock, though, you would see the light following a longer, zig-zag path. If the speed of light is indeed the same in both frames of reference, it should appear to you that the light in the moving frame takes longer to travel the zig-zag path from light to detector than the light in the ground-based clock. Consequently,

An atomic clock at the National Institute of Standards and Technology in Boulder, Colorado. Precise clocks like this run measurably more slowly when strapped into a speeding jet aircraft.

time dilation A phenomenon in special relativity in which moving clocks appear to tick more slowly.

from your point of view on the ground, the moving clock must tick more slowly. The two clocks are identical, but *the moving clock runs slower.* This surprising phenomenon is known as **time dilation,** and it is an essential consequence of relativity.

Remember that each observer regards the clocks in his or her own reference frame as completely normal, while all other clocks appear to be running slower. Thus, paradoxically, while we observe the spaceship clock as slow, observers in the speeding spaceship see the Earth-based clock moving and, as above, believe that the Earth-based clock is running more slowly than theirs.

Relativity's prediction of time dilation can be tested in a number of ways. Scientists have actually documented relativistic time dilation by comparing two extremely accurate atomic clocks, one on the ground and one strapped into a jet aircraft. Even though jets travel at a paltry hundred-thousandth of the speed of light, the difference in the time recorded by the two clocks can be measured.

Time dilation can also be observed with high-energy particle accelerators that routinely produce unstable subatomic particles (see Chapter 15). The normal half-life of these particles is well known. When accelerated to near the speed of light, however, these particles last much longer because of the relativistic slowdown in the decay rates.

Thus, although the notion that moving clocks run slower than stationary ones violates our intuition, it seems to be well documented by experiment. Why, then, aren't we aware of this effect in our daily lives? To answer that question, we have to ask how big an effect time dilation is. How much do moving clocks slow down?

The Size of Time Dilation

In general, we have tried to talk about science in familiar terms and stay away from inaccessible formulas and language in this book. But we have now run into a rather fundamental question that requires simple mathematical calculations. In this section, you'll be able to follow the kind of thought process that Einstein used when he first formulated his revolutionary theory.

Consider the two identical light clocks in Figure 16–5, one moving at a velocity, v, and one stationary on the ground. Each clock has a

Table 16–1   •   Symbols for Deriving Time Dilation Description

Symbol	Description
v	The velocity of the moving light clock relative to the ground
d	The distance between the clock's light and mirror
t_{GG}	Time for one tick (ground clock, ground observer)
t_{MG}	Time for one tick (moving clock, ground observer)
t_{GM}	Time for one tick (ground clock, moving observer)
t_{MM}	Time for one tick (moving clock, moving observer)
c	The speed of light, a constant.

light-to-mirror separation distance of d. The various symbols used are summarized in Table 16–1.

The notation for the time it takes for light to travel the distance d from the light to its opposite mirror—that is, one "tick" of the stationary clock—is a little trickier, because we have to keep track of which clock we're looking at *and* from which reference frame we're looking. We will use two subscripts: the first subscript will tell us if the clock is on the ground (G) or moving (M), and the second subscript will indicate if the observer is on the ground or moving. Thus, t_{GG} is the time for one tick of the ground-based clock as noted by a ground-based observer, and t_{MG} is the time for one tick of the moving clock from the point of view of this ground-based observer. According to the principle of relativity, all observers see clocks in their own reference frames as normal. The ground-based observer sees the ground-based clock behaving exactly as the moving observer sees the moving clock. In equation form, this idea is stated as

$$t_{GG} = t_{MM}$$

As ground-based observers, we are interested in determining the relative values of t_{GG} and t_{MG}—what we see as "ticks" of the stationary versus the moving clocks. In the stationary ground-based frame of reference, one tick is simply the time it takes light to travel the distance d:

$$\text{time} = \frac{\text{distance}}{\text{speed}}$$

Substituting values for the light clock into this equation, we have

$$\text{time for one tick} = \frac{\text{light-to-mirror distance}}{\text{speed of light}}$$

or

$$t_{GG} = \frac{d}{c}$$

where c is the standard symbol for the speed of light (3×10^8 m/s).

We argued that, to the observer on the ground, it appears that the light beam in the moving clock travels on a zig-zag path as shown in Figure 16–6, and that this made the moving clock appear to run more slowly. In what follows, we will show how to take such an intuitive statement and convert it into a precise mathematical equation. We begin by labeling the dimensions of our two clocks.

The moving clock travels a horizontal distance of $v \times t_{MG}$ during each of its ticks. In order to determine the value of t_{MG}, we must first determine how far light must travel in the moving clock as seen by the observer on the ground. As illustrated in Figure 16–4, we know the lengths of the two shortest sides of a right triangle. One side has length d, representing the vertical distance between light and mirror (a distance, remember, that is the same in both frames of reference). The other side is $v \times t_{MG}$, which corresponds to the distance traveled by the moving clock as observed in the stationary frame of reference. The distance traveled by the moving light beam in one tick is represented by the hypotenuse of this right triangle, and is given by the Pythagorean theorem:

Figure 16–6
Light clocks with dimensions labeled. Both the stationary clock (*a*) and the moving clock (*b*) have light-to-mirror distance *d*. During one tick the moving clock must travel a horizontal distance $v \times T_{MG}$.

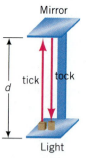

(*a*) Stationary light clock

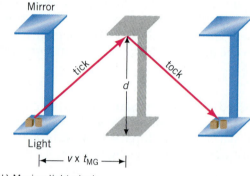

(*b*) Moving light clock

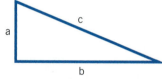

Figure 16–7
The Pythagorean theorem states that the lengths, *a*, *b*, and *c* of a right triangle are related by the formula ($a^2 + b^2 = c^2$), where *c* is the longest side. This equation is useful in the calculation of time dilation.

▶ **In words:**

The square of the length of a right triangle's long side equals the sum of the squares of the lengths of the other two sides (see Figure 16–7).

▶ **In symbols:**

$$(\text{distance light travels})^2 = d^2 + (vt_{MG})^2$$

We can begin to simplify this equation by taking the square roots of both sides.

$$\text{distance light travels} = \sqrt{d^2 + (vt_{MG})^2}$$

Applying this theorem to our light clock, the square of the distance light travels during one tick equals the sum of the squares of the light-to-mirror distance and the horizontal distance the clock moves during one tick.

We've got a lot of unknowns here, but we can eliminate some by substitution. Remember, time equals distance divided by velocity. Therefore, the time it takes light to travel this distance, t_{MG}, is given by the distance, $\sqrt{d^2 + (vt_{MG})^2}$, divided by the velocity of light, *c*:

$$t_{MG} = \frac{\sqrt{(d^2 + v^2 t_{MG}^2)}}{c}$$

Next, we must engage in a bit of algebraic manipulation. First, square both sides of this equation:

$$t_{MG}^2 = \frac{d^2}{c^2} + \frac{v^2 t_{MG}^2}{c^2}$$

But we saw previously that $t_{GG} = d/c$, so, substituting

$$t_{MG}^2 = t_{GG}^2 + \frac{v^2 t_{MG}^2}{c^2}$$

Dividing both sides by t_{MG}^2 gives

$$\frac{t_{MG}^2}{t_{MG}^2} = \frac{t_{GG}^2}{t_{MG}^2} + \frac{v^2 t_{MG}^2/c^2}{t_{MG}^2}$$

or,

$$1 = \left(\frac{t_{GG}}{t_{MG}}\right)^2 + \left(\frac{v}{c}\right)^2$$

Finally, regrouping yields:

$$t_{MG} = \frac{t_{GG}}{\sqrt{1 - (v/c)^2}}$$

This equation expresses in mathematical form what we said earlier in words; moving clocks appear to run slower. It tells us that t_{MG}, the time it takes for one tick of the moving clock as seen by an observer on the ground, is equal to the time it takes for one tick of an identical clock on the ground *divided by a number less than one*. Thus the time required for a tick of the moving clock will always be greater than that for a stationary one.

The expression $\sqrt{1 - (v/c)^2}$ is a number, called the *Lorentz factor*, that appears repeatedly in relativistic calculations. In the case of time dilation, the Lorentz factor arises from an application of the Pythagorean theorem.

An important point to notice: if the velocity of the moving clock is very small compared to the speed of light, the quantity $(v/c)^2$ becomes very small and the Lorentz factor is almost equal to one. In this case, the time on the moving clock is equal to the time on the stationary one, as our intuition demands that it should be. Only when speeds get very high do the effects of relativity become important.

Science by the Numbers

How Important Is Relativity?

To illustrate why we aren't aware of relativity in our daily lives, let's calculate the size of the time dilation for a clock in a car moving at 70 km/h (about 50 mph).

First, we must convert the familiar speed in km/h to a speed in meters per second so we can compare it to the speed of light. There are $60 \times 60 = 3600$ seconds in an hour, so a car traveling 70 km/h is moving at a speed of

$$70 \text{ km/h} = \frac{70{,}000 \text{ m}}{3600 \text{ s}}$$
$$= 19.4 \text{ m/s}$$

For this speed, the Lorentz factor is

$$\sqrt{1 - (19.4/300{,}000{,}000)^2} = 0.9999999999999999$$

Thus the passage of time for a stationary and speeding car differs by only one part in the sixteenth decimal place.

To get an idea of how small the difference is between the ground clock and the moving clock in this case, we can note that if you watched the moving car for a time equal to the age of the universe, you would observe it running *10 seconds slow* compared to your ground clock.

For an object traveling at 99% of the speed of light, however, the Lorentz factor is:

$$\sqrt{1 - (v/c)^2} = \sqrt{1 - (0.99)^2}$$
$$= \sqrt{0.0199}$$
$$= 0.1411$$

The value of time dilation, 0.1411, is about $\frac{1}{7}$. In this case, you would observe the stationary clock ticking about seven times as fast as the moving one. In other words, the ground clock would tick about seven times while the moving clock ticked just once.

This numerical example illustrates a very important point about relativity. Our intuition and experience tell us that the exterior clock on our local bank doesn't suddenly slow down when we view it from a moving car. Consequently, we find the prediction of time dilation to be strange and paradoxical. But all of our intuition is built up from experiences at very low velocities—none of us has ever moved at an appreciable fraction of the speed of light. For the everyday world, the predictions of relativity coincide precisely with our experience. It is only when we get into regions near the speed of light, where that experience isn't relevant, that the "paradoxes" arise. ●

Space Travel and Aging

While human beings do not now experience the direct effects of time dilation in their day-to-day lives, some day they might. If we ever develop interstellar space travel with near-light speed, then time dilation may wreak havoc with family lives (and genealogists' records).

Imagine a spaceship that accelerates to 99% of the speed of light and goes on a long journey. While 15 years seem to pass for the crew of the ship, more than a century may go by on Earth. The space explorers could return 15 years older than when they left, but biologically younger than their great grandchildren! Friends and family would all be long-since dead.

If we ever enter an era of extensive high-speed interstellar travel, people may drift in and out of other people's lives in ways we can't easily imagine. Parents and children could repeatedly leapfrog each other in age, and the notion of "close relatives" could take on complex twists in a society with widespread relativistic travel.

Distance and Relativity

Many results from relativistic calculations, like the time dilation example above, run counter to our intuition. These results can be derived using

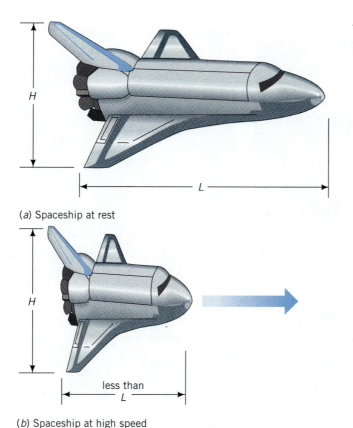

(a) Spaceship at rest

(b) Spaceship at high speed

Figure 16–8
A spaceship in motion appears to contract in length, L, along the direction of motion. The height, H, and width of the ship do not appear to change, however.

procedures similar to (but more complicated than) the one we just gave for working out time dilation. In fact, using procedures like those above, Einstein showed that *moving yardsticks must appear to be shorter than stationary ones in the direction of motion* (see Figure 16–8).

The equation that relates the ground-based observer's measurement of a stationary object's length, L_{GG}, to that observer's measurement of the length of an identical moving object, L_{MG}, is

$$L_{MG} = L_{GG} \times \sqrt{1 - (v/c)^2}$$

You should recognize the term on the right side of this equation as the Lorentz factor, which we derived from our study of light clocks. The equation tells us that the length of the moving ruler can be obtained by multiplying the length of the stationary ruler by a number less than one, and thus must appear shorter. This phenomenon is known as **length contraction.**

Note that the height and width of the moving object do not appear to change; only the length along the direction of motion. Thus a moving basketball takes on the flattened appearance of a pancake at speeds near those of light.

Length contraction is not just an optical illusion. While relativistic shortening isn't something most people observe on a daily basis, the effect is real. Physicists who work at particle accelerators inject "bunches" of particles into their machines. As these particles approach light speed,

length contraction The phenomenon in relativity that moving objects appear to be shorter than stationary ones in the direction of motion.

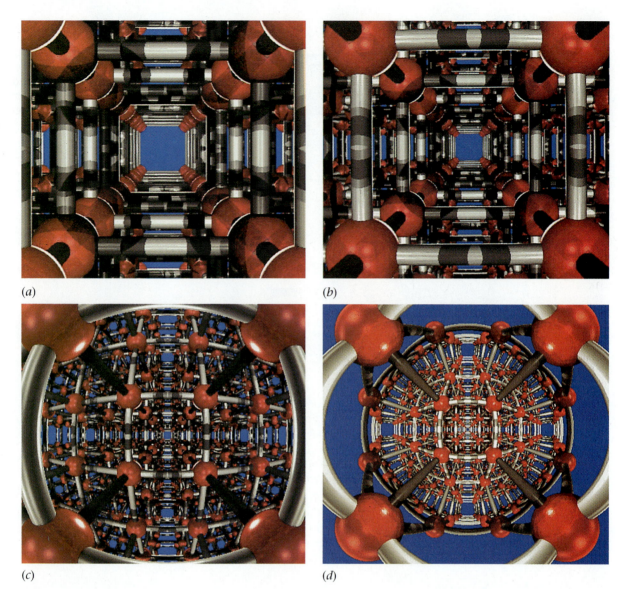

(a)

(b)

(c)

(d)

This series of four computer-generated images shows the changing appearance of a network of balls and rods as it moves toward you at different speeds. (*a*) At rest—the normal view. (*b*) At 50% of light speed, the array appears to contract. (*c*) At 95% of light speed, the lattice has curved rods. (*d*) At 99% of the speed of light, the network is severely distorted.

the bunches are observed to contract according to the Lorentz factor. This effect must be compensated for in the design of all particle accelerators.

What About the Train and the Flashlight?

Now that we understand a little more about how relativity works, we can go back and unravel the paradox we discussed earlier in this chapter—how do both an observer on the ground and an observer on the train see light from a flashlight moving at the same speed?

Recall that velocity is the distance traveled divided by the time required. Since both length and time appear to be different for different observers, it should come as no surprise that the velocities appear different as well. The Newtonian rules that tell us how to add velocities (like the velocity of the light and the train) are thus more complicated at relativistic speeds. Our intuition that tells us that we can simply add velocities (as in the example of the train and the ball) is valid at small velocities, but breaks down for objects moving near the speed of light. For those objects, a more complex addition has to be done.

The more complex addition takes into account the differences in distances and times in different frames of reference. It yields the following result: If we let u be the velocity of a moving frame of reference (for example, the train), and v be the velocity of something moving in that frame as measured in that frame (for example, the speed of the baseball or light beam), then V, the velocity of the baseball or light beam as measured by an observer on the ground, will be:

$$V = \frac{u + v}{1 + uv/c^2}$$

Notice that when u and v are very small compared to the velocity of light (as they would be for the example of the train and the baseball), the term uv/c^2 in the denominator of the fraction can be neglected. In this case, the equation becomes

$$V = u + v$$

In other words, when velocities are small compared to light, the equations of relativity tell us to just go ahead and add the velocities, precisely as our intuition dictates. This is another example in which relativity reduces to ordinary Newtonian results for small velocities.

In the case of the flashlight beam, however, the velocity we have called v, which is the speed of the light beam as measured by someone on the train, will be c, the speed of light. In this case,

$$
\begin{aligned}
V &= \frac{u + c}{1 + uc/c^2} \\
&= \frac{u + c}{1 + u/c} \\
&= \frac{c(u + c)}{(u + c)} \\
&= c
\end{aligned}
$$

In other words, *both the observer on the ground and the observer on the train see the flashlight beam moving at the same velocity.* This result violates our intuition, but it follows from the principle of relativity.

Developing Your Intuition

Simultaneity

Relativistic calculations often violate our intuition. At extreme velocities, events don't match up with our ideas about how the world should operate. This fact supplies us with a large number of seeming paradoxes to sharpen our intuition about what the world looks like at velocities near the speed of light. Here's a famous one called "The Train and the Tunnel."

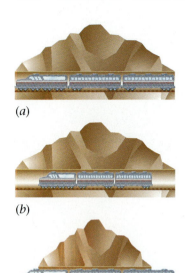

(a)

(b)

(c)

Figure 16–9
The train and the tunnel paradox. When both are at rest, they are the same length (*a*). From the point of view of the gatekeeper, the train's length should shrink when it is moving, and thus fit inside the tunnel (*b*). From the point of view of the engineer, however, the tunnel should shrink and the train should stick out at both ends (*c*).

A railroad track runs through a tunnel in a mountain. At the ends of the tunnel are two gates, which can be closed by a switch (Figure 16–9). When the train is standing still, it is exactly as long as the tunnel. The engineer of a train and the gatemaster make a bet. The engineer will get the train moving toward the tunnel, and the gatemaster will try to trap the train in the tunnel by closing the two gates at the same time. The engineer reasons as follows: In terms of relativity, I am standing still and the tunnel is moving toward me. As it approaches, it will shrink and be shorter than the train, and it will be impossible for the train to be trapped. The gatemaster, on the other hand, reasons as follows: The train will be moving toward me, and will therefore be shorter than the tunnel. I will, therefore, be able to trap it by closing the two doors.

The engineer backs up the train and comes barreling toward the tunnel. Who wins the bet?

To resolve this paradox, ask how events occurring in the gatemaster's frame of reference look to the engineer. In the Earth frame of reference, the gatemaster waits until the shrunken train is completely inside the tunnel, then closes both gates simultaneously. The engineer, however, is in a moving frame of reference. In that frame, it appears that the front gate closes first (because the engineer is moving toward it), and the back gate closes later. The engineer claims that the gatemaster cheated by closing the front gate before the train got to it, and then waited to close the back gate until the end of the train had passed through. In the end, neither one wins the bet, because relativity dictates that events that appear simultaneous in one frame will not be so in another. ●

Mass and Relativity

Perhaps the most far-reaching consequence of Einstein's theory of relativity was the discovery that mass, like time and distance, is relative to one's frame of reference. So far, we have addressed two strange conclusions of Einstein's theory of relativity:

1. Clocks run fastest for stationary objects; moving clocks slow down. As the speed of light is approached, time slows down and approaches zero.

2. Distances are greatest for stationary objects; moving objects shrink in the direction of motion. As the speed of light is approached, distances shrink and approach zero.

Einstein showed that a third consequence followed from his principle:

3. Mass is lowest for stationary objects; moving objects become more massive. As an object's velocity approaches the speed of light, its mass approaches infinity.

Einstein showed that if the speed of light is a constant in all reference frames, which must follow from the central assumption of the theory of relativity, then an object's mass depends on its velocity. The faster an object travels, the greater its mass, and the harder it is to deflect from its course. If a ground-based observer measures an object's stationary or "rest" mass, m_{GG}, then the apparent mass, m_{MG}, of that object moving at velocity v is

$$m_{GG} = m_{MG} \times \sqrt{1 - (v/c)^2}$$

Once again the Lorentz factor comes into play. As we observe an object approach the speed of light, its mass appears to us to approach infinity.

This property of mass leads to the common misperception that relativity predicts that nothing can travel faster than the speed of light. In fact, the only thing we can conclude is that nothing that is now moving at less than the speed of light can be accelerated to or past that speed. It also says that, should there exist objects already moving faster than light, they could not be decelerated to speeds of less than c, and that the only objects that travel at the speed of light (such as photons) are those that have zero rest mass.

Mass and Energy

Time, distance, and mass—all quantities that we can easily measure in our homes or laboratories—actually depend on our frame of reference. But not everything in nature is so variable. The central tenet of relativity is that natural laws must apply to every frame of reference. Light speed is constant in all reference frames in accord with Maxwell's equations. Similarly, the first law of thermodynamics—the idea that the total amount of energy in any closed system is constant—must hold no matter what the frame of reference. Yet here, Einstein's description of the universe seems to be flawed. He claimed that the observed mass depends on your frame of reference. But, in that case, kinetic energy, which is defined as mass times velocity squared, could not follow the conservation of energy law. In Einstein's treatment, an object appears to have different amounts of energy depending on its frame of reference. Where does the extra energy come from?

Conservation of energy appeared to be violated because we missed one key form of energy in our equations: mass. In fact, Einstein was able to show that the amount of energy contained in any mass turns out to be the mass times a constant.

▶ **In words:**

All objects contain a rest energy (in addition to any kinetic or potential energy), which is equal to the object's rest mass times the speed of light squared.

▶ **In word equation form:**

rest energy = rest mass $\times$ (speed of light)^2

▶ **In symbols:**

$$E = mc^2$$

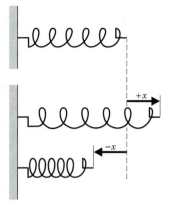

Figure 16–10
A compressed or stretched spring has a greater mass because it has elastic potential energy.

This familiar equation has become an icon of our nuclear age, for it defines a new form of energy. It says that mass can be converted to energy, and vice versa (Figure 16–10). Furthermore, the amounts of energy involved are prodigious (because the constant, the speed of light squared, is so large). A handful of nuclear fuel can power a city; a fist-sized chunk of nuclear explosive can destroy it.

Until Einstein traced the implications of special relativity, the nature of mass and its vast potential for producing energy was hidden. Now, a significant fraction of electrical energy in the United States is produced in nuclear reactors that constantly confirm the predictions of Einstein's theory (see Chapter 14).

GENERAL RELATIVITY

Special relativity is a fascinating and fairly accessible subject, requiring nothing more than an open mind and a knowledge of basic algebra. General relativity, which deals with all reference frames including accelerating ones, is much more challenging in its full rigor. While the details are tricky, you can get a pretty good feeling for Einstein's general theory by thinking about the nature of forces.

The Nature of Forces

Begin by imagining yourself in a completely sealed room that is accelerating at exactly one "*g*," the Earth's gravitational acceleration. Could you devise any experiment that would reveal one way or the other if you were on Earth or accelerating in deep space?

If you dropped this book on Earth, the force of gravity would cause it to fall to your feet (see Figure 16–11*a*). If you dropped the book in an accelerating spaceship, however, Newton's first law tells us that it will keep moving with whatever speed it had when it was released (see Figure 16–11*b*). The floor of the ship, still accelerating, will therefore come up to meet it. To you, standing in the ship, it appears that the book falls, just as it does if you are standing on Earth.

Figure 16–11
If you dropped a book in an accelerating spaceship (*a*), the book would stay in place and the floor would accelerate up to meet it (*b*). From your frame of reference in the spaceship, it would be impossible to tell the difference between this situation and one in which the book was dropped on the surface of the Earth.

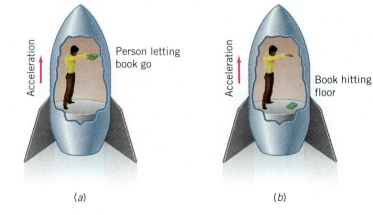

(*a*)   (*b*)

From an external frame of reference, these two situations would involve very different descriptions. In the one case, the book falls due to the force of gravity; in the other, the spaceship accelerates upward to meet the free-floating book. But no experiment you could devise *in your reference frame* could distinguish between acceleration in deep space and the force of the Earth's gravitational field.

In some profound way, therefore, gravitational forces and acceleration are equivalent. Newton saw this connection in a way, for his laws of motion equate force with an accelerating mass. But Einstein went a step further by recognizing that what we call gravity versus what we call acceleration is a purely arbitrary decision, based on our reference frame. Whether we think of ourselves as stationary on a planet with gravity, or accelerating on spaceship Earth, makes no difference in the passage of events.

Although this connection between gravity and acceleration may seem abstract, your experiences demonstrate it is true. Have you ever been in an elevator and felt momentarily heavier when it started up or momentarily lighter when it started down? If so, you know that the feeling we call "weight" can indeed be affected by acceleration.

The mathematical details required to demonstrate this notion of the equivalence of acceleration and gravity are complicated, but a simple analogy can help you visualize the difference between Einstein's and Newton's views of the universe. In the Newtonian universe, forces and motions can be described by a ball rolling on a perfectly flat surface with neatly inscribed grid lines (see Figure 16–12a). The ball rolls on and on, following a line exactly, unless an external force is applied. If, for example, a large mass rests on the surface, the rolling ball will change its direction and speed; it will accelerate in response to the force of gravity. Thus, for Newton, motion occurs along curving paths in a flat universe.

The description of that same event in general relativity is very different. In that case, as shown in Figure 16–12b, we would say that the heavy object distorts the surface. Peaks and depressions on the surface influence the ball's path, deflecting it as it rolls across the surface. For Einstein, the ball moves in a straight line across a curved universe, as opposed to the ball curving across a flat surface.

Given these differing views, Newton and Einstein would present very different descriptions of physical events. Newton would say, for example, that the Moon orbits the Earth because of an attractive gravitational force between the two bodies. Einstein, on the other hand, would say that space has been warped in the vicinity of the Earth-Moon system, and this warping of space governs the Moon's motion. In the relativistic view, space deforms around the Sun, and planets follow the curvature of space like marbles rolling around in the bottom of a curved bowl.

We now have two very different ways of thinking about the universe. In the Newtonian universe, forces cause objects to accelerate. Space and time are separate dimensions that are experienced in very different ways. This view more closely matches our everyday experience of how the world seems to be. In Einstein's universe, objects move according to distortions in space, while the distinction between space and time depends on your frame of reference.

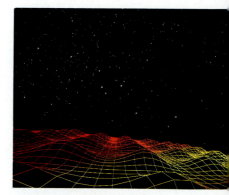

A computer-generated image of a gravity field reveals masses as wells on an otherwise flat grid.

Figure 16–12
Newtonian and Einsteinian universes treat the motion of rolling balls in different ways. In the Newtonian scheme (*a*), a ball travels in uniform motion unless acted upon by a force; motion occurs along curved paths in a flat universe. In the Einsteinian universe (*b*), a ball's mass distorts the universe; it moves in a straight path across a curved surface.

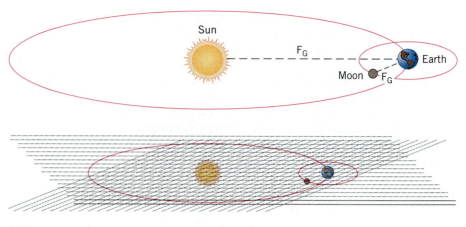

(*a*) Newtonian universe: gravitational forces in a "flat" universe

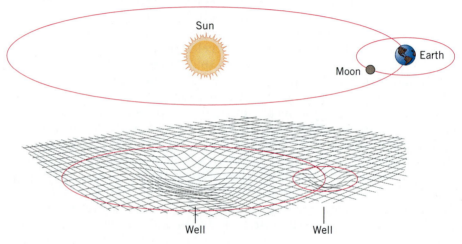

(*b*) Einstein's universe: motion in a curved universe.

Predictions of General Relativity

The mathematical models of Newton and Einstein are not just two equivalent descriptions of the universe. They lead to slightly different quantitative predictions of events. In three specific instances, the predictions of general relativity have been confirmed.

1. The Gravitational Bending of Light. One consequence of Einstein's theory is that light bends as it travels along the warped space near strong gravitational centers such as the Sun. Einstein predicted the exact amount of deflection that would occur near the Sun, and his prediction was confirmed by precise measurements of star positions during a solar eclipse in 1919 (see Figure 16–13). This observation, more than any other single event, established Einstein's international reputation and led to the widespread acceptance of his theory of general relativity. Today, these measurements are made with much more precision by measuring the deflection of radio waves emitted by distant quasars (see Chapter 22).

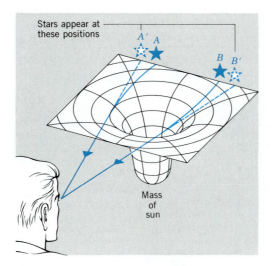

Stars appear at these positions

A' A

B B'

Mass of sun

Figure 16–13
Einstein's explanation for the change in two stars' angular separation during an eclipse. The mass of the Sun curves the space-time around it. The bent paths make stars that are really located at *A* and *B* appear to be located at *A'* and *B'*.

2. Planetary Orbits. In Newton's solar system, the planets adopt elliptical orbits, with long and short axes that remain pointed in the same directions, except for the gravitational influence of other planets. Einstein's calculations predict nearly elliptical orbits, but ones in which the axes advance slightly from orbit to orbit. In Einstein's theory, for example, the axis of the orbit of innermost planet Mercury was predicted to advance by 43 seconds of arc per century due to relativistic effects—a small perturbation superimposed on much larger effects due to the other planets. Einstein's prediction almost exactly matched the observed shift in Mercury's orbit.

3. The Gravitational Redshift. The theory of relativity predicts that as a photon (a particle of electromagnetic radiation) moves up in a gravitational field, it must lose energy in the process. The speed of light is constant, so this energy loss is manifest as a slight decrease in frequency (a slight increase in wavelength, known as the *Doppler effect*, as described in Chapter 8). Thus, lights on the Earth's surface will appear slightly redder than they do on Earth if they are observed from space. By the same token, a light shining from space to the Earth will be slightly shifted to the blue end of the spectrum. Careful measurements of laser light frequencies have amply confirmed this prediction of relativity.

Recent advances in highly sensitive electronics are now providing more opportunities for researchers to measure the predictions of general relativity. One of the most intriguing tests will be conducted as part of a satellite mission in the mid-1990s. Meticulously machined quartz spheres will be set into rotation and carefully measured. According to general relativity, these spheres should develop a small wobble as they rotate in the Earth's gravitational field. Sensitive electronics will detect any perturbations of this sort.

As scientists get better and better at making precise measurements, more and more tests of the extremely small differences between Newtonian and relativistic predictions of physical events will be made. A few years ago, for example, a group of scientists at institutions in the Washing-

Figure 16–14
In a proposed test of general relativity, a laser beam would be directed from the University of Maryland (*a*) to a mirror on the top of the National Cathedral (*b*) and reflected to the Naval Research Laboratory (*c*). In this figure, the White House is at location (*d*).

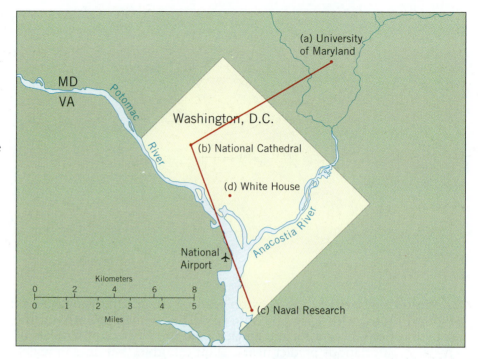

ton, D.C., area proposed an experiment in which light from a laser would be sent out over the city from the University of Maryland (Figure 16–14). The light would be reflected from a mirror on top of the National Cathedral, and would be received by detectors at the nearby Naval Research Laboratory. They predicted that because of the Earth's rotation, there should be a minute difference in travel time between light traveling east and light traveling west; the theory of relativity predicts an additional, smaller difference as well. With extremely accurate clocks and short enough laser pulses, these kinds of differences can be measured with almost as much accuracy as instruments in a satellite.

Who Can Understand Relativity?

Einstein's general theory of relativity was extraordinary, but when first introduced it was difficult to grasp, in part because it relied on some complex mathematics that were unfamiliar to many scientists at the time. Furthermore, while the theory made specific predictions about the physical world, most of those predictions were difficult to test. Soon after the theory's publication, it became conventional wisdom that only a handful of geniuses in the world could understand it.

Einstein made one very specific prediction, however, that could be tested. His proposal that the strong gravitational field of the Sun would bend the light coming from a distant star was new and quantitatively different from other theories. The total eclipse of the Sun in 1919 gave scientists the chance to test Einstein's prediction. Sure enough, the apparent position of stars near the Sun's disk was shifted by exactly the amount he had predicted.

Around the world, front-page newspaper headlines trumpeted Einstein's success. He became an instant international celebrity and his theory of relativity became a part of scientific folklore. Attempts to explain the revolutionary theory to a wide audience began almost immediately.

Thus, while only a few scientists may have grasped the main ideas of general relativity in 1915, when the full theory was first unveiled, that certainly is not true today. The basics of special relativity are taught to tens of thousands of college freshmen every year, while hundreds of graduate students in astronomy and physics explore general relativity in its full mathematical splendor.

If this subject intrigues you, you might want to read some more, watch TV specials or videos about relativity, or even sign up for one of those courses!

THINKING MORE ABOUT RELATIVITY

Was Newton Wrong?

The theory of relativity describes a universe about which Isaac Newton never dreamed. Time dilation, contraction of moving objects, and mass as energy play no role in his laws of motion. Curved space-time is alien to the Newtonian view. Does that mean that Newton was wrong? Not at all.

In fact, all of Einstein's equations reduce *exactly* to Newton's laws of motion, *at speeds significantly less than the speed of light*. This equivalence was shown specifically for time dilation in the Science by the Numbers section in this chapter. Newton's laws, which have worked so well in describing our everyday world, fail only when dealing with extremely high velocities or extremely large masses. Thus Newton's laws represent an extremely important special case of Einstein's more general theory.

Science often progresses in this way, with one theory encompassing previous valid ideas. Newton, for example, merged discoveries by Galileo of Earth-based motions and Kepler's laws of planetary motion into his unified theory of gravity. And some day, Einstein's theory of relativity may be incorporated into an even grander view of the universe.

▶ Summary

Every observer sees the world from a different *frame of reference*. Descriptions of actual physics events are different for different observers, but the *theory of relativity* states that all observers must see the universe operating according to the same laws. Since the speed of light is built into Maxwell's equations, this principle requires that all observers must see the same speed of light in their frames of reference.

Special relativity deals with observers who are not accelerating with respect to each other, while *general relativity* deals with observers in any frame of reference. In regard to special relativity, simple arguments lead to the conclusion that moving clocks appear to tick more slowly than stationary ones; this phenomenon is known as *time dilation*. Furthermore, moving objects appear to get shorter in the direction of motion, which is the phenomenon of *length contraction*. Finally, moving objects become more massive than stationary ones, and an equivalence exists between mass and energy, as expressed by the famous equation, $E = mc^2$.

General relativity begins with the observation that the force of gravity is connected to acceleration, and describes a universe in which heavy masses warp the fabric of space-time and affect the motion of other objects. There are three classical tests of general relativity—the bending of light rays passing near the Sun, the changing orientation of the orbit of Mercury, and the redshift of light passing through a gravitational field.

▶ Key Equations

Time dilation:
$$\frac{t_{GG}}{t_{MG}} = \sqrt{1 - (v/c)^2}$$

Length contraction:
$$\frac{L_{MG}}{L_{GG}} = \sqrt{1 - (v/c)^2}$$

Mass effect:
$$\frac{m_{GG}}{m_{MG}} = \sqrt{1 - (v/c)^2}$$

Rest mass:
$$E = mc^2$$

▶ Review Questions

1. What is a frame of reference?

2. Give examples of two frames of reference you were in today.

3. Give an example of a situation that can be presented from two different frames of reference.

4. What is the central idea of Einstein's theory of relativity?

5. What is the difference between special and general relativity?

6. What is time dilation?

7. How fast does something have to be moving for time dilation to be appreciable?

8. What is the Lorentz factor?

9. When is the Lorentz factor most commonly used?

10. According to an observer on the ground, how does the length of a moving object compare to the length of an identical object on the ground?

11. According to an observer on the ground, how does the mass of a moving object compare to the mass of an identical object on the ground?

12. What is the relation between the mass of an object and its energy?

13. How can we say that gravitational forces and acceleration are equivalent?

14. How might it be possible for a child to be older than a parent? What phenomenon is at work?

15. Explain three tests of general relativity. Are these the only tests possible?

16. Describe length contraction. How does it work?

▶ Fill in the Blanks

Complete the following paragraph with phrases from the list.

frame of reference	special relativity
general relativity	theory of relativity
length contraction	time dilation

We all observe the universe from a particular _____. The assumption that all observers moving at constant speed see the same laws of physics leads to _____, while including all observers, accelerating or not, leads to _____. Some consequences of the _____ are that moving clocks slow down (_____) and moving objects appear shortened (_____).

▶ Discussion Questions

1. Imagine arriving by spaceship at the solar system for the first time. Identify three different frames of reference that you might choose to describe the Earth.

2. Did Einstein disprove Newton's laws of motion? Explain your answer.

3. In Chapter 3, we talked about the idea that Newton's work was profoundly in tune with the time in which he lived. In what sense might you say that relativity is in tune with the twentieth century?

4. The twentieth century has been called the age of relativism, where each person has his or her own ethical system and no set of values is absolute. Do you agree? Does the theory of relativity imply that no values are absolute?

5. Which theory, Newton's or Einstein's, is more consistent with the way you view the universe? Please explain your answer.

6. Answer the following questions about frames of reference.

  a. From the frame of reference of the Earth, describe the motion of the Moon.

  b. From the frame of reference of the Sun, describe the motion of the Moon.

  c. From the frame of reference of the center of our galaxy, describe the motion of the Moon.

7. What is the significance of the word relativity in Einstein's description of the fundamental laws of motion?

8. How does the principle of relative motion relate to the principle that every observer must experience the same natural laws?

9. You and your friend buy two identical watches. You depart and one day you see your friend traveling relative to you at 25% of the speed of light. Is your friend's watch running faster or slower than your watch? Does your friend agree with you? Explain.

10. What is the main idea behind the fact that the speed of light is the same for all reference systems?

11. How can the equivalence between energy and mass be used to predict that light will bend around massive objects such as the Sun?

12. In your own words, explain why nothing can travel faster than the speed of light.

13. How is the "warping of space" around a massive object an equivalent but different description of the gravitational force generated by that object?

▶ Problems

1. While running at 15 mph directly toward Patrick, Lisa passes a basketball to him. Patrick is stationary and receives the ball moving at a speed of 35 mph. At what speed does Lisa throw the ball?

2. Paula is traveling on her bicycle at a speed of 10 mph when a car passes her. From her frame of reference, she estimates that the car is going 45 mph toward her and 45 mph away from her. What is the speed of the car from the frame of reference of someone on the ground?

3. Rafael was traveling overseas on the Concorde, moving at Mach 2 (700 m/s), when a radio signal was received from the Paris control tower informing him of a major area of turbulence over the English Channel. Rafael received a second message confirming this report from the New York control tower. What speed did Rafael measure for the radio message sent from Paris? From New York?

4. Astronomers have routinely observed distant galaxies moving away from the Milky Way galaxy at speeds over 10% of the speed of light (30,000 km/s). At what speed does this distant light reach astronomers?

5. Joshua is traveling by a building at 150,000 km/s, moving along the width of the building. Joshua measures the building to be 50 ft wide and 100 ft tall. What is the height and width of the building, relative to a person standing at rest by the building?

6. A certain molecule will naturally vibrate at 50 Hz in your laboratory. Assume that the molecule is moving at 250,000 km/s relative to your laboratory, directly across your line of sight.

   a. Will you measure a different natural vibration?
   b. If so, will it be larger than, equal to, or smaller than the laboratory frequency (in Hz)?
   c. If so, calculate the new frequency.

7. Jim is riding on a cosmic train at 200,000 km/s as he passes Joannie, who is sitting next to the train track. A ball, traveling in the same direction as Jim, reaches Joannie as soon as she sees Jim pass her. She measures the speed of the ball to be 150,000 km/s. What speed does Jim record for the ball?

8. Jim was riding on a cosmic train at 250,000 km/s as he passes Joannie, who is sitting next to the train track. A moonbeam, traveling in the opposite direction as Jim, reaches Joannie as soon as she sees Jim pass her. She measures the speed of the moonbeam to be 300,000 km/s. What speed does Jim record for the moonbeam?

9. Calculate the speed that an interplanetary spaceship must travel so that an observer at rest can measure a 3m wide window on the space ship as 1.5 m.

10. You are traveling 80 km/h in your car when you throw a ball 50 km/h. What is the ball's apparent speed to a person standing by the road when the ball is thrown:

   a. straight ahead?
   b. sideways?
   c. backward?

11. Calculate the Lorentz factor for objects traveling at 1%, 50%, and 99.9% of the speed of light.

12. What is the apparent mass of a 1-kg object that has been accelerated to 99% of light speed?

13. If a moving clock appears to be ticking half as fast as normal, at what percent of light speed is it traveling?

14. Draw a picture illustrating how a spaceship passing the Earth might look at 1%, 90%, and 99.9% of light speed.

▶ Investigations

1. Read a biography of Albert Einstein. What were his major scientific contributions? For what work did he receive the Nobel Prize?

2. Take a bathroom scale into an elevator in a tall building, stand on it, and record your weight under acceleration and deceleration. Why does the scale reading change?

3. Read the novel *Einstein's Dreams* by Alan Lightman. Each of the chapters explores different time-space relationships. Which chapters teach you something about Einstein's theory of relativity?

4. Investigate the influence of Einstein's theory of relativity on twentieth-century art and philosophy.

5. Write a short story with a plot that centers on an example of space travel and aging.

▶ Additional Reading

Lightman, Alan. *Einstein's Dreams*. New York: Pantheon, 1993.

Pais, Abraham. *Subtle is the Lord: The Science and Life of Albert Einstein*. New York: Oxford University Press, 1982.

17 THE EARTH AND OTHER PLANETS

THE EARTH, ONE OF NINE PLANETS THAT ORBIT THE
SUN, FORMED 4.5 BILLION YEARS AGO FROM A GREAT
CLOUD OF DUST.

Looking at the Night Sky

Walk outside and look at the sky tonight just after sunset. Chances are that you will find two or three particularly bright objects that stand out among the stars, even in the haze and illumination of a city. They don't seem to twinkle like the stars, but shine steadily. If you look at them through binoculars, they appear as small disks.

If you look at the same bright objects on successive nights, you'll notice that over a period of weeks or months, they seem to wander among the stars, never appearing in the same place two nights in a row. The Greeks called these objects "wanderers," or planets, and assigned them the names of the gods. In the evening and morning, for example, you are likely to see Venus, the goddess of love, and swift-moving Mercury, the messenger of the gods; and the night sky is often dominated by Jupiter, the Roman name for the king of the gods.

Today we know that those disks of light in the sky are objects that are in many ways similar to our own planet, the Earth. They remind us that we are part of a system that includes not only the Earth, but the Sun and the other planets as well. Our rocket-powered space probes have visited all of them and landed on two—Mars and Venus. Visionaries talk of the day when science fiction will become reality and human beings will work and live on these other planets. In every sense of the word, the planets are the next frontier.

THE SUN AND ITS PLANETS

The Copernican revolution radically altered human perceptions of our place in the universe (see Chapter 3). Rather than occupying what was assumed by many to be the center of creation, the Earth became just one of a number of planets orbiting the Sun. The **solar system,** which includes the Sun, the planets and their moons, and all other objects gravitationally bound to the Sun, displays several distinctive characteristics. Describing these features, and explaining in detail how they came to be, remains one of the main challenges faced by scientists today.

solar system   The Sun, the planets and their moons, and all other objects bound by gravitation to the Sun.

Features of the Solar System

How can we deduce the origin and present state of the solar system? Until recently, all of our observations of the Sun and planets have been made from the surface of the Earth. While we have been able to see points of light moving in the skies, we have not been able to translate that information into a vivid picture of a dynamic system.

Humans have been studying the solar system for thousands of years, making observations and proposing models. Ancient scholars recorded, among other things, the changing positions of the brightest planets, such as Venus and Jupiter. The development of the telescope by Galileo (see Chapter 3) led to the discovery of many new, faint objects, including dozens of moons and other small bodies. Our present understanding of the solar system, therefore, demonstrates the cumulative effect of centuries of observation.

As astronomers gathered data on the solar system, they noticed striking regularities regarding the orbits of planets and the distribution of mass. These patterns provide the key to understanding the evolution of our home planet, Earth.

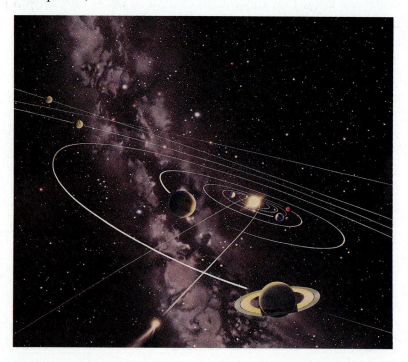

Figure 17–1
The solar system.

Planetary Orbits.   Think about what Newton's laws tell us regarding satellites orbiting a central body. A satellite can go in any direction: east to west or west to east, around the equator, or north to south over the poles. There are no constraints regarding the orientation of the orbit, and planets could orbit any which way around the Sun. Yet in our solar system we see two very curious features, as illustrated in Figure 17–1:

- All the planets move in orbits in the same direction around the Sun, and this direction is the same as that of the rotation of the Sun.

- All the orbits of the planets are in more or less the same plane. The solar system resembles a bunch of marbles rolling around on a single flat dish.

Scientists puzzled over why the solar system is arranged in this orderly way, and they wondered what these features tell us about its history.

The Distribution of Mass.   You could imagine a solar system in which mass was evenly distributed, with all planets more or less the same size and same chemical composition. However, our solar system is not that way at all (Figure 17–2). Instead:

- Approximately 99% of the matter that makes up the solar system is contained within the Sun. The remaining 1% of the matter forms the planets and other objects in orbit.

- There are two separate kinds of planets. Near the Sun, in what we call the "inner" solar system, we find planets like the Earth—relatively small, rocky, high-density worlds. These are called the **terrestrial planets,** and include *Mercury, Venus,* the Earth, *Mars,* and (although it isn't really a planet) the *Earth's Moon.* Farther out from the Sun, in the "outer" solar system, we find huge worlds made up primarily of liquids and gases. We call them "gas giants," or **Jovian planets,** and they are *Jupiter, Saturn, Uranus,* and *Neptune. Pluto,* the outermost planet, is something of an anomaly, being small and rocky. The planets of the solar system and some of their characteristics are listed in Appendix D.

terrestrial planets Those relatively small, rocky, high-density planets located in the inner solar system: Mercury, Venus, Earth, the Earth's Moon, and Mars.

Jovian planets Huge worlds also known as "gas giants," located in the outer solar system, where the effects of solar heat and the solar wind are reduced. These planets are made primarily of frozen liquids and gases such as hydrogen, helium, ammonia, and water, with atmospheres of nitrogen, methane, and other compounds: Jupiter, Saturn, Uranus, Neptune.

Figure 17–2
Most of the mass in the solar system is in the Sun, and most of the rest is in the Jovian planets.

asteroids Small, rocky objects, which circle the Sun like miniature planets.

asteroid belt A collection of small, rocky planetesimals, which never managed to collect into a single planet. They are located in a circular orbit between Mars and Jupiter.

comet An object, usually found outside the orbit of Pluto, composed of chunks of materials such as water ice and methane ice embedded with dirt; may fall toward the Sun, if its distant orbit is disturbed, and become visible in the night sky.

- Interspersed with the planets are many other kinds of objects. All the planets except for the innermost two, Mercury and Venus, are orbited by one or more *moons*. While some moons are little more than boulders a few kilometers across, others are much larger, and Saturn's largest moon, Titan, is about the same size as Mercury. Although Saturn has the most dramatic *rings*, composed of millions of tiny moons, rings are known to exist around all of the Jovian planets. Small, rocky **asteroids** that circle the Sun like miniature planets are found primarily in orbits between Mars and Jupiter, in what is called the **asteroid belt,** although some have orbits that cross the Earth's. Far beyond Pluto, we find a swarm of icy **comets,** with compositions something like a "dirty snowball." Occasionally, one is jostled loose from its orbit and becomes part of the realm of the planets, becoming visible in the night sky.

These regularities in the distribution of the solar system's mass give us important insights as to how the system was formed. These insights, in turn, lead to more theories and more observations of the heavens.

Science in the Making

The Discovery of Pluto

Five of the planets (in addition to the Earth) are visible to the naked eye and have been known from ancient times—Mercury, Venus, Mars, Jupiter, and Saturn. The three most distant planets—Uranus, Neptune, and Pluto—were discovered after the telescope was invented. In fact, the existence of Neptune and Pluto was predicted before they were observed because of their gravitational effects. Neptune causes significant shifts in the orbit of Uranus, for example, which led to its discovery in 1846.

The most recent discovery of a planet occurred on February 18, 1930, when Clyde Tombaugh, a Kansas farm boy employed as a technician at Lowell Observatory, in Flagstaff, Arizona, uncovered convincing evidence for the existence of Pluto. As a teenager, Tombaugh had built a small telescope, using parts from an old cream separator to make its stand. He sketched the surface of Mars and sent his illustrations to Lowell Observatory, which was then in the process of observing the red planet. The sketches, made with a small amateur's instrument, corresponded so well to what astronomers at Lowell were seeing through their state-of-the-art telescope that Tombaugh received a job offer by return mail.

Clyde W. Tombaugh, the discoverer of Pluto.

At Lowell, he was given the task of doing a systematic search of the skies for what was then called "Planet X." The founder of the observatory, American astronomer Percival Lowell, had predicted the existence of such a planet on the basis of some questionable data on variations in the orbit of Neptune. Tombaugh's task was straightforward, if tiring. As the Earth swept around in its orbit, he took photographs of each section of the sky, then took another photograph a few days later. The two photographs were then put into a machine that would show first one photograph, then the other, in an eyepiece. Any object

The discovery photos for Pluto reveal the shift of one point of light between January 23, 1930 (left), and January 29, 1930 (right). The white arrows point to Pluto.

that had moved between the time of the two photographs would appear to jump back and forth as the photographs were "blinked," while stars would remain stationary.

The main problem Tombaugh faced is that the plane of the solar system is littered with asteroids, each of which could show up as a moving light on such photos. Therefore, Tombaugh wasn't just looking for something out there that moved; there were plenty of such objects. He had to find a point of light that moved as much as Kepler's laws tell us that a planet beyond Neptune would move in a few days (see Chapter 3). That is exactly what Tombaugh found on that day in February, some 10 months into his search. The new "Planet X" was called Pluto.

After becoming one of only three human beings to have discovered a new planet, Tombaugh went back to college. Much to his surprise, he was not allowed to take introductory astronomy. "They cheated me out of four hours!" he says.

And his old telescope? When one of the authors asked him a few years ago whether he was going to donate it to the Smithsonian Institution, the 83-year-old astronomer replied, "They want it, but they can't have it. I'm not through using it yet!" ●

THE NEBULAR HYPOTHESIS

The modern theory of the solar system's formation was first proposed by French mathematician and physicist Pierre Simon Laplace (1749–1827). His was a simple idea, and one that takes into account many of the distinctive characteristics of the solar system, including the rotation of the Sun, the orbits of the planets, and the distribution of mass into several large objects and many smaller ones. The key to his hypothesis is the effect of gravitational force on a cloud of dust and gas.

Dust, Gas, and Gravity

According to the model, called the *nebular hypothesis*, long ago (we now estimate about 4.5 billion years ago) a large cloud of dust and gas floated in space in the region now occupied by the solar system (Figure 17–3a). Such dust and gas clouds, called **nebulae,** are common throughout our galaxy, the Milky Way. They typically contain more than 99% hydrogen

nebulae Dust and gas clouds, common throughout the Milky Way galaxy, rich in hydrogen and helium.

Figure 17–3
As the nebula that formed the solar system collapsed, it began to rotate and flatten into a disk. The stages in solar system formation include (a) a slowly rotating nebula, (b) a flattened disk with massive center, (c) planets in the process of birth represented as mass concentrations in the nebula, and (d) the solar system.

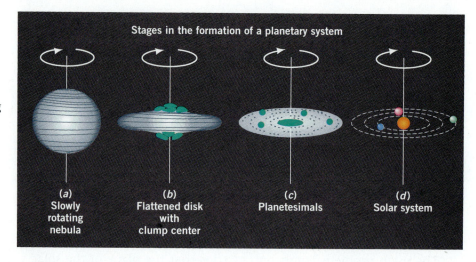

Stages in the formation of a planetary system

(a) Slowly rotating nebula

(b) Flattened disk with clump center

(c) Planetesimals

(d) Solar system

and helium, with lesser amounts of all the other naturally occurring elements.

Under the influence of gravity, the nebula slowly, inexorably, started to collapse on itself. As it did so, it began to rotate faster and faster. In a sense, the collapsing, rotating cloud can be compared to an ice skater beginning a spin. Going into the spin she will have her arms extended out, but as the spin progresses she pulls her arms in and spins faster and faster—sometimes so fast that you can barely make out her features. In the same way, as the nebula from which the solar system was formed began to collapse, it too began to spin faster and faster.

This spin had several consequences for the formation of the Sun and planets. First of all, some of the material in the outer parts of the cloud began to spin out into a flat disk. This common consequence of fast rotation is familiar to anyone who has watched a pizza maker creating a flat disk of dough by spinning a mass overhead. The rapid rotation in the solar system flattened the material into a pancake shape. In fact, if you imagine the solar system at this stage of its formation as a large pancake with a big lump in the middle, you'll have a pretty good idea of the situation. The big lump represents the material that eventually became the Sun, and the material in the thin flattened disk eventually became the planets and the rest of the solar system (see Figure 17–3b).

The flattening of the nebula into a rotating disk explains another feature of the solar system. The planets had to form in this rotating disk, and hence their eventual orbits had to lie close to the disk's plane. Thus the fact that all planetary orbits lie near the same plane is a simple consequence of the solar system's rapid rotation as the nebular cloud contracted.

An ice skater in a spin twirls faster as she pulls her arms into her body. Similarly, the solar nebula began to spin faster and faster as mass was pulled toward the central region.

The formation of stars, such as the Sun, is discussed in more detail in Chapter 21, so here we will confine our attention to what happened in the outer, thin regions of the disk. In any conglomeration of matter such as the spinning disk, purely by chance, matter is more densely collected in some regions than others. These regions exert a stronger gravitational force than their neighbors, so that nearby matter tends to gravitate toward them. Once the nearby matter is pulled in by gravity, the concentration of matter at that point becomes even greater, and will pull in even more material. As more material accumulates, solid grains start to stick together.

A consequence of this gravitational force is the rapid breakup of the disk into small objects called *planetesimals*, which range in size from boulders to masses several kilometers in diameter. Once this occurs, the process of gravitational attraction continues at a grander scale, in which larger objects capture smaller ones and continue growing (Figure 17–3c).

The Formation of Planets

About the time that this consolidation of planetesimals was taking place, the material at the center of the solar system, which included more than 99% of the nebula's original mass, began to turn into a star. Light energy began to radiate away from the Sun, and temperature differences began to develop in the disk. Those parts nearest the Sun became hot, while those farther out warmed only a little. Furthermore, as the Sun began to burn, an intense *solar wind*, made up of particles ejected from the Sun, blew outward. This wind was strongest for the inner planets. As a result, the inner and outer solar systems developed differently. In the warm inner system, materials such as water, hydrogen, and helium were in gaseous form, while farther out they were frozen into solids.

Thus everyday physical processes as familiar as boiling water and making ice explain one of the crucial facts about the solar system. The terrestrial planets—Mercury, Venus, Earth, and Mars—were formed from materials that remain solid at high temperatures (see Figure 17–3d). Consequently, they are small, rocky, high-density worlds.

Farther out in the solar system, we find the Jovian planets—Jupiter, Saturn, Uranus, and Neptune. The composition of those planets is essentially the same as the material concentrated in the original nebula, in that they contain large amounts of hydrogen and helium. These planets were formed from material that remained solid (or at least liquid) because of their lower temperature so far from the Sun. Consequently, they have a markedly different chemical composition from the terrestrial planets closer to the Sun (see Figure 17–4).

Figure 17–4
Because regions closer to the Sun were warmer than regions farther away, the terrestrial planets formed from different materials than the Jovian planets. This effect explains the differences in composition between the two types of planets.

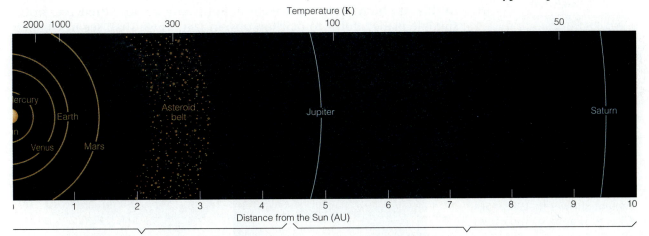

High-temperature condensates of iron-nickel alloys, silicate and oxide minerals. Planets and moons are rocky with metallic cores.

Low-temperature condensates of ices of water, methane, and ammonia. Planets are gassy, moons sheathed in water ice.

The structure of the Jovian planets is layered, like that of the terrestrial planets, but they do not have a distinct solid surface like the Earth and Moon. Moving down from space into the body of Jupiter or Saturn would be a strange experience. You would move through progressively denser and denser layers of clouds and then pass imperceptibly into a layer where the gases change into liquids because of the high pressure. In fact, landing on Jupiter would be more like landing on a giant ice cream sundae than landing on the planet Earth or the Moon.

Just as any construction site has a pile of leftover materials lying around when a structure is completed, so too does the solar system have its "scrap pile." These leftovers that never got taken up into planets are in the form of rocky asteroids and icy comets that still orbit the Sun.

Developing Your Intuition

Forming Planets

A dynamic frontier of space research is the search for planets orbiting other stars. Under what circumstances do planets form in distant solar systems? What properties would a star have to have for there to be no terrestrial-type planets? What properties would lead to no Jovian-type planets?

The presence of the two kinds of planets around the Sun, we have argued, depends on the differences in temperature between the inner and outer solar system when the planets were forming. In order for there to be no terrestrial-type planets, the temperature in the inner solar system would have had to be low enough so that substances such as helium and methane would be retained by the planets. Thus, for the solar system to have no terrestrial-type planets, the Sun would have to be much cooler and less energetic than it is.

Similarly, for there to be no Jovian-type planets, temperatures at the orbit of Neptune would have to be high enough to vaporize helium, methane, and other volatiles. This would require that the Sun be much hotter than it is. And although the possibility of life developing in this sort of system might seem high (if only because there are more suitable planets for it to develop on), in fact, as we shall see in Chapter 21, very bright stars tend to have very short life spans. Thus life probably wouldn't have time to develop in this particular hypothetical star system. ●

(*a*) (*b*)

(*a*) A bright fireball observed during the Perseid meteor shower in 1988. (*b*) Chemical analysis indicates that this meteorite, discovered in Antarctica, came from Mars.

EVOLUTION OF THE EARTH AND MOON

The collapse of the solar nebula into the Sun and planets was the first step in the evolution of the solar system. Following the initial accretion of planets, each object evolved in its own distinctive way.

The Formation and Early History of the Earth

Once the countless thousands of planetesimals were formed, the formation of planets followed quickly. As planetesimals moved through their orbits, they collected smaller planetesimals through the process of gravitational attraction. Then these larger planetesimals collided and coalesced into the beginnings of a planet. As the process of accumulation went on, the growing planet gradually swept up all the debris that lay near its orbit.

If you had been standing on the surface of the newly forming Earth during this stage, you would have seen a spectacular display. A steady rain of debris left over from the initial period of planetary formation fell to the surface, constantly adding mass to the Earth. During this period, called the **great bombardment,** the large amounts of kinetic energy carried by the shower of stones were converted into heat, which was added to the newly forming planet. According to some theories, much of the Earth's surface would have glowed bright red from this accumulating heat, and each large impact would have been accompanied by a spectacular splash of molten rock. Although the addition of material to the Earth has slowed considerably since the beginning, it has not stopped. For example, every time you see a **meteor,** often called a shooting star, you are seeing an object roughly the size of a grain of sand that is being added to our planet. Scientists estimate that the mass of the Earth grows by about 20 metric tons (20,000 kilograms, or 2×10^7 g) per day by accretion of material falling from space.

When the nebular hypothesis was first proposed in the eighteenth century, there seemed little chance that any direct observational evidence could be found to support it. In 1992, however, astronomers using the Hubble Space Telescope (see Chapter 21) were able to detect thick masses of dust encircling newborn stars in a region of space called the Orion Nebula. It appears that in these cases we are seeing a distant solar system in the process of being born. Such observations give us a measure of confidence in our model of how planets come into existence.

In regions like the Orion Nebula, new stars (and presumably new planetary systems) are forming today.

great bombardment An event following the initial period of planetary formation in which meteorites showered down on planets, adding matter and heat energy.

meteor A piece of interplanetary debris that hits the Earth's atmosphere and forms a bright streak of light from friction with atmospheric particles: often called a shooting star.

Science by the Numbers

The Growth of the Earth

The Earth's mass is approximately 6×10^{27} g. Adding 2×10^7 grams per day, how many years would it take to double the Earth's mass?

▶ **Reasoning:** First, we have to calculate how many grams are added to the Earth each year by multiplying the daily added mass by the number of days in a year. The Earth's total mass, divided by the mass added each year, gives us the time it would take to double the present mass.

▶ **Solution:** First we determine the yearly mass added to the Earth:

$$\text{mass added per year} = (365 \text{ days/year}) \times (2 \times 10^7 \text{ g/day})$$
$$= 730 \times 10^7 \text{ g/year}$$
$$= 7.3 \times 10^9 \text{ g/year}$$

Then, divide the Earth's total mass by the mass added every year:

$$\text{number of years} = \frac{6 \times 10^{27} \text{ g}}{7.3 \times 10^9 \text{ g/year}}$$
$$= 0.82 \times 10^{18} \text{ years}$$
$$= 8.2 \times 10^{17} \text{ years}$$

This number is the time (in years) that would be required to double the mass of the Earth at its present rate of growth. This immense period of time—nearly a billion-billion years—is vastly greater than the lifetime of our planet, which is a "paltry" 4.5 billion years. From this calculation we can see that the total amount of mass now being added to the Earth is trivial. ▲

The Layered Structure of the Earth

Each time another planetesimal hit the early Earth during the great bombardment, all of its kinetic and potential energy was converted into heat. That heat diffused rapidly through the planet. The Earth glowed red hot and reached temperatures of thousands of degrees in its deep interior. Eventually, scientists postulate, the Earth either melted completely, or else it was heated to high enough temperatures so that it was very soft all the way through. Heavy, dense materials (such as iron and nickel) sank under the force of gravity toward the center of the planet, while lighter, less dense materials floated to the top. As a result of this process, called **differentiation**, the present-day Earth has a very definite layered structure (as shown in Figure 17–5).

What happened to the Earth long ago is similar to what happens to a mixture of oil and water when it is shaken up and then allowed to stand. Eventually, under the influence of gravity, the lighter oil will float to the top and the heavier water will sink to the bottom. In the same way, when the Earth underwent differentiation, it separated into layers of different densities.

At the center of the Earth, with a radius of about 3400 kilometers (2000 miles), is the **core**, made primarily of iron and nickel metal. Temperatures at the center of the Earth are believed to exceed 5000°C, a temperature generally hot enough to melt those metals. However, pressures at the Earth's center are so high (about 3.5 billion grams per square centimeter, or almost 50 million pounds per square inch) that the iron-nickel inner core is solid (see the Technology section). A little farther out from the Earth's center, the pressures are somewhat lower, so that the outer region of the iron-nickel core is a liquid.

The metal core is surrounded by a thick layer of material, the **mantle**, which is rich in the elements oxygen, silicon, magnesium, and iron. Whereas metallic bonding predominates in the iron-rich core, the mantle

differentiation The process by which heavy, dense materials (for example, iron and nickel) sank under the force of gravity toward the molten center of the planet, while lighter, less dense materials floated to the top, resulting in a layered structure of the present-day Earth.

core in geology: heaviest elements of the Earth's mass, primarily iron and nickel, concentrated at the center with a radius of about 3400 km (2000 miles).

mantle The thick layer—rich in oxygen, silicon, magnesium, and iron, and containing most of the Earth's mass—overlying the metal core of the Earth.

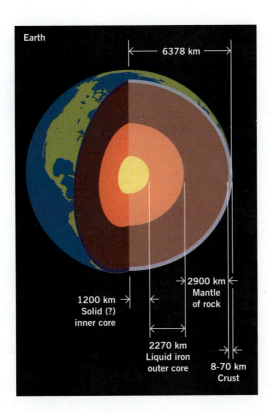

Figure 17-5
The layered Earth. The principal layers, which differ in chemical composition and physical properties, are the core, mantle, crust, and atmosphere. When looked at in detail, each of these layers is composed of smaller layers.

features minerals with primarily ionic bonds between negatively charged oxygen ions and positively charged silicon, magnesium, and other ions. Mantle rocks are similar in composition to some familiar surface rocks, but the atoms in these high-pressure materials are packed together much more densely.

The very outer layer of the Earth is known as the **crust,** which is made up of the lightest materials. The crust's thickness ranges from less than 10 kilometers (about 6 miles) in parts of the oceans, to as much as 70 kilometers (about 45 miles) beneath parts of the continents. The crust is the only layer of the solid Earth with which human beings have had contact, and it remains the source of almost all the rocks and minerals that we use in our lives.

You might be wondering how scientists know so much about parts of the Earth that no human being has ever seen. In the next chapter, we will talk about *seismology,* a branch of science that has provided (among other things) our present picture of the Earth's interior.

crust A thin layer at the surface of the Earth formed from the lightest elements, ranging in thickness from 10 km (6 miles) in parts of the ocean to 70 km (45 miles) beneath parts of the continents.

Technology

Producing High Pressure

The force of gravity, which causes the layers of the Earth to pull inward, results in immense internal pressures that exceed 3 million times the atmospheric pressure at the planet's center. What effects does this pressure have on rocks and minerals? Researchers have learned to sustain laboratory pressures greater than those at the center of the Earth, and are providing surprising answers.

Synthetic diamonds, produced at high temperature and pressure, find widespread use as abrasives for cutting and polishing. *Inset:* Individual diamond crystals, ×100.

Of all the materials that form deep within the Earth, none holds more fascination than diamond, the high-pressure form of carbon. Not only does diamond provide magnificent gemstones, but as the hardest known substance, it also serves as the most efficient abrasive for machining the tough metal parts of modern industry. Until the mid-1950s, diamond was available only from a few natural sources. Then in 1954, scientists at General Electric Company discovered how to manufacture diamond by duplicating the extreme temperatures and pressures that exist hundreds of kilometers deep. The researchers squeezed black carbon between the jaws of a massive metal vise and heated their sample with a powerful electrical current. Although the first experiments yielded only a fraction of a gram of diamond, giant factories (including a major GE plant in Worthington, Ohio) now produce dozens of *tons* of diamond every year. This output has greatly exceeded the total amount of diamond mined since Biblical times.

Studying the Earth helped us understand how diamond was made, but now scientists use diamond to learn how the Earth was made. The highest sustained laboratory pressures available today are obtained by clamping together two tiny pointed anvils of diamond. Samples squeezed between the diamond-anvil faces have been subjected to pressures of several million kilograms per square centimeter; pressures significantly greater than at the center of the Earth. At such extreme conditions, crustal rocks and minerals compress to new, dense forms with compact atomic structures occupying less than half their original volumes. Dramatic changes in chemical bonding are also observed, with many ionically bonded compounds transforming to metals at high pressure (see Chapter 11). ●

Researchers attain high pressures, equivalent to those inside the Earth and other planets, by using a diamond anvil cell. Looking through such diamond cells you can observe pressurized samples such as this high pressure ice crystal formed at room temperature by squeezing water.

Example 17–1: The Volume and Mass of the Earth's Core

What fraction of the Earth's volume is taken up by the core? What fraction of the Earth's mass is in the core?

▶ **Reasoning 1:** We must solve this problem in two parts. First, we need to compare the volume of the Earth's core, a sphere with a 3500-km radius, to the volume of the entire Earth, with approximately a 6400-km radius. The volume of a sphere is given by the formula

$$\text{sphere volume} = \tfrac{4}{3} \times \pi \times \text{radius}^3$$

where the constant π is approximately 3.14.

▶ **Solution 1:** The ratio of the core's volume to the entire Earth's volume is thus

$$\frac{\text{vol}_{\text{core}}}{\text{vol}_{\text{Earth}}} = \frac{\tfrac{4}{3} \times 3.14 \times 3500^3}{\tfrac{4}{3} \times 3.14 \times 6400^3}$$

Note that the numerical terms $\tfrac{4}{3}$ and 3.14 cancel out on the right side of this equation, so:

$$\frac{\text{vol}_{\text{core}}}{\text{vol}_{\text{Earth}}} = \frac{3500^3}{6400^3}$$
$$= \frac{4.3 \times 10^{10}}{2.6 \times 10^{11}}$$
$$= 0.17, \text{ or } 17\%.$$

Thus the core accounts for about 17% of the Earth's volume, while the outer layers—the crust and mantle—account for the other 83%.

▶ **Reasoning 2:** The second part of this problem considers the mass of the core compared to the mass of the entire Earth. The average density of the highly compressed iron core is about 10 g/cm^3, which is slightly less than the density of lead. This value is almost twice as great as the 5.5 g/cm^3 average density of the entire Earth, which is composed mostly of rocks that are much less dense than iron.

▶ **Solution 2:** The mass of an object equals its volume times its density. Thus the mass of the Earth is given by

$$\text{mass}_{\text{Earth}} = \text{vol}_{\text{Earth}} \times \text{density}_{\text{Earth}}$$

Similarly,

$$\text{mass}_{\text{core}} = \text{vol}_{\text{core}} \times \text{density}_{\text{core}}$$

The relative mass of the core, $\text{mass}_{\text{core}}/\text{mass}_{\text{Earth}}$, is thus given by

$$\frac{\text{mass}_{\text{core}}}{\text{mass}_{\text{Earth}}} = \frac{\text{vol}_{\text{core}} \times \text{density}_{\text{core}}}{\text{vol}_{\text{Earth}} \times \text{density}_{\text{Earth}}}$$

But we calculated above that $\text{vol}_{\text{core}}/\text{vol}_{\text{Earth}} = 0.17$, and we know that the average values of core and Earth densities are 10 and 5.5 g/cm³, respectively. Thus

$$\frac{\text{mass}_{\text{core}}}{\text{mass}_{\text{Earth}}} = \frac{0.17 \times (10 \text{ g/cm}^3)}{5.5 \text{ g/cm}^3}$$
$$= 0.31, \text{ or } 31\%$$

The core represents slightly less than a third of our planet's mass. ▲

The Formation of the Moon

The origin of the Moon, the only large body in orbit around the Earth, poses one of the oldest puzzles in planetary science. The Moon is one of the largest bodies in the solar system. It is larger than the planets Mercury and Pluto, and almost as large as Mars. The Moon's density and chemical composition, which were studied in detail by the astronauts of the Apollo lunar missions in the late 1960s and 1970s, are quite different from the composition of the Earth as a whole. However, the density and composition are remarkably similar to those of the Earth's mantle.

The mystery of the Moon is this: How could it have formed near the Earth, and yet have a composition so unlike the whole Earth?

Various attempts have been made to solve this problem. Some astronomers argued that the Moon was formed elsewhere in the solar system in a region of low-density, mantle-like material and was later captured by the Earth's gravitational pull. Other scientists suggested that the Moon was somehow thrown out of a rapidly rotating Earth after differentiation of the mantle had taken place. In fact, in the early twentieth century, astronomers often pointed to the Pacific Basin as the likely "birth scar" of the Moon. Recent computer models of planetary motions, however, prove that neither of these theories is likely to be true. It turns out to be very difficult for a planet to capture a large body in a near circular orbit, and even harder for a planet to throw off a moon-sized chunk.

The current theory (and one that seems to the authors to have a reasonable chance of being right) is called the "big splash." In this theory,

Figure 17–6
A computer simulation of the "big splash." The impact is shown in (a). Both bodies have metal cores (red and pink) and rocky mantles (brown and green). A plume of material spreads out after the impact (b) and (c), and eventually the Moon accretes out of the material thrown into orbit.

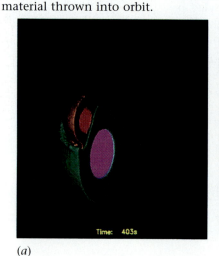
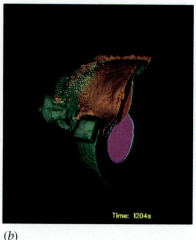
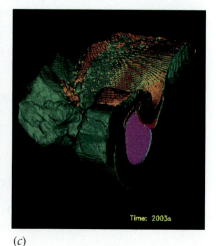

Time: 403s

Time: 1204s

Time: 2003s

(a)

(b)

(c)

it is believed that the Earth underwent differentiation as described earlier, but while it was still in a formative state, it was struck by an object about the size of Mars. This catastrophic collision blew a huge quantity of molten mantle rocks out into orbit around the Earth. The Moon then formed from this material (see Figure 17–6). This theory accounts for the Moon's unusual composition, density, and size. Meanwhile, the great crater in the Earth created by this epic impact was quickly filled by molten rock and eventually weathered away, so no evidence of it appears today on Earth.

The Behavior of the Other Planets

The natural processes that occurred during the Earth's formation affected the other planets as well. Mercury, Mars, and the Moon, for example, display surface cratering suggesting that large chunks of rock bombarded all of these bodies late in their formation. The Earth undoubtedly was heavily cratered over 4 billion years ago, but all evidence of those early craters has been weathered away. On Mercury and the Moon, which have no atmosphere and therefore no wind or water to cause weathering, there has been nothing to affect the craters, and so they are still there.

The early bombardment may also have affected other characteristics of the terrestrial planets. The direction of Venus's rotation, for example, is opposite that of the Earth. (Planets revolve around the Sun, but rotate about their axes.) The Earth's axis of rotation, furthermore, is tilted at 23 degrees to the plane of its orbit, while Uranus has its axis of rotation close to the plane of its orbit; this is a full 90 degrees from an upright orientation. Current thinking is that these differences resulted from random collisions with large objects, perhaps hundreds of kilometers in diameter, that occurred near the end of planetary formation. You might expect that the details of these late-stage collisions were different for each planet. Thus the nebular hypothesis not only explains how it is that the planets all have their orbits in one plane and move in the same direction around the Sun, but also allows us to explain why the rotations of individual planets can be so different.

The Evolution of Planetary Atmospheres

The Earth didn't always have the kind of atmosphere it has today. In fact, scientists now suggest that originally it had no atmosphere at all, and that once an atmosphere formed, its chemical composition evolved gradually to that of its present state. The question of how the Earth's atmosphere evolved is extremely important, because this history goes hand-in-hand with understanding the origin and evolution of life on our planet.

The early Earth may well have collected some gases through gravitational attraction. During the early formation of the Sun, however, the intense solar wind would have blown off any atmosphere the Earth had accumulated. Thus, for all intents and purposes, we can think of the first stage in the evolution of the Earth's atmosphere as corresponding to an airless ball of hot (or even molten) rock.

An artist's impression of the Earth's surface during the great bombardment.

Because there is no weather on the Moon, craters like these, formed during the great bombardment, still survive. This photograph shows the Apollo 12 lunar module preparing to land.

During the period of cooling that followed the great bombardment, large amounts of water vapor, carbon dioxide, and other gases were released from deep within the solid Earth. Countless volcanoes and fissures belched steam and other gases into the newly forming atmosphere. This process, called *outgassing*, resulted in a completely new atmosphere surrounding the solid Earth.

Outgassing, which was violent and rapid early in the Earth's history, still occurs on a small scale. We tend to think of volcanic eruptions as involving the flow of red-hot lava, but if you look at pictures of eruptions, you will probably see large clouds of smoke and steam that accompany the glowing lava flows. Even today, more than 4.5 billion years after the planet's formation, volcanoes release large amounts of gases from the interior of the Earth.

Scientists are engaged in a spirited debate about the composition of the Earth's early atmosphere. One recent theory claims that this atmosphere consisted primarily of methane (CH_4), ammonia (NH_3), carbon dioxide (CO_2), hydrogen (H_2), and water (H_2O). Presumably, the same sort of outgassing occurred on the other terrestrial planets. For a time, the atmosphere was probably too hot for water to condense from a gas to a liquid, but, eventually, atmospheric temperatures dropped and torrential rains began to fill the ocean basins.

Once a planet has acquired an atmosphere as a result of outgassing, there are two important ways that its atmosphere can evolve and change. The simplest is *gravitational escape*. The molecules in an atmosphere heated by the Sun may move sufficiently fast so that a large percentage of them can actually escape the gravitational pull of their planet. The Moon, Mercury, and Mars are examples of bodies that had denser atmospheres early in their history, but lost much of these gases through gravitational escape long ago. Most of the light elements such as hydrogen and helium were presumably lost in the same way from the Earth, but the heavier gases such as carbon dioxide and water vapor remained because they were too heavy to escape the Earth's gravitational force.

A second cause of atmospheric change is the effect of living things. To the best of our knowledge, such changes occur only on the Earth. We know of no life anywhere else in the solar system (although some scientists argue that there may have been life at one time on Mars). By the time the Earth was 1 billion years old, photosynthetic organisms had evolved to use the Sun's energy to power the chemical reactions essential for life. In photosynthesis, carbon dioxide and water are taken into the structures of living things, and oxygen is given off as a waste product. As the number of living things on the planet increased, the amount of free oxygen increased as well. Today, oxygen comprises about 20% of the atmosphere.

We tend to think of oxygen as a benign and beautiful substance, but from a chemical point of view it's really rather nasty stuff. As we saw in Chapter 9, oxygen reacts violently with many materials (think of fire burning or the explosion of hydrogen or gasoline). In fact, some scientists have called the production of oxygen by living things the Earth's first global pollution event. Most scientists now agree that you can tell whether a planet has life on it simply by looking at its atmosphere with a spectrometer (see Chapter 9) for the tell-tale signature of oxygen.

THE INNER SOLAR SYSTEM

All of the planets of the inner solar system (with the exception of the Earth's Moon) formed in a process much like that of the Earth. They experienced a period of bombardment from space debris and, as a result, went through the process of differentiation (either total or partial) early in their history. We'll look at each of the terrestrial planets one by one.

Mercury and the Moon

Mercury and the Moon share many characteristics. Because of their low masses, the Moon and Mercury very quickly lost whatever atmospheres they might have acquired through outgassing. Consequently, there is no weather or erosion on these worlds, and the marks of the early bombardment can still be seen on craters on their surfaces. These worlds also cooled rapidly, so that they are now thought of as cold worlds whose internal features have largely been frozen into place. They have no active volcanoes or earthquakes.

Mercury is the smallest planet, only 5% the mass of the Earth. It has a "year" 88 Earth days long and a "day" 56 Earth days long. It does have a tenuous atmosphere, consisting of sodium and potassium ions baked off the rocks. Mercury's interior is frozen, with a large metal core producing a magnetic field. (See Figure 17–7.) Recently, telescopes have detected evidence for tiny amounts of ice at the poles (in regions shadowed from the Sun). The surface temperature ranges from −200 to +500 °C, depending on whether the area in question is exposed to the Sun or deep space.

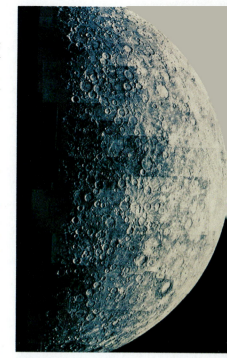

Mercury.

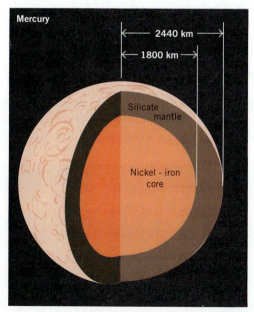

Figure 17–7
A theoretical model of the interior of Mercury. Like the Earth, Mercury has a layered structure, but seems to have a proportionately larger metal core.

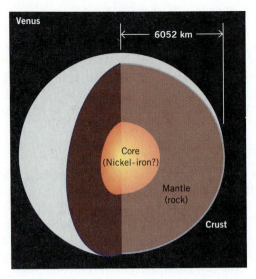

Figure 17–8
A theoretical model for the interior of Venus.

The surface of Venus shows a lack of ocean basins.

Venus

Venus, on the other hand, is a planet only a little smaller than the Earth. (Its mass is 82% that of our planet.) It probably has a metal core much like that of the Earth (see Figure 17–8). However, its slow rotation (its "day" is 243 Earth days long) is thought to be responsible for its lack of a magnetic field. The *Magellan* spacecraft arrived at Venus in 1990 and used onboard radar to make a detailed map of the Venusian surface. There seem to be active volcanoes on the planet and evidence of volcanic activity in the past. Some regions, called highlands, appear to be caused by the upward welling of hot material from the interior. Yet, unlike Earth, Venus shows no planetwide geological activity (see Chapter 18).

The atmosphere of Venus is largely carbon dioxide, with clouds of sulfuric acid, so infrared radiation from the surface tends to be trapped. This gives rise to a *runaway greenhouse effect* that raises the temperature far above what it would be if there were no atmosphere. Surface temperatures are about 450 °C. Venus rotates slowly, with its day almost equal to its year. It has no satellites and no detectable magnetic field.

The surface of Venus is shrouded in clouds, but the *Magellan* spacecraft produced radar images of the surface. In this computer-generated view of a Venusian volcano, the vertical relief has been greatly exaggerated.

Earth

The largest terrestrial planet, Earth is the only known abode of life in the solar system (or, for that matter, in the universe). It has an atmosphere of roughly 78% nitrogen and 21% oxygen and has had liquid water on its surface since early in its history. The interior, as we have seen, is differentiated and there is planetwide geological activity. It has a magnetic field.

Mars

Mars, the most distant terrestrial planet from the Sun, is a small planet (only 11% the mass of the Earth). The red color of Mars comes from the presence of iron compounds on its surface. Its atmosphere, mostly carbon dioxide and nitrogen, is very thin (equivalent to the Earth's atmosphere at an altitude of 40 kilometers) and temperatures at its surface rarely exceed 0 °C. Its polar ice caps are divided into two layers—an upper layer of frozen carbon dioxide (dry ice) and a lower layer of water ice.

Because of the thin atmosphere, Mars displays little weather or erosion. Consequently, many ancient craters mark the planet's surface. Extinct volcanoes are observed on Mars but, as with Venus, scientists see no evidence of planetwide geological activity. Like Mercury and the Moon, the interior of Mars is probably entirely solid at this time (see Figure 17–9).

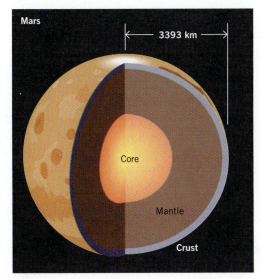

Figure 17–9
A theoretical model of the interior of Mars.

Photos of the Martian surface taken by the Mariner spacecraft showing evidence that there was once running water on the Martian surface. The photo shows an area about 20 km across.

In 1976, the *Viking 1* and *2* spacecrafts landed on Mars. The most important result from onboard experiments was a total lack of evidence for the existence of life on the planet. Since Mars was considered to be the planet most likely to contain life, this result has led to the widespread belief that the Earth is probably the only abode of living things in the solar system.

Photo of the surface of
Mars taken by the
Viking 2 lander.

Probably the most unexpected result that came back from space probes was the existence of long channels on the Martian surface. These channels show that earlier in its existence Mars had liquid water flowing on its surface. This discovery has led some scientists to suggest that life may have developed on Mars, only to become extinct when water vanished from the surface.

THE OUTER SOLAR SYSTEM

Space probes launched from Earth in the 1970s and 1980s have visited most of the outer planets. As a result, we have a pretty good idea of what the outer solar system is like. The distances to the giant outer planets are immense. The closest Jovian planet, Jupiter, orbits at five times the distance between the Earth and the Sun—more than 800,000,000 kilometers away. Saturn is twice that distance from the Sun, while Uranus and Neptune are several billion kilometers away. That far out in the system, the Sun looks like a small marble in the sky, and its warming effects are feeble indeed. Compounds that are normally gases on the Earth, such as carbon dioxide, nitrogen, and methane, are found in liquid or even solid form as we move to the frigid outer solar system. The Jovian planets are fairly simple in structure, although we don't know much about many of their details (see Figures 17–10, 11, and 12).

Jupiter

Jupiter.

The largest planet in the solar system, Jupiter, has outer layers composed primarily of gases. Underneath these layers, scientists expect to find layers of liquid and metallic hydrogen, and possibly a small solid core at the center. The planet rotates rapidly (its day is about 10 hours long), and has bands of rapidly moving gases in the visible part of the atmosphere. A great storm (one that could easily swallow the entire Earth) called the Great Red Spot has been raging on its surface for at least 300 years. Jupiter has a magnetic field and a large collection of moons and rings.

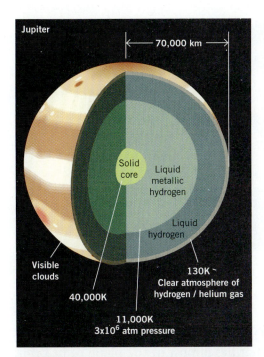

Figure 17–10
A theoretical model of the interior of Jupiter. Most of the planet's volume is highly compressed hydrogen and helium.

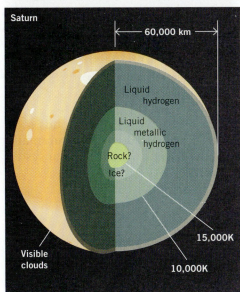

Figure 17–11
A theoretical model for the interior of Saturn.

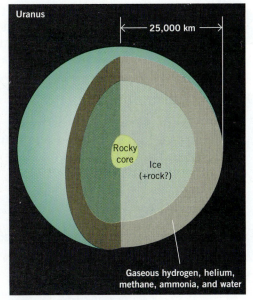

Figure 17–12
A theoretical model for the interior of Uranus.

Saturn

The most striking feature of Saturn is its magnificent system of rings, made from billions of bits of debris in orbit around the planet. Constructed along the same lines as Jupiter, Saturn has a density less than that of water—if you could find an ocean large enough, the planet would float! Saturn has numerous moons, including Titan, the largest moon (except for Earth's) in the solar system.

Saturn.

Uranus.

Uranus

The first planet to be discovered after the development of the telescope, Uranus has an axis of rotation that lies almost in the plane of its orbit. Because of large amounts of methane in its upper atmosphere, it appears bluish to the eye. It is the least massive of the Jovian planets (only 15 times the mass of the Earth). It has a magnetic field of unknown origin, a collection of small satellites, and a system of thin, dark rings.

Neptune.

Neptune

The most distant of the Jovian planets, Neptune has high clouds in its hydrogen-helium atmosphere. It has two large moons and several smaller ones. Triton, its largest moon, is one of the few moons in the solar system with an atmosphere.

Pluto/Charon

The outermost planet, Pluto, remains something of a mystery. While it can be studied through a telescope, it has yet to be visited by a space

Pluto and Charon. (a) The best ground-based image; (b) a photograph taken with the Hubble Space Telescope.

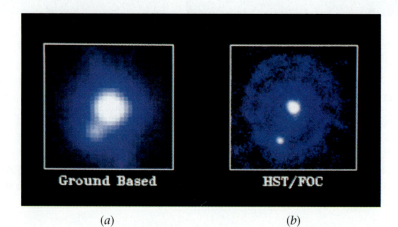

Ground Based   HST/FOC

(a)   (b)

probe. Pluto is a small planet, only about 0.2% the mass of the Earth, and it has a moon called Charon that is only slightly smaller than the planet. Pluto has a thin atmosphere consisting of methane. It made its closest approach to the Sun in 1989, and scientists expect methane to freeze into "snow" and coat the surface as the planet moves farther out.

Pluto's orbit is highly elliptical and, for part of each 250-year-long trip around the Sun, it is actually inside the orbit of Neptune. Because it is a small, rocky planet, unlike the neighboring gas giants, some planetary scientists argue that Pluto actually formed as a captured comet or asteroid, rather than a planet in the usual sense of the word. Indeed, the origin of Pluto is very much an open question in planetary astronomy.

Moons

Astronomers are deeply interested in the amazingly varied moons that encircle Jupiter, Saturn, Uranus, and Neptune. They have found dozens of such moons, although it is a matter of semantics whether something can be reasonably called a "moon" when it appears to be just a large orbiting rock.

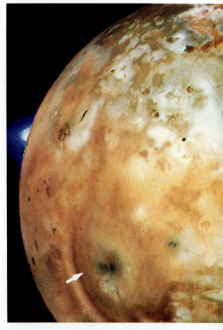

Io, a moon of Jupiter, is the only moon in the Solar System with active volcanoes. The arrow points to an eruption. The heat needed to drive the volcanoes comes from the effects of Jupiter's gravity on the body of Io.

Each of these moons can be thought of as a small laboratory that sheds some light on the formation of terrestrial planets. For example, the moon, Io, which circles close to Jupiter, is the only moon in the solar system known to have active volcanoes. Scientists think that the flexing and twisting of the moon's body due to the powerful gravitational forces of Jupiter produces the energy to drive those volcanoes.

Saturn's moon, Titan, is one of the largest moons in the solar system; it is about the same size as the planet Mercury. While we don't know the exact nature of its surface, it may have frigid lakes of methane (CH_4) and ethane (C_2H_6). Furthermore, simple chemical reactions in its atmosphere may have produced a layer of organic materials there. Thus this moon may reveal something about how organic materials accumulate, and about how cellular life developed on our own planet.

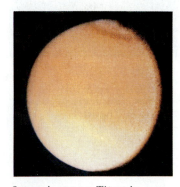

Saturn's moon Titan is, next to the Earth's Moon, the largest satellite in the Solar System. It has an atmosphere whose pressure at the surface is greater than that of Earth's.

A false-color close-up of the rings of Saturn, showing many gaps and rings. Working out the motion of particles in the ring remains a challenge for scientists.

Science in the Making

The Voyager Space Probes

On August 20, 1977, a rocket blasted off from Cape Canaveral, Florida, to launch a small space probe on its course. Sixteen days later, another rocket did the same (see Figure 17–13). These two probes, called *Voyager 1* and *2*, spent the next 15 years moving past the planets of the outer solar system, providing scientists with their first close look at the Jovian planets and their moons.

Each spacecraft had 10 scientific instruments on board, each designed to measure a different aspect of the deep-space environment. The ones that had the greatest public impact were the cameras that took pictures of moons, rings, and planets, but measurements were also made of magnetic fields, cosmic ray abundances, and infrared and ultraviolet radiation. Taken together, the *Voyager* probes produced a good deal of the detailed information about the outer solar system contained in this chapter.

One interesting feature of space flights of this type is that by the time a spacecraft has been in flight for a few years, all of its instrumentation has become obsolete. In the 10 years between its encounters with Jupiter and Neptune, for example, significant changes were made in the way *Voyager 2* analyzed and transmitted data, and in the way the data was received on Earth. As a result of these changes, the rate of data transmission during the Neptune flyby in 1989 was not significantly different from that of the Jupiter encounter in 1979, despite the greater technical difficulties of collecting and sending data so far from the Sun.

Among their discoveries, the *Voyagers* found new rings around all the Jovian planets, recorded a volcanic eruption on Io, tripled the number of known moons around Uranus, and clocked record winds on the surface of Neptune. Today, both *Voyagers*, together with a couple of earlier space probes called *Pioneers*, have moved outside of Pluto's orbit into interstellar space. *Voyager 2* is expected to keep returning data until its plutonium power supply runs down, sometime around 2020. By that time, it will have traveled far beyond the limits of the solar system, thereby becoming the first human-made object to have broken free from all the influences of the Sun. ●

Figure 17–13
Voyager 2 passed within 4,800 kilometers of Neptune in August 1989.

Asteroids, Comets, and Meteors

As the solar system formed, not all the material in the planetary disk was taken up into the bodies of the planets and moons. Even after hundreds of millions of years of accumulation and bombardment, a lot of debris is still floating in space. This debris comes in two main forms: asteroids and comets. These objects mimic the compositional differences of the terrestrial and Jovian planets.

Asteroids   Asteroids, like the inner terrestrial planets, are small rocky bodies in orbit around the Sun. These rocky chunks range from small boulders to objects many kilometers in diameter. Most asteroids are found in a broad, circular asteroid belt between Mars and Jupiter (see Figure

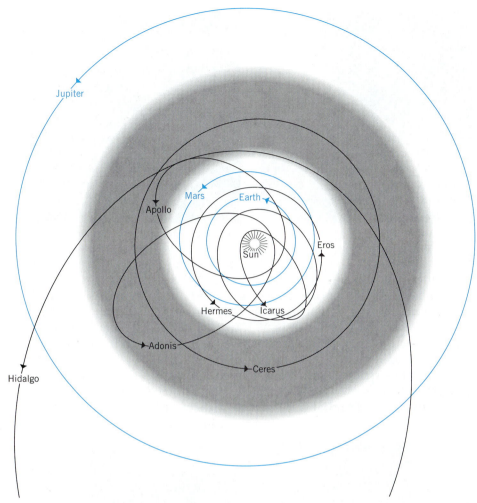

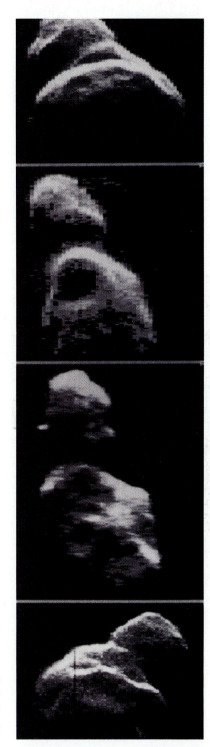

Figure 17–14
Asteroids move in orbits around the Sun. Some of the orbits cross that of the Earth, and may someday collide with us.

17–14). This belt is thought to be a collection of planetesimals that never managed to collect into a single planet. The most likely explanation is that the nearby planet Jupiter had a disrupting gravitational effect. In addition, many asteroids possess orbits that cross that of the Earth, and they produce occasional large impacts on our planet.

Comets Comets can best be thought of as "dirty snowballs." Unlike rocky asteroids, comets consist of chunks of material such as water ice and methane ice in which a certain amount of solid, rocky material or dirt is embedded. Also unlike asteroids, most of the comets in the solar system circle the Sun far outside the orbit of Pluto. Billions of them are to be found in a region called the *Oort cloud* (named after Dutch astronomer Jan Oort, 1900–1992, who suggested its existence). Occasionally, when comets in the cloud collide with each other, one will be deflected so that it falls toward the Sun. When this happens, the increasing temperature of the inner solar system begins to boil off materials, and we see a large "tail," blown away from the Sun by the solar wind, that reflects light to us.

A radar image of the asteroid Toutatis reveals two separate pieces rotating around each other.

The asteroid Gaspara. (*a*) Photographed by the Galileo spacecraft from a distance of 5300 km; (*b*) shown (top) in a photo montage with the moons of Mars, Deimos (lower left) and Phobos (lower right) at the same scale.

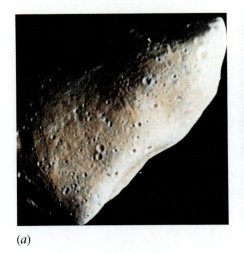

(*a*)

(*b*)

Sometimes a comet will be captured and fall into a regular orbit around the Sun. The most famous of these periodic comets is Halley's Comet (see Chapter 4), which returns to the vicinity of the Earth about every 76 years. Halley's return in 1910 was quite spectacular, because the comet passed near the Earth when it was at its highest temperature and therefore had its largest and most spectacular tail. The return in 1986 was much less spectacular because the comet was on the far side of the Sun when it was at its brightest. The next predicted return in 2061, unfortunately, will probably be just as unspectacular.

Comets are distinguished by their bright gaseous tails, which stream away from the Sun.

Meteoroids, Meteors, and Meteorites *Meteoroids* are small pieces of ancient space debris in orbit around the Sun. Occasionally one of these bits, perhaps the size of a sand grain, will fall into the Earth's atmosphere, where it becomes briefly visible as a *meteor*. Most meteors burn completely until they are microscopic particles of ash that slowly, imperceptibly, rain down on Earth. The meteors' bright streaks of light record the path of this burning. Occasionally, if the object is big enough so that only the outer surface burns, a piece of rock may actually reach the Earth's surface. Any such rock that has fallen to Earth from space is called a **meteorite**.

meteorite The fragment of a meteor that hits the Earth.

The 1,200-meter-wide Meteor Crater in Arizona formed from a collision about 20,000 years ago. The meteorite was about the size of a small car.

Historical drawing of a meteor shower in 1833. Though probably exaggerated, an intense shower can produce hundreds of meteors per hour.

Meteor showers are a set of spectacular annual events in the night sky. During a shower, you can see bright streaks in the sky every minute or so. They occur when the Earth collides with clouds of small debris that travel around the orbits of comets. Some of these clouds may be comets that have been broken up by the gravitational pull of one of the planets. Table 17–1 lists some of the most spectacular meteor showers.

Meteorites are extremely important in the study of the solar system because they represent the material from which the system was originally made. Scientists study them intensively, both to get a notion of how and when the Earth was made, and to learn what kinds of things human beings will find when they leave the Earth to explore the rest of the solar system.

Table 17–1 • Major Meteor Showers

Name	Date of Maximum	Hourly Rate of Meteors
Quadrantid	January 3	30
Perseid	August 12	40
Geminid	December 14	55

Environment

Meteorites and the Environment

We tend to think of environmental change as being a relatively slow process, with small changes accumulating over long periods of time. The truth is that the sudden impact of large bodies on the Earth, from the great bombardment down to the present, have had an enormous effect on living things on our planet.

During the great bombardment, for example, impacts of large objects could well have wiped out life on the planet. Suppose, for a moment, that oceans had formed and life had begun to develop 4 billion years ago or more. Suppose, further, that a good-sized asteroid—one the size of the state of Ohio, for example—hit the Earth. Such an asteroid would carry enough energy to vaporize the oceans and wrap the planet in a cloud of steam for a thousand years. It would, in the most literal sense, sterilize the Earth. We don't know if there were forms of life on the Earth before our own ancestors. If there were, all traces of them would have been wiped away by the impact.

In more recent times, the sudden spectacular demise of the dinosaurs 65 million years ago is now generally thought to have been the result of a meteorite hitting in the region of the Yucatan peninsula of Mexico. In the early 1980s, Luis and Walter Alvarez, a father and son team from the University of California at Berkeley, found evidence that an asteroid approximately 10 km across had struck the Earth at about the time of the great extinction. The impact would have thrown enough dust and dirt into the atmosphere to block sunlight from reaching the surface for several months, wiping out about two thirds of all the species in existence at the time.

Compared to the impact of an asteroid hitting the Earth, the environmental problems we are now experiencing, such as global warming discussed in Chapter 19, are minor. But you needn't worry. By one estimate, we aren't due for another dinosaur-style impact for another 11 million years or so. ●

THINKING MORE ABOUT
PLANETS

Human Space Exploration

Since astronaut Neil Armstrong walked on the surface of the Moon in 1969, the scientific community has debated the question of how best to explore the solar system. The question is whether future missions to the planets should carry people or only machines.

Those who advocate using machines for space exploration point to the enormous technical difficulties involved in providing a safe habitat for human beings in the harsh environment of space. Why, they ask, should we make the tremendous effort to put a human being on the surface of Mars,

(*a*) Neil Armstrong left his footprint (*b*) when he walked on the Moon in July 1969.

(*a*)

(*b*)

for example, when we can learn just as much by sending instrument packages and robots controlled from the Earth?

On the other side of the issue, scientists advocating space exploration by astronauts argue that no machine has the flexibility and ingenuity of a human being. They note that no matter how well designed a machine might be, when it is millions of kilometers from Earth, things can go wrong, and only a trained astronaut can salvage the mission. They point out that even a mammoth project such as the Hubble Space Telescope needed astronauts to replace the optical systems after the main mirror was built incorrectly. Besides, they argue, if one goal of the space program is to establish human colonies on other bodies in the solar system, you can't do that with machines.

What do we hope to learn from our studies of the solar system? Is colonization of the rest of the solar system the real long-term goal of the space program? How much extra effort (and taxpayer's dollars) is it worth spending to put people instead of machines on the surface of Mars?

▶ Summary

The Earth is believed to have formed approximately 4.5 billion years ago along with the Sun and other planets in our *solar system* from a *nebula*, which is a large dust cloud rich in hydrogen and helium. As that dust cloud began to contract due to gravitational forces, it also began to rotate and flatten out into the disk that now defines the planetary orbits. More than 99% of the original nebula's mass concentrated at the center, which became the Sun.

Gradually, the matter in the flat disk began to form clumps under its own gravitational forces. The largest of these masses swept up more and more debris as they orbited the early Sun, and they began to define a string of planets. *Terrestrial planets*, the four nearest the Sun, were subjected to high temperatures and strong solar winds, so that lighter gases such as hydrogen and helium were swept out into space. Thus the inner four planets, Mercury, Venus, Earth, and Mars, are dense, rocky places with relatively low content of gaseous elements.

The Earth's formation was probably typical of these planets. After the principal mass of the Earth had been collected together, additional rocks and boulders showered down in the *great bombardment*, adding matter and heat energy to the planet. Dense iron and nickel separated from lighter material by the process of *differentiation* and sank to the center to form a metallic *core*. Most of the Earth's mass concentrated in the thick *mantle*, while the lightest elements formed a thin *crust*. The Moon, Earth's only large satellite, may have formed when a planet-sized body hit the Earth early in its history.

The Solar System's outer *Jovian planets*, which are Jupiter, Saturn, Uranus, and Neptune, are quite different from the inner planets. Lying beyond the strong effects of solar heat and wind, they accumulated large amounts of gases, such as hydrogen, helium, ammonia, and water. These outer planets are giant balls of ice, with thick atmospheres and great frigid oceans of nitrogen, methane, and other compounds that are gases on Earth. Beyond the Jovian planets lies Pluto, a rocky body that is the smallest planet. All of the planets except Mercury and Venus, the two closest to the Sun, have moons in orbit.

Interspersed with the planets and their moons is a large number of other types of objects. Small, rocky *asteroids*, most of which are concentrated in an *asteroid belt* between Mars and Jupiter, circle the Sun like miniature planets. Far outside the solar system, swarms of "dirty snowballs" called *comets* are concentrated in the *Oort cloud*. If a comet's distant orbit is disturbed, it may fall toward the Sun and create a spectacular display in the night sky. When a piece of interplanetary debris hits the Earth's atmosphere, it creates a *meteor*, or shooting star, which burns up with a fiery trail. Occasionally, a meteor fragment will hit the Earth and become a *meteorite*.

▶ Review Questions

1. Describe the characteristics of terrestrial planets.

2. Name the terrestrial planets.

3. Describe the characteristics of Jovian planets.

4. Name the Jovian planets.

5. What is the nebular hypothesis? How does it explain the orbits of the planets?

6. What are planetesimals, and what role do they play in the formation of planets?

7. Why are the terrestrial planets different in composition from the Jovian planets?

8. Describe the process of differentiation.

9. How has the process of differentiation affected the Earth?

10. What is the Earth's core and what is it made of?

11. How does the high pressure in the core affect chemical bonds?

12. What is the Earth's mantle and what is it made of?

13. What is the composition of the Earth's crust?

14. Describe the "big splash."

15. What was the result of the great bombardment?

16. What is outgassing? Is it still going on today?

17. Describe two ways that a planet's atmosphere can evolve.

18. What is an asteroid? Where are most asteroids found?

19. What is the difference between a meteor and a meteorite?

20. What is a comet? Where are most comets found?

21. How could asteroids and comets affect life on Earth?

22. Which planets have moons?

23. How does gravitational escape affect a planet's atmosphere?

24. How does life affect a planet's atmosphere?

25. What evidence suggests that the Moon was once part of the Earth?

▶ Fill in the Blanks

Complete the following paragraphs with words and phrases from the list.

asteroids	mantle
asteroid belt	meteors
comets	meteorite
core	moons
crust	nebula
differentiation	Pluto
great bombardment	solar system
Jovian planets (Jupiter, Saturn, Uranus, Neptune)	terrestrial planets (Mercury, Venus, Earth, Mars)

The Sun, the planets, comets and asteroids make up the _____. The _____ are small and rocky, while the _____ are large and gaseous. The outermost planet, _____, is an anomaly. Many planets have _____ in orbit around them, and each of these has different characteristics. _____ are debris left over from the formation of the solar system, and are concentrated in the _____. _____ are mostly found in orbit outside the farthest planets, but occasionally fall into the inner solar system.

The solar system formed from a cloud of gas and dust known as a _____. Early in history their planets underwent a period known as the _____, when debris rained onto their surfaces. The continuation of this process can be seen today with _____, which are objects entering the Earth's atmosphere and burning up. If such a body survives to reach the ground, it is called a _____.

Early in its history, the Earth melted and underwent _____, so that its interior now consists of an innermost _____ surrounded by a _____, with a thin outer layer known as the _____.

▶ Discussion Questions

1. What distinctive characteristics of the Earth make it suitable for life?

2. How has life altered the chemistry of the Earth's atmosphere?

3. Why are meteorites important to our study of the solar system?

4. Review the history of studies of the solar system. What new technologies have enhanced our ability to make observations of objects in the solar system?

5. What factors might influence the development of life on other planets?

6. What sources of data might help us to determine more about how the Earth's Moon formed?

7. How does Clyde Tombaugh's work fit into the scientific method?

8. Do you think that we should continue to explore space? Why or why not?

9. Do you think colonization on another planet is an achievable goal?

10. How does the process of photosynthesis affect the Earth's atmosphere today? What is its role?

11. Classify qualitatively, relative to the Earth (i.e., larger, smaller, faster, slower, etc.) the nine major planets according to their size, mass, density, surface temperatures, size of their orbits, position relative to the Sun, average orbital speeds and periods, rotation rates and directions, and number of natural moons.

12. What patterns do astronomers know about the physical position and the direction of the planets' orbits and rotations that suggest a common formation with the Sun?

13. What patterns exist between the average density of the terrestrial planets, meteorites, and Earth and Moon rocks that may suggest a common origin?

14. List the major steps and the resulting environment (relative temperatures, compositions, positions of the nebular material) for the formation of the solar system using a nebular hypothesis. Include in this list the general features that such steps and environments help explain.

15. Explain outgassing and how it relates to the formation and evolution of the Earth's atmosphere. Is this process still occurring today? Is outgassing a process that is seen only on the Earth, or has it also occurred on other planets or planetary-like objects? Explain.

16. Outline the development of the Earth's atmosphere over time.

17. Classify asteroids and comets according to their location in the solar system, their orbits, and their composition.

18. How are meteors, meteoroids, and meteorites related to each other? Are they closer in origin to comets or asteroids?

19. What is the origin of the initial high temperatures in the Earth's core that resulted in its differentiation?

▶ Problems

1. Calculate and compare the relative size of the Moon's orbit around the Earth and the diameters of the Sun and Jupiter.

2. Since the Moon is in the same planetary environment as the Earth, calculate the amount of time it would take to double its mass, using the current rate of accretion of matter for the Earth.

3. How many Earths (consider only the volumes) can fit into the Sun? Into Jupiter?

4. Calculate the fraction of the Earth's volume and mass taken up by the mantle. Use 4.5 g/cm^3 for the average density of the mantle.

5. If the present Earth were formed in 1 billion years via early bombardment,
   a. what is this bombardment rate assuming that all of the present Earth's mass accreted during this time?
   b. how does this rate compare with the current rate of accretion by the Earth?

6. What is the average orbital speed of the Earth in km/s? In mph? (Assume a perfect circular orbit.)

7. What is the average orbital speed of the Moon around the Earth in km/s? In mph? (Assume a perfect circular orbit.)

8. Given the diameters of the planets in Table 17–1, what are the relative volumes of Earth, Mercury, and Jupiter? From the same table, what are the relative masses?

9. From the values of mass in Table 17–1, calculate the densities (mass divided by volume) for Earth, Mercury, and Jupiter. Why do you think the relative masses and densities are different?

10. How many asteroids 10 km in diameter would be required to make a planet about the size of Mars? Neglect the effects of compression in the planet's interior.

11. If the average thickness of the Earth's crust is 10 km, what fraction of the solid Earth's total volume is in the crust?

▶ Investigations

1. Investigate the history of unmanned planetary probes. What are the names, dates, and target planets of these probes? What countries sponsored them? What kind of data did they return? Are any planetary missions now under way?

2. Read a history of the Apollo missions to the Moon. What theories about the Moon's origins prevailed before these missions? What new data changed theories about the origin of the Moon?

3. How would you respond to the following argument: No one was present when the Earth was formed, so how can scientists talk about the details of the formation process?

4. Listen to *The Planets*, a suite for orchestra by the British composer Gustav Holst. In what ways do the musical descriptions of each planet reflect the physical characteristics of that planet? What sources of inspiration other than scientific data did Holst use in creating these pieces? Which planets did he omit and why?

5. There are many more meteor showers than the ones listed in Table 17–1. Find out which ones may be coming in the next month or two and plan a meteor-watching party. Document what you see.

6. Scientists are attempting to document the paths of asteroids with Earth-crossing orbits. Investigate this research and comment on the probability that a large asteroid might hit the Earth. Should we increase funding for asteroid monitoring?

7. Research NASA's plans for upcoming space probes. Do you think your representatives and senators will vote to approve funding? Do you think they should?

8. Investigate the various ideas that have been proposed to establish a permanent base on the Moon. What are the principal technical difficulties in such a venture? How much might it cost?

9. Investigate the various ideas that have been proposed for a human mission to Mars. How long would such a mission take? What are some of the principal technical difficulties to be overcome? How much might such a mission cost?

10. Discuss the pros and cons of international cooperation in space exploration. Should the United States attempt such exploration alone?

11. Read about the mission of Apollo 13 or see the 1995 movie based on it. Explain how the astronauts were returned to Earth after an accident caused them to lose most of their oxygen.

▶ Additional Reading

Comins, Neil F. *What If the Moon Didn't Exist?* New York: Harper-Collins, 1993.

Greeley, Ronald. *Planetary Landscapes*, 2d ed. New York: Chapman and Hall, 1993.

Hartmann, William K. *Moons and Planets*, 3d ed. Belmont, Calif.: Wadsworth, 1993.

Sagan, Carl and Ann Druyan. *Comet.* New York: Random House, 1985.

18 THE DYNAMIC EARTH

THE ENTIRE EARTH IS STILL CHANGING, DUE TO THE
SLOW CONVECTION OF SOFT, HOT ROCKS DEEP
WITHIN THE PLANET.

A Construction Site

Think about the last time you passed a new construction site after a rain shower. You could probably see little valleys, freshly carved by water running over the bare earth, and shallow pools where fine-grained material collected in layers. Such features are small-scale examples of soil that is eroded by rain. If the construction workers had dug a deep pit, you might have noticed different layers of earth and rock freshly exposed; these layers represent sediments deposited by water long ago.

Now think about the last time you were at the beach when the wind was blowing. Did you feel a lot of sand and grit in the air? The wind is constantly moving soil at the Earth's surface.

These sorts of small-scale changes in the Earth's surface are mirrored by much more dramatic large-scale changes. When Mount St. Helens, a volcano in Washington State, erupted in 1980, the entire side of a mountain was blown away and many square kilometers of forest were flattened. Large earthquakes in our lifetime will change the course of rivers and destroy villages and towns.

Perhaps the most extraordinary changes of all are ones that we cannot easily perceive in human time scales without sophisticated instruments. For example, the most permanent features of the Earth, such as the continents and ocean crust, are constantly changing. The continent of North America is almost a meter farther from Europe than it was when you were born; this movement is the result of forces deep within the Earth.

No feature on the Earth's surface, no desert or broad plain, no mountain or ocean, is permanent. Every feature is constantly changing and evolving into something different.

THE CHANGING EARTH

How do we know that the Earth's surface is changing if we don't often observe dramatic changes in the landscape? You can make a simple estimate, based on familiar observations of streams, that will convince you of the Earth's dynamic nature. Mountains appear to be permanent features on Earth, yet, as you will see in the following section, mountains must wear away in times much shorter than the estimated 4.5-billion-year age of the Earth. You will also see that continents are not as permanent as they may appear to be.

Science by the Numbers

The Life Span of a Mountain

While there is no such thing as a "typical" mountain, for the purposes of this rough estimate think of a mountain as a rectangular mass about 2 km high, 4 km long, and 4 km wide (that's about 1.2 mi by 2.5 mi by 2.5 mi) (Figure 18–1). The volume of this mid-sized mountain is

$$\text{volume} = \text{length} \times \text{height} \times \text{width}$$
$$= 2 \text{ km} \times 4 \text{ km} \times 4 \text{ km}$$
$$= 32 \text{ km}^3$$

Expressed in cubic meters, that is

$$2000 \text{ m} \times 4000 \text{ m} \times 4000 \text{ m} = 3.2 \times 10^{10} \text{ m}^3.$$

Think about a stream running down a mountainside. Such a stream carries a certain amount of sand and dirt. You can see this because the stream has a sandy bottom and, as the stream flows, it moves the sand downstream. You might also see gravel and boulders in the stream—evidence that, from time to time, heavy rains cause much more violent movement of material down the mountainside. This rock and sand are a result of erosion that occurs higher up on the mountain. In fact, the existence of the stream means that the mountain is constantly being eroded away.

You can estimate how long a mountain might survive by examining signs of erosion in a stream. Suppose, for example, that four principal streams run off the sides of the mountain, and that each stream

Figure 18–1
Pike's Peak near Colorado Springs, Colorado, may be approximated as a 2 × 2 × 4 km rectangular block of rock.

carries an average of one tenth of a cubic meter of earth per day off that mountain (the actual amount would vary from day to day depending on the kind of rock, the amount of water flowing down hill, and other factors). One tenth of a cubic meter per day is not very much material; it corresponds to a pile of sand, dirt, and gravel about 50 cm (a foot and a half) on a side. (Such a pile could fit under an ordinary kitchen chair.) The next time you are near a stream, put your hand down into the water and feel the sediment that is being constantly transported.

Over a period of a year, the four streams might thus remove

$$4 \text{ streams} \times 0.1 \text{ m}^3/\text{stream-day} \times 365 \text{ days/year} = 146 \text{ m}^3/\text{yr}$$

Every year, therefore, close to 150 m³ of material, or about six dump trucks full, could be removed from a mountain as a result of the normal processes of erosion.

If the mountain streams remove about 150 m³ each year, then the lifetime of the mountain can't be much longer than the volume of the mountain, divided by the volume lost each year:

$$\frac{3.2 \times 10^{10} \text{ m}^3}{150 \text{ m}^3/\text{yr}} = 0.0213 \times 10^{10} \text{ years}$$
$$= 213{,}000{,}000 \text{ years.}$$

This estimate, while very rough and not directly applicable to any specific mountain, tells us that under normal circumstances, mountains won't last more than a few hundred million years. According to our estimate, mountains will disappear as a result of erosion and will become low, rounded hills in time periods much shorter than the Earth's age of 4.5 billion years. ●

The Case of the Disappearing Mountains

If all mountains eventually erode, why are there any mountains left on Earth? We must conclude that if the erosion of mountains takes a few hundred million years, then any mountain that existed when the Earth first formed 4.5 billion years ago would have been worn away long ago,

Stream erosion can carve spectacular scenery, as in the canyons of the Green and Colorado rivers in Utah.

Steep slopes and angular peaks characterize young mountains such as the 65-million-year-old Rocky Mountains in Alberta, Canada, photographed from Mt. Rae (*a*). The 400-million-year-old Smoky Mountains in North Carolina display the rounded character of older mountains (*b*).

(*a*)

(*b*)

and the Earth's surface should by now be smooth and featureless. The only explanation consistent with the fact that mountains still exist on the surface of the Earth is that mountains are continuously formed.

Thus everyday observations of erosion and some simple arithmetic leads us to a startling conclusion: tremendous forces must be acting on the Earth, creating new mountain chains as the old ones are worn away. Although mountains seem to symbolize eternal solidity to human beings, they are, in fact, transitory. Geologists who map the distribution and ages of rocks have shown that the Appalachian Mountains were formed a few hundred million years ago, and the Rocky Mountains about 60 million years ago.

Newton's laws of motion (Chapter 4) tell us that nothing happens unless a force acts. What forces in the Earth could create entire mountain ranges? Until recently, this question remained one of the greatest puzzles in geology.

Volcanoes and Earthquakes: Evidence of Forces in the Earth

Most geological processes like mountain building and erosion are slow by human standards, and take thousands or even millions of years to effect noticeable change. But volcanoes and earthquakes may transform a landscape in an instant, thus revealing the tremendous energy stored in our dynamic planet.

Figure 18–2
A cross section of a volcano reveals a magma chamber, which stores molten rock, and a system of pipes, cracks, and vents that lead to the surface. The terms in black refer to the kinds of rocks formed from cooled magma. Xenoliths are fragments of the original rocks encased in this cooled magma.

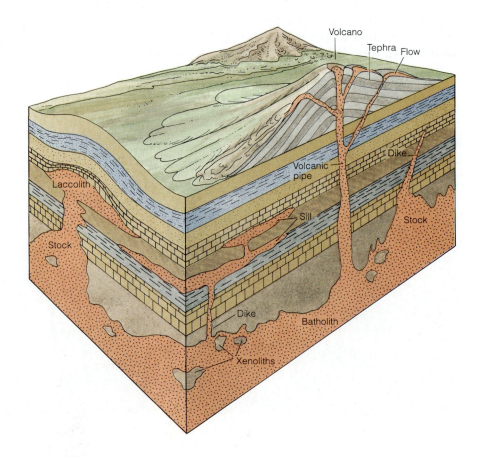

Volcanic eruptions provide the most spectacular process by which new mountains are formed. In a typical **volcano,** subsurface molten rock called *magma,* concentrated in the Earth's upper mantle or lower crust, breaks through to the surface of the Earth, as shown in Figure 18–2. This breakthrough may be sudden, giving rise to the kind of dramatic events we saw when Mount St. Helens exploded in 1980, or it may be relatively slow, with a stately surface flow of molten rock, known as *lava.* In both cases, however, magma eventually breaches the surface and hardens into new rock.

Earthquakes occur when rock suddenly breaks along a more-or-less flat surface, called a *fault.* Have you ever stretched a thick, strong elastic band, only to have it snap back painfully against your hand? When this happened, you gradually added elastic potential energy to the band, and that energy was suddenly released and converted to violent kinetic energy. The same thing happens in the Earth when rock under stress suddenly snaps. This sudden release of energy is called an **earthquake.**

When a rock suddenly breaks, tremendous amounts of potential energy are released. The two sides of the fault can't fly apart like an elastic band, so the energy is transmitted in the form of a sound wave, or *seismic wave* (see Chapter 8). These waves, traveling at speeds of several kilometers per second, cause the ground to rise and fall like the surface of the ocean. Normally "solid" ground sways and pitches in a motion that can cause severe damage to buildings and other structures. Furthermore, if an earthquake occurs under or near a large body of water, these violent motions of the ground can transfer energy into great waves of water that can devastate low-lying coastal areas. Such waves are called *tsunamis* (pronounced "sue-Nah-me"), a Japanese term for harbor or bay waves (see Figures 18–3 and 18–4).

Great earthquakes, such as the one that killed 20,000 people in India in September 1993, are fortunately relatively rare. But smaller earthquakes, barely noticeable to the average person, occur every day by the thousands. Earthquakes were once rated on the *Richter scale,* named after the United States geologist Charles Richter, who devised it. Today, earthquakes are measured by a related scale that refers to the amount of ground motion that would be measured by an instrument a fixed distance from the center of the earthquake. The scale is such that each increase of 1 unit corresponds to 10 times more ground motion, and 30 times more energy released. Thus an earthquake that measures 7 will have 100 times

volcano Places where subsurface molten rock breaks through to the surface of the Earth to form dramatic short-term changes in the landscape.

earthquake Disturbance caused when stressed rock on Earth suddenly snaps, converting potential energy into released kinetic energy.

Recent volcanic eruptions. (*a*) Mount Saint Helens in Washington State erupted in a violent explosion on July 22, 1980. (*b*) Puu O'O on the flanks of Hawaii's most active volcano, Kilauea, erupted periodically throughout most of the 1980s (this photograph was taken in 1983). (*c*) Mount Pinatubo in the Philippines erupted in 1991, spreading a thick blanket of ash over thousands of square kilometers.

(*a*)

(*b*)

(*c*)

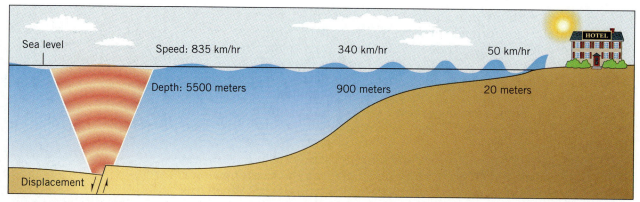

Figure 18–3
A tsunami, or harbor wave, is generated by a displacement of the ocean floor. It builds in height as it approaches a shoreline.

more ground motion and 900 times more energy released than one that measures 5, and so on.

Earthquakes that measure around 5 on this scale will be felt by most people, but will do little damage in areas with well-constructed buildings. An earthquake between 6 and 7 will do considerable damage to buildings, and a magnitude 8 earthquake will level large areas. In October 1989, a 7.1-magnitude earthquake that occurred in the San Francisco Bay area destroyed many buildings and caused sections of elevated freeways to collapse. A similar large earthquake destroyed tens of thousands of buildings and killed more than 5000 people in Kobe, Japan, in January 1995, and resulted in perhaps $100 billion in damage. Even so, these events were relatively minor compared to "The Big One," a magnitude-8 or greater earthquake that must inevitably strike California some time in

Figure 18–4
It takes a tsunami about 5 hours to get from Alaska to Hawaii, 15 hours to get from Chile to Hawaii, and so on. The dots indicate the origins of tsunamis in the past 50 years, with the size indicating the severity of the damage caused. How much time does it take for a tsunami to travel from Tokyo to Hawaii?

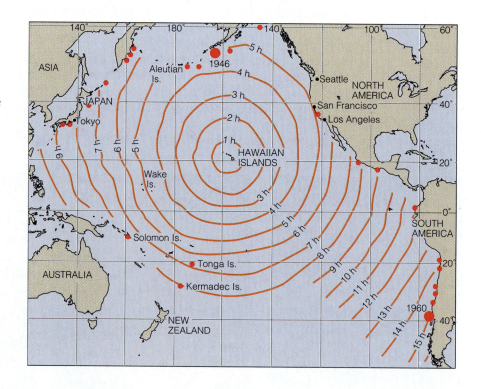

the coming decades. No earthquakes greater than 9 on the Richter scale have ever been recorded, probably because no rocks can store that much energy before they let go.

The puzzle remains, however: Where does all the energy that powers volcanoes and earthquakes come from? To answer this question, we need to know more about the structure of the Earth.

The Movement of the Continents

Think about a large map of the world. In your mind, move North and South America eastward toward Eurasia and Africa. Have you ever noticed how the two seem to fit together (Figure 18–5)?

This jigsaw puzzle pattern was one of the first indications that the Earth's surface is not static, but in a state of constant flux. English states-man and natural philosopher Francis Bacon (1561–1626) pointed out this fact in 1620. It wasn't until the beginning of the twentieth century, however, that anyone took the parallel coastlines seriously enough to question the origin of this pattern. In 1912, a German meteorologist named Alfred Wegener (1880–1930) proposed that the Earth's continents are in motion. The reason that the Americas fit so well into the coastline of Europe and Africa, he suggested, is that they were once joined and have since been torn apart.

Wegener's theory, called *continental drift*, was dismissed by most earth scientists. He amassed some geological evidence to back it up, such as the matching locations of distinctive rock formations on opposite sides of the Atlantic Ocean. But most of this evidence was fragmentary and unconvincing to other scientists. Indeed, some of Wegener's arguments later turned out to be wrong. Furthermore, he failed to provide a reason-able mechanism by which continents could move. For most of the twenti-

In January 1995, a magnitude-7.2 earthquake caused extensive damage and loss of life in the city of Kobe, Japan.

Figure 18–5
A map of the world's continents reveals the similar shapes of coastlines and continental shelves on the two sides of the Atlantic Ocean.

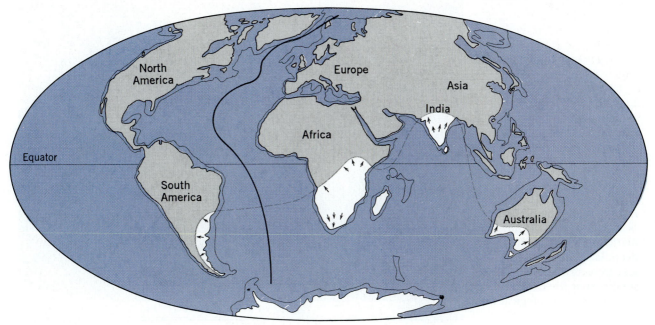

eth century, continental drift was regarded as a far-fetched exercise in theory, and few geologists paid much attention to it. Beginning in about 1960, however, geologists and oceanographers obtained new evidence to support one aspect of Wegener's notion, the idea that the continents are not fixed.

The discovery that continents do indeed move depended on a combination of three different kinds of newly acquired geological evidence. These new data included topographic profiles of the ocean floors, maps of rock magnetism, and information on rock ages.

Figure 18–6
A topographic map of ocean floors reveals dramatic mountain chains, deep flat planes, canyons, and trenches. These features suggest that the ocean basins are active geological regions.

A photo of deep sea vents taken from the submarine research vessel ALVIN reveals astonishing, hitherto unknown life forms.

1. Ocean Floors.   When the contours of the ocean floor were mapped in the years following World War II, oceanographers discovered remarkable, unsuspected features. Most scientists thought that the deep ocean bottoms were simply flat plains, which had been passively collecting the sediments that had gradually eroded off the continents. Instead, they found steep-walled canyons and lofty undersea mountains, indicating that the ocean floor is as dynamic and changing as the continents themselves (Figure 18–6). The longest mountain range on Earth, for example, is not on a continent, but in the middle of the Atlantic Ocean. Called the Mid-Atlantic Ridge, this feature extends from Iceland in the North Atlantic to the South Sandwich Islands in the South Atlantic. Similar ridges, also called rises, were found on the floors of all the Earth's oceans. These ridges are sites of continuous geological activity, including numerous earthquakes, volcanoes, and lava flows. Oceanographers have now mapped more than 85,000 km of ocean ridges.

2. Magnetic Reversals.   To understand the nature of magnetic data—the second kind of evidence that pointed to continental motion— you have to recall that the Earth has a magnetic field of the dipole type, with north and south magnetic poles. For reasons that are not fully understood, this field flips direction sporadically over time, something like an electromagnet in which the direction of current in the coils changes occasionally (see Chapter 7). Well over 300 reversals of the Earth's magnetic field have been recorded in ancient rocks, spanning about 200 million years. During recent episodes of field reversals, what is now the north magnetic pole of the Earth was located somewhere near the South Pole in Antarctica, while the south magnetic pole was somewhere near the North Pole.

When lava flows out of a fissure in the Earth, it contains small crystals of natural iron oxides, including the naturally magnetic mineral magnetite. These bits of iron ore act as tiny magnets, and, since the rock is still in a fluid state, their magnetic dipoles are free to turn around and align

Figure 18–7
Magnetic measurements of the seafloor in the late 1950s and early 1960s revealed patterns of magnetic stripes running nearly parallel to the Vancouver and Washington State coastlines.

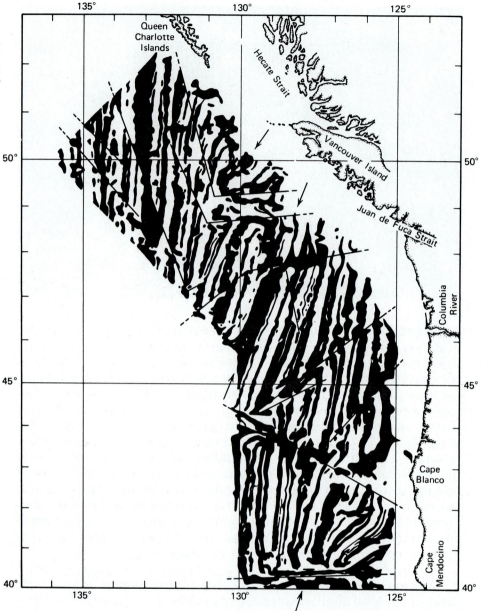

themselves in a north-south direction parallel to the Earth's magnetic field. Think of them as small compass needles embedded in the fluid rock. Once the rock hardens, however, the bits of magnetite are frozen in place; they can no longer move. Thus the volcanic rock carries within it a memory of where the magnetic north pole was when the rock solidified. If we examine the tiny compass needles, we can tell whether the magnetic field of the Earth was as it is today, or whether it was reversed. The discipline devoted to the study of this sort of effect is called *paleomagnetism*, and it came into maturity in the early 1960s.

In the mid-1950s, an oceanic exploration ship named *Pioneer* began taking magnetic measurements near the ocean floor off the coast of Washington State. At first the data made little sense, but as more and more

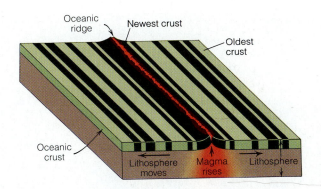

Figure 18–8
Magnetic stripes that parallel ocean ridges must form as new magma wells up from the fissure and pushes out to the sides. In this cross-sectional view, older rocks lie farther from the ridge.

profiles were obtained, a pattern of stripes on the ocean floor began to emerge. It seemed that there were parallel strips of rock in which the magnetic direction in neighboring strips alternated. We show some of the original data in Figure 18–7. Within the next few years, through the mid-1960s, earth scientists collected more and more data of this type, and it soon became clear that the Mid-Atlantic Ridge, the East Pacific Rise, and many other places on the ocean floor show the same pattern of alternating magnetic stripes.

How could this configuration of magnetic stripes be explained? In an environment in which the orientations of north and south magnetic poles were alternating back and forth over time, the only way to get the observed striping is shown in Figure 18–8. The seafloor must have been getting wider, or *spreading*, as molten rock came from deep within the Earth and erupted through volcanic fissures on the seafloor.

Over time, the rock at the surface today will be pushed aside as the continents move apart and as more magma comes up to take its place. The compass needles in this newer rock will point to the location of the magnetic north pole when they reach the surface. Each new batch of magma will solidify and lock in the current magnetic orientation, despite any further reversals. Each time the Earth's magnetic field reverses, the dipole direction of the natural magnets changes and gets locked into the rock. Thus, over long periods of time, we would expect to see alternating bands of magnetic orientation—exactly the kind of zebra-striped pattern of alternating north and south compass orientations shown in Figure 18–7.

3. Rock Dating. The conclusions based on magnetic data were reinforced by studies of the age of volcanic rocks in the oceanic crust. Volcanic rocks contain radioactive isotopes that can be used for dating (see Chapter 14). Rocks near the Mid-Atlantic Ridge and other similar features were found to be quite young, a few million years old or less. Rocks collected successively farther away from the ridge proved to be correspondingly older.

New Evidence for Moving Continents

The new data on the topography of the ocean floor, as well as the magnetic properties and ages of its rocks, suggested to many scientists that the width of the Atlantic Ocean must be increasing yearly as the seafloor spreads. The continents of Eurasia and North America, it appears, are

By using radio telescopes like these, astronomers in Europe and North America were able to document the slow separation of their two continents.

moving apart. Could the hypothesis that this distance is increasing be tested by an experimental measurement?

Up until the late 1980s, no one was able to measure the motion of continents directly. The breakthrough came from an unexpected source: radio astronomy. Astronomers in North America and Europe trained radio telescopes on distant quasars (see Chapter 21). By measuring the times of arrival of the crests of the same radio wave, astronomers were able to get very accurate measurements of the distance between their telescopes. By repeating these sorts of experiments over a period of a few years, they were able to measure exactly the separation of the continents. Their work shows that North America is separating from Europe at the rate of about 5 cm (about 2 inches) a year.

Science by the Numbers

The Age of the Atlantic Ocean

If continents are constantly moving at speeds up to several centimeters per year, what would a map of the Earth's surface have looked like millions of years ago? We can make an educated guess by using present rates of movement and "reversing the tape," so to speak. We know, for example, that the floor of the Atlantic Ocean is spreading at about 5 cm/yr. Assuming that the rate of spreading has remained more or less constant, how old is the Atlantic Ocean?

To calculate the answer, we need to apply the equation for travel time:

$$\text{time of travel} = \frac{\text{distance}}{\text{speed}}$$

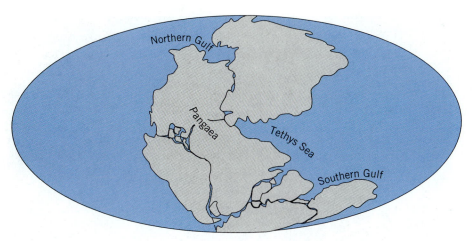

Figure 18–9
A map of the ancient continent of Pangaea reveals how the positions of present-day continents appeared 200 million years ago.

(You use the same relationship every time you estimate the driving time for a vacation.) The speed in this equation is the spreading rate, 5 cm/yr, and the present width of the Atlantic Ocean is approximately 7000 km, or 7×10^8 cm. The Atlantic Ocean began to open, therefore,

$$\text{time} = \frac{7 \times 10^8 \text{ cm}}{5 \text{ cm/yr}}$$

$$= 1.4 \times 10^8 \text{ yr}$$

$$= 140 \text{ million yr ago}$$

Think about what the Earth must have looked like 140 million years ago, during the age of the dinosaurs. Europe and the Americas would be joined together and there would be no Atlantic Ocean. In fact, by retracing the wandering paths of continents, earth scientists have discovered that 200 million years ago what we now call the continents of North America, South America, Eurasia, and Africa were once locked together in one giant continent called Pangaea (Figure 18–9). A globe of that ancient Earth would be all but unrecognizable, with a huge land mass on one side, and a giant ocean on the other.

By the same token, 100 million years from now the Earth's continents will have moved thousands of kilometers and appear completely different from today's configuration. ●

PLATE TECTONICS: A UNIFYING VIEW OF THE EARTH

The compelling model of the dynamic Earth that has emerged from studies of paleomagnetism, rock dating, and other data is called **plate tectonics.** "Tectonics" is related to the word "architect" and has to do with building or putting things together. Plate tectonics refers to a picture of the world in which many of the Earth's large-scale surface features and related phenomena result from the motions of the relatively thin, outermost layers of brittle rock.

plate tectonics The model of the dynamic Earth that has emerged from studies of paleomagnetism, rock dating, and much other data. A theory that explains how a few thin, rigid tectonic plates of crustal and upper mantle materials are moved across the Earth's surface by mantle convection.

Figure 18–10
The major plates of the Earth with their directions of motion. The lengths of the arrows indicate the relative speeds of plate motions.

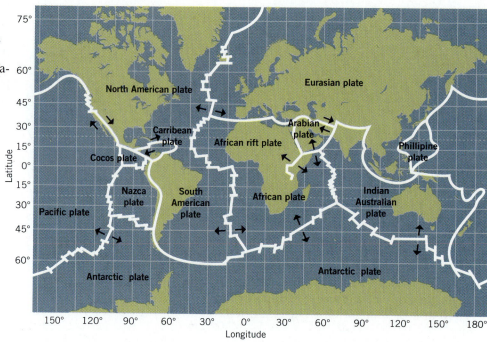

tectonic plate   One of a dozen large, and some smaller, sheets of moving rock forming the surface of the Earth.

The central idea of the plate tectonics theory is that the surface of the Earth is broken into about a dozen large pieces (as well as a number of smaller ones), called **tectonic plates** (see Figure 18–10). Each plate is a rigid, moving sheet of rock up to 100 km (60 miles) thick, composed of the crust and part of the upper mantle. Oceanic plates have an average 8- to 10-kilometer thickness of dense rock known as *basalt* capping the mantle rock. Continental plates have, on average, an additional 35-kilometer thickness of lower-density rock such as *granite* capping the basalt and mantle rock.

Tectonic plate boundaries are not the same as those of the continents and oceans. Some plates have continents on all or part of their surface, while some are covered only by oceanic crust. Most of the North American continent, for example, rests on the 8000-kilometer wide North American plate, which extends from the middle of the Atlantic Ocean to the edge of the Pacific plate on the west coast of North America.

About one quarter of the Earth's surface is covered by continent; the rest is ocean. On time scales of millions of years, the plates shift about on the planet's surface, carrying the continents with them like passengers on a raft. Thus it is not the motion of the continents themselves that is fundamental to understanding the dynamics of the Earth, but the constant motion of the underlying plates. Continental motion (what Wegener called continental drift) is just one manifestation of that plate motion.

The Convecting Mantle

Nothing happens without a force, but what force could possibly be large enough to move not only the continents, but also the plates of which they are a part? According to current theory, continents move as a result of the force generated by **mantle convection** deep within the Earth. These epic motions are driven by our planet's internal heat energy.

At the Earth's surface, we are seldom aware of the prodigious amount of heat energy stored deep underground. We normally think of caves and cellars as cool places—much cooler, on the average, than the surface. But miners are well aware of the heat constantly generated in the Earth. Once a mine penetrates more than a few hundred meters deep, the temperature begins to rise. This increase in temperature, which averages several degrees each thousand meters, can lead to some surprising engineering problems. At the Homestake Gold Mine in the Black Hills of South Dakota, for example, engineers planning to open shafts about 3 kilometers below the surface expect to find rock temperatures there well over 100°F. In such an environment, mining engineers have to air condition the shafts before miners can be allowed to go into them.

Two sources of energy contribute to the Earth's interior heat. Much of this heat energy is left over from the gravitational potential energy released during the great bombardment and differentiation of the mantle and core as the Earth formed (see Chapter 17). The decay of uranium and other radioactive elements (see Chapter 14), which are fairly common throughout the Earth's core and mantle, provide a second important source of heat energy. These elements decay over time and produce energetic, fast-moving decay products—particles that collide with atoms in surrounding rocks and give up their energy as heat. A surprising amount of internal heat can be generated in this way. If you take a chunk of granite, for example, and could suddenly release all the heat that could be generated by its radioactive elements, you could melt the rock. Deep in the Earth, all this heat energy softens rocks to the point that they can flow slowly like hot taffy.

In Chapter 6 we saw that heat energy, once generated, must move spontaneously toward cooler regions. The heat generated in the interior of the Earth, for example, must eventually be radiated into space. If the Earth were somewhat smaller, or if only small amounts of heat were generated, the energy could be carried to the surface entirely by *conduction*, which is the movement of heat by atomic collisions (see Chapter 6). This process operates in the Moon and Mars, for example. In the case of the Earth, however, there is too much internal heat to be carried away by conduction alone. Rocks in the mantle have been heated to the point where they are able to flow, and deep within the mantle of the Earth immense *convection cells* are set up, as shown in Figure 18–11. In a sense, the rocks of the mantle behave like water in a boiling pot.

There are, of course, very important differences between the Earth's mantle and a boiling pot; in particular, the time scales differ. Solid rock does not flow very far on time scales comparable to human lifetimes. But give soft, hot rocks a few hundred million years and they can move large distances. Earth scientists now estimate that the Earth's convection cells go through a full cycle, with hot rocks rising from the lower mantle, cooling near the surface, and falling back to be replaced by other warmer rocks, on a time scale of about 200 million years. In effect, the Earth behaves something like a giant spherical stove, with the burners on the inside and the circulating rocks bringing the heat to the surface.

mantle convection The convective motion deep within the Earth, driven by internal heat energy, that moves continents and plates of which they are a part.

Figure 18–11
Inside the Earth, warm rock rises and cool rock sinks over hundreds of millions of years. The resulting convection cells are responsible for tectonic motion.

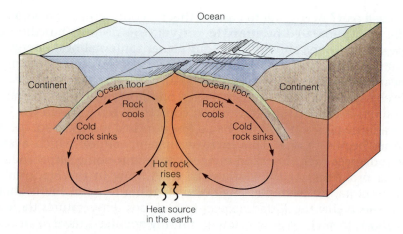

Near the surface of the Earth, the partially molten convecting mantle rocks encounter the comparatively thin plates that cover the surface in a relatively cool, brittle layer. Along the oceanic ridges, the brittle plates fracture and basaltic lava erupts, initiating seafloor spreading and plate motion. The plates move along with the convection cells underneath them, more or less as a film of oil on boiling water would move along on top of the water. The plates shift around, crash into each other, join each other, and are split apart, in a constant dance. And on top of these plates, floating along like a thin scum with no control whatsoever over their destiny, are the continents, the places that we call the "solid Earth."

Science in the Making

Reactions to Plate Tectonics

At first, when geologists observed the magnetic stripes on the ocean's floor, rock ages, and other seemingly odd data, they tried to find an explanation that did not require the creation of new seafloor. These geologists had been taught about a "fixed Earth" model in their college courses, and that was the model upon which they based their work. Most of these scientists were not prepared to accept the radical new idea—that the continents might actually move—without a fight.

Some experts questioned the statistical significance of the magnetic data, and they wondered whether the distinctive striped patterns weren't the result of some as yet unknown effect (such as small electric currents running around the ocean floor). Others simply tried to ignore the whole thing. But as geologists and oceanographers collected more data, and as new measurements of magnetic patterns and more precise rock ages were obtained, the evidence for seafloor spreading simply became overwhelming.

Less than a decade from the time of the first puzzling data from the *Pioneer* expedition, geologists had, for the most part, accepted a theory that radically changed many of the central principles of their discipline. This dramatic change in perspective brings us back to a point we made in Chapter 1. Good scientists will eventually accept the implications of their observations, whether those implications violate preconceived ideas or not. A scientist can't look at data without some preconceptions. Few scientists, for example, took the original continental drift arguments of Wegener seriously because, in part, no obvious mechanism could cause entire continents to move. But as more and more data supported the increasingly convincing plate tectonics model, the majority of earth scientists readily changed their notions as the data demanded it. The fact that the fixed Earth, which was one of the most revered and widely accepted geological theories, could be abandoned in the space of a few years when confronted by powerful contrary evidence indicates that the scientific method works.

The theory of plate tectonics caused a revolution in the earth sciences. For the first time, scientists had a single, coherent picture of our planet that included the oceans, the land masses, the planet's deep interior, and the interconnections between all these systems. There was a time when geologists of different specialties, including women and men who studied the ocean currents or ancient fossils, ore deposits or the planet's deep interior, had little to say to each other. That situation is no longer true. All of these different disciplines now share a common view of the Earth, and are now able to share ideas and gain breadth and depth from their mutual interactions. It may be that your college or university no longer has a department of geology, but instead has a department of earth sciences, or environmental sciences. These changes in name are more than just bureaucratic shufflings. They represent a very real change in the way scientists view and study the Earth. ●

Plate Boundaries

The boundaries between the Earth's tectonic plates are active sites that determine much of the geological character of the surface. Three main types of boundaries separate the Earth's tectonic plates: *divergent*, *convergent*, and *transform*.

1. Divergent Plate Boundaries. We have already seen one aspect of plate motion when we talked about seafloor spreading at the Mid-Atlantic Ridge, where new plate material is formed. We can now understand how such a spreading feature arises. In Figure 18–12, we show what happens when plates happen to lie above a zone where magma comes

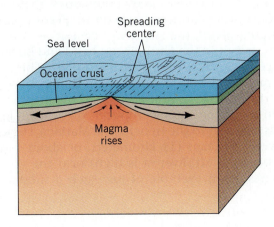

Figure 18–12
A divergent plate boundary defines a line along which new plate material is formed from molten magma.

A satellite photo of the great Rift Valley. The narrow body of water defines a divergent plate boundary at which new plate material is being created and plates are moving out to either side.

to the surface. Not only does the volcanic action form a chain of mountains, but the motion of the magma also pushes the two adjoining plates farther and farther away from each other. The newly erupted molten material cools to rock and becomes new plate material. As the brittle tectonic plates crack and separate, shallow earthquakes of relatively low energy occur. This mechanism drives the seafloor spreading that gave us our first indication of the nature of continental motion. Such a spreading zone of crustal formation is called a **divergent plate boundary.**

When a divergent plate boundary occurs on the ocean floor, the seafloor spreads, basalt lava erupts from the newly created fissures, the two plates are pushed apart, and any continents that might be located on other portions of those plates are pushed apart as well. Eurasia and North America, for example, are separating right now at the rate of about 5 centimeters per year; consequently, the Atlantic Ocean is getting wider. Note that old divergent plate boundaries, such as the Mid-Atlantic Ridge, are always located in the middle of an ocean.

New spreading centers, on the other hand, may begin anywhere, even in the middle of a continent. If a continent happens to be sitting above what will eventually become a divergent plate boundary, then the continent itself will literally be torn apart. The Great East African Rift Valley, which extends south from Ethiopia, along the east-coastal interior of central Africa to the coast of Mozambique, and north into Israel and Syria, is a modern-day example of this motion. Millions of years in the future, the western part of Africa may be separated from the eastern part by an ocean. In fact, at the point where the Great Rift Valley crosses the African coast, this sea is already beginning to form. Look at the Dead Sea, Red Sea, and Gulf of Aqaba and you will see this rift in progress.

2. Convergent Plate Boundaries. The Earth is not getting any larger. If new material pushes tectonic plates apart at places such as the oceanic ridges, then old plate material must be pushed together and taken into the Earth somewhere else. A place where two plates are coming together is called a **convergent plate boundary.**

At most convergent plate boundaries, one plate sinks beneath another to form a **subduction zone.** The plate that is subducted, or "taken beneath," in this way sinks down to rejoin the mantle material from which it came, as illustrated in Figure 18–13.

Earth scientists observe three broadly different kinds of surface features associated with convergent plate boundaries; they are all shown in Figure 18–13. First, if no continents are on the leading edge of either of the two converging plates, the result will be a *deep ocean trench*. As one plate penetrates deep into the Earth, it can bend and buckle the adjacent plate to produce a deep furrow in the ocean floor. This subducting slab may partially melt as it penetrates into the hot mantle; the resulting magma slowly rises toward the surface and, if the magma penetrates the overlying ocean crust, continuing eruptions of lava will build a chain of volcanic islands adjacent to the trench. The Mariana Trench, the deepest point in the world's oceans (11 kilometers, or about 7 miles, deep—2 kilometers deeper than Mt. Everest is high) near the volcanic coastline of the Philippines, is an example of just such a subduction zone and volcanic terrain.

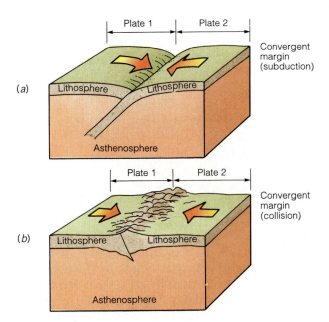

Figure 18–13
Convergent plate boundaries may display a variety of surface features, depending on the distribution of continental material. (*a*) When neither tectonic plate carries continental material at the convergent boundary, then a subduction zone is formed. A deep ocean trench and chain of island volcanoes are often created in this case. (*b*) When both tectonic plates carry continental material at the boundary, the continental materials buckle and fold to form nonvolcanic mountains.

If continents ride on top of both converging plates, they will collide. Continental material will be compressed together like crumpled cloth and will be pushed up to form a high, jagged mountain chain. The Himalayas, for example, which began to form about 30 million years ago, are still growing taller at about 1 centimeter per year as the once separate Indian subcontinent collides with Asia. Similarly, the Alps mark the point at which the Italian peninsula became joined to Europe. All of these geological processes involve the production of mid-continental mountain chains.

Finally, if a continent rests at the leading edge of only one of the two colliding plates, the denser oceanic plate will subduct beneath the continent. Just as in the case where two oceanic plates converge, a deep trench may form a short distance offshore, while the continental material may be crumpled into a coastal mountain range. In addition, the subducting tectonic slab may partially melt; as a result, the magma will rise to form a chain of volcanic mountains parallel to the coast. The Andes Mountains of South America and the Cascade Mountains of the northwestern United States are spectacular examples of this phenomenon.

The oceanic plate material on the Earth renews itself roughly every 200 million years. The ocean crust is constantly being replaced by the process of seafloor spreading and subduction, while the lower-density continental plate material experiences no subduction. This process explains why, while some rocks on ancient continents that formed billions of years ago are still preserved there, no rocks on the ocean floor are older than about 200 million years.

3. Transform Plate Boundaries. The third kind of boundary between plates occurs when one plate scrapes past the other, with no new plate material being produced. This kind of plate contact is called a **transform plate boundary;** it is shown in Figure 18–14. The most famous transform boundary (and the only active plate boundary in the

transform plate boundary
The type of boundary between plates that occurs when one plate scrapes past the other, with no new plate material being produced; for example, California's San Andreas fault.

The San Andreas Fault in California marks the transform boundary between the North American plate and the Pacific plate.

Figure 18–14
A transform plate boundary, showing the relative motion of the adjacent plates. The terms "asthenosphere" and "lithosphere" refer to sections of the crust and mantle.

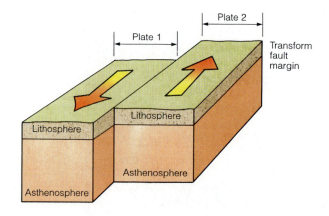

continental United States) is the San Andreas Fault in California. At the San Andreas Fault, the Pacific plate is moving northwestward with respect to the North American plate at the rate of several centimeters a year.

The process by which two plates slide past each other is not smooth. Over time, the motion of the plates compresses and strains rocks at the boundary. Friction normally prevents the stressed rocks from moving, but periodically the rocks simply break, moving as much as several meters in one sudden burst. When they do so, an earthquake occurs. No mountain building or volcanism is associated with transform boundaries.

Developing Your Intuition

Mountain Building

The Urals are a relatively low mountain chain that runs north-south in the center of the Eurasian landmass. They are often cited as the dividing line between Europe and Asia. Given what you know about plate tectonics, can you make a guess as to how (and when) these mountains were formed?

Midcontinental mountain chains are often the result of the collision between two continents riding on plates above a converging boundary. It is reasonable, then, to guess that the Ural Mountains must mark the spot where Europe and Asia joined together. Since the mountains are low and worn down, you could guess that they were formed more than a hundred million years ago, and have been eroding ever since.

In fact, the Urals were formed by the collision of Europe and Asia some 250 million years ago. ●

Environment

The Effect of Plate Motion on Human Beings

Human beings differ from apes in one important way; we walk upright on two legs, rather than swinging along on our knuckles or walking on all fours. Most students of human evolution now believe that it was the upright posture that freed our hands from walking and stimulated the development of our brains. And Richard Leakey, a well-known paleoanthropologist, has suggested that our upright posture might have indirectly resulted from the movement of the Earth's tectonic plates.

His argument is as follows: 30 million years ago, most of eastern Africa was covered by a lush jungle. No fewer than 20 different species of apes, including our ancestral species, flourished in that environment and were especially well adapted to living in trees. When a divergent boundary started to pull the continent apart along the East African Rift Valley, the environment started to change. The forest began to disappear, to be replaced first by open plains dotted with stands of trees, and finally, 3 million years ago, by the savannah that exists there today. Most of the apes became extinct long before our 3-million-year-old ancestors, but Leakey argues that walking upright and being able to get from one forest "island" to another would have been a distinct advantage in that sort of environment. The result, according to Leakey, is that today there are only three kinds of descendants from those apes in Africa: gorillas, chimpanzees, and human beings. ●

The Geological History of North America

The epic movements of tectonic plates provided earth scientists with a new way of thinking about the history of the Earth's surface. Earth scientists now use our understanding of plate tectonics to tell us something about the formation of our own continent. The oldest parts of the North American continent are in northeastern Canada. Here we find large geological formations of rocks, several billion years old, that form the core of the continent. Over hundreds of millions of years, land was added to this continent by tectonic activity. Most of the western part of the United States, as shown in Figure 18–15, is made up of small chunks of land called *terranes*. These are masses of rock several hundred kilometers across. Originally, these terranes were large islands in the Pacific Ocean. They were carried toward the North American continent by plate activity and added to the mainland as tectonic plates converged. The hills near Wichita, Kansas, for example, are old mountains that once marked the addition of a large South American terrane onto what then comprised North America. (The idea that Wichita might have once had ocean-front property is one of the strange discoveries that comes out of plate tectonics.)

The Appalachian Mountains, which at one time may have rivaled the Himalayas in majesty, were formed over a period from about 450 to 300 million years ago when the continents that are now Eurasia and Africa collided with the continent that is now North America. A series of long folds and fractures in the surface rocks formed the present-day Appalachians. This process explains, for example, why roads in the mountainous regions of the eastern part of the United States tend to run from southwest to northeast; they follow the mountain valleys that were created by erosion of these folded rocks. Thick wedges of sediments eroded off these mountains, forming the Coastal Plains of eastern North America. This erosion also contributed to the sediments of the Great Plains.

The dramatic geological features of the western United States record a great variety of mountain-building events. The Rocky Mountains formed approximately 60 million years ago from a broad warping, and subsequent

Figure 18–15
A geological map of the western United States. Each different color represents a different terrane that "docked" on the continent as a result of tectonic activity.

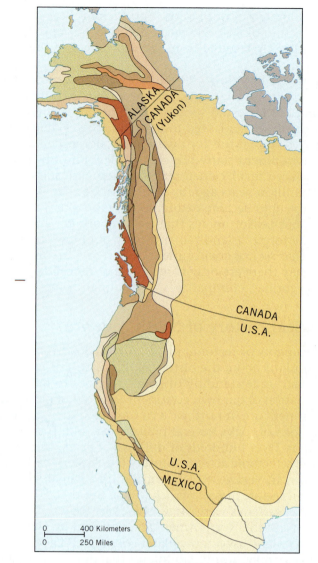

folding and fracturing, of continental material. The Colorado Plateau, comprising parts of the states of Colorado, Arizona, and New Mexico, experienced a more gentle uplift, as rivers incised features such as the Grand Canyon. The Sierra Nevada in California formed more recently when molten rock pushed up a huge block of sediments. These processes of uplift and erosion continue to this day in many places around the world.

VOLCANOES AND EARTHQUAKES

As you are seeing, plate tectonics provides us with a dynamic picture of the Earth. Plates are slowly but continually moving over the hot, partially molten rocks in the mantle: they crash together, rip apart, and scrape by each other. In the process, tall mountain chains are uplifted and then worn down, while wide ocean basins are opened and closed as continents come together and split apart. Nothing on the Earth is permanent, because heat continuously flows from the hot interior to the cooler surface, and mantle convection provides the primary mechanism for that heat transfer. The scale of these ongoing processes is so vast that it is nearly outside our ability to comprehend. Occasionally, however, we are reminded of the power of geological processes.

For thousands of years, humans have realized that the Earth's most violent events—volcanoes and earthquakes—do not occur randomly. Earthquakes are common in California and Alaska, but extremely rare in Kansas or Florida. Volcanoes are commonplace in Hawaii and the Pacific Northwest, but never appear in New York or Texas. Why should this be? Plate tectonics provides an answer.

Plates and Volcanism

The global distribution of volcanoes may be understood in terms of the principles of plate tectonics. Volcanoes are common in three geological situations: along divergent plate boundaries, near convergent plate boundaries, or above places called "hot spots."

1. Divergent Plate Boundaries. New crustal rocks are formed along the volcanic ridges of divergent plate boundaries. New basalt plate material spreads out at the rate of a few centimeters per year along about 85,000 kilometers of oceanic ridges around the world.

2. Convergent Plate Boundaries. Volcanoes are also common near subduction zones, except where two continental tectonic plates collide. As water-rich crustal material plunges into the mantle, it becomes hotter and may partially melt. This magma, which is highly mobile fluid rock, rises to the surface to form chains of volcanoes, typically about 200 kilometers inland from the line of subduction. The Ring of Fire, a dramatic string of volcanoes that borders much of the Pacific Ocean, is a direct consequence of plate subduction (see Figures 18–16 and 17).

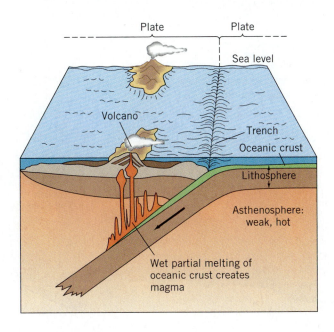

Figure 18–16
Volcanoes form above a subduction zone when heated plate material partially melts. The hot magma rises through the overlying crust to form a chain of volcanic islands.

Figure 18–17
The "Ring of Fire" around the Pacific basin marks the location of volcanoes which result from the motion of tectonic plates, specifically subduction zones. The Ring of Fire is also called the Andesite Line because of the type of igneous rocks formed from its volcanic magma. (The volcano names printed in black erupt basaltic magma, not andesitic magma.)

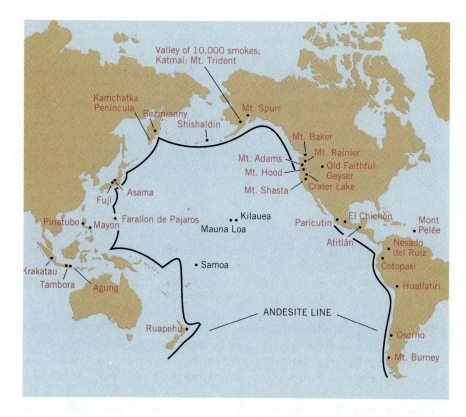

Figure 18–18
Each dot represents a hot spot on the Earth's surface. Unlike earthquakes and volcanoes, these dots are not clustered on plate boundaries. Why do you suppose they aren't?

The volcanoes that form the Cascade Mountain chain along the northwestern coast of the United States (including Mount Rainier and Mount St. Helens) are striking examples of the processes associated with subduction of an oceanic plate beneath a continent. Frequent dramatic eruptions of similar volcanoes in Central America, Japan, and the Philip-

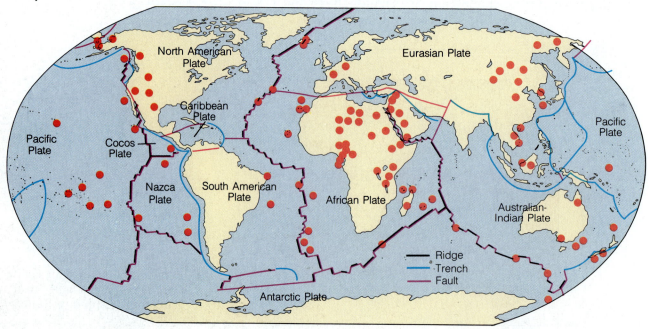

pines point to other places where subduction and volcanism occur in tandem.

3. Hot Spots. Finally, **hot spots** are a dramatic type of volcanism indirectly associated with plate tectonics. Earth scientists recognize dozens of hot spots around the world, in places such as Hawaii, Yellowstone National Park, and Iceland (see Figure 18–18). At these locations, large, isolated chimney-like columns of hot rock rise to the surface, more or less like bubbles coming to the surface in water being heated on a stove. These columns, also known as mantle plumes, originate in the lower mantle or even at the core-mantle boundary.

hot spot A dramatic type of volcanism indirectly associated with plate tectonics; large isolated chimney-like columns of hot rock, or mantle plumes, rising to the surface in certain places of the Earth; for example, Yellowstone National Park, Iceland, and Hawaii.

(*a*) An example of a chain of volcanoes forming at a convergent boundary. The snow-capped peaks are volcanoes in Ecuador, produced by the subduction of the Nazca plate beneath the South American plate. (*b*) A chain of volcanoes in Washington and Oregon extends above a subduction zone, where the Juan de Fuca Plate plunges beneath the North American Plate.

(*a*)

(*b*)

Figure 18–19
The Hawaiian Islands stretch along a northwest-southeast line that reveals the northwesterly motion of the Pacific Plate over a fixed hot spot. As the Pacific plate moves, new volcanic islands are created to the southeast, while older islands (ages in millions of years) erode away.

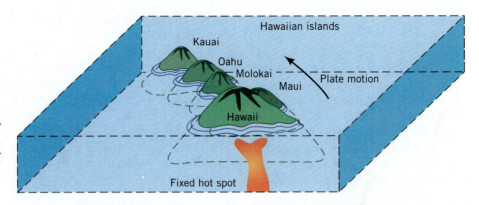

On a geological time scale, sources of hot spots are relatively stationary, so if a tectonic plate slowly moves over the fixed hot spot, the result will be a chain of volcanoes. The Hawaiian Islands, for example, are the result of a series of basaltic lava eruptions (see Figure 18–19). These islands were created one at a time as the Pacific Plate moved above the localized hot spot. The present-day volcano Kilauea on the "big island" of Hawaii, the site of most of the island chain's active volcanism, is directly over a hot spot. The volcanic islands to the northwest are progressively older and smaller owing to erosion. This distribution of islands reveals that the motion of the Pacific Plate is also toward the northwest. In fact, a series of eroded underwater peaks that were islands millions of years ago stretches hundreds of kilometers farther to the northwest. In several million years, the most northwesterly of the present Hawaiian islands will also erode beneath the waves. However, new volcanic activity has already begun on the ocean floor southeast of Kilauea, promising new islands that will replace the old.

Earthquakes

Stress builds up in brittle rock for several reasons. Heated rock expands and cooling rock contracts, changes that cause a solid formation to warp and distort. Rock may also become stressed in response to changes in pressure, as overlying mountains wear away or new layers of sediment weigh down. And, of course, stress builds up to extreme levels as two tectonic plates attempt to move past each other at a transform plate boundary.

Earthquakes may be felt near any plate boundary. Minor, shallow earthquakes occur near divergent plate boundaries as two oceanic plates move apart. Stronger earthquakes, including "deep-focus" earthquakes originating more than 100 km down, occur near subduction zones. Many of the most destructive shocks in Japan are of this type. In the United States, earthquakes at the transform plate boundary along the San Andreas Fault in California receive the most attention because of the fault's unusual activity, length, and proximity to major population centers. There are, however, occasional earthquakes in the middle of plates (in southeastern Missouri, for example) whose origins are not fully understood (see Figure 18–20).

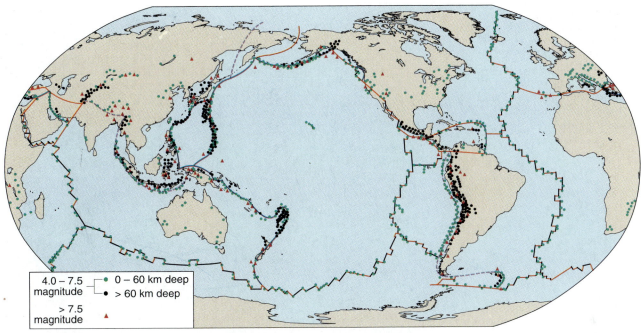

Figure 18–20
Each dot marks the location of an earthquake of magnitude greater than 4.0. Note how the dots are clustered at the boundaries of the tectonic plates. Why should this be so?

Seismology: Exploring the Earth's Interior with Earthquakes

Scientists who study earthquakes discovered that these violent events provide the best means for exploring the deep interior of our planet. Geologists can't obtain samples from much more than the outer 10 kilometers or so of rock layers. Everything we know about the interior of the Earth deeper than a few kilometers has to be obtained by indirect means. The science of **seismology,** which is the study and measurement of vibrations within the Earth, is dedicated to deducing our planet's inner structure.

The basic idea of seismology is simple. When an earthquake or explosion occurs, waves of vibrational energy called *seismic waves* (see Chapter 8) move outward through the rocks. Some of these waves travel through the center of the Earth, some move along the surface, and still others bounce off layers deep within the planet.

There are two principal types of seismic waves. *Compressional waves,* also called *longitudinal waves,* are like sound: the molecules in the rock move back and forth in the same direction as the wave. *Shear waves,* on the other hand, are transverse waves, like water waves in which the molecules move up and down perpendicular to the direction of wave motion. (See Chapter 8 for a review of wave motion.) These kinds of waves travel through rock at different velocities depending, among other things, on the rock type, its temperature, and the pressure. After a major earthquake, scientists at laboratories around the world record the inten-

seismology The study and measurement of vibrations within the Earth, dedicated to deducing our planet's inner structure.

A geologist uses a seismograph to detect sound waves passing through the Earth.

Figure 18–21
Seismic waves passing through the Earth can take a variety of paths. The speeds of *P* (compressional) and *S* (shear) waves will differ depending on the type of rock, its temperature, and the pressure.

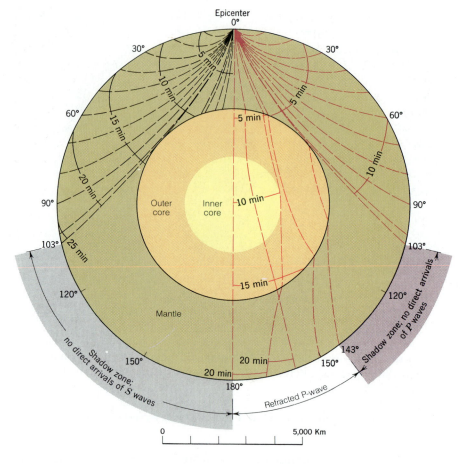

sity and time of the arrival of the various kinds of waves—both those that pass along the surface and those that pass through the interior of the Earth (see Figure 18–21). By comparing the arrival behavior of waves from the same earthquake at many different sites on the Earth, a computer can construct a picture of the material through which those waves passed.

The picture of the Earth we gave in Chapter 17, in which we discussed the solid and liquid core, the layered mantle, and the thin brittle crust,

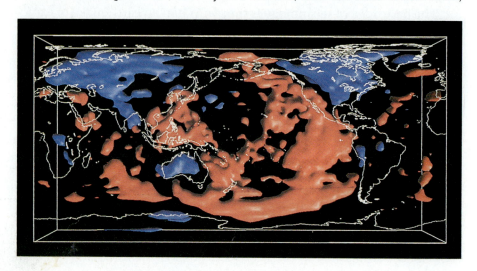

Figure 18–22
A seismic tomograph provides a picture of the Earth's deep interior, based on millions of individual measurements of seismic wave velocities.

came from studies of seismic waves. As more and more seismic data are collected, and ever-faster computers permit new ways to process those data, a new branch of earth science called *seismic tomography* is enabling geophysicists to obtain astonishing three-dimensional pictures of the Earth's interior (Figure 18–22). We are now able to document the basic movements of cold subducting slabs, hot upwelling mantle plumes, and the convection cells that drive plate tectonics.

The Design of Earthquake-Resistant Buildings

Most people who die in earthquakes do so not because of the violent shaking, but because of falling buildings. Over the past few decades, structural engineers have learned a great deal about how to design buildings so that they do not collapse during an earthquake. The basic problem they face is how to design a building so that it maintains its integrity when the ground on which it sits moves. There are two general solutions to this problem: make it flexible, or make it rigid.

The first strategy, widely used in designing tall buildings, is best typified by a tree bending in the wind. The idea is that a building can be designed with a specially reinforced steel skeleton so that it will bend and vibrate as the Earth shakes, but come back to its original orientation without damage when the quake stops.

The second approach is widely used in individual houses and apartment buildings. The idea is to construct the building so that it tosses about on the moving earth like a ship on the ocean. The main preoccupation of the engineers is to guarantee that the building's corners are maintained at 90-degree angles, no matter how much it moves around. This rigidity is achieved by reinforcing corners and rigid connections to the foundation and roof. Sometimes this rigidity can be obtained simply by covering all the walls of the building with plywood sheets before the outer siding is installed. ●

(*a*) After an earthquake near Santa Cruz, California, plywood sheathing was used to strengthen new construction. (*b*) Office building construction in San Francisco must conform to strict codes to limit earthquake damage.

(*a*)

(*b*)

THINKING MORE ABOUT PLATE TECTONICS

Earthquake Prediction

One of the most important roles scientists can play in modern society is to provide warnings of natural disasters such as earthquakes, volcanic eruptions, and large storms. In the summer of 1992, for example, early warnings allowed residents of southern Florida to escape the path of Hurricane Andrew. The storm caused billions of dollars of property damage, but surprisingly few lives were lost. Similarly, volcanic eruptions, which are usually preceded by numerous small earthquakes, can also be predicted with some certainty. The successful prediction of Mount Pinatubo's 1991 eruption in the Philippines gave thousands of people living on the flanks of the long-dormant volcano time to flee.

Predicting earthquakes is another story. Earthquakes occur because of a gradual, but predictable and measurable, buildup of stresses in the Earth's crust. However, the sudden failure of rock is not so predictable. You are familiar with this situation if you've ever gradually bent a pencil to the breaking point. It's easy to predict that a bent pencil will eventually break, but it's very hard to say exactly at what point the failure will occur. Much of the scientific effort related to predicting earthquakes centers on the careful documentation of stress buildup in earthquake-prone areas. But these measurements alone tell us little about the exact time when failure will occur.

In their efforts to predict earthquakes, many scientists search for "precursor events," which are measurable phenomena that precede the quake. There is much anecdotal evidence that such events may indeed occur. Folklore has it that some domesticated animals become highly agitated before a strong earthquake, or that changes are seen in the flow of well water or hot springs. Similarly, scientists have noticed changes in the regularity of geysers and have recorded swarms of minor quakes just prior to some big earthquakes. Nevertheless, no reliable methods for predicting earthquakes have yet been devised. At best, we can reliably predict the probability that a large earthquake might occur in a given area during a period of a few decades.

Ironically, if we do develop a way to predict the timing of earthquakes precisely, a whole new series of problems could arise. Suppose you could predict with 80% certainty that there would be a major earthquake in the Los Angeles Basin sometime in the next 30 days. Would you announce it? What would happen in the Los Angeles Basin if you made such an announcement? What would happen if you made the announcement and the earthquake never occurred? What would happen if you didn't make the announcement and the earthquake did occur?

▶ Summary

The surface of the Earth is constantly changing. Mountains are created and worn away, while entire continents slowly shift, opening up oceans and closing them again. *Plate tectonics*, a relatively new theory that explains how a few relatively thin, rigid *tectonic plates* of crustal and upper mantle material are moved across the Earth's surface by *mantle convection*, provides a global context for these changes. According to this theory, plates move over a partially molten, underlying section of the Earth's mantle, like rafts on an ocean, in response to the convection of pliable mantle rocks.

Three different kinds of observations—the geological features of ocean floors, parallel stripes of magnetic rocks situated symmetrically about volcanic ridges on the ocean floor, and the ages of these rocks— have provided direct evidence that new crust is being created at *divergent plate boundaries*. Meanwhile, old crust returns to the mantle in *subduction zones*, where plates converge. *Convergent plate boundaries* in the ocean create deep trenches and associated volcanic islands. When an oceanic plate subducts beneath a plate carrying a continent, an offshore trench and chain of continental volcanoes parallel to shore result. The collision of two plates that carry continents at their margins produces mountain ranges of crumpled continental material. *Transform plate boundaries* occur where two plates scrape by each other.

Most *volcanoes* form near plate boundaries, either along the volcanic ridges of diverging plates, or above subducting plates. Other volcanoes such as the Hawaiian Islands form above *hot spots* originating in the Earth's mantle. *Earthquakes* occur when stressed rock ruptures. Earthquakes may be felt at all plate boundaries. The only plate boundary in the United States, California's San Andreas Fault, is a transform boundary in which one block of crust moves horizontally past the opposing block. The science of *seismology*, which documents the passage of earthquake-generated sound waves through the Earth, is providing new insights into the dynamic processes that drive plate tectonics.

▶ Review Questions

1. What evidence suggests that mountain ranges are not permanent features on the Earth's surface?

2. What evidence suggests that Europe, Africa, and North and South America were once joined?

3. What pieces of evidence pointed to the process of seafloor spreading, or diverging plates?

4. Hot mantle rocks are sometimes compared to a pot of soup. Why?

5. Identify three kinds of plate boundaries.

6. Explain how mountain ranges such as the Alps and the Himalayas formed as a result of plate motions.

7. How do mountains form at divergent plate boundaries?

8. How do mountains form at convergent plate boundaries?

9. Why does a ring of volcanoes and earthquakes surround much of the Pacific Ocean?

10. What North American mountain range may have been the tallest in the world approximately 300 million years ago?

11. State some direct evidence for the movement of tectonic plates.

12. What is the Richter scale and where might you read about it?

13. How do the ages of rocks on the ocean floor help support the theory of seafloor spreading?

14. Describe magma and the process it undergoes when a volcano erupts.

15. What is lava and where is it found?

16. What is a fault and when does it occur?

17. What is seismology, and how is it used?

18. What are the two kinds of seismic waves?

19. Identify three places where volcanoes occur.

20. Why are earthquakes difficult to predict?

▶ Fill in the Blanks

Complete the following paragraph with words and phrases from the list.

convergent plate bound-
 aries
plate tectonics
seismology
divergent plate bound-
 aries
earthquakes
hot spots
mantle convection
subduction zones
tectonic plates
transform plate bound-
 aries
volcanoes

The _____ theory says the Earth's surface is made of regions called _____, which are in constant motion. The motion of the surface is driven by _____ deep underground. At _____, new crust is being created, while at _____ old crust is taken back into the earth's interior. At _____, plates slide by each other without new material being created or destroyed. Eruptions of magma from _____, and violent motions of surface rocks called _____, are consequences of this tectonic activity. Volcanoes are formed at _____, near converging boundaries, and above _____, as in the Hawaiian Islands. The branch of science devoted to studying the interior of the earth through the use of earthquake-generated waves is called _____.

▶ Discussion Questions

1. Identify two volcanic mountain chains that bear a close relationship to plate boundaries.

2. What plate do you live on? How many adjacent plates are there? What kinds of boundaries do you find to the north, south, east, and west?

3. The continent of Antarctica has rocks with plant and animal fossils that suggest the Antarctic climate was once temperate. Explain at least two different ways in which these warm-climate fossils might have ended up in what are now polar regions. How might you test your hypotheses?

4. Volcanic islands, including the Azores, the Canaries, and Iceland, lay scattered across the Atlantic Ocean. If you were to date the rocks on these and other Atlantic islands, what pattern do you predict you would find?

5. Geology has been called an integrated science, because it calls on several scientific disciplines to help explain features and processes of the Earth. Explain how geologists have used other sciences to answer the following questions:

a. How old is a piece of rock?

b. How is heat transferred from the Earth's deep interior to the surface?

c. How does the Earth's magnetic field change over time?

d. What is the structure of the Earth's interior?

e. What is the topography of the seafloor?

6. At some convergent plate boundaries, deep ocean trenches lie a short distance from tall mountains. Why are these two contrasting features related to each other?

7. Popular media sometimes describe the California coast as poised to "slide into the ocean." Based on your understanding of plate tectonics, is this a plausible occurrence? Why or why not?

8. Should scientists announce if they believe there is a 50% chance of an earthquake in your area during the next month? Why or why not?

9. Can you think of islands other than the Hawaiian Islands that may have formed as a result of volcanoes? Explain.

10. What is Dr. Leakey's theory about the development of human intelligence? Do you think this theory is reasonable?

11. What is magma and how is it related to volcanoes?

12. How are faults and earthquakes related?

13. What are the general properties of a tectonic plate?

14. What are the differences between the properties of oceanic and continental plates?

15. What are the sources of energy for the Earth's interior heat?

16. Why do geologists think that convection probably is not prevalent in the interiors of the Moon and Mars? Is there any evidence for plate tectonics on the Moon and Mars?

17. Classify the three different plate boundaries according to the following:

a. Their location on the surface of the Earth.

b. Their geologic products, such as mountain building, earthquakes, and so forth.

c. The location of these products, such as earthquakes, mountains, volcanoes, and so on, relative to the plate boundaries.

18. Draw a diagram of a subduction zone and locate the position of deep ocean trenches, earthquakes, mountains, and other features. Draw a lateral view (across the subduction zone) and a corresponding aerial view as seen from a satellite directly overhead.

19. How does the presence of a subduction zone help explain these facts?

a. Earthquakes are usually found on only one side of an ocean trench.

b. Shallow earthquakes are found closer to ocean trenches.

c. Deep earthquakes and volcanoes are found farther away from the ocean trenches.

20. The San Andreas Fault is an example of what type of plate boundary?

21. Use a diagram or outline to demonstrate how the Hawaiian Island chain was formed by a hot spot over the Pacific plate.

22. Classify the two different kinds of seismic waves according to the direction of motion by the rocks they move, and the direction of propagation of the seismic wave.

▶ Problems

1. If the African Rift Valley opens up at the rate of 5 cm per year, how long will it take before a body of water 1000 km wide divides the African continent?

2. Estimate the probable lifetime of your favorite mountain. (*Hint:* Get its dimensions from a map, then estimate its volume.)

3. The Mississippi River will transport tons of silt per hour past New Orleans. What is the origin of this material?

4. The big island of Hawaii has two young shield volcanoes, Mauna Loa and Mauna Kea, that are approximately 8700 m above the ocean floor, each nearly 4200 m above sea level. Assume that each volcano has a base 20 km × 20 km at sea level.

a. How long will it take to erode these volcanoes down to sea level?

b. How long will it take to erode these volcanoes down to the ocean floor?

5. The Pacific Plate is moving at an average speed of 8.1 cm per year at the Hawaiian Islands. The center of the island of Oahu is about 300 km from the center of the big island of Hawaii. If you assume that the big island just recently formed and the same hot spot is responsible for the Hawaiian island chain, did the island of Oahu form?

6. Using the Richter scale, compare two earthquakes, one of magnitude 4.0 and the second of magnitude 7.0. Which one released more energy? By what ratio?

7. What is the relative difference in energy released by a 7.2-magnitude and a 5.2-magnitude earthquake?

8. Miners often find the temperature in deep mines quite hot. If the change in temperature in the earth's crust is 10°C per km, how deep does one have to go before

a. You double the temperature at the Earth's surface?

b. You reach the temperature of boiling water? (Assume that you start at the surface of the Earth at 20°C.)

9. About 50 million years ago, the Indian subcontinental plate was about 2500 km from its present location. How fast (in cm per year) was the Indian subcontinental plate moving before it collided with the Asian plate? Is your answer reasonable?

▶ Investigations

1. Examine original sources related to Wegener's continental drift theory. Why was this theory rejected by the majority of earth scientists in the 1920s? Compare and contrast the major features of the continental drift theory with plate tectonics theory.

2. How did ancient civilizations explain the occurrence of earthquakes and volcanoes?

3. Does plate tectonics operate on any other terrestrial planet in our solar system? Why or why not?

4. In your library, examine newspaper reports of major volcanoes and earthquakes during the past 20 years (each student could take one year). Plot these events on a world map. Do you see any obvious geographic patterns? How do the locations of these events relate to the plate boundaries shown in Figure 14–8?

5. A few years ago someone predicted an earthquake on the New Madrid Fault in Missouri, the site of a destructive shock in 1812. He was widely believed and schoolchildren were trained in what to do in case of an earthquake. The earthquake did not occur, and still has not occurred. What evidence did the amateur scientist use to make his case? How would you analyze that evidence?

▶ Additional Reading

Glen, William. *The Road to Jamarillo*. Stanford, Calif. Stanford University Press, 1982.

McPhee, John. *Basin and Range*. New York: Farrar, Strauss, and Giroux, 1981.

————. *In Suspect Terrain*. New York: Farrar, Strauss, and Giroux, 1983.

————. *Rising from the Plains*. New York: Farrar, Strauss, and Giroux, 1986.

Parker, Ronald. *Inscrutable Earth*. New York: Scribners, 1984.

19

CYCLES OF
THE EARTH

ALL MATTER ABOVE AND BENEATH THE SURFACE OF
THE EARTH MOVES IN CYCLES.

A
RANDOM
WALK

Through the Day

Every place we go, every day of our lives, we experience Earth's cycles. A day may start out clear and sunny, with mild temperatures and gentle breezes. But, suddenly, the wind picks up as a line of dark clouds presses in from the west. With the wind comes the first threatening rumble of thunder, flashes of distant lightning, and spurts of rain. A heavy downpour and powerful gusts of wind follow, as thunder booms and lightning illuminates the grey sky around us. Trees sway, windows rattle, and limbs come crashing down at the height of the storm.

Within an hour, skies are sunny again, though the temperature is noticeably cooler. The air has a fresh, clean smell, and the skies seem more deeply blue than before. In the span of a few hours we have experienced one of Earth's many cycles—the atmospheric cycle called weather.

Day and night, summer and winter, life and death—these and many other cyclical changes characterize our dynamic planet. As the author of the Book of Ecclesiastes wrote 3000 years ago, as he looked at the world around him:

> One generation passeth away, and another generation cometh: but the earth abideth forever;
> The sun also ariseth, and the sun goeth down, and hasteth to the place where he arose;
> The wind goeth toward the south, and turneth about unto the north; it whirleth about continually, and the wind returneth again according to his circuits;
> All the rivers run into the sea; yet the sea is not full; unto the place from whence the rivers come, thither they return again . . .
> The thing that hath been, it is that which shall be; and that which is done is that which shall be done: and there is no new thing under the sun.

Bauxite, an important ore of aluminum, is mined in Venezuela.

CYCLES SMALL AND LARGE

Think about the last time you drank a can of soda. What did you do with the aluminum can when you were through? You may have taken the time to place the can in a recycling bin. Alternatively, you may have tossed it into a trash container, or even by the side of the road. Does it really matter? To answer this question, we have to think about where the aluminum atoms end up.

Recycling Earth's Atoms

The atoms that make up the Earth, with the exception of a few radioactive isotopes (see Chapter 14), will last forever. A single aluminum atom, for example, will appear in many different guises during its lifetime. It may form part of a swirling lava flow in which it is tightly bonded to oxygen atoms. It may then be incorporated with those oxygen atoms into a solid rock. As the rock weathers away, the atom may be concentrated with other aluminum atoms in soil, which is mined in great open pits. Giant smelters separate the aluminum atom from its oxygen neighbors, in a process that consumes prodigious amounts of energy, to produce the aluminum metal that made your soda can.

Once discarded, the atom may be recycled into new cans, or it may go back to the soil where it once again bonds to oxygen. The Earth has a vast, though finite, number of aluminum atoms. The advantage to recycling aluminum is that we save all the energy that went into finding and mining the aluminum-rich deposit, as well as the energy that was required to break aluminum-to-oxygen bonds. But no matter what you do with your can, the aluminum atoms are still part of the Earth.

The Earth also has a finite number of carbon atoms. Every time you breathe, you participate in the grand cycle by which carbon moves through the environment (see Figure 19–1). You take in carbon atoms when you eat food, which stores chemical energy in the bonds between its atoms. You release that energy by *oxidizing* the carbon-bearing molecules in your cells (see Chapter 12); carbon dioxide is produced as a by-product, carried to your lungs by your blood, and returned to the environment when you breathe out. Some molecules of carbon dioxide enter the ocean and are added to the shells of molluscs or the skeletons of microscopic organisms. When an organism dies, these hard parts sink to the ocean bottom, where, in the form of calcium carbonate, they are turned into limestone.

Carbon atoms in limestone can remain locked up for hundreds of millions of years until the rock is weathered and the carbon is released into the atmosphere. From the atmosphere, carbon atoms may be taken into plants and incorporated into plant tissue through the process of photosynthesis. Someday, you may consume those carbon atoms when you eat an apple or a piece of bread, and thus start the whole process again.

A single atom of carbon, in other words, may have gone through many different chemical reactions during the 4.5-billion-year life of the planet. These reactions will continue as long as there is life on Earth; the atom will probably never leave the planet. For all practical purposes, the

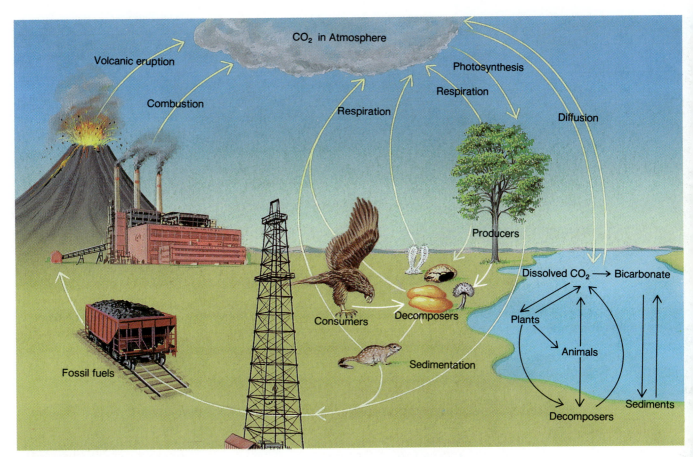

Figure 19–1
The carbon cycle. Carbon cycles through the different parts of the Earth's eco-system. It is found in carbon dioxide in the air, then taken into plants to be-come part of the plant's structure. If the plant is eaten, the carbon may be re-turned to the air through respiration or become part of the animal's tissue. When the animal dies, the carbon dioxide returns to the air.

Earth is an enormous closed system: aluminum, carbon, and all the other elements play a never-ending game of musical chairs, cycling for billions of years through plants and animals, the atmosphere and oceans, and other parts of our dynamic planet. To understand our planet, therefore, we must view it not as a collection of isolated objects, but as a vast series of interconnected cycles.

Environment

The Problem of Urban Solid Waste

The fact that atoms never disappear has led to a growing problem in urban America. Garbage (so-called solid waste) is generated at an enor-mous rate in American cities today. New York City alone adds 17,000 tons of solid waste daily to its landfill on Staten Island. Environmental

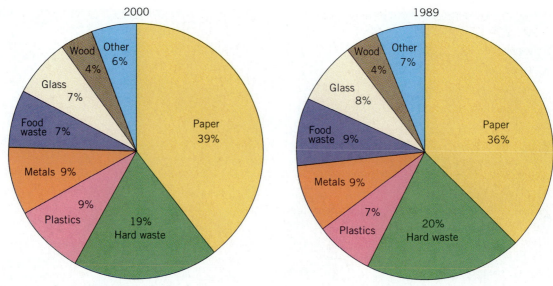

Figure 19–2
The percentage of different kinds of trash in urban landfills in 1989 (right) and projected percentages for 2000 (left). Not much progress, is there?

engineers estimate that at these current rates, every individual American will generate solid waste equivalent in volume to the Statue of Liberty in only about 5 years (see Figure 19–2).

To make matters worse, in most modern landfills the normal process of breakdown and decay, which turns carbon and nitrogen compounds into soil, is slowed enormously. In a landfill, solid waste is dumped on the ground and compacted, then covered with a layer of dirt, then another layer of compacted waste, then another layer of dirt, and so on. The materials in such a landfill are cut off from air and water, and the bacteria that normally work to decompose the waste cannot thrive. Archaeologists digging into landfills have discovered, for example, that newspapers from the 1950s are still readable after having been buried for 40 years! This means that, unlike an ordinary garden compost pile in which materials are quickly broken down by the action of bacteria, the landfill is more of a burial site than a recycling site (see Figure 19–3).

Figure 19–3
The cross section of a modern landfill, with a plastic liner to keep chemicals from seeping into the ground and monitoring wells to keep track of purity of groundwater. (The vadose zone is an area of water or solutions in the crust above the groundwater level.)

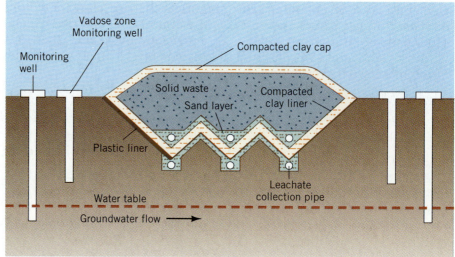

Fresh Kills landfill, a modern landfill in New York, accepts about 15,000 tons of waste per day.

The landfill at Sussex City, New Jersey, uses state-of-the-art technology with liners.

Recycling has become crucial in the United States in the 1990s because all the obvious places to dump solid waste near major cities are becoming full, and no replacement sites are feasible. Even though recycling newspapers is not always a paying business proposition in the short term, most municipalities realize that it's a lot cheaper to pay whatever little is needed to recycle newspapers than to find a new waste disposal site in which to dump them. ●

The solid waste produced in the United States during one year is enough to build a 500-foot wide solid wall across the Grand Canyon.

Science by the Numbers

Trash

How much solid waste is produced in the United States annually? Engineers estimate that the average American is responsible for about 40 tons (80,000 lb) of trash annually. This total includes everything from disposable containers, newspapers, and mail-order catalogs, to old automobiles and appliances, as well as the industrial wastes necessary to manufacture the things we buy. What is the total volume of this waste? Well-compacted trash weighs perhaps 80 lb/ft^3, which is somewhat denser than water, but less dense than rock. The volume of 40 tons, therefore is equivalent to a volume of

$$\frac{80,000\,\text{lb}}{80\,\text{lb/ft}^3} = 1000\,\text{ft}^3$$

That's enough compacted trash to fill two large dump trucks for every man, woman, and child in the United States every year. Thus 250,000,000 Americans produce a total annual volume of trash of

$$250,000,000 \text{ people} \times 1,000 \text{ ft}^3/\text{person} = 2.5 \times 10^{11} \text{ ft}^3$$

That's almost 2 mi^3 of trash every year, enough to build a solid 500-ft-wide wall across the Grand Canyon at its widest and deepest point. ●

Technology

The Science in Recycling

Because land that can be used for dumps near big cities is growing ever more scarce, and because the environmental costs of using materials once and then throwing them away are steadily growing, governments have recently begun to pay more attention to recycling. Not only does every recycled plastic milk jug or sheet of paper mean less material in landfills, but it also means that less petroleum is taken from the ground and fewer trees are cut down.

But recycling is not simple. A great deal of science and engineering has to be done before even the simplest materials can be reused. In addition, different processes are usually required to recycle different materials. Recycling different kinds of plastics, for example, requires different kinds of chemical reactions, and processes that work for plastic soft drink bottles will not necessarily work for ketchup containers. (You may have noticed small numbers located inside the recycle symbol on plastic products; these numbers specify which kind of plastic was used.) Each type of material that is to be recycled poses its unique problems for the engineer.

Let's consider white paper as an example. The average office worker generates about 250 lb of high-grade paper waste per year, and many offices around the country have paper recycling programs. The first step in this process is simple: the paper is ground up into a pulp and added to water to make a liquid, called a *slurry*. Ink particles from typewriters and pens rise to the surface of the slurry and can be skimmed off, leaving a material that can be added to fresh pulp to make new paper. As we pointed out in Chapter 7, however, copying machines and laser printers work by melting bits of carbon mixed with resins onto the paper. That type of ink makes heavier particles when the paper is ground up, and those particles sink to the bottom along with the paper fibers. Until quite recently, such paper could only be recycled into items such as cardboard or tissue paper, for which color quality is not important.

The new technology for dealing with this problem involves adding substances called *surfactants* to the pulp. The molecules in these substances are shaped so that they bind to the heavier ink particles on one end, and to bubbles of gas on the other. Once the molecules are attached to the ink, various kinds of gas are bubbled through the slurry. The surfactants and their load of ink rise to the surface with the bubbles and can be skimmed off, leaving clean paper fibers for reuse.

The national recycling effort will involve hundreds of different processes such as this. Each will be geared to a specific material, and each will do its part to conserve energy and material resources. ●

The Three Great Cycles of Earth Materials

Two central ideas frame our understanding of our constantly changing planet:

> **Earth materials, both above and beneath the surface, move in cycles.**

and,

> **A change in one cycle affects the others.**

The bottoms of many plastic containers carry a number that indicates the type of plastic, and thus the exact process to be used in recycling.

Paper recycling greatly reduces the amount of solid waste that we generate.

The histories of aluminum and carbon described earlier are just two examples of the many ways atoms move among the Earth's atmosphere, oceans, and interior. All of these processes are part of the Earth's three great cycles:

1. The Atmospheric Cycle: including the processes of the weather, the seasons, and the global climate;
2. The Water Cycle, or Hydrological Cycle: including the movement of water molecules among oceans, rivers, clouds, and other repositories; and,
3. The Rock Cycle: including the formation and weathering of rocks at and beneath the solid surface of the planet (see Chapter 20).

Some Earth processes—weather for example—vary by the hour, while the events that form and alter rocks may take hundreds of millions of years. But whatever the time scale, all of these processes illustrate a central theme in the earth sciences: our planet's atoms are constantly moving and recycling. Above the Earth, the gases in the atmosphere flow in the great atmospheric cycles of weather, the seasons, and global climate. Water moves from rivers to oceans to glaciers and to clouds as it takes part in the Earth's dynamic hydrological cycle. And the solid surface of the Earth itself slowly forms, alters, erodes away, and forms again in the stately rock cycle (see Chapter 20).

Many of the Earth's cycles are driven by the tendency of heat to spread out—to flow from hot to cold in what we described as the second law of thermodynamics (see Chapter 6). The Earth has two primary sources of heat energy: the Sun, and its own geothermal processes, each of which drives its own thermal cycles. More heat energy from the Sun falls at the equator than at the poles, and heat transfer by convection moves gases in the atmosphere and water in the oceans in the great cycles that control weather and climate. Similarly, heat energy stored in the Earth's core and mantle drives the convection cycles that move tectonic plates.

Thus Earth's cycles reflect the most basic properties of matter and energy. These cycles may be studied at many levels, from an atomic scale for individual elements such as aluminum or carbon, to the global cycles involved in the movements of water and the atmosphere. We find it especially useful to consider the Earth in terms of the three most familiar cycles around us: the cycles of air and water, which are reviewed in this chapter, and the rock cycle, which is considered in the next.

THE ATMOSPHERIC CYCLE

atmospheric cycle The circulation of gases near the Earth's surface, including the short-term variations of weather and the long-term variations of climate.

The Earth's atmosphere and oceans play the most important role in distributing heat across the surface of the planet. Within the atmosphere are chemical cycles involving nitrogen, oxygen, and carbon, which are intimately bound up with the presence of living things on the Earth. The circulation of gases near the Earth's surface, including both the daily variations of weather and the longer patterns of climate, is called the **atmospheric cycle.**

The General Circulation of the Atmosphere

The atmosphere circulates in vast currents of air that cover the globe from the equator to the poles. This circulation of air is powered by the energy of the Sun, as air in the tropics is heated and rises. If the Earth did not rotate, we would expect to have a situation like the one shown in Figure 19–4. Warm air would rise at the equator, cool off, and sink at the poles. This pattern of flow is the familiar convection cell we saw in Chapter 6. Such a pattern arises whenever a fluid is heated nonuniformly in a gravitational field.

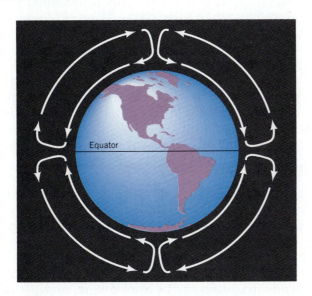

Figure 19–4
If the Earth did not rotate, the circulation of the atmosphere would take place in convection cells that create high-altitude winds from equator to pole, and low-altitude winds from pole to equator.

If the Earth did not rotate, then prevailing winds in the Northern Hemisphere would flow from north to south. In fact, they do nothing of the kind. The weather patterns in North America move, in general, from west to east; meteorologists call this pattern *prevailing westerlies*. The Earth's rotation breaks the north-south atmospheric convection cell that would exist in the absence of rotation into three cells as shown in Figure 19–5. In addition, the rotation "stretches out" the shape of the air circula-

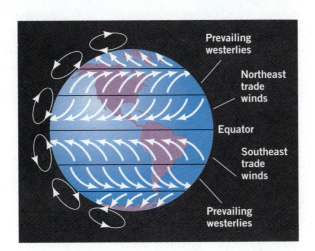

Figure 19–5
Atmospheric convection on the rotating Earth, showing the principal air circulation cells. Compare this diagram with the satellite photograph of the Earth. (*See page 537.*)

Jupiter's atmosphere displays complex banding and turbulence that result from high-velocity winds.

tion pattern in each cell. In the cell near the equator, the winds at the surface tend to blow from east to west; these air currents are the so-called trade winds that drove sailing ships from Europe to North America. In midlatitude zones, the Earth's rotation causes the winds to blow from west to east, creating regions in which weather patterns also usually move from west to east. Finally, in the Arctic and Antarctic, the winds blow once again from east to west.

Similar patterns of atmospheric motion can be seen on all the planets in the solar system that have atmospheres. In some cases—for instance, the planet Jupiter—the rapid rotation of the planet and the atmospheric dynamics produce more than three convection cells. Jupiter, in fact, has no fewer than 11.

Because of the force of gravity, the Earth's atmosphere gets thinner as altitude increases. Half of the mass of the atmosphere is below an altitude of 5.5 km (3.4 miles), and 99% is below 32 km (20 miles). The upper atmosphere is called the *stratosphere*, while the region near the ground is called the *troposphere*. You occasionally see these terms used in discussions of the weather.

Weather and Climate

weather   Daily changes in rainfall, temperature, amount of sunshine, and other variables, resulting partly from the general circulation in the atmosphere, and partly from local disturbances and variations.

climate   The average weather conditions of a place or area over a period of years.

The word **weather** refers to daily changes in rainfall, temperature, amount of sunshine, and other familiar variables that are part of our everyday lives. The term **climate,** on the other hand, refers to trends in the weather longer than the cycle of seasons. Thus a city like Seattle may have hot, sunny weather on a particular day, but have a generally moderate, rainy climate. The weather in a given region results partly from the general circulation in the atmosphere, and partly from local disturbances and variations caused by the presence of mountain ranges, bodies of water, or other features.

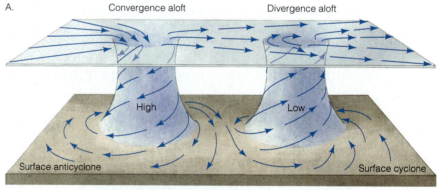

A.   Convergence aloft   Divergence aloft
High   Low
Surface anticyclone   Surface cyclone

B. Surface isobars as they appear on a map

100.4 kPa
100 kPa   100.8 kPa
H
100 kPa
L
99.2 kPa
99.6 kPa

Figure 19–6
This three-dimensional diagram shows how air moves in curved paths from regions of high pressure to regions of low pressure. (Anticyclones and cyclones are systems of winds that rotate around the centers of highs and lows, respectively.)

Figure 19–7
A satellite photo of a low-pressure system that is centered over Ireland and is moving toward the European continent. Notice the counterclockwise rotation of the air.

The most important variations underlying the daily weather are differences in atmospheric pressure, which relates to the amount of air actually present at a given point. Air, like any other fluid, can slosh around. There will be regions of higher pressure, where more air than average accumulates, and regions of lower pressure as well. When low-pressure areas develop in the general flow, air from the surrounding regions tends to rush in to fill up those lows. In the Northern Hemisphere, the rotation of the Earth forces the air to flow in a generally clockwise direction around an area of high pressure, and in a counterclockwise direction around an area of low pressure. (It is just the reverse of the Southern Hemisphere.) This flow gives rise to the circular patterns we often see in satellite weather photographs (see Figures 19–6 and 19–7).

Air in low-pressure areas rises, cooling as it does so. At some altitude, the temperature falls to the point where water vapor condenses and clouds form. Low-pressure areas are thus often characterized by extensive cloud cover and precipitation. In high-pressure areas, air moves downward and becomes warmer and relatively drier. High-pressure areas, therefore, are associated with sunny and clear weather. A typical weather map will show two or three high- or low-pressure areas across a continent such as North America at any given time, and it is the constant interplay and change among these patterns as they move across the continent that produces the daily changes of weather.

The place where two masses of air of different temperature meet is called a *front*. We experience the passage of a front as a change in the

Figure 19–8
When warm air encounters a cold mass, it is forced upward, clouds form, and the weather turns rainy.

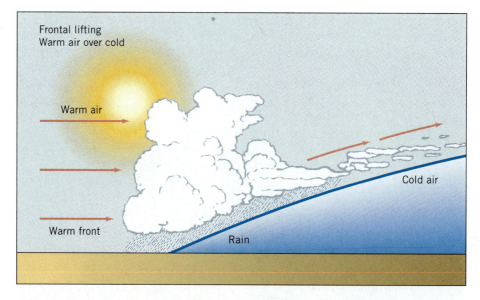

Frontal lifting
Warm air over cold

Warm air

Cold air

Warm front

Rain

weather. For example, Figure 19–8 illustrates what happens when a cold air mass is overtaken by a warm one. The warm air is lifted up over the cold, and, as a result, clouds begin to form. Standing on the ground, you would first see high, thin clouds, as shown. The clouds would gradually thicken until it began to rain. Have you ever noticed this sequence of events where you live?

Figure 19–9
The jet stream is a fast-moving, high-altitude air current above North America. (*a*) The jet stream often follows a relatively straight path, with minor undulations. (*b*) Strongly developed undulations may pull a mass of cold arctic air to the south.

A. Jet stream with small undulations

Cold air

Jet stream

Warm air

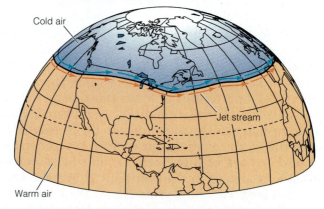

B. Strongly-developed undulations pull a trough of cold air south

Cold air

Warm air

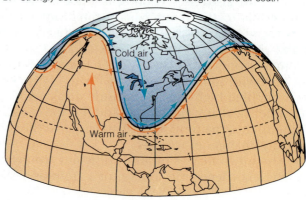

Another important determinant of the weather in North America is the position of the **jet stream.** This high-altitude stream of fast-moving winds marks the boundary between the northern polar cold air mass and the warmer air of the temperate zone. It generally shifts north and south on an annual basis, but bends in the jet stream can prevent low-pressure areas from forming (a process that can produce drought) and can allow cold air to spill down into the temperate zones (a process nicknamed the Alberta Clipper or Siberian Express), creating cold waves (see Figure 19–9). You sometimes see the jet stream referred to in weather reports as an explanation for unusual weather patterns.

jet stream A high-altitude stream of fast-moving winds that marks the boundary between the northern polar cold air mass and the warmer air of the temperate zone.

Developing Your Intuition

The Coriolis Force

Why should winds rotate in different directions in the Northern and Southern hemispheres? The reason has to do with the rotation of the Earth.

Imagine someone standing at the North Pole and throwing a ball to someone standing at the equator. While the ball is in flight, the rotation of the Earth will carry the catcher to the east, and the ball will fall to his left. From his point of view, the path of the ball has curved to the west (i.e., toward his left). Using Newton's first law (see Chapter 3), he would argue that a force was acting on the ball to cause this deviation from straight-line motion. Like the centrifugal force we discussed in Chapter 4, this force arises from the motion of the accelerated frame of reference. It is called the Coriolis force, after the French engineer and mathematician Gaspard Coriolis (1792–1843).

Can you see that the force will bend the ball's track to the right in the Southern Hemisphere? ●

Common Storms and Weather Patterns

Many kinds of severe weather conditions affect our world. Some of these have caused devastation and tragedy, while others are merely inconvenient. In the following section, we review a few types of violent weather.

Tropical storms are severe storms that start as low-pressure areas over warm ocean water. They draw energy from the warm water, and grow in great rotating patterns called cyclones that are hundreds of kilometers in diameter (see Figure 19–10). Tropical storms can cause serious damage when they strike land because of the high winds they generate, and the fact that the low-pressure area at their center causes an increase in sea level, called a *storm surge.* Once over land, these storms lose their source of energy and eventually die out.

Tropical storms that begin in the Atlantic Ocean off the coast of Africa and affect North America are called *hurricanes;* those that begin in the North Pacific are called *typhoons.* Until recently, these violent weather systems often hit unprotected coastlines with little warning. Today, weather satellites spot and track tropical storms long before they approach land.

Figure 19–10
A satellite weather photo reveals Hurricane Andrew approaching the Florida coast on August 24, 1992.

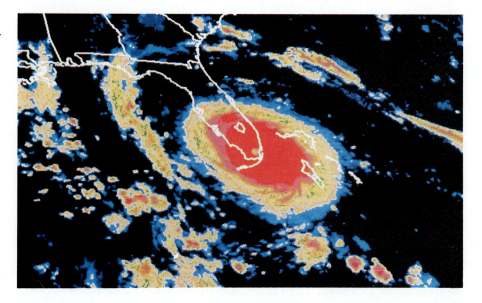

Tornadoes are rotating air funnels some tens to hundreds of meters across. Tornadoes descend from storm clouds to the ground, causing intense damage along the narrow path where the funnel touches the ground. Although their damage is usually limited to a small area, tornadoes are the most violent weather phenomenon known, with air speeds in excess of 500 kilometers (about 300 miles) per hour.

Monsoons include any wind systems on a continental scale that reverse their direction seasonally. These winds arise primarily from seasonal variations in relative temperatures over land and sea. Land cools off and warms up faster than the ocean when the seasons change, and, since atmospheric pressure increases over colder areas, these winds tend to blow from cold (higher pressure) to warm (lower pressure) regions. Thus, in summer monsoons come from the ocean and bring rain, and in winter from the land and bring drought. Monsoon systems occur across eastern Africa and southern Asia, northern Australia, and the Gulf Coast of the United States.

A tornado in North Dakota.

El Niño is a weather cycle in the Pacific basin—a cycle that recurs every four to seven years. The name (which means "Christ Child") comes from the fact that the phenomenon, when it happens, usually begins around Christmas time. El Nino can cause severe storms and flooding all along the western coast of the Americas, and drought from Australia to India.

El Niño is an example of a coupling between two of the Earth's cycles, in this case the atmospheric and water cycles, for it results from an interaction of winds and ocean currents. Here's what happens: during normal years (see Figure 19–11) the normal easterly winds produce an upwelling of cold water along the coast of South America. The cold ocean temperature chills the air above it, so that it does not rise and form clouds. In this situation, there is rain in the western Pacific (around Indonesia), but not in the east.

When the easterly winds weaken, however, the eastern ocean begins to warm and the upwelling slows. This situation allows the air above the ocean to warm and rise and the rain to move eastward. These changes, in turn, affect atmospheric pressures in the Pacific in a way that further weakens the easterlies and strengthens the growing El Niño.

In this way, the El Niño (or Southern Oscillation, as it is sometimes caused) results from a delicate interplay between the sea and the air. Changes in the easterlies induce changes in sea surface temperatures, which induce changes in rainfall and induce further changes in the winds, and so on. Scientists are now using computer models to predict the onset of the El Niño as they would other kinds of weather changes.

Figure 19–11
During normal years (*a*) cold upwelling water in the eastern Pacific prevents the formation of clouds. When an El Niño event begins (*b*), the easterlies weaken, the eastern Pacific warms, and rainfall moves eastward. This cycle repeats every 4 to 7 years.

Wind

Indonesia

South America

(*a*)

Wind

Indonesia

South America

(*b*)

A computerized radar weather map shows areas of cloud cover and precipitation.

Technology

Doppler Radar

Radar has been a staple tool for weather forecasters for decades. In fact, you probably see radar maps of local weather conditions on TV every night. The way radar works is as follows: Microwaves are sent out from a central antenna. When they encounter objects such as raindrops, snowflakes, or ice in the air, the waves are reflected back. Each kind of material produces a distinct pattern of reflected waves because its density is different from that of the surrounding air. Using these reflection patterns, a map of local storm activity can be assembled.

Ordinary radar, however, cannot detect winds, even winds of high velocity. The reason is that the density of moving air is usually not very different from stationary air, and the two produce the same reflection pattern. But in many situations, the detection of air currents is important. Near airports, for example, sudden changes in wind direction create violent air turbulence called *wind shear*, an extremely dangerous condition that we need to be able to detect.

Doppler radar is designed to detect motions of the air by analyzing reflected microwaves for their frequency as well as their intensity (see Chapter 8). First, the difference between the emitted and reflected frequencies is calculated. An analysis of the Doppler effect then yields the velocity of the object from which the wave was reflected. In this way, high winds and atmospheric turbulence can be detected at a safe distance. ●

Understanding Climate

While the position of the jet stream and the creation of high- and low-pressure zones dominates daily weather patterns, overall climate depends

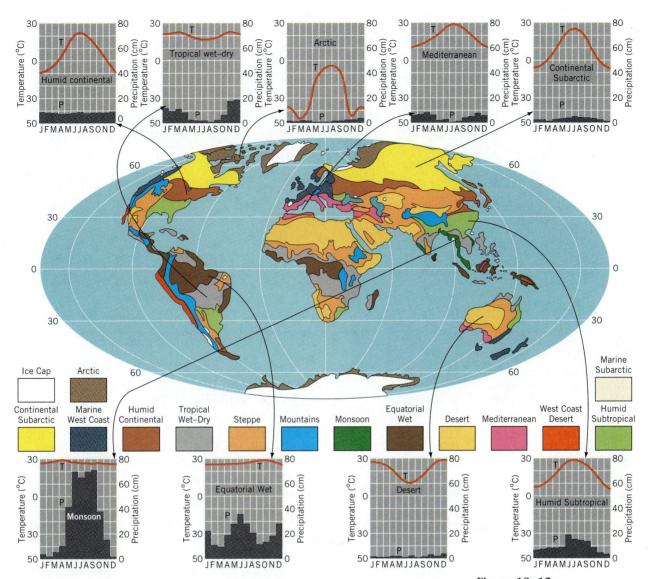

Figure 19–12
Rainfalls and climates of the world.

on long-range features of the Earth's surface (see Figure 19–12). Factors dictating climate include the distribution of heat due to the stabilizing temperature of oceans, and the presence of mountains, which alter the direction and intensity of air currents. The climate is also extremely sensitive to the amount of sunlight that falls on our planet and the amount of heat that is radiated back to space.

At the moment, our best attempts at predicting long-term climate depend on complex computer models of the atmosphere called *global circulation models* (GCMs). In a typical GCM, the computer splits the world's surface into squares several hundred kilometers on a side, and slices the atmosphere into about 10 vertical compartments (see Figure 19–13). In each of these little boxes, the laws of motion and thermodynamics are used to calculate the amount of heat that flows in and out, how much water vapor comes out of the air, and so on. The computer balances the inflow and outflow from all of the boxes in the atmosphere

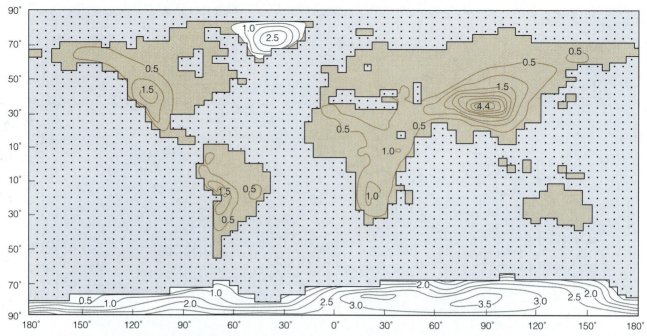

Figure 19–13
This is a typical view of the world as drawn by a computer running a GCM. How close to the real world is it? (The numbers indicate elevation in thousands of meters.)

and predicts long-term climate trends. Our current models are still rather crude (they have a great deal of difficulty taking into account the effects of clouds or the ocean, for example), but they represent the best attempts to date in understanding what affects the Earth's climate. These models also play a critical role in discussing various types of ecological changes (see the section, Global Warming, later in this chapter).

THE HYDROLOGICAL CYCLE

Water plays a vital role in the unique chemistry of the Earth's outer layers. Water saturates the air, falls to the ground as precipitation, moves through a complex system of rivers and streams, and is stored for long periods in underground reservoirs, oceans, and ice. Water shapes the surface of our planet, and it provides the only known medium in which life can begin. The combination of processes by which water moves from repository to repository near the Earth's surface is called the **hydrological cycle.**

The total amount of water on the Earth's surface has stayed roughly the same since it formed. Water first reached the surface during the outgassing of the young, volcano-covered Earth (see Chapter 17). When the planet's surface temperature finally fell below 100°C, this water condensed into liquid form and began to fill the ocean basins. A small amount of water is lost from the Earth each year, as ultraviolet rays from the Sun break up water molecules high in the atmosphere, freeing hydrogen atoms. These atoms may escape into space because of their small mass. According to one estimate, the amount lost is no more than one or two Olympic-sized swimming pools of water per year. Thus, for all intents and purposes, we can treat the Earth as if it has had a fixed amount of water for billions of years.

hydrological cycle The combination of processes by which water moves from repository to repository near the Earth's surface.

Most of the world's ice is locked up in the Antarctic ice sheets.

Reservoirs of Water

The Earth boasts several major water repositories (see Table 19–1). In addition to oceans, lakes, and rivers, significant amounts of water are locked into the Earth's polar ice caps and glaciers, which are bodies of ice that form in regions where snowfall exceeds melting. **Ice caps** are layers of ice that form at the north and south polar regions of the Earth. **Glaciers** are large bodies of ice that slowly flow down a slope or valley under the influence of gravity. Approximately 96% of glaciers (by volume) occur in Antarctica and Greenland, while the rest are widely scattered in mountainous areas throughout the world. All of these places where water occurs tap into the same central supply. During its lifetime, a given molecule of water will cycle through many different kinds of bodies, over and over again (see Figure 19–14).

The hydrological cycle with which most of us are familiar involves the short-term, back-and-forth transfer of water molecules between the oceans and the land. Water evaporates off the surface of the ocean, forms into clouds, falls as rain on the land, and then returns to the oceans via rivers and streams. Most terrestrial life depends upon this simple cycle.

ice cap Layers of ice that form at the north and south polar regions of the Earth.

glacier Large body of ice that slowly flows down a slope or valley under the influence of gravity; found primarily in Greenland and Antarctica.

Table 19–1 • Distribution of Water on the Earth

Location	Volume of Water (in millions of km³)
Ocean	1370
Glaciers and ice sheets	25
Groundwater	5.1
Lakes and rivers	0.5
Atmosphere	0.015

Figure 19–14
Water of the Earth's surface cycles through clouds, rainfall, oceans, lakes and rivers, and groundwater.

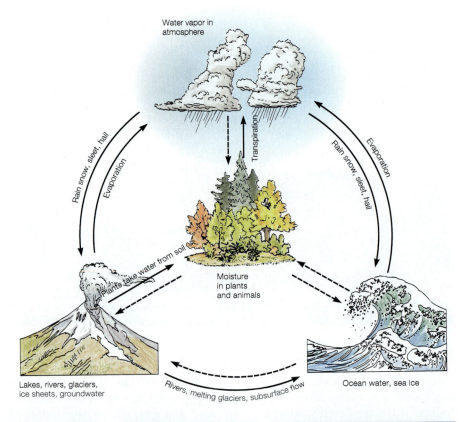

Common types of clouds include (a) cumulus clouds; (b) cumulonimbus clouds; and (c) altocumulus clouds.

(a)

(b)

(c)

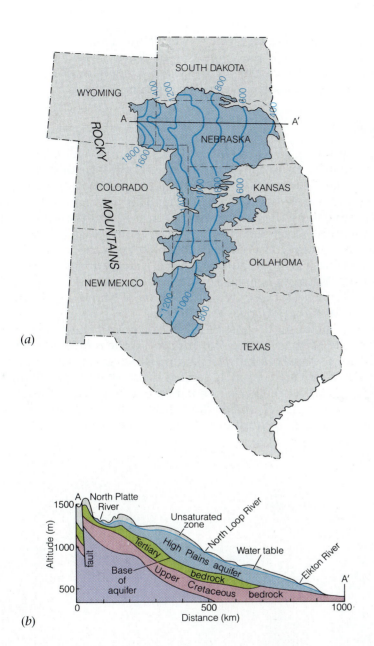

Figure 19–15
An aquifer is a rock layer that stores water and can be tapped by a well. The High Plains aquifer of the central United States provides much of the water used for irrigation in this region. (*a*) A map of the aquifer with contour lines showing the depth to the water level (in meters). (*b*) A cross section of the aquifer reveals the slope of the water-bearing rock formations.

A vast part of this cycle remains unseen, however. Some of the water that falls on the continents does not immediately return to the ocean; rather it seeps into the Earth to become **groundwater.** Groundwater may move into large *aquifers*, which are bodies that act as underground storage tanks of water. By some estimates, more than 98% of the world's fresh water is stored as groundwater. Water typically percolates into the ground and fills the tiny spaces between grains of sandstone and other porous rock layers. These layers of water-saturated rock are often bounded by impermeable materials such as clay, which keeps the water from seeping away (see Figure 19–15). Humans tap into aquifers when they drill wells to supply water for cities and agricultural use.

One problem with using aquifers as a water supply is that it may take many thousands of years to fill them, but only a few years to drain them.

groundwater   Fresh water from the surface that typically percolates into the ground and fills the tiny spaces between grains of sandstone and other porous rock layers.

For example, when farmers in the central United States take water from the Ogallala (High Plains) aquifer, one of the great underground reservoirs, they are, in effect, mining the water. Underground water is not a renewable resource, except over very long time scales.

Ice Ages

ice age A period of several million years during which glaciers have repeatedly advanced and retreated, causing radical changes in climate and influencing human evolution.

The fixed amount of water on the Earth has had many important geological consequences. From time to time, much of the Earth's water supply becomes locked into glaciers that advance across land from the poles. Such a period is called an **ice age.** When this happens, the amount of water available to fill the ocean basins decreases, and the sea level drops. During the most recent advance of glaciers some 20,000 years ago, for example, the eastern coast of the United States was about 250 kilometers farther east than it is today, and to the west a land bridge made it possible to walk from Alaska to Siberia. This land bridge provided a route that is thought to have been taken by the ancestors of Native Americans when they moved into the Americas from Siberia.

The periodic onset of massive glaciers is one of the most intriguing features of the Earth's climate. We are now in the middle of such an ice age, a period of several million years during which glaciers have repeatedly advanced and retreated. In fact, geologists say that we are living in an *interglacial period*, between two major advances of glaciers. About 20,000 years ago, massive glaciers spread down from eastern and central Canada, covering a good deal of northern North America. They then receded to

Figure 19–16
The extent of the most recent North American glaciation, approximately 20,000 years ago.

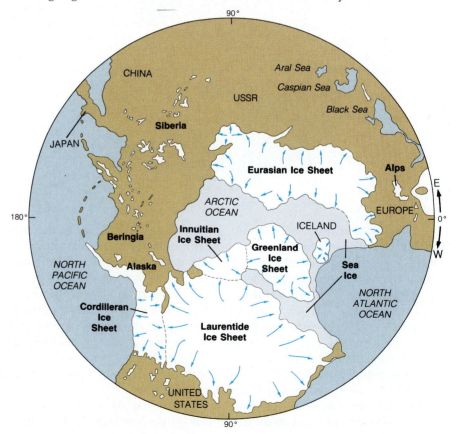

Greenland about 10,000 years ago, as illustrated in Figure 19–16. Glaciers have come and gone many times, and it appears that the periods of cyclic glaciation like the one in which we now live have occurred relatively often during the past 2 million years of the Earth's history.

The occurrence of the present ice age, and the resulting cycles of glaciation, may have had a profound influence on human evolution. In fact, some scholars have argued that the constant radical changes in climate during recent human history is responsible for the development of technology. They point out, for example, that the emergence of modern human beings and the disappearance of Neanderthals in Europe about 35,000 years ago coincided with one of these glacial periods. Thus you might say that the fact that you are sitting in a heated or air-conditioned room, reading a book produced on a modern printing press, is ultimately due to the fact that ice ages are part of the Earth's climate.

Milankovitch Cycles

The primary causes of periodic glaciations were first explained by Milutin Milankovitch, a Serbian civil engineer, in the early part of the twentieth century. His theory was that the relation between the Earth and the Sun is affected by a number of variations in the Earth's rotation and orbit. These variations cause slight changes in the amount of solar radiation absorbed by the Earth, as shown in Figure 19–17.

The easiest of the orbital effects to understand is the behavior of the Earth's axis of rotation. If you have ever watched a child's top spinning, you may have noticed how the top spins rapidly around its axis while the axis itself describes a lazy circle in space (Figure 19–17). This circular motion is called *precession*. In the same way that a top may precess every second or two, the Earth's axis, which is tilted at an angle of 23 degrees, precesses once every 26,000 years. While this is occurring, the axis of the Earth's elliptical orbit around the Sun also shifts due to the gravitational

Figure 19–17
The precession of the Earth's axis (*a*). Small changes in the tilt of the axis (*b*), and changes in the shape of the orbit induced by the other planets (*c*), contribute to recurring ice ages.

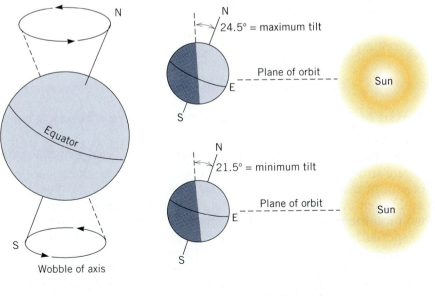

(*a*) Precession of the equinoxes
(period = 23,000 years)

(*b*) Tilt of the axis
(period = 41,000 years)

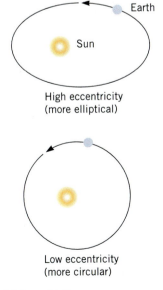

(*c*) Eccentricity
(dominant period =100,000 years)

effects of the other planets. The net effect of these two motions is that the axis of the Earth gradually changes its orientation. The axis tilts the Northern Hemisphere more toward the Sun for half of a 23,000-year cycle, and the Southern Hemisphere more toward the Sun for the other half. Today, the Earth's axis is in such a position that the Northern Hemisphere is tilted away from the Sun during its winter months (the period in which the Earth is actually closest to the Sun). About 11,500 years from now, however, the situation will be reversed, and the Northern Hemisphere will be tilted toward the Sun during its winter.

Other orbital effects involve a slow change in the angle of the axis of rotation (it rocks back and forth by about a degree every 41,000 years), and a small change induced in the shape of the Earth's orbit by other planets (a cycle that takes about 100,000 years).

Milankovitch proposed that the global climate changes in cycles (now called *Milankovitch cycles*) when all of these orbital effects reinforce each other. If we find ourselves in a period of decreasing solar energy absorption and increased precipitation, more snow will fall in the winter and it will stay on the ground longer. Snow and ice reflect sunlight, so this extra snow and ice further cools the Earth and more snow falls and stays on the ground even longer. Thus a decline in absorption of sunlight may trigger a sequence of events that can eventually lead to glacial advance. By the same token, a period of increased absorption of sunlight will result in warmer periods during which glaciers will tend to retreat.

Other factors important in controlling the extent and distribution of glaciers include the distribution of land masses (which changes because of tectonic motion), and the location of mountain chains (which alters wind and precipitation patterns). Recent evidence also suggests that the amount of volcanic dust and gases in the atmosphere can cause short-term changes in global temperatures. Finally, some scientists suspect that the energy output of the Sun is cyclical, and that variations in the Sun's energy output impose a cyclic variation in the Earth's temperature.

Many scientists think that we may experience a new phase of glacial advance within the next 10,000 years. This cooling trend may be offset for a time, however, by global warming, potentially resulting from an enhanced greenhouse effect (see below).

Science in the Making

Milutin Milankovitch Decides on His Life's Work

Milutin Milankovitch didn't seem to be headed toward a career in the sciences. Trained in Vienna as a civil engineer, he designed reinforced-concrete structures in central Europe for a few years before becoming a professor of mathematics in his native Belgrade just before World War I. Swept up in the nationalistic movements that were then (as now) prominent in the Balkans, he became friends with some poets who specialized in writing patriotic verse. One evening, drinking coffee to celebrate a new book of verse, Milankovitch and his friends came to the attention of a banker at a neighboring table. The banker was so taken with the poems that he bought 10 copies on the spot.

With the new money in their pockets, the friends started to celebrate with wine. After the first bottle, Milankovitch says in his journal,

"I looked back on my earlier achievements and found them narrow and limited." By the third bottle, he had decided to "grasp the entire universe and spread light to its farthest corners."

He then methodically set apart a few hours each day to study and interpret climate records. Even during World War I, when he served as an engineer on the Serbian general staff and became a prisoner of war, he kept up his work. A member of the Hungarian Academy of Sciences arranged for him to have a desk at the academy after he gave his word of honor that he would not try to escape, and a good deal of the work described in his text was done under those conditions. The final work, published in 1920, was quickly recognized and accepted by the scientific community. ●

Ocean Currents: The Physical Circulation of the Oceans

The Earth's oceans, an important part of the hydrological cycle, are in constant motion (see Figure 19–18). At one time, scientists viewed the oceans as little more than passive containers into which the Earth's water flowed, but it would be very misleading to think of them as a series of gigantic bathtubs. In fact, we now know that the Earth's oceans, even more than the atmosphere, are key in distributing heat across the surface of the planet, and thus in determining climate.

Currents are like rivers of moving water within the larger ocean. Each ocean basin, such as the North Atlantic, has a surface current. Some surface currents carry warm waters from the equator, where a large amount of heat energy from the Sun is absorbed, toward the cooler poles. At the same time, other surface currents carry cold water from the poles back to the equator to be heated and cycled again. These great *gyres* (rhymes with "tires"), as they are called, have a profound effect on the weather of the land past which they flow. In the North Atlantic, for

current A riverlike body of moving water in an ocean basin.

Figure 19–18
Ocean currents play a major role in redistributing the Earth's heat.

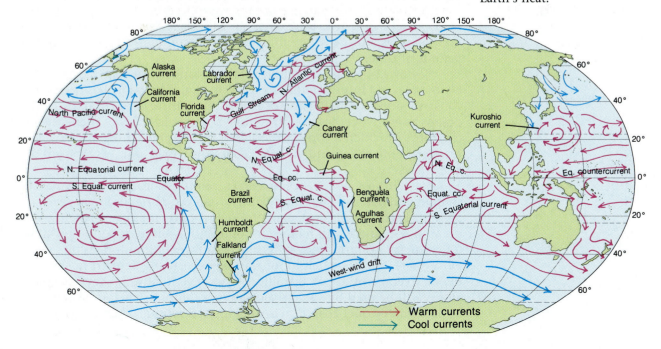

example, the Gulf Stream current originates in the Caribbean and flows past the eastern coast of the United States. It comes near England, making the British Isles much warmer than you might expect them to be based on their latitude, which is farther north than Maine. Indeed, the city of London possesses a mild climate even though it is at about the same latitude as Canada's Hudson Bay.

A much colder current, on the other hand, flows along the western coast of Europe back to the tropics to complete the cycle. In each of the Earth's great ocean basins, a similar set of currents can be seen. In the North Pacific, the Kuroshio current carries warm waters from the tropics past the east coast of Japan, while the California current carries cold water back down toward the equator along the western coast of North America.

In addition to this relatively rapid circulation of water at the surface of the ocean, we find deeper three-dimensional circulation. When the effects of surface currents and wind along a coast act in such a way as to push surface water away from the land, colder water from the depths rises to create an area of upwelling. The waters along the coast of California display this phenomenon, which does much to explain why the ocean there is so cold. On a larger scale, water from the Arctic and Antarctic, which is both very cold and salty because so much fresh water is removed to form ice, sinks to the bottom of the sea and rolls sluggishly toward the equator, producing the slow, deep currents that characterize much of the bottom currents in the world's ocean basins.

Chemical Cycles in the Oceans

Just as water in the ocean is in constant physical motion, so too are the chemicals that make it salty. It used to be thought that the oceans were simply passive receptacles for materials washed into them from the land. The old belief was that rivers flowing into the ocean carried dissolved minerals and salts with them, and when water evaporated, these minerals were left behind so that the oceans were becoming increasingly salty with time. In this view, the extremely salty Dead Sea in Israel was always pointed to as a place where minerals had accumulated over the longest periods of time.

We now know that the salinity of the ocean has not changed appreciably over several hundred million years. In fact, it probably has been fairly constant since soon after the oceans formed. Instead of thinking of the oceans as passive receptacles, it is more accurate to think of them as large test tubes in which a constant round of chemical reactions occur. These reactions both affect and are affected by the input of minerals from the world's rivers.

The saltiness of the ocean comes primarily from the presence of sodium and chlorine, but many other dissolved minerals are in seawater as well. Each chemical element follows a different set of reactions in ocean water, and it will remain in solution for different periods of time. Calcium, for example, may enter the ocean when limestone is dissolved. Once there, a given calcium ion can be expected to remain in solution for an average of about 8 million years. Eventually, however calcium atoms will be used by some sea creature to build its shell or skeleton or be precipitated out of solution in a chemical reaction. When the organism

Table 19–2 • Some Typical Residence Times for Elements in the Ocean

Element	Concentration (parts per million)	Residence Time (million years)
Sodium	10,800	260
Calcium	413	8
Chlorine	19,400	infinite
Gold	0.00005	0.042
Potassium	387	11
Copper	0.003	0.05

dies, the shell will sink to the bottom, and the calcium atom may once more be incorporated in limestone. Many millions of years later, that limestone may be lifted up by tectonic forces and exposed at the Earth's surface, causing the calcium atom to be carried back to the sea through erosion or dissolution.

The average length of time that any given atom will stay in the ocean water before it is removed by a chemical reaction is called the *residence time* (see Table 19–2). The element sodium (one of the ions that gives seawater its distinctive taste) enters the ocean after having been dissolved out of various kinds of rocks. A sodium ion will stay in suspension on average for about 260 million years before it is incorporated again into various kinds of clays and muds on the ocean bottom. Once so incorporated, it can go through the same cycle of uplift and erosion as limestone.

Most chloride ions, on the other hand, will stay in the ocean forever. Have you ever gone to the beach and noticed that your skin was salty at the end of the day, even if you didn't go in the water? Bubbles of sea foam burst, evaporate, and spew little bits of salt (sodium chloride, or NaCl) into the air. This salt is carried by the wind and may be suspended for some time before it sticks to your skin. Sailors know that a rain in the open ocean is often salty, and rains that fall near coasts also contain a fair amount of sodium and chlorine. These salty rainfalls often give crops raised in coastal areas a distinctive taste. The flavor of artichokes grown in fields south of San Francisco are one example of such a phenomenon, and the unique taste of meat from the sheep that graze on the coastal meadows of Brittany in France is another. The chlorine that leaves the ocean in this way is quickly returned in rain or by coastal streams.

Only when a body of salt water evaporates, forming dry salt deposits such as those surrounding the Great Salt Lake in Utah, is chlorine removed from its water environment for any appreciable length of time. But even these deposits, slowly buried and compressed, could eventually (in perhaps a few hundred million years) rise to the surface as salt domes, where they would weather away, returning the salt once again to the sea.

To summarize chemical cycles in the oceans: The Earth's rivers continuously transport all sorts of elements into the sea. Each of these elements resides for a certain amount of time in the ocean and then is removed by one sort of reaction or another. The supply of every kind of atom is being constantly renewed, and the oceans may never be any more salty than they are today.

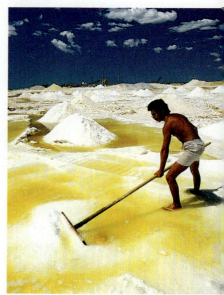

A worker in Brazil harvests salt by evaporating sea water. Most of the world's salt supply is obtained in this manner.

Developing Your Intuition

Element Residence Times in Your Body

Do you ever wonder why you need to eat sources of minerals such as iron and calcium throughout your life? Once you are grown, after all, you have all the atoms you need to stay alive.

It turns out that your body constantly recycles atoms, just like the oceans. Some of your body's cells—for example, the lining of the intestines—are replaced every few days. You need fresh supplies of carbon, oxygen, and nitrogen daily to help replace these cells. Red blood cells last much longer, on average 120 days, but you need a regular intake of iron to produce these important cells, as well. Failure to digest enough iron can lead to anemia, a condition characterized by fatigue due to insufficient red blood cells.

Atoms in your bones and tendons last much longer—a decade or more, on average. However, even those atoms are constantly being replaced. Gradual loss of calcium atoms in bone, for example, is of special concern in older people who may not consume enough replacement calcium. Osteoporosis, a disease in which bones become weak and brittle, may result from this calcium deficiency. A similar mysterious loss of bone calcium affects astronauts who spend more than a few days in the weightless environment of space.

Some harmful elements, including lead, mercury, and other so-called heavy metals, do not easily recycle once taken into the body, because we have no effective biological process to remove them. Their average residence times are much longer than a human lifetime. Concentrations of these atoms can thus build up over time and may result in sickness or even death. Lead poisoning remains a serious health issue, and you may hear about efforts in your community to test drinking water for high concentrations of lead that are found in old plumbing systems. ●

Comparison of (*a*) normal and (*b*) calcium deficient bone in electron microphotographs.

(*a*)

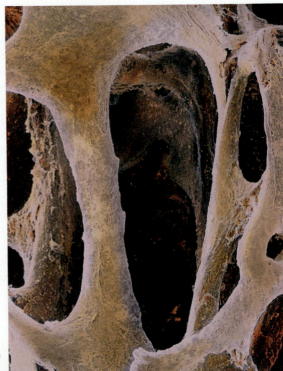

(*b*)

Science by the Numbers

The Ocean's Gold

We've said that every element can be found in seawater. If this is so, how much gold is there, say, in a cubic kilometer of sea water? Could we get rich recovering this gold?

According to Table 19–1, gold is present in the ocean at a concentration of 0.00005 parts per million. A concentration of 1 part per million corresponds to 1 milligram (mg) of a solid dissolved in a liter (l) of water. A liter is a volume measurement equal to a thousandth of a cubic meter (m^3). The total amount of gold in a cubic meter of sea water, therefore, is 0.00005 mg/l times 1000 l/m cubed:

$$(0.00005 \text{ mg/l}) \times (1000 \text{ l/m}^3) = 0.05 \text{ mg/m}^3$$

Each gram contains 1000 mg, so

$$(0.05 \text{ mg/m}^3) \times (1 \text{ g/1000 mg}) = 0.00005 \text{ g/m}^3$$

A km^3 contains 1000 m × 1000 m × 1000 m = 10^9 m^3, so the total amount of gold in a km^3 of sea water is

$$(10^9 \text{ m}^3) \times (0.00005 \text{ g/m}^3) = 5 \times 10^4 \text{ gm}$$

Every cubic kilometer of ocean water holds about 50,000 grams (about 100 pounds) of gold. At $300.00 an ounce, that much gold is worth about half a million dollars.

The total amount of gold dissolved in the world's oceans is vast, but there is no known economical way to extract and concentrate these riches. The equipment and energy required to process that much water would cost far more than the value of any gold recovered. ●

THE GLOBAL ENVIRONMENT

All living things are dependent on the **environment,** which includes the atmosphere, oceans, and land on which we rely for our existence. As we have seen, the Earth is a dynamic planet, constantly changing through its interlocking cycles. But human activities can alter and accelerate these changes, thereby altering nature's cycles.

The most serious environmental problems result from **pollution,** which is the introduction by humans of potentially harmful elements and compounds into the atmosphere, oceans, and land. As the human population grows, demand for energy and material goods increases. The more we consume, the more we are likely to cause large-scale environmental problems. We will examine three of these problems—the degradation of the ozone layer, acid rain, and the greenhouse effect. All of these problems are serious, but their solutions entail different levels of national and international commitment. Each of these problems will need to be resolved in order for our modern industrial society to keep functioning.

global environment The combination of atmosphere, oceans, and land on which all living things rely for their existence.

pollution The introduction by humans of potentially harmful elements and compounds into the atmosphere, oceans, and land.

The Ozone Problem

Although the Sun gives off most of its radiation in visible light, a certain amount of that radiation comes in the form of the *ultraviolet radiation* from the higher-energy part of the spectrum (see Chapter 8 for a review of ultraviolet and other forms of *electromagnetic radiation*). Ultraviolet radiation can be very damaging to living organisms; indeed, it is routinely used to sterilize equipment in hospitals. The surface of the Earth must be shielded in some way from the Sun's ultraviolet rays for life to exist on land.

ozone A molecule made up of three oxygen atoms, instead of the usual two, that absorbs ultraviolet radiation.

Ozone, a molecule made up of three oxygen atoms instead of the usual two, absorbs ultraviolet radiation (see Figure 19–19). If enough ozone molecules exist in the atmosphere, they will protect life at the Earth's surface by absorbing most of the Sun's ultraviolet radiation and keeping it from reaching the ground. In fact, a protective shield of ozone formed high in the Earth's atmosphere several hundred million years ago, and it was only after this shield had formed that life could move from the oceans onto land.

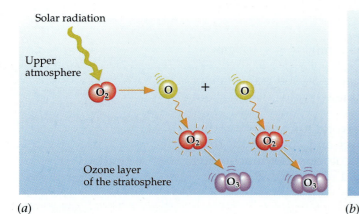
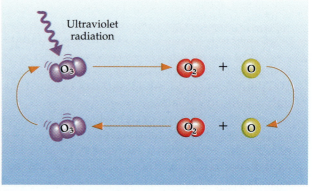

(a)

(b)

Figure 19–19
(*a*) Photons from the Sun break up oxygen molecules, a process necessary for the formation of the ozone layer. (*b*) In the upper atmosphere, oxygen atoms cycle between molecular oxygen (O_2) and ozone (O_3).

Scientists detect ozone in the atmosphere using several techniques. One is simply to fly specialized aircraft into the region where ozone is common and collect samples. For the past decade or so, this kind of sampling has been done routinely by organizations such as the National Oceanic and Atmospheric Administration (usually called "Noah," after its acronym NOAA) and its counterparts in other countries. Another way to detect ozone is to measure characteristic spectral lines given off by the ozone molecule (see Chapter 9). These measurements can be made from satellites, aircraft, balloons, or by ground-based observers.

Measurements reveal that ozone is a *trace gas* that constitutes less than one molecule in a million in the Earth's atmosphere; its concentration varies significantly with height in the atmosphere (see Figure 19–20). While ozone is found at every altitude (you are breathing a small amount even as you read this), most is concentrated in a thick layer some 30 kilometers up, although even there the concentrations are very small. This region of enhanced ozone concentration is called the **ozone layer** (see Figure 19–20). Most of the ultraviolet radiation is absorbed in this layer, but it should not be thought of as anything analogous to a cloud bank in the sky.

ozone layer A region of enhanced ozone (O_3) 30 to 45 kilometers (about 20 to 30 miles) above the Earth's surface where most of the absorption of the sun's ultraviolet radiation occurs.

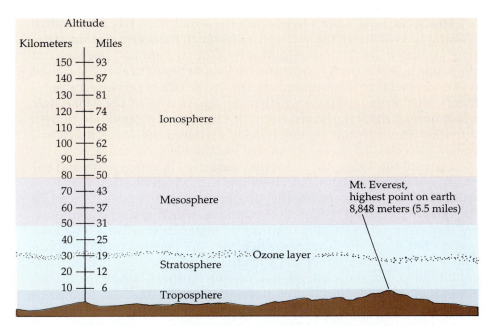

Figure 19–20
The ozone layer. Although ozone is found everywhere in the atmosphere, even at ground level, it is concentrated in a layer some 20 miles above the surface. The labels on the right are the standard terms scientists use to describe different levels of the atmosphere.

In 1985, British scientists working in Antarctica noticed that during the Antarctic spring (roughly the months of September through November) the amount of ozone in the ozone layer dropped significantly. Later studies from satellites and ground-based experiments confirmed these results. During this period of the year, the concentration of ozone above Antarctica falls by different amounts in different years. This phenomenon was promptly dubbed the **ozone hole** (see Figure 19–21).

ozone hole A volume of atmosphere above Antarctica during September through November in which the concentration of the trace gas ozone has declined significantly.

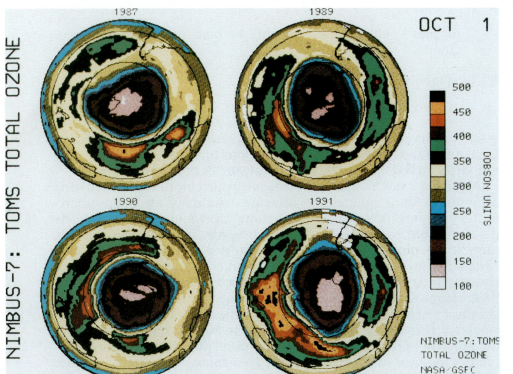

Figure 19–21
The ozone hole over the South Pole. The dark region right over the pole is where ozone concentrations are significantly lower than normal; the bright band represents normal concentrations. This pattern persists only for a few months during the Southern Hemisphere spring.

The ozone hole is not a hole where the atmosphere has deteriorated. It is simply a volume of the atmosphere in which the concentration of the trace gas ozone has declined significantly. Once the annual appearance of the ozone hole was firmly established, scientists began to ask what caused it and whether it could be prevented. Due to the lack of previous measurements, they were uncertain whether the appearance of the ozone hole was a normal natural phenomenon, or one that was caused by human activities. This scientific inquiry quickly narrowed the causes down to a widely used class of chemicals known as *chlorofluorocarbons* (*CFCs*).

Up until the 1950s, the fluids that were used in refrigerators and air-conditioners were rather nasty chemicals such as ammonia. If they were released into the atmosphere, they would cause a noticeable decrease in the quality of the air. Chlorofluorocarbons, which replaced ammonia, are very stable and generally nonreactive materials that last a long time and do not break down readily when they are released into the atmosphere. The introduction of CFCs made possible the great boom in air-conditioning that has made so much of the southern part of the United States more liveable during the summer months. CFCs also seemed to be an environmental boon. And unlike ammonia and the other chemicals they replaced, CFCs did not present an immediate danger in the case of accidental leaks.

We now know that CFCs do, in fact, present a very real danger to the Earth's ozone layer. Over periods of time that range into the decades, molecules of CFCs work their way into the upper regions of the atmosphere, where they can be broken apart by high-energy sunlight, not available in the lower atmosphere. The chlorine atoms that are freed in this way act as a catalyst in a reaction that can be written as follows:

▶ **In words:**

(ozone + chlorine + sunlight) become (ordinary oxygen + chlorine)

▶ **In symbols:**

$$2O_3 + Cl + sunlight \rightarrow 3O_2 + Cl.$$

While this reaction proceeds very slowly, each chlorine atom liberated from a CFC can, over time, destroy millions of ozone molecules before it is safely locked into another chemical species in the atmosphere.

Over most of the Earth, the effect of this chlorine is not striking because new ozone molecules are being created all the time. In the Antarctic, however, a number of unusual circumstances come together to create the ozone hole. For one thing, during the months immediately preceding the hole, no sunlight falls in the Antarctic region of the Earth. This period of darkness leads to the appearance of high clouds made entirely of ice crystals; these are known as *polar stratospheric clouds*. Crystals of ice in these clouds provide sites on which chlorine atoms undergo a series of chemical reactions which proceed up to the final step before chlorine atoms are released. As soon as energy in the form of sunlight returns in the Antarctic spring, the destruction of ozone proceeds very quickly because of the large amounts of chlorine. The ozone is destroyed in a matter of days or weeks, and the ozone hole results.

Clouds like these form over Antarctica during the winter. Made from ice crystals, they are the site of many of the chemical reactions that produce the ozone hole.

A comparison of Los Angeles during a clear day (top) and on a smoggy day (bottom).

You might think that the disappearance of the ozone shield in the Antarctic spring would not be a major environmental problem. After all, almost no life exists on the Antarctic continent (and very few people go sunbathing there). The real danger of the ozone hole, however, is that it is caused by chemical reactions that could have long-term effects on the entire ozone layer. Not only has the average size of the ozone hole grown larger over the past decade, but some recent measurements suggest that the ozone layer has been depleted by a few percent worldwide.

In 1986, an international congress meeting in Montreal produced a treaty by which all the industrial nations of the world agreed first to limit, then to eliminate, their production of CFCs. This decision triggered a lot of activity in major chemical companies, where researchers are looking to find replacement substances. The main thrust of this work is to develop compounds that, although more expensive than CFCs, perform the same functions without releasing chlorine to the environment. Many of these substitute compounds contain bromine instead of chlorine. In 1992, the reduction of CFCs was proceeding so quickly that the target date for elimination of most CFCs was set at 1996. Recent estimates predict a return of ozone levels to normal early in the twenty-first century.

We view the environmental problems posed by the ozone hole as an example of a serious environmental concern, but one that has a relatively straightforward solution. Scientists have clearly established the cause of the problem. The effects of the ozone hole, while serious, are not totally devastating. And, finally, the cost of solving the problem is relatively low. The problem of ozone depletion appears to be well on its way to being solved, and represents a major environmental victory.

Acid Rain and Urban Air Pollution

Burning, which is the chemical reaction of rapid *oxidation* (see Chapter 12), inevitably introduces chemical compounds into the atmosphere. Carbon dioxide and water vapor, for example, are always released when

Figure 19–22
The amount and sources of atmospheric sulfur dioxide (*a*) and nitrogen oxides (*b*) in the United States.

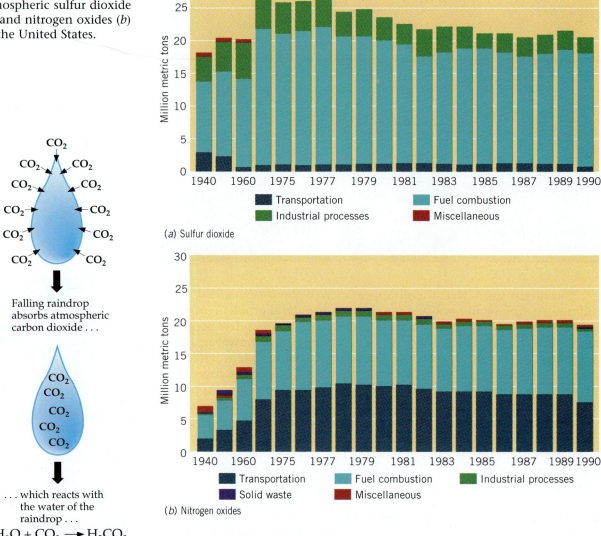

(*a*) Sulfur dioxide

(*b*) Nitrogen oxides

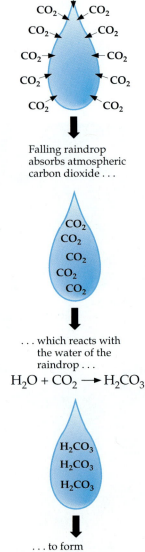

Falling raindrop absorbs atmospheric carbon dioxide . . .

. . . which reacts with the water of the raindrop . . .

$$H_2O + CO_2 \longrightarrow H_2CO_3$$

. . . to form carbonic acid.

Figure 19–23
The process by which normal rainfall becomes slightly acidic.

hydrocarbons are burned (see Chapter 13). But burning produces three other significant sources of pollution: nitrogen oxides, sulfur compounds, and hydrocarbons (see Figure 19–22).

1.   Nitrogen oxides. Whenever the temperature of the air is raised above approximately 500°C, nitrogen in the air combines with oxygen to form what are called NO_X compounds. Two important examples are nitrogen oxide (NO) and nitrogen dioxide (NO_2). The x subscript refers to the fact that these compounds have different numbers of oxygen atoms in them.

2.   Sulfur compounds. Petroleum and coal-based fossil fuels usually contain small amounts of sulfur, either as a contaminant or as an integral part of their structure. The result is that chemical combinations of sulfur and oxygen, particularly sulfur dioxide (SO_2), are released into the atmosphere as well.

3. Hydrocarbons. The long-chain molecules that make up hydrocarbons are seldom burned perfectly in any real-world situation. As a result, a third class of pollutants, composed of bits and pieces of unreacted hydrocarbon molecular chains, enters the atmosphere.

The emission of NO_X compounds, sulfur dioxide, and hydrocarbons gives rise to a number of serious environmental problems. One of these that has immediate consequences for urban residents is air pollution. Sunlight hitting nitrogen compounds and hydrocarbons in the air triggers a set of chemical reactions that, in the end, produce ozone. And, while ozone in the stratosphere is essential to life on Earth, ozone at ground level is a caustic, stinging gas that can cause extensive damage to the human respiratory system. This "bad ozone" is a major product of modern urban air pollution associated with photochemical *smog*, which is the brownish stuff that you often see over major cities during the summer.

Urban air pollution is a serious problem, but it is also an immediate and transitory one. If the air quality in a city declines, people know it immediately. While the arrival of a thunderstorm or stiff winds can quickly solve the problem for a particular city, such acts of nature merely move the problem somewhere else.

Nitrogen and sulfur compounds in the air cause long-term problems that may not have an immediate effect on the place where the emissions occur. These airborne compounds interact with water in the air to form tiny droplets of nitric and sulfuric acid. (The latter is the type of acid normally used in automobile batteries.) When it rains, these droplets of acid wash out and become, in effect, a rain of dilute acid rather than of water. This phenomenon is known as **acid rain.** (We should note that rain is normally slightly acidic because carbon dioxide dissolves in raindrops to make a weak solution of carbonic acid; see Figure 19–23. The term acid rain, therefore, refers to the considerable extra acidity produced by human activities.)

You can see one effect of this sort of acid rain in cities. Many of the great historical monuments in European cities are made from limestone, which is particularly susceptible to the effect of acid. Over the years, the acid rain simply dissolves the surfaces of buildings and monuments (see Figure 19–24).

acid rain A phenomenon that occurs when nitrogen and sulfur compounds in the air interact with water to form tiny droplets of nitric and sulfuric acid, which makes raindrops more acidic than normal.

Figure 19–24
The effects of acid rain. Over a period of 60 years, this sandstone statue on a castle in Germany has been completely destroyed.

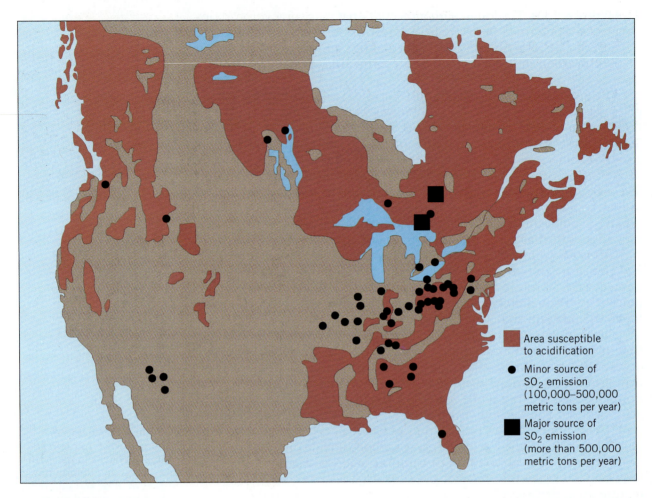

Figure 19–25
Dark areas represent places where the problem of acid rain is most acute.

Area susceptible to acidification

● Minor source of SO_2 emission (100,000–500,000 metric tons per year)

■ Major source of SO_2 emission (more than 500,000 metric tons per year)

Trees near Lake Tahoe, California, have been damaged by a combination of ozone pollution, acid rain, and road salt.

In the middle part of the twentieth century, tall smokestacks were built to deal with the local effects of acid rain and other kinds of pollution in the United States. The effect of these smokestacks was to put pollutants high enough in the atmosphere to be taken away by prevailing winds. But, in keeping with the fact that atoms and molecules never disappear, that approach didn't really solve the problem; it merely displaced it. The nitrogen and sulfur compounds emitted in smokestacks of the Midwest simply fell as acid rain on the more distant forests of New England and Canada (see Figure 19–25).

The government's response to urban air pollution has centered on reducing the levels of emissions associated with the burning of fossil fuels. In California, for example, by 1998, 2% of the cars sold in the state will have to be electric cars that are emission-free (although the plants that generate the electricity to run the cars still produce pollution). Because of the strength of the California automobile market, these laws will probably lead to the rapid improvement of electrical car design. At the same time, large facilities such as power plants, which emit huge amounts of pollutants, will be required to install complex engineering devices known as *scrubbers*. Scrubbers remove sulfur compounds from the smokestack before they become part of the atmosphere.

(a)

(b)

Acid rain and air pollution are examples of moderate environmental problems. We understand the problems and consequences of pollution, and we know what has to be done to prevent it. The costs of dealing with these problems, however, are considerably higher than the costs of reversing the depletion of the ozone layer. Political and economic questions become very important. How much are we willing to pay for clean air? How much is the preservation of a mountaintop in New England worth compared to the jobs that might be lost by closing down an outmoded factory in the Midwest? These are not easy questions, nor are they questions that can be answered by science alone.

Smokestacks equipped with an electrostatic precipitator create an electric field that attracts ash and soot and collects them before they can pollute the atmosphere (a). When the precipitator is turned off (b), thick clouds rise from the stacks.

The Greenhouse Effect

The temperature at the Earth's surface is determined in large measure by the extent to which gases in the atmosphere absorb out-going infrared radiation (see Chapter 8). The atmosphere is largely transparent to the Sun's incoming visible and ultraviolet radiation, which warms the surface, but it is somewhat opaque to the infrared (heat) energy that radiates out into space (see Figure 19–26). Thus, like a greenhouse, the atmosphere traps some of the Sun's energy and raises the temperature of the Earth. Without the atmosphere, the Earth's average temperature would be about $-20°C$, compared to its present average of about $+20°C$. The temperature increase that results from atmospheric trapping of heat energy is known as the **greenhouse effect.**

greenhouse effect   A global temperature increase caused by the fact that the Earth's atmospheric gases trap some of the Sun's infrared (heat) energy before it radiates out into space.

Solar radiation passes through the greenhouse glass and is converted to heat, which is trapped within the greenhouse.

Solar radiation passes through the atmosphere and is converted to heat, which is trapped within the planetary atmosphere.

Solar radiation

Atmospheric shell of CO_2 and other greenhouse gases

Panes of greenhouse glass

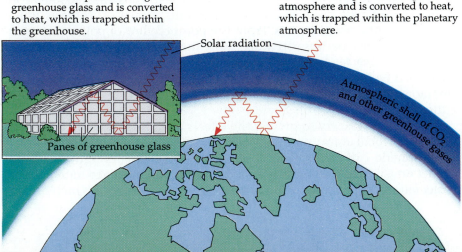

Figure 19–26
The greenhouse effect. Just as the Sun's energy passes through the glass of a greenhouse and becomes trapped inside as heat, so too does the atmosphere act as a greenhouse to warm up the Earth.

Many atmospheric gases contribute to the greenhouse effect. Most newspaper accounts emphasize the role of carbon dioxide, certainly an important greenhouse gas, but CO_2 accounts for only about 10% of the total infrared absorption. Water vapor, especially in clouds, is the dominant greenhouse gas, while the trace gases methane and CFCs, which make up less than a few millionths of the atmosphere, are molecule-for-molecule the most efficient infrared absorbers.

When people talk about the greenhouse effect these days, they are usually concerned about possible increases in average global temperatures. These temperature increases could result from the increasing concentration of human-generated carbon dioxide and other industrial gases in the atmosphere. While smokestack emissions can be scrubbed to reduce the sulfur and nitrogen compounds that cause acid rain, it is not possible to remove carbon dioxide created when fossil fuels are combined with oxygen. The only way to release energy from fossil fuel, after all, is to liberate chemical potential energy by the production of water and carbon dioxide.

Figure 19–27
Measurements of atmospheric CO_2 concentrations made at the Mauna Loa Observatory in Hawaii, reveal both an annual cycle and a gradual increase.

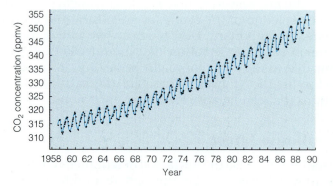

The graph in Figure 19–27 shows the amount of carbon dioxide in the Earth's atmosphere, as measured on a high mountaintop observatory in Hawaii during the past four decades. This graph shows that since the detailed measurements were begun, the amount of carbon dioxide has increased steadily year by year. The small wiggles in the graph correspond to the annual cycle in which leaves take in carbon dioxide for photosynthesis in the spring, and return carbon dioxide to the atmosphere in the fall. (In a sense, the wiggles in this graph show the Earth breathing.)

Without question, burning fossil fuels has increased the amount of carbon dioxide in the Earth's atmosphere. All scientists agree on this fact, and all scientists also agree that carbon dioxide absorbs infrared radiation. Once we get past these two basic statements, however, there is much disagreement among experts. From the point of view of policy, the questions that must be answered are: Will the increased concentration of carbon dioxide lead to an increase in the Earth's temperature? If so, what will be the social and economic consequences? These questions are extraordinarily difficult to answer based on currently available scientific knowledge.

Scientists who try to predict the effects of carbon dioxide concentrations on global warming make their calculations for the case in which atmospheric carbon dioxide doubles. In 1990, the Intergovernmental Panel on Climate Change attempted to make the most accurate possible calculations, based on the combined judgment of the world's meteorological community on this question. Their prediction was that a doubling of CO_2 would lead to a global warming of 1°C to 5°C by the mid twenty-first century.

Scientists are engaged in many ongoing arguments regarding the nature and consequences of the greenhouse effect. Many scientists mistrust the validity of all calculations undertaken so far. Some fear more catastrophic levels of warming will occur. Others doubt that warming will take place at all. The reasons for these uncertainties stem from the fact that our only way of describing the Earth's atmosphere is through the global circulation models described earlier in this chapter. These models are, at best, imperfect ways of predicting changes in climate. For one thing, they break the atmosphere into unrealistic, uniform chunks several hundred miles on a side. Such a coarse-grained look at the atmosphere cannot realistically deal with effects of clouds that are typically only a few miles on a side.

Clouds play a critical role in controlling global temperatures. More clouds in the atmosphere reflect more sunlight into space. You experience this phenomenon whenever a cloud comes between you and the Sun; you immediately feel cooler than you did when the cloud wasn't there. If increased carbon dioxide increases the temperature of the atmosphere slightly, one consequence might be increased evaporation of water from the ocean, and, hence, an increased formation of clouds. Thus some scientists argue that clouds produce an automatic feedback that counteracts the effects of carbon dioxide. Indeed, when a group at the British meteorological office tested their predictions using global circulation models and the ordinary description of clouds, versus a prediction using a somewhat more realistic description of how clouds interact with solar radiation, they found that the predicted warming increment dropped from 5°C to 2.5°C.

Another important effect that is difficult to incorporate into global circulation models is that of the world's oceans. A constant interplay takes place between water and the atmosphere at the ocean's surface, and carbon dioxide moves into and out of the ocean all the time. In fact, the amount of carbon dioxide locked in the oceans and seafloor sediments is much greater than that stored in the atmosphere. Even small changes in the way that oceans interact with atmospheric carbon dioxide can have huge effects on the world's climate. In addition, as we saw earlier, ocean currents are instrumental in spreading heat around the surface of the Earth. Small changes in those currents could have enormous effects on the Earth's climate.

Many unknown effects might, in the ways just described, mitigate the effects of added carbon dioxide in the atmosphere and lower the warming due to the greenhouse effect. It is also possible, however, that unknown effects might work in the opposite way and increase the warming. At the moment, we simply do not understand the workings of the atmosphere well enough to know.

Why do so many conflicting arguments arise about the effects of additional greenhouse gases on the climate? The main reason is that the Earth has experienced a large greenhouse effect for billions of years. What we are trying to predict with our models is not whether a greenhouse effect will occur, but rather how the existing greenhouse effect will be modified by a small change in atmospheric composition. What we need to know is the likelihood of a relatively small change in the existing greenhouse effect due to the addition of human-made pollutants to the atmosphere. Such a minor change in the global climate is extremely difficult to predict, and it requires a much more detailed understanding of the way the Earth works than we have at the present time.

The range of the possible consequences of greenhouse warming is also the subject of much debate. As a general rule, it appears that for every half degree Celsius of greenhouse warming, a global line of a given temperature will move about 100 miles northward in the northern hemisphere. Thus, for 2°C warming, temperatures in Washington, D.C., will be comparable to those in Atlanta, and temperatures in Minneapolis will be comparable to those in St. Louis. Depending on how large you expect the greenhouse warming to be and how quickly you expect it to happen, the effects on the Earth's biosphere and ecosystems can be large or small. The total warming in the Northern Hemisphere after the last ice age, for example, was about 5°C, and took place over a period of several thousand years. We know from studies of pollen deposited in the bottom of lakes that this warming, large as it was, was sufficiently gradual that plant populations were able to adapt and migrate north with the retreat of the glaciers. More recently, studies of the northern Atlantic Ocean have indicated that there have been periods in which the temperature in that region has changed by 5°C over a much shorter period, perhaps as little as a few decades. No known ecological disasters appear to be associated with these events. The predictions of consequences of greenhouse warming, should it occur, are thus also surrounded with a great deal of uncertainty.

THINKING MORE ABOUT CYCLES

Global Warming

In the opinion of the authors, the greenhouse effect seems to be the most difficult and the most potentially alarming of the many environmental problems that face the global ecosystem. On the one hand, it is the most difficult to model because the effect of adding carbon dioxide to the atmosphere is uncertain and the cost of doing something about it is very high. The idea that you can take the world economy, which runs almost entirely on fossil fuels, and change it over to other sources of energy in a very short period of time is unrealistic. In the past, it has taken many decades to make similar changes in a society's energy use and consumption. Figure 19–28 shows transitions from wood to coal and from coal to oil and gas in the United States. As you can see, it takes 30 to 50 years for a new fuel to work its way into the economy. If the more disastrous predictions of greenhouse warming are true, in about 50 years, the warming will already have occurred and it will be too late to do anything about it.

The best scientific estimates now indicate that if warming occurs, it will not become evident for several decades, far beyond the planning horizon of corporations, governments, and other major institutions in any society. Human beings find it difficult to suffer real hardship in the present to prevent an uncertain event from happening in the future. While we are waiting for more definitive answers to these questions, we need to decide how to answer questions such as this: Am I willing to give up driving my car three days a week because a possibility exists that for global warming will take place sometime during the lifetime of my grandchildren?

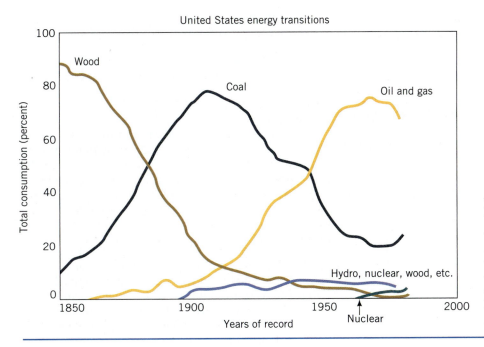

Figure 19–28
U.S. energy transitions. In the past, it has taken 30 to 50 years to make the transition from one type of fuel to another. We could expect a transition to solar energy, for example, to take about this long.

► Summary

Matter that forms the outer layers of the Earth follows many cycles, driven by the energy of the Sun and the Earth's inner heat. The *atmospheric cycle* of the *weather* redistributes solar energy from the warmer regions near the equator to higher latitudes through the development of global convection cells of air. The prevailing westerly flow of weather across North America marks one of these large cells, while the *jet stream* delineates the boundary between this flow and the contrary cell to our north. The *climate*, in contrast to weather, varies much more slowly in response to ocean circulation, the Sun's energy, the positions of continents and mountain ranges, and other relatively fixed conditions.

The *hydrological cycle* traces the path of water as it evaporates from the oceans, falls back to Earth as rain, and forms lakes, rivers, *ice caps*, *glaciers*, and *groundwater* reservoirs. During unusually cold climatic periods, more water falls as snow, creating a white reflective blanket that further reduces the amount of absorbed solar radiation. This situation, if prolonged, can lead to an *ice age*, during which ocean levels drop significantly and great sheets of ice cover the land at high and middle latitudes. Temperatures in these high latitudes may be moderated, however, by ocean *currents* that are important in redistributing heat energy at the Earth's surface.

Human actions are causing changes in the Earth's global *environment*. The use of *chlorofluorocarbons* (*CFCs*) during the past several decades, for example, is having a pronounced effect on *ozone*, a molecule of three oxygens found as a trace gas in the *ozone layer* of the upper atmosphere. Ozone provides important protection on Earth from the Sun's harmful ultraviolet radiation, but chlorine atoms from CFCs hasten the breakdown of ozone molecules, thus creating a growing *ozone hole*.

Burning untreated coal and other fossil fuels releases sulfur and nitrogen compounds into the atmosphere—chemicals that contribute to *air pollution* and *acid rain*. Dealing with acid rain will require cleaning up emissions in automobiles and generating plants.

Carbon dioxide, a necessary product of all combustion of carbon-based fuels, adds to the atmosphere's store of infrared-absorbing gases, and thus contributes to the *greenhouse effect*. At present, scientists are not able to predict the consequences of these global changes with much certainty.

► Review Questions

1. What happens to matter during a cyclical process?

2. What is the difference between weather and climate? Describe the weather and climate of your area.

3. What is the prevailing direction of the jet stream? How does it influence your weather?

4. What are the principal repositories of water in the Earth?

5. How does the atmosphere distribute heat across the Earth's surface?

6. What are the differences between ice caps and glaciers?

7. What factors might cause glaciers to advance from polar areas to more temperate zones?

8. How do ocean currents affect local climate?

9. Why do we call groundwater in most areas a "non-renewable resource"?

10. What is ozone? Why is it important to life on Earth?

11. What is the ozone layer? Where is it located?

12. What is the ozone hole? How have governments responded to the discovery of the ozone hole?

13. Why is ozone part of the urban air pollution problem?

14. What is acid rain? What steps would have to be taken to solve the acid-rain problem?

15. What is the greenhouse effect?

16. Why are predictions of global warming so uncertain?

17. What steps would have to be taken to reduce the severity of the greenhouse effect?

18. Explain how the rotation of the Earth affects climate.

19. What is a prevailing westerly, and where is one found?

20. What is an ice age?

21. Are we in an ice age now?

22. What are the important characteristics of a hurricane?

23. What are the important characteristics of a tornado?

24. What are the important characteristics of a monsoon?

25. What are the important characteristics of a typhoon?

26. Who was Mulitin Milankovitch and what did he do?

27. What is the purpose of the global circulation model?

28. What is El Niño, and why is it so named?

29. What is the danger posed by CFCs and how is this problem being addressed?

30. How do gyres alter weather patterns on land?

31. What topographical features influence local weather?

▶ Fill in the Blanks

Complete the following paragraphs with words and phrases from the list.

acid rain	hydrological cycle
atmospheric cycle	ice ages
climate	ice caps
currents	jet streams
front	ozone hole
glaciers	ozone layer
greenhouse effect	residence times
groundwater	weather

The motion of the air around the Earth constitutes the _____. This cycle is involved in short-term effects such as _____ and long-term effects such as _____. When air masses of different temperatures meet, they form a _____ and the weather changes. _____, which are fast-moving currents of air that divide arctic from temperate air masses, also play a major role in the weather.

The movement of water around the Earth constitutes the _____. Ocean _____ redistribute heat on the surface. Elements spend different _____ in the ocean before they are removed by chemical reactions. Most fresh water resulting from rainfall is stored as _____, although some is frozen and taken into _____ and _____. Periods when glaciers advance are known as _____.

Three major environmental problems are _____, resulting from sulfur and nitrogen compounds in the air; the destruction of the _____ by chemical reactions catalyzed by chlorine, which gives rise to the annual _____ over the Antarctic; and the _____, resulting from increasing amounts of carbon dioxide and other gases in the atmosphere.

▶ Discussion Questions

1. The thickness of polar ice caps depends critically on the location of continents; much thicker ice can accumulate on land than in water. Where is the only polar continent now?

2. How do oceans redistribute the Earth's heat? How does the atmosphere accomplish this? Do rocks redistribute heat? Which global cycle do you think is most efficient in transferring heat? Why?

3. You can often distinguish surface ocean currents because their color is different from the surrounding ocean. Why do you think this is so?

4. What environmental changes might result from a global warming of 2°C? What countries might be most affected?

5. How did you affect your environment today? How did it affect you?

6. What would the political, social, and economic consequences be in your community if serious steps were taken to reduce acid rain or air pollution?

7. Where do electrical cars obtain their energy? Is it true that they do not generate pollution because they have batteries?

8. Describe the climate where you live. Describe this in relation to the Earth's rotation.

9. Retell the carbon cycle from the perspective of a carbon atom.

10. What are the two primary heat sources of the Earth that drive atmospheric, meteorological, and geological cycles? Which heat source drives which cycle? Explain.

11. Why would the prevailing winds in the Southern

Hemisphere come from the South Pole if the Earth did not rotate?

12. Outline the prevailing wind directions according to latitude for the Earth. Start at the equator and continue in steps of 10 degrees in latitude north, then south.

13. Compare and contrast the circulation pattern, cloud coverage, relative density and temperature of the air, and any other significant feature between high and low pressure areas in the Northern and Southern hemispheres.

14. The El Niño weather cycle affects Pacific Basin countries such as Australia and India, and the countries of Central and South America, as well as the southwestern United States. It is a result of a combination of a meteorological and water cycle. Outline this complicated process and its effects on the weather on Pacific Coast states such as California.

15. Retell the Milankovitch cycles in your own words.

16. What are some of the consequences of the ocean currents in the North Atlantic? In the North Pacific?

17. Using the residence times for sodium, chlorine, calcium, and gold given in Table 19–1, retell a typical chemical cycle in the ocean from the point of view of those atoms.

18. How are oceanic chemical cycles responsible for the constant salinity of the oceans over long geologic time scales?

19. What are some of the consequences of acid rain on buildings, trees, lakes, and other natural or human-made structures?

▶ Problems

1. How much copper is in a cubic kilometer of seawater? (*Hint:* Refer to Table 19–1.)

2. How much trash have you generated in your lifetime, assuming you have produced 40 tons per year? What is the volume of that trash if compacted?

3. Assuming that there is a global warming of 1°C, what kind of weather will your town have after the warming has taken place? (*Hint:* Find a town to the south of you that has that sort of weather now.)

4. The average office worker generates 250 pounds of paper per year. The density of this paper is about 1.2 g/cm³. How large a volume is this?

5. Marco decided to start a recycling center for his physical science class. He surveyed the class of 50 students and they decided to store crushed aluminum cans for the semester. A crushed aluminum can is approximately 1 cm high with a circular base of radius 2.5 cm. The instructor gave Marco a storage room that is 2 m high and 2.5 m × 2.5 m square. How many aluminum cans can Marco's class store over the course of a semester (120 days)? Is this a reasonable project for the class?

6. Stacy and Rebecca decided that it would be a wonderful summer project to obtain 1 kg of gold from sea water, just to have it for their life's savings. How much seawater (in gallons) must Stacy and Rebecca process to obtain this amount of gold? Is this reasonable on their part?

▶ Investigations

1. Look at some weather maps in your local newspaper over a period of several weeks and determine if looking at weather patterns to the west of your location is a good predictor of your weather. Why should this be so?

2. Where does your water come from at your college? Is the water processed or treated in any way? How long is that source of water expected to last? What alternatives exist if that supply is totally depleted?

3. Investigate the biological cycle of calcium in your body. Where in your body are calcium atoms used? How often are they replaced? How much calcium do you need to consume each day? What are the best food sources of this element?

4. What are the "doldrums" and how do they form? What role have they played in poetry and literature?

5. Do commercial jet airliners fly through the ozone layer? What effects might this have?

6. How are solid wastes such as plastics handled in your community? Does that method represent a long-term or short-term solution to the problem? What other options are available to your community?

7. Write a short story chronicling the passage of a carbon atom through 10 different stages.

8. What is an environmental impact statement? Are there any projects in your area that have required environmental impact statements? Where should they be required?

9. What happens to your garbage? Is there a recycling program in your town? What materials can be recycled?

10. Read about the "pea soup fogs" that used to plague London. What caused them? How many people died during them?

11. Investigate the history of the climate in your area during the past 10,000 years. How do scientists determine the climate from so long ago?

▶ Additional Reading

Bascom, Willard. *Waves and Beaches*. New York: Doubleday, 1964.

Trefil, James. *A Scientist at the Seashore*. New York: Scribners, 1984.

Trefil, James. *Meditations at Sunset*. New York: Scribners, 1987.

van Andel, Tjered. *Tales of an Old Ocean*. New York: Norton, 1974.

20 HISTORICAL GEOLOGY

ROCKS REVEAL THE HISTORY OF THE EARTH'S
LANDFORMS AND THE EVOLUTION OF LIFE.

Changes

Look around your school's campus. It probably hasn't changed much in the past few years. You see many of the same roads and buildings that were there 50 years ago. The living things around campus haven't changed much, either: the same trees, fields, lawns, and even a few professors have been around for decades. And, of course, the rivers, lakes, mountains, plains, and other landforms that surround your school look pretty much as they did when humans first settled your area.

Although our physical surroundings seem permanent, this sense of permanence is an illusion. Rocks at the Earth's surface tell stories of sweeping changes. The place where you are now sitting was probably, at one time, deep under the water of an ocean or sea. At another time, it may have been part of a great mountain range or a parched, dry desert. The seemingly unchangeable life around you appeared vastly different in the past, as well. Ancient rocks reveal the remains of creatures far stranger than anything imagined in myths and fables.

Our planet has a magnificent, epic history, but one that is not easily told. To decipher that tale we must turn to the awesome testimony of the rocks.

THE ROCK CYCLE

rock cycle   An ongoing cycle of internal and external Earth processes by which rock is created, destroyed, and altered.

igneous rock   The first rock to form on a cooling planet, solidified from hot, molten material; intrusive or extrusive (volcanic).

volcanic rock   Extrusive igneous rock that solidifies on the surface of the Earth.

intrusive rock   Igneous rock that cools and hardens underground.

When the Earth first formed, there were no rocks. Over 4 billion years ago, the *great bombardment* released prodigious amounts of energy as swarms of meteorites crashed into the growing planet, converting gravitational potential energy into kinetic energy and heat (see Chapter 17). That heat produced a molten ball orbiting the Sun. There were no continents, no oceans, and no atmosphere. Only when the bombardment subsided and the Earth began to cool did rocks appear. First, as the temperature dropped below the melting point of the surface rocks, the outer crust of the Earth must have gradually solidified like the first layer of ice on a pond in winter. Then, when surface temperatures dropped below the boiling point of water, the first rains must have fallen. Together, these two events began the **rock cycle,** which is the cycle of internal and external Earth processes by which rock is created, destroyed, and altered.

Igneous Rocks

Igneous rocks, which solidify from a hot liquid, were the first to appear on the Earth. They come in two principal types. **Volcanic rocks** (also known as *extrusive rocks*) solidify on the Earth's surface in volcanic eruptions (see Figure 20–1). Volcanoes are by far the most spectacular of all rock-forming events, as red-hot fountains and glowing rivers of lava add to the slopes of the growing cone-shaped mountain. The most common variety of volcanic rock is *basalt*, a dark, even-textured rock rich in oxides of silicon, magnesium, iron, calcium, and aluminum. Basalt makes up most of the rock in Hawaii, as well as most of the new material formed at mid-ocean ridges. Other volcanoes feature rocks richer in silicon; if these magmas mix with a significant amount of water or other volatile (easily vaporized) substance, the volcanic rock can become the frothy rock *pumice*.

Igneous rocks that harden underground are called **intrusive rocks.** Dark-colored basalt often exploits underground cracks near volcanoes to form layers or sheets of igneous rock. The Palisades on the Hudson River near New York City formed in this way. *Granite* is perhaps the most common intrusive rock in the Earth's crust. It is lighter in color and

Figure 20–1
Igneous rocks. (*a*) Devil's Tower in Wyoming represents the neck of a would-be volcano that never reached the surface. The surrounding sediments have subsequently eroded away. (*b*) Granite from a quarry in Barre, Vermont, solidified from magma deep underground.

(*a*)

(*b*)

Lava flow from a Hawaiian volcano in 1983 is an example of the formation of extrusive igneous rock.

density than basalt. Hard, durable granite, with its attractive pale grayish or pinkish color and speckled array of light and dark minerals, makes an ideal ornamental building stone. New England is particularly famous for its many fine granite quarries.

Igneous rocks are still being formed on Earth; for example, new plate material is formed at diverging boundaries (see Chapter 18) or in active volcanoes. In other places, such as the Yellowstone National Park region, *hot springs* and *geysers* reveal hidden sources of underground heat and may indicate places where intrusive igneous rocks are forming today.

An actual-size photograph of granite reveals the uniform texture of its grains.

The Mississippi River Delta is an active site where new sediments are being deposited.

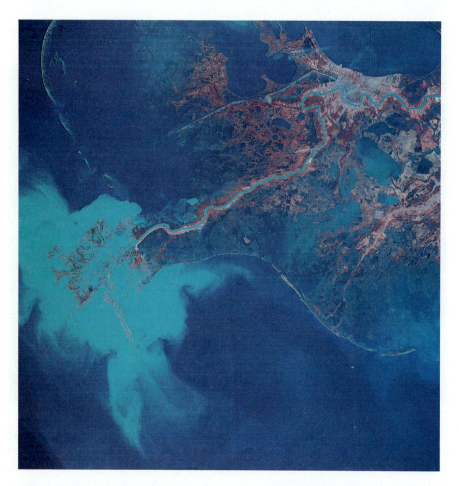

Sedimentary Rocks

When the first rains began to fall on the first igneous rocks, the process of weathering began. Small grains washed off the recently hardened volcanic rocks, flowed down through streams and rivers into the ocean, and were deposited on the seafloor when the fast-moving waters of the river met the slower currents of the ocean. Over time, layers of this sediment accumulated, especially at the mouths of rivers near the shores of the Earth's new oceans. As more and more sediment collected, these layers became thicker and thicker. In many places on the Earth right now, such as the Mississippi River delta that extends into the Gulf of Mexico, layers of sediment may reach several kilometers in thickness.

As the first sediments were buried deeper and deeper, they were subjected to the high temperatures and pressures of the Earth's interior. In addition, water flowed through the layers of sediments, dissolving and redepositing gluelike chemicals, which can be compared to the crusty deposits that can build up on an ordinary faucet when water drips continuously. The net result of all of these processes—pressure, heat, and the effects of mineral-laden water flowing between the grains—was to weld the bits of sediment together into new layered rocks. This kind of rock, called **sedimentary rock,** is made up of grains of material worn off previous rocks (see Figure 20–2). Other common sedimentary rocks, including salt deposits, may form from layers of chemical precipitates.

sedimentary rock A type of rock that is formed from layers of sediment produced by the weathering of other rock or by chemical precipitation.

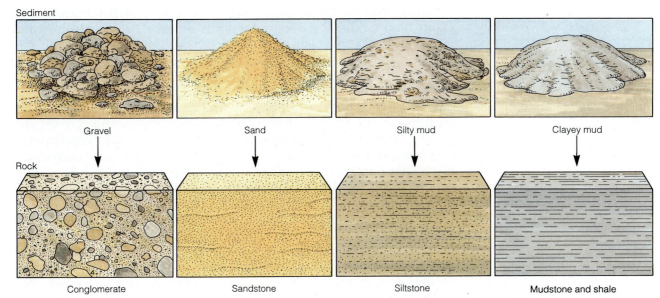

Sediment

Gravel   Sand   Silty mud   Clayey mud

Rock

Conglomerate   Sandstone   Siltstone   Mudstone and shale

Figure 20–2
The origin of different
kinds of sedimentary rocks.

While uniform sedimentary rocks can form at the base of a single
mountain or cliff, the collection of grains often comes together from
many different places. The grains in a single fragment of sedimentary
rock being formed in the Mississippi River delta, for example, may have
come from a cliff in Minnesota, a valley in Pennsylvania, and a mountain
in Texas. Similarly, sediments deposited near the mouth of the Colorado
River carry bits of history from much of the North American West. Deltas
inevitably contain particles from all the rocks in their rivers' drainage area.

As you travel across the United States, you will encounter many
common varieties of sedimentary rock. They are easy to spot in highway
road cuts and outcroppings of rock because of their characteristic layered
appearance, like the pages of a book or a many-layered cake (Figure 20–3).
Sandstone forms mostly from sand-sized grains of quartz (silicon dioxide,
or SiO_2), the most common mineral at the beach, and from other hard
mineral and rock fragments. Many sandstones represent ancient beaches,
deserts, or streambeds—places where concentrations of sand are found
today. Sandstones usually feel rough to the touch, and you can just barely
see the individual grains that have been cemented together.

Figure 20–3
A spectacular example of
sedimentary rocks in Utah.
The different colored bands
correspond to layers of different kinds of materials being deposited on the floor
of a long-vanished ocean.

Shales and *mudstones* form from sediments that have much finer grains than sand. These rocks commonly accumulate beneath the calm waters of lakes or in deep ocean basins, places often teeming with life. These organisms, both large and small, die and are buried in muddy ooze, where they may eventually form into fossils that provide us with much information about the evolution of life on Earth.

Limestone, another distinctive type of sedimentary rock, forms from an accumulation of the calcium carbonate ($CaCO_3$) skeletons of sea animals or from a chemical precipitate of calcium carbonate. Some types of limestone form from a gradual rain of microscopic debris or broken shells, while others represent a coral reef that spread across the floor of a shallow sea (see Figure 20–4). Like shales and mudstones, limestones commonly bear fossils.

Sedimentary rocks that originally formed at the bottom of the ocean are often folded and uplifted by the motions of the Earth's tectonic plates. Indeed, it is not at all unusual to see sedimentary rocks in mountain passes thousands of meters above the ocean, or in the middle of continents thousands of kilometers from the nearest open water. For example, dramatic exposures of reef limestone form the cliffs of Lookout Mountain high above Chattanooga, Tennessee, as well as the magnificent peaks of the Canadian Rockies in Alberta and British Columbia.

One of the best places to gain an appreciation of sedimentary rocks is at the beach. If you pick up a handful of sand, you will notice that each grain is different. Some are dark or deeply colored, others are white or transparent. Some have sharp, angular edges, while others are smooth and worn down. Each of these grains of sand was once part of a rock in a drainage of a river that feeds into the ocean. As the rock weathered away, each grain was chipped off and carried to the sea by wind and water. Eventually, the grains of sand you hold in your hand will form into solid rock—sandstone that will be subjected to the forces of plate tectonics. That sandstone may someday be uplifted to an altitude well above sea level, where the grains will be weathered again, thus starting the whole cycle over. Thus each grain of sand in your hand may have made the trip from rock to beach to sandstone many times in its existence.

Figure 20–4
The Great Barrier Reef in Australia is a site of active limestone formation.

Figure 20–5
The Green Mountains in New Hampshire are made of metamorphic rocks—rocks that have been altered by intense pressure and temperature deep within the Earth and then uplifted and eroded into mountains.

The story of the grain of sand provides a good model for the way materials move about the surface of the Earth. The atoms of rocks, just like those of the air, water, or your body, as we saw in Chapter 19, are always shifting around, but the atoms being recycled are always the same.

metamorphic rock   Igneous or sedimentary rock that is buried and transformed by the Earth's intense internal temperature and pressure.

Metamorphic Rocks

Sedimentary rocks may be buried deep within the Earth and subjected to intense pressure and heat. There, they will be turned into yet another kind of rock, transformed by the Earth's extreme conditions into **metamorphic rock**(Figure 20–5). If shales or mudstones are buried like this, they may eventually turn into brittle, hard *slates*, which are the kind of rock that school blackboards were made from in the past. Even higher temperatures and pressure can transform slates into spectacularly banded rocks, called *schists* and *gneisses* (pronounced "nices"), which often boast fine crystals of garnets and other high-pressure minerals. Highway road cuts and outcrops of these metamorphic rocks can look like an intensely

(*a*) A grain of garnet grew in gneiss, a metamorphic rock, at high pressure and temperature. The garnet contains small inclusions of minerals from an earlier rock. (*b*) An actual-size photograph of metamorphic gneiss displays a characteristic banding and alignment of mineral grains.

(*a*)

(*b*)

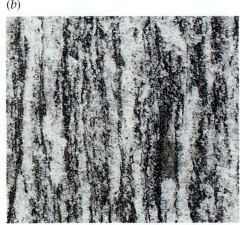

Quartzite is a metamorphic rock that consists almost entirely of recrystallized sand grains. The arrow points to faint traces of an original rounded grain.

folded cloth, or a giant cross-section of swirled marble cake. Sandstones, when exposed to high temperature and pressure, also metamorphose, recrystallizing to a durable rock in which the original sand grains fuse into a solid mass known as *quartzite*.

The Story of Marble

Of all the metamorphic rocks, none tells a more astonishing tale than *marble*, a rock of extraordinary beauty. If you ever travel the roads of Vermont, chances are you will pass an outcrop or road cut of distinctive greenish-white cast—a rock with intricate bands and swirls. These marbles take a high polish and have been prized for centuries by sculptors and architects. But no works of humans can match the epic process that formed the stone.

Most marbles began as limestone, which are rocks that originate primarily from the skeletal remains of sea life. Over the ages, limestone was deeply buried, crushed under the weight of many kilometers' worth of sands, shales, and more limestones in an ancient sea. But no ocean or sea can last forever on our dynamic planet. An ancient collision of the Eurasian and North American plates compressed and deformed this ocean basin, crumpling the layered rock into tight folds. During the intense pressure and high temperatures associated with the converging tectonic plates, the limestones were metamorphosed to the marble that we use today. Then the buckled and contorted sedimentary formations were uplifted to high elevations when the Appalachian Mountains formed. Many millions of years of erosion and uplift have exposed these ancient rocks on the Earth's surface, where they have gradually weathered away and started the cycle again. And humans, in a futile quest for immortality, quarry the marble for their monuments and tombstones and other transient reminders of Earth's incessant change.

The Lincoln Memorial is one of many famous monuments carved from marble, a metamorphosed limestone.

Igneous, sedimentary, and metamorphic rocks all participate in the rock cycle (see Figure 20–6). Igneous rocks, once formed, can be weathered to form sedimentary rocks, and can themselves undergo metamorphism. Layers of sedimentary rocks can be transformed into metamorphic rocks, as well. All three kinds of rocks can be *subducted* into the Earth's mantle (see Chapter 18), partially melted, and reformed as new igneous rocks. Thus the rock cycle never ceases.

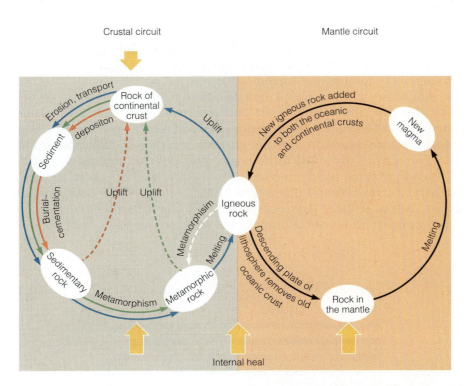

Crustal circuit   Mantle circuit

Figure 20–6
The rock cycle. The igneous rocks can be weathered to form sedimentary rocks which can, through the action of heat and pressure, become metamorphic rocks. At any stage in the cycle, rocks can be subducted and returned to the Earth's interior to start the cycle again.

Science in the Making

James Hutton and the Discovery of "Deep Time"

Near the town of Jedburgh, Scotland, there exists a curious cliff that reveals vertical layers of rock overlain by horizontal layers (see Figure 20–7). How could such a sequence have occurred? In the final decades

Figure 20–7
James Hutton recognized the immense spans of time required to form this spectacular outcrop of rock at Siccar Point, Berwickshire, in Scotland. Vertical sedimentary rocks are capped by horizontal layers.

of the eighteenth century, Scottish scientist James Hutton (1726–1797), a man who is often called "the father of modern geology," studied this remarkable cliff and realized that he was seeing the result of an incredibly long period of geological turmoil.

Knowing what you now know about sedimentary rocks, you will realize that this cliff represents the end product of a long chain of events (Figure 20–8). First, a series of sedimentary rocks was laid down in the usual horizontal fashion, one flat layer on top of another. Then, some tectonic activity disrupted those layers, breaking and folding them until they were tilted nearly vertically. As the result of still more tectonic activity, the rocks found themselves at the bottom of an ocean, and another layer of sedimentary rocks formed on top of them. Finally, an episode of uplift and erosion brought the rocks to our view.

Figure 20–8
This series of sketches shows several stages in the history of the Jedburgh outcrop. (*a*) Layers of sediment were gradually deposited in water. (*b*) Those sediments, deeply buried, were compressed and tilted during tectonic activity. (*c*) Uplift brought those tilted sediments to the surface, where they were partly eroded. (*d*) The rocks subsided, and a new cycle of sedimentation began.

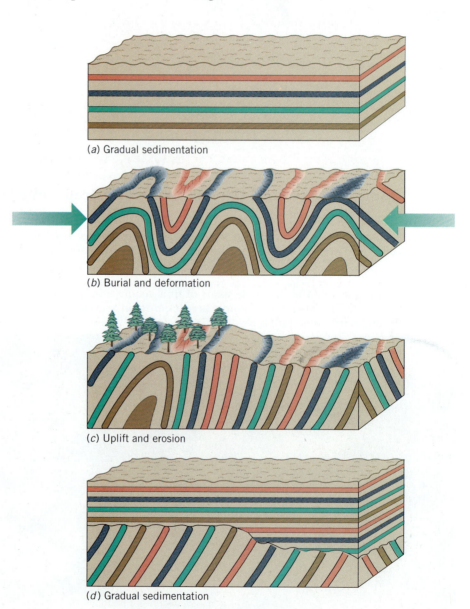

(*a*) Gradual sedimentation

(*b*) Burial and deformation

(*c*) Uplift and erosion

(*d*) Gradual sedimentation

Hutton, having deduced the complex history of the Jedburgh rocks, realized that geological forces must have been operating for a very long time. Each step of the formation process—gradual sedimentation, burial, folding, uplift, more sedimentation, and so on—required countless generations, based on observations of ongoing geological processes. In the words of nature writer John McPhee, Hutton had discovered "deep time." In order for a formation like the one at Jedburgh to exist, the Earth had to have existed not for thousands of years or even hundreds of thousands of years, but for many millions of years.

Today, we know that the age of the Earth is calculated in billions of years, and the existence of structures like this are not surprising. At the time of its interpretation by James Hutton, however, the rocks at Jedburgh provided a totally new insight into the inconceivable antiquity of our planet.

In the words of Hutton, the testimony of the rocks offered

"No vestige of a beginning, no prospect of an end." ●

Developing Your Intuition

The Interdependence of Earth Cycles

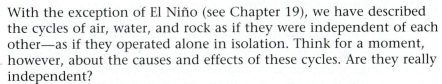

With the exception of El Niño (see Chapter 19), we have described the cycles of air, water, and rock as if they were independent of each other—as if they operated alone in isolation. Think for a moment, however, about the causes and effects of these cycles. Are they really independent?

We saw from the description of El Niño that the interaction of wind and ocean currents affects weather and rainfall. The amount of rainfall in a given location affects the rate of erosion and thus the amount of sediment being deposited in deltas. The deposition rate, in turn, affects the amount of sedimentary rock being formed. In this way, the atmospheric and water cycles affect the rock cycle. Similarly, the breakdown of rock is essential to the formation of soils in which

A symmetrical fan of sediments forms where a stream enters Death Valley, California, from a steep mountain canyon.

plants grow. The presence of plants, in turn, affects the absorption of sunlight at the Earth's surface and thus the energy balance that controls the movement of the winds and ocean currents. Thus, although each of the three cycles operates on a different time scale, they constantly influence each other.

Finally, over hundreds of millions of years, the global cycle of plate tectonics, which controls the distribution of the Earth's mountains and oceans, influences all other cycles. ●

THE EVOLUTION OF LIFE

evolution   The process by which new life forms arise and change.

The diversity of life on our planet is enormous. You could well have seen hundreds of different kinds of plants and animals in the last 24 hours, whether you took note of them or not. Biologists estimate that there may be more than 10 million species on our planet right now. Examining the records of the past, we find that life has existed on our planet for billions of years. The process by which new life forms arise and change is called **evolution.**

Evidence for Evolution: The Fossil Record

Today the vast majority of scientists accept evolution as a fact. Like every scientific theory, however, the theory of evolution must be supported by experimental and observational data. A tremendous body of scientific literature is devoted to this subject. Much of the evidence is biological: all living things share the same biochemical and genetic mechanisms, for example. But the most compelling and unambiguous evidence for evolution as a historical fact comes from fossils.

When a plant or animal dies, the remains usually decay. A tree will rot, the carcass of an animal will be torn apart by scavengers and dispersed, a crab shell will be broken up by the action of the surf. Occasionally, however, an organism escapes this fate and is preserved, typically by being buried in sediments and sealed off. In this situation, the hard parts of the organism may remain underground for long periods of time.

fossil   Evidence in rock for past life. Usually a replica in stone of the original organism, created when calcium and other atoms in the hard parts of the buried organism are replaced by minerals in the water flowing through the surrounding area.

As time goes by, two things may happen. First, the material around the organism may go through the rock cycle and be turned into sedimentary rock. Second, minerals in the water flowing through the surrounding area may gradually replace the calcium and other atoms in the buried hard parts, thus creating a **fossil,** which is a replica in stone of the original organism.

When we think of fossils, we usually think of large dinosaur skeletons that are displayed in a museum. Such fossils are the rock replicas we have just described. The term is also used to refer to other kinds of record of past life, such as the imprint of a leaf on mud that changes into rock, or an insect preserved in ancient tree sap, called *amber.*

fossil record   A term that refers to all of the fossils that have been found, catalogued, and studied since human beings first began to study them in a systematic way.

The term **fossil record** refers to all of the fossils that have been found, catalogued, and studied since human beings first began to examine them in a systematic way in the early part of the nineteenth century. This fossil record gives us the best indication of how different organisms came to be what they are. The fossil record of horses, for example, includes a sequence of animals beginning with ones about the size of a cat some

(a)

(b)

50 million years ago, and changing through many intermediate forms up to modern times. Throughout this sequence of fossil mammals, you can see gradual transitions from a small quick animal to a large grazing one.

The fossil record also contains some examples of actual changes in species. In order to do this, the fossil record has to be very complete, with many thousands of years of continuous sediments. Such continuity is rare, but in some instances the actual transitions from one species to another can be documented.

Even so, the major problem with the fossil record is that it is very incomplete. It is estimated that only one species (*not* one individual) out of every 10,000 early life forms is actually represented in the fossil record. Thus, in interpreting the past, we must always be aware that we are dealing with a very small and select sample of what was actually there. This sample is strongly biased toward organisms that were more likely to have been buried soon after death. Therefore, we have a much better record of corals and clams that lived on the continental shelf than we do of insects that flew around primeval forests. Nevertheless, the fossil record was the first (and for a long time the only) evidence that backed up the notion that life is constantly changing and evolving.

Three key ideas have emerged from studying fossils. First, the older the rocks, the more their animal and plant fossils differ from modern forms. Mammals in 5-million-year-old rocks are not terribly different from today's fauna, but few species from 50 million years ago would be familiar, and 150 million years ago dinosaurs rather than mammals were dominant. Similar patterns occur in shells, plants, fish, and all other forms. Often the earlier forms appear to combine characteristics of later organisms. Ancient insects preserved in amber, for example, show some forms that may be intermediate between ants and wasps. Early mammals, similarly, have general mammalian characteristics, but few of the specialized structures that have evolved in such creatures as bats, whales, tigers, and rabbits.

Fossils also display general trends in overall complexity of form. All known fossils from before about 570 million years ago are either single-celled organisms or simple invertebrates such as jellyfish. Marine invertebrates with hard parts, such as mollusks, corals, and crustacea, dominate the record for the next 200 million years or so. Very simple land animals and plants appear next, followed by flowering plants and a much greater

Fossil trilobites. (*a*) This 500 million year old specimen from Utah is typical of the first animals to develop a hard outer coating. (*b*) By about 400 million years ago, some trilobite species had developed bizarre arrays of spines, as in this specimen from Morocco.

(a)

(b)

(c)

Fossils occur in many different forms. (a) A bony fish from Utah; (b) dinosaur footprints; (c) a 38-million year old mosquito in amber; (d) ammonites, an ancient relative of the squid.

(d)

variety of large land animals. This long-term trend toward increasing complexity of organization is consistent with all theories of evolution.

Finally, the fossil record proves beyond a doubt that most species that have lived on Earth have died out and are now **extinct.** Scientists estimate that for every species on the planet today, as many as 999 species have become extinct at some time in the past. They estimate that the average lifetime of a species in the fossil record is about 3 million years. In this view, species, like individuals, are born, live out their lives, and die. This fact alone indicates that some natural mechanism must exist to produce new species as the old disappear.

extinction   The disappearance of a species on Earth.

Two Stages of Evolution

The Earth started out as a hot, lifeless ball of molten rock (see Chapter 17). The first rocks were formed when the planet cooled, but even then the Earth looked nothing like it does today. Water filled the ocean basins, but no fish swam in it and no algae floated on it. All of the millions of different life forms that would some day develop were absent in this early stage.

The transition from a lifeless planet to one that teems with living things came in two stages. The first stage involved the appearance of the first living cell from the lifeless chemical compounds that existed in the early Earth. Chemical evolution was governed by the laws of chemistry and physics.

The second stage in the evolution of life saw the gradual transformation of a single living cell into a wide diversity of complex living things, by a new dynamic process called natural selection. This mechanism for evolution, not proposed until the middle of the nineteenth century, recognized that all living things must compete for a limited supply of resources.

Chemical Evolution: The Miller-Urey Experiment

Chemical evolution is the process by which life arose from nonlife. But how can one start with the simple chemical compounds that were most likely present in the Earth's early atmosphere and wind up with a living cell? This relatively new area of research is one in which there are still many unknowns. Perhaps the most important experiment relating to chemical evolution was performed in 1953 by Stanley Miller (b. 1930) and Harold Urey (1893–1981) at the University of Chicago. The novel apparatus of the **Miller-Urey experiment,** which attempted to reproduce conditions of the early Earth, is sketched in Figure 20–9.

Based on analysis of gases that are released by volcanoes today, scientists have argued that the Earth's early atmosphere contained water vapor, hydrogen (H_2), methane (CH_4), and ammonia (NH_3). Miller and Urey mixed these materials together in a large flask in order to mimic the Earth's early atmosphere. Then, realizing that powerful lightning would have laced the turbulent atmosphere of the early Earth, they caused electric sparks to jump between electrodes in the flask. Could life have arisen in such an environment, they wondered?

After a period of a couple of weeks, Miller and Urey noticed that the liquid in the flask had become cloudy and had started to turn dark brown. Analysis revealed that this brownish liquid contained a large number of amino acids, one of the basic building blocks of life (see Chapter 12). Miller and Urey did not create life in the laboratory, but they did demonstrate that complex molecules essential to life could arise from natural processes.

By the 1950s, scientists had found that amino acids, the building blocks of proteins, could be generated quite easily by natural processes in the atmosphere of the Earth. Since that time, it has been found that energy sources such as ultraviolet radiation (from the Sun) and heat (supplied by volcanoes, for example) will also produce amino acids. In subsequent experiments at the University of Chicago and other labora-

chemical evolution The process by which simple chemical compounds present in Earth's early atmosphere became an organized reproducing cell.

Miller-Urey experiment An experiment performed in 1953 by Stanley Miller and Harold Urey, which showed that a combination of gases, believed to be present in the early atmosphere, and a series of electric sparks, simulating the lightning on the early Earth, will produce amino acids, a basic building block of life.

Figure 20–9
The Miller-Urey experiment. Several of the chemical compounds thought to have been present on the early Earth were mixed and subjected to electrical discharges. Within a few weeks, amino acids had formed.

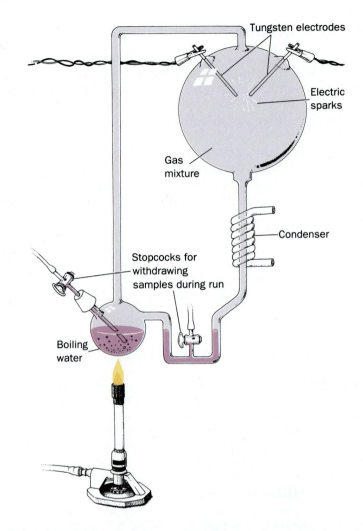

tories, scientists have used modified Miller-Urey devices to make other organic molecules essential to life, including lipids and bases, as well as complex substances such as long protein chains.

Some scientists suggest that this scenario has important implications for the early Earth. For perhaps several hundred million years, they say, the amino acids and other molecules created by the Miller-Urey process were concentrated in the ocean, producing a rich broth, sometimes called the *primordial soup*. Additional amino acids may have been added to the early oceans by other possible sources, as well. For example, amino acids have been found in meteorites. Thus meteorites could have added to, or even provided a substitute for, the Miller-Urey process in the atmosphere. However it happened, it seems clear that the enrichment of organic chemicals in the Earth's early oceans required nothing more than normal chemical processes. This small piece of the chemical evolution puzzle seems to be well understood.

The First Living Cell

Our greatest gap in the evolutionary story comes next. How was it possible for the countless molecules, floating in random patterns in the ocean, to become organized into a living cell? While mechanisms such as condensation polymerization (see Chapter 12) can join simple organic molecules together, sunlight tends to break these bonds apart. Where could large clusters of molecules have formed near enough to the Sun's energy to accumulate concentrations of organic material, but far enough away from direct sunlight to avoid destruction?

Over the years, there have been many conjectures about how the first cell might have formed, but it is safe to say that we do not, at present, understand how this process occurred. Here are some of the current theories.

- The *tidal pool theory* suggests that the first cell formed in an environment shielded from the sun's damaging ultraviolet radiation. Perhaps a cell developed in the protected water a few meters under the surface of a tidal pool, or perhaps the cell formed between the layers of certain kinds of clay. One implication of the tidal pool theory is that life could develop only on planets with large moons, which cause significant tides. Since current thinking is that the formation of a large moon is an unlikely situation, it would imply that life is very rare in the universe.

- The *RNA world theory* holds that nucleic acids (see Chapter 11) developed first and made possible the subsequent development of the cell. This theory follows from the recent discovery that some types of nucleic acid molecules called RNA can themselves act as enzymes for chemical reactions.

- The authors' favorite theory has been nicknamed *the primordial oil slick*. The theory is based on the fact that processes of the Miller-Urey type can produce lipids as well as amino acids. Based on this theory, the early ocean would have had huge numbers of lipid globules on

One theory of the development of the first life form involves tidal pools like this. Each high tide brings new water into the pool, and subsequent evaporation concentrates substances in the water.

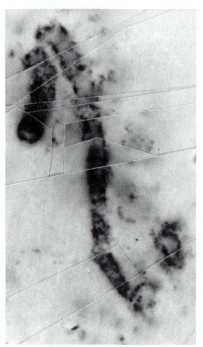

Figure 20–10
The earliest evidence of life, showing clusters of bacteria that lived about 3.5 billion years ago. At least 11 different kinds of cells were identified, so the earliest life must have appeared considerably earlier.

its surface, each enclosing a sample of the organic molecules present in the water. Each of these bubbles can be thought of as a tiny chemistry experiment. Eventually, one bubble happened to have the right mix of chemicals to reproduce itself, and the first cell was born.

Which of these theories will survive when scientists finally work out the origin of the first cell? There is no way of knowing that at the present time; in fact, a completely different theory may be required. What we do know is that however the first cell originated, it appeared very soon after the oceans formed.

The Age of Life on Earth

Whatever chemical processes led to the first reproducing organism, we know that those processes had to take place rapidly. We know that the Earth, like all the other planets, went through the period of the great bombardment (see Chapter 17). During this period, large chunks of debris fell onto the planets from space, bringing with them enormous amounts of energy. The impact of an asteroid several hundred miles across would heat the Earth enough to boil the oceans. As a result, the planet would literally be sterilized, and any life that might have developed before the impact would have been wiped out. Therefore, the process that led to the ancestors of all present life on Earth could not have begun until after the last big impact. The best estimate for this date is about 4 billion years ago.

Recent discoveries have made it clear that by about 3.5 billion years ago, life was not only present on Earth, but flourishing. In 1993, William Schopf of the University of California at Los Angeles discovered evidence for colonies of primitive bacteria in rocks from that period; these rocks were dated using radiometric techniques (see Chapter 14). Schopf's evidence is pretty clear. When he takes thin slices of these old rocks and looks at them under a microscope, he sees the imprints left behind by at least 11 different varieties of primitive one-celled organisms (see Figure 20–10).

These 3.5-billion-year-old fossil organisms, found in sediments in what is now Australia, are similar to modern green pond scum. Scientists estimate that it must have taken hundreds of millions of years for life to develop from the first cell to the more complex organisms observed in Australia. The age of life on Earth, then, must be placed within a well-constrained interval of time. The origin of life can be no older than the time of the last big impact, nor can it be younger than the appearance of fossils. This interval extended roughly from 4.0 to 3.5 billion years ago, and the origin of life may well have occurred in the earlier part of that interval.

Think about the unique status of the very first cell on Earth. That cell, unlike life today, had no competition for resources, so it did not have to be particularly efficient in using the chemicals found in its environment. There were no predators, and no other life forms to compete for the abundant stock of organic molecules that enriched the early ocean. Once the first cell formed, it would have been able to multiply rapidly.

Science by the Numbers

Cell Division

The first cell was a microscopic organism, but it may not have taken long for that first bit of life to spread great distances around the globe. To get a feel for this process, imagine how long it would take to fill up the Mediterranean Sea starting with a single cell that divides once a day, assuming all cells survive and continue to divide.

To figure this out, we must estimate the volume of an ordinary bacterium and compare it with the volume of the Mediterranean Sea. A typical bacterium is about one thousandth of a centimeter across, so its volume is approximately

$$(\tfrac{1}{1000} \, cm)^3 = 10^{-9} \, \frac{cm^3}{bacteria}$$

A recent world atlas gives the surface area of the Mediterranean Sea as about 2.5 million km², with an average depth of 1.4 km, for a total volume of

$$2{,}500{,}000 \, km^2 \times 1.4 \, km = 3{,}500{,}000 \, km^3 \, (3.5 \times 10^9 \, km^3)$$

The question boils down to how many times would you have to double a $10^{-9} \, cm^3$ bacterium to make $3.5 \times 10^6 \, km^3$. To make this calculation easier, we convert cubic kilometers to cubic centimeters:

$$1 \, km = 10^5 \, cm$$
$$1 \, km^3 = 10^{15} \, cm^3$$

The total volume of the Mediterranean in cubic centimeters is

$$(3.5 \times 10^6 \, km^3) \times (10^{15} \, cm^3/km^3) = 3.5 \times 10^{21} \, cm^3$$

How many bacteria would it take to fill this volume? We divide the immense volume of the Mediterranean Sea by the tiny volume of a single bacterium:

$$\frac{3.5 \times 10^{21} \, cm^3}{10^{-9} \, cm^3/bacteria} = 3.5 \times 10^{30} \, bacteria$$

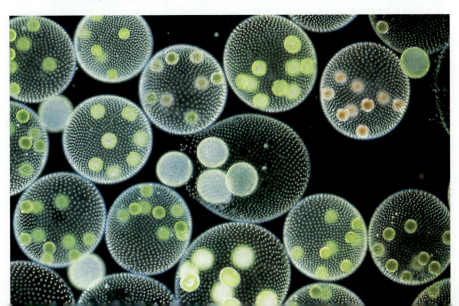

The colonial one-celled organism, *Volvox*, arises from repeated division of an original cell.

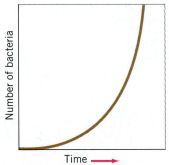

Figure 20–11
The number of bacteria grows rapidly, in what scientists call an exponential curve. In a matter of months, a single cell whose descendents divide once a day could easily populate a large ocean.

natural selection The mechanism by which nature can introduce wide-ranging changes in living things over long periods of time by modifying the gene pool of a specific species.

Starting with a single bacterium on the first day, there would be two on the second day, four on the third, eight on the fourth, and so on. After about three weeks, there would be more than a million bacteria, which is only about one thousandth of a cubic centimeter's worth. But day by day, the number would increase geometrically. After two months, there would be more than 10^{18} bacteria; after three months, 10^{27} individuals, occupying more than 10,000 km^3. And in just 10 days more, which is just 100 days after the first cell began to divide, the Mediterranean Sea would be completely filled with bacteria (see Figure 20–11).

Naturally, no body of water could be "completely filled" with bacteria. Early life probably did not spread this fast, nor was the process this regular and predictable. But the implication is clear. While it may have taken hundreds of millions of years for the first cell to evolve, the descendants of that first cell could have taken over the world's oceans relatively quickly. ●

Natural Selection and the Development of Complex Life

Once we get beyond how the first cell formed, our understanding of how life developed becomes much more detailed and precise. Our knowledge is largely due to the work of one man, the British naturalist Charles Darwin (1809–1882). His book *On the Origin of Species*, published in 1859, is arguably one of the most influential books ever written, and certainly one of the most influential books about science. In it, he presented a theory about how living things evolve. Darwin's theory was influential and controversial because it identified a simple mechanism for evolving complex multicellular life forms from single-celled life. His entire theory is built around the process of **natural selection,** by which life evolves through competition.

Artificial Selection.   Darwin introduced his controversial theory of natural selection by describing the selective breeding of animals, a process very familiar to his British readers. Farmers have known for millennia that the way to get bigger fruit, healthier plants, or animals with more meat on them is to breed the best individuals. To obtain large potatoes, for example, a farmer should plant only the eyes from the largest potatoes in any given crop. Over long periods of time, this practice will provide a variety of potato that is significantly improved from the original. Since human choice, not nature, drives this process, Darwin called it artificial selection. Artificial selection explains how animals as diverse as longhorn and Angus cattle, or Chihuahuas and Great Danes, can be bred from the same ancestral stock.

Natural Selection.   Darwin reasoned that competition among living things in nature could, given sufficient time, result in changes far greater than those of artificial selection. Two basic premises govern natural selection:

1. Every population contains hereditary diversity; that is, the individual members of the population possess a range of characteristics. Some

Contrasting breeds of dogs, including (*a*) St. Bernard, (*b*) Dalmation, and (*c*) Shih Tzu, illustrate the changes possible with artificial selection.

(*a*)

(*b*)

(*c*)

are able to run a little faster than others, some have a slightly different color than others, and so forth.

2. Many more individuals are born than can possibly survive. Therefore, those characteristics that make it more probable that a given member of a population will live long enough to reproduce will tend to be passed on to a greater percentage of the population's subsequent generations.

Let's look at a hypothetical example to see how this works. Suppose an island supports a certain number of birds, and suppose that the environment of this island is such that having a color that blends in with the local vegetation makes it easier for those birds to avoid their predators. Just by chance, some members of that population will have colors that match the colors of local leaves and trees better than others.

Better-camouflaged birds will be less likely to be eaten by predators, and will, therefore, be more likely to survive to adulthood and mate. You would expect, then, that the particular hereditary traits that give this advantage will be more likely to be passed to the next generation. If this process goes on for a long period of time, the entire population would eventually begin to share those advantageous traits for feather color. Thus nature "selects" those characteristics that will be propagated in any given species.

Natural selection, it should be remembered, is neither as controlled nor as rapid as artificial selection. It is always possible that birds who do not carry the selected gene will, in some generations, be more successful at mating than those who do. Over the long haul, however, the selective advantage granted by color will win. Thus Darwin envisioned natural selection as a process that operates over long periods of time to produce gradual changes in populations, not a process that can explain short-term variations in a few individual traits.

Since Darwin's time, the theory of natural selection has been expanded and developed to the point where it now unifies all of biology. A biochemist working on cells, a zoologist studying the organisms of a tropical lake, and a geneticist investigating DNA all share the central ideas of Darwin's theory, and hence will have a common vocabulary and a common way of attacking problems.

Figure 20–12
Light and dark versions of the peppered moth. The population shifted toward dark colors when coal burning became widespread in central England, then started to shift back when environmental controls were instituted.

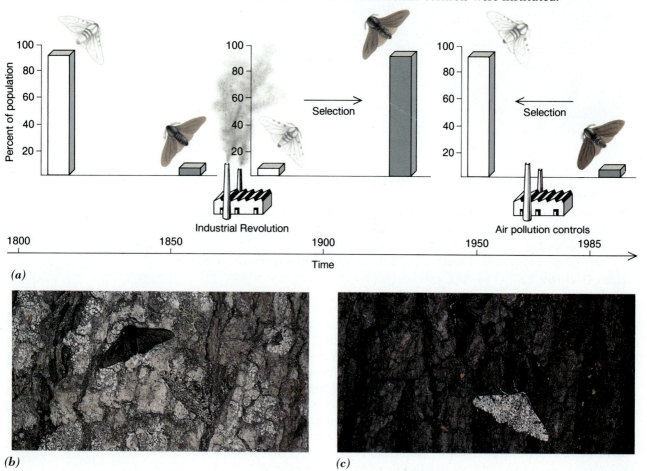

Environment

The Peppered Moth

About the time that Darwin's book came out, a rather extraordinary example of the power of natural selection appeared in his native England. The Midlands area of England in the nineteenth century had become a vast industrial belt. Factories poured smoke into the air (this was before the time of environmental awareness) and covered much of the countryside with dark soot. The peppered moth, a small moth that lived in this region, featured black and white wing patterns that blended in well with the light-colored lichen that had grown on trees in this area for centuries. At the time, lighter-colored moths predominated, though an occasional dark one was seen. While this lighter coloration gave the moth protection from predators in an environment that did not include soot, the common, lighter moths stood out clearly on trees darkened by industrial activity. Over a period of less than a century, the selective pressure on the moth population shifted in favor of darker individuals (see Figure 20–12). By the turn of the century, most of the peppered moths to be found were almost black. Scientists point to this rapid shift in moth coloration as a modern example of natural selection in operation.

Charles Darwin (1809–1882)

Today's scientists can watch this same process in reverse. Since environmental controls have been adopted in England, the soot emissions from Midlands factories have been cut significantly. As Darwin would have predicted, the peppered moth population has started to lighten again, as the original mossy tree trunks become more common. ●

Science in the Making

The Reception of Darwin's Theory

Charles Darwin formulated the basic outline of his theory of natural selection in 1838, but he waited more than 20 years to publish his findings. This delay was not simple procrastination. He realized that the central precepts of his theory would cause a furor. Eventually, learning that another British naturalist, Alfred Russell Wallace, had developed similar ideas, Darwin hastened to get *On the Origin of Species* into print.

Written in accessible prose and published in a widely available edition, Darwin's theory evoked intense reactions among those who could not accept his view of an old Earth. Naturally, some theologians denounced the book for its denial of a miraculous creation and the relatively short chronology demanded by a literal reading of the Bible. More disturbing to Darwin was the reaction of the majority of readers who embraced the "theory of evolution" as scientific evidence for God's hand in the progress of nature. They used Darwin's theory to support the idea of man's moral and spiritual superiority. These readers seized Darwin's discovery as a shining example of God's wisdom and beneficence. Some intellectuals of the late nineteenth century even went so far as to cite Darwin in their defense of an economically and socially stratified society. They claimed that the most "fit" individuals were those who rose to the top of the economic ladder.

The "progression" from chimpanzee to human is often used as an icon of Darwinian evolution. Darwin, however, did not see natural selection as leading inevitably from lower forms to humans.

Ironically, Darwin never intended his theory to suggest the idea of inevitable "progress" in nature, only inevitable change. Indeed, Darwin never even used the word "evolution," a word that connotes improvement, in his book. Nor did he address the question of human origins in *On the Origin of Species*. Far from being guided by a divine hand, he saw natural processes as violent and amoral, a constant struggle for survival in which the ability to reproduce fertile offspring was the only measure of success. He observed successful natural strategies that his contemporaries would have viewed as repulsive in any moral sense—species whose females devour their mates, species whose offspring eat each other until just a few survive, and parasites and predators that kill without thought in the frantic quest for energy to survive. To Darwin, human ascendancy seemed an evolutionary accident rather than a divine plan, and he saw no sign of God in the brutal process of natural selection. Nevertheless, in his own concluding words:

> There is grandeur in this view of life, with its several
> powers, having been originally breathed by the Creator into
> a few forms or into one; and that, whilst this planet has
> gone cycling on according to the fixed law of gravity, from
> so simple a beginning endless forms most beautiful and most
> wonderful have been, and are being, evolved. ●

THE HISTORY OF LIFE ON EARTH

The 3.5-billion-year history of life on Earth, from the first living cell to students reading textbooks, is a record of continuous change among organisms and their environment. Before the development of radiometric dating in this century (see Chapter 14), scientists knew about the existence of many varieties of fossils. They could see from the relative positions of sedimentary layers that some fossils were older than others, but they had no way of attaching an age to the changes they could see in the fossil record. Nevertheless, several important landmarks in the process of evolution were used as boundaries in the delineation of past times.

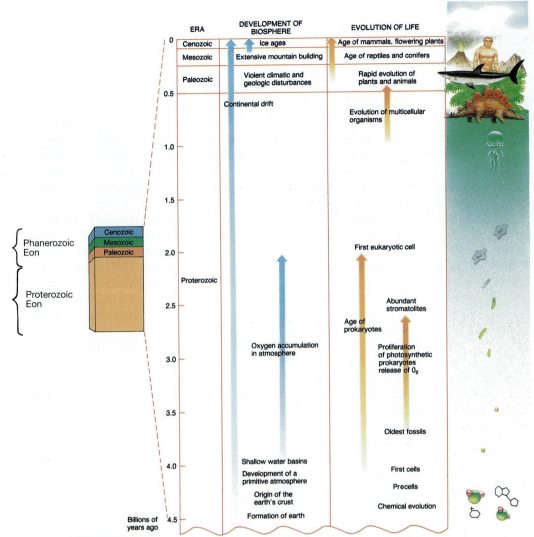

Figure 20–13
The geological time scale, with representative living things illustrated.

In the nineteenth century, scientists were not aware of fossil bacteria, or even fossils of ancient soft-bodied organisms. To them, the evidence seemed to indicate that life suddenly appeared at the beginning of what they called the Cambrian period (beginning, as we know now, about 570 million years ago, when fossils of hard-bodied organisms first appeared). The period before the Cambrian was, therefore, called the *Proterozoic* ("before life") era. Next was the *Paleozoic* ("old life") era, from about 570 to 245 million years ago. This era saw a marvelous diversification of life, including the appearance of fish, amphibians, land plants and animals, and rudimentary forms of reptiles. The third great era (245 to 65 million years ago) was the *Mesozoic* ("middle life"), also known the age of dinosaurs, when the major vertebrate life forms on Earth were large reptiles. Finally, the *Cenozoic* ("new life") era began with the extinction of the dinosaurs some 65 million years ago and continues to the present day. This is the time when mammals proliferated and began to dominate the Earth. The human species evolved at the very end of the Cenozoic.

Throughout this long and intricate process of change, the principle of natural selection was always at work, shaping and molding life forms (see Figure 20–13).

A mutant multi-toed cat.

The Proterozoic Era: Single Cells and the Earth's Atmosphere

As soon as the first cell split into two competing individuals, natural selection began to operate. In that early environment, where the first cells were surrounded by energy-rich molecules and very few neighbors, competition was not severe. Before long, however, natural processes called *mutations* introduced new traits into living things. Some cells developed differently from others, and some of those differences improved the efficiency with which the cells acquired food. Over time, beneficial mutations came to be shared throughout the entire population through the process of natural selection.

At this early stage, just as in today's life forms, the vast majority of mutations caused differences that were not beneficial. Nonbeneficial mutations died out quickly, and only beneficial mutations remained. This process is a little like our view of movies from the 1930s and 1940s. A great many poor films were made in those days, but we rarely see them anymore. What we remember, and preserve, are the most successful films, such as *Citizen Kane* and *Casablanca*. In the same way, only the "greatest hits" of all the mutations survived into the future.

As the first cells replicated and spread around the Earth and occupied most of the oceans, some cells wound up in different environments than others. Some, for example, were in tropical waters, while others were near the poles. Some were in deep oceans, while others were near the shore. Each of these environments exerted slightly different pressures on the cells. An adaptation that might be very advantageous in the tropics, for example, might not be advantageous near the poles, and vice versa. The driving force of natural selection, coupled with the fact that many different environments exist on our planet, produced many very different living things. Most scientists thus believe that the creation of new and diverse organisms began very early in the history of life.

Single-celled organisms. (*a*) A prokaryotic cell has no nucleus; (*b*) a eukaryotic cell has a well-defined, membrane-enclosed nucleus.

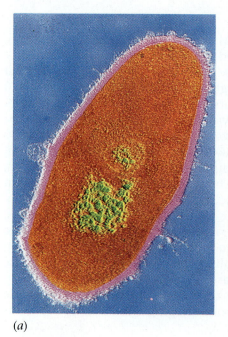

(*a*)

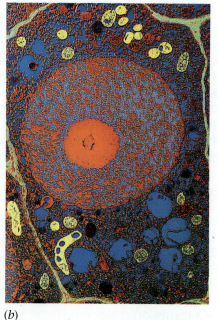

(*b*)

Our knowledge of this early period of life is limited by the fact that we have very little in the way of hard physical evidence that pertains to it. The fossil record does not preserve many single-celled organisms. This is not to say, however, that we have no record at all. As Figure 20–10 shows, we do have examples of fossil bacteria, and even a few cases of fossil bacteria caught in the act of dividing.

The best guess as to what went on until about a billion years ago is that the new living things spread around the world and differentiated, driven by natural selection. We suspect that, in the early part of this evolutionary process, all life consisted of simple, one-celled organisms, called *prokaryotes*, that produce oxygen as a byproduct of photosynthesis. To an outside observer, the Earth would have looked remarkably sterile. No life at all existed on land, but the margins of the oceans were covered with collections of scum, which was taking in carbon dioxide and returning oxygen to the atmosphere. Ever so gradually, life altered the composition of the Earth's atmosphere, replacing its carbon dioxide with oxygen. Some of this oxygen dissolved in the surface layers of the ocean; this change was essential in supporting many kinds of sea life.

About a billion years ago, single-celled life became more complex, as some smaller cells found that they did better living inside their larger neighbors than they could do on their own, as illustrated in Figure 20–14. This change led to cells, called *eukaryotes*, with an internal structure called the *nucleus* (not to be confused with the nucleus of the atom). These cells, like their neighbors and ancestors the prokaryotes, remained as single-celled organisms.

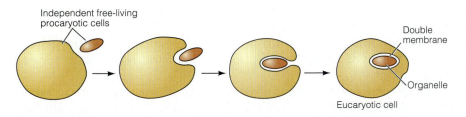

Independent free-living procaryotic cells

Double membrane

Organelle

Eucaryotic cell

Figure 20–14
One way that eukaryotes might have developed would have been for two free-living cells to come together as shown. If they were more likely to survive together than apart, natural selection would have favored this development.

Sometime during this period, cells began to come together to form large colonies. At first, these assemblages were probably nothing more than clumps of single-celled organisms living next to each other. Later, however, they developed into larger bodies. Indeed, by about 600 million years ago, the seas were probably full of large multicellular animals and plants, some of which resembled modern jellyfish. The stage was set for one of the most important developments in the history of life, the evolution of the skeleton.

The Paleozoic Era: When Life Got Hard

About 570 million years ago, a crucial development took place in living systems. By a process that we don't fully understand, but which probably involved a new enzyme that converted calcium in ocean water into shell material, some animals began to grow hard shells. Protective shells provided a tremendous advantage, so the seafloor was soon teeming with many different kinds of hard-shelled animals. As always happens when a useful evolutionary characteristic appears, there was a great deal of

(*a*)

Living fossils. (*a*) The chambered nautilus is a close relative of animals that thrived during the age of dinosaurs; (*b*) a 145-million year old fossil of a horseshoe crab is virtually unchanged from the modern animal.

(*b*)

competition and experimentation among living things as they tried to find the best type of outer shell, body design, and metabolism for each environment.

From the scientist's point of view, one of the most important aspects of this development was that, for the first time, living things left large numbers of fossils. In fact, for most of the nineteenth and twentieth centuries, before scientists found the fossils that indicated the presence of primitive forms of life, it looked as if life suddenly exploded at the beginning of this period. Many new creatures, including distinctive many-legged animals called trilobites, and other now-extinct forms, crowd the Paleozoic fossil record. This sudden change of life on Earth, therefore, is often referred to as the *Cambrian explosion*. (Geologists refer to the time during which skeletons developed as the *Cambrian period*, after "Cambria," the old Roman name for Wales, where rocks from this period were first studied).

Following the development of shells, the last half billion years or so has been a period of enormous growth in both the complexity and diversity of living things. A short summary of major developments is given in Table 20–1. Among the most important events of the Paleozoic era were the appearance of fish and other vertebrates, and the gradual adaptation of plants and animals to land. By the close of the Paleozoic, approximately 245 million years ago, lush jungles teaming with insects and other animals covered much of the Earth's continents.

Pikaia

An artist's conception of what life in the early Cambrian period, just after the development of skeletons, might have looked like.

Table 20–1 • Major Steps in the Evolution of Life

Time (millions of years ago)	Event
4000–3500	First cell (prokaryotes)
1000	Cells with nuclei (eukaryotes)
700	Multicellularity
570	Animals with shells
about 500	Early fishes
440	Land plants
360	Early trees
350	Reptiles
245	Mass extinction
200	Early birds and mammals
150	Flowering plants
65	Mass extinction Early primates (human ancestors)
3	Humanlike primates

Evidence from the fossil record, however, reveals a dramatic change in life on Earth at the 245-million-year boundary. Suddenly, for reasons that are still a matter of intense debate, perhaps 90% of all life forms, including the trilobites and many other large groups of animals, disappeared. Within a few million years, a mere blink in geological time, countless new organisms evolved to take their place.

The Mesozoic Era: The Age of Dinosaurs

The Earth at the beginning of the Mesozoic era would have appeared strange, indeed. There was no Atlantic Ocean. The continents we now know as Europe and North America were locked together, having collided to form the mighty Appalachians, which at the time was perhaps the tallest mountain range on the planet. Geologists call this immense land mass *Pangaea*, and recognize it as just one brief, passing stage in the continuing process of plate tectonics (see Chapter 18).

The feature of life in the Mesozoic that gets the most publicity today is the dramatic success of reptiles, which came to dominate the lands, seas, and skies with a wonderful diversity of dinosaurs and other forms. Ferocious carnivores such as *Tyrannosaurus rex* and immense herbivores such as *Appatosaurus* (also known as *Brontosaurus*) receive the most attention from the popular press and movies, but hundreds of smaller reptiles have been found in the fossil record.

The Mesozoic era featured important changes in marine life, as well. Coiled shells called *ammonites*, which are closely related to modern squids, appear in most ocean sedimentary rocks of the era, while mollusks, corals, and fish were also common. On land, common plants included ferns, conifer trees, and the ginkgo (virtually unchanged to this day). The evolution of flowering plants during the middle of the Mesozoic marked an important change in the character of land plants, and greatly influenced the evolution of insects and other animal life.

The Mesozoic era ended abruptly 65 million years ago. Dinosaurs, ammonites, and many other forms of life disappeared in a geological instant. This sudden catastrophe, and the even larger event that marked the end of the Paleozoic era, now appear to be just two of many mass extinctions that have altered Earth's history.

The movie *Jurassic Park* used computer animation to bring dinosaurs like the *Tyrannosaurus* back to "life." During the Mesozoic era, life on Earth was dominated by reptiles like these dinosaurs.

Mass Extinctions and the Rate of Evolution

Under normal circumstances, the rate of extinction seems to be such that roughly 10% to 20% of the species at any given time will be extinct in a matter of 5 or 6 million years. The fossil record shows, however, that not all extinctions are "normal." Rare catastrophic events in the past have caused large numbers of species to become extinct suddenly. These events are called **mass extinctions** (see Figure 20–15).

By "large numbers of species," we mean anywhere from 30% to 90% of the species alive at the time. By "suddenly," we mean a time too short to be resolved by standard geological techniques. The extinction may have taken place over a period of a few tens of thousands of years, or over a single weekend.

The best-known mass extinction is the one in which the dinosaurs perished some 65 million years ago, at the end of the Mesozoic era. In that particular extinction, about two thirds of all living species disappeared. For some organisms, such as ocean plankton, this number may have climbed as high as 98%. But the extinction at the end of the Mesozoic was neither the largest, nor the most recent mass extinction. About 245 million years ago, at the end of the Paleozoic, about 90% of existing species disappeared in a single extinction event. A somewhat milder extinction, which wiped out 30% of existing species, appears to have taken place about 11 million years ago. In fact, geologists who study the past history of life in more detail distinguish as many as 11 of these mass extinctions.

One of the most interesting explanations for how these mass extinctions may have occurred was put forward in 1980 by the father-and-son team of Walter Alvarez (a geologist) and Luis Alvarez (a Nobel laureate in physics). Based on evidence they accumulated, they suggested that the impact of a large asteroid killed off the dinosaurs and other life forms. Such an impact would have raised a dust cloud that would have blocked out sunlight for several months. This catastrophe would have been such a shock to the world ecosystem that it is a wonder anything survived at all.

mass extinction Rare and catastrophic events in the past that have caused large numbers of species to become extinct suddenly.

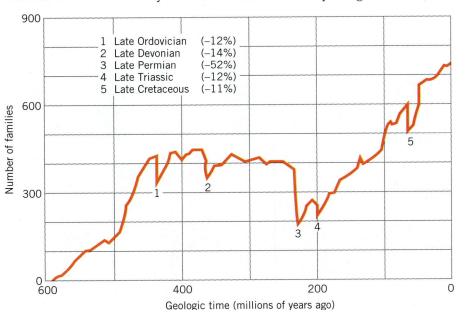

Figure 20–15
The diversity of life on Earth has not increased steadily, but has gone through a series of sharp changes. This graph of marine animals in the fossil record is indicative of the overall growth in families and of extinctions. The extinction of the dinosaurs is indicated by the event labeled 5.

Figure legend:
1 Late Ordovician (–12%)
2 Late Devonian (–14%)
3 Late Permian (–52%)
4 Late Triassic (–12%)
5 Late Cretaceous (–11%)

y-axis: Number of families (0, 300, 600, 900)
x-axis: Geologic time (millions of years ago) (600, 400, 200, 0)

Most scientists today accept that an asteroid hit the Earth 65 million years ago, at the end of the Cretaceous period, and agree that it was at least partly responsible for the mass extinction. This conclusion was bolstered in 1992, when a crater over 100 miles across was discovered buried under the surface near the Yucatan Peninsula in Mexico. Less certain is the role that other factors played in these events. The world ecosystem was under a great deal of stress at that time because of relatively rapid changes in climate and the recent creation of mountain chains, both of which were in the process of altering habitat.

Mass extinctions illustrate an important point about the history of life on our planet. Evolution is not a smooth, gradual process through time. There are times when sudden changes (such as those in the mass extinctions) are followed by rapid evolution, as new species develop to take the place of those that disappeared. After the extinction of the dinosaurs, for example, the number of species of mammals increased dramatically.

Scientists continue to debate about the rate of evolution. There are two extremes in the debate. One is the *gradualism* hypothesis, which holds that most change occurs as a result of the accumulation of small adaptations. The second hypothesis is referred to as *punctuated equilibrium*, which holds that changes usually occur in short bursts separated by long periods of stability. These theories are shown schematically in Figure 20–16. It now appears that both of these extremes, and probably any rate of evolution in between, have occurred at some time in the Earth's past.

The Cenozoic Era: The Age of Mammals

If you could visit the ancient Earth during the Paleozoic or Mesozoic eras, you would immediately notice striking differences. During the early

Figure 20–16
In gradualist theories, changes in organisms accumulate slowly over time to produce new species (*a*). Punctuated equilibrium holds that new species are formed suddenly, then persist relatively unchanged for long periods of time (*b*). Examples of both kinds of evolution, and many intermediate processes, can be found in the fossil record.

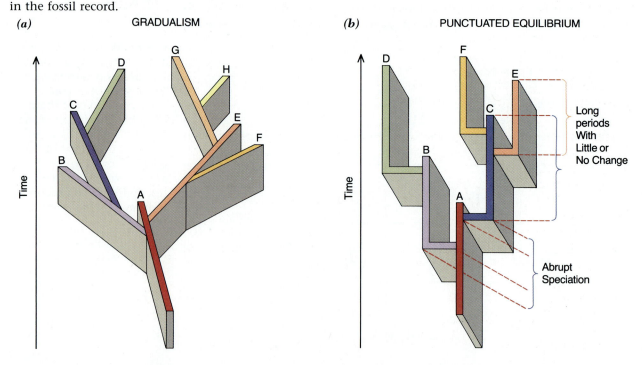

Paleozoic, you would find a landscape devoid of life. Later in the era, the vegetation, insects, and other animals would appear strange and alien. The Paleozoic oceans, too, were populated with strange creatures such as trilobites and armor-plated fish. Life during the Mesozoic era would seem a bit more familiar to you, except for the dinosaur lurking around the bend or the flying reptile overhead. By the Cenozoic era, however, life was starting to look much more as it does today. Flowers, insects, birds, and mammals—some rather similar to today's plants and animals—occupied many of the same settings that they do today.

The mass extinction of dinosaurs created opportunities for new life forms to exist. For example, agile tree dwellers, swift grassland herbivores, and powerful carnivores started to populate the Earth. Many of these "niches" in the environment were taken over, and are still dominated, by mammals. Mammals are warm-blooded animals that usually give live birth and who nourish their young by milk from special glands in the female. Though small, primitive mammals were present throughout most of the Mesozoic era, it was not until the demise of the dinosaurs that mammals diversified to an extent unparalleled by any other episode in the history of vertebrate life.

The oceans, too, saw swift changes, as fish took over the predatory opportunities left by the demise of ammonites. But perhaps the most dramatic change in the history of life on Earth, at least from its potential impact on the global environment, occurred only recently, in the last few million years. That change was the development of human beings.

The Evolution of Human Beings

The evolution of our own species, **Homo sapiens**, is no different in principle from that of any other species. The subject, nevertheless, holds an intrinsic interest for us that the evolution of, say, the mayfly, does not. Fossils suggest that our branch of the family tree broke off from that of other primates about 8 million years ago. At that time, the branch that would eventually be human beings split off from the branch that led to the great apes. A possible human family tree is sketched in Figure 20–17. Unfortunately, rocks created in the period 3 to 8 million years ago are rather rare on the Earth's surface, so it is difficult to fill in this gap. Nevertheless, some progress is being made.

Australopithecus.   The oldest known hominid is a fossil found in 1994 in Ethiopia. Called ***Australopithecus*** *ramidus* ("southern ape, root of humanity"), it appears to be midway between later, more human, fossils and the great apes. The best known humanlike fossils are bones of *Australopithecus afarensis* ("southern ape from the Afar Triangle region of Ethiopia"). Better known as *Lucy*, after the name given by paleontologists to a near-complete skeleton of the species, this fossil was discovered relatively recently, in 1974. The name arose because the paleontologists celebrated their discovery around a campfire while playing tapes of the Beatles' song "Lucy in the Sky With Diamonds." Radiometric dating established Lucy's age as about 3.5 million years. While *A. afarensis* is not a human, it is closer to us than to any other primate.

Lucy and her family walked erect, but had brains about the size of a modern chimpanzee's. They were also rather small (the adults probably

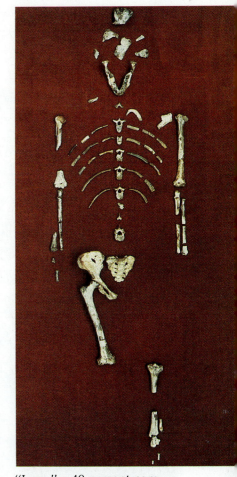

"Lucy," a 40 percent complete skeleton of *Australopithecus afarensis*, lived 3.5 million years ago in what is now northern Africa. This skeleton proved that hominids of this period walked erect.

Homo sapiens   The single species that includes all branches of the human race; recognized in fossils as old as 200,000 years.

Australopithecus   The first hominid, a primate closer to humans than any other; lived approximately 4.5 million years ago, walked erect, and had a brain about the size of that of a modern chimpanzee.

Figure 20–17
A chart of human evolution. The earliest hominid appeared about 3 million years ago. For most of the intervening period, there was more than one kind of "human" on the Earth. The disappearance of Neanderthals 35,000 years ago left *Homo sapiens* as the sole hominid on Earth.

weighed no more than 60 to 80 pounds), and may well have been covered with hair. If you saw an example of *A. afarensis*, the only thing that might make you think it is related to modern human beings would be its upright gait.

Scientists used to think that the development of large brains made human beings special, and that the brain's development led to upright walking. In fact, the evolutionary story seems to be the other way around. Primates first walked upright (freeing the hands for use), then the large brain developed. Some scientists have suggested that it was the evolutionary advantage bestowed by hand-eye coordination that provided the competitive edge for *Australopithecus*, which led to the large brain.

Homo habilus. Following Lucy, the line of humanlike australopithecines developed larger and larger brains. About 2 million years ago, however, the first true humans, *Homo habilus* or "man the tool-maker," appear in the fossil records of East Africa. *H. habilus* was larger than Lucy, and had a larger brain. More importantly, *H. habilus* fossils are found with crude stone tools, so the association of human beings with toolmaking started with this species.

Homo erectus. Shortly after the appearance of *H. habilus*, another member of our genus, *Homo erectus* ("man the erect"), appeared. *H. erectus* fossils are found not only in East Africa, but in Asia and the Middle East. Many of the famous fossil humans whom you may have heard of, including Java Man and Peking Man, are examples of this species. *H. erectus* lived at the same time as some of the later australopithecines, and survived until about 1.5 million years ago. *H. erectus* was the first in the line of human ancestors to use fire.

Homo sapiens. Fossils that we recognize as anatomically modern humans, belonging to the species *Homo sapiens*, begin to appear in rocks that are about 200,000 years old. A major debate has emerged among paleontologists as to whether anatomically modern humans began in Africa and spread to the rest of the world, or whether they emerged at several different points around the world. This question is still the subject of serious discussion and, unless major new finds of fossil hominids are unearthed, the debate is not likely to be resolved soon.

Neanderthal Man. About the same time, yet another type of human being, the so-called Neanderthal man, appeared on the scene. We commonly use the term "neanderthal" to denote something stupid. The term reflects conclusions drawn from questionable early studies of Neanderthal fossils, which stated that this species walked stooped over, with knuckles swinging, and had the thick brow ridge we associate with gorillas. In fact, these early suggestions were based on the study of a single skeleton of an old man who had a severe case of arthritis.

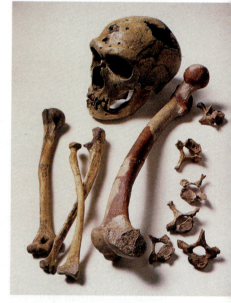

Bones of Neanderthal man are nearly indistinguishable from those of modern man.

Modern studies on other fossils reveal that Neanderthals, although not identical to modern human beings, were not all that different. They tended to be short, with thick, powerful arms and legs, and a skull that is much more elongated and pulled forward than that of modern *Homo sapiens*. However, Neanderthals had a large brain, on the average 10% larger than that of modern humans. Recently discovered fossil sites reveal that Neanderthals had a complex social structure. They cared for elderly and infirm members of their tribe, and they performed burials with ritual—a fact that suggests the presence of both religion and language. Overall, Neanderthals were not too different from their contemporaries among the anatomically modern humans.

Several major mysteries and controversies surround Neanderthal. The first puzzle is how closely Neanderthals were related to modern human beings. Were they, as some scientists claim, merely a subspecies of *Homo sapiens*? Scientists who adopt this view classify Neanderthal as *Homo sapiens neanderthalensis*. On the other hand, the traditional view (and the view of many modern scientists) is that Neanderthal, although our nearest relative, was a separate species. People who hold this view classify Neanderthal as *Homo neanderthalensis*, to indicate that it was a different species from modern humans. This debate will continue as more fossil evidence becomes available.

The next great mystery pertaining to Neanderthals is: what happened to them? In Europe, where the fossil record is most complete, it appears that Neanderthals flourished until 35,000 years ago, and then disappeared rather suddenly. Their disappearance coincided with the entry into Europe of modern *Homo sapiens*. Several very different theories have been put forward to explain Neanderthals' disappearance. Some suggest Neanderthals were wiped out by the invading members of our own species in what might be described as a prehistoric instance of genocide. Other scientists have proposed that Neanderthals intermarried with the invaders so that a certain percentage of the genes of modern human beings are Neanderthal in origin. In order for this to be true, Neanderthal would, by definition, have been a subspecies of (as opposed to a separate species from) *Homo sapiens*. Finally, some have suggested that Neanderthal couldn't compete with the more technologically advanced newcomers, and died out. In this case Neanderthal was not wiped out by acts of war, but was simply moved away from the desirable settlement locations, and eventually disappeared. This situation would be an example of the displacement of one species by another, a common phenomenon in the history of life.

From this description of the descent of modern human beings, one important point emerges. In the past, many different beings could be classified as humanlike. For whatever reason, none of them survived to this day, except ourselves. The branch of the family tree leading to human beings, in other words, has been extensively pruned by the processes of natural selection. This fact, we believe, made it easy for people in the nineteenth century to discount or misinterpret Darwin, and to believe that the human race was special and not related to the rest of our planet's interwoven web of life.

THINKING MORE ABOUT HISTORICAL GEOLOGY

Creationism

Opposition to the idea of an immensely old Earth and Darwin's theory of natural selection did not end in the nineteenth century. As recently as March of 1981, the legislature of the state of Arkansas passed a law requiring that public schools teach the Biblical story of the creation of the Earth along with the theory of evolution. The law was contested in the courts, and the federal courts eventually ruled that this law was an attempt to impose religious beliefs in the public schools, something expressly forbidden by the United States Constitution.

What is creationism, and why does it have such a strong hold in modern America? Three central beliefs of creationism, all based on a literal reading of the Bible, are:

1. The Earth and the universe were created a short time ago, perhaps 6,000 to 10,000 years in the past.

2. All life forms were created by God in a miraculous act, in essentially their modern forms.

3. The present disrupted surface of the Earth and the distribution of fossils are primarily the consequence of a great catastrophic flood.

From the creationist's point of view, almost all of the diversity we see among modern life forms has existed from the start, and very little subsequent change in life on Earth has occurred.

The debate between evolutionists and creationists illustrates a conflict between two differ-

The 1960 movie *Inherit the Wind* dramatized the debate between creationists and Darwin's theory of evolution by natural selection. Spencer Tracy (left) and Frederick March (right) acted the roles of the protagonists.

ent "ways of knowing" (see Chapter 1). Evolutionists point to the observational evidence of fossils and living things, as reviewed in this chapter, to support their view of an ancient Earth with gradually changing life forms. Creationists, with equal conviction, accept the Bible as their authority for a relatively young Earth with essentially unchanging life. This conflict cannot be resolved in any simple way.

Science and religion provide us with different (often complementary) ways of understanding our place in the universe. The theory of evolutionism is based on observations of the natural world, and is subject to change, based on new observations—the process we have called science. Creationism is based on faith, which cannot be revised based on observations—creationism is a religious belief.

To what extent do you think that parents should have the right to decide what scientific theories and ideas are presented in schools? To what extent do you think that opposing scientific and religious views should be taught? Should the passionately held views of creationists be given special consideration?

▶ Summary

The solid materials that make up the Earth's crust are subject to the *rock cycle*. The first rocks to form on the cooling planet were *igneous rocks*, which are formed from hot, molten material. *Volcanic*, or *extrusive*, igneous rocks solidify on the surface, while *intrusive* igneous rocks cool underground. The first igneous rocks were subjected to weathering by wind and rain, which eventually produced layers of sediment and the first *sedimentary rocks*. Sandstones, shales, limestones, and other sedimentary rocks have been deposited in ocean basins, layer upon layer, in sequences that often are many kilometers thick. Igneous and sedimentary rocks were subsequently buried and transformed by the Earth's internal temperature and pressure to form *metamorphic rocks*. Each of the three major rock types—igneous, sedimentary, and metamorphic—can be converted into the others by the ongoing processes of the rock cycle.

Fossils provide evidence that life on Earth has undergone the process of *evolution*. The *fossil record*, though incomplete, reveals that the appearance and *extinction* of new life forms has been going on for at least 3.5 billion years. Life on Earth evolved in two stages. The first period of *chemical evolution* was characterized by the gradual buildup of organic chemicals in the primitive oceans. The *Miller-Urey experiment* showed that simple compounds including water, methane, ammonia, and hydrogen, subjected to electrical sparks or some other energy source, combine to make the building blocks of life: amino acids, lipids, and other molecules. Through a sequence of events not yet well understood, a primitive but complex self-replicating chemical system developed. All subsequent life evolved from that first cell.

The first cell, free from competitors, quickly multiplied in the nutrient-filled oceans. As the oceans became crowded and competition for resources increased, a new stage of evolution—*natural selection*—began. The theory of natural selection, introduced by Charles Darwin in his 1859 monograph *Origin of Species*, recognized that every species exhibits variations in traits, and that some traits enhance an individual's ability to survive and produce offspring. Just as breeders develop new varieties of animals by selecting desirable traits artificially, so too, Darwin argued, nature selects traits through the struggle for survival. In this way, over immense spans of time, new species arise. Geological time is divided into Proterozoic, Paleozoic, Mesozoic, and Cenozoic eras, depending on the kinds of fossils found from the period when the rocks formed.

Extinction is a continuous process; however, there have been a number of catastrophic episodes of *mass extinction*, when most species disappeared in a brief time interval. Asteroid impacts may account for some of these events.

Human evolution can be traced back approximately 3.5 million years to *Australopithecus*, a hominid that walked erect but had a brain about the size of a chimpanzee's. Modern humans, who are of the species *Homo sapiens*, are recognized in fossils as old as 200,000 years.

▶ Review Questions

1. Why were igneous rocks the Earth's first rocks?

2. What are the three principal kinds of rocks, and how do they form?

3. If you were driving past a large road cut through rock, what features might you observe to tell you its origin?

4. Describe the process of evolution.

5. What are the two stages of evolution of life on Earth?

6. What is chemical evolution? How does it differ from natural selection?

7. Describe the Miller-Urey experiment. Why are its results important?

8. What is the primordial soup? What role might it have played in the origin of life?

9. Where is the greatest gap in our knowledge of the evolution of life?

10. What is natural selection? Give examples of natural selection at work.

11. How does natural selection differ from artificial selection?

12. State the two basic facts that govern the operation of natural selection.

13. How old are the earliest known fossils?

14. What is the Cambrian explosion?

15. What is the fossil record? How does it support the theory of evolution?

16. What is a mass extinction? Give an example. What are some possible reasons for mass extinctions?

17. When did the first members of the hominid family appear on the Earth? Who was Lucy?

18. Which came first in human evolution, large brains or upright posture?

19. Give an example of a theory used to explain the disappearance of Neanderthal.

20. What is creationism?

21. What might hot springs and geysers indicate in regions such as Yellowstone National Park?

22. What are some differences between sandstones, shales, and mudstones?

23. What occurs if shales and mudstones are subjected to high pressure and high temperature?

24. Describe what is meant by "deep time."

▶ Fill in the Blanks

Complete the following paragraphs with words and phrases from the list.

Australopithecus	mass extinctions
chemical evolution	metamorphic rocks
evolution	Miller-Urey
extinction	natural selection
fossils	Neanderthal
fossil record	rock cycle
Homo sapiens	sedimentary rocks
Homo erectus	volcanic or extrusive
igneous rocks	rocks
intrusive rocks	

Rocks on the Earth's surface continuously change form in a process known as the _____. Molten material hardens to form _____, which can be broken down and carried by water to lakes and oceans, where they form _____. If these rocks are buried and subjected to heat and pressure, they may again change form to become _____. Molten rocks flowing on the surface are called _____, while those flowing into cracks underground are called _____.

Life on Earth developed by the process of _____. Dead organisms may become _____, and the _____ is one piece of evidence to support the theory of evolution. At first, life developed from inorganic materials in a process known as _____. The _____ experiment was a key event in our understanding of this first stage of evolution. Once life appeared, its development was governed by the principle of _____, which describes how the environment affects organisms over long periods of time. When a life form disappears from the fossil record, we say that _____ has occurred. There seem to be events in the past, known as _____, when large numbers of organisms perished suddenly.

The earliest "humans" were known as _____, and are now extinct. Our family tree includes _____, the first humans to have fire, and _____, who was either a close relative or a subspecies of modern humans, or _____.

▶ Discussion Questions

1. Describe three places where you might find volcanic rocks forming today.

2. Describe three places where you could watch sedimentary rocks forming today.

3. Where would you have to go to watch metamorphic rocks form?

4. Discuss the possible connection between plate tectonics and the constant process of natural selection. How might a study of fossils and plate motions be used to test Darwin's theory?

5. All scientific theories must be able to make testable predictions. What are some testable predictions that follow from Darwin's theory of natural selection?

6. In what ways were Neanderthals different from modern humans? In what ways were they similar?

7. Why do most scientists think all life evolved from a single cell? What evidence do we have to support this hypothesis? What alternative hypotheses can you propose?

8. Do you think that intelligent life is an inevitable consequence of natural selection? Why or why not?

9. Fossils are usually found in sedimentary rocks. Why aren't they likely to be found in igneous rocks? What biases might this introduce into the fossil record?

10. Some diseases, such as adult-onset diabetes and Parkinson's, begin to affect individuals only in middle age. Why hasn't natural selection eliminated individuals with these diseases?

11. How fast are species disappearing from the Earth today? What is the main reason for these extinctions?

12. Some people have argued that the progress of modern medicine has stopped the workings of natural selection for human beings. What basis might there be for such an argument? Should this argument be taken into account in formulating public policy? Why or why not?

13. Classify the three major rock types according to their place of origin, original compositions, and one or two distinguishing features.

14. What is the difference between intrusive and extrusive igneous rocks?

15. Classify the three different sedimentary rocks—sandstone, shales, and limestone—according to the place and composition of origin.

16. Retell in your own words, the story of the origin of a piece of marble.

17. What is the meaning of the fossil record? What is one of the major problems with the fossil record?

18. List three key ideas that have emerged from the study of fossils. How have these ideas contributed to our understanding of evolution?

▶ Problems

1. If the average lifetime of a species is about 3 million years, what percentage of the age of the Earth does an average species exist?

2. In the Science by the Numbers section, demonstrate the result that in two months there will be 10^{18} bacteria.

▶ Investigations

1. What kind of rock underlies your campus? Are there any rock outcrops nearby? If so, was your area ever under a large body of water?

2. The three kinds of rocks—igneous, sedimentary, and metamorphic—are described in this chapter as being quite distinct, yet some Earth scientists have engaged in an intense debate about the origins of certain rocks—for example, granites—that formed at high temperature deep within the Earth. Some scientists claim that these rocks are igneous, while others say they are metamorphic. How could such a debate arise, and how could it be resolved? (*Hint:* Think about making taffy.)

3. Read *The Panda's Thumb* by Stephen J. Gould. Discuss the book in the context of evidence for evolution.

4. Consult your geology department and find out locations of the nearest fossil-bearing rocks. Visit a fossil location and collect a variety of samples. What kinds of life forms did you find, and what kind of environment did they live in? How old are these fossils? What living organisms most resemble these fossils? What kinds of rocks were they found in?

5. Read accounts of one of the recent trials on evolution versus creationism. What legal arguments were advanced on each side? How did the courts rule?

6. Investigate the concept of social Darwinism. What was this doctrine? What government policies did it encourage? When was it in fashion?

7. Read an account of Charles Darwin's visit to the Galapagos Islands while he was on the *Beagle*. What did he see there that led him to formulate the theory of evolution by natural selection?

8. Read an account of the development of the hypothesis that an asteroid was responsible for the extinction of the dinosaurs, such as *Nemesis*, by David Raup. What role did the chemistry of the element iridium play in this hypothesis? Does the history of this idea support our argument in Chapter 1 that scientists must believe the data that results from their observations?

9. Read an account of the Scopes "Monkey Trial," or see a movie based on it, such as *Inherit the Wind*. How was the conflict between science and religion portrayed in these writings and movies? Is such a conflict inevitable?

10. How might life evolve on another planet? How might the process vary on a different world? What aspects of the processes should be the same?

▶ Additional Reading

Darwin, Charles. *On the Origin of Species.* Cambridge, Mass.: Harvard University Press, 1975 (reprint of 1859 edition).

Johanson, Donald E., and Maitland, Edey A. *Lucy— Beginnings of Human Kind.* New York: Simon and Schuster, 1981.

Mushat Frye, Roland, ed. *Is God a Creationist?* New York: Scribners, 1981.

Rudwick, Martin. *The Meaning of Fossils.* New York: Science History Publications, 1976.

Ward, Peter. *The End of Evolution.* New York: Bantam, 1994.

21 THE STARS

THE SUN AND OTHER STARS USE NUCLEAR FUSION REACTIONS TO CONVERT MASS INTO ENERGY. EVENTUALLY, WHEN A STAR'S NUCLEAR FUEL IS DEPLETED, THE STAR MUST BURN OUT.

A Shopping Trip

Think about the remarkable range of elements you can buy at your local shopping center. The hardware store stocks aluminum siding, copper wire, and iron nails. The drugstore sells iodine for cuts, zinc and calcium compounds as dietary supplements, and perhaps tanks of oxygen for patients with breathing difficulties. The jeweler displays rings and necklaces of silver, gold, and platinum set with diamonds, which are a pure form of carbon. And your local electronics dealer offers an amazing assortment of audio and video equipment, which are able to operate because of integrated circuits fabricated from silicon, perhaps doped with small amounts of aluminum or phosphorus (see Chapter 13).

Where did all these elements come from? Were they always part of the universe, or was there a time in the history of the universe when most of them weren't around? It may come as a surprise to you that the answers to questions like these lie in studies of the stars.

ENERGY AND THE STARS

astronomy   The study of objects in the heavens.

star   Any object like our Sun that forms from giant clouds of interstellar dust and generates energy by fusion.

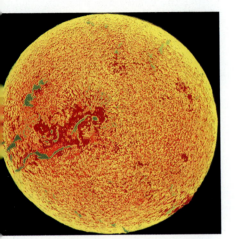

The Sun. The fact that the Sun emits energy means that it must have a finite lifetime.

Astronomy, the study of objects in the heavens, is perhaps the oldest science. When you examine the night sky, the most striking features are the thousands of visible stars. Each **star** is an immense fusion reactor in space. The Sun, the nearest star to Earth, is just one of countless trillions of stars in our universe.

You have already learned enough about the way the universe works to understand some of the implications of your nighttime view of the stars. The fact that you can see the stars, for example, means that they are emitting electromagnetic radiation (see Chapter 8). Every star you see sends huge amounts of radiant energy into space in every direction. Your eye intercepts a tiny fraction of that energy and converts it into the image you see.

The laws of nature, including the laws of thermodynamics that describe the behavior of energy, apply everywhere in the universe. If a star sends energy out into space, then that star must have a source of energy. Furthermore, since every star is a finite object, it must have a finite store of energy. This simple observation leads to perhaps one of the most profound insights about the universe:

> **All stars have a beginning and an ending.**

In other words, the magnificent display that we see in the night sky is a temporary phenomenon. It has lasted a long time—from the time that stars first formed until the present, a matter of perhaps 15 billion years. The view we see in the heavens will not last forever, however. Each star will eventually run out of energy.

Science by the Numbers

The Sun's Total Energy Budget

To understand the energy budget of stars, we must first determine how much energy a star releases. If you go outside on a warm day, you can feel the Sun's heat on your body. In fact, if you measured the sunlight, you would find that a square meter on the Earth's surface receives an average of about 750 watts (750 J/s)—just about enough energy to light your desk and heat your room on a cool evening. Above the Earth's atmosphere, away from the absorbing effects of air and dust, a square meter would receive about twice that much, or 1.5 kilowatts.

Imagine an immense sphere centered on the Sun with the surface passing through the Earth (see Figure 21–1). The Earth's orbit would be a line around the sphere's circumference. None of the Sun's radiant energy can disappear while in transit between the Sun and this imaginary sphere, so the total amount of energy going through the sphere is exactly the same as the total amount of energy generated by the Sun. Every single square meter on this sphere will receive the same amount of energy as a square meter located above the Earth's atmosphere.

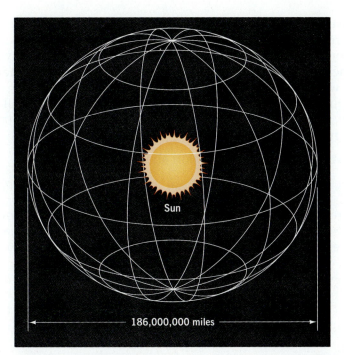
186,000,000 miles

Figure 21–1
An imaginary sphere centered about the Sun with a radius equal to the Sun-Earth distance. Each square meter of this sphere would receive the same amount of energy each second as a one-meter-square panel above the Earth's atmosphere. The total energy passing through the sphere is the same as the total number of square meters in the sphere (its area) multiplied by the energy passing through each square meter.

To find the Sun's total energy output, first we must calculate the total area of this immense sphere with radius equal to the Sun-to-Earth distance, which is about 150 million km (1.5×10^{11} m), or 93 million miles. The surface area of a sphere is given by

$$\text{area} = 4 \times \pi \times \text{radius}^2$$

where π is the constant 3.14. Thus we can calculate how many square meters there are on the surface of this imaginary sphere:

$$\begin{aligned}
\text{area} &= 4 \times 3.14 \times (1.5 \times 10^{11} \text{ m})^2 \\
&= 4 \times 3.14 \times 2.25 \times 10^{22} \text{ m}^2 \\
&= 28.3 \times 10^{22} \text{ m}^2 \\
&= 2.83 \times 10^{23} \text{ m}^2
\end{aligned}$$

The total energy of the Sun is this area times the energy per square meter at the distance of the sphere:

$$\text{Sun's total energy} = (\text{energy per meter}^2) \times (\text{sphere's total area})$$

or

$$\begin{aligned}
\text{energy} &= (1.5 \text{ kW/m}^2) \\
&\quad \times (2.83 \times 10^{23} \text{ m}^2) \\
&= 4.24 \times 10^{23} \text{ kW}
\end{aligned}$$

This number is astronomical (in every sense of the word). In one second, the Sun produces 1000 times the annual energy budget of the United States. ●

Example 21–1:   Moving Closer: the Sun's Energy and Venus

What is the energy from the Sun per square meter on the planet Venus, which is about 108 million km from the Sun?

▶ **Reasoning:** The total amount of energy from the Sun is the same for an imaginary sphere at any radius. Thus the total energy passing the orbit of Venus at a radius of 108 million km is exactly the same as the energy passing the orbit of the Earth at a radius of 150 million km, which we just calculated to be 4.24×10^{23} kw. The area of the imaginary sphere at Venus's orbital radius is much smaller than a similar sphere at the Earth's orbital radius, so Venus receives more energy per square meter.

▶ **Solution:** From the equation above,

Sun's total energy = (energy per meter^2) × (sphere's total area)

where, as before, the area of a sphere at the orbit of Venus will be $4 \times \pi \times r^2$, where r is the distance between Venus and the Sun. Therefore:

$$\text{energy per meter}^2 \text{ at Venus} = \frac{\text{Sun's total energy}}{\text{sphere's total area at Venus}}$$

$$= \frac{4.24 \times 10^{23} \text{ kW}}{4 \times 3.14 \times (1.08 \times 10^{11} \text{ m})^2}$$

$$= \frac{4.24 \times 10^{23} \text{ kW}}{4 \times 3.14 \times 1.17 \times 10^{22} \text{ m}^2}$$

$$= \frac{4.24 \times 10^{23} \text{ kW}}{14.7 \times 10^{22} \text{ m}^2}$$

$$= 2.89 \text{ kW/m}^2$$

Thus Venus, which is about three fourths of the Earth's distance to the Sun, receives twice the energy per square meter. ▲

The Sun's Energy Source: Fusion

Once people understood the concept of conservation of energy (see Chapter 5), two key questions about the stars naturally arose: (1) What is their energy source? (2) How can they continue to burn, emitting huge amounts of energy into space, yet remain seemingly unchanged for such long periods of time?

In the nineteenth century, several scholars attempted to explain the energy source of the Sun and other stars. One astronomer, for example, calculated how long the Sun could burn if it were composed entirely of anthracite coal, the best fuel available at that time. (The answer turns out to be about 10,000 years.) The Sun's energy source also figured in the famous debate at the end of the nineteenth century regarding the age of the Earth (discussed in Chapter 5).

Today, we understand that the Sun is indeed "burning" a fuel, but that fuel is hydrogen, which burns by means of nuclear fusion (see Chapter 14). The stars are born in the depths of space, in nebular clouds of gases and other debris (see Chapter 17). In such a cloud, matter happens to be more concentrated in some regions than in others. These regions of higher mass concentration attract their neighbors by the ordinary force of gravity, and clumps of matter come together, some eventually building up to very large masses.

As a new star begins to form, and as more and more mass pours into it from the surrounding region, the pressure and temperature at the center begin to climb. The new star will be made primarily of hydrogen gas, since that is the most common material in the universe. As a star forms and heats up, the electrons are torn from the hydrogen and other atoms, creating a *plasma* made up primarily of protons (the nucleus of the hydrogen atom) and electrons.

Normally, positively charged protons would repel each other (see Chapter 7). As matter accumulates in the new star, however, the protons move faster as they are pressed closer and closer together under tremendous gravitational pressure. Eventually, they acquire enough energy to overcome the electrical repulsion between them and they start to fuse.

The fusion process in the Sun's core does not take place all at once, with four particles suddenly coming together to make a helium-4 nucleus (see Figure 21–2). Instead, it takes place in three steps:

STEP 1: P + P ⟶ D + e⁺ + neutrino

Two protons come together to form a deuterium nucleus (an isotope of hydrogen made up of one proton and one neutron), a positron (e^+, the antiparticle of the electron as described in Chapter 15), and a neutrino.

STEP 2: D + P ⟶ ³He + photon

Another proton collides with the deuterium produced in the first step to form ³He, the isotope of helium that has two protons and one neutron in its nucleus. A photon in the form of an energetic gamma ray is also produced.

STEP 3: ³He + ³He ⟶ ⁴He + 2 protons + photon

Two ³He nuclei collide to form ⁴He, two protons, and a photon (another gamma ray).

The net effect of this three-step process, called *hydrogen burning*, is that four protons are converted into a ⁴He nucleus with a few extra particles thrown in. As we saw in Chapter 14, the sum of the masses of all the particles produced in this reaction amounts to *less* than the mass of the original four protons. The lost mass has been converted into energy—the nuclear energy that powers the Sun and eventually radiates out into space.

How long could the Sun burn hydrogen at its present rate? If you simply add up all the hydrogen in the Sun and ask how long it could last, the answer turns out to be something like 75 billion years. Actually, no star ever consumes all of its hydrogen in this way. The hydrogen burning is generally confined to a small region in the center of the star called the *stellar core*. The best current estimate of the total lifetime of the Sun is about 11 billion years; in other words, the Sun is almost halfway through its hydrogen-burning phase.

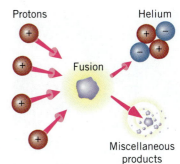

Figure 21–2
The net result of the three fusion steps outlined in the text is that four protons have produced a helium nucleus, some miscellaneous products, and energy corresponding to the mass difference between the initial and final states.

Science in the Making

The Solar Neutrino Problem

Our explanation of the Sun's energy source is a theory that is subject to experimental verification. In fact, a great deal of observational evi-

This solar neutrino apparatus, located a mile underground in the Homestake gold mine in South Dakota, receives only about a third of the expected number of neutrinos from the Sun.

dence supports the notion that the hydrogen burning reactions just outlined account for the Sun's energy. One crucial piece of evidence, however, seems to indicate that we may know less about the interior of the Sun than we think.

Since the early 1970s, a large experiment located a mile underground in a gold mine in Lead, South Dakota, has been returning results that are puzzling, to say the least. The purpose of this experiment is to detect the neutrinos that are expected to be given off in the Sun's nuclear reactions. Because they are small and electrically neutral, almost all of these neutrinos are expected to pass right through the outer layers of the Sun, and most of the time they pass right through the Earth as well. Occasionally, however, a neutrino will interact with an atom in the apparatus of this experiment, so that it can be detected. Thus the experiment provides us with a "telescope" that can see right down to the center of the Sun. If we understand what reactions are going on there, then we should be able to predict how many neutrinos will be observed.

When these measurements are made, scientists find only about one third to one half of the number of neutrinos they had expected to find. In the late 1980s, additional experiments of this type were begun in Europe and in the former Soviet Union, and these experiments seem to bear out the South Dakota results. The missing one half to two thirds of the expected neutrinos are known as the *solar neutrino problem.*

One possible explanation of the solar neutrino problem is tied to the nature of elementary particles such as neutrinos, and the fact that there are three different kinds of neutrinos (see Chapter 15). According to this idea, the predicted number of ordinary neutrinos is produced in the Sun, but while they're in transit to the Earth, some of them are converted into the two other kinds of neutrinos (the ones associated with the mu and tau particles, as described in Chapter 15). Such transitions between neutrino types are possible in some versions of modern unified field theories. If neutrinos change spontaneously in this way as the neutrino stream travels to Earth, only a fraction of the original particles will be ordinary neutrinos capable of interacting with ordinary atoms in an experimental apparatus. Such a solution to the problem would be enormously satisfying to astrophysicists, but only time will tell if it is the correct one. ●

Technology

Dyson Spheres

Science fiction fans like to speculate about how an extremely advanced technological civilization would use the energy from a star. One approach to the question is to note that most of the energy generated by stars is radiated into empty space and, from the point of view of planet dwellers, wasted. Physicist Freeman Dyson has suggested that one way to avoid this energy loss would be to dismantle a large planet and convert its material into a sphere surrounding a star—a hypothetical structure now called a Dyson sphere. A Dyson sphere would not only intercept and use all of a star's energy, but it could support a tremendous population. Such an immense structure would have an internal surface area comparable to more than 500 million Earthlike planets, and it would be uniformly heated in eternal daylight.

A star enclosed in a Dyson sphere would radiate its energy in the visible and higher energies, as usual, but all that energy would be absorbed by the sphere, which would be warmed and radiate the energy into space as infrared radiation. Scientists have suggested that one way to search for advanced extraterrestrial civilizations would be to look for the characteristic infrared signal. ●

THE ANATOMY OF STARS

Stars are much more than uniform balls of gas. They have a complex and dynamic interior structure that is constantly changing and evolving. The Sun is the only star that is near enough to influence the Earth. It is, in fact, the only star for which we have the kind of detailed knowledge that allows us to talk about how the various parts function. In this sense, the Sun does not only give life on our planet, but it also gives knowledge. Without a star close by to study, we would probably not know much about the billions of stars that are farther away.

The Structure of the Sun

The scientific community considers the Sun to be a rather ordinary star. Its stellar core comprises about 10% of the Sun's total volume. This core is the Sun's furnace, where nuclear fires rage, and energy generated in the core streams out from the center. This heat energy is then transferred outward by two primary means. Deep within the Sun, the energy transfer takes place largely through the collisions of the high-energy particles, including protons and positrons, that are generated in the core's nuclear reactions (see Figure 21–3).

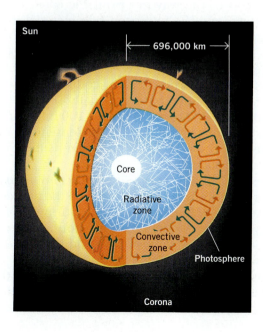

Figure 21–3
A cutaway view of the Sun. In the radiative region, energy is transferred by collisions, then by convection in the convective region.

Figure 21–4
The Sun's chromosphere and corona become visible during a total eclipse. This halo of incandescent plasma is normally blotted out by light from the Sun's main disk.

Figure 21–5
The magnetic field of the Earth is swept out into a long tail by the solar wind.

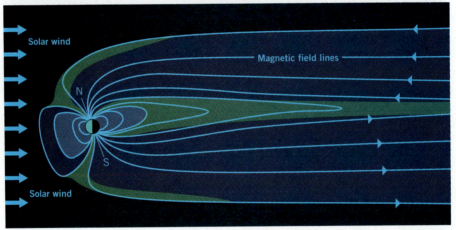

About four fifths of the way out from the core, the energy transfer mechanism changes, and the hydrogen-rich material in the Sun begins to undergo large-scale convection (see Chapter 5). This outer region, comprising the upper 200,000 kilometers (about 125,000 miles) of the Sun, is called the *convection zone*. Thus energy is brought from the core to the surface in a stepwise process—first by collisions, then by convection.

The only part of the Sun that we actually see is a thin outer layer. We can peer perhaps 150 kilometers (about 100 miles) into the Sun; any deeper and the stellar material becomes too dense to be transparent. The outer part of the Sun, the part that actually emits most of the light that we see, is called the *photosphere*.

The Sun does not actually have a sharp outer boundary, but gradually becomes thinner and thinner farther away from the surface. These gaseous layers are not usually visible from the Earth. During a total eclipse of the Sun, however, when the Moon passes in front of the Sun, the Sun's spectacular halo—called the *chromosphere* and the *corona*—may become visible for a few minutes (see Figure 21–4).

The Sun constantly emits a stream of particles, called the **solar wind,** into space. These particles, which consist primarily of ions (electrically charged atoms) of hydrogen and helium, blow by the Earth all the time.

solar wind A stream of charged particles—mainly ions of hydrogen and electrons—emitted constantly by the Sun into the space around it.

Northern lights.

Because the particles are charged, they affect the magnetic fields of the planets, compressing the fields on the "upstream" side and dragging them out on the "downstream" side (see Figure 21–5). The particles are also affected by the Earth's magnetic field. The interaction of the solar wind with the outer reaches of the Earth's atmosphere gives rise to the *aurora borealis*, or *northern lights*.

Thus we see that the flow of energy out from the Sun is complex. Beginning with the conversion of mass in fusion reactions, energy slowly percolates outward, first in collisions and later in great convection cells under the solar surface. It takes a few tens of thousands of years for the energy to work its way to the photosphere; however, it takes only eight minutes to cover the distance between the Sun and the Earth.

Once the sunlight reaches our planet, a portion of it is converted, by the process of photosynthesis in plants, into the chemical energy of large molecules. This energy then becomes the primary source of food for all living things on the planet.

Environment

The Human Eye

Our eyes take advantage of the sunlight of the environment in which they evolved. In Chapter 7, we saw that of all the possible waves in the electromagnetic spectrum, the human eye can detect only the small interval of wavelengths between red and violet. One reason the human eye is made this way is that the Earth's atmosphere is transparent to these wavelengths, so it is possible for the waves to travel long distances through the air. But another aspect of the human eye has to do not with the Earth, but with the Sun. Because of the fusion reactions at the Sun's core, the temperature of the outer part of the Sun is quite high—about 5500°C (for reference, the melting point of gold is about 1065°C). Every object above absolute zero radiates electromagnetic waves, and both the total amount of energy and the wavelengths of the radiation depend on the body's temperature.

In Figure 21–6, we show the amount of energy that the Sun radiates at each wavelength, with the visible spectrum represented by the vertical lines. We see that the Sun's peak output of energy is in the middle of the visible spectrum. Thus the wavelengths between red and violet are visible for two reasons—the air is transparent to them, and the Sun, our main source of light, emits the greatest proportion of its energy in that form.

Science fiction writers often use this fact when they portray imaginary beings from other planets. Creatures from the environment of planets around stars cooler than the Sun are often portrayed as having large eyes, so that they can absorb more of the scarcer photons. For the reasons cited above, however, humans don't require such large optical collectors. ●

Figure 21–6
The Sun's peak output of energy is in the middle of the visible spectrum, with lesser amounts of energy emission at different wavelengths.

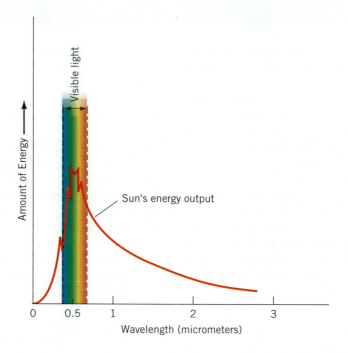

MEASURING THE STARS WITH TELESCOPES AND SATELLITES

telescope A device that focuses and concentrates radiation from distant objects; used by astronomers to collect and analyze radio waves, microwaves, light, and other radiation.

Our only source of data pertaining to distant stars is electromagnetic radiation, which is energy streaming through space at 3×10^8 meters per second. Part of this radiation is visible, but much of it is at wavelengths not detectable by the eye. To collect and analyze radio waves, microwaves, light, and other radiation, astronomers have devised a variety of **telescopes,** which are devices to focus and concentrate electromagnetic radiation from distant objects.

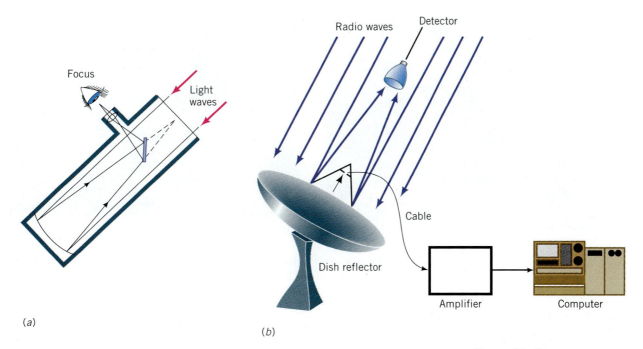

(a)

(b)

Figure 21–7
Schematic diagrams of telescopes. In an optical telescope (a) light strikes a curved mirror and is focused on a light-sensitive detector such as the eye or a piece of film. In a radio telescope (b) radio waves from space strike a curved metal dish that focuses the waves onto an antenna. Signals are amplified and processed by computer.

Earth-Based Telescopes

The first telescopes, developed more than 500 years ago, had a series of lenses that focus light. Most modern astronomers work with reflecting telescopes (see Figure 21–7a), which use a large curved mirror to collect and focus light to produce an image of the object being studied.

A new generation of reflecting telescopes employ a series of small, independently controlled, lightweight mirrors, instead of a single mirror made of a solid block of glass. Taken together, the series of mirrors produce an image. The Keck telescope on Mauna Kea in Hawaii is the biggest of these new-age instruments, and was put into operation in 1992. Other things being equal, the larger the telescope, the more useful it is, because the more light it is able to collect and the fainter the objects it is able to detect.

It was not until this century that astronomers took advantage of the invisible portions of the electromagnetic spectrum. In the 1930s, they built radio receivers that did for radio waves what the reflecting telescope did for light waves (see Figure 21–7b). Today, large radio telescope facilities can be found all around the world.

Orbiting Observatories

Except for visible light and radio waves, the atmosphere of the Earth is largely opaque to the electromagnetic spectrum. Most infrared and all ultraviolet radiation, long-wavelength microwaves, X-rays, and gamma rays entering the top of the atmosphere are absorbed long before they can reach instruments at the surface. The use of satellite observatories during the last half of the twentieth century has ushered in a golden age

(*a*)

(*b*)

(*a*) Several telescopes take advantage of the excellent conditions at the summit of Hawaii's Mauna Kea. (*b*) The Very Large Array in Socorro, New Mexico, features a Y-shaped array of radio telescopes. The effect of many small telescopes is equivalent to one very large telescope.

of astronomy. In this period, we have been able to put instruments in orbit far above the atmosphere and, for the first time, record and analyze the entire flood of electromagnetic radiation coming to us from the cosmos. Some particularly important orbiting observatories are:

Cosmic Background Explorer (COBE)   The Cosmic Background Explorer measured the presence of microwave radiation that is present as background noise in every direction of the skies. COBE's data have played a significant role in our understanding of the early history of the universe (see Chapter 22). The satellite ceased operation in 1994.

A view of the Milky Way taken by the Cosmic Background Explorer (COBE). The dust and gas that clog the central plane of the galaxy are transparent to microwaves, so we can "see" to the center of the galaxy something we could not do with visible light.

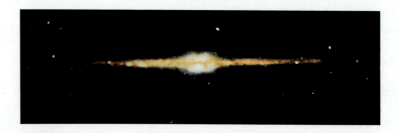

Infrared Astronomical Satellite (IRAS)   Launched in 1983 by the United States, the United Kingdom, and the Netherlands, IRAS functioned for nine months and gave us our first broad view of infrared radiation in the universe. Since every object in the universe, including the infrared telescope itself, emits some infrared radiation, parts of IRAS had to be cooled with liquid helium to within a few degrees of absolute zero in order for the telescope to not "see itself." IRAS photos gave us new views of the dust clouds that clog the central disks of galaxies, and of regions in the Milky Way where new stars are forming. IRAS stopped functioning only when it ran out of liquid helium; as one commentator remarked, it was "good to the last drop."

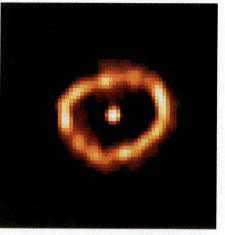

Comparison of images of an exploding star, Nova Cygni 1992, taken by the Hubble Space Telescope before (left) and after (right) repairs were made.

Astronaut F. Story Musgrave helped repair the Hubble Space Telescope (HST) during a space shuttle mission in December 1993. The quality of HST images improved dramatically following the repairs. (See comparison photos at left).

Hubble Space Telescope (HST)   Launched in 1990 from the United States, HST is a reflecting telescope with a 2.4–meter mirror that gives unparalleled resolution in the visible and ultraviolet wavelengths. Despite the fact that flaws in the manufacturing of the main mirror prevented some planned observations from being made and garnered a great deal of negative publicity, HST made an impressive series of discoveries even before a new optical system was installed by astronauts in late 1993. Among its accomplishments were detailed photographs of storm systems on Saturn, and discoveries of details that explain how stars form in dust clouds. Perhaps the most striking HST photos are those showing the impact of comet Shoemaker-Levy with Jupiter in July 1994.

The collision of Comet Shoemaker-Levy with Jupiter produced dramatic flares. This false-color photograph was taken with infrared cameras by Hawaii's Keck Telescope.

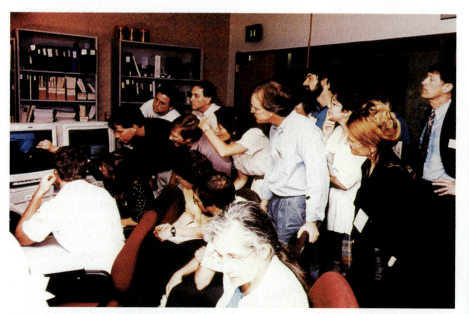

NASA's control room was the focus of intense activity during the collision of Comet Shoemaker-Levy with Jupiter in July 1994.

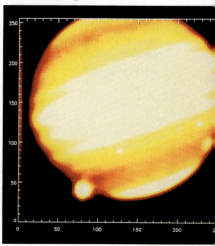

Roentgen Satellite (ROSAT)   Launched in 1990 by the United States, United Kingdom, and Germany, ROSAT is the latest in a series of satellites equipped to detect X-rays. X-rays are emitted by energetic processes, hence an X-ray map of the sky indicates where violent, high-energy events are occurring. Over a period of years, ROSAT will complete an X-ray survey of the entire sky.

Gamma Ray Observatory (GRO)   Launched in 1991, GRO detects the highest-energy end of the electromagnetic spectrum—gamma rays—which are emitted by black holes, exploding stars, and some active distant galaxies called *quasars* (see Chapter 22). GRO and HST are the first two telescopes in the National Aeronautic and Space Administration's (NASA's) Great Observatories Program, whose ultimate goal is to have permanent orbiting observatories monitoring all parts of the electromagnetic spectrum. Two such observatories now in the planning stage at NASA are SIRTF (Space Infrared Telescope Facility) and AXAF (Advanced X-Ray Astronomy Facility).

THE ZOOLOGY OF STARS

Knowing how one star, our own Sun, works should help us understand the other stars we see in the sky. When you look at the stars in the night sky, one of the first things you notice is that they don't all look the same. Some are very bright; some are barely visible. Some seem to have a reddish color; some are almost blue. Part of these differences in appearance arise because the stars are all different distances from the Earth. An unusually bright star located far away will appear dim to us, while an average star located nearby might appear very bright. The brightness of a star is related to the amount of energy the star is producing. Astronomers often refer to the total energy emitted by a star as its *luminosity*.

apparent magnitude   The brightness a star appears to have when it is viewed from the Earth.

absolute magnitude   The brightness a star appears to have when it is viewed from a standard distance.

Differences in the appearances of stars due to distance are taken into account by distinguishing between a star's **apparent magnitude,** which is the brightness it appears to have when viewed from the Earth, and its **absolute magnitude,** which is the brightness it would have if viewed from a standard distance. The brightest stars in apparent magnitude are said to be first magnitude, the next brightest second magnitude, and so on. The human eye can see down to sixth magnitude, and telescopes routinely detect stars down to the magnitude 23.

Differences in the stars' appearance, however, cannot be attributed entirely to distance; there are many varieties of stars themselves. Some stars shine a thousand times brighter than the Sun, while others are a thousand times dimmer. Some stars contain 40 times more mass than the Sun, while others have much less. As we shall see, we can bring order to this tremendous diversity of stars by recognizing that the behavior of every star depends primarily on just two factors: its total mass, and its age.

The Astronomical Distance Scale

When we look at the sky, we see what appears to be a flat (two-dimensional) display; all stars look equally distant. To add depth to the third dimension of this picture, we must find distances to the stars. Astrono-

THE ZOOLOGY OF STARS **625**

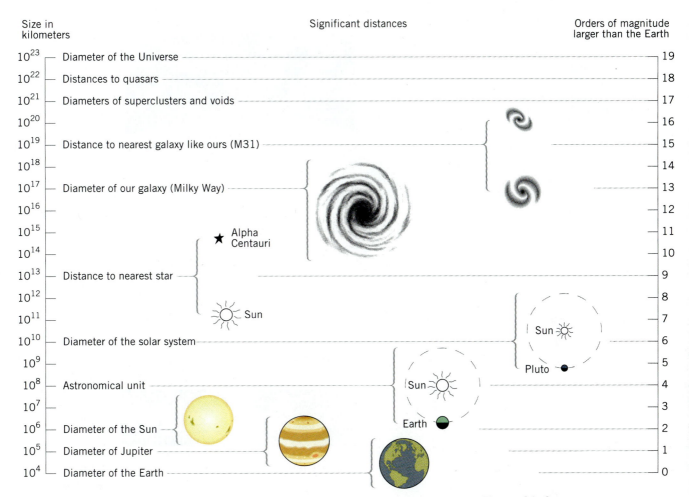

Size in kilometers	Significant distances	Orders of magnitude larger than the Earth
10^{23} — Diameter of the Universe		19
10^{22} — Distances to quasars		18
10^{21} — Diameters of superclusters and voids		17
10^{20}		16
10^{19} — Distance to nearest galaxy like ours (M31)		15
10^{18}		14
10^{17} — Diameter of our galaxy (Milky Way)		13
10^{16}		12
10^{15}	Alpha Centauri	11
10^{14}		10
10^{13} — Distance to nearest star		9
10^{12}	Sun	8
10^{11}		7
10^{10} — Diameter of the solar system	Sun	6
10^{9}	Pluto	5
10^{8} — Astronomical unit	Sun	4
10^{7}		3
10^{6} — Diameter of the Sun	Earth	2
10^{5} — Diameter of Jupiter		1
10^{4} — Diameter of the Earth		0

Figure 21–8
A scale of astronomical sizes, shown in absolute size and in relation to the size of the Earth.

light-year The distance light travels in one year, 10 trillion km (about 6.2 trillion miles).

mers customarily measure these great distances in **light years** (abbreviated ly), which is the distance light travels in one year, or about 9.5 trillion kilometers (about 6.2 trillion miles). Alternatively, they may use a unit called the *parsec*, a distance of about 3.3 light-years. This convenient measure roughly corresponds to the average distance between nearest neighbor stars in our galaxy. (See Figure 21–8 for a scale of astronomical sizes.)

In practice, no single method can be used to find the distance to every star. Just as you might use a ruler, a tape measure, and a surveyor's tape to measure successively larger distances, astronomers measure distances to stars with a series of "yardsticks," each appropriate to a particular distance scale.

For short distances (up to a few hundred light-years), several different methods involving simple geometry can often be used. For nearby stars, for example, the angle of sight to the star measured at opposite ends of the Earth's orbit (see Figure 21–9) can be used to work out the distance. Navigators on the Earth's surface use a similar method, called *triangulation*, to determine the positions of ships.

For greater distance, a standard type of star called a *Cepheid variable* is used. These stars, the first of which was discovered in the constellation of Cepheus, show a regular behavior of steady brightening and dimming

Figure 21–9
The triangulation of stellar distances. By measuring the angle of sight to a given star from two points of known separation, we can determine the star's distance from us.

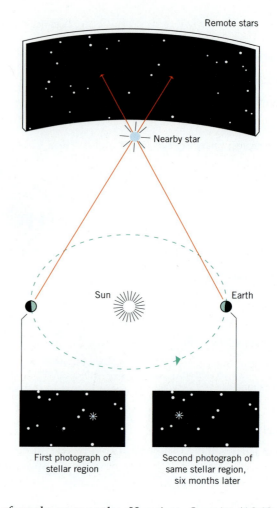

Remote stars

Nearby star

Sun

Earth

First photograph of stellar region

Second photograph of same stellar region, six months later

Annie Jump Cannon (left) and Henrietta Swan Leavitt (right) contributed important studies of the spectroscopy of stars at the Harvard College Observatory.

over a period of weeks or months. Henrietta Leavitt (1868–1921) of Harvard College Observatory showed that the absolute magnitude of these stars is related to the time it takes for them to go through a dimming-brightening-dimming sequence. Thus we can watch a Cepheid variable change in brightness for several months and deduce how much energy it is pouring into space. This measurement, together with a knowledge of how much energy we actually receive, tells us how far away it is.

In Figure 21–10, we show a simplified version of the relation between the period of brightening and luminosity for one type of Cepheid variable. The general rule, as you can see, is that the longer the period we observe, the greater the luminosity of the star.

Example 21–2: Distance to a Cepheid Variable

Suppose that we observe a Cepheid variable to have a period of about 100 days, and suppose that the amount of light we get from that star corresponds to an energy flow of about 2×10^{-14} W/m² at the location of our telescope. How far is that star from the solar system?

▶ **Reasoning:** We first use Figure 21–10 to find the luminosity of the star in question. Then, as in the Science by the Numbers section in

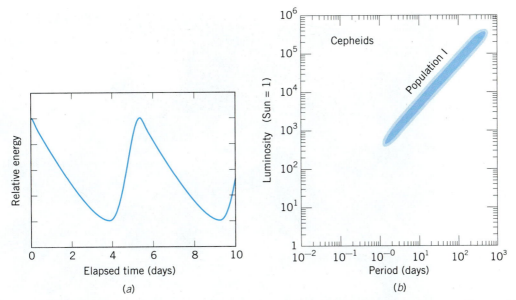

Relative energy

0   2   4   6   8   10
Elapsed time (days)

(a)

10^6

10^5

Cepheids

10^4

Luminosity (Sun = 1)

Population I

10^3

10^2

10^1

1

10^{-2} 10^{-1} 10^{-0} 10^1 10^2 10^3
Period (days)

(b)

Figure 21–10
As time goes by, the luminosity of a hypothetical Cepheid variable star increases and then falls (a). The length of time it takes for the cycle to occur is related to the luminosity of a particular class of Cepheid variables as shown in (b).

this chapter, all of the energy emitted by the star has to pass through a sphere whose center is at the star, and whose surface passes through the solar system. We calculate how big the radius of that sphere has to be to get the observed energy flow. The radius of the imaginary sphere will then be the same as the distance to the star.

▶ **Solution:** From Figure 21–10, a star with a 100–day period will have a luminosity about 150 times that of the Sun. Using the Science by the Numbers section of this chapter, we find that the luminosity (L) of the star is

$$L = 150 \times 4.24 \times 10^{23} \text{ kW}$$
$$= 6.4 \times 10^{28} \text{ W}$$

The amount of energy per second flowing through a sphere of radius R is

$$\text{energy} = \frac{L}{4\pi R^2}$$

so that

$$R^2 = \frac{L}{4\pi \text{ energy}}$$
$$= \frac{6.4 \times 10^{28} \text{ W}}{4\pi (2 \times 10^{-14} \text{ W/m}^2)}$$
$$= 2.6 \times 10^{41} \text{ m}^2$$

so that R, the distance to the star, is:

$$R = 5 \times 10^{20} \text{ m} = 2600 \text{ light-years}$$

If you can see an individual Cepheid variable, then you can always use this method to find the distance to it. The method of Cepheid variables can be used for distances of many millions of light-years. In fact, this was a crucial ingredient in the birth of modern cosmology, as we shall see in Chapter 22. ▲

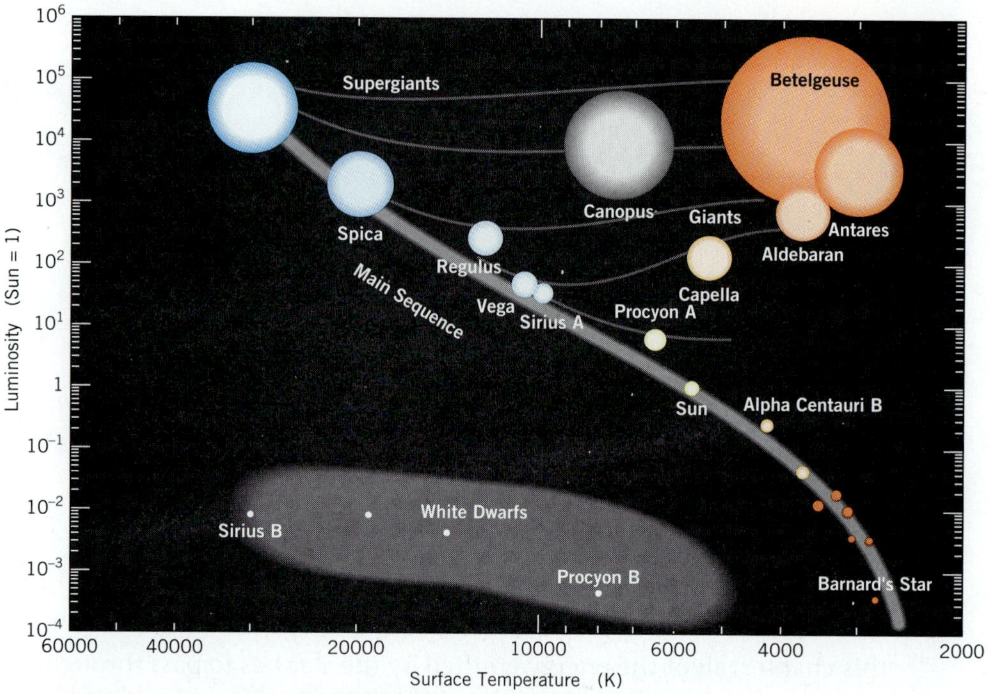

Figure 21–11
A Hertzsprung-Russell diagram plots a star's temperature versus its energy output. Stars in the hydrogen-burning stage, including the Sun, lie along the main sequence, while red giants and white dwarfs represent subsequent stages of stellar life.

main-sequence star A star that derives energy from the fusion reactions of hydrogen burning; found on the Hertzsprung-Russell (H-R) diagram within a bandlike pattern.

red giant A very large star that emits a lot of energy but whose surface is very cool and therefore appears somewhat reddish in the sky; found on the upper right corner of the H-R diagram.

The Hertzsprung-Russell Diagram

Early in this century, two astronomers, Ejnar Hertzsprung of Denmark and Henry N. Russell of the United States, independently discovered a way to find order among the diversity of stars. The product of their work, called the *Hertzsprung-Russell (H-R) diagram*, is a simple graphical technique widely used in astronomy. It works like this: On the graph's vertical axis, astronomers plot the amount of energy given off by a star, as measured by estimating the star's distance and brightness. On the graph's horizontal axis, they plot the star's temperature, as determined by its spectrum (see Chapter 8). Each star has its own characteristic combination of energy and temperature, and so it appears as a single point on the Hertzsprung-Russell diagram. The Sun, for example, is a point on the H-R diagram, along with other stars, in Figure 21–11.

When stars are plotted this way, the majority (the ones that are like the Sun) fall on a band of **main-sequence stars** that stretches from the upper left to the lower right in the diagram. Most stars conform to this trend, from very hot stars emitting a lot of energy, down to relatively cool stars emitting less energy. All of these main-sequence stars are in the hydrogen-burning phase of their lives, and their energy is derived from the fusion reactions that we described earlier in this chapter.

Two additional clumpings of stars appear in the H-R diagram. One clumping, in the upper right corner, corresponds to stars that emit a lot of energy but whose surfaces are very cool. These stars must be very large so that the low temperature (and low-energy emission of each square foot of surface area) is compensated for by the large surface area. Stars of this type are called **red giants,** and they often do appear somewhat reddish in the sky.

Another grouping of stars appears in the lower left corner of the H-R diagram. These stars, called **white dwarfs,** have very low emission of energy, but very high surface temperatures; they are very small and very hot.

Both red giants and white dwarfs play crucial roles in the life cycle of stars, as you shall see shortly.

THE DEATH OF STARS

At some time in the distant future, the core of the Sun will run out of hydrogen fuel, as indeed the core of every star must. In order to understand the spectacular events that occur when a star dies, we have to understand a little more about the life of stars.

One way to look at the life of a star such as the Sun is to think of it as continually battling against the force of gravity. From the moment the Sun's original gas cloud started to contract, the force of gravity has acted on every particle, forcing it inward and trying to make the entire

white dwarf A star that has a very low emission of energy but very high surface temperature; found on the lower left corner of the H-R diagram.

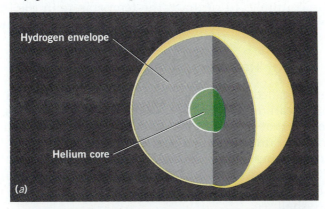

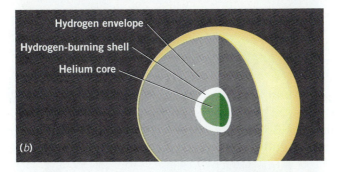

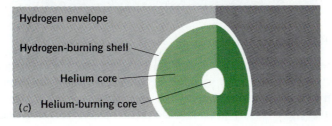

Figure 21–12
(*a*) As a main-sequence star burns hydrogen, the helium accumulates as "ash" in its core. (*b*) When the hydrogen in a star's core starts to give out, the star begins to contract. The temperature in the interior increases, and hydrogen begins to burn in a shell around the core. (*c*) When the temperature in the core gets high enough, the helium itself can burn, as this close-up of the star's interior shows.

structure collapse on itself. When the nuclear fires ignited in the core of the Sun 4.5 billion years ago, gravity was held at bay. The increase in temperature in the center raised the pressure in the star's interior and balanced the inward pull of gravity. But this balance can be only a temporary state of affairs. The Sun can stave off the inward tug of gravity only as long as it has hydrogen to burn. When hydrogen fuel in the core is depleted, the amount of energy generated in the core will decrease, and gravity will begin to take over. The core will then begin to contract, even as the hydrogen-rich outer envelope starts to expand.

This dramatic situation will have two effects. First, the temperature in the region immediately surrounding the core will begin to rise. Any remaining hydrogen in that region that had not been burned, because the temperature had been too low, will begin to burn. Thus a hydrogen-burning shell will begin to form around the extinguished core. The second effect is that helium, the "ash" of hydrogen burning, will have become concentrated in the Sun's core. The Sun then begins to look like the layers of an onion, with a core of helium surrounded by a hydrogen-burning layer (see Figure 21–12). As the pressure continues to increase, and the Sun continues to contract, the temperature in the interior will eventually get hot enough so that the helium itself will begin to undergo nuclear fusion reactions. The net reaction will be:

$$^4He + {}^4He + {}^4He \longrightarrow {}^{12}C$$

This process is called *helium burning*, in which the helium in the core burns to make carbon. Figure 21–13 summarizes this process.

The notion that the ashes of one nuclear fire serve as fuel for the next is central to understanding the life and death of stars. With stars like the Sun, the temperature never gets high enough to ignite the carbon, so the helium burning is the final energy-producing stage. In more massive stars, this process of successive burning cycles can go on for quite a while, as we shall see.

Figure 21–13
The "burning" of three helium nuclei to form a carbon nucleus.

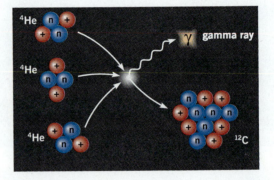

The Death of the Sun

The Sun will burn at more or less its present size and temperature for billions of years more. But, in its final stages, our star will undergo dramatic changes. When the core burns out, the hydrogen-burning shells surrounding the central region will be pulled in. When this occurs, there will be a temporary collapse, which will increase the amount of energy generated by fusion. This increased energy will cause the surface of the Sun to balloon out. At its maximum expansion, the dying Sun will extend out past the present orbit of Venus. Because the solar wind will increase during this period, removing material, the Sun's mass will decrease. One surprising consequence of this change in the Sun's mass is that the planets will move outward.

In the end, as the Sun expands, only Mercury will actually be swallowed. During this phase of its life, the Sun will emit its energy through a much larger surface than it has now, and that surface will appear cooler than at present; it would appear to glow red hot to our eyes. In fact, our Sun will become a red giant, and the helium in the Sun's core will burn to produce an inner core primarily of carbon.

As carbon accumulates in the core, a slow collapse will ensue until some other force intervenes. In the case of the Sun, that force will come from the Pauli exclusion principle. Recall from Chapter 9 that this principle dictates that no two electrons can occupy the same state. As the core starts to collapse, its electrons will be compressed into a smaller and smaller volume. They will reach the point (what we called the "full parking lot") where they can no longer be pushed together. At this point, the Pauli principle will take over and the electrons' intrinsic need for "elbow room" will control the Sun's size. The collapse will stop because the electrons cannot be pushed together any closer.

Astronomers call this countervailing force exerted by electrons the *degeneracy pressure*. They think of it as a permanent outward force on every particle in the star, which cancels the inward force of gravity.

When the Sun reaches this stage, it will be small, probably about the size of the Earth, and it will no longer generate energy. It will be very hot and will take a long time to cool off. During this phase, the temperature of each part of the Sun's surface will be very high, but, because the Sun will be so small, the total amount of radiation coming from it will not be very large. It will be, in other words, a white dwarf (see Figure 21–14). The carbon that is the end product of helium burning will remain locked in the white dwarf, and will not be returned to the cosmos.

Figure 21–14
A comparison among the sizes of the Sun, the Earth, and Sirius B. The latter is a white dwarf of approximately the same mass as the Sun.

One way of thinking about the life cycle of the Sun is to look at the path that it traces out in the H-R diagram. As a nebular cloud, the Sun had a low temperature and low energy. It entered the H-R diagram from the lower right corner. By the time it had become a full-fledged star, it had moved to the main sequence, where it sits now. Six billion years from now, when the Sun runs out of hydrogen, it will start to move into

Figure 21–15
The life cycle of the Sun on a Hertzsprung-Russell diagram. The Sun started hydrogen burning in its core more than 4.5 billion years ago on the main sequence (at point *1*), and it will remain near that point on the diagram for several billion years more. As the hydrogen in the core is consumed, however, a short period of helium burning (point *2*) will move the Sun's position on the diagram rapidly upward toward the red giant stage (point *3*). Once the helium is consumed, the nuclear fusion reactions will cease and gravitational contraction will cause the Sun to heat up (point *4*). Eventually the Sun will shrink to a white dwarf (point *5*).

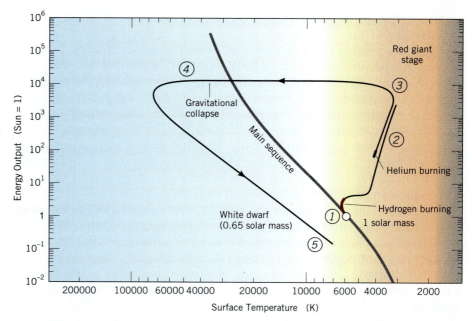

the upper right corner of the H-R diagram, as a red giant. It will then move quickly down into the lower left corner to become a white dwarf. Any other star like the Sun will trace out a similar path on the H-R diagram (see Figure 21–15).

Environment

The Role of the Sun

Now that we understand the life story of the Sun, from its birth in a contracting dust cloud to its death as a cooling white dwarf, there is one point that should be noted. During the billions of years that the Sun is on the main sequence, the total amount of energy it gives off stays roughly constant. Since it first entered the main sequence 4.5 billion years ago, the amount of energy generated by the Sun has increased by only about 30%. This long-term stability has important implications for the development of life on this planet.

Had life not developed on the Earth, the gradual warming of the Sun over the eons would have eventually caused the oceans to boil, filling the atmosphere with water vapor. Since water vapor is a greenhouse gas (see Chapter 19), this process would have turned the Earth into a somewhat cooler version of Venus. It would have been the ultimate environmental catastrophe. Fortunately, as we discussed in Chapter 19, while the Sun was warming, the composition of the Earth's atmosphere was changing. As it turned out, this change was just enough to compensate for the increased brightness of the Sun, but not enough to cool the Earth to the point where it entered a permanent ice age.

Had the Earth's distance from the Sun been a few percent less than it is, the change in atmosphere would not have been enough, the oceans would have boiled, and all life would have been lost. Had the distance been a few percent greater than it is, the oceans would have frozen solid and, again, all life would have disappeared. Mars may be an example of what such a frozen world would be like. In fact, scientists calculate that there is a very narrow band around a star, called the *continuously habitable zone*, in which a planet can exist for billions of years with liquid water on its surface. Only planets in this zone can support life. ●

The Death of Very Large Stars

Stars up to about eight times the mass of the Sun will carry out the life cycle of main sequence to red giant to white dwarf much as the Sun does. Such stars will have different lifetimes, depending on their size, but essentially the same life history.

One of the paradoxes of astronomy is that larger stars—those with the most hydrogen fuel—have the shortest lifetimes. This paradox arises because the largest stars have to burn hydrogen at a prodigious rate in order to overcome the intense force of gravity. Thus a star 10 times as massive as the Sun may live for a very short time—perhaps only 30 million years—compared to the Sun's 11-billion-year span. At the same time, stars much less massive than the Sun may live for hundreds of billions of years, rather than the Sun's 11 billion.

Large stars die differently from the Sun. For these stars, the pressure exerted by gravity is high enough so that the helium in the core not only burns to carbon, but the carbon can also undergo fusion reactions to produce oxygen, magnesium, silicon, and other larger nuclei. For such a star, the successive collapses and burnings will produce a layered, onion-like structure such as that shown in Figure 21–16.

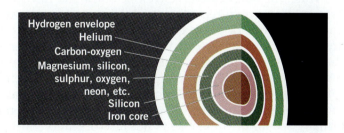

Figure 21–16
The interior of a large star displays concentric shells of fusion reactions, yielding progressively heavier elements toward the core.

In fact, this chain of nuclear burning goes on until iron, the element with 26 protons, is produced. As we noted in Chapter 14, iron is the most tightly bound nucleus: it requires energy to break the iron nucleus apart (nuclear fission), and energy to add more protons and neutrons to it (nuclear fusion). Thus it is impossible to extract energy from iron by any kind of nuclear reaction.

The cores of large stars will eventually fill up with iron "ash," and, no matter how high the pressure and temperature get, iron simply will not burn to produce a countervailing force to gravity. In fact, the iron core builds up until the force of gravity becomes so great that even the degeneracy pressure of the electrons cannot prevent collapse. At the incredible pressures and temperatures at the center of the star, the elec-

Figure 21–17
In a supernova, the core of a large star collapses to a neutron star, and shock waves blow the outer envelope of the star into space.

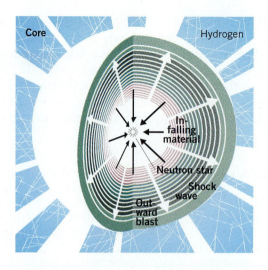

A false-color image of supernova 1987A, taken by the Hubble Space Telescope, reveals an expanding cloud of debris.

supernova   A stupendous explosion of a star, which increases its brightness hundreds of millions of times in a few days; results from the implosion of the core of a massive star at the end of its life.

neutron star   A very dense, very small star, usually with a high rate of rotation and a strong magnetic field; the core remains of a supernova, held up by the degeneracy pressure of neutrons.

trons actually combine with protons inside the iron nuclei, forming neutrons—a process that is the exact opposite of radioactive beta decay (see Chapter 14). Within a second or so, all of the protons in the iron nuclei are turned into neutrons, and all of the electrons disappear. At this point, the core of the star begins a catastrophic collapse. The collapse will go on until another force appears on the scene to counteract gravity. In this case, the force may be provided by the degeneracy pressure of the neutrons, which, like electrons, are subject to the Pauli exclusion principle.

The core collapses so fast that it falls inward beyond the point where the degeneracy pressure of the neutrons can balance gravity (see Figure 21–17). Like an acrobat jumping on a trampoline, the star's falling matter first bounces inward, then rebounds as the neutrons exert a counterpressure. Meanwhile, the outer gaseous envelope of the star has suddenly lost its support, and it begins a free-fall toward the interior of the star. When the collapsing envelope of dense gas meets the rebounding core of neutrons, intense shock waves are set up in the star, and the entire outer part of the star literally explodes. Someone watching this event from a distance will see a sudden brightening of the star in the sky, usually in a matter of a day or so. We call this dramatic event a **supernova.** Supernovas probably happen about every 30 years in the Milky Way galaxy. We don't normally see all of these events because of intervening dust, but we do see them in neighboring galaxies.

During the explosion, intense shock waves tear back and forth across the exploding star, raising the temperature enough to form all of the chemical elements in the periodic table. In a complex set of collisions, some of the nuclei up to iron that have been created by the successive fusion reactions soak up neutrons and undergo beta decay (see Chapter 14) to form nuclei up to uranium and beyond. All elements beyond iron are created in the short-lived maelstrom of the supernova explosion.

For a while, a supernova is surrounded by a cloud of ejected material. This expanding cloud dissipates into interstellar space, leaving behind the core of neutrons that was created in the collapse. A star that is being held up by degeneracy pressure of neutrons is called a **neutron star.**

Developing Your Intuition

Looking for Extraterrestrials

In many science fiction stories, extraterrestrials populate planets around stars that appear very bright in the Earth's sky—Rigel and Deneb, for example. Is this scenario reasonable?

Most stars that appear bright to us have a very high luminosity. We know this means that the stars are burning hydrogen at prodigious rates. But massive stars that burn hydrogen quickly have short lifetimes—sometimes only in the tens or hundreds of millions of years—after which they become supernovas. It took simple cells a billion years to develop on Earth, and intelligent life over 4 billion years. Planets around big stars just don't have time for this to happen, so science fiction writers should look elsewhere for their extraterrestrials. ●

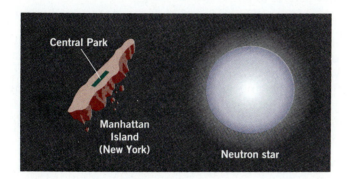

Figure 21–18
A neutron star is smaller than a typical city. Here one is shown in relation to the island of Manhattan in New York City.

Neutron Stars and Pulsars

A neutron star is, in essence, a giant nucleus—incredibly dense and very small. A typical neutron star might be 10 miles across—enough to fit within the city limits of even a moderate-sized metropolis (see Figure 21–18). Several significant things happen when a large star shrinks down into something the size of a city. For one thing, the rate of rotation of the star goes up substantially: just as an ice skater increases her spin when she pulls in her arms, a star rotates faster and faster as it contracts. In fact, some neutron stars in our galaxy rotate 1000 times a second. You can compare this to the sedate motion of the Sun, which rotates once every 26 days.

Neutron stars do not give off much light, and would probably have gone undetected if some of them didn't exhibit unusual behavior in the radio part of the spectrum. The reason for this behavior can be understood if you follow the collapse that leads to the neutron star. As the star collapses, the strength of its magnetic field increases. If a normal star has a dipole field (see Chapter 7), for example, then during the collapse, the field lines are dragged in with the material of the star so the field becomes much more concentrated and intense. Some neutron stars in our galaxy possess fields as much as a trillion times that of the magnetic field at the surface of the Earth.

Jocelyn Bell Burnell detected the first pulsars in 1967.

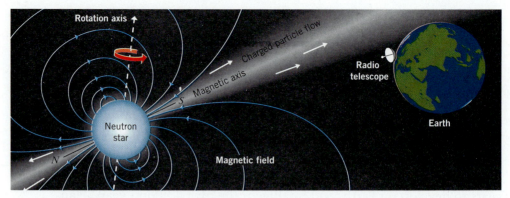

Figure 21–19
A schematic diagram of a pulsar reveals its two key attributes—rapid rotation and an intense magnetic field. This combination of traits produces a pulsing lighthouse-like pattern of energetic radiation.

pulsar   A neutron star in which fast moving particles speed out along the intense magnetic field lines of the rotating star, giving off electromagnetic radiation that we detect as a series of pulses of radio waves.

These two effects—a strong magnetic field and rapid rotation—may combine to produce a special kind of neutron star, which astronomers call a **pulsar.** Fast-moving particles speed out along the intense magnetic field lines of the rotating neutron star, and these accelerating particles give off electromagnetic radiation, as shown in Figure 21–19. Most of this radiation is in the radio range, so the neutron star's signal is seen primarily with radio telescopes.

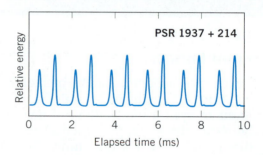

Figure 21–20
A graph showing regular radio pulses from a rapid pulsar. The pulses are about a thousandth of a second apart, indicating that the star is rotating five hundred times each second.

Figure 21–21
The Crab Nebula represents the remnants of a supernova that was observed in 1054 A.D. Today, the stellar material is being returned to space, where it may eventually form part of new stars. At the center of the nebual is a pulsar.

One way of thinking about a pulsar is to imagine it as being somewhat like a searchlight in the sky. Radio waves are continuously emitted along an axis that goes between the north and south magnetic poles of the neutron star, and this line describes a circle in space as the neutron star rotates. If you are standing in the line you will see a burst of radio waves every time the north or south pole of the pulsar is pointing toward you, and nothing when it's not. You will, in other words, see a series of pulses of radio waves. The signature of a pulsar in the sky is a series of regularly spaced pulses, typically some tens to thousands per second (see Figure 21–20). The pulsar represents one possible end state of a supernova. All pulsars are neutron stars, although all neutron stars are probably not seen as pulsars by earthbound astronomers.

We know of several pulsars that are the remnants of previous supernovas. The Crab Nebula (see Figure 21–21), a supernova seen from Earth in A.D. 1054, contains at its core one of the first pulsars discovered. Likewise, Supernova 1987A (see the following section) is also expected to reveal a pulsar when all the dust clears.

Our current theories of stellar evolution say that stars more than 10 times as massive as the Sun will go through the supernova process we've just described, and eject large amounts of heavy elements into space.

Science in the Making

The Discovery of a Planetary System

In 1994, Alexander Wolszczan at Penn State University announced the discovery of the first planetary system outside of our own. What is most surprising about the discovery, given our description of the violent nature of supernovae, is that the planetary system circles a pulsar.

This particular pulsar (called PSR B1257+12) rotates about 160 times per second. If a planet is orbiting the pulsar, then the gravitational force exerted by the planet will pull the pulsar toward it, and there will be a slight shift in the arrival times of the radio pulses. If the pulsar is moving away from us, the crests will arrive farther apart than average, and the opposite will be true if it is moving toward us, simply because the pulsar itself has moved in the 0.06 second between pulses. (This effect is similar to the Doppler effect discussed in Chapter 8).

After monitoring the pulsar for several years, Wolszczan was able to gather enough data to unravel the effects of three objects in orbit around the pulsar—two planets about three times the size of the Earth, and one about the size of the Moon. If this claim holds up, it will be the first successful detection of another planetary system.

How could a pulsar, the product of a supernova, have planets? The best guess is that the star that exploded to form PSR B1257+12 was once part of a double star system, and that the disk from which the planets were formed (see Chapter 17) came from the breakup of the companion after the supernova. ●

Alexander Wolszczan discovered the first planetary system outside our own.

Supernova 1987A

On February 23, 1987, a supernova was seen to explode in the Large Magellanic Clouds, a small galaxy-like structure near the Milky Way galaxy. Although the supernova was 170,000 light years from the Earth, it caused a great stir in science because it was the first supernova to be observed with modern observatories, including satellites. It was seen by large neutrino detectors on the Earth, by X-ray and gamma-ray observatories above the atmosphere, and by hundreds of ordinary telescopes on the Earth itself. Since it was the first supernova observed in 1987, it was given the name 1987A.

1987A was a classic supernova, and perhaps the biggest surprise to come out of the experience was that there were so few surprises. The intricate theories of nuclear reactions that take place in those incredibly complex few hours when a star explodes were largely confirmed. In the

This pair of photos was taken (a) on February 23, before, and (b) on February 26, after the explosion of Supernova 1987A.

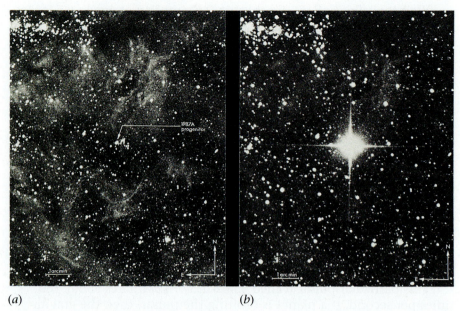

(a)   (b)

Figure 21–22
Light emitted by a star normally escapes into space (a), but as the star collapses, the space around it is warped and the path of emitted light begins to bend (b). Eventually, when a black hole forms (c), light is pulled back and cannot escape.

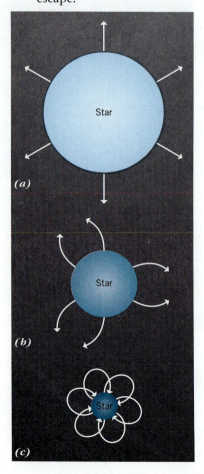

years that followed the first appearance of 1987A, astronomers have observed a sequence of particles characteristic of the decay of different nuclei that we believe are created in supernovas. In each case, the predictions of the theory seem to be matched by what actually happened.

Black Holes

Occasionally, a large star may die in a way that does not lead to the formation of a pulsar. If a star is large enough—perhaps 30 times as massive as the Sun—there may be processes, as yet only imperfectly understood, by which even the degeneracy pressure of neutrons is overcome and the star collapses. The result is the ultimate triumph of gravity, a **black hole.** A black hole is an object so dense, a mass so concentrated, that nothing, not even light, can escape from its surface (see Figure 21–22).

We do not know how often black holes are formed in the galaxy. Indeed, it comes as a surprise to most people to learn that no widely

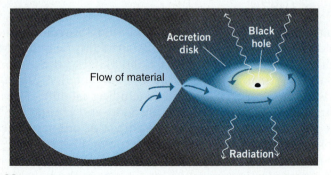

Figure 21–23
One way to look for a black hole is to try to detect X-rays emitted as material from a companion star falls into one.

accepted black hole with the mass of a single large star (a *stellar black hole*) has been detected in our galaxy. There is good evidence, however, for much larger black holes, equal in mass to many thousands of stars, at the center of this and other galaxies. Black holes are difficult to detect because you can't see them. Several promising candidates—objects that could very well be black holes—have been spotted, but none you could point to and say, "Yes, this is beyond doubt an example of a black hole."

The search for stellar black holes concentrates on double star systems in which one star has evolved into a black hole. The idea is that even though we can't see the black hole itself, we can see its effect on its partner. The most striking thing to look for is material falling into the black hole (see Figure 21–23). The enormous gravitational energy released in the process is partially converted into X-rays and gamma rays, which can be detected by orbiting observatories.

black hole Formed at the death of a very large star, an object so dense, with a mass so concentrated, that nothing— not even light—can escape from its surface.

THINKING MORE ABOUT STARS

The Generation of Chemical Elements

Almost everything you see about you was made in a supernova. Your body, for example, is made primarily from elements that formed in some distant exploding star more than 4.5 billion years ago. We say this because, as we shall see in Chapter 22, the universe began its life with only light elements—hydrogen, helium, and small amounts of lithium. These elements formed the first stars and were processed in the first stellar nuclear fires. In stars like the Sun, elements heavier than helium may be made, but they remain in that star and never return to the cosmos.

In large stars, however, all the elements up to uranium (the element with 92 protons) and beyond are made and spewed back into the interstellar medium in the titanic explosions we call supernovas. These heavy elements enrich the surrounding galaxy, and when new stars are formed, these new elements are incorporated into them. The Sun, which formed fairly late in the history of our own galaxy, thus incorporated many heavy elements that had been made in previous supernovas.

In fact, you can think of the history of our

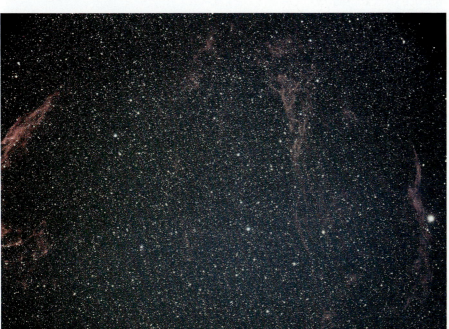

Bright reddish wisps in the constellation Cygnus define a circle, which suggests the location of an ancient supernova. Supernova material, enriched by the addition of heavy elements, is being returned to the interstellar medium.

galaxy as one of successive and cumulative enrichment by nuclear processing in large, short-lived stars. These stars, with lifetimes as short as tens of millions of years, take the original hydrogen in the galaxy and convert it into heavier elements. Thus we expect that older, smaller stars that have been shining since the early history of the universe will have fewer heavy elements than relatively young stars like the Sun—a prediction that is borne out by astronomical observations.

Think about what this means as you look around you. All the objects in your life—this book, your clothes, even your skin and bones—all are made of atoms that formed in the hearts of giant stars long ago.

▶ Summary

Astronomy, the study of objects in the heavens, has given us much knowledge about the nature and origins of *stars*. Stars like our own Sun form from giant clouds of interstellar dust—clouds that gradually collapse under the force of gravity. This collapse subjects the star's atoms, primarily hydrogen, to tremendous temperatures and pressures. The life of a star is a continuous struggle against this gravitational force.

Conditions of extreme temperature and pressure deep inside a star cause its hydrogen core to undergo nuclear fusion reactions, burning to create helium and heat energy. Ignition of these nuclear fires creates an outward flow of particles, called the *solar wind*. The fusion reaction proceeds in three steps, in which (1) two protons come together to form deuterium, (2) a proton and a deuterium nucleus come together to form helium-3, and (3) two helium-3 nuclei fuse to make helium-4. This energy creates the pressure that balances the force of gravity pulling the star inward.

Stars that are burning hydrogen to produce energy are said to be *main-sequence stars*. Larger stars burn hotter and emit more energy, while smaller stars are cooler and radiate less energy. Main-sequence stars are found in a simple bandlike pattern on a *Hertzsprung-Russell diagram*, which graphs a star's energy output versus its temperature.

We study stars with *telescopes*, instruments that gather and focus electromagnetic radiation. Earth-based telescopes detect visible and radio waves, while orbiting observatories detect all other regions of the electromagnetic spectrum. A star's brightness is measured in magnitude. The brightness of the star as seen from earth is its *apparent magnitude*, which depends on both the amount of energy it produces and its distance from us. The *absolute magnitude* is a star's brightness when viewed from a fixed distance, and depends only on its energy output. The most powerful telescopes can detect stars that are hundreds of millions of *light-years* away.

When a star like the Sun consumes most of its core hydrogen, a helium-rich central region remains. The star once again begins to collapse under gravity, and internal temperatures rise again. Hydrogen burning begins in shells outside the core, while the core's helium may also combine in nuclear fusion reactions to form carbon. These new nuclear processes may briefly cause a star like the Sun to expand and become a *red giant*, a star whose outer layers glow red hot. Eventually, however, nuclear fuel must be exhausted. Gravity will dominate, and the carbon-rich star will collapse to a very small, very hot *white dwarf*.

Stars at least eight times larger than the Sun may evolve beyond hydrogen and helium burning. If temperatures and pressure are high enough, carbon can undergo additional nuclear reactions to form elements as heavy as iron, the ultimate nuclear ash. Once iron is formed, however, there can be no more energy produced by these reactions and burning will cease. The sudden extinguishing of a star causes a catastrophic gravitational collapse and rebound—a *supernova*—in which the star literally explodes and spews all the chemical elements into the heavens. A dense, spinning *neutron star* or *pulsar* may be the only remnant of the original star. The largest stars may collapse into a *black hole*—an object so massive that not even light can escape its gravitational pull.

▶ Review Questions

1. What is a star?

2. Why do stars emit energy in the form of radiation?

3. What are the major layers of the Sun?

4. What two properties of stars do scientists plot on a Hertzsprung-Russell diagram? Why do they choose these properties?

5. In what ways is the Sun a typical star?

6. What is the solar wind?

7. Describe two kinds of telescopes. In what ways are they similar?

8. What are the advantages of placing a telescope in orbit?

9. Describe two ways to determine the distance to another star.

10. Describe hydrogen burning. Where does it take place?

11. Why must the Sun eventually die? What changes will the Sun undergo before it dies?

12. Why do large stars have shorter lives than smaller stars?

13. Why won't the Sun become a supernova or a black hole?

14. How are supernovas and neutron stars related to each other?

15. How are neutron stars and pulsars related to each other?

16. If iron is the ultimate nuclear ash, where do elements heavier than iron come from?

17. Why is it difficult to detect a black hole from Earth?

▶ Fill in the Blanks

Complete the following paragraphs with words and phrases from the list.

absolute magnitude	pulsars
apparent magnitude	red giant
astronomy	solar wind
black hole	stars
light-years	supernovas
main-sequence star	telescopes
neutron star	white dwarf

_____ is the branch of science devoted to studying the heavens. _____ are large bodies of gas that derive energy from fusion reactions and radiate it into space. The brightness of a star as seen from Earth is called its _____, while the brightness seen from a standard distance is called its _____. We observe stars with Earth-based _____ and orbiting observatories. Typically, the distance between stars is measured in _____, the distance light travels in a year.

The Sun is a _____ star. It emits particles called the _____. It will eventually become a _____ and then a _____. Larger stars will become _____, and their remnants will be highly compact _____. A rotating star of this type may appear as a _____. If the star is massive enough, its end state may be a _____.

▶ Discussion Questions

1. Why do we see stars only at night? Do they shine during the day?

2. How might you determine the age of a star from an Earth-based telescope? What measurements might you make?

3. In the science fiction movie classic *Star Wars*, Han Solo speaks the following line: "Fast? This is the ship that made the Kessel run in under 12 parsecs. She's fast enough for you, old man." What grade would Captain Solo receive in Astronomy 101? Why?

4. What was happening on Earth when the energy in the sunlight falling outside your window today was first produced in fusion reactions at the Sun's core? Were the Pyramids built yet? Did people live in cities? Were Neanderthals still around?

5. Given the connection between the Sun and the human eye, speculate about the sorts of stars that might have been found in solar systems from which different fictional extraterrestrials are supposed to have come. Would the Klingon and Vulcan stars in *Star Trek* have been much different from our own? Why or why not?

6. How can we talk about the evolution of stars over billions of years when human beings have been observing stars for only a few thousand years?

7. How does the principle of conservation of energy apply to a supernova?

8. Would a star surrounded by a Dyson sphere be visible at night? How would you go about searching for such a star? What kind of instruments would you need to detect it?

9. Most stars we see are on the main sequence. Stars spend most of their lives consuming their initial stock of hydrogen. Is there a connection between these two statements? If so, what?

10. What are some differences between a larger star and a smaller star?

11. What are some differences between a star and a planet?

12. Why do all stars eventually have an ending? Retell this insight using the basic principles of thermodynamics.

13. What is meant by "hydrogen burning" in the Sun, and why does it have to take place in the center of the Sun?

14. Outline the major layers of the Sun according to their distance from the center of the Sun. List the relative size, temperature, density, and one principal characteristic for each layer.

15. What are the advantages and disadvantages of placing a telescope in orbit around the Earth? On the surface of the Moon?

16. How is the apparent magnitude of a star related to the absolute magnitude of the same star?

17. Why is the period-luminosity relation for Cepheid variable stars significant?

18. How is hydrogen burning related to the main sequence of stars?

19. Outline the evolution of a star similar to the Sun from its main-sequence phase to its white dwarf stage. For each significant step, briefly describe and distinguish the physical conditions in the interior of the star, and those properties that are observed on the surface of the star.

20. Compare and contrast the evolutionary processes and stages between a low-mass star (like the Sun), and a high-mass star with nearly 10 times the Sun's mass.

▶ Problems

1. Mercury, the closest planet to the Sun, lies at an average distance of 58 million km (about 36 million miles) from the Sun. How much energy falls on a square meter of Mercury's sunlit surface? What does this number imply about the possibility of life surviving on the planet?

2. If the Sun shines for 11 billion years at about its present energy output, how much total energy will the star send into space during its lifetime?

3. How much energy would fall on a one-square-meter detector above the atmosphere of Mars (2.3×10^{11} m from the Sun)? Of Jupiter (7.8×10^{11} m)? Pluto (5.9×10^{12} m)?

4. How far away is Alpha Centauri, the nearest star? How long would it take to get there at a speed of 2000 mph (the speed of a fast jet plane)?

5. Calculate the total amount of solar energy that reaches the sunlit side of the Earth's surface every second. (*Hint:* Assume that the Earth is a perfect sphere.)

6. Calculate the total amount of energy that reaches the sunlit side of the surface of the planet Mars every second. Mars is approximately 1.5 times more distant from the Sun than the Earth.

7. Jeanne did not believe the estimate that if the Sun were completely composed of anthracite coal, the Sun could last for only about 5000 years if it radiated continuously at its present luminosity. Verify this calculation for Jeanne by answering the following questions:

  a. What is the total amount of energy available from the Sun if it is composed entirely of anthracite coal? The heat of combustion of coal is 3.3×10^7 J/kg.

  b. Calculate the total time that the Sun can radiate at its present rate of 4.24×10^{23} kw.

8. Julio thought that if the Sun were composed of methane gas, it could last for a considerably longer of time since methane has one of the highest heats of combustion for known fuels. The heat of combustion for methane is 1.3×10^4 kcal/kg.

  a. How long would the Sun continue to radiate at

its present rate if it were composed entirely of methane gas?

  b. From your answers to this question and the previous one, state whether geological time scales are consistent with the age of the Sun if it were composed entirely of coal or natural gas. If not, provide a plausible alternative explanation.

9. Rank the following celestial objects according to their apparent brightness, the brightest first.

Object	Apparent magnitude
Full Moon	−12.6
Sun	−26.8
Sirius, brightest star in sky	−1.4
Venus, maximum brightness	−4.4
Pluto, maximum brightness	+14.9
Faintest star observed by eye	+6.0
Brightest quasar	+12.8
Faintest star, Hubble Telescope	+30.0

10. Calculate the number of kilometers and miles in one parsec.

11. Polaris, the North Star, is a Cepheid variable with a period of nearly 4 days. The amount of light from Polaris that reaches your telescope is 3.8×10^{-23} w/m². How far is Polaris from the solar system?

12. The Large Magellanic Cloud (LMC) is one of the nearest galaxies to the Milky Way galaxy (and the solar system). If the distance to the LMC is 150,000 light-years, what is the amount of light that reaches our telescopes (in w/m²) for a Cepheid variable with a 100–day period?

13. The pulsar in the Crab Nebula has a rotational period of 0.033 seconds, one of the shortest known periods for a pulsar.

  a. What is its rotational frequency in hertz?

  b. How many times in one second does it rotate?

  c. If a spot on the neutron star producing this frequency was at the equator, how fast is this spot moving? (Assume that the neutron star is a sphere with a radius of 10 km.)

  d. Compare this speed to the speed of light.

14. Some neutron stars in our galaxy have rotational frequencies of 0.001 Hz.

  a. How many times in one second will these fast-rotating neutron stars rotate?

  b. If the spot on a neutron star producing this frequency is at the equator, how fast is this spot moving? (Assume that the neutron star is a sphere with a radius of 10 km.)

  c. Compare this speed to the speed of light.

▶ Investigations

1. Locate some stars in the sky and find out their apparent magnitudes. (You might want to start with some familiar stars such as those in the Big Dipper.)

2. The Crab Nebula is the remains of a supernova event that was sighted almost 1000 years ago, and must have been visible as a brilliant object for several days. What cultures left a record of this astronomical event? How did they explain what they saw?

3. You can set up an analog to the astronomical distance scale by using two "yardsticks"—a ruler and a tape measure, for example—to measure distances. Measure the dimensions of your classroom this way. How would you make sure that distances on each "yardstick" were the same? Does this exercise suggest a way for astronomers to check the consistency of their distance scale?

4. The nineteenth-century American poet Walt Whitman (1819–1892) wrote the following poem about astronomy.

When I Heard the Learn'd Astronomer

When I heard the learn'd astronomer,
When the proofs, the figures, were ranged in columns before me,
When I was shown the charts and diagrams, to add, divide, and measure them,
When I sitting heard the astronomer where he lectured with much applause in the lecture-room,
How soon unaccountable I became tired and sick,
Till rising and gliding out I wander'd off by myself,
In the mystical moist night air, and from time to time,
Look'd up in perfect silence at the stars.

Although Whitman was unimpressed by the facts and figures of the "Learn'd Astronomer," astronomers of the past century have changed the way we think about our place in the universe. In this respect, how does science complement poetry? How do poetry and astronomy differ as ways of understanding why we are here? How would you answer the poet today?

▶ Additional Reading

Ferris, Timothy. *Coming of Age in the Milky Way*. New York: William Morrow, 1988.

Goldsmith, Donald. *Supernova! The Exploding Star of 1987*. New York: St. Martin's Press, 1989.

———. *The Astronomers*. New York: St. Martin's Press, 1991.

Sullivan, Walter. *Black Holes*. New York: Doubleday, 1979.

Trefil, James. *Space Time Infinity*. Washington, D.C.: Smithsonian Books, 1985.

22 COSMOLOGY

THE UNIVERSE BEGAN BILLIONS OF YEARS AGO IN THE
BIG BANG, AND IT HAS BEEN EXPANDING EVER SINCE.

The Campfire

One of life's most pleasant experiences is to sit outside around a campfire as dusk falls, watching the flames leap and dance. If you observe
closely, you may notice that the color of the coals in the fire changes,
depending on how hot the fire is. They are ordinarily red, but in a roar-
ing blaze they may actually become white. Then, as the fire starts to go
out and people start to unroll their sleeping bags and get ready for
bed, the coals glow a dull red and, eventually, stop glowing altogether.

Even when the coals aren't glowing, they give off energy in the
form of infrared radiation, which you can feel if you put your hands
close to the fire. You may still be able to feel the radiation the next
day as the embers are cooling. It was this phenomenon that led
twentieth-century scientists to a completely new understanding of the
structure and history of the universe in which we live.

THE NATURE OF THE COSMOS

Milky Way A collection of about 100 billion stars forming the galaxy of which the Sun is a part.

galaxy A large assembly of stars (between millions and hundreds of billions of them), together with gas, dust, and other materials, that is held together by the forces of mutual gravitational attraction.

cosmology The branch of science that is devoted to the study of the structure and history of the entire universe.

On any given night you can see several thousand points of light in the sky, almost all of which lie within a collection of about 100 billion stars we call the **Milky Way** galaxy. A **galaxy** is a large assembly of stars (between millions and hundreds of billions of them), gas, dust, and other materials. Galaxies are held together by the forces of mutual gravitational attraction.

Copernicus, Kepler, and Newton had shown that the Earth is not at the center of the universe. Although we knew that the Earth was only one of several planets circling the Sun, it was not until the first part of the twentieth century that we had to shed the belief that our own Milky Way was the only large collection of stars in the universe. In the last 70 years or so, we have come to understand that the collection of stars we call home is only one of many billions of similar collections in the sky. The more we learn about our universe, the less unique our home planet Earth seems to be.

In 1924, a young American astronomer named Edwin Hubble established beyond a shadow of a doubt that the Milky Way is just one of a countless number of galaxies in the universe. In doing so, he set the tone for a century of progress in the new branch of science called **cosmology,** which is devoted to the study of the structure and history of the entire universe.

Science in the Making

Edwin Hubble and the Birth of Modern Cosmology

In 1919, Edwin Hubble (1889–1963) went to work at Mount Wilson Observatory near Los Angeles. Mount Wilson was then a lonely observatory far from any major city. Today it has been engulfed by the Los Angeles metropolitan area, but in those days it afforded astronomers a chance to look at the sky through clear, unpolluted air. This observatory had what was then the world's largest telescope, with a mirror that measured 100 inches, nearly 3 meters, across (see Figure 22–1).

Edwin Hubble led an extraordinary life and excelled at an astonishing variety of endeavors. The Missouri-born Hubble became an honor student at the University of Chicago, where he lettered in track and played on a championship basketball team. He was also a talented amateur boxer, and at one point had to choose between continuing his education and becoming a professional prize fighter. He opted to take a Rhodes Scholarship, and studied law at Oxford. He practiced law for a while in Kentucky, decided he didn't like it, and returned to Chicago to study for a Ph.D. in astronomy. He got his degree at about the time that the United States entered World War I, and enlisted in the infantry. By the end of the war, he had risen to the rank of major and, at age 30, was ready to begin the career that was to change forever our view of humanity's place in the universe.

By the time Hubble joined the astronomers at Mount Wilson, the debate about the nature of the universe we live in was almost a century old. Scientists had known for a long time that the Sun is one of a large collection of stars in the Milky Way. But the question that re-

Figure 22–1
Edwin Hubble (1889–1953) at the 100-inch telescope of California's Mount Wilson Observatory.

mained was whether this collection is just one of many "island universes" in the vastness of space, or is the Milky Way all there is? Throughout the latter part of the nineteenth century and the early part of the twentieth century, the debate raged among astronomers. The controversy continued for one simple reason: while it was evident that stars in the Milky Way are no more than about 100,000 light-years away, no one could provide an accurate measurement of the distance to any object that might lie outside the Milky Way.

The debate centered around objects called *nebulae*—fuzzy structures in the sky that could not be seen clearly with the telescopes available at that time. Nebulae were long thought to be dust clouds or large collections of stars, but no one was sure whether they were in the Milky Way or outside of it. The new 100–inch telescope at Mount Wilson allowed Hubble to see individual Cepheid variable stars (see Chapter 21) in some nebulae (something no one had been able to do before), thus he was able to measure distances to them. It turned out that the distance to the nearest one, the Andromeda nebula, was some 2 million light-years—far outside the bounds of the Milky Way. Thus, with a single observation, Hubble established one of the most important facts about the universe we live in: it is made up of countless galaxies, of which the Milky Way is but one. ●

Kinds of Galaxies

The Milky Way is a rather typical *spiral galaxy*. As shown in Figure 22–2, it is a flattened disk about 100,000 light-years across, with a central bulge known as a *nucleus*. Bright regions in the disk, known as *spiral arms*, mark

(a)

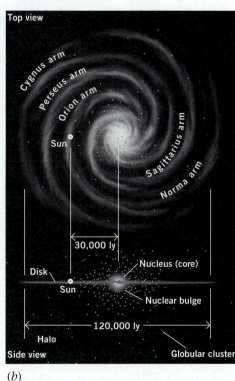

(b)

Figure 22–2
The Milky Way Galaxy. (*a*) A photograph of a small portion of the galaxy shows a tiny fraction of the 100 billion stars it contains. (*b*) A more detailed diagram shows the spiral arms and the location of the Sun.

Figure 22–3
A typical spiral galaxy, with a bright core and spiral arms where new stars are forming.

Figure 22–4
A typical elliptical galaxy. This one is known as M84 and is located in the constellation Virgo.

A false-color image of radio emissions from a quasar reveals a jet of material streaming away from the nucleus.

areas where new stars are being formed. About 75% of the brighter galaxies in the sky are of this spiral type (see Figures 22–2 and 22–3).

Another class of galaxies, known as *ellipticals*, exhibit a shape something like a football (see Figure 22–4). The brightest elliptical galaxies tend to have more stars than spiral galaxies do, and comprise about 20% of bright galaxies. In addition to the relatively large and bright elliptical and spiral galaxies, the universe is littered with small collections of stars known as *irregular galaxies* and *dwarf galaxies*. Most astronomers suspect that these two types—given that they are faint and therefore difficult to detect—are probably the most common galaxies in the universe.

Larger elliptical and spiral galaxies and smaller irregular and dwarf galaxies can be thought of as quiet, homey places, where the process of star formation and death proceed in an orderly fashion. But a small number of galaxies, perhaps 10,000 among the billions known, are quite different and are referred to collectively as *active galaxies*. The most spectacular of these unusual objects are the *quasars* (for quasi-stellar radio sources). Quasars are wild, explosive, violent objects, where as-yet-unknown processes pour vast amounts of energy into space each second from an active center no larger than the solar system. Astronomers suggest that the only way to generate this kind of energy is for the center of a quasar to be occupied by an enormous black hole (with masses, in some cases, millions of times greater than that of the Sun), and for the energy to be generated by huge amounts of mass falling into this center.

Because they are so bright, quasars are the most distant objects we can see in the universe.

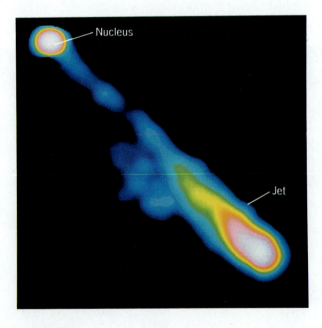

THE REDSHIFT AND HUBBLE'S LAW

Hubble's recognition of galaxies other than our own Milky Way wasn't the end of his discoveries. When he looked at the light from nearby galaxies, he noticed (as had others) that the distinctive colors emitted by

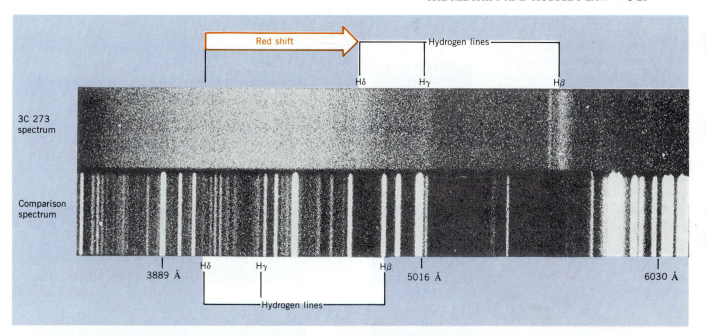

different elements seemed to be shifted toward the red (long-wavelength) end of the spectrum, compared to light emitted by atoms on Earth. Hubble interpreted this **redshift** as an example of the Doppler effect (see Chapter 8). This is the same phenomenon that causes the sound of a car whizzing past to change its pitch. Hubble's observation meant that distant galaxies are moving away from the Earth. Furthermore, Hubble noticed that the more distant a galaxy, the faster it moves away from us (Figures 22–5 and 22–6).

On the basis of measurements of a few dozen nearby galaxies, Hubble suggested that a simple relationship exists between the distance of an object from the Earth, and that object's speed away from the Earth. Comparing two galaxies, one twice as far away from the Earth as the other, the farther galaxy moves away from us twice as fast. This statement, which has been amply confirmed by measurements in the subsequent half-century, is now called **Hubble's law.** Hubble's law says

▶ **In words:**

The farther away a galaxy is, the faster it recedes

▶ **In equation form:**

galaxy's velocity = (Hubble's constant) × (galaxy's distance)

▶ **In symbols:**

$v = H \times d$

Hubble's law tells us that we can determine the distance to galaxies by measuring the redshift of the light we receive from them, whether we can make out individual stars in them or not. Astronomers continue to debate the exact value of Hubble's constant of proportionality, but most experts agree that it is between 50 and 100 kilometers per second per

The spectrum of a quasar shows a dramatic red shift of three characteristic hydrogen lines.

redshift An increase in the wavelength of the radiation received from a receding celestial body as a consequence of the Doppler effect; a shift toward the long-wavelength (red) end of the spectrum.

Hubble's law The law relating the distance to a galaxy, d, and the rate at which it recedes from Earth as measured by the red shift: $v = Hd$.

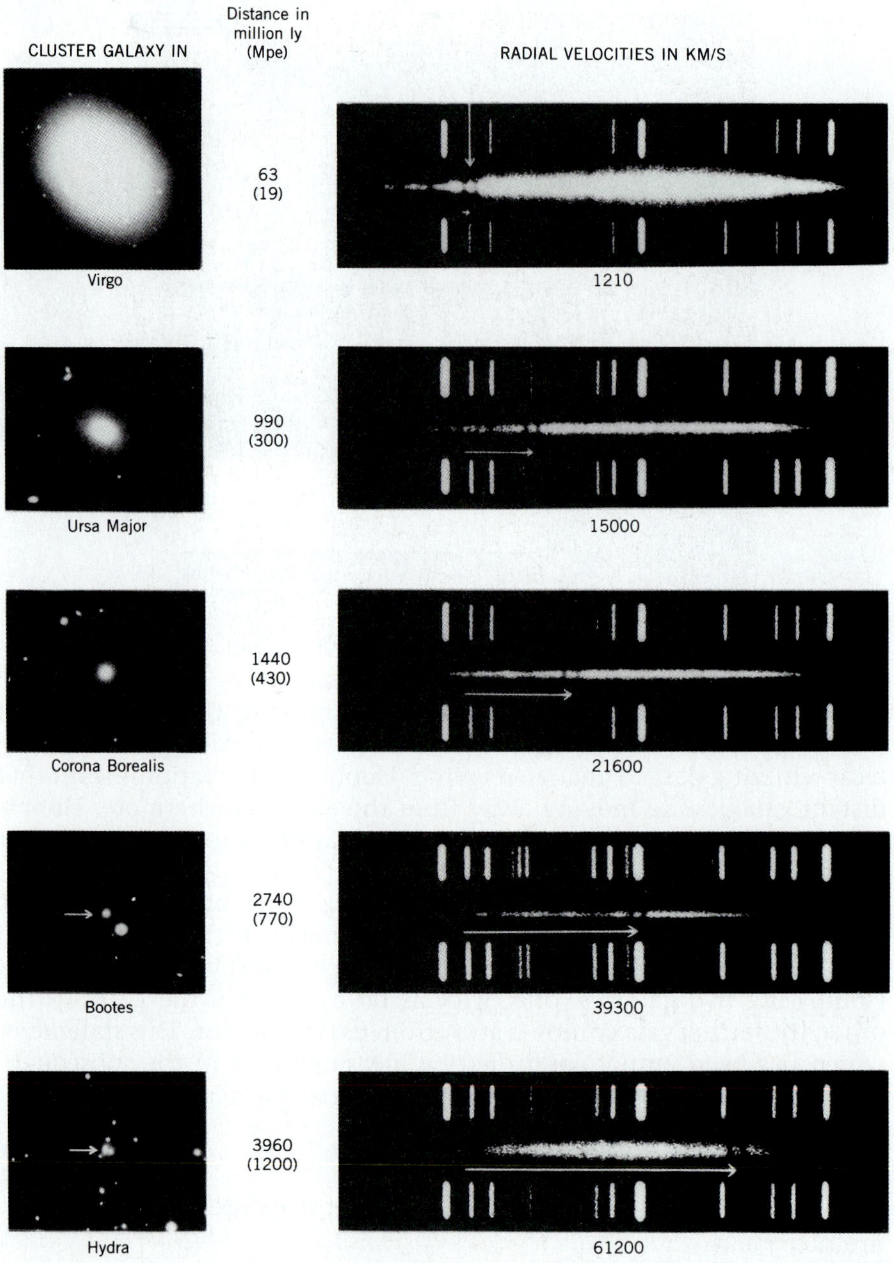

Figure 22–5
Photographs of galaxies as seen through a telescope (on the left), with spectra of those galaxies (on the right). The double dark lines in the spectra, characteristic of the calcium atom, are shifted farther to the right (toward the red) the farther away the object is. Thus more-distant galaxies are traveling away from us at higher velocities. This phenomenon was used by Edwin Hubble to derive his law.

CLUSTER GALAXY IN	Distance in million ly (Mpe)	RADIAL VELOCITIES IN KM/S
Virgo	63 (19)	1210
Ursa Major	990 (300)	15000
Corona Borealis	1440 (430)	21600
Bootes	2740 (770)	39300
Hydra	3960 (1200)	61200

megaparsec (a megaparsec, abbreviated Mpc, is a million parsecs, or 3.3 million light-years). In this view of the cosmos, the redshift becomes the final "ruler" in the astronomical distance scale (see Chapter 21).

One way of interpreting Hubble's constant is to notice that if a galaxy were to travel from the location of the Milky Way to its present position with velocity v, then the time it would take to make the trip would be

$$\text{time} = \frac{\text{distance}}{\text{velocity}}$$
$$= \frac{d}{v}$$

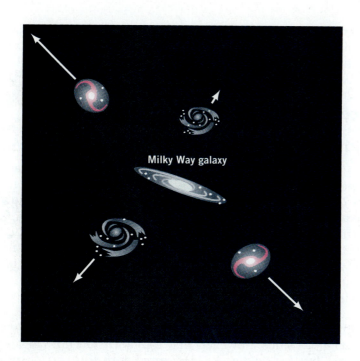

Figure 22–6
Illustration of Hubble expansion. The more distant a galaxy is from the Earth, the faster it moves away from us (as shown by the lengths of the arrows).

Milky Way galaxy

Substituting for *v* from Hubble's law,

$$\text{time} = \frac{d}{(H \times d)}$$

$$= \frac{1}{H}$$

Thus the Hubble constant provides a rough estimate of the time that the expansion has been going on and, hence, of the age of the universe. A Hubble constant of 50 km/s/Mpc corresponds to an age of the universe of about 16 billion years, while a constant of 100 km/s/Mpc corresponds to an age of about 8 billion years.

Example 22–1:   The Distance to a Receding Galaxy

Astronomers discover a new galaxy and determine from its redshift that it is moving away from us at approximately 100,000 km/s (about a third of the speed of light). Approximately how far away is this galaxy? Assume an intermediate value of 75 km/s/Mpc for the Hubble constant.

▶ **Reasoning:** According to Hubble's law, a galaxy's distance equals its velocity divided by the Hubble constant.

▶ **Solution:**

$$\text{distance (in Mpc)} = \frac{\text{velocity (in km/s)}}{\text{Hubble's constant (in km/s/Mpc)}}$$

$$= \frac{100,000 \text{ km/s}}{75 \text{ km/s/Mpc}}$$

$$= \frac{100,000}{75} \text{ Mpc}$$

$$= 1333 \text{ Mpc}$$

Remember, a parsec equals about 3.3 ly, so this galaxy is more than 4 billion light years away. The light that we observe from such a distant galaxy began its trip about the time that our solar system was born. ▲

Science by the Numbers

Analyzing Hubble's Data

In his original sample, Hubble observed 46 galaxies, but was able to determine distances to only 24. Some of his data are given in Figure 22–7.

How does one go about analyzing data such as these? One way is to make a graph. In this case, the vertical axis is the velocity of recession of the galaxy, based on measurements of redshift, and the horizontal axis is the distance to the galaxy, based on measurements of cepheid variables. In Figure 22–7, we show the data as originally plotted by Hubble.

Looking at the data, the general trend of Hubble's law becomes clear—the farther you go to the right (i.e., the farther away the galaxies are), the higher the points (i.e., the faster the galaxies are moving away). You also notice, however, that the points do not fall on a straight line, but are scattered. Confronted with this sort of situation, you can do one of two things. You can assume that the scattering is due to experimental error, and that more precise experiments will verify that the points fall on a straight line; or you can assume that the scatter is a real phenomenon and try to explain it.

Hubble took the first alternative; he assumed that the distance to a galaxy was *directly proportional* to the speed of the galaxies (see Chapter 2 for a discussion of proportionality). So the only problem left was to find the line about which experimental error was scattering his data. This is usually done by finding the line for which the sum of the distance between the line and each data point is smaller than for any other line. In effect, you find the line that comes closest to all the data points. The slope of this line, which measures how fast the velocity increases for a given change in distance, is the best estimate of Hubble's constant. ●

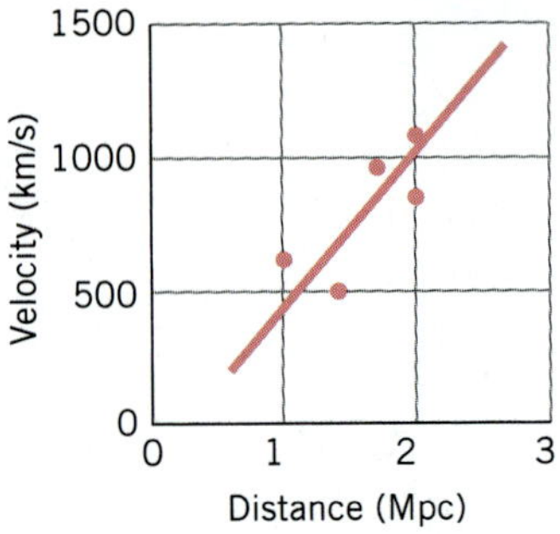

Distance to galaxy (Mpc)	Velocity (km/s)
1.0	620
1.4	500
1.7	960
2.0	850
2.0	1090

Figure 22–7
Hubble's distance-versus-velocity data for five galaxies.

THE BIG BANG

Hubble's law reveals something extraordinary about our universe: it is expanding. Nearby galaxies are moving away from us, and more-distant galaxies are moving away even faster. The whole universe is blowing up like a balloon. This startling fact, in turn, leads us to perhaps the most amazing discovery of all. If you look at our expanding universe today and imagine moving backward in time (think of running a videotape in reverse), you can see that at some point in the past the universe must have started out as a very small object. In other words, Hubble's observations lead us to a startling conclusion about the universe:

> **The universe began at a specific point in the past, and it has been expanding ever since.**

This picture of the universe—that it began from a small, dense collection of matter and has been expanding ever since—is called the **big bang theory.** This theory constitutes our best idea of what the early universe was like.

big bang theory The idea that the universe began at a specific point in the past and has been expanding ever since.

Think how different the big bang theory of the universe is from that of the Greeks or the medieval scholars, or even the great scientists of the nineteenth century whose work we have studied. To them, the Earth went in stately orbit around the Sun, and the Sun moved among the stars, but the collection of stars you can see at night with your naked eye or with a telescope was all that there was. Suddenly, with Hubble's work, the universe grew immeasurably. Our own galaxy became just one of at least 100 billion galaxies in a universe in which galaxies are flying away from each other at incredible speeds. It is a vision of a universe that began at some time in the distant past, and may end at some time in the future.

Developing Your Intuition

A Different Universe

What would Hubble have seen if galaxies in the universe were scattered at random, but there was no overall expansion?

In general, we would expect the galaxies to move around with respect to each other. In fact, you could probably think of the motion of these galaxies as being analogous to the movement of atoms in a gas. Looking into such a collection, you would expect to see some galaxies moving toward you and some moving away. If the universe had been organized this way, therefore, Hubble would have seen roughly as many spectra blueshifted (i.e., moving toward us) as redshifted (i.e., moving away from us). The failure to see many blueshifted galaxies is a tip-off that there is something ordered about the motion of galaxies (see Figure 22–8).

Furthermore, in such a randomly organized universe, it would make no difference whether a galaxy was nearby or far away. Hubble would have seen exactly the same distribution of velocities for close, bright galaxies and the most distant ones he could measure. ●

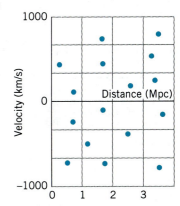

Figure 22–8
If the universe had been different, Hubble might have seen data like this rather than the points shown in Figure 22–7. The negative values of the velocity in this hypothetical diagram would represent galaxies moving toward us, rather than receding.

The Large-Scale Structure of the Universe

The Milky Way is part of a group of galaxies known as the *Local Group*, made up of ourselves, the Andromeda galaxy, and perhaps a dozen small "suburban" galaxies (see Figure 22–9). The Andromeda galaxy is actually visible to the unaided eye from the Earth; on a clear summer night it appears as a fuzzy patch of light in the northeast. The Local Group, in turn, is part of the *Local Supercluster*, a collection of galaxies about 100 million light-years across.

We now know that literally billions of galaxies populate the universe, each galaxy a collection of billions of stars. Most galaxies seem to be clumped together into *groups* and *clusters*, many of which are, in turn, grouped into larger collections called *superclusters* containing thousands of galaxies.

Figure 22–9
Here's one idea of how the Milky Way and its neighbors would look to someone in another galaxy.

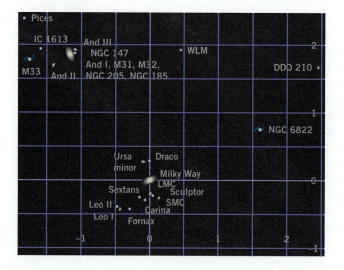

In the 1980s, astronomers began making "redshift surveys" of the sky. In these surveys, they not only plotted the locations of distant galaxies, but they also measured their redshifts so we would know how far away they are. In this way, they constructed a full, three-dimensional picture of the distribution of matter in the cosmos. With the results from these observations, led primarily by the team of Margaret Geller and John Huchra at the Harvard-Smithsonian Astrophysical Observatory, astronomers put together a picture of the universe that is very different from what they originally expected. Instead of finding galaxies scattered more or less at random through space, they found that galaxies are collected into large structures that run for billions of light-years across the sky. In fact, to get an excellent picture of the structure of the universe imagine taking a knife and slicing through a big pile of soapsuds. The result will give you a structure in which large empty spaces are surrounded by soap film. In exactly the same way, matter in the universe seems to be concentrated in superclusters on the surfaces of large empty areas called *voids* (see Figure 22–10). Attempts to understand the reason for this very complex structure in the universe remain one of the major preoccupations of modern cosmology.

A cluster of galaxies in the constellation Virgo.

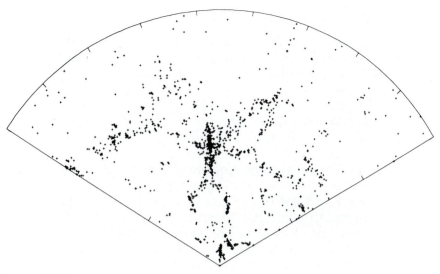

The Andromeda Galaxy is a spiral galaxy like our own Milky Way, and is one of our nearest galactic neighbors at two million light years.

Figure 22–10
The large-scale structure of the universe. This figure shows a thin slice of the universe, with each galaxy represented as a point (the Milky Way lies at the lower point in this figure). Matter is concentrated in superclusters, with large, relatively empty areas called voids between them.

The Large Magellanic Cloud is a small, irregular galaxy near the Milky Way galaxy.

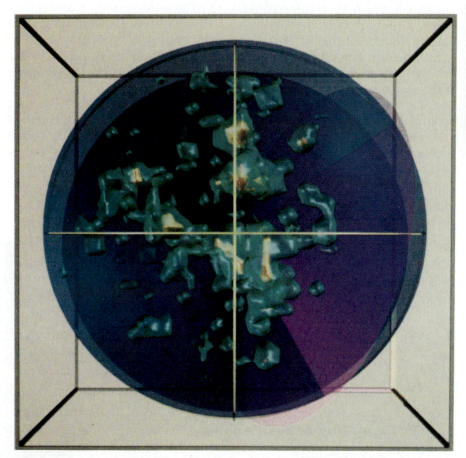

A computer simulation of the distribution of galaxies in space. Concentrations of galaxies (in orange) are separated by voids.

Some Useful Analogies

The big bang picture of the universe is so important that we should spend some time thinking about it. Many analogies can be used to help us picture what the expanding universe is like, and we'll look at two. Be forewarned, however: none of these analogies is perfect. If you pursue any of them far enough they fail, because none of them captures the entirety and complexity of the universe in which we live. And yet each of the analogies can help us understand aspects of that universe.

1. The Raisin-Bread Dough Analogy One standard way of thinking about the big bang is to imagine the universe as being analogous to a huge vat of rising bread dough in a bread factory (see Figure 22–11). If raisins scattered through that dough represent galaxies, and if you're standing on any one of those raisins, then you would look around you and see other raisins moving away from you. You could watch as a nearby raisin moved away because the dough between you and it is expanding. A nearby raisin wouldn't be moving very fast, because there isn't much expanding dough between you and that raisin. A raisin three times as far away, however, would move away faster—three times faster, in fact, because three times as much dough lies between you and that more distant raisin.

The raisin-bread dough analogy is very useful because it makes it easy to visualize how everything could seem to be moving away from us, with objects that are farther away moving faster. If you stand on any raisin in the dough, all the other raisins look as though they're moving away. This analogy, therefore, explains why the Earth seems to be the center of the universe. It also explains why *every* point appears to be at the center of the universe.

But the expanding-dough analogy fails to address one of the most commonly asked questions about the Hubble expansion: What is outside the expansion? A mass of bread dough, after all, has a middle and an outer surface; some raisins are closer to the center than others. But we believe the universe has no surface, no outside and inside, and no unique central position. In this regard, the surface of an expanding balloon provides a better analogy.

Figure 22–11
The raisin-bread dough analogy of the expanding universe. As the dough expands, all raisins move apart from each other—the farther apart the raisins, the faster the distance increases.

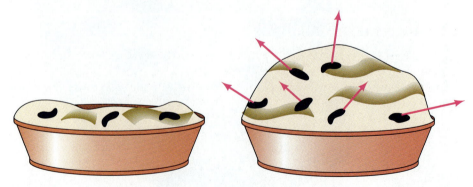

2. The Expanding-Balloon Analogy Imagine that you live on the surface of a balloon in a two-dimensional universe. You would be absolutely flat, living on a flat-surfaced universe (similar to the way we are three dimensional, living in a three-dimensional universe). Evenly spaced points cover the balloon's surface, and one of these points is your home. As the inflating balloon expands, you observe that every other point moves away from you—the farther the point, the faster it moves away along the surface of the balloon (see Figure 22–12).

Where is the edge of the balloon? What are the "inside" and "outside" of the balloon in two dimensions? The answers, at least from the perspective of a two-dimensional being on the balloon's surface, are that every point appears to be at the center, and the universe has no edges, no inside, and no outside. The two-dimensional being experiences one continuous, never-ending surface. We live in a universe of higher dimensionality, but the principle is the same: our universe has no center, and no inside versus outside.

The balloon analogy is useful because it can help us visualize an answer to a paradoxical question that is often asked about the expanding universe: What is it expanding *into*? If you think about being on the balloon, you realize that you could start out in any direction and keep traveling. You might come back to where you started, but you would never come to an end. In that continuous, two-dimensional world, there is no such thing as an "into." The surface of a balloon is an example of a system that is bounded (in two dimensions), but that has no boundaries.

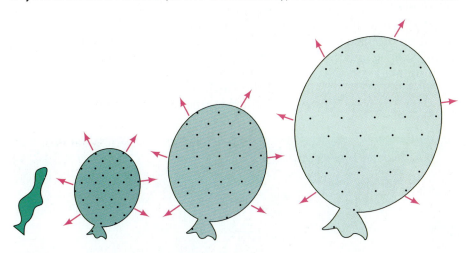

Figure 22–12
The expanding-balloon analogy of the universe. All points on the surface of the expanding balloon move away from each other—the farther apart the points initially, the faster they move apart.

Evidence for the Big Bang

In Chapter 1, we discussed how every scientific theory must be tested, and must have experimental or observational evidence backing it up. While the big bang theory provides a comprehensive picture of what our universe might be like, is there sufficient observational evidence to support it? Three important pieces of observational evidence make the big bang idea extremely compelling to scientists.

1. The Universal Expansion   Edwin Hubble's observation of universal expansion provided the first strong piece of evidence for the big bang theory. If the universe began from a compact source and has been expanding, then we would expect that the expansion would continue today. The fact that such an expansion is occurring serves as evidence for a big bang event in the past. This evidence, however, is not conclusive.

Many other theories of the universe have incorporated an expansion, but not a specific beginning in time. During the 1940s, for example, scientists proposed a theory called the *steady state universe*. In this universe, galaxies move away from each other, but new galaxies are constantly being formed in the spaces that are being vacated. Thus the steady state model describes a universe that is constantly expanding and forming new galaxies, but with no trace of a beginning.

Because of the possibility of this kind of theory, the universal expansion, in and of itself, does not compel us to accept the big bang theory.

2. The Cosmic Microwave Background   In 1964, Arno Penzias and Robert W. Wilson, two scientists working at Bell Laboratories in New Jersey, used a primitive radio receiver to scan the skies for radio signals. Their motivation was a simple one. They worked during the early days of satellite broadcasting, and they were measuring microwave radiation to document the kinds of background signals that might conceivably interfere with radio transmission. They found that, whichever way they pointed their receiver, they heard a faint hiss in their apparatus. There seemed to be microwave radiation falling onto the Earth from all directions. We now call this radiation the **cosmic microwave background radiation.**

cosmic microwave background radiation Microwave radiation, characteristic of a body about 3K, coming to Earth from all directions. This radiation is evidence for the big bang.

At first Penzias and Wilson suspected that this background noise might be an artifact—a fault in their electronics, or even interference caused by droppings from a pair of pigeons that had nested inside their funnel-shaped microwave antenna. However, a thorough testing and cleaning made no difference. A constant influx of microwave radiation of wavelength 7.35 cm flooded the Earth from every direction in space. The scientists were left to ask: Where is this radiation coming from?

Arno Penzias and Robert W. Wilson in front of their microwave receiver in New Jersey.

In order to understand the answer to their question, you need to remember that every object in the universe that is above the temperature of absolute zero emits some sort of radiation (see Chapter 8). As we saw in the opening Random Walk, a coal on a fire may glow white hot and emit the complete spectrum of visible electromagnetic radiation. As the fire cools, it will give out light that is first concentrated in the yellow, then orange, and eventually dull red. Even after it no longer glows with visible light, you can tell that the coal is giving off radiation by holding out your hand to it and sensing the infrared or heat radiation that still pours from the dying embers. As the coal continues to cool, it will give off wavelengths of longer and longer radiation.

One way to think about the cosmic microwave background, then, is to imagine that you are inside a cooling coal on a fire. You would find radiation coming uniformly from every direction, and the wavelengths of that radiation would move progressively from white to orange to red light and eventually all the way down to microwave as the coal cooled.

In 1964, a group of theorists at Princeton University in New Jersey (not far from Bell Laboratories where Penzias and Wilson worked) pointed out that if the universe had indeed begun at some time in the past, then today it would still be giving off electromagnetic radiation. In fact, the best calculations at the time indicated that the radiation would be in the microwave range, characteristic of an object at a few degrees above absolute zero. When Penzias and Wilson contacted these theorists, the reason they couldn't get rid of the microwave signal became obvious. Their microwave signal was real, and it was evidence for the big bang itself. For their discovery, Penzias and Wilson shared the Nobel Prize in 1978.

We said before that it is possible to imagine theories, such as the steady state theory, in which the universe is expanding but has no beginning. However, it is difficult to imagine a universe that does not have a beginning but that produces the kind of microwave background we're talking about. Thus Penzias and Wilson's discovery put an end to belief in the steady state theory.

In 1989, the satellite observatory called the Cosmic Background Explorer, or COBE (see Chapter 21), measured the microwave background to extreme levels of accuracy, as shown in Figure 22–13. The purpose of this measurement was to see, in great detail, whether the predictions of the big bang theory about the nature of the cosmic microwave background radiation were correct.

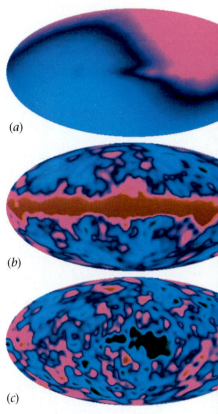

(a)

(b)

(c)

Data on strength of 2.7-Kelvin cosmic microwave radiation as taken by COBE. (a) observed microwave data; (b) what remains after correcting for the Sun's motion; (c) what remains after correcting for microwave emissions from the Milky Way.

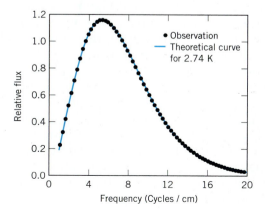

Figure 22–13
Data taken by COBE on the strength of cosmic microwave radiation as a function of wavelength. The dots are data, and the line shows the radiation you would expect from a body at 2.74 K.

These data established beyond any doubt that we live in a universe where the average temperature is just 2.7 degrees above absolute zero, or 2.7 kelvin (see Chapter 6). This finding reaffirmed the validity of the big bang theory in the minds of most scientists.

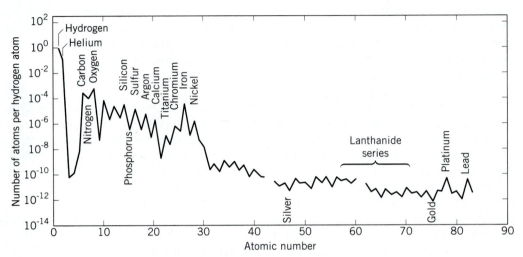

Figure 22–14
The abundance of chemical elements in the universe, on a scale where the abundance of hydrogen is 1.

3. The Abundance of Light Elements The third important piece of evidence for the big bang theory comes from studies of the abundance of light nuclei in the universe (see Figure 22–14). As we'll see at the end of this chapter, for a short period in the early history of the universe, atomic nuclei were formed from elementary particles. Cosmologists believe that the only nuclei that could have formed in the big bang are isotopes of hydrogen, helium, and lithium (the first three elements, with one, two, and three protons in their nuclei, respectively). All elements heavier than lithium were formed later in stars, as discussed in Chapter 21.

The conditions necessary for this formation were twofold. First, matter had to be packed together densely enough for there to be many collisions that could produce a fusion reaction. Second, the temperature had to be high enough for those reactions to occur, but not so high that nuclei created by fusion would be broken up in subsequent collisions. In an expanding universe, the density of matter will decrease rapidly because of the expansion. Thus nuclei form in a very narrow window of opportunity. Calculations based on density and collision frequency, together with known nuclear reaction rates, make rather specific predictions about how much of each isotope could have been made before matter spread too thinly. Thus the cosmic abundances of elements such as deuterium (the isotope of hydrogen with one proton and one neutron in its nucleus), helium-3 (the isotope of helium with two protons and one neutron), and helium-4 (with two protons and two neutrons) comprise another test of our theories about the origins of the universe.

In fact, studies of the abundances of these isotopes find that they agree quite well with predictions made in this way. The prediction for the primordial abundance of helium-4 in the universe, for example, is that it cannot have exceeded 25%. Observations of helium abundance are quite close to this prediction. If the abundance of helium were significantly higher than this, or if the abundance fell below 20% or so, the theory would no longer match observations.

THE EVOLUTION OF THE UNIVERSE

Our vision of an expanding universe leads us to peer back in time, to the early history of matter and energy. What can we say about the changes that must have taken place during the past 16 billion years?

Some General Characteristics of an Expanding Universe

Have you ever pumped up a bicycle tire with a hand pump? If you have, you may have noticed that after you've run the pump for awhile, the barrel gets very hot. This phenomenon occurs because all matter heats up when it is compressed.

This increase in temperature is true of the universe as well. A universe that is more compressed and denser than the one in which we live would also be hotter on average. In such a universe, the cosmic radiation background would correspond to a temperature much higher than 2.7 K (where it is today), and the wavelength of the background radiation would be shorter than 7.35 cm.

When the universe was younger, it must have been much hotter and denser than it is today. This principle guides our understanding of how the universe evolved. In fact, the big bang theory we have been discussing is often called the "hot big bang" to emphasize the fact that the universe began in a very hot, dense state and has been expanding and cooling ever since.

In Chapter 11, we saw that changes of temperature often correspond to changes of state in matter. If you cool water, for example, it eventually turns into ice at the freezing point. In just the same way, modern theories

Figure 22–15
The sequence of "freezings" in the universe since the big bang. The earliest freezings involve the splitting of forces, while later freezings involve forms of matter.

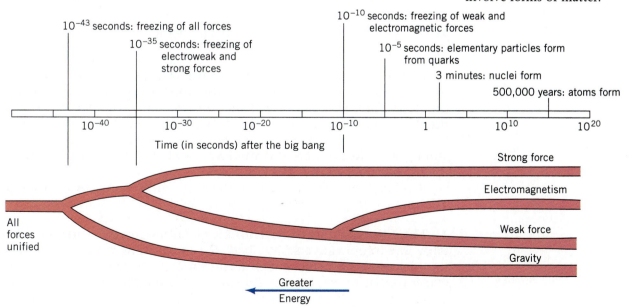

10^{-43} seconds: freezing of all forces

10^{-35} seconds: freezing of electroweak and strong forces

10^{-10} seconds: freezing of weak and electromagnetic forces

10^{-5} seconds: elementary particles form from quarks

3 minutes: nuclei form

500,000 years: atoms form

10^{-40} 10^{-30} 10^{-20} 10^{-10} 1 10^{10} 10^{20}

Time (in seconds) after the big bang

Strong force

Electromagnetism

All forces unified

Weak force

Gravity

Greater Energy

claim, as the universe cooled from its hot origins, it went through changes of state very much like the freezing of water. We will refer to these dramatic changes in the fabric of the universe as *freezings*, even though they are not actually changes from a liquid to a solid state.

A succession of freezings dominates the history of the universe. Six distinct episodes occurred, and each had its own unique effect on the universe in which we live. Between each pair of freezings was a relatively long period of steady and rather uneventful expansion. Once we understand these crucial transitions in the history of the universe, we will have come a long way toward understanding why the universe is the way it is.

Let's look at the transitions in order, from the earliest (when the universe was only 10^{-43} second old) to the most recent, as summarized in Figure 22–15. The first set of freezings involves the unification of forces we discussed in Chapter 15, while the last set involves the coming together of particles to form more complex structures such as nuclei and atoms.

10^{-43} Second: The Freezing of All Forces

The freezings that marked the very early stages of the universe do not involve particles at all, but rather the unification of forces we described in Chapter 15. Cosmologists calculate that the first freezing after the beginning of the universe took place at about 10^{-43} second. (This is really a small number—0.0000000000000000000000000000000000000000001 second!) Before this time, there was only a single unified force. At 10^{-43} second, gravity split off from the strong-electroweak force, so there were two fundamental forces acting in nature.

We cannot reproduce in our laboratories the unimaginably high temperatures that existed at this freezing, and we do not have successful theories that describe the unification of gravity with the other forces. Thus this earliest freezing remains both the theoretical frontier and the limit of our knowledge about the universe at the present time.

10^{-35} Second: The Freezing of the Electroweak and Strong Forces

Unified field theories that describe the behavior of all matter and forces (see Chapter 15) tell us that before the universe was 10^{-35} second old, the strong force was unified with the electroweak force. At 10^{-35} second, the strong force split off from the electroweak force. That is, before this time there were only two fundamental forces acting in the universe (the strong-electroweak force, and the force of gravity), but after this time, there were three.

Two important events are associated with the freezing at 10^{-35} second. They are:

1. The Elimination of Antimatter Antimatter (see Chapter 15) is fairly rare in our universe. We had to wait until the twentieth century until scientists were able to identify antimatter, and we have compelling evidence that no large collections of antimatter exist anywhere in our universe. We have already landed spaceships on the Moon, Mars, and Venus, for example. If any of those bodies had been made of antimatter, those spaceships would have been annihilated in a massive burst of gamma rays. As they did not, we conclude that none of those planets is made of antimatter.

By the same token, the solar wind, composed of ordinary matter, is constantly streaming outward from the Sun to the farthest reaches of the solar system. If any objects in the solar system were made of antimatter, the protons in the solar wind would be annihilated with that body, and we would see evidence of it. The entire solar system, therefore, is made of ordinary matter. By the same type of argument, scientists have been able to show that our entire galaxy is made of ordinary matter, and that no clusters of galaxies anywhere in the observable universe are made of antimatter.

The question, then, is this: If antimatter is indeed simply a mirror image of ordinary matter, and if antimatter appears in our theories on an equal footing with ordinary matter, as it does, then why is there so little antimatter in the universe?

The unified field theories give us an explanation of this rather striking feature of the cosmos. In our laboratories, we find one instance of a particle that decays preferentially into matter; that is, a particle whose decay products more often contain matter than antimatter. This particle is called the K_L^0 ("K-zero-long"), one of the many heavy particles such as protons and neutrons that were discovered in the latter part of the twentieth century.

If you take the theories that are successful in explaining this laboratory phenomenon and extrapolate them to the very early universe, you find that they predict that, during the freezing at 10^{-35} second, about 100,000,001 protons were formed for every 100,000,000 antiprotons. In the maelstrom that followed, all the antiprotons were annihilated by contact with protons, leaving only a sea of intense radiation and leftover protons to mark their presence (see Figure 22–16). From this collection of leftover protons, all the matter in the universe (including the Earth and its environs) was made. This discovery allowed physicists who study particles at high energies to explain the very puzzling absence of antimatter in the universe, and in the 1980s led to an enormous burst of interest in the evolution of the early universe.

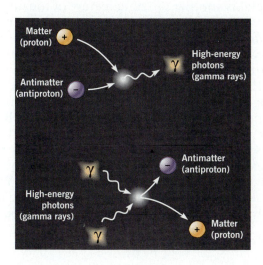

Figure 22–16
Matter and antimatter can annihilate to make photons.

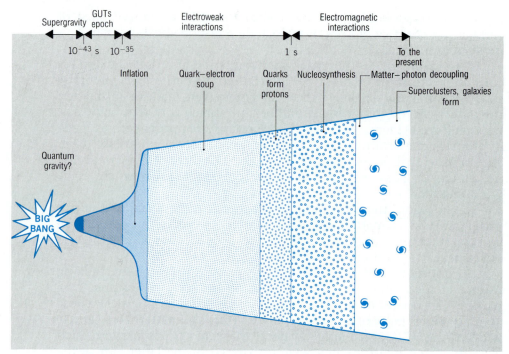

Supergravity | GUTs epoch | Electroweak interactions | Electromagnetic interactions

10^{-43} s | 10^{-35} | 1 s | To the present

Inflation | Quark–electron soup | Quarks form protons | Nucleosynthesis | Matter–photon decoupling | Superclusters, galaxies form

Quantum gravity?

BIG BANG

Figure 22–17

The evolution of the universe through the freezings as discussed in the text. Note the rapid expansion associated with the inflationary period.

2. Inflation According to the unified field theories, the freezing at 10^{-35} second was accompanied by an incredibly rapid (but short-lived) increase in the rate of expansion of the universe. This short period of rapid expansion is called *inflation*, and theories that incorporate this phenomenon are called *inflationary theories*.

One way to think about inflation is to remember that changes in volume are often associated with changes of state. Water, for example, expands when it freezes, which explains why water pipes may burst open when the water freezes in very cold weather. In the same way, scientists argue, the universe underwent a period of very rapid expansion during the period when the strong force froze out from the electroweak. Roughly speaking, at this time the universe went from being much smaller than a single proton to being about the size of a grapefruit (see Figure 22–17).

Inflation explains another puzzling feature of the universe. We have repeatedly observed that the cosmic microwave background is remarkably uniform. The temperatures associated with microwaves coming from one region of the sky differ from those coming from another region by no more than 1 part in 1000. But calculations based on a uniform rate of expansion say that different parts of the universe would not have been close enough together to have established such a uniform temperature.

The inflationary theory helps resolve this problem. Before 10^{-35} second, all parts of the universe were in contact with each other because the universe was much smaller than we would have guessed based on a uniform rate of expansion. There was time to establish equilibrium before inflation took over and increased the size of the universe. The temperature equilibrium, established early, was preserved through the inflationary era and is seen today in the uniformity of the microwave background.

Thus the coming together of the theories of elementary particle physics and the study of cosmology has produced solutions to long-standing problems and questions about the universe.

10^{-10} Second: Freezing the Weak and Electromagnetic Forces

At 10^{-10} second (that's one ten-billionth of a second), the weak and electromagnetic forces were unified. In other words, before 10^{-10} second, there were only three fundamental forces operating in the universe. These were the strong, gravitational, and electroweak forces. After 10^{-10} second, the full complement of four fundamental forces was present.

The 10^{-10}-second time also marks another milestone in our discussion of the evolution of the universe. The modern particle accelerators of high-energy physics can just barely reproduce the incredible concentration of energy associated with that event. This means that from this point forward, it is possible to have direct experimental checks of the theories that describe the evolution of the universe.

10^{-5} Second: The Freezing of Elementary Particles

Up to this point, the matter in the universe had been in its most fundamental form—quarks and leptons (see Chapter 15). The remaining freezings involve the coming together of those basic particles to create the matter we see around us. At 10^{-5} second (10 microseconds), the first of these events occurred, when elementary particles were formed out of quarks.

In other words, before 10^{-5} second, matter existed in the form of independent quarks and leptons, what we referred to in Chapter 15 as the "dots and dashes" of the universe. After this time, matter existed in the form of hadrons and leptons—that is, the ordinary elementary particles we see in our laboratories today. The universe composed of these particles kept expanding and cooling until nuclei and atoms formed.

Three Minutes: The Freezing of Nuclei

Three minutes marks the age at which nuclei, once formed, could remain stable in the universe. Before this time, if a proton and a neutron came together to form deuterium, the simplest nucleus, then the subsequent collisions of that nucleus with other particles in the universe would have been sufficient to knock that nucleus apart. Before 3 minutes, matter existed in the form of elementary particles only. Those elementary particles could not come together to form nuclei.

At 3 minutes, a short burst of nucleus formation occurred, as we discussed earlier. Thus, from 3 minutes on, the universe was littered with nuclei, which formed part of the plasma that was the material of the early universe. Before that time, the universe consisted of a sea of high-energy radiation whizzing around between all the various species of elementary particles we discussed in Chapter 15.

Before 1 Million Years: The Freezing of Atoms

The most recent transition occurred gradually, between the time the universe was a few hundred thousand and a million years old. Before this time, the background temperature of the hot, dense universe was so great that electrons could not remain in shells around the nucleus to form atoms. Even if an atom formed by chance, its subsequent collisions were sufficiently violent that the atom could not stay together. Thus all of the universe's matter was in the form of a plasma, which you may recall from Chapter 11 is a hot fluid mixture of electrons and simple nuclei. During this period, matter in the form of atoms simply could not exist.

This freezing marks an extremely important point in the history of the universe, because it is a point at which radiation such as light was no longer locked into the material of the universe. You know from your own experience that light can travel long distances through the atmosphere (which is made of atoms). But light cannot travel freely through a plasma, which quickly absorbs light and other forms of radiation. Thus, when atoms formed, the universe became transparent and radiation was released. It is this radiation, cooled and stretched out, that we now see as the cosmic microwave background.

The formation of atoms marks an important milestone for another reason. Before this event, if clumps of matter started to form (under the influence of gravity, for example), they would absorb radiation, then be blown apart. This means that there must have been a "window of opportunity" for the formation of galaxies. If galaxies are made of ordinary matter, they couldn't have started to come together out of the primordial gas cloud until atoms had formed, about 500,000 years after the big bang. By that time, however, the Hubble expansion had spread matter out so thinly that the ordinary workings of the force of gravity would not have been able to make a universe of galaxies, clusters, and superclusters. Known as the "galaxy problem," this puzzle remains the great riddle that must be answered by cosmologists.

Another way of stating this problem is to compare the "clumpiness" of matter in the universe with the smoothness of the cosmic background radiation. The background radiation seems to be the same no matter which way you look in the universe. This uniformity argues that the universe had a smooth, regular beginning. How can this statement be reconciled with the lumpy structure of galaxies and clusters of galaxies we see when we look at the present distribution of matter?

DARK MATTER AND RIPPLES AT THE BEGINNING OF TIME

Scientists' picture of the early history of the universe may seem complicated and complete, but significant gaps exist in our understanding. Some of these gaps were closed by the development of the *unified field theories* (see Chapter 14) and the inflationary scheme of the universe. Scientists still have difficulties explaining the existence of galaxies, clusters, and superclusters, however.

It now appears that the impressive array of luminous objects in the sky constitutes less than 10% of the matter in the universe—perhaps a good deal less than 10%. The other 90% (or more) of matter exists in forms that we cannot see, but whose effects we can measure. This mysterious material is called **dark matter.**

The easiest place to see evidence for dark matter is in galaxies such as our own Milky Way. Far out from the stars and spiral arms that we normally associate with galaxies, we can still see a diffuse cloud of hydrogen gas. This gas gives off radio waves, so we can detect its presence and its motion. In particular, we can tell how fast this gas is rotating. When we do these sorts of measurements, a rather startling fact emerges. In Chapter 3, we saw that Kepler's laws implied that any object orbiting around a central body under the influence of gravity will travel more slowly the farther out it is. The distant planet Jupiter, for example, moves more slowly in orbit around the Sun than does the Earth. Similarly, you would expect that when hydrogen molecules are very far away from the center of a galaxy, these more distant atoms would move more slowly than those closer in. Even though we can observe these hydrogen atoms at distances three times and more the distance from the center of the Galaxy to the ends of the spiral arms, no one has ever measured the predicted slowing down (see Figure 22–18).

dark matter Material that exists in forms that cannot be seen and constitutes 90% of the matter of the universe.

NGC 2998

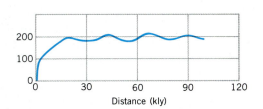

Distance (kly)

Figure 22–18
Rotation curve of a spiral galaxy as measured by astronomer Vera Rubin. This galaxy is observed to rotate much faster than expected based on its visible mass.

The only way to explain this phenomenon is to say that those hydrogen atoms are still in the middle of the gravitational influence of the Galaxy. This means that the luminous matter—the bright stars and spiral arms—are not the only things that are exerting gravitational forces. Something else, something that makes up at least 90% of the mass of the galaxy and that extends far beyond the stars, exerts a gravitational force and affects the motion of the hydrogen we observe. Studies of many galaxies show the same effect, and scientists are now convinced that at least 90% of the universe is made of this mysterious dark matter. Scientists also find evidence that dark matter exists in between galaxies, in clusters, and in other places in the universe.

Dark matter is strange, indeed. It does not interact through the electromagnetic force. If it did, it would absorb or emit photons, and it wouldn't be "dark" in the sense we're using the term here. Yet, because we know that it exerts a gravitational attraction, we can conclude that this unseen "stuff" must be a form of matter—but matter that interacts with ordinary matter only through the gravitational force. Detecting dark matter, and finding out what it is, remains a very active research field today.

The existence of dark matter might help us understand a key event in the early history of the universe, because dark matter could have formed into clumps *before atoms formed*. In the first several hundred

 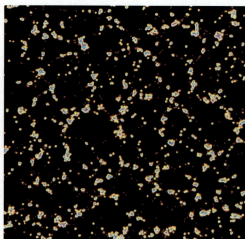 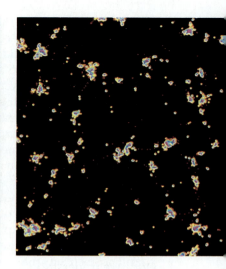

A supercomputer provides a simulation of the clumping of matter in the universe that produced large-scale structures. False color represents density: blue the highest and red the lowest.

thousand years of the big bang, photons blew apart collections of ordinary luminous matter that were trying to form galaxies. Light would not have affected the clumping of dark matter, however. Therefore, when atoms began to clump together about 500,000 years after the big bang, they found themselves in a universe in which concentrations of dark matter already existed. The force of gravity would have pulled luminous matter into these clusters, to form what we now see as galaxies. Thus, if dark matter exists, and if it formed clumps early in the history of the universe, the problem of the uneven distribution of matter is solved.

In 1992, important new results from the COBE satellite supported the notion of dark matter. Examining the microwave background in great detail, astronomers found evidence that some regions in the sky emit microwave background radiation at a slightly higher temperature than the surrounding regions. These "ripples at the beginning of time" were not very large, and they were found to correspond to temperature increases of much less than a fraction of a degree. Nevertheless, they could have been emitted only from regions that were more dense and slightly hotter than their neighbors. These ripples appear to mark the beginning of the collection of luminous matter immediately after the formation of atoms. Thus these recent data support both the existence of dark matter in the early universe, and its role in producing the structure we see in the sky.

The Cosmic Background Explorer (COBE) produced this map of microwave radiation from the entire sky. Blue indicates regions that are 0.01% cooler than average, whereas red indicates 0.01% warmer regions. These "ripples at the beginning of time" suggest that dark matter pulled visible matter into clumps shortly after atoms started to form.

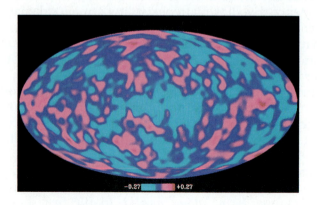

In 1993, astronomers announced the discovery of one kind of dark matter, called *MACHOS* (*ma*ssive *c*ompact *h*alo *ob*jects). These bodies form a swarm of large dark objects, each a fraction of the size of the Sun, that orbit the Milky Way.

The study of dark matter will continue to be important in the coming years. Astronomers still seek to find out the nature of this invisible material and document its distribution. Keep reading and watching the news and you'll be sure to find information about the latest discoveries.

THINKING MORE ABOUT COSMOLOGY

The Future of the Universe

Once we understand the big bang, a simple question comes to mind. Will the expansion we see going on around us today continue in the future, or will all those outgoing galaxies someday slow down, reverse their motion, and fall together in an event that astronomers, half in jest, call the "big crunch"? (See Figure 22–19.) This question involves one of the most fundamental inquiries we can make, for it involves nothing less than the future of the universe itself.

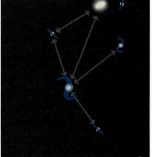

Figure 22–20
The Hubble expansion carries galaxies farther apart from each other, and as this happens the gravitational force between galaxies weakens (but never vanishes completely).

Present	Distant future	Fireball

Figure 22–19
One possible end for the universe, if there is enough mass in it, is a reversal of the Hubble expansion into a "Big Crunch."

This question can be answered (at least in principle) by observing the universe today. The only force we know that is capable of reducing the speed of a receding galaxy is gravity (see Figure 22–20). If the universe holds enough matter, even those quasars at the edge of the observable universe will someday come falling back in. So the question of the fate of the universe comes down to the question of whether the universe contains enough matter to exert that force.

If the universe has enough matter to reverse the expansion, then we say the universe is *closed*. If it doesn't, then we say the universe is *open*, and it will go on expanding forever. The boundary between these two, in which the expansion slows and just comes to a halt after infinite time has passed, is called a *flat* universe.

If you count all the luminous matter (the stuff we can actually see), then we observe only about 1% of the matter needed to close the universe. Dark matter adds significantly to this total. So far, astronomers have found perhaps 20% to 30% of the matter required, and the search for more continues.

One feature of the debate about the future of the universe that you might want to think about is this: The inflationary theories that explain the behavior of the universe at 10^{-35} second also predict that exactly enough matter exists for the universe to be flat. Some theoretical astrophysicists have taken this to mean that the universe *must be* flat, and that observational astronomers should work harder to find the rest of the dark matter. Observers, on the other hand, say that if the universe is open, that's all there is to it.

How much faith do you think should be put in theoretical predictions of this sort? How hard should observers look for the "missing" matter?

► Summary

Early in the twentieth century, Edwin Hubble made two extraordinary discoveries in the area of *cosmology*, the study of the structure and behavior of the universe. First, he demonstrated that our home, the collection of stars known as the *Milky Way*, is just one of countless *galaxies* in the universe, each containing billions of stars. By measuring the *redshift* of galaxies, he also discovered that these distant objects are moving away from each other. According to *Hubble's law*, the farther away a galaxy is, the faster it is moving away. This relative motion implies that the universe is expanding.

One theory that accounts for universal expansion is the *big bang*, which is the idea that the universe began at a specific moment in time and has been expanding ever since. Evidence from the *cosmic microwave background radiation* and the relative abundances of light elements, in addition to the expansion, support the big bang theory.

At the moment of creation, all forces and matter were unified in one unimaginably hot and dense volume. As the universe expanded, however, a series of six "freezings" led to the universe we see today. Freezings at 10^{-43} second, 10^{-35} second, and 10^{-10} second caused a single unified force to split into the four forces we observe today: the gravitational, strong, electromagnetic, and weak forces. At that early stage of the universe, when all matter and energy were contained in a volume no larger than a grapefruit, matter was in its most elementary form of quarks and leptons.

At 10^{-5} second, the quarks bonded together to form heavy nuclear particles such as protons and neutrons. Subsequent freezings saw these particles first fuse to nuclei at 3 minutes, and ultimately join with electrons to form atoms at 500,000 years. Stars, which formed from those atoms, then could begin the processes that provided all the other chemical elements.

The search for *dark matter*, which is mass that we cannot see with our telescopes, is a frontier in cosmological research that may help us determine if the universe will continue expanding forever.

► Review Questions

1. What is cosmology? How does cosmology differ from astronomy?

2. What is a galaxy? How does a galaxy differ from a star?

3. How are galaxies distributed in the universe?

4. How did Edwin Hubble discover that there are galaxies in the universe other than the Milky Way?

5. Describe Hubble's law. How did Hubble discover it?

6. What is the big bang theory? What kinds of evidence support it?

7. Why is interstellar matter composed mostly of hydrogen, helium, and lithium?

8. What event occurred when the universe was 500,000 years old?

9. What event occurred when the universe was 3 minutes old?

10. What event occurred when the universe was 10^{-5} second old?

11. What event occurred when the universe was 10^{-10} second old?

12. What event occurred when the universe was 10^{-35} second old?

13. What event occurred when the universe was 10^{-43} second old?

14. What is the significance of the discovery of cosmic microwave background radiation by Penzias and Wilson?

15. What are "ripples" in the universe, and why are they important?

16. What is "dark matter" and what evidence exists for it?

17. Describe what spiral arms are.

18. What are the characteristics of elliptical galaxies?

19. What are dwarf galaxies composed of?

20. Describe the major characteristics of quasars.

21. What were the contributions of Geller and Huchra?

22. Describe groups, clusters, and superclusters.

23. What was the steady state theory of the universe?

24. Describe inflationary theories and how they explain the expansion of the universe.

25. What is the difference between an open and closed universe?

► Fill in the Blanks

Complete the following paragraphs with words and phrases from the list.

big bang theory

cosmic microwave
background radiation

cosmology

dark matter

galaxies

Hubble's law

Milky Way

redshift

_____ is the branch of science devoted to understanding the origin and structure of the universe. We know that the matter of the universe is clumped into _____, and that these clumps are receding from each other. Our own galaxy is called the _____.

_____ states that the farther a galaxy is from us, the faster it is receding. We detect this motion by noticing the _____ of the light that reaches us. The idea that the universe began in a hot, compressed state and has been expanding ever since is called the _____. The discovery of the _____ is one of the main pieces of experimental evidence for this theory. Theorists believe that over 90% of the material in the universe is _____, which does not interact with electromagnetic radiation.

11. Because of the expansion of the universe, the Sun and our Milky Way galaxy are getting larger. Explain why this statement may be true or false.

12. Order the major components of the universe from the Earth to superclusters of galaxies alphabetically. Include in your list of objects the Moon and planets, the Earth, the Sun, asteroids, the solar system, the Milky Way galaxy, the Andromeda galaxy, the Local Group of galaxies, and finally groups, clusters, and superclusters of galaxies.

13. What are the similarities and differences between the steady state and big bang cosmological models?

14. Outline the major events for the current big bang model of the evolution of the universe.

15. What is meant by "freezings" in the explanations of the current cosmological model?

16. What are the fates of an open, closed, and flat universe?

17. How is "dark matter" related to an open, closed, or flat universe?

▶ Discussion Questions

1. Why does the Earth seem to be at the center of the Hubble expansion?

2. If the universe is closed, describe the results that some future scientist like Edwin Hubble would obtain when he or she looked through a telescope during the period of contraction. Would other galaxies be visible? Would they display a redshift?

3. Why was the steady state theory of the universe abandoned? How does this episode fit into the discussion of the scientific method in Chapter 1?

4. How will we know whether the universe is open or closed? What measurements or observations could answer this question?

5. Louis Pasteur once said that "chance favors only the mind that is prepared." Apply this saying to the discoveries of Edwin Hubble, and of Penzias and Wilson.

6. Some advances in our knowledge have been made possible through better equipment, like Hubble's discoveries at Mount Wilson. What other major discoveries in cosmology relied on technical improvements in existing apparatus?

7. List the three classes of galaxies and describe one important characteristic of each class.

8. How does a quasar differ from a nearby galaxy?

9. How is the slope of the velocity-distance relationship for distant galaxies related to the Hubble constant?

10. How is the age of the universe determined from the Hubble constant?

▶ Problems

1. Assuming a Hubble constant of 75 km/s/Mpc, what is the approximate velocity of a galaxy 100 Mpc away? 1000 million Mpc away?

2. If a galaxy is 500 Mpc away, how fast will it be receding from us?

3. An observer on one of the raisins in our bread-dough analogy measures distances and velocities of neighboring raisins. The data look like the following:

Distance (cm)	Velocity (cm/hr)
0.5	1.02
0.9	2.00
1.4	2.90
2.1	4.05
3.0	5.90
3.4	7.10

Plot these data on a graph and use the plot to estimate a "Hubble constant" for the raisins.

4. From the data in Problem 3 and the graph you drew, estimate the time that has elapsed since the dough started rising. Estimate the largest and smallest values of this number consistent with the data.

5. In the inflationary period, some theories say that

the scale of the universe increased by a factor of 10^{50}. Suppose your height were to increase by a factor of 10^{50}. How tall would you be? Express your answer in light-years and compare it to the size of the observable universe.

6. Suppose a proton (diameter about 10^{-13} cm) were to inflate by a factor of 10^{50}. How big would it be? Convert the answer to light-years and compare it to the size of the observable universe.

7. How fast will a galaxy 5 billion light-years from Earth be moving away from us? What fraction of the speed of light is this?

▶ Investigations

1. The Milky Way is actually a band of stars that, seen from Earth in the summer months, stretches all the way across the sky. Given what you know about galaxies, why do you suppose that our own galaxy appears this way to us? Who was the first natural philosopher to figure this out?

2. Will the Andromeda galaxy, which is in the constellation of Andromeda, be above the horizon tonight? If so, go out and try to spot this galaxy. What does it look like? Examine the galaxy with binoculars or a telescope.

What features do you see?

3. Investigate the concept of the "Great Attractor." How does the existence of such an object fit in with the concept of the Hubble expansion? How would you modify the raisin-bread dough analogy to put in the Great Attractor?

4. Investigate the cosmologies of other societies. How do they explain the origins of the universe?

5. What agencies or organizations fund cosmological research? In particular, investigate the role of the Carnegie Institution of Washington in Edwin Hubble's research.

▶ Additional Reading

Dressler, Alan. *Voyage to the Great Attractor: Exploring Intergalactic Space*. New York: Knopf, 1994.

Ferris, Timothy. *The Red Limit*. New York: William Morrow, 1977.

Hawking, Stephen. *A Brief History of Time*. New York: Bantam, 1988.

Silk, Joseph. *The Big Bang*. New York: W.H. Freeman, 1989.

Trefil, James. *The Dark Side of the Universe*. New York: Scribners, 1988.

APPENDIX A

QUANTITATIVE THINKING

The Grocery Store

Think about the last time you went shopping at the supermarket. You may not have noticed, but during that trip you were confronted with quite a variety of mathematical problems. Probably, your first priority was to keep a mental note of the total cost of your purchases so you didn't overspend. Each time you put an item into your shopping cart, you added its cost to the running total. But that was only the beginning of your quantitative thinking.

When you selected some green beans or grapes from the produce section, you had to multiply the weight of your purchase by the price per pound to compute its cost. In another part of the store you had to decide on the most economical purchase: Is it better to buy 24 ounces of detergent for $2.75 or 40 ounces for $4.00? (Many supermarket chains help you by posting "unit pricing" information.)

You were also confronted with many different units of measurement at the store. You purchase milk by the gallon, soda by the liter, and vanilla extract by the fluid ounce. Coffee beans are sold by the pound, exotic spices by the gram, plastic wrap by the square foot, and trash bags by their thickness in mils (thousandths of an inch). For many items, more than one unit of measurement is given: cereal boxes list contents in ounces *and* in grams; beverages may be labeled in gallons *and* in liters. You probably have a general idea about the relative amounts or sizes of these units, but products destined for international commerce must be precisely labeled in both English and SI units (see Chapter 2). Without unit labeling, and our ability to convert easily from one unit of measurement to another, modern industry and commerce would grind to a halt.

Science also depends on quantitative thinking. In this appendix, we review some of the most common facts that you will need in dealing with mathematical problems.

A-1

CONVERSION FACTORS

Recall from Chapter 2 that we usually group together into a *system of units* fundamental quantities like length, mass, time, and temperature. The two common systems of units in the United States are the English system, with the foot, pound, and second, and the metric system (the Systéme Internationale or SI), with the meter, kilogram, and second.

Units in the SI system are based on multiples of ten. Thus, the centiliter is 1 one-hundredth the volume of a liter, the milliliter 1 one-thousandth, and so on. In the same way, the units kilometer, kilogram, and kilowatt specify 1000 meters, grams, and watts, respectively. The English system is not quite as logical: it's hard to see rhyme or reason in a system in which twelve inches equal one foot, three feet equal one yard, and 1760 yards equal one mile.

In science, as in life, you must frequently convert from one system of units to another. You already have experience in converting inches into feet or minutes into hours, but many other conversions aren't as simple, because the conversion factors aren't integers. Tables of some of the most useful and important conversion factors follow.

Here is the way to use the tables. If, for example, you want to convert a known distance of 5.14 kilometers (an SI unit) into a distance in miles (an English unit), look at the table with length conversion factors from SI to English units and find the appropriate factor. In this case, it is 0.6214. Then multiply 5.14 kilometers by this factor:

$$(5.14 \text{ kilometers}) \times (0.6214 \text{ miles/kilometer}) = 3.19 \text{ miles}$$

All conversion factors work the same way.

Length Conversions from SI to English Units

To get:	Multiply:	By:
inches	meters	39.4
feet	meters	3.281
yards	meters	1.094
miles	kilometers	0.6214

Length Conversions from English to SI Units

To get:	Multiply:	By:
meters	inches	0.0254
meters	feet	0.3048
meters	yards	0.9144
kilometers	miles	1.609

Mass Conversions from SI to English Units

To get:	Multiply:	By:
pounds	kilograms	2.205
ounces	kilograms	35.27
troy ounces	kilograms	32.2
pounds	newtons*	0.2248
ounces	newtons	3.599
Troy ounces	newtons	3.281

* Recall that the weight of a 1-kilogram mass is equivalent to a force of 9.806 newtons.

Mass Conversions from English to SI Units

To get:	Multiply:	By:
kilograms	pounds	0.4535
kilograms	ounces*	0.0283
kilograms	troy ounces*	0.0311
newtons	pounds	4.448
newtons	ounces	0.2778
newtons	troy ounces	0.3048

* One pound contains 16 ounces, but only 12 troy ounces.

Area Conversions from SI to English Units

To get:	Multiply:	By:
square inches	square meters*	1.55×10^3
square feet	square meters	10.76
square yards	square meters	1.196
acres	square kilometers*	2.47×10^{-4}
square miles	square kilometers	0.3861

* One square kilometer contains 10^6 square meters.

Area Conversions from English to SI Units

To get:	Multiply:	By:
square meters	square inches	6.45×10^{-4}
square meters	square feet	0.0929
square meters	square yards	0.8361
square meters	acres	4.05×10^3
square kilometers	square miles	2.590

Volume Conversions from SI to English Units

To get:	Multiply:	By:
gallons (US)	liters*	0.2198
gallons (Canadian)	liters	0.2639
cubic feet	liters	0.0353
cubic inches	liters	60.97
fluid ounce (US)	liters	33.78
fluid ounce (UK)	liters	35.21
cubic yards	cubic meters*	1.308

* Recall that a liter is the volume of a cube 10 centimeters on a side. Therefore, 1 liter equals 1000 cubic centimeters, or 0.001 cubic meters.

Volume Conversions from English to SI Units

To get:	Multiply:	By:
liters	gallons (US)	4.55
liters	gallons (Canadian)	3.79
liters	cubic feet	28.3
liters	cubic inches	0.0164
liters	fluid ounce (US)	0.0296
liters	fluid ounce (UK)	0.0284
cubic meters	cubic yards	0.7646

Temperature Conversion Equations for Degrees Fahrenheit (°F), Degrees Celsius (°C), and Kelvins (K)

Temperature conversion equations are similar to conversion factors, but they include extra terms. These terms arise from the fact that 0° is different in each of the three scales.

$$°F = (1.8 \times °C) + 32$$
$$°F = (1.8 \times K) - 492$$
$$°C = (°F - 32)/1.8$$
$$°C = K - 273$$
$$K = (°F + 492)/1.8$$
$$K = °C + 273$$

Units of Force, Energy, and Power

Once the basic units of mass, length, time, and temperature have been defined, the units of other quantities like force and energy follow. Recall that the energy units we have defined in the text are:

joule: a force of 1 newton acting through 1 meter

foot-pound: a force of 1 pound acting through 1 foot

calorie: energy required to raise the temperature of 1 gram of water by 1 degree Celsius

British Thermal Unit, or BTU: energy required to raise the temperature of 1 pound of water by 1 degree Fahrenheit

kilowatt-hour: 1000 joules per second for 1 hour

Furthermore, power units are:

watt: 1 joule per second

horsepower: 550 foot-pounds per second

Energy and Power Conversion from SI to English Units		
To get:	Multiply:	By:
BTUs	joules	0.00095
calories	joules	0.2390
kilowatt-hours	joules	2.78×10^{-7}
foot-pounds	joules	0.7375
horsepower	watts	0.00134

Energy and Power Conversion from English to SI Units		
To get:	Multiply:	By:
joules	BTUs	1055
joules	calories	4.184
joules	kilowatt-hours	3.6 million
joules	foot-pounds	1.356
watts	horsepower	745.7

WORKING WITH NUMBERS

A few mathematical skills are essential to describe the behavior of the physical world. We review some of these skills in the following sections.

Powers of Ten Notation

Powers of ten notation allows us to write very large or very small numbers conveniently, in a compact way. Any number can be written by following three rules:

1. Every number is written as a number between 1 and 10 followed by 10 raised to a power, or an exponent.
2. If the power of ten is positive, it means "move the decimal point this many places to the right."

3. If the power of ten is negative, it means "move the decimal point this many places to the left."

Thus, using this notation, five trillion is written as 5×10^{12}, instead of 5,000,000,000,000. Similarly, five trillionths is 5×10^{-12}, instead of 0.000000000005.

Multiplying or dividing numbers with powers of ten requires special care. If you are multiplying two numbers, such as 2.5×10^3 and 4.3×10^5, you multiply 2.5 and 4.3, but you add the two exponents:

$$(2.5 \times 10^3) \times (4.3 \times 10^5) = (2.5 \times 4.3) \times 10^{3+5}$$
$$= 10.75 \times 10^8$$
$$= 1.075 \times 10^9$$

When dividing two numbers, such as 4.3×10^5 divided by 2.5×10^3, you divide 4.3 by 2.5, but you subtract the denominator exponent from the numerator exponent:

$$(4.3 \times 10^5)/(2.5 \times 10^3) = (4.3/2.5) \times 10^{5-3}$$
$$= 1.72 \times 10^2$$
$$= 172$$

Significant Figures and Rounding

One of the most popular characters in the classic *Star Trek* episodes was Mr. Spock, who had amazing computational abilities. When asked about how long it would take to arrive at their next destination, he would answer, "Approximately 39 minutes, 18.3 seconds," even though an answer of "About 40 minutes" would have sufficed.

Spock's reply is a humorous example of reporting too many significant figures. You may be faced with similar situations in your daily activities. For example, you are riding in your car at about 55 miles per hour, and you have to drive 17 miles to work. How long will it take you? The travel time is calculated by dividing the distance to be driven by your speed:

$$17 \text{ miles}/55 \text{ miles per hour} = 17/55 \text{ hours}$$
$$= 0.309090909090 \text{ hours}$$
$$= 18.5454545455 \text{ minutes}$$

Each digit in this answer is called a "significant figure." In this example, we have reported our estimated travel time to be 12 significant figures. Is that wrong?

In general, you should not report more significant figures than are contained in the measurements you use. Your estimates of speed and distance, for example, have only two significant figures, 55 miles per hour and 17 miles, and the estimated travel time *can't* be more accurate than two significant figures. You have to discard the last 10 digits, so a better estimate would be "18 or 19 minutes."

The process of discarding extra digits is called "rounding off." When a calculation doesn't come out exactly evenly, you will have to round off the last significant figure to the nearest value. Follow these two rules:

1. If the first digit to be discarded is 0, 1, 2, 3, or 4, just omit the extra digits. Thus 18.45454 would round off to 18.
2. If the first digit to be discarded is 5, 6, 7, 8, or 9, add 1 to the last significant digit, and omit the rest. Thus, 18.54545 would round off to 19.

Units Cancellation and Solving Equations

Equations in scientific calculations, as well as those you use in everyday life, usually require canceling units. Consider the equation previously cited for travel time. To estimate travel time in hours, you must divide distance in miles by average speed in miles per hour. Note that each of the three factors in the equation—time, distance, and speed—have different units. For the equation to work out properly, these units must cancel:

$$\frac{\text{distance (\cancel{miles})}}{\text{speed (\cancel{miles}/hour)}} = \text{time (hours)}$$

When you divide miles by miles per hour, the miles units cancel, giving an answer in hours, as required. If you are asked for a solution in SI units, you may have to convert the answer using the conversion factor tables.

As you attempt to solve science problems, pay close attention to the units. You may find it helpful to follow three basic steps any time you have to work out such a problem.

1. Based on the information given in the problem, and the answer requested, select the appropriate equation. For example, if you are asked to calculate the force of gravity between two objects, you will be given the masses of the two objects and the distance between them. With this information, you can apply Newton's equation for gravitation force:

$$\text{force} = [G \times (\text{first mass}) \times (\text{second mass})]/(\text{distance})^2,$$

where G is the universal gravitational constant.

2. Identify the quantities to be substituted in the equation, making sure you convert each of these quantities into the appropriate SI units by using the conversion tables. Mass should be in kilograms and distance in meters, for example. Also remember that many constants include units; the universal gravitational constant, G, is in units of $N\text{-}m^2/kg^2$. Thus, in our example, if we have two masses of 1 kg at a separation of 1 m, we obtain by substitution into Newton's equation:

$$\begin{aligned}
\text{force (in N)} &= [(6.673 \times 10^{-11}\, N\text{-}m^2/kg^2) \times (1\text{ kg}) \\
&\quad \times (1\text{ kg})]/(1\text{ m})^2 \\
&= [(6.673 \times 10^{-11}\, N\text{-}\cancel{m^2}/\cancel{kg^2}) \times \cancel{kg^2}]/\cancel{m^2} \\
&= 6.673 \times 10^{-11}\, N
\end{aligned}$$

3. Always check your answer by making sure that all the units on the right side of the equation cancel to give the desired unit on the left. In this case the m^2 and kg^2 terms cancel, leaving the correct unit for force, N.

A Warning! Sometimes units, at first glance, don't appear to cancel. Think about Newton's second law of motion which defines force as mass times acceleration:

$$\text{force (in N)} = \text{mass (in kg)} \times \text{acceleration (in m/s}^2)$$

The right side of the equation gives an answer in kg-m/s^2, while the left side is supposed to be in newtons. Units don't seem to match up until you remember that a newton *is defined* as 1 kg-m/s^2! The following table lists the definitions for several commonly used units.

Definitions of Common Units

Force: 1 newton = 1 kg-m/s^2

Frequency: 1 hertz = 1 s^{-1}

Energy, work: 1 joule = 1 N-m = 1 m^2-kg/s^2

Power: 1 watt = J/s = 1 m^2-kg/s^3

Calculating Areas and Volumes

The following lists define the area and volume of simple geometrical forms in terms of their dimensions.

The area of a square is the edge length squared.

The area of a rectangle is the length times the height.

The area of a right triangle is half the product of the two shorter sides.

The area of a triangle equals half the base times the height.

The area of a circle is π times the radius squared.

The area of the surface of a sphere is 4 π times the radius squared.

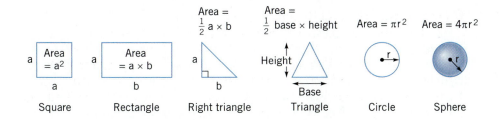

The volume of a cube is the edge length cubed.

The volume of a rectangular solid is the product of the three edge lengths.

The volume of a sphere is $\frac{4}{3}\pi$ times the radius cubed.

The volume of a right circular cylinder is π times the radius squared times the height.

The volume of a right circular cone is $\frac{1}{3}\pi$ times the base radius squared, times the height.

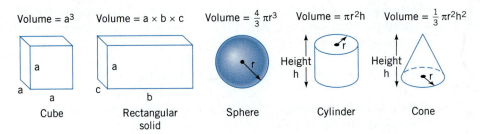

Volume = a^3 Volume = $a \times b \times c$ Volume = $\frac{4}{3}\pi r^3$ Volume = $\pi r^2 h$ Volume = $\frac{1}{3}\pi r^2 h^2$

Cube   Rectangular   Sphere   Cylinder   Cone
                   solid

APPENDIX B

THE GEOLOGIC

TIME SCALE

A
RANDOM
WALK

The Abandoned Farmhouse

As you drive through the countryside, you may come across an abandoned farmhouse, its windows boarded up, its roof open to the elements, its only inhabitants animals and insects. The surrounding fields, once cultivated, may have been overtaken by the prairie grasses that were there long before the farmhouse was built. Within a very short time span, perhaps as short as 25 years, that farmhouse may be razed and a subdivision built on the same site. Over a longer time span, the same site may have been at the bottom of a prehistoric ocean, then lifted to the top of a mountain, and finally eroded to the level at which the farmhouse was built. All things change over time.

All things also affect others as they change. In the following pages, we show you the *geologic time scale*, a chronological arrangement of geologic time units as currently understood. In addition, we show you the major steps in evolution that were made possible, in part, by the conditions that existed at every step of that time scale.

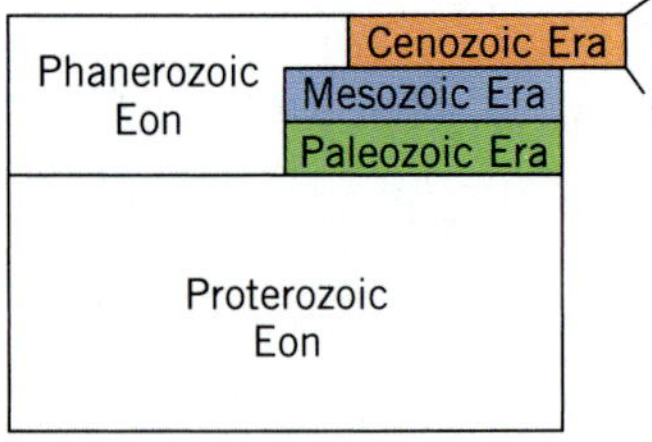

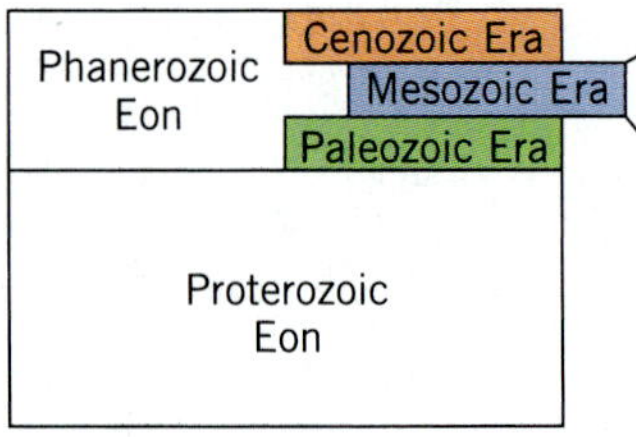

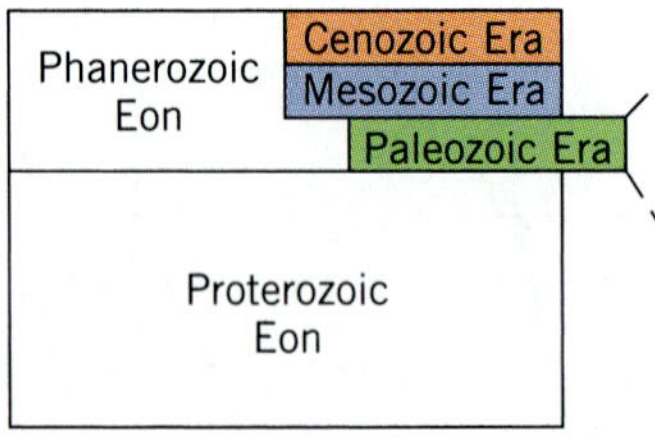

PERIOD	EPOCH	PLANT EVOLUTION
Quaternary	Holocene	Repeated glaciation
	Pleistocene	
Tertiary	Pliocene	Decline of forests, spread of grasslands
	Miocene	
	Oligocene	
	Eocene	Explosive radiation of flowering plants
	Paleocene	
Cretaceous		First flowering plants
Jurassic		Forests of gymnosperms and ferns over most of the Earth
Triassic		Gymnosperms dominant
Permian		Widespread extinction Decline of nonseed plants
Carboniferous Pennsylvanian		Gymnosperms appear
		Widespread forests of giant club moss trees, horsetails and tree fern — create vast coal deposits
Mississippian		
Devonian		First seed plants Development of vascular plants: club mosses and ferns
Silurian		First vascular plants First land plants
Ordovician		
Cambrian		Algae dominant

Time markers (Millions of years ago): 2.5, 65, 135, 195, 240, 285, 375, 420, 450, 520, 570

Phanerozoic Eon
  Cenozoic Era
  Mesozoic Era
  Paleozoic Era
Proterozoic Eon

Millions of years ago

ANIMAL EVOLUTION	MAJOR GEOLOGICAL EVENTS
Appearance of *Homo sapiens* First use of fire Appearance of *Homo erectus*	Worldwide glaciations Linking of North America and South America
Appearance of hominids Appearance of first apes All modern genera of mammals present In seas, bony fish abound Rise of mammals First placental mammals	Opening of Red Sea Formation of Himalayan Mountains Collision of India with Asia Separation of Australia and Antarctica Opening of Norwegian Sea and Baffin Bay
Dinosaurs extinct Modern birds	Formation of Alps Formation of Rocky Mountains
First birds Age of dinosaurs	
Explosive radiation of dinosaurs First dinosaurs First mammals Complex arthropods dominant in seas First beetles	Opening of Atlantic Ocean
Widespread extinction Appearance of mammal-like reptiles Increase of reptiles and insects Decline of amphibians	Final assembly of Pangaea
Early reptiles First winged insects Increase of amphibians	Formation of coal deposits
Amphibians diversify into many forms First land vertebrates — amphibians	
Golden Age of fishes First land invertebrates — land scorpions	
First vertebrates — fishes Increase of marine invertebrates	
Trilobites dominant Explosive evolution of marine life	

	Plant Evolution	Animal Evolution	Major Geological Events
1,000			
1,500			Formation of early super-continent
2,000			
2,500			
3,000	Early algae	Early bacteria	
3,500			
4,000			Oldest Earth rocks
4,500			Oldest Moon rocks / Heavy meteorite bombardment
4,600			Formation of the Earth

Millions of years ago

Time	Eon*
	Proterozoic
Precambrian	Archean
	Hadean

* No further subdivisions into eras or periods are in common use.

APPENDIX C

SELECTED

PHYSICAL

CONSTANTS AND

ASTRONOMICAL

DATA

Avogadro's number
6.02×10^{23}/mol

Charge on electron
1.6×10^{-19} C

Electron mass
$m_e = 9.10939 \times 10^{-31}$ kg

Gravitational constant
$G = 6.673 \times 10^{-11}$ N · m²/kg²

Planck's constant
$h = 6.62608 \times 10^{-34}$ J · s

Proton mass
$m_p = 1.6726 \times 10^{-27}$ kg
$= 1836.1\ m_e$

Speed of light in a vacuum
$c = 2.9979 \times 10^8$ m/s

Astronomical unit
$AU = 1.4959789 \times 10^{11}$ meters

Hubble's constant
$H \sim 20$ km/s/Mly

Light-year
$ly = 9.46053 \times 10^{15}$ meters
$= 6.324 \times 10^4$ AU

Mass of Sun
$M_{sun} = 1.989 \times 10^{30}$ kg

Radius of Sun
$R_{sun} = 6.96 \times 10^5$ km

Mass of Moon
$M_{moon} = 7.348 \times 10^{22}$ kg

Radius of Moon
$R_{moon} = 1.738 \times 10^3$ km

APPENDIX D

PROPERTIES OF

THE PLANETS

Planet	Length of Day	Distance from Sun (millions of km)	Length of Year (Earth year)	Average Radius (km)	Radius (Earth radii)	Mass (kg)	Mass (Earth masses)
Mercury	58.65 days	57.9	0.24	2,439	0.38	3.30×10^{23}	0.0562
Venus	243.01 days (retrograde)	108.2	0.615	6,052	0.95	4.87×10^{24}	0.815
Earth	23 h 56 min 4.1 s	149.6	1.000	6,378	1.00	5.974×10^{24}	1.000
Mars	24 h 37 min 22.6 s	227.9	1.881	3,397	0.53	6.42×10^{23}	0.1074
Jupiter	9 h 50.5 min	778.4	11.86	71,492	11.19	1.899×10^{27}	317.9
Saturn	10 h 14 min	1424	29.46	60,268	9.45	5.68×10^{26}	95.1
Uranus	17 h 14 min (retrograde)	2872	84.01	25,559	4.01	8.66×10^{26}	14.56
Neptune	16 h 3 min	4499	164.8	25,269	3.96	1.03×10^{26}	17.24
Pluto	6.39 days (retrograde)	5943	248.6	1,140	0.18	1.1×10^{22}	0.0018

APPENDIX E

STAR CHARTS

FOR THE

NORTHERN

HEMISPHERE

INDEX TO STAR MAPS
Choose the date and time; the appropriate map is then specified.

Date	8 P.M.	9 P.M.	10 P.M.	11 P.M.	12 P.M.	1 A.M.
Jan 05		1		2		3
Jan 20	1		2		3	
Feb 05		2		3		4
Feb 20	2		3		4	
Mar 05		3		4		5
Mar 20	3		4		5	
Apr 05		4		5		6
Apr 20	4		5		6	
May 05		5		6		7
May 20	5		6		7	
Jun 05		6		7		8
Jun 20	6		7		8	
Jul 05		7		8		9
Jul 20	7		8		9	
Aug 05		8		9		10
Aug 20	8		9		10	
Sep 05		9		10		11
Sep 20	9		10		11	
Oct 05		10		11		12
Oct 20	10		11		12	
Nov 05		11		12		1
Nov 20	11		12		1	
Dec 05		12		1		2
Dec 20	12		1		2	

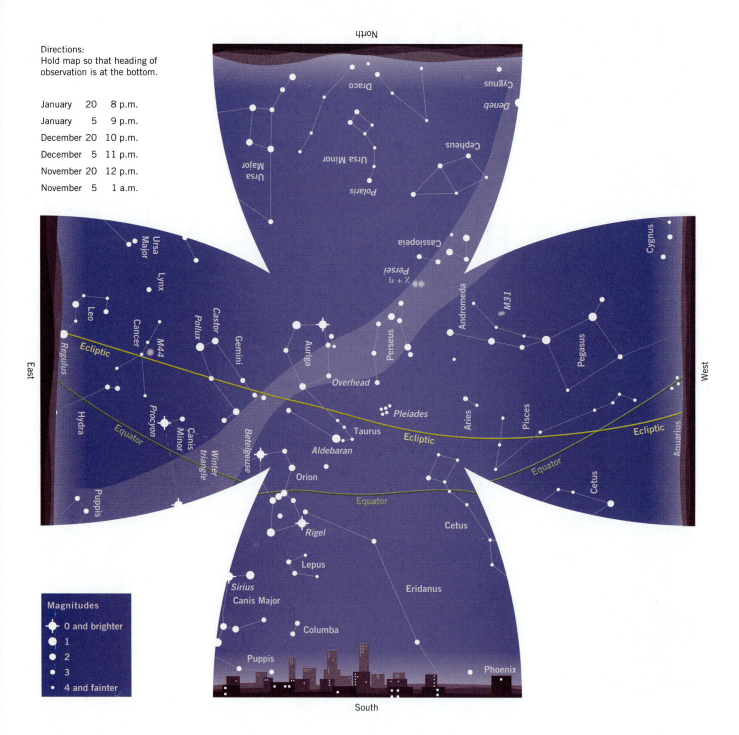

Directions:
Hold map so that heading of
observation is at the bottom.

January	20	8 p.m.
January	5	9 p.m.
December	20	10 p.m.
December	5	11 p.m.
November	20	12 p.m.
November	5	1 a.m.

Magnitudes
● 0 and brighter
● 1
● 2
● 3
· 4 and fainter

Map 1

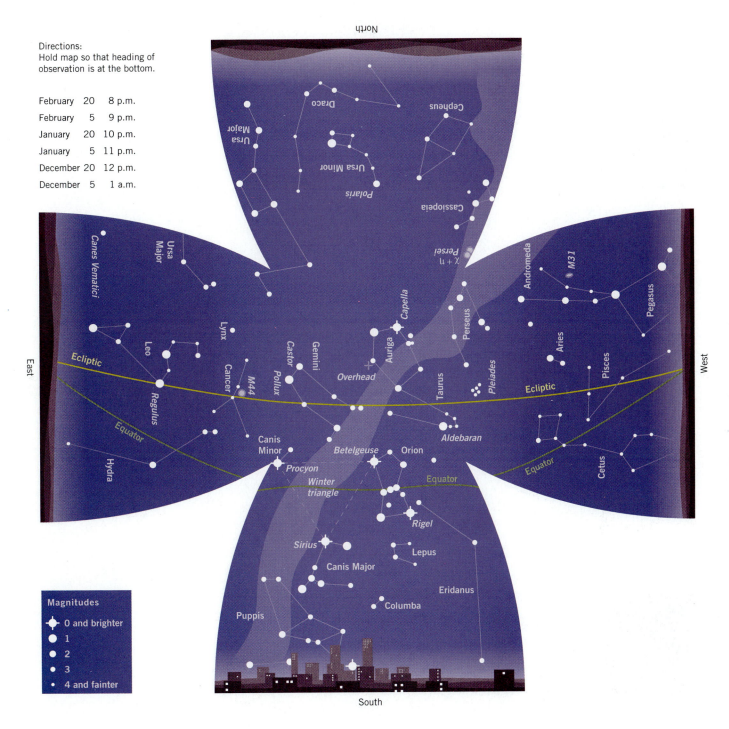

Map 2

Directions:
Hold map so that heading of
observation is at the bottom.

March	20	8 p.m.
March	5	9 p.m.
February	20	10 p.m.
February	5	11 p.m.
January	20	12 p.m.
January	5	1 a.m.

Magnitudes
- 0 and brighter
- 1
- 2
- 3
- 4 and fainter

Map 3

Directions:
Hold map so that heading of
observation is at the bottom.

April	20	8 p.m.
April	5	9 p.m.
March	20	10 p.m.
March	5	11 p.m.
February	20	12 p.m.
February	5	1 a.m.

North

East

West

South

Magnitudes
- 0 and brighter
- 1
- 2
- 3
- 4 and fainter

Map 4

Directions:
Hold map so that heading of
observation is at the bottom.

May	20	8 p.m.
May	5	9 p.m.
April	20	10 p.m.
April	5	11 p.m.
March	20	12 p.m.
March	5	1 a.m.

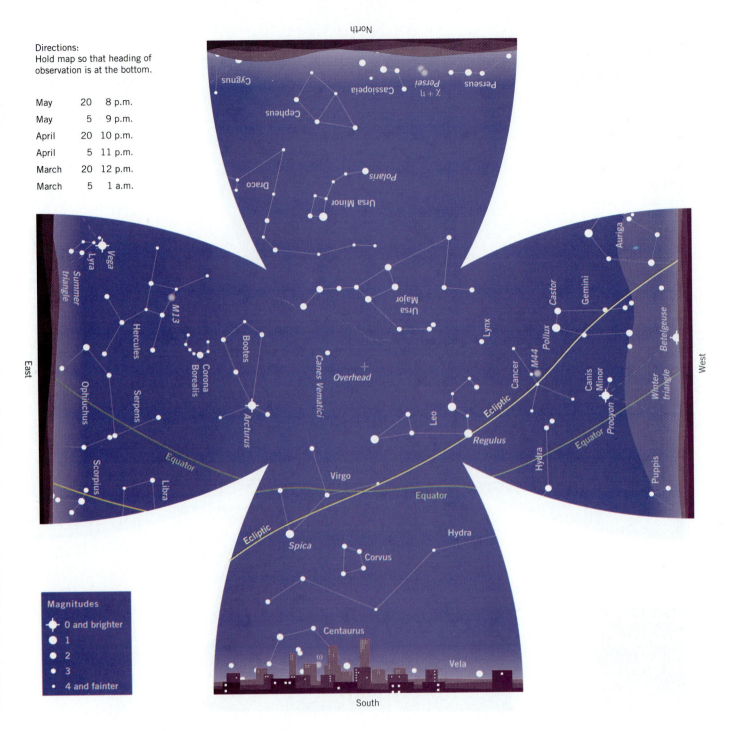

Map 5

Directions:
Hold map so that heading of observation is at the bottom.

June	20	8 p.m.
June	5	9 p.m.
May	20	10 p.m.
May	5	11 p.m.
April	20	12 p.m.
April	5	1 a.m.

Magnitudes
- 0 and brighter
- 1
- 2
- 3
- 4 and fainter

Map 6

Directions:
Hold map so that heading of
observation is at the bottom.

July	20	8 p.m.
July	5	9 p.m.
June	20	10 p.m.
June	5	11 p.m.
May	20	12 p.m.
May	5	1 a.m.

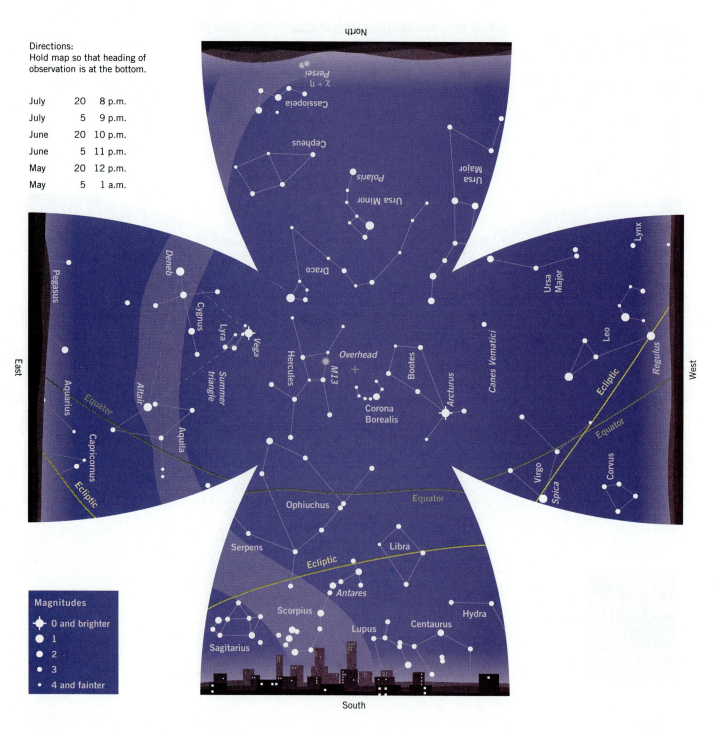

Magnitudes

✦ 0 and brighter
● 1
● 2
● 3
• 4 and fainter

Map 7

Directions:
Hold map so that heading of observation is at the bottom.

August	20	8 p.m.
August	5	9 p.m.
July	20	10 p.m.
July	5	11 p.m.
June	20	12 p.m.
June	5	1 a.m.

North

East

West

South

Overhead

Summer triangle

Ecliptic

Equator

Equator

Ecliptic

Andromeda
M31
Pegasus
Pisces
Aquarius
Capricornus
Deneb
Cygnus
Lyra
Vega
Altair
Aquila
Draco
Hercules
M13
Serpens
Ophiuchus
Serpens
Capricornus
Sagittarius
Antares
Scorpius
Lupus
Libra
Corona Borealis
Bootes
Arcturus
Serpens
Libra
Virgo
Spica
Hydra
Leo
Canes Venatici
Ursa Major
Polaris
Ursa Minor
Cepheus
Cassiopeia
χ + η Persei

Magnitudes
✦ 0 and brighter
● 1
● 2
• 3
· 4 and fainter

Map 8

Directions:
Hold map so that heading of
observation is at the bottom.

September 20	8 p.m.
September 5	9 p.m.
August 20	10 p.m.
August 5	11 p.m.
July 20	12 p.m.
July 5	1 a.m.

Magnitudes

- ✦ 0 and brighter
- ⬤ 1
- ● 2
- • 3
- · 4 and fainter

Map 9

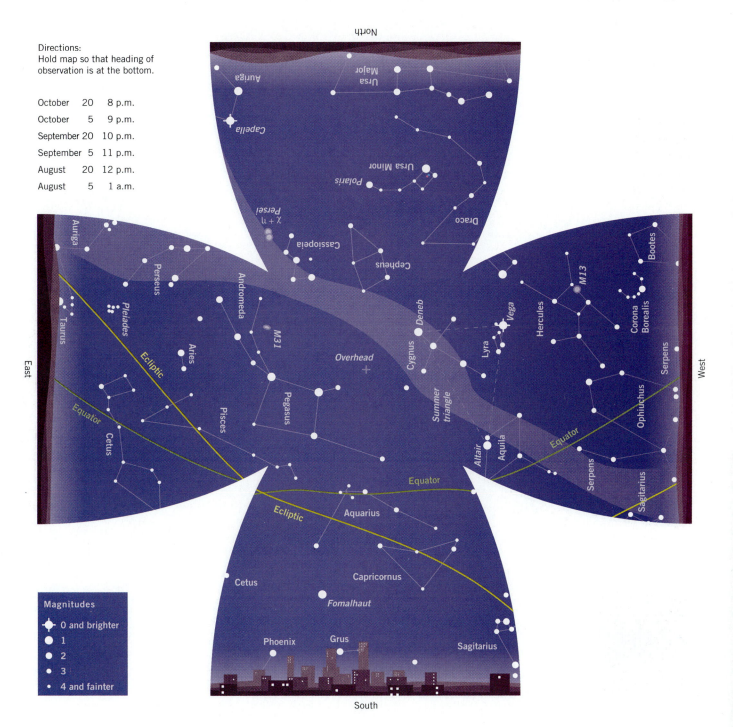

Directions:
Hold map so that heading of
observation is at the bottom.

October	20	8 p.m.
October	5	9 p.m.
September	20	10 p.m.
September	5	11 p.m.
August	20	12 p.m.
August	5	1 a.m.

East

West

South

Magnitudes
- 0 and brighter
- 1
- 2
- 3
- 4 and fainter

Map 10

Directions:
Hold map so that heading of
observation is at the bottom.

November 20   8 p.m.
November 5   9 p.m.
October   20   10 p.m.
October   5   11 p.m.
September 20   12 p.m.
September 5   1 a.m.

Magnitudes

✦ 0 and brighter
● 1
● 2
• 3
· 4 and fainter

Map 11

North

Directions:
Hold map so that heading of
observation is at the bottom.

December 20	8 p.m.
December 5	9 p.m.
November 20	10 p.m.
November 5	11 p.m.
October 20	12 p.m.
October 5	1 a.m.

Ursa Major

Draco

Ursa Minor

Polaris

Cepheus

Cassiopeia

χ + η Persei

Vega

Lyra

Deneb

Cygnus

Summer triangle

Altair

Lynx

Cancer

M44

Pollux

Gemini

Canis Minor

Procyon

Capella

Auriga

Perseus

M31

Andromeda

Pegasus

Aquarius

Equator

Ecliptic

Capricornus

East

West

Betelgeuse

Winter triangle

Orion

Aldebaran

Taurus

Pleiades

Overhead

Aries

Ecliptic

Pisces

Equator

Sirius

Canis Major

Lepus

Rigel

Cetus

Eridanus

Magnitudes

- 0 and brighter
- 1
- 2
- 3
- 4 and fainter

Fomalhaut

Phoenix

South

Map 12

THE CHEMICAL ELEMENTS

Table of Atomic Weights

Elements by Name, Symbol, Atomic Number, and Atomic Weight
(Atomic weights are given to four significant figures for elements below atomic number 104.)

Name	Symbol	Atomic Number	Atomic Weight	Name	Symbol	Atomic Number	Atomic Weight
Actinium	Ac	89	227.0	Erbium	Er	68	167.3
Aluminum	Al	13	26.98	Europium	Eu	63	152.0
Americium	Am	95	243.1	Fermium	Fm	100	257.1
Antimony	Sb	51	121.8	Fluorine	F	9	19.00
Argon	Ar	18	39.95	Francium	Fr	87	223.0
Arsenic	As	33	74.92	Gadolinium	Gd	64	157.2
Astatine	At	85	210.0	Gallium	Ga	31	69.72
Barium	Ba	56	137.3	Germanium	Ge	32	72.59
Berkelium	Bk	97	247.1	Gold	Au	79	197.0
Beryllium	Be	4	9.012	Hafnium	Hf	72	178.5
Bismuth	Bi	83	209.0	Hahnium	Ha	105	—
Boron	B	5	10.81	Hassium	Ha	108	—
Bromine	Br	35	79.90	Helium	He	2	4.003
Cadmium	Cd	48	112.4	Holmium	Ho	67	164.9
Calcium	Ca	20	40.08	Hydrogen	H	1	1.008
Californium	Cf	98	252.1	Indium	In	49	114.8
Carbon	C	6	12.01	Iodine	I	53	126.9
Cerium	Ce	58	140.1	Iridium	Ir	77	192.2
Cesium	Cs	55	132.9	Iron	Fe	26	55.85
Chlorine	Cl	17	35.45	Krypton	Kr	36	83.80
Chromium	Cr	24	52.00	Lanthanum	La	57	138.9
Cobalt	Co	27	58.93	Lawrencium	Lr	103	260.1
Copper	Cu	29	63.55	Lead	Pb	82	207.2
Curium	Cm	96	247.1	Lithium	Li	3	6.941
Dysprosium	Dy	66	162.5	Lutetium	Lu	71	175.0
Einsteinium	Es	99	252.1	Magnesium	Mg	12	24.30

Name	Symbol	Atomic Number	Atomic Weight	Name	Symbol	Atomic Number	Atomic Weight
Manganese	Mn	25	54.94	Ruthenium	Ru	44	101.1
Meitnerium	Mt	109	—	Rutherfordium	Rf	104	—
Mendelevium	Md	101	256.1	Samarium	Sm	62	150.4
Mercury	Hg	80	200.6	Scandium	Sc	21	44.96
Molybdenum	Mo	42	95.94	Seaborgium[1]	Sg	106	—
Neodymium	Nd	60	144.2	Selenium	Se	34	78.96
Neon	Ne	10	20.18	Silicon	Si	14	28.09
Neptunium	Np	93	237.0	Silver	Ag	47	107.9
Nickel	Ni	28	58.69	Sodium	Na	11	22.99
Nielsbohrium	Ns	107	—	Strontium	Sr	38	87.62
Niobium	Nb	41	92.91	Sulfur	S	16	32.07
Nitrogen	N	7	14.01	Tantalum	Ta	73	180.9
Nobelium	No	102	259.1	Technetium	Tc	43	98.91
Osmium	Os	76	190.2	Tellurium	Te	52	127.6
Oxygen	O	8	16.00	Terbium	Tb	65	158.9
Palladium	Pd	46	106.4	Thallium	Tl	81	204.4
Phosphorus	P	15	30.97	Thorium	Th	90	232.0
Platinum	Pt	78	195.1	Thulium	Tm	69	168.9
Plutonium	Pu	94	239.1	Tin	Sn	50	118.7
Polonium	Po	84	210.0	Titanium	Ti	22	47.88
Potassium	K	19	39.10	Tungsten	W	74	183.8
Praseodymium	Pr	59	140.9	Uranium	U	92	238.0
Promethium	Pm	61	144.9	Vanadium	V	23	50.94
Protactinium	Pa	91	231.0	Xenon	Xe	54	131.3
Radium	Ra	88	226.0	Ytterbium	Yb	70	173.0
Radon	Rn	86	222.0	Yttrium	Y	39	88.91
Rhenium	Re	75	186.2	Zinc	Zn	30	65.39
Rhodium	Rh	45	102.9	Zirconium	Zr	40	91.22
Rubidium	Rb	37	85.47				

[a] Names of elements 104, 105, and 107 to 109 have been endorsed by a committee of the American Chemical Society. An international chemistry organization recommends different names for elements 104 to 108.

[1] Proposed symbol and name

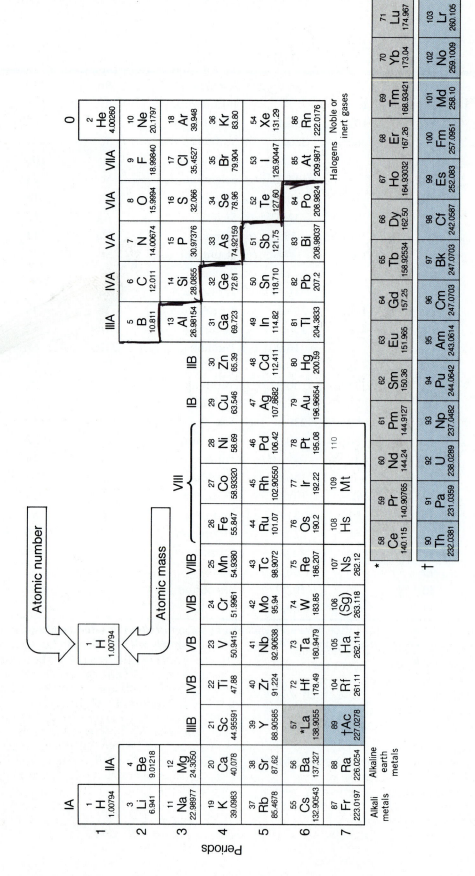

Names of elements 104, 105, and 107 to 109 have been endorsed by a committee of the American Chemical Society.
An international chemistry organization recommends different names for elements 104 to 108.

ANSWERS TO FILL-IN-THE-BLANK QUESTIONS AND SELECTED ODD-NUMBERED PROBLEMS

Chapter 1

Fill-in-the-Blank Answers: science, scientific method, observation, experiment, hypothesis, scientist, theory, law of nature, basic research, applied research, research and development (R&D), technology, peer review, National Academy of Science, National Institute of Health, National Science Foundation.

3A. The speed increases every 2 s by 10 mph until the car reaches 50 mph (10 s after your mother started) when the car continued at a constant speed of 50 mph for 6 s. After 16 s from the start, the car decreases its speed by 15 mph every 2 s; **B.** As the length of the string doubles, the square of the period of oscillation doubles (a direct relationship); **C.** The maximum heart rate of a human being decreases with age. There is a decrease by 10 beats per min for every increase in 10 years of age; **D.** The amount of gasoline pumped into a car increases by 2 gallons every 2.5 s; **E.** The altitude of the sun increases from 19° above the horizon in January and reaches its maximum altitude at 62° on June 21, and then decreases following the same curved pattern to its minimum altitude from June 21 to December; **F.** The world population increases over time, gradually at first, then rapidly toward the latter part of this century (an exponential relationship). A second description is that the world population increases by a factor of two about every 100 years.

A-31

Chapter 2

Fill-in-the-Blank Answers: scalar, vector, English system, metric system or international system (SI), foot, meter, second, kilogram, pound.

1. 43 mph; **3.** 3.7 minutes; **5a.** 70,000; **b.** 0.0007; **c.** 6,410,000; **d.** 0.00000641; **7a.** The acceleration increases by 0.85 m/s^2 for every increase in 5° of inclination; (the acceleration a is directly proportional to the inclination angle i, i.e., $a = k \times i$). **b.** The time of descent t decreases with every increase in inclination angle i (an inverse proportionality where $t = k \times 1/i = k/i$); **9.** The brightness of a light bulb (B) decreases with an increase in the distance from the bulb (d). In fact, the brightness of the bulb decreases as the square of the distance (an inverse-square relationship where $B = k \times 1/d^2$); **11.** P.J. expends 360 Cal running every Sunday; 720 Cal walking every Monday, Wednesday, and Friday; 990 Cal every Tuesday and Thursday playing volleyball. Total Cal expended = 2,070 Cal. Cal/day = 296 Cal/day (including Saturday); **13a.** 1.5×10^{11}; **b.** 4.3×10^3 ft; **c.** 2.3×10^{-5} m; **d.** 9.2×10^{-8} s; **e.** 7.4×10^{-2} g; **f.** 6.17×10^5; **g.** 4.3×10^{-5} breweries; **15a.** 1.1×10^{16}; **b.** 9.3×10^{-15}; **c.** 3.0×10^{-6}; **d.** 4.9×10^2

Chapter 3

Fill-in-the-Blank Answers: astronomy, Kepler's laws of planetary motion, mechanics, speed, velocity, acceleration, acceleration due to gravity (g), law of compound motion.

1. 273 ft/s = 0.05 mi/s. Since the car changes direction (oval) it accelerates; **3.** The hare and the tortoise will meet (be at the same position) 20 s after the gun goes off; **5.** The rock will take 0.63 s to hit the ground and will land 12.8 m from the point (horizontal position) it was dropped; **7.** 9.9×10^4 persons;

9.

Object	Period (years)
Earth	1.00
Mars	1.87
Pluto	248
Halley's Comet	75.7

11. Fastest to slowest: A, B, E, C, D; **13.** A plot of the square of the period against the cube of the averages distances gives a straight line with a slope of 1; **15a.** 1.7×10^8 years or 170 million years; **b.** km; **17.** 30.5 hours difference; **19.** −2 mph/s.

21.

Time (s)	Velocity (m/s)	Distance (m)
0.0	0.0	0.0
1.0	9.8	4.9
2.0	19.6	19.6
3.0	29.4	44.1
3.5	34.3	60.0
4.5	44.1	99.2

23. 44.1 m; **25.** Neglecting air resistance, they both will reach the ground at the same time; **27a.** 7.9 km/s. **b.** 84.5 min.

Chapter 4

Fill-in-the-Blank Answers: Newton's laws of motion, force, mass, force, momentum, conservation of momentum, Newton's law of universal gravitation, gravitational constant, weight, mass, centripetal force.

1a. 60 kg-m/s; **b.** 100 kg-m/s; **c.** 700 kg-m/s; **d.** 100 kg-m/s; **3.** For a 120 lb person the weight in newtons would be 535 N; **5.** The same; **7.** 1.4 times longer; **9.** Ocean liner (greatest), mosquito (lowest); **11a.** 2/3 m/s^2; **b.** 3 m; **c.** 2 m/s; **13.** 20 ft/s; **15a.** Both experience the same force (Newton's third law); **b.** The insect experiences the greatest acceleration; **c.** Both experience the same change in momentum; **17a.** 2,250 kg-m/s; **b.** 2,250 kg-m/s; **c.** 4.4 m/s; **19.** Force = 9.8 N on earth and 1.6 N (9.8 N/6) on the moon. Use 1738 km for the radius of the moon. **21.** Reduce all forces by 1/6 for the moon and 1/2.6 for Mars; **23.** 1.0 km/s; **25.** 1.1×10^{30} N.

Chapter 5

Fill-in-the-Blank Answers: work, energy, power, kinetic energy, potential energy, thermal energy, first law of thermodynamics, conservation law.

1. Assume a 55 kg (about 120 lb) person. **a.** 1.6×10^3 J; **b.** 2.2×10^5 J; **c.** 135 flights of stairs; **3.** Assume the kinetic energy (KE) for a car moving at 10 mph equals 1.0;

KE	Speed	KE	Speed
1.0	10 mph	9.0	30 mph
4.0	20 mph	36.0	60 mph

5a. 3.24×10^7 J; **b.** $0.90; **7a.** 120 J; **b.** 0.4 m/s; **9.** 1,900 ft-lbs or 2500 J; **11.** 9.0×10^5 J; **13.** 25 min;

15. 624 Cal; **17.** $1.35; **19a.** 26.5 m/s or 59 mph;
b. 17,375 J; **21a.** 6.39×10^{11} J; **b.** 145 km; **c.** 639
gigawatt plants.

Chapter 6

Fill-in-the-Blank Answers: temperature, absolute
zero, heat capacity, heat transfer, conduction, convection cell, radiation, second law of thermodynamics,
efficiency, entropy.

1a. 50%; **b.** 58%; **c.** Not worth the extra expenses to
work in the extreme cold; **3a.** 322 K; **b.** 233 K;
c. 6644 K; **d.** 2 K; **5a.** −193°C; **b.** 27°C; **c.** 5727°C;
d. 272°C; **7a.** 7.4×10^5 J or 176 Kilocalories; **b.** The
air; **c.** 3.5 kg. (Use 330 KJ/kg for the heat of fusion.);
9a. 1.2×10^6 J; **b.** 1.1×10^6 J; **c.** 0.38°C or 0.68°F;
11. 0.27°C; **13a.** 1.47×10^6 J; **b.** 17 W; **c.** For a 150
lb person using 2250 Cal (see Chapter 5) this is about
16%.

Chapter 7

Fill-in-the-Blank Answers: static electricity, electric charge, negative charge, positive charge, Coulomb's law, electric field, magnetic force, poles (north
and south), magnetic field, electric current, electricity, electric generator, electromagnet, electric motor,
alternating current (AC), direct current (DC), electric
circuit, ampere, volts, electrical resistance, ohms.

1a. No, no current will flow through bulbs B and C;
b. Yes, current flows through bulbs A and C; **c.** Yes,
current flows through bulbs A and B; **3.** 100 ohms;
5. 6.25×10^{18} electrons; **7a.** 5.25 W; **b.** 18,900 J;
9. 7500 W; **11.** c; **13.** The current is 14.3 amps. The
lights will not go off since the current does not exceed 15 amps. **15.** 36 days.

Chapter 8

Fill-in-the-Blank Answers: wave, wavelength, frequency, frequency, hertz, interference, constructive
interference, destructive interference, Doppler effect,
electromagnetic wave or radiation, speed of light (*c*),
electromagnetic spectrum, radio waves, microwaves,
infrared radiation, visible light, ultraviolet radiation,
x-rays, gamma rays.

1. 55 Hz; **3.** 3.16×10^9 or about 100 years; **5.** 0.5 Hz;
7. 3 ft/s; **9.** 1870 m; **11.** 3.07 Hz; **13a.** Should be
less. If the velocity remains constant but the distance
to the wall increases, the frequency must decrease;
b. 1.4 Hz; **c.** 42 claps.

Chapter 9

Fill-in-the-Blank Answers: chemical elements,
atom, electrons, nucleus, molecules, Bohr atom, quantum leap or jump, photon, spectrum, spectroscopy,
laser, periodic table of the elements, Pauli exclusion
principle, electron shell.

1. 10 lines; **3a.** 158 g; **b.** 20 g; **c.** 44 g; **d.** 56 g.

Chapter 10

Fill-in-the-Blank Answers: quantum mechanics,
uncertainty principle, probability, wave function,
wave-particle duality.

1. 6.63×10^{-32} m; **3.** 6.63×10^{-8} m or 0.06 microns;
5a. 1.0 cm; **b.** 2.0 cm; **7.** 5 cm.

Chapter 11

Fill-in-the-Blank Answers: chemical bonds, ionic
bond, covalent bond, metallic bond, hydrogen
bonds, van der Waals forces, states of matter, gas,
plasma, liquid, solid, crystals, glasses, plastics, polymers, changes of state.

3. 37,500 cal to get all 500 g to 100°C; 270,000 cal to
boil the 500 g; 307,500 cal total energy.

5.

Atomic Number	Type of Bonding	Atomic Number	Type of Bonding
1	Ionic	17	Ionic
2	Neutral	18	Neutral
6	Ionic	19	Ionic
7	Ionic	20	Ionic
8	Ionic	26	Ionic
10	Neutral	29	Ionic
11	Ionic	47	Ionic
13	Ionic	79	Ionic
14	Ionic	82	Ionic
16	Ionic	86	Neutral

7. 4.5 gm/m³; **9.** 21.7% (See problem 6); **11.** 9600
cal; **13.** 0°C with 21.9 g of ice and 998.1 g of liquid
water; **15a.** 2.43×10^5 cal or 1.0×10^6 J; **b.** 280 W.

Chapter 12

Fill-in-the-Blank Answers: chemical reactions, oxidation, reduction, acids, bases, pH, polymerization,

hydrocarbons, amino acids, proteins, enzymes, carbo-hydrates, sugar, DNA, nucleic acid, double helix, lipids.

3a. 2; **b.** 2; **c.** 3; **d.** 2, 3; **e.** 2, 1, 2; **f.** 3, 2; **g.** 2, 6, 2; **5a.** acid; **b.** acid; **c.** base; **d.** acid. 790% isooctane, 10% n-heptane.

Chapter 13

Fill-in-the-Blank Answers: composite materials, conductors, insulators, electrical resistance, semiconductors, superconductors, ferromagnet, domains, doping, diode, transistor, microchip, bits, byte, computer.

1a. 4.5 Megabytes/picture or 1 megaword/picture; **b.** 1 gigabits/sec; **5.** 4.5 bytes/word; **7a.** twice the capacity; **b.** 3.1×10^5 words on a high density disk and 1.5×10^5 words on a normal disk; **9a.** 93.6 s; **b.** 23.4 s; **11.** Color TV, 31 megabytes/s; black and white TV, 10 megabytes/s.

Chapter 14

Fill-in-the-Blank Answers: protons, neutrons, atomic number, isotopes, strong force, radioactivity or radioactive decay, alpha decay, beta decay, gamma radiation, half-life, radiometric dating, fission, fusion, nuclear reactor.

1a. Element = Oxygen, Atomic No. = 8, Atomic Mass = 16, Charge = −2; **b.** Oxygen, 8, 17, −2; **c.** Fluorine, 9, 17, −1; **d.** Oxygen, 8, 16, −1; **3.** There are many heavy isotopes of Cobalt; **5.** Between 3 and 4 half-lifes (15–20 days) or approximately 17.5 days. **7.** Step 1 Element = Thorium, Atomic Mass = 234, At. No. = 90; Step 2 Protactinium, 234, 91; Step 3 Uranium, 234, 92; Step 4 Thorium, 230, 90; **9.** No, it does not; **11.** 10 Standard X-rays; **13.** $1.5 \times 10{-}10$ J; **15.** 6.5×10^{11} J; **17a.** Element = Hydrogen, Atomic Mass =3, Charge = −1; **b.** Carbon, 14, +1; **c.** Aluminum, 27, 0; **d.** Cobalt, 60, −2; **e.** Xenon, 131, +3; **f.** Lead, 230, −5; **g.** Thorium, 232, 0.

Chapter 15

Fill-in-the-Blank Answers: high energy physics or elementary particle physics, cosmic rays, particle accelerators, synchrotrons, linear accelerators, hadrons, leptons, antiparticle or antimatter, positron, quarks, gauge particles, unified field theory.

1. Negative, neutral; **3.** +1; **5.** Atomic mass = 0, Atomic Number = 0, Charge = +1.

Chapter 16

Fill-in-the-Blank Answers: frame of reference, special relativity, general relativity, theory of relativity, time dilation, length contraction.

1. 20 mph; **3.** Both at the speed of light; **5.** Height = 100 ft, Width = 57.7 ft; **7.** $-c/8$; assumes that the ball and Jim are traveling in the same direction; **9.** 0.87 c; **11a.** 0.99995; **b.** 0.867; **c.** 0.0447; **13.** 0.867 c.

Chapter 17

Fill-in-the-Blank Answers: Solar System, terrestrial planets, Jovian planets, Pluto, moons, asteroids, asteroid belt, comets, nebula, great bombardment, meteors, meteorite, differentiation, core, mantle, crust.

1. Sun's radius is 1.8 times the radius of the moon's orbit, which is 5.4 times the radius of Jupiter; **3.** Volume of the Sun is 1.3×10^6 times the volume of the Earth. Volume of Jupiter is 1,400 times the volume of the Earth; **5.** 5.98×10^{16} kg/yr or 8.2 million times larger than the present rate; **7.** 2,290 mph; **9.** Density of Earth is 1.02 times the density of Mercury. Density of Earth is 4.15 times the density of Jupiter. Interiors are different; **11.** 0.5%.

Chapter 18

Fill-in-the-Blank Answers: plate tectonics, tectonic plates, mantle convection, divergent plate boundaries, convergent plate boundaries, transform plate boundaries, volcanoes, earthquakes, subduction zones, hot spots, seismology.

1. 20 million years; **3.** The midwest and central plains states; **5.** 3.7×10^6 yrs; **7.** 2 magnitude difference = 900 times the energy; **9.** 5.0 cm/yr. This is reasonable.

Chapter 19

Fill-in-the-Blank Answers: atmospheric cycle, weather, climate, front, jet streams, hydrological cycle, currents, residence times, groundwater, glaciers, ice caps, ice ages, acid rain, ozone layer, ozone hole, greenhouse effect.

1. 3000 kg; **3.** Must collect about 330 cans per school day per student; unreasonable.

Chapter 20

Fill-in-the-Blank Answers: rock cycle, igneous rocks, sedimentary rocks, metamorphic rocks, volcanic or extrusive rocks, intrusive rocks, evolution, fossils, fossil record, chemical evolution, Miller–Urey experiment, natural selection, extinction, mass extinctions, *Australopithecus*, *Homo erectus*, Neanderthal, *Homo sapiens*.

1. 0.07%.

Chapter 21

Fill-in-the-Blank Answers: astronomy, stars, apparent magnitude, absolute magnitude, telescopes, light years, main sequence star, solar wind, red giant, white dwarf, supernovas, neutron stars, pulsar, black hole.

1. 10.0 Kw/m^2; **3.** Mars; 0.64 Kw/m^2; Jupiter 0.05 Kw/m^2; Pluto; 0.00097 Kw/m^2.

Chapter 22

Fill-in-the-Blank Answers: cosmology, galaxies, Milky Way, Hubble's law, red shift, big bang theory, cosmic microwave background radiation, dark matter.

1a. 7500 km/s; **b.** 7.5×10^{10} km/s; Impossible: this exceeds the speed of light! **3.** 2.1 (cm/hour)/cm; **5.** Assume 1.8 m height = 1.9×10^{-16} LY. A 10^{50} inflation gives 1.9×10^{34} LY; much larger than the observable universe; **7.** 115,000 km/s or 0.38 c.

GLOSSARY

AAAS (pronounced "triple-A S") See *American Association for the Advancement of Science.*

absolute magnitude The brightness a star appears to have when it is viewed from a standard distance. (Ch 21)

absolute zero The temperature, 0 K, at which no energy can be extracted from atoms; the coldest attainable temperature, equal to $-273.16°C$ or $-459.67°F$. (Ch 6)

absorption A process by which light energy is converted into some other form, usually heat energy. See also *transmission* and *scattering.* (Ch 8)

absorption line A dark line in an absorption spectrum that corresponds to the absorbed wavelength of light. (Ch 9)

absorption spectrum A characteristic set of dark lines used to identify a chemical element or molecule from the photons absorbed by atoms or molecules. (Ch 9)

AC See *alternating current.*

acceleration The amount of change in velocity divided by the time it takes the change to occur. Acceleration can involve changes of speed, changes in direction, or both. (Ch 3)

acceleration due to gravity A constant numerical value for the acceleration that all objects experience at the Earth's surface, determined by measuring the actual fall rate in a laboratory: $9.8 m/s^2 = 32 ft/s^2$. (Chs 3, 4)

acid Any material that when put into water produces positively charged hydrogen ions (i.e., protons) in the solution; for example, lemon juice and hydrochloric acid. (Ch 12)

acid rain A phenomenon that occurs when nitrogen and sulfur compounds in the air interact with water to form tiny droplets of nitric and sulfuric acid, which makes raindrops more acidic than normal. (Ch 19)

addition polymerization The formation of a polymer in which the basic building blocks are simply joined end to end; for example, polyethylene. (Ch 12)

air pollution A serious environmental problem, with immediate consequences for urban residents, from the emission of NO_x compounds, sulfur dioxide, and hydrocarbons into the atmosphere. (Ch 19)

alkali metals Elements that are highly reactive, such as lithium, sodium, and potassium; listed in far left-hand column of the periodic table of elements. These elements possess one valence electron. (Ch 9)

alkaline earth metals Elements that combine with oxygen in a one-to-one ratio and form colorless solid compounds with high melting temperatures. Listed in the second column in the periodic table of elements: beryllium, magnesium, calcium, and others. These elements possess two valence electrons. (Ch 9)

alkane A family of molecules, based on the methane molecule, that burns readily and is used as fuels. (Ch 12)

alloy The combination of two or more chemical elements in the metallic state; for example, brass (a mixture of copper and zinc) or bronze (an alloy of copper and tin). (Ch 11)

alpha decay The loss by the nucleus of a large and massive particle composed of two protons and two neutrons. (Ch 14)

alpha particle A particle, made of two protons and two neutrons, emitted during alpha decay, used by Ernest Rutherford in a well-known experiment in which the nucleus was discovered. (Ch 9)

alternating current (AC) A type of electric current, commonly used in household appliances and cars, in which charges periodically alternate their direction of motion. (Ch 7)

AM See *amplitude modulation.*

American Association for the Advancement of Science (AAAS) One of the largest scientific societies, representing all branches of the physical, biological, and social sciences. AAAS is a strong force in establishing science policy and promoting science education. (Ch 1)

amino acid The building block of protein, incorporating a carboxyl group (COOH) at one end, an amino group (NH_2) at the other end, and a side group (which varies from one amino acid to the next). (Ch 12)

amino group A group of atoms of nitrogen and hydrogen (NH_2) that forms one end of an amino acid. See *carboxyl group.* (Ch 12)

ampere (amp) A unit of measurement for the amount of current (number of charges) flowing in a wire or elsewhere per unit time. (Ch 7)

amplifier A device that takes a small current and converts it into a large one to do work. (Ch 13)

amplitude The height of a wave crest above the undisturbed level of the medium. (Ch 8)

amplitude modulation (AM) A process that a radio station uses to vary the amplitude of a radio wave signal being transmitted, usually between 530 and 1600 kilohertz. After transmission, the signal is converted to sound by the radio receiver. (Ch 8)

angular momentum The property of rotating objects which tends to make them keep rotating. (Ch 4)

annihilation The most efficient and violent energy conversion process known in the universe; occurs when a particle collides with its antiparticle, completely converting both masses to energy. (Ch 15)

antimatter Particles that have the same mass as their matter twins, but with an opposite charge and opposite magnetic characteristics. (Ch 15)

apparent magnitude The brightness a star appears to have when it is viewed from the Earth. (Ch 21)

applied research The type of research performed by scientists with specific and practical goals in mind. This research is often translated into practical systems by large-scale research and development projects. (Ch 1)

aquifer An underground body of stored water, often a layer of water-saturated rock bounded by impermeable materials such as clay. (Ch 19)

artificial intelligence   A field based on the idea that computers might someday be able to perform all functions of thought that we normally think of as being distinctly human. (Ch 13)

artificial selection   The process of conscious breeding for specific characteristics in plants and animals. (Ch 20)

asteroid belt   A collection of small, rocky planetesimals, which never managed to collect into a single planet. They are located in a circular orbit between Mars and Jupiter. (Ch 17)

asteroids   Small, rocky objects, which circle the Sun like miniature planets. (Ch 17)

astronomy   The study of objects in the heavens. (Ch 21)

atmospheric cycle   The circulation of gases near the Earth's surface, including the short-term variations of weather and the long-term variations of climate. (Ch 19)

atom   Fundamental building blocks for all matter. A particle having a nucleus and electrons in orbit; the smallest representative sample of an element. (Ch 9)

atomic number   The number of protons in the nucleus, which determines the nuclear charge and therefore the chemical identity of an atom. (Ch 14)

atomism   The hypothesis that for each chemical element there is a corresponding species of indivisible objects called atoms. (Ch 9)

Australopithecus   The first hominid, a primate closer to humans than any other; lived approximately 4.5 million years ago, walked erect, and had a brain about the size of that of a modern chimpanzee. (Ch 20)

basalt   A dense, dark, even-textured volcanic rock forming the oceanic plates; rich in oxides of silicon, magnesium, iron, calcium, and aluminum. (Ch 20)

base   A class of corrosive material that when put into water produces negatively charged hydroxide ions; usually tastes bitter and feels slippery. (Ch 12)

base pair   One of four possible bonding combinations of the bases adenine, thymine, guanine, and cytosine on the DNA molecule: AT, TA, GC, and CG. (Ch 12)

basic research   The type of research performed by scientists who are interested in simply finding out how the world works: in knowledge for its own sake. (Ch 1)

battery   A device that converts stored chemical energy into kinetic energy of charged particles (usually electrons) running through an outside wire. (Ch 7)

beta decay   A kind of radioactive decay in which a particle such as the neutron spontaneously transforms into a collection of particles that includes an electron. (Ch 14)

big bang theory   The idea that the universe began at a specific point in the past and has been expanding ever since. (Ch 22)

bit   Binary digit: a unit of measurement for information equal to "yes-no" or "on-off." (Ch 13)

black hole   Formed at the death of a very large star, an object so dense, with a mass so concentrated, that nothing—not even light—can escape from its surface. (Ch 21)

blueshift   The result of the Doppler effect on light waves, when the source of light moves toward the observer: light wave crests bunch up and have a higher frequency. (Ch 8)

Bohr atom   A model of the atom, developed by Niels Bohr in 1913, in which electrons exist only in allowable orbits, located at fixed distances from the center of an atom. In these orbits, the electrons maintain fixed energy for long periods of time, without giving off radiation. (Ch 9)

boiling   A change of state from liquid to gas caused by the increase in temperature or decrease in pressure, which speeds up the vibration of individual molecules of the liquid, allowing them to break free and form a gas. (Ch 11)

Brownian motion   A phenomenon that describes the rapid, random movements caused by atomic collisions of very small objects suspended in a liquid. (Ch 9)

byte   In a computer, a group of eight switches storing eight bits of information; the basic information unit of most modern computers. (Ch 13)

c   A constant, 300,000 kilometers per second (about 186,000 miles per second), equal to the product of the wavelength and frequency of an electromagnetic wave; the speed of light. (Ch 8)

calorie   A common unit of energy defined as the amount of heat required to raise 1 gram of room-temperature water 1 degree Celsius in temperature. (Ch 5)

Cambrian explosion   The sudden change in life on Earth, well documented in the fossil record, when hard-bodied organisms first appeared about 570 million years ago. (Ch 20)

Cambrian period   The time, beginning about 570 million years ago, when animals first began to develop shells and skeletons. (Ch 20)

carbohydrate   A class of modular molecules, which are made of clusters of sugar molecules (carbon, hydrogen, and oxygen), and which form part of the solid structure of plants and play a central role in how living things are supplied with energy. (Ch 12)

carboxyl group   A group of atoms of carbon, hydrogen, and oxygen (COOH) that forms one end of an amino acid string. See *amino group*. (Ch 12)

central processing unit (CPU)   The part of a computer in which transistors store and manipulate relatively small amounts of information at any one time. (Ch 13)

centrifugal force   The illusory force produced by an object moving in a circle; the force is directed outward, away from the center of the circle. (Ch 4)

centripetal force   The force needed to keep an object moving in a circle; the force is directed inward, toward the center of the circle. (Ch 4)

Cepheid variable   A type of star with a regular behavior of brightening and dimming, whose period is related to the star's luminosity. Cepheid variables are used to calculate distances out to many millions of light-years. (Ch 21)

ceramic   A crystalline solid that includes a broad class of hard, durable solids, including bricks, concrete, pottery, porcelain, and numerous synthetic abrasives. (Ch 11)

CFCs   See *chlorofluorocarbons*.

chain reaction   The process in which nuclei undergoing fission produce neutrons that will cause more splitting, resulting in the release of large amounts of energy. (Ch 14)

change of state   Transition between the solid, liquid, and gas states caused by changes in temperature and pressure. The processes involved are freezing and melting (for solids and liquids), boiling and condensation (for liquids and gases), and sublimation (for solids and gases). (Ch 11)

chaos A field of study modeling systems in nature that can be described in Newtonian terms but whose futures are, for all practical purposes, unpredictable; for example, the turbulent flow of water or the beating of a human heart. (Ch 4)

chemical bond The name given to the attraction resulting from the redistribution of electrons that leads to a more stable configuration between two or more atoms—particularly by filling the outer electron shells—that holds the two atoms together. The principal kinds of chemical bonds are ionic, covalent, and metallic. (Ch 11)

chemical evolution The process by which simple chemical compounds present in Earth's early atmosphere became an organized reproducing cell. (Ch 20)

chemical potential energy The type of energy that is stored in the chemical bonds between atoms, such as the energy in flashlight batteries or detergents. (Ch 5)

chemical reaction The process by which atoms or smaller molecules come together to form large molecules, or by which larger molecules are broken down into smaller ones; involves the rearrangement of atoms in elements and compounds, as well as the rearrangement of electrons to form chemical bonds. (Ch 12)

chlorofluorocarbons (CFCs) A class of stable and generally nonreactive chemicals widely used in refrigerators and air-conditioners until the late 1980s. See *ozone hole*. (Ch 19)

cholesterol An essential component of the cell membrane synthesized by the body from saturated fats in the diet; in high levels, can cause fatty deposits that clog arteries. (Ch 12)

chromosphere One of the Sun's outer layers, visible for a few minutes as a spectacular halo during a total eclipse of the Sun. (Ch 21)

climate The average weather conditions of a place or area over a period of years. (Ch 19)

closed system A type of system in which matter and energy are not freely exchanged with the surroundings; also known as an isolated system. (Ch 5)

closed universe A universe in which the expansion will someday reverse, because the universe holds enough matter to exert a strong enough gravitational force to reverse the motion of receding galaxies. (Ch 22)

cluster A collection of galaxies. (Ch 22)

COBE See *Cosmic Background Explorer.*

combustion A rapid combination with oxygen, producing heat and flame. (Ch 12)

comet An object, usually found outside the orbit of Pluto, composed of chunks of materials such as water ice and methane ice embedded with dirt; may fall toward the Sun, if its distant orbit is disturbed, and become visible in the night sky. (Ch 17)

composite material A combination of two or more substances in which the strength of one of the constituents is used to offset the weakness of another, resulting in a new material whose strengths are greater than any of its components; for example, plywood and reinforced concrete. (Ch 13)

compressional wave One of two principal types of seismic waves in which the molecules in the rock move back and forth in the same direction as the wave; a longitudinal wave. (Ch 18)

compressive strength A material's ability to withstand crushing. (Ch 13)

computer A machine that stores and manipulates information. (Ch 13)

condensation A change of state from gas to liquid caused by the decrease in temperature or pressure of the gas, which slows down the vibration of individual molecules of the gas, allowing them to form a liquid. (Ch 11)

condensation polymerization A chemical reaction often used to manufacture plastics and other polymers and which, in the body, occurs during the formation of a peptide bond. (Ch 12)

condensation reaction The formation of a polymer in which each new polymer bond releases a water molecule as the ends of the original polymer molecules link up. (Ch 12)

conduction The movement of heat by collisions between vibrating atoms or groups of atoms (molecules); one of three mechanisms by which heat moves. (Ch 6)

conduction electron An electron in a material that is able to move in an electric field. (Ch 13)

conductor A material capable of carrying an electric current; any material through which electrons can flow freely. (Ch 13)

conservation law Any statement that says that a quantity in nature does not change. (Ch 5)

conservation of momentum A law which states that the total momentum of a system does not change unless an outside force is applied.

constructive interference A situation in which two waves act together to reinforce or maximize the wave height at the point of intersection. (Ch 8)

continuously habitable zone The narrow area around a star in which a planet can revolve for billions of years with liquid water at its surface. Only planets in this zone can support life. (Ch 21)

convection The transfer of heat by the physical motion of masses of fluid: dense, cooler fluids (liquids and gases) descend in bulk and displace rising warmer fluids, which are less dense; one of three mechanisms by which heat moves. (Ch 6)

convection cell A region in a fluid in which heat is continuously being transferred by a bulk motion of heated fluid from a heat source to the surface of the fluid, where heat is released. The cooled fluid then sinks and the cycle repeats. (Ch 6)

convection zone The outer region of the Sun, comprising the upper 200,000 km (about 125,000 mi) where the dominant energy transfer mechanism changes from collision to convection. (Ch 21)

convergent plate boundary A place where two tectonic plates are coming together. (Ch 18)

core (*a*) geology: the central spherical region of the Earth, composed primarily of dense iron and nickel metal with a radius of about 3400 km (2000 mi). (*b*) astronomy: a small region in the center of a star where hydrogen burning is generally confined. (Ch 17)

corona One of the Sun's outer layers, visible for a few minutes as a spectacular halo during a total eclipse of the Sun. (Ch 21)

Cosmic Background Explorer (COBE) An orbiting observatory that measures microwave radiation that is present as background noise in every direction of the skies. (Ch 21)

cosmic microwave background radiation Microwave radiation, characteristic of a body about 3K, coming to Earth from all directions. This radiation is evidence for the big bang. (Ch 22)

cosmic rays Particles (mostly protons) that rain down continuously on the atmosphere of the Earth after being emitted by stars in our galaxy and in others. (Ch 15)

cosmology The branch of science that is devoted to the study of the structure and history of the entire universe. (Ch 22)

coulomb (pronounced "koo-loam") The unit for measuring the magnitude of an electric charge. (Ch 7)

Coulomb's law An empirically derived rule that states "The magnitude of the electrostatic force between any two objects is proportional to the charges of the two objects, and inversely proportional to the square of the distance between them." (Ch 7)

covalent bond A chemical bond in which neighboring molecules share electrons in a strongly bonded group of at least two atoms. (Ch 11)

cpu See central processing unit.

critical mass The minimum number of radioactive atoms needed to sustain a nuclear chain reaction to the point that large amounts of energy can be released. (Ch 14)

crust A thin layer at the surface of the Earth formed from the lightest elements. It ranges in thickness from 10 km (6 mi) in parts of the ocean to 70 km (45 mi) beneath parts of the continents. (Ch 17)

crystal A solid in which atoms occur in a regularly repeating sequence. Crystal structure is described by first determining the size and shape of the repeating boxlike group of atoms, and then recording the exact type and position of every atom that appears in the box. (Ch 11)

current A riverlike body of moving water in an ocean basin. (Ch 19)

cyclone A great rotational pattern in the atmosphere, hundreds of kilometers in diameter, which can draw energy from the warm oceanic waters and create low-pressure tropical storms. (Ch 19)

cyclotron The first of the particle accelerators, for which Ernest Lawrence won the 1939 Nobel Prize in physics. (Ch 15)

dark matter Material that exists in forms that cannot be seen and may constitute 90% or more of the matter of the universe. (Ch 22)

DC See direct current.

decay chain A series of decays, or radioactive events, ending with a stable isotope. (Ch 14)

deep ocean trench A surface feature associated with convergent plate boundaries in which no continents are on the leading edge of either of the two converging plates, and one plate penetrates deep into the Earth. (Ch 18)

degeneracy pressure A permanent outward force, exerted by electrons or neutrons, that cancels the inward force of gravity in some stars. (Ch 21)

dendrite One of a thousand projections on each nerve cell in the brain through which nerve signals move; each is connected to different neighboring nerve cells in the brain. (Ch 13)

deoxyribonucleic acid (DNA) A strand of nucleotides with alternating phosphate and sugar molecules in a long chain, and with base molecules adenine, guanine, cytosine, and thiamine at the side. The nucleotide strand bonds with a second nucleotide strand to make a molecule with a ladderlike double helix shape. DNA stores the genetic information in a cell. (Ch 12)

depolymerization The breakdown of a polymer into short segments. (Ch 12)

destructive interference A situation in which two waves intersect in a way that decreases or cancels out the wave height at the point of intersection. (Ch 8)

differentiation The process in planetary formation by which heavy, dense materials (for example, iron and nickel) sink under the force of gravity toward the molten center of a planet, while lighter, less dense materials float to the top, resulting in the layered structure of the present-day Earth. (Ch 17)

diffuse scattering A process by which light waves are absorbed and re-emitted in all directions by a medium such as clouds or snow. (Ch 8)

diode An electronic device that allows electric current to flow in only one direction. (Ch 13)

dipole field The magnetic field that always arises from the two poles of a magnet. (Ch 7)

direct current (DC) A type of electric current in which the electrons flow in one direction only; for example, in the chemical reaction of a battery. (Ch 7)

distillation A process by which engineers separate the complex mixture of petroleum's organic chemicals into much purer fractions. (Ch 12)

divergent plate boundary A spreading zone of crustal formation; a place where neighboring tectonic plates move away from each other. (Ch 18)

Divine Calculator An eighteenth-century idea, proposed by Pierre Simon Laplace, which stated that if the position and velocity of every atom in the universe is known, with infinite computational power, the future position and velocity of every atom in the universe could be predicted. (Ch 4)

DNA See deoxyribonucleic acid.

doping The addition of a minor impurity to a semiconductor. (Ch 13)

Doppler effect The change in frequency or wavelength of a wave detected by an observer because the source of the wave is moving. (Ch 8)

double bond The type of covalent bond formed when two electrons are shared by one atom. (Ch 11)

double helix The twisted double strand of nucleotides that forms the structure of the DNA molecule. (Ch 12)

earthquake Disturbance caused when stressed rock on Earth suddenly snaps, converting potential energy into released kinetic energy. (Ch 18)

efficiency The amount of work you get from an engine, divided by the amount of energy you put in; a quantification of the loss of useful energy. (Ch 6)

El Niño A weather cycle in the Pacific Basin that recurs every four to seven years and can cause severe storms and flooding all along the western coast of the Americas, as well as drought from Australia to India. (Ch 19)

elastic limit The point at which a material stops resisting external forces and begins to deform permanently. (Ch 13)

elastic potential energy The type of energy that is stored in a flexed muscle, a coiled spring, or a stretched rubber band. (Ch 5)

electric charge An excess or deficit of electrons on an object. (Ch 7)

electric circuit An unbroken path of material that carries electricity and consists of three parts: a source of energy, a closed path, and a device to use the energy. (Ch 7)

electric current A flow of charged particles, measured in amperes. (Ch 7)

electric field The force that would be exerted on a positive charge if it were brought to a position near a charged object. Every charged object is surrounded by an electric field. (Ch 7)

electric generator A source of energy producing an alternating current in an electric circuit through the use of electromagnetic induction, which may be located in a power plant many miles from the energy's ultimate use or under the hood of a car. (Ch 7)

electric motor A device that operates by supplying current to an electromagnet to make the magnet move and generate mechanical power. The motor usually employs permanent magnets and rotating loops of wire inside the poles of these magnets. (Ch 7)

electrical conductivity A number that measures the ease with which a material allows electrons to flow. The inverse of electrical resistance. (Ch 13)

electrical potential energy The type of energy that is found in a battery or between two wires at different voltages; energy associated with the position of a charge in an electrical field. (Ch 5)

electrical resistance The quantity, measured in ohms, that represents how hard it is to push electrons through a material. High-resistance wires are used when electron energy is to be converted into heat energy. Low-resistance wires are used when energy is to be transmitted from one place to another with minimum loss. (Chs 7, 13)

electricity A force, more powerful than gravity, that moves objects both toward and away from each other, depending upon the charge. (Ch 7)

electromagnet A device that produces a magnetic field from a moving electrical charge. (Ch 7)

electromagnetic induction A process by which a changing magnetic field produces an electric current in a conductor, even though there is no other source of power available. (Ch 7)

electromagnetic radiation See *electromagnetic wave.*

electromagnetic spectrum The entire array of waves, varying in frequency and wavelength, but all resulting from an accelerating electric charge; includes radio waves, microwaves, infrared, visible light, ultraviolet, X-rays, gamma rays, and others. (Ch 8)

electromagnetic wave A form of radiant energy that reacts with matter by being transmitted, absorbed, or scattered. A self-propagating wave made up of electric and magnetic fields fluctuating together. A wave created when electrical charges accelerate, but requiring no medium for transfer; electromagnetic radiation. (Ch 8)

electromagnetism A term used to refer to the unified nature of electricity and magnetism. (Ch 7)

electron Tiny, negatively charged particles that circle around a positively charged nucleus of an atom. (Ch 9)

electron shell A specific energy level in an atom that can be filled with a predetermined number of electrons. (Ch 9)

electrostatic charge The type of electric charge that doesn't move once it has been placed on an object, and the forces exerted by such charges. (Ch 7)

electroweak force A force resulting from the unification of the electromagnetic and weak forces. (Ch 15)

element Materials made from a single type of atom. Elements cannot be broken down any further by chemical means. (Ch 9)

elementary-particle physics The study of particles that are believed to comprise the basic building blocks of the universe; for example, the particles that make up the nucleus, and particles such as the electron. See *high-energy physics.* (Ch 15)

elementary particles Particles that make up the nucleus, together with particles such as the electron; the basic building blocks of the universe. (Ch 15)

ELF radiation Extremely low-frequency waves associated with the movement of electrons to produce the alternating current in household wires. (Ch 8)

emission lines Lines of specific wavelengths or colors emitted by atoms caused by electron transitions to lower states of energy. (Ch 9)

emission spectrum A characteristic set of emission lines that corresponds to the wavelengths of light emitted by atoms.

endothermic A chemical reaction in which the final energy of the electrons in the reaction is greater than the initial energy; energy must be supplied to make the reaction proceed. (Ch 12)

energy The ability to do work; the capacity to exert a force over a distance. A system's energy can be measured in joules or foot-pounds. (Ch 5)

entropy The thermodynamic quantity that describes the degree of randomness of a system. The greater the disorder or randomness, the higher the statistical probability of the state, and the higher the entropy. (Ch 6)

environment The combination of atmosphere, oceans, and land on which all living things rely for their existence. (Ch 19)

enzyme A molecule that facilitates reactions between two other molecules, but which is not itself altered or taken up in that overall reaction. (Ch 12)

eukaryote An advanced single-celled organism and all multicelled organisms that are made from cells that contain a nucleus. (Ch 20)

evolution The process by which new life forms arise and change. (Ch 20)

excited state All energy levels of an atom above the ground state. (Ch 9)

exothermic A chemical reaction in which the final energy of the electrons is less than the initial energy, and therefore energy is given off in some form. (Ch 12)

experiment The manipulation of some aspect of nature to observe an outcome. (Ch 1)

extinction The disappearance of a species on Earth. (Ch 20)

extrusive rock See *volcanic rock.*

fault A fracture in a rock along which movement occurs. (Ch 18)

ferromagnetism The property of a few materials in nature, such as iron, cobalt, and nickel metals, in which the individual atomic magnets are arranged in a nonrandom manner, lined up with each other into small magnetic domains to produce an external magnetic field. (Ch 13)

field The force—magnetic, gravitational, or electric—that would be felt at a particular point. For example, forces exerted by one object that would be felt by another object in the same region. (Ch 7)

first law of thermodynamics The law of the conservation of energy. In an isolated system the total amount of energy, including heat energy, is conserved. (Ch 5)

fission A reaction that produces energy when heavy radioactive nuclei split apart into fragments that together have less mass than the original isotopes. (Ch 14)

flat universe A model of the future of the universe in which the expansion slows and comes to a halt after infinite time has passed. (Ch 22)

fluorescence A phenomenon in which energy contained in ultraviolet wavelengths is absorbed by the atoms in some materials and partially emitted as visible light. (Ch 8)

FM See *frequency modulation*.

foot-pound The amount of work done by a force of one pound acting through one foot. The unit of energy in the English system. (Ch 5)

force A push or pull that, acting alone, causes a change in acceleration of the object on which it acts. (Ch 4)

fossil Evidence in rock for past life. Usually a replica in stone of the original organism, created when calcium and other atoms in the hard parts of the buried organism are replaced by minerals in the water flowing through the surrounding area. (Ch 20)

fossil fuel Carbon-rich deposits of ancient life that burn with a hot flame and have been the most important energy source for 150 years. Examples include coal, oil, and natural gas. (Ch 5)

fossil record A term that refers to all of the fossils that have been found, catalogued, and studied since human beings first began to study them in a systematic way. (Ch 20)

frame of reference The physical surroundings from which a person observes and measures the world. (Ch 16)

freezing A change of state from liquid to solid caused by the decrease in temperature or change in pressure of the liquid, which slows the vibration of individual molecules and forms the solid structure. (Ch 11)

frequency The number of wave crests that go by a given point every second. A wave completing one cycle (sending one crest by a point every second) has a frequency of 1 hertz (Hz). (Ch 8)

frequency modulation (FM) A process by which information is transmitted by varying the frequency of a signal. After being transmitted, the signal may be converted to sound by circuits in the receiver. (Ch 8)

fusion A process in which two nuclei come together to form a third, larger nucleus. When this reaction combines light elements to make heavier ones, the mass of the final nucleus is less than the mass of its constituent parts. The "missing" nuclear mass can be converted into energy. (Ch 14)

G See *gravitational constant*.

g The acceleration of any object due to the force of gravity at the Earth's surface; equal to 9.8 m/s^2 or 32 ft/s^2.

galaxy A large assembly of stars (between millions and hundreds of billions of them), together with gas, dust, and other materials, that is held together by the forces of mutual gravitational attraction. (Ch 22)

gamma radiation A kind of radioactivity involving the emission of energetic electromagnetic radiation from the nucleus of an atom, with no change in the number of protons or neutrons in the atom. (Ch 14)

gamma ray The highest-energy wave of the electromagnetic spectrum, with wavelengths between the size of an atom and the size of a nucleus, less than one trillionth of a meter; normally emitted in very high-energy nuclear particle reactions. (Chs 8, 14)

Gamma Ray Observatory (GRO) An orbiting observatory that detects the highest-energy end of the electromagnetic spectrum, gamma rays. One of the first permanent orbiting observatories launched by the National Atmospheric and Space Administration's Great Observatories Program for monitoring all parts of the electromagnetic spectrum. (Ch 21)

gas Any collection of atoms or molecules that expands to take the shape of and fill the volume available in its container. (Ch 11)

gauge particle A particle that produces one of the fundamental forces that hold everything together when exchanged by two other particles; examples include photons, gravitons, gluons, and W and Z particles. (Ch 15)

GCM See *global circulation models*.

general relativity The second and more complex of two parts of Einstein's theory of relativity, which applies to all reference frames, whether or not those frames are accelerating relative to each other. (Ch 16)

glacier Large body of ice that slowly flows down a slope or valley under the influence of gravity; found primarily in Greenland and Antarctica. (Ch 19)

glass A solid with predictable local environments for most atoms, but no long-range order to the atomic structure. Compared to crystals, glass lacks the repeating unit of atoms. (Ch 11)

global circulation models (GCMs) Complex computer models of the atmosphere that are the best attempts to date to predict long-term climate and to discuss various types of ecological changes such as global warming. (Ch 19)

glucose An important sugar ($C_6H_{12}O_6$) in the energy cycle of living things; figures prominently in the energy metabolism of every living cell. (Ch 12)

gluon A massless gauge particle, confined to the interior of particles, that mediates the strong force holding quarks together. (Ch 15)

gneiss The metamorphic rock formed from slate under extreme temperature and pressure. (Ch 20)

gradualism A hypothesis that holds that most evolutionary change occurs as a result of the accumulation of small adaptations. (Ch 20)

granite A lower-density rock, capping the mantle rock and forming much of the continental plates. (Ch 20)

gravitational constant (G) A universal constant that expresses the exact numerical relation between the masses of two objects and their separation, on the one hand, and the force between them on the other; equal to 6.67×10^{-11} N-m^2/kg^2. (Chs 3, 4)

gravitational escape One way that a planet's atmosphere

can evolve and change: molecules in the atmosphere heated by the Sun may move sufficiently fast so that appreciable fractions of them can escape the gravitational pull of their planet. (Ch 17)

gravitational potential energy   Energy associated with the position of a mass in a gravitational field. The gravitational potential energy of an object equals its weight (the force of gravity exerted by the object) times its height above the ground. (Ch 5)

graviton   The gauge particle of gravity. (Ch 15)

gravity   A force that acts on every object in the universe. (Ch 4)

great bombardment   An event following the initial period of planetary formation in which meteorites showered down on planets, adding matter and heat energy. (Ch 17)

greenhouse effect   A global temperature increase caused by the fact that the Earth's atmospheric gases trap some of the Sun's infrared (heat) energy before it radiates out into space. (Ch 19)

GRO   See *Gamma Ray Observatory.*

ground state   The lowest energy level of an atom. (Ch 9)

groundwater   Fresh water from the surface that typically percolates into the ground and fills the tiny spaces between grains of sandstone and other porous rock layers. (Ch 19)

gyre   A circulation of water at the surface of the ocean, transporting warm water from the equator toward the cooler poles, and cold water from the poles back to the equator to be heated and cycled again. (Ch 19)

hadron   Particles, including the proton and neutron, that are made from quarks and are subject to the strong force. (Ch 15)

half-life   The rate of radioactive decay measured by the time it takes for half of a collection of isotopes to decay into another element. (Ch 14)

heat (thermal energy)   A measure of the quantity of atomic kinetic energy contained in every object. (Ch 5)

heat capacity   A measure of the ability of a material to absorb heat energy, defined as the constant of proportionality between an amount of heat and the change of temperatures that this heat produces in the material. (Ch 6)

heat transfer   The process by which heat moves from one place to another, through conduction, convection, and/or radiation. (Ch 6)

helium burning   The final energy-producing stage of many stars in which the temperature in the interior becomes so hot that the helium itself will begin to undergo nuclear fusion reactions to make carbon. The net reaction will be: $^4He + {^4He} + {^4He} \longrightarrow {^{12}C}$. (Ch 21)

hertz (Hz)   The unit of measurement for the frequency of waves; one wave cycle per second. (Ch 8)

Hertzsprung-Russell (H-R) diagram   A simple graphical technique widely used in astronomy to plot a star's temperature (determined by its spectrum) versus the star's energy output (measured by its energy and brightness). (Ch 21)

high-energy physics   The field of study of elementary particles and their properties. See *elementary particle physics.* (Ch 15)

high-grade energy   Sources of energy that can be used to produce very high-temperature reservoirs; for example, petroleum and coal. (Ch 6)

high-temperature reservoir   Any hot object from which

energy is intended to do work. Within the cylinder of a gasoline engine is a high-temperature reservoir. (Ch 6)

hole   The absence of an electron; in a silicon crystal, for example, a hole is left behind after a conduction electron is shaken loose. (Ch 13)

Homo erectus   ("man the erect") The species of modern human's genus who first walked erect and learned to use fire; disappeared about 500,000 years ago. (Ch 20)

Homo habilis   The first member of the genus of modern humans, who appeared about 2 million years ago in East Africa; distinguished by a larger brain and stone tools. (Ch 20)

Homo sapiens   The single species that includes all branches of the human race; recognized in fossils as old as 200,000 years. (Ch 20)

horsepower   Equal to 550 foot-pounds per second in the English system of measurement; commonly used to assess the power of engines and motors. (Ch 5)

hot spot   A dramatic type of volcanism indirectly associated with plate tectonics; large isolated chimney-like columns of hot rock, or mantle plumes, rising to the surface in certain places of the Earth; for example, Yellowstone National Park, Iceland, and Hawaii. (Ch 18)

HST   See *Hubble Space Telescope.*

Hubble Space Telescope (HST)   A reflecting telescope, launched in 1990, with a 2.4-m mirror designed to give unparalleled resolution in the visible and ultraviolet wavelengths. Manufacturing flaws in the main mirror were corrected by astronauts in late 1993. (Ch 21)

Hubble's law   The law relating the distance to a galaxy, d, and the rate at which it recedes from Earth as measured by the redshift: $V = Hd$. (Ch 22)

hurricane   Tropical storms having winds in excess of 120 km/h (75 mph) that begin in the Atlantic Ocean off the coast of Africa and affect North America. (Ch 19)

hydrocarbon   A chainlike molecule from a chemical compound of carbon and hydrogen, which provides the most efficient fuels for combustion, with only carbon dioxide and water as products. (Ch 12)

hydrogen bond   A bond that may form when a polarized hydrogen atom links to another atom by a covalent or ionic bond. (Ch 11)

hydrogen burning   A three-step process, generally confined to a small region in the center of a star, in which four protons are converted into a ^4He nucleus, two protons, and a photon. (Ch 21)

hydrogenation   The addition of hydrogen atoms into the carbon chains of polyunsaturated products; a process that eliminates the carbon-carbon double bonds. (Ch 12)

hydrological cycle   The combination of processes by which water moves from repository to repository near the Earth's surface. (Ch 19)

hypothesis   A tentative guess about how the world works, based on a summary of experimental or observational results and phrased so that it can be tested by experimentation. (Ch 1)

Hz   See *hertz.*

ice age   A period of several million years during which glaciers have repeatedly advanced and retreated, causing radical changes in climate and influencing human evolution. (Ch 19)

ice cap   Layers of ice that form at the north and south polar regions of the Earth. (Ch 19)

igneous rock   The first rock to form on a cooling planet, solidified from hot, molten material; intrusive or extrusive (volcanic). (Ch 20)

inertia   The tendency of a body to remain in uniform motion; the resistance to change. (Ch 4)

inflation   A short period of rapid expansion of the universe, which, according to the grand unified theories, accompanied the "freezing" at 10^{-35} second. (Ch 22)

inflationary theories   Theories that incorporate the phenomenon of inflation at a specified time after the big bang. (Ch 22)

Infrared Astronomical Satellite (IRAS)   An orbiting observatory launched in 1983 by the United States, the United Kingdom, and Netherlands to view infrared radiation in the universe. IRAS is no longer functioning. (Ch 21)

infrared energy   A form of electromagnetic radiation that travels from a source to an object, where it can be absorbed and converted into the kinetic energy of molecules. (Ch 6)

infrared radiation   Wavelengths of electromagnetic radiation that extend from a millimeter to a micron; felt as heat radiation. (Ch 8)

insulator   A material that will not conduct electricity. (Ch 13)

integrated circuit   A microchip made of hundreds or thousands of transistors specially designed to perform a specific function. (Ch 13)

interference   When waves from two different sources come together at a single point, they interfere with each other; the observed wave amplitude is the sum of the amplitudes of the interfering waves. (Ch 8)

international system (SI)   An internationally recognized system of units based on multiples of ten. (Ch 2)

intrusive rock   Igneous rock that cools and hardens underground. (Ch 20)

inversely proportional   The relationship between two measurable quantities so that if one quantity increases, the other quantity decreases, and vice versa, by a constant proportion. (Ch 2)

ion   An atom that has an electrical charge, from either the loss or gain of one or more electrons. (Ch 9)

ionic bond   A chemical bond in which the electrostatic force between two oppositely charged ions holds the atoms in place; often formed as one atom gives up an electron while another receives it, lowering chemical potential energy when atom shells are filled. (Ch 11)

ionization   Stripping away one or more of an atom's electrons to produce an ion. (Ch 14)

IRAS   See *Infrared Astronomical Satellite*.

isolated system   See *closed system*.

isomer   A molecule that contains the same atoms as another molecule, but has a different structural arrangement. (Ch 12)

isotope   Two nuclei are isotopes if they have the same number of protons but a different number of neutrons. (Ch 14)

jet stream   A high-altitude stream of fast-moving winds that marks the boundary between the northern polar cold air mass and the warmer air of the temperate zone. (Ch 19)

joule   The amount of work done when you exert a force of one newton through a distance of one meter. (Ch 5)

Jovian planets   Huge worlds also known as "gas giants," located in the outer solar system where the effects of solar heat and the solar wind are reduced. These planets are made primarily of frozen liquids and gases such as hydrogen, helium, ammonia, and water, with atmospheres of nitrogen, methane, and other compounds: Jupiter, Saturn, Uranus, Neptune. (Ch 17)

Kepler's laws of planetary motion   Three mathematical laws of planetary motion derived by German mathematician Johannes Kepler. (1) All planets orbit the Sun in an elliptical path. (2) A planet's velocity will increase as it moves closer to the Sun. (3) The farther a planet is from the Sun, the longer its year will be. (Ch 3)

kilowatt   A commonly used measurement of electrical power equal to 1000 watts and corresponding to the expenditure of 1000 joules per second. (Ch 5)

kinetic energy   The type of energy associated with moving objects: the energy of motion. Kinetic energy is equal to the mass of a moving object times the square of that object's velocity, multiplied by $\frac{1}{2}$. (Ch 5)

laser   An instrument that uses a collection of atoms, energy, and mirrors to emit photons that have wave crests in exact alignment. The instrument name is the acronym for *l*ight *a*mplification by *s*timulated *e*mission of *r*adiation. (Ch 9)

lava   Molten rock discharged by volcanoes that flows across the Earth's surface and eventually hardens into new rock. (Ch 18)

law of compound motion   Motion in one direction has no effect on motion in another. (Ch 3)

law of nature   An overarching statement of how the universe works, following repeated and rigorous observation and testing of a theory or group of related theories to show that the theory seems to apply everywhere in the universe. (Ch 1)

length contraction   The phenomenon in relativity that moving objects appear to be shorter than stationary ones in the direction of motion. (Ch 16)

lepton   A particle (such as the electron, muon, and neutrino) that participates in the weak, but not the strong, interaction. (Ch 15)

light   A form of electromagnetic wave to which the human eye is sensitive. Light travels at a constant speed in a vacuum and needs no medium for transfer. (Ch 8)

light-year   The distance light travels in one year, approximately 10 trillion km (about 6.2 trillion miles). (Ch 21)

limestone   A sedimentary rock formed from the calcium carbonate ($CaCO_3$) skeletons of sea animals, shells, and coral. (Ch 20)

linear accelerator   A device for making high-velocity particles, which relies on a long, straight vacuum tube into which particles are injected to ride an electromagnetic wave down the tube. (Ch 15)

lipid   An organic molecule that is insoluble in water. At the molecular level, lipids form the cell membranes that separate living material from its environment. Lipids are also an extremely efficient storage medium for energy; for example, fat in foods, wax in candles, and grease for lubrication. (Ch 12)

liquid   Any collection of atoms or molecules that has no fixed shape but maintains a fixed volume. (Ch 11)

liquid crystal   A synthesized substance, used in digital displays, that is formed from very long molecules that may adopt a very ordered arrangement even in the liquid form. (Ch 11)

load   The location in a circuit where the useful work is done, such as the filament in a light bulb or the heating element of a dryer. (Ch 7)

longitudinal wave   A kind of wave in which the motion of the medium is in the same direction as the wave movement; pressure wave or sound wave. (Ch 8) Also, one of two principal types of seismic waves in which the molecules in rock move back and forth in the same direction as the wave; see *compressional wave.* (Ch 18)

Lorentz factor   A number, equal to the square root of $[1 - (v/c)^2]$, that appears in relativistic calculations and is an indication of the magnitude of change in time and scale. (Ch 16)

low-temperature reservoir   The ambient atmosphere into which the waste heat generated by an engine is dumped; for example, from a cylinder in a gasoline engine to the atmosphere. (Ch 6)

luminosity   The total energy radiated by a star. (Ch 21)

magma   Subsurface molten rock, concentrated in the upper mantle or lower crust, which can breach the surface and harden into new rock. (Ch 18)

magnetic field   A collection of lines that map out the direction that compass needles would point in the vicinity of a magnet. (Ch 7)

magnetic force   The force exerted by magnets on each other. (Ch 7)

magnetic monopole   A hypothetical single isolated north or south magnetic pole, proposed in some theories of matter but not yet located through experimentation. (Ch 7)

magnetic potential energy   The type of energy stored in a magnetic field. (Ch 5)

magnetism   A fundamental force in the universe. (Ch 7)

main-sequence star   A star that derives energy from the fusion reactions of hydrogen burning; found on the Hertzsprung-Russell (H-R) diagram within a bandlike pattern. (Ch 21)

mammal   One of a group of vertebrates made up of individuals that are warm-blooded, have hair, and whose females nurse their young. (Ch 20)

mantle   The thick layer—rich in oxygen, silicon, magnesium, and iron, and containing most of the Earth's mass—overlying the metal core of the Earth. (Ch 17)

mantle convection   The convective motion deep within the Earth, driven by internal heat energy, that moves continents and plates of which they are a part. (Ch 18)

marble   A metamorphic rock that began as limestone and was subjected to intense pressure and high temperatures. (Ch 20)

mass   The amount of matter contained in an object, independent of where that object is found. (Ch 4)

mass extinction   Rare and catastrophic events in the past that have caused large numbers of species to become extinct suddenly. (Ch 20)

mass number   A quantity related to the number of neutrons plus the number of protons, which determines the mass of an isotope. (Ch 14)

Maxwell's equations   Four fundamental laws of electricity and magnetism. (1) Coulomb's law: like charges repel and unlike charges attract. (2) Magnetic monopoles do not exist in nature. (3) Magnetic phenomena can be produced by electric effects. (4) Electrical phenomena can be reproduced by magnetic effects. (Ch 7)

mechanics   The branch of science that deals with the motions of material objects and the forces that act on them; for example, a rolling rock or a thrown ball. (Ch 3)

meltdown   The most serious accident that can occur at a nuclear reactor, in which the flow of water to the fuel rods is interrupted and the enormous heat stored in the central part of the reactor causes the fuel rods to melt. (Ch 14)

melting   A change of state from solid to liquid caused by an increase in temperature or pressure of the solid, which increases the vibration of individual molecules and breaks down the structure of the solid. (Ch 11)

metal   Characterized by a shiny luster and ability to conduct electricity, an element or combination of elements in which the sharing of a few electrons results in a more stable electron arrangement. (Ch 11)

metallic bond   A chemical bond in which electrons are redistributed so that they are shared by all the atoms in the material as a whole. (Ch 11)

metamorphic rock   Igneous or sedimentary rock that is buried and transformed by the Earth's intense internal temperature and pressure. (Ch 20)

meteor   A piece of interplanetary debris that hits the Earth's atmosphere and forms a bright streak of light from friction with atmospheric particles: often called a shooting star. (Ch 17)

meteor showers   A set of spectacular, regularly occurring events in the night sky, caused by the collision of the Earth with clouds of small debris that travel around the orbits of comets. (Ch 17)

meteorite   The fragment of a meteor that hits the Earth. (Ch 17)

metric system   An internationally recognized system of measurement comprised of units based on multiples of 10; standard units include the meter as the unit of length, the kilogram as the unit of mass, and the second as the unit of time. See *international system* (SI). (Ch 2)

microchip   A complex array of *p*- and *n*-type semiconductors, which may incorporate hundreds or thousands of transistors in one integrated circuit. (Ch 13)

microwave   Electromagnetic waves, with wavelengths ranging from approximately 1 meter to 1 millimeter, which are used extensively for line-of-sight communications and cooking. (Ch 8)

Milky Way   A collection of about 100 billion stars, forming the galaxy of which the Sun is a part. (Ch 22)

Miller-Urey experiment   A demonstration of possible steps in chemical evolution, performed in 1953 by Stanley Miller and Harold Urey, which showed that a combination of gases, believed to be present in the early atmosphere, and a series of electric sparks, simulating the lightning on the early Earth, will produce amino acids, a basic building block of life. (Ch 20)

moderator   In a nuclear reactor, the fluid whose function is to slow down neutrons that leave the fuel rods. (Ch 14)

molecule   A cluster of atoms that can be isolated; the basic constituent of many different kinds of material. (Chs 9, 11)

momentum   The product of the mass of an object moving in a straight line times its velocity. (Ch 4)

monosaccharide   An individual sugar molecule. (Ch 12)

mono-unsaturated   A lipid with one carbon-carbon double bond, and thus two fewer hydrogen atoms than a fully saturated lipid. (Ch 12)

monsoon   Any wind system on a continental scale that seasonally reverses its direction because of seasonal variations in relative temperatures over land and sea. (Chs 6, 19)

Moon   The Earth's only satellite, which may have formed when a planet-sized body hit the Earth early in its history. (Ch 17)

mudstone   A sedimentary rock formed from sediments that are much finer-grained than sand. (Ch 20)

mutation   A change in the genetic material of a parent that is inherited by the offspring. (Ch 20)

N   See *newton*.

n-type semiconductor   A type of conductor formed from doping, which has a slight excess of mobile negatively charged electrons. (Ch 13)

National Academy of Science   A nationally recognized association of scientists, elected to membership by their peers to provide professional advice for the government on policy issues ranging from environmental risks and natural resource management to education and funding for science research. (Ch 1)

National Institutes of Health   A federal agency that provides funding for basic and applied research in medicine and biology. (Ch 1)

National Science Foundation   A federal agency that funds American scientific research and education in all areas of science. (Ch 1)

natural selection   The mechanism by which nature can introduce wide-ranging changes in living things over long periods of time by modifying the gene pool of a specific species. (Ch 20)

Neanderthal man   A type of human with a large brain who lived until 35,000 years ago in groups with a complex social structure; either a separate species of the genus *Homo* or a subspecies of *Homo sapiens*. (Ch 20)

nebulae   Dust and gas clouds, common throughout the Milky Way galaxy, rich in hydrogen and helium. (Ch 17)

nebular hypothesis   A model that explains the formation of the solar system from a large cloud of gas and dust floating in space 4.5 billion years ago. This cloud collapsed upon itself under the influence of gravity and began to spin faster and faster, eventually forming the planets and the rest of the solar system along a thin flattened disk of matter surrounding a central star. (Ch 17)

negative electric charge   An excess of electrons on an object. (Ch 7)

neutrino   A subatomic particle, emitted in the decay of the neutron, which has no electric charge, travels at the speed of light, and has no rest mass. (Ch 15)

neutron   A type of subatomic particle, located in the nucleus of the atom, which carries no electrical charge, but has approximately the same mass as the proton; one of two primary building blocks of the nucleus. (Ch 14)

neutron star   A very dense, very small star, usually with a high rate of rotation and a strong magnetic field; the core remains of a supernova, held up by the degeneracy pressure of neutrons. (Ch 21)

newton (N)   The unit of force. The force required to give an acceleration of 1 m/s² to a mass of 1 kg. (Ch 4)

Newton's law of universal gravitation   Between any two objects in the universe there is an attractive force (gravity) that is proportional to the masses of the objects and inversely proportional to the square of the distance between them. In other words, the more massive two objects are, the greater the force between them will be, and the farther apart they are, the less the force will be. (Ch 4)

Newton's laws of motion   Three basic principles, expressed as laws, that govern the motion of everything in the universe, from stars and planets to cannonballs and muscles. *The first law* states that a moving object will continue moving in a straight line at a constant speed, and a stationary object will remain at rest, unless acted on by an unbalanced force. *The second law* states that the acceleration produced on a body by a force is proportional to the magnitude of the force and inversely proportional to the mass of the object. *The third law* states that for every action there is an equal and opposite reaction. (Ch 4)

noble gases   Elements listed in the far right-hand column of the periodic table of elements, including helium, argon, and neon, which are odorless, colorless, and slow to react. (Ch 9)

nonrenewable resources   Resources such as coal and petroleum, which are forming at a much slower rate than they are being consumed. (Ch 5)

nuclear reactor   A device that controls fission reactions to produce energy when heavy radioactive nuclei split apart. (Ch 14)

nucleic acid   A molecule originally found in the nucleus of cells that carries and interprets the genetic code; includes DNA and RNA. (Ch 12)

nucleotide   A molecule that is the basic element from which all DNA and RNA are built; formed from a sugar, a phosphate group, and one of four bases (adenine, guanine, cytosine, and thymine or uracil). (Ch 12)

nucleus   (*a*) The very small, compact object at the center of an atom; made up primarily of protons and neutrons. (*b*) A prominent structure in the interior of a cell that contains the cell's genetic material—the DNA—and controls the cell's chemistry. (Chs 9, 20)

observation   The act of observing nature without manipulating it. (Ch 1)

ohm   A unit of measurement for the electrical resistance of a wire. (Ch 7)

oil shale   A form of fossil fuel in which petroleum is dispersed through solid rock. (Ch 5)

Oort cloud   A region beyond the orbit of Pluto that contains billions of comets circling the Sun; the reservoir for new comets. (Ch 17)

open system   A type of system within which an object can exchange matter and energy with its surroundings. (Ch 5)

open universe   A model of the future of the universe in which the expansion will continue forever because the universe lacks the matter to exert a gravitational force to slow receding galaxies. (Ch 22)

organic chemistry   The branch of science devoted to the study of carbon-based molecules and their reactions. (Chs 11, 12)

organic molecules Molecules that contain carbon, whether or not they come from living systems. (Ch 12)

outgassing Release of gases from nongaseous materials; extrusion of gases from the body of a planet after its formation. (Ch 17)

oxidation A chemical reaction in which an atom such as oxygen accepts electrons while combining with other elements; examples include rusting of iron metal into iron oxide, and animal respiration. (Ch 12)

oxides Chemical compounds that contain oxygen, like most common minerals and ceramics. (Ch 13)

ozone A molecule made up of three oxygen atoms, instead of the usual two, that absorbs ultraviolet radiation. (Ch 19)

ozone hole A volume of atmosphere above Antarctica during September through November in which the concentration of the trace gas ozone has declined significantly. (Ch 19)

ozone layer A region of enhanced ozone (O_3) 30 to 45 kilometers (about 20 to 30 miles) above the Earth's surface where most of the absorption of the sun's ultraviolet radiation occurs. (Ch 19)

p-type semiconductor A type of conductor formed from doping, which has a slight deficiency of electrons, resulting in mobile positively charged holes. (Ch 13)

paleomagnetism The field devoted to the study of remnant magnetism in ancient rock recording the direction of the magnetic poles at some time in the past. (Ch 18)

parsec A unit of measurement equal to about 3.3 light-years, which roughly corresponds to the average distance between nearest neighbor stars in our galaxy. (Ch 21)

particle accelerator A machine such as a synchrotron or linear accelerator that produces particles at near light speeds for use in the study of the fundamental structure of matter. (Ch 15)

Pauli exclusion principle A statement that says no two electrons can occupy the same state at the same time. (Ch 9)

peer review A system by which the editor of a scientific journal submits manuscripts considered for publication to a panel of knowledgeable scientists who, in confidence, evaluate the manuscripts for mistakes, misstatements, or shoddy procedures. Following the review, if the manuscript is to be published, it is returned to the author with a list of modifications and corrections to be completed. (Ch 1)

peptide bond A connection between two atoms that remains after the hydrogen (H) at one end of an amino acid and the hydroxyl (OH) from the end of another amino acid combine, releasing a water molecule (H_2O). The process is identical to the condensation polymerization reaction. (Ch 12)

periodic table of the elements An organizational system, first developed by Dmitri Mendeleev in 1869, now listing more than 110 elements by atomic weight (in rows) and by chemical properties (in columns). The pattern of elements in the periodic table reflects the arrangement of electrons around their nuclei. (Ch 9)

petroleum Thick black liquid found deep underground and derived from many kinds of transformed molecules of former life. (Ch 12)

phospholipid The class of molecules that form membranes in cells. Lipids have a long, thin structure with a carbon backbone and a phosphate group at one end of the molecule. (Ch 12)

photoelectric effect A phenomenon that occurs when photons strike one side of a material and cause electrons of that material to be emitted from the opposite side. This effect is observed in cameras, CAT scans, and fiber optics. (Ch 10)

photon A particle-like unit of electromagnetic radiation, emitted or absorbed by an atom when an electrically charged electron changes state; the form of a single packet of electromagnetic radiation. (Ch 9)

photosphere The gaseous layers of the Sun's outer part, which emit most of the light we see. (Ch 21)

Planck's constant (h) A constant named after the German physicist Max Planck is the central constant of quantum physics; equal to 6.63×10^{-34} joule-seconds in SI units. (Ch 10)

planetesimal Small objects, which range in size from boulders to several miles across, formed from the accretion of solid material during the formation of the planets. (Ch 17)

plasma A state of matter existing under extreme temperatures in which electrons are stripped from their atoms during high-energy collisions, forming an electron sea surrounding positive nuclei. (Chs 11, 21)

plastic A solid composed of intertwined chains of molecules called polymers, with an ability to be molded or formed into virtually any desired shape: clear film, dense casting, strong fiber, colorful molding. (Ch 11)

plate A rigid moving sheet of rock up to 100 km (60 mi) thick, composed of the crust and part of the upper mantle. See *plate tectonics*. (Ch 18)

plate tectonics The model of the dynamic Earth that has emerged from studies of paleomagnetism, rock dating, and much other data. A theory that explains how a few thin, rigid tectonic plates of crustal and upper mantle materials are moved across the Earth's surface by mantle convection. (Ch 18)

polar molecule Atom clusters with a positive and negative end; exerts electrical force on neighboring atoms. Water is a polar molecule. (Ch 11)

polarization The subtle electron shift from negative to positive that takes place when the electrons of an atom or a molecule are brought near a polar molecule such as water, resulting in a bond caused by the electrical attraction between the negative end of the polar molecule and the positive side of the other molecule. (Ch 11)

poles The two opposite ends of a magnet, named north and south, which repel a like magnetic pole and attract an unlike magnetic pole. (Ch 7)

pollution An excess of a substance generated by human activity and present in an undesired location. (Ch 19)

polymer A material composed of extremely long and large molecules that are formed from numerous smaller molecules, like links in a chain, with predictable repeating sequences of atoms along the chain. (Ch 11)

polymerization A reaction that includes all chemical reactions that form long strands of polymer fibers by linking small molecules. (Ch 12)

polypeptide   A bonded chain of amino acids. See *peptide bond*. (Ch 12)

polysaccharide   A molecule that is the result of many sugar molecules strung together in a chain; for example, starch and cellulose. (Ch 12)

polyunsaturated   A type of lipid that forms when two or more kinked "double bonds" between carbon atoms are in the molecule. See *unsaturated*. (Ch 12)

positive electric charge   A deficiency of electrons on an object. (Ch 7)

positron   The positively charged antiparticle of the electron. (Ch 15)

potential energy   The energy a system possesses if it is capable of doing work but is not doing work now. Types of potential energy include magnetic, elastic, electrical, and chemical. (Ch 5)

power stroke   The downward motion of a piston in a gasoline engine, in which the actual work is done and the energy released by combustion is translated into the motion of the car. (Ch 6)

power   The rate at which work is done, or the rate at which energy is expended. The amount of work done, divided by the time it takes to do it. Power is measured in watts in the metric system, horsepower in the English. (Ch 5)

precession   The circular motion of the spinning axis of the Earth in space, which causes the tilt of the Northern Hemisphere to change on a 23,000-year cycle. (Ch 19)

precipitation   A chemical reaction that is the reverse of a solution reaction, producing a solid that separates from very concentrated solutions. (Ch 11)

prediction   A guess about how a particular system will behave, followed by observations to see if the system did behave as expected within a specified range of situations. (Ch 1)

prevailing westerly   Wind in the midlatitude zones that blows primarily from west to east, causing weather patterns to move from west to east. (Ch 19)

primary structure   The simplest of the four stages of the organization of the amino acids in a protein molecule; the exact order of amino acids along the protein string. (Ch 12)

primordial soup   A rich broth of the amino acids and other molecules created by the Miller-Urey process and concentrated in the early oceans for several hundred million years. (Ch 20)

probability   The likelihood that an event will occur or that an object will be in one state or another; how nature is described in the subatomic world. (Ch 10)

prokaryote   A type of primitive cell in which the DNA is coiled together but not separated in the nucleus. Prokaryotes constitute the kingdom Monera, including all cells that do not have a nucleus. (Ch 20)

protein   A molecule which can consist of hundreds of amino acids and thousands of atoms formed in a chain structure. Proteins function as enzymes and direct the cell's chemistry. (Ch 12)

proton   One of two primary building blocks of the nucleus, with a positive electric charge of +1 and a mass $(1.6726430 \times 10^{-24}$ g) approximately equal to that of the neutron. (Ch 14)

pseudoscience   A kind of inquiry, falling in the realm of belief or dogma, that includes subjects that cannot be proved or disproved with a reproducible test. The subjects include creationism, extrasensory perception (ESP), unidentified flying objects (UFOs), astrology, crystal power, and reincarnation. (Ch 1)

publication   A peer-reviewed paper written by a scientist or group of scientists to communicate the results of their research to a larger audience. A publication will include the technical details of the methodology, so that the research can be reproduced, and a concise statement of the results and conclusions. (Ch 1)

pulsar   A neutron star in which fast-moving particles speed out along the intense magnetic field lines of the rotating star, giving off electromagnetic radiation that we detect as a series of pulses of radio waves. (Ch 21)

pumice   Frothy volcanic rock rich in silicon from magmas that mix with a significant amount of water or other volatile substance. (Ch 20)

pumping   The process in a laser that adds energy to the system from the outside to return atoms continuously to their excited states so that more coherent photons can be produced. (Ch 9)

punctuated equilibrium   A hypothesis that holds that evolutionary changes usually occur in short bursts separated by long periods of stability. (Ch 20)

quantized   Whenever energy or another property of a system can have only certain definite values, and nothing in between those values, it is said to be quantized. (Ch 10)

quantum jump   See *quantum leap*.

quantum leap   A process by which an electron changes energy levels without ever traversing any of the positions between the original and final levels; also known as quantum jump. (Ch 9)

quantum mechanics   The branch of science that is devoted to the study of the motion of objects that come in small bundles, or quanta. (Ch 10)

quark   (*pronounced "quork"*)   The truly fundamental building block of the hadrons: a particle that has fractional electric charge and cannot exist alone in nature. (Ch 15)

quartzite   A durable rock in which the original sand grains of sandstone, under high temperature and pressure, recrystallize and fuse into a solid mass. (Ch 20)

quasar   Quasi-stellar radio source. Objects in the universe, where as-yet-unknown processes pour vast amounts of energy into space each second from an active center no larger than the solar system; the most distant objects known. (Ch 22)

quaternary structure   The joining of separate protein chains, each with its own secondary and tertiary structure. (Ch 12)

R&D   See *research and development*.

radiation   (*a*) The particles emitted during the spontaneous decay of nuclei. (Ch 14) (*b*) The transfer of heat by electromagnetic radiation; the only one of the three mechanisms of heat transfer that does not require atoms or molecules to facilitate the transfer process. (Ch 6)

radio wave   Part of the electromagnetic spectrum that ranges from the longest waves—wavelengths longer than the Earth—to waves a few meters long. (Ch 8)

radioactive decay   The process of spontaneous change of unstable isotopes. (Ch 14)

radioactivity The spontaneous release of energy by certain atoms, such as uranium, as these atoms disintegrate. The emission of one or more kinds of radiation from an isotope with unstable nuclei. (Chs 5, 14)

radiometric dating A technique based on the radioactive half-lives of carbon-14 and other isotopes that is used to determine the age of materials. (Ch 14)

radon A colorless, odorless inert gas that can cause an indoor pollution problem when it undergoes radioactive decay. (Ch 14)

red giant A very large star that emits a lot of energy but whose surface is very cool and therefore appears somewhat reddish in the sky; found on the upper right corner of the H-R diagram. (Ch 21)

redshift An increase in the wavelength of the radiation received from a receding celestial body or any other light source as a consequence of the Doppler effect; a shift toward the long-wavelength (red) end of the spectrum with a corresponding decrease in frequency. (Chs 8, 22)

reduction A chemical reaction in which electrons are transferred from an atom to other elements, resulting in a gain in electrons for the material being reduced; for example, smelting of metal ores and photosynthesis. (Ch 12)

reductionism The quest for the ultimate building blocks of the universe. An attempt to reduce the seeming complexity of nature by first looking for an underlying simplicity and then trying to understand how that simplicity gives rise to the observed complexity. (Ch 15)

reflection A process by which light waves are scattered at the same angle as the original wave; for example, from the surface of a mirror. (Ch 8)

refraction A process by which light slows down or changes direction as it passes through matter. See also *absorption* and *transmission*. (Ch 8)

relativity See *theory of relativity*.

reproducible A criterion for the results of an experiment. In the scientific method, observations and experiments must be reported so that anyone with the proper equipment can verify the results. (Ch 1)

research and development A kind of research, aimed at specific problems, usually performed in government and industry laboratories. (Ch 1)

residence time The average length of time that any given atom will stay in ocean water before it is removed by some chemical reaction. (Ch 19)

rock cycle An ongoing cycle of internal and external Earth processes by which rock is created, destroyed, and altered. (Ch 20)

Roentgen Satellite (ROSAT) Launched in 1990 by the United States, United Kingdom, and Germany as the latest in a series of satellites equipped to detect X-rays. (Ch 21)

ROSAT See *Roentgen Satellite*.

salts Molecules formed by the neutralization of an acid and a base. (Ch 12)

sandstone A sedimentary rock formed mostly from sand-sized grains of quartz (silicon dioxide) and from other hard mineral and rock fragments. (Ch 20)

saturation A fully bonded carbon atom in a lipid. In a straight lipid chain, every carbon atom bonds to two adjacent carbons along the chain and two hydrogen atoms on the sides. (Ch 12)

scalar A quantity that can be described by a single number, such as weight, length, or time. (Ch 2)

scattering A process by which electromagnetic waves may be absorbed and rapidly re-emitted; can be diffuse scattering or reflection. (Ch 8)

schist The metamorphic rock formed from slate under extreme temperature and pressure. (Ch 20)

science The discipline that uses the scientific method to ask and answer questions about the physical world. (Ch 1)

scientific method A continuous process used to collect observations, form and test hypotheses, make predictions, and identify patterns in the physical world. (Ch 1)

scientists Women and men who work in one or more areas of science. (Ch 1)

second law of thermodynamics Any one of three equivalent statements: (1) Heat will not flow spontaneously from a colder to a hotter body; (2) It is impossible to construct a machine that does nothing but convert heat into useful work; (3) The entropy of an isolated system always increases. (Ch 6)

secondary structure Shapes taken by the string of amino acids that make up the primary structure of a protein. (Ch 12)

sedimentary rock A type of rock that is formed from layers of sediment produced by the weathering of other rock or by chemical precipitation. (Ch 20)

seismic tomography A new branch of earth science that is enabling geophysicists to obtain astonishing three-dimensional pictures of the Earth's interior. (Ch 18)

seismic wave The form through which an earthquake's energy is transmitted; may cause the Earth to rise and fall like the surface of the ocean. (Chs 8, 18)

seismology The study and measurement of vibrations within the Earth, dedicated to deducing our planet's inner structure. (Ch 18)

semiconductor Materials that will conduct electricity but do not conduct it very well; neither a good conductor nor a perfect insulator; for example, silicon. (Ch 13)

shale A sedimentary rock formed from sediments that are much finer-grained than sand. (Ch 20)

shear strength A material's ability to withstand twisting. (Ch 13)

shear wave One of two principal types of seismic waves, in which the molecules move perpendicular to the direction of the wave motion; a transverse wave. (Ch 18)

single bond The type of covalent bond formed when only one electron is shared. (Ch 11)

slate A brittle and hard metamorphic rock, formed from shale or mudstone. (Ch 20)

solar system The Sun, the planets and their moons, and all other objects bound by gravitation to the Sun. (Ch 17)

solar wind A stream of charged particles—mainly ions of hydrogen and electrons—emitted constantly by the Sun into the space around it. (Ch 21)

solid All materials that possess a fixed shape and volume, with chemical bonds that are both sufficiently strong and directional to preserve a large-scale external form. (Ch 11)

solutes Any element or compound that has been dissolved in a liquid. (Ch 11)

sound wave Sound is a longitudinal wave that is created by a vibrating object and transmitted only through the motion of molecules in a solid, gas, or liquid. The energy of the sound wave is associated with the kinetic energy of those molecules. (Ch 5)

special relativity The first of two parts of Einstein's theory of relativity, dealing with reference frames that do not accelerate. (Ch 16)

specific heat A measure of heat capacity per unit mass of a material, defined as the quantity of heat required to raise the temperature of 1 g of that material by 1°C. (Ch 6)

spectroscopy The study of emission and absorption spectra of materials to discover the chemical makeup of that material; a standard tool used in almost every branch of science. (Ch 9)

spectrum The characteristic signal from the total collection of photons emitted by a given atom that can be used to identify the chemical elements in a material; the atomic fingerprint. (Ch 9)

speed The distance an object travels divided by the time that it takes to travel that distance. (Ch 3)

speed of light (c) The velocity at which all electromagnetic waves travel in a vacuum, regardless of their wavelength or frequency; equal to 300,000 km/s (about 186,000 mi/s). (Ch 8)

spreading The widening of the seafloor, as magma comes from deep within the Earth and erupts through fissures on the seafloor. (Ch 18)

standard model Theories, supported by experimental evidence, that predict the unification of the strong force with the electroweak force. (Ch 15)

star Any object like our Sun that forms from giant clouds of interstellar dust and generates energy by fusion. (Ch 21)

states of matter Different modes of organization of atoms or molecules, which result in properties of gases, plasmas, liquids, or solids. (Ch 11)

static electricity A manifestation of the electric force, caused by the transfer of electrons between objects: often observed as lightning, or as sparks produced when walking across a wool rug on a cold day. (Ch 7)

steady state universe A model that describes a universe that is constantly expanding, constantly forming new galaxies, but with no trace of a beginning. (Ch 22)

strength The ability of a solid to resist changes in shape; directly related to chemical bonding. (Ch 13)

strong force The force responsible for holding the nucleus together; one of the four fundamental forces in nature. This force operates between particles in the nucleus over extremely short distances and also between quarks to hold elementary particles together. (Ch 14)

subduction zone A region of the Earth where plates converge and old crust returns to the mantle. (Ch 18)

sublimation The direct transformation of a solid to a gaseous state, without passing through the liquid state. (Ch 11)

sugar The simplest of the carbohydrates: common sugars contain five, six, or seven carbon atoms arranged in a ringlike structure. (Ch 12)

superclusters Large collections of clusters and groups of thousands of galaxies. (Ch 22)

superconductivity The property of some materials to exhibit the complete absence of any electrical resistance, usually when cooled to within a few degrees of absolute zero. (Ch 13)

supernova A stupendous explosion of a star, which increases its brightness hundreds of millions of times in a few days; results from the implosion of the core of a massive star at the end of its life. (Ch 21)

surface tension The strong attractive force between adjacent molecules in a liquid that causes the liquid's surface to resist disruption. (Ch 11)

synchrotron A particle accelerator in which magnetic fields are increased as particles become more energetic, keeping them moving on the same track. (Ch 15)

system A part of the universe under study and separated from its surroundings by a real or imaginary boundary. (Ch 5)

system of units A method of measuring objects using a common set of standard quantities. (Ch 2)

technology The application of the results of science to specific commercial or industrial goals. (Ch 1)

tectonic plate One of a dozen large, and some smaller, sheets of moving rock forming the surface of the Earth. (Ch 18)

telescope A device that focuses and concentrates radiation from distant objects; used by astronomers to collect and analyze radio waves, microwaves, light, and other radiation. (Ch 21)

temperature A quantity that reflects how vigorously atoms are moving and colliding in a material. (Ch 6)

tensile strength A material's ability to withstand pulling apart. (Ch 13)

terrane A mass of rock as much as several hundred kilometers across, found in most of the western part of the United States, which was once a large island in the Pacific Ocean and carried toward the North American continent by plate activity. (Ch 18)

terrestrial planets Those relatively small, rocky, high-density planets located in the inner solar system: Mercury, Venus, Earth, the Earth's Moon, and Mars. (Ch 17)

tertiary structure The complex folding of a protein, caused by the cross-linking of chemical bonds from side groups in the amino acid chain. (Ch 12)

theory A description of the world that covers a relatively large number of phenomena and has met many observational and experimental tests. A conclusion based upon observations of nature. (Ch 1)

theory of relativity An idea that the laws of nature are the same in all frames of reference and that every observer must experience the same natural laws. (Ch 16)

thermal conductivity The ability of a material to transfer heat energy from one molecule to the next by conduction. When thermal conductivity is low, as in wood or fiberglass insulation, the transfer of heat is slowed down. (Ch 6)

thermodynamics The study of the movement of heat; the science of heat, energy and work. (Ch 5)

time dilation A phenomenon in special relativity in which moving clocks appear to tick more slowly than stationary ones. (Ch 16)

tornado The most violent weather phenomenon known: a rotating air funnel some tens to hundreds of meters across, descending from storm clouds to the ground, causing intense damage along the path where the funnel touches the ground. (Ch 19)

total momentum The sum of all the momenta in a system. (Ch 4)

trace gas A gas that constitutes less than one molecule in a million in the Earth's atmosphere; for example, ozone. (Ch 19)

transform plate boundary The type of boundary between tectonic plates that occurs when one plate scrapes past the other, with no new plate material being produced; for example, California's San Andreas Fault. (Ch 18)

transistor A device that sandwiches *p* and *n* semiconductors in an arrangement that can amplify an electrical current running through it; a device that played an essential role in the development of modern electronics. (Ch 13)

transmission A process by which light energy passes through matter. See also *absorption* and *refraction*. (Ch 8)

transverse wave A kind of wave in which the motion of the wave is perpendicular to the motion of the medium on which the wave moves; pond ripples. (Ch 8)

triangulation A geometrical method used to measure the distances to the nearest stars up to a few hundred light-years away: the angle of sight to the star is measured at opposite ends of the Earth's orbit and the distances are worked out. (Ch 21)

tropical storm A severe storm that starts as a low-pressure area over warm ocean water and, while drawing energy from the warm water, grows and rotates in great cyclonic patterns hundreds of kilometers in diameter. (Ch 19)

tsunami A great wave, which can devastate low-lying coastal areas, occurring when the energy of an earthquake under or near a large body of water is transferred through the water. (Ch 18)

typhoon Tropical storms that begin in the North Pacific. (Ch 19)

ultraviolet radiation High-frequency wavelengths, shorter than visible light, ranging from 400 nanometers to 100 nanometers. (Ch 8)

uncertainty principle The idea quantified by Werner Heisenberg in 1927, that at a quantum scale, the precise location and velocity of an object can never be known at the same time, because quantum-scale measurement affects the object being measured. Specifically, "The error or uncertainty in the measurement of an object's position, times the error or uncertainty in that object's velocity, must be greater than a constant, h, divided by the object's mass." (Ch 10)

unified field theory The general name for any theory in which fundamental forces are seen as different aspects of the same force. (Ch 15)

uniform motion The motion of an object if it travels in a straight line at a constant speed. All other motions involve acceleration. (Ch 3)

unsaturation A lipid chain in which two carbon atoms are linked by a kinked double bond. These carbon atoms are each bonded to one hydrogen atom, rather than the usual two.

valence The combining power of a given atom, determined by the number of its outermost electrons. (Ch 11)

valence electron An outer electron of an atom that can be exchanged or shared during chemical bonding. (Ch 11)

van der Waals force A net attractive but weak force resulting from the polarization of electrically neutral atoms or molecules. (Ch 11)

vector A quantity measured with two numbers, indicating magnitude and direction, such as velocity, acceleration, or momentum. (Ch 2)

velocity The distance an object travels divided by the time it takes to travel that distance, including the direction of travel. The velocity of a falling object is proportional to the length of time that it has been falling. (Ch 3)

visible light Electromagnetic waves with a wavelength that can be interpreted by receptors in the brain: wavelengths range from 700 nanometers for red light to 400 nanometers for violet light. (Ch 8)

voids Large, empty areas of the universe around which superclusters of galaxies assemble. (Ch 22)

volcanic rock Extrusive igneous rock that solidifies on the surface of the Earth. (Ch 20)

volcano Places where subsurface molten rock breaks through to the surface of the Earth to form dramatic short-term changes in the landscape. (Ch 18)

voltage The pressure produced by the energy source in a circuit, measured in volts. (Ch 7)

watt A unit of power that is the expenditure of 1 joule of energy in 1 second. (Ch 5)

wave A traveling disturbance that carries energy from one place to another without requiring matter to travel across the intervening distance. (Ch 8)

wave mechanics Another term for quantum mechanics indicating the dual (wave and particle) nature of quantum objects. (Ch 10)

wavelength The distance between crests, the highest points of adjacent waves. (Ch 8)

weather Daily changes in rainfall, temperature, amount of sunshine, and other variables, resulting partly from the general circulation in the atmosphere, and partly from local disturbances and variations. (Ch 19)

weight The force of gravity on an object located at particular point. (Ch 4)

white dwarf A star that has a very low emission of energy but very high surface temperature; plots on the lower left corner of the H-R diagram. (Ch 21)

wind shear Violent air turbulence created from sudden downdrafts, which can cause an extremely dangerous condition near airports. (Ch 19)

work A force that is exerted times the distance over which it is exerted; measured in joules in the metric system, in foot-pounds in the English. (Ch 5)

X-rays High-frequency and high-energy electromagnetic waves that range in wavelength from 100 nanometers to 0.1 nanometers, used in medicine and industry. (Ch 8)

CREDITS

PHOTOS

Chapter 1 *Opener:* Ken Biggs/Tony Stone Images/New York, Inc. *Page 3:* Henry Groskinsky/Peter Arnold, Inc. *Page 5:* NASA/Ressmeyer/Corbis. *Pages 7 and 8:* Granger Collection. *Page 10:* Bettmann Archive. *Page 13 (top left):* Roger Ressmeyer/Corbis. *Page 13 (top center):* Dennis Hallinan/FPG International. *Page 13 (top right):* Steve Niedorf/The Image Bank. *Page 13 (bottom left):* Andy Sacks/Tony Stone Images/New York, Inc. *Page 13 (bottom right):* Kaluzny/Thatcher/Tony Stone Images/New York, Inc. *Page 15:* Mark Simon/Black Star. *Page 16:* Courtesy Lisa Passmore.

Chapter 2 *Opener:* Thompson & Thompson/Tony Stone Images/New York, Inc. *Page 28:* Larry Ulrich/Tony Stone Images/New York, Inc. *Page 29 (top):* Frank Siteman/Stock, Boston. *Page 29 (bottom):* Courtesy Ford Motor Company. *Page 31:* Wolfgang Bayer/Bruce Coleman, Inc. *Page 32:* John Henley/The Stock Market. *Page 33 (top):* Peter Lamberti/Tony Stone Images/New York, Inc. *Page 33 (bottom):* Aaron Haupt/Stock, Boston. *Page 34:* D. Brewster/Bruce Coleman, Inc. *Page 35:* Courtesy GM/Saturn. *Page 36:* Dr. Tony Brain/Science Photo Library/Photo Researchers. *Page 38:* Robert E. Daemmrich/Tony Stone Images/New York, Inc. *Page 39:* Peter Lerman. *Page 40:* Granger Collection. *Page 41:* Courtesy Cincinnati Dept. of Transportation. *Page 43 (bottom left):* Courtesy National Institute of Standards and Technology. *Page 43 (right):* Courtesy Bureau Internationale des Poids et Mesures, France. *Page 44 (top left):* David W. Hamilton/The Image Bank. *Page 44 (top right):* Doug Armand/Tony Stone Images/New York, Inc. *Page 44 (bottom):* Joan Iaconetti/Bruce Coleman, Inc.

Chapter 3 *Opener:* Roger Ressmeyer/Corbis. *Page 52:* Jerry Schad/Photo Researchers. *Page 54:* Ric Ergenbright/Tony Stone Images/New York, Inc. *Page 57:* Granger Collection. *Page 58:* Granger Collection. *Page 59 (center):* Bettmann Archive. *Page 59 (top right):* Granger Collection. *Page 60:* Foundation Saint-Thomas, Strasbourg, France. *Page 63:* Scala/Art Resource. *Page 64 (top):* Granger Collection. *Page 64 (margin):* Art Resource. *Figure 3-15:* © Estate of Harold Edgerton, courtesy of Palm Press, Inc. *Page 73 (bottom):* James Sugar/Black Star. *Page 75 (top):* Ken Edward/Photo Researchers. *Page 75 (bottom):* Courtesy Ringling Brothers and Barnum & Bailey Circus. *Page 75 (margin):* Richard Megna/Fundamental Photographs.

Chapter 4 *Opener:* Barrett & MacKay/Masterfile. *Page 84:* Painting by Sir Godfrey Kneller, National Portrait Gallery, London. *Page 85 (bottom):* Michael Melford/The Image Bank. *Page 85 (margin):* Ch. Petit/Photo Researchers. *Page 87:* Duomo Photography, Inc. *Figure 4-1:* Courtesy Mercedes-Benz of North America. *Page 88 (bottom):* Courtesy NASA. *Page 89:* Mitchell Layton/Duomo Photography, Inc. *Page 90:* Henry Groskinsky/Peter Arnold, Inc. *Page 91:* Milton Heiberg/Photo Researchers. *Page 92:* H. R. Bramaz/Peter Arnold, Inc. *Page 95:* Grapes/Michaud/Photo Researchers. *Page 97:* Courtesy NASA/Johnson Space Center. *Page 99:* David Madison/Duomo Photography, Inc. *Page 100:* Courtesy NASA. *Figure 4-9:* Bettmann Archive. *Page 104:* Astronomical Society of the Pacific. *Page 105:* Sylvie Chappaz/Vandystadt/Photo Researchers.

Chapter 5 *Opener:* George Hunter/Tony Stone Images/New York, Inc. *Figure 5-1:* Jerry Ohlinger's Movie Material Store. *Page 113:* Granger Collection. *Page 114:* David Madison/Duomo Photography, Inc. *Figure 5-3:* John P. Kelly/The Image Bank. *Page 115 (bottom):* Granger Collection. *Page 116:* Craig J. Brown/Gamma Liaison. *Page 119:* Runk/Schoenberger/Grant Heilman Photography. *Figure 5-4:* The New York Historical Society. *Figure 5-5(a):* W.J. Scott/H. Armstrong Roberts. *Figure 5-5(b):* The Image Bank. *Figure 5-5(c):* Ken Straiton/The Stock Market. *Page 123:* Granger Collection. *Page 124:* Greg Stott/Masterfile. *Figure 5-7:* © Estate of Harold Edgerton, courtesy of Palm Press, Inc. *Figure 5-8:* Shahn Kermani/Gamma Liaison. *Page 129:* Steve Leonard/Black Star. *Figure 5-9:* The Royal Society, London. *Page 132:* McCutcheon/Visuals Unlimited. *Page 133 (top):* Nicholas Devore III/Bruce Coleman, Inc. *Page 133 (bottom):* Ken Graham/Bruce Coleman, Inc. *Page 136:* Bob Wallace/Stock, Boston.

Chapter 6 *Opener:* Michael Gadomski/Bruce Coleman, Inc. *Page 142 (top):* Al Sattenwhite/The Image Bank. *Figure 6-1(a):* David Leah/Tony Stone Images/New York, Inc. *Figure 6-1(b):* Bruce Forster/Tony Stone Images/New York, Inc. *Figure 6-1(c):* Damir Frkovic/Masterfile. *Page 144 (left):* AP/Wide World Photos. *Page 144 (right):* Bob Daemmrich/The Image Works. *Page 146:* Christel Hesse/Tony Stone Images/New York, Inc. *Figure 6-3:* Charles Thatcher/Tony Stone Images/New York, Inc. *Page 149 (top):* Fritz Goro. *Figure 6-5:* Astrid & Hanns-Frieder Michler/Science Photo Library/Photo Researchers. *Page 151 (top left):* Art Wolfe/Tony Stone Images/New York, Inc. *Page 151 (top right):* Galen Rowell/Mountain Light Photography, Inc. *Page 151 (bottom):* Tom Bledsoe/DRK Photo. *Figure 6-6:* F. Stuart Westmorland/Tony Stone Images/Seattle. *Page 153:* Dale Durfee/Tony Stone Images/New York, Inc. *Page 160:* Granger Collection.

Chapter 7 *Opener:* Richard Kaylin/Tony Stone Images/New York, Inc. *Page 168:* Peter Menzel/Tony Stone Images/Seattle. *Figure 7-2:* Painting by Benjamin Wes, c. 1805, Mraud, Mrs. Wharton Sinkler Collection, Philadelphia Museum of Art. *Page 170 (left):* George Loeher/The Image Bank. *Page 171:* Granger Collection. *Page 176:* Courtesy Colchester and Essex Museum. *Page 177 (bottom):* Jeff Greenberg/Visuals Unlimited. *Page 177 (margin):.* Blakwill-D. Maratez/Visuals Unlimited. *Page 179 (top):* Granger Collection. *Page 179 (bottom):* Jerry Ohlinger's Movie Material Store. *Page 181:* From original painting by C. W. Eckersberg. *Page 182:* Steve Elmore/The Stock Market. *Page 186:* Mark Green/FPG International. *Page 187:* Courtesy Smithsonian Institution. *Page 188:* Courtesy Southern California Edison Co. *Page 193:* Courtesy of the International Cadmium Association.

Chapter 8 *Opener:* Chad Ehlers/Tony Stone Images/New York, Inc. *Figure 8-3:* Oli Tennent/Tony Stone Images/New York, Inc. *Page 206:* J. Cochin/The Image Bank. *Page 207:* Bob Krist/Tony Stone Images/New York, Inc. *Page 209:* Stephen Dalton/Animals Animals. *Figure 8-7:* Manfred Cage/

Researchers. *Page 418:* Courtesy Fermi National Accelerator Laboratory. *Page 420:* The Kobal Collection. *Page 421 (left):* Courtesy Carl Anderson, California Institute of Technology. *Page 421 (right):* Courtesy Lawrence Berkeley Laboratory. *Figure 15-3:* Dan McCoy/Rainbow. *Figure 15-3(a):* Courtesy Brookhaven National Laboratory. *Page 427:* Courtesy National Archives. *Page 429:* Science Photo Library/Photo Researchers.

Chapter 16   *Opener:* Charles Thatcher/Tony Stone Images/ New York, Inc. *Page 439:* Courtesy Dr. Helen Ghiradella, Dept. of Biological Sciences, SUNY, Albany. *Page 441:* National Institute of Standards and Technology, Boulder Laboratories, U. S. Department of Commerce. *Page 448:* 1989 Hsiung. *Page 453:* Imtek Imagineering/Masterfile.

Chapter 17   *Opener:* Courtesy NASA. *Figure 17-1:* Trans. #1307, Photo by Helmut Wimmer, Dept. of Library Services, American Museum of Natural History. *Page 464:* Courtesy Stephen Shawl. *Page 465:* Lowell Observatory Photograph. *Page 466:* Bob Martin/Tony Stone Images/New York, Inc. *Page 468 (left):* Courtesy of Steve Traudt, Synergistic Visions. *Page 468 (right):* Courtesy NASA. *Page 469:* Jim Riffle/ Astroworks, Corp. *Page 472 (top):* Courtesy GE Super Abrasives. *Page 472 (bottom):* Van Valte. *Page 474 (center):* By Chesley Bonestell/Space Art International/© Life Picture Sevice. *Page 474 (right):* Courtesy NASA. *Figure 17-6:* Courtesy M. E. Kipp, Sandia National Laboratories, and H. J. Melosh, University of Arizona. *Page 477:* Astronomical Society of the Pacific. *Pages 478, 479, 480, 482, 483 (bottom), and 485:* Courtesy NASA. *Page 483 (top):* Image created by A. S. McEwen, Arizona State University and the U.S. Geological Survey. Courtesy Jet Propulsion Laboratory and NASA. *Figure 17-13:* JPL/NASA. *Figure 17-32:* Dennis DiCicco/Sky and Telescope Magazine. *Page 486:* Courtesy NASA. *Page 487 (left):* Breck Kent/Earth Scenes. *Page 487 (right):* Courtesy American Museum of Natural History Library. *Page 488 (left):* Astronomical Society of the Pacific. *Page 488 (right):* Courtesy NASA.

Chapter 18   *Opener:* G. Brad Lewis/Tony Stone Images/New York, Inc. *Figure 18-1:* David A. Rosenberg/Tony Stone Images/Seattle. *Page 495 (top):* David Hiser/Tony Stone Images/New York, Inc. *Page 495 (left):* J. A. Kraulis/ Masterfile. *Page 495 (right):* Carr Clifton/Tony Stone Images/ Seattle. *Page 497 (left):* Ralph Perry/Tony Stone Images/ Seattle. *Page 497 (center):* Richard A. Cooke, III/Tony Stone Images/Seattle. *Page 497 (right):* Gamma Liaison. *Page 499:* Iwasa/Sipa Press. *Figure 18-6:* Marie Tharp. *Page 501:* Dudley Foster/Woods Hole Oceanographic Institution. *Page 504:* Roger Ressmeyer/Corbis. *Page 510:* Courtesy NASA. *Page 511:* Dewitt Jones Prod./Tony Stone Images/ Seattle. *Page 517 (top):* Ric Ergenbright. *Page 517 (bottom):* Loren McIntyre/Woodfin Camp & Associates. *Page 519:* Andrew Rafkind/Tony Stone Images/New York, Inc. *Figure 18-22:* Photo by Paul Morin/data by Zhang & Tanimoto. *Page 521 (left):* Lee Foster/FPG International. *Page 521 (right):* Martin/ Prism Rogers/FPG International.

Chapter 19   *Opener:* Phil Schofield/Tony Stone Images/New York, Inc. *Page 528:* Brian J. Skinner. *Page 531 (top):* SABA. *Page 531 (bottom):* Ray Pfortner/Peter Arnold, Inc. *Page 532:* Chad Ehlers/Tony Stone Images/New York, Inc. *Page 533 (top):* Richard Megna/Fundamental Photographs. *Page 533*

(bottom): Stephen Kline/Bruce Coleman, Inc. *Page 536:* Courtesy NASA. *Figure 19-7:* National Oceanic and Atmospheric Administration/DLR-FRG/Roger Ressmeyer/ Corbis. *Figure 19-10:* NOAA/Photo Researchers. *Page 540 (bottom):* Impact Photos. *Page 542:* Walter Bibkow/The Image Bank. *Page 545:* Ralph A. Clevenger/West Light. *Page 546 (top):* Tom Bean/DRK Photo. *Page 546 (bottom):* S. Nielsen/Bruce Coleman, Inc. *Page 553:* Luiz Claudio Marigo/Peter Arnold, Inc. *Page 554:* Professor P. Motta/ Science Photo Library/Photo Researchers. *Figure 19-21:* Courtesy NASA. *Page 559 (top margin):* Jim Mendenhall. *Page 559 (top):* Philippe Plailly/Science Photo Library/Photo Researchers. *Page 559 (bottom margin):* Jim Mendenhall. *Figure 19-24:* Courtesy Westfälisches Amt für Denkmalpflege. *Page 562:* Ray Pfortner/Peter Arnold, Inc. *Page 563:* Courtesy Environmental Elements Corporation.

Chapter 20   *Opener:* Bill Brooks/Masterfile. *Figure 20-1(a):* Michael J. Howell/ProFiles West, Inc. *Figure 20-1(b):* Robert Frerck/Odyssey Productions. *Page 573 (top):* J. D. Griggs/ USGS. *Page 573 (bottom):* William Sacco. *Page 574:* Courtesy NASA. *Figure 20-3:* Tom Bean/DRK Photo. *Figure 20-4:* David Ball/Tony Stone Images/New York, Inc. *Figure 20-5:* David Muench/Tony Stone Images/Seattle. *Page 577 (bottom left):* Brian J. Skinner. *Page 577 (bottom right):* William Sacco. *Page 578 (top):* Craig Johnson. *Page 578 (bottom):* Bill Ross/H. Armstrong Roberts, Inc. *Figure 20-7:* Dr. K. Roy Gill. *Page 581:* Martin Miller. *Page 583 and 584 (top left):* A. Kerstitch/ Visuals Unlimited. *Page 584 (top right):* Francois Gohier/ Photo Researchers. *Page 584 (bottom left):* Visuals Unlimited. *Page 584 (bottom right):* Don Fawcett/Visuals Unlimited. *Page 587:* Norbert Wu/Peter Arnold, Inc. *Figure 20-10:* BPS/ Terraphotographics. *Page 589:* Kim Taylor/Bruce Coleman, Inc. *Page 591:* Gerard Lacz © Natural History Photographic Agency. *Figure 20-10:* Courtesy Prof. Lawrence Cook, University of Manchester. *Page 593:* Courtesy of the American Museum of Natural History. *Page 594:* from Stephen Jay Gould, *Wonderful Life*, NY, 1989, W.W. Norton and Company. *Page 596 (top):* A. J. Copley/Visuals Unlimited. *Page 596 (bottom left):* Dr. Jeremy Burgess/Photo Researchers. *Page 596 (bottom right):* CNRI/Photo Researchers. *Page 598 (left):* Tom McHugh/Tony Stone Images/Seattle. *Page 598 (right):* Tom Stack & Associates. *Page 598 (Inset):* Courtesy K.W. Barthel, Museum beim Solenhofer Aktienverein, Germany. *Page 600:* The Kobal Collection. *Page 603:* Courtesy the Institute of Human Origins. *Page 605:* John Reader/Science Photo Library/Photo Researchers. *Page 607:* Photofest.

Chapter 21   *Opener:* Anglo Australian Telescope Board. *Page 612:* Science Photo Library/Photo Researchers. *Page 616:* Courtesy Brookhaven National Laboratory. *Figure 21-4:* National Center for Atmospheric Research. *Page 619:* Johnny Johnson/Earth Scenes/Animals Animals. *Page 622 (top left):* Richard Winscoat/Douglas Peebles Photography. *Page 622 (top right):* Courtesy National Radio Astronomy Observatory. *Pages 622 (bottom) and 623 (top right):* Courtesy NASA. *Page 623 (top and bottom left):* Space Telescope Science Institute. *Page 623 (bottom right):* California Association for Research in Astronomy/Science Photo Library/Photo Researchers. *Page 626:* Courtesy Harvard College Observatory. *Page 634:* Courtesy NASA and ESA. *Page 635:* Courtesy Royal Observatory, Edinburgh. *Figure 21-21:* Courtesy NASA. *Page 637:* Courtesy Cornell

University. *Page 638 (top):* Courtesy NASA. *Page 639:* J. Riffle/Astroworks, Corp.

Chapter 22   *Opener:* Bill Brooks/Masterfile. *Figure 22-1:* Courtesy AIP Emilio Segre Visual Archives/Hale Observatories. *Figure 22-2:* Dennis DiCicco/Sky and Telescope Magazine. *Figure 22-3:* David Malin/Anglo Australian Telescope Board. *Page 648 (bottom margin):* Courtesy NASA. *Page 648 (bottom):* Courtesy of Richard Davis, Jodrell Bank; observations with the Merlin array. *Page 650:* Courtesy M. Schmidt. *Figure 22-13:* Courtesy of Palomar Observatory, California Institute of Technology. *Page 654:* Courtesy KPNO/NOAO. *Page 655 (top right):* Courtesy Hale Observatories. *Page 655 (bottom left):* Courtesy Brent Tully, Institute for Astronomy, University of Hawaii. *Page 655 (bottom right):* (c) 1984 Royal Observatory, Edinburgh, Anglo-Australian Telescope Board. *Page 658:* Astronomical Society of the Pacific. *Page 659:* Courtesy NASA. *Figure 22-18:* Courtesy V. Rubin. *Page 668 (top):* Courtesy J.A. Tyson/AT & T Bell labs. *Page 668 (bottom):* Courtesy of NASA/Goddard Space Flight Center.

LINE ART

Figure 16-13 (p. 455): Adapted from a diagram by C. Misner, K. Thorne, and J. Wheeler.

Figure 17-4 (p. 467): Based on Michael Zeilik, *Astronomy: The Evolving Universe,* 6th, ed., 1991 (New York, John Wiley & Sons, Inc.), Fig. 12-23, p. 247.

Figure 17-9 (p. 479): Based on a model by J. S. Lewis.

Figure 19-12 (p. 543): Modified from W. M. Marsh and J. Dozier, *Landscape,* p. 131, copyright © 1981, John Wiley & Sons. Reprinted with permission of John Wiley & Sons, Inc.

Figure 19-13 (p. 544): Reprinted by permission from M. E. Schlesinger (1983), A review of climate model simulations of CO_2-induced climatic change, *Report No. 41,* Oregon State University, Climatic Research Institute, Fig. 3; as adapted by permission from M. E. Schlesinger and W. L. Gates (1981a), Preliminary analysis of four general circulation model experiments on the role of the ocean in climate, *Report No. 25,* Oregon State University, Climatic Research Institute.

Figure 19-22 (p. 560): Council on Environmental Quality, 1991.

Figure 19-25 (p. 562): After "How Many More Lakes Have to Die?" *Canada Today,* 12, No. 2, 1981.

Figure 19-27 (p. 564): C. D. Keeling, et al. A three-dimensional model of atmospheric CO_2 transport based on observed winds: 1. Analysis of observational data, *AGU Monograph 55,* 1989, Figure 16. Copyright by the American Geophysical Union.

Figure 19-28 (p. 567): J. Fowler, *Energy and the Environment,* 2nd. Ed., 1984 (New York: McGraw-Hill, Inc.), p. 82. Reproduced with the permission of McGraw-Hill, Inc.

Figure 20-9 (p. 586): S. L. Miller and L. E. Orgel, *The Origins of Life on Earth,* 1974 (Englewood, NJ: Prentice-Hall), p. 84.

Figure 20-14 (p. 597): Adapted by permission from G. Brum, L. McKane, and G. Karp, *Biology: Exploring Life,* 2nd ed., 1994 (New York: John Wiley & Sons, Inc.), Fig. 5-24 (a and b).

Figure, p. 599: Based on a drawing by Virge Kask in Derek

E. G. Briggs (1991), "Extraordinary Fossils," *American Scientist,* vol. 79, no. 2, p. 133.

Figure 20-15 (p. 601): D. M. Raup, and J. J. Sepkoski, Jr. (1982), "Mass Extinctions in the Marine Fossil Record," *Science* 215:1501-2.

Figure 21-3: Courtesy Paul DiMare.

Figure 21-17: Adapted from a diagram by J. C. Wheeler.

Figure 21-20: Based on observations by D. C. Backer, S. R. Kulkarni, C. Heiles, M. M. Davis, and W. M. Goss.

Figure 21-22: Adapted from a diagram by W. Kaufmann.

Figure 22-13: Adapted from a NASA diagram.

Figure 22-17: Adapted from a diagram by J. Burns.

The following figures were reprinted from Robert Robbins, William Jefferys, and Stephen Shawl, *Discovering Astronomy,* 3rd Ed. Copyright © 1995 John Wiley & Sons. Reprinted by permission of John Wiley & Sons, Inc.: 3-3, 21-8, 21-12, 21-14, 21-16, 21-18, 21-23, 22-9, 22-14, 22-19, 22-20, E-1, E-2, E-3, E-4, E-5, E-6, E-7, E-8, E-9, E-10.

The following figures were reprinted from Michael Zeilik, *Astronomy: The Evolving Universe,* 7th Ed. Copyright © 1994 John Wiley & Sons. Reprinted by permission of John Wiley & Sons, Inc.: 3-7, 3-8, 3-9, 3-19, 17-7, 17-8, 17-11, 17-12, 17-14, 21-10, 21-13, 22-16.

The following figures were reprinted from Carl Snyder, *The Extraordinary Chemistry of Ordinary Things,* 2nd Ed. Copyright © 1995 John Wiley & Sons. Reprinted by permission of John Wiley & Sons, Inc.: 4-6, 9-1, 14-3, 14-6, 14-10, 14-14, 19-19, 19-20, 19-23.

The following figures were reprinted from Brian Skinner and Stephen Porter, *The Blue Planet: An Introduction to Earth Systems Science,* Copyright © 1995 John Wiley & Sons. Reprinted by permission of John Wiley & Sons, Inc.: 17-2, 18-4, 18-11, 18-15, 18-17, 19-6, 19-14, 19-17, 20-6.

The following figures were reprinted from Brian Skinner and Stephen Porter, *The Dynamic Earth: An Introduction to Physical Geology,* 3rd Ed. Copyright © 1995 John Wiley & Sons. Reprinted by permission of John Wiley & Sons, Inc.: 5-11, 14-1, 18-18, 18-20, 20-2.

The following figures were reprinted from Gil Brum, Larry McKane, and Gerry Karp, *Biology: Exploring Life,* 2nd Ed. Copyright © 1994 John Wiley & Sons, Inc. Reprinted by permission of John Wiley & Sons, Inc.: 11-6, 11-9, 12-2, 20-16.

The following figures were reprinted from Michael Zeilik, *Conceptual Astronomy,* Copyright © 1993 John Wiley & Sons, Inc. Reprinted by permission of John Wiley & Sons, Inc.: 16-13, 21-3, 22-17, 22-18.

The following figures were reprinted from David Halliday, Robert Resnick, and Jearl Walker, *Fundamentals of Physics,* 4th Ed. Copyright © 1993 John Wiley & Sons. Reprinted by permission of John Wiley & Sons, Inc.: 4-3, 7-19, 9-17, 13-2.

The following figures were reprinted from John Cutnell and Kenneth Johnson, *Physics,* 3rd. Ed., Copyright © 1995 John Wiley & Sons, Inc. Reprinted by permission of John Wiley & Sons, Inc.: 3-17, 15-1, 16-10.

The following figures were reprinted from Daniel Botkin and Edward Keller, *Environmental Science: Earth as a Living Planet,* Copyright © 1995 John Wiley & Sons, Inc. Reprinted by permission of John Wiley & Sons, Inc.: 19-2, 19-3a.

INDEX

Page numbers in italics represent illustrations.